Electronics

A systems approach

Electronics

A systems approach

Neil Storey

University of Warwick, Coventry

2nd edition

An imprint of Pearson Education

Harlow, England · London · New York · Reading, Massachusetts · San Francisco · Toronto · Don Mills, Ontario · Sydney
Tokyo · Singapore · Hong Kong · Seoul · Taipei · Cape Town · Madrid · Mexico City · Amsterdam · Munich · Paris · Milan

Pearson Education Limited
Edinburgh Gate
Harlow
Essex CM20 2JE
England

and Associate Companies throughout the World.

Visit us on the World Wide Web at:
http://www.pearsoneduc.com

Cover designed by The Senate, London
and printed by The Riverside Printing Co. (Reading) Ltd
Text design by Sally Castle
Typeset in 10/12pt Times by 25
Printed and bound in Great Britain by Biddles Ltd, Guildford and King's Lynn

ISBN 0-201-17796-X

Library of Congress Cataloging-in-Publication Data
Storey, Neil.
Electronics : a systems approach / Neil Storey. -- 2nd ed.
p. cm. -- (Electronic systems engineering series)
Includes index
ISBN 0-201-17796-X (alk. paper)
1. Electronic systems. 2. Electronic circuits. I. Title
II. Series
TK7870.S857 1998
621.381--dc21 98-5086
CIP

10 9 8 7 6 5 4 3
04 03 02 01 00 99

To Jill, Jonathan and Emma

Contents

Preface

When the first edition of this book was published it represented a very novel approach to the teaching of electronics. Up to that time most texts in this area had adopted a decidedly 'bottom-up' approach to the subject. They often started by looking at semiconductor materials and worked their way through diodes, transistors and biasing, before looking at simple amplifier circuits. Eventually, several chapters later, they might actually look at the uses of the circuits being considered.

The first edition of *Electronics: A Systems Approach* pioneered a new approach to the teaching of electronics by explaining the uses and required characteristics of circuits before embarking on detailed analysis. This aids comprehension and makes the process of learning much more interesting. Another innovation within the first edition was that it provided a unified treatment of both analogue and digital electronics within a single volume, allowing common ground between these two areas to be developed. While many texts are still resolutely device centred, it is pleasing to see that several more enlightened authors are now following a more systems oriented approach.

One of the great misconceptions concerning this approach is that it is in some way less rigorous in its treatment of the subject. It seems that some instructors believe that any book that does not start with several pages of complex mathematics does not do justice to the teaching tradition. The systems approach does not define the depth to which a subject is studied but only the order and manner in which the material is presented. Many students *will* need to look in detail at the operation of electronic components and to understand the physics of its materials; however, this material will be more easily absorbed if the characteristics and uses of the components are understood first.

One of the many advantages of a systems approach is that it allows a single text to be used by a wide variety of students. *All* engineers and scientists should have an appreciation of the basics of electronics since it is an essential enabling technology across a broad spectrum of disciplines. Presenting the information in a 'top-down' manner makes it easier to assimilate for all readers. For the student who is destined to be an electronic specialist this route provides the information in the most accessible order, thus aiding comprehension. For the non-specialist the information is presented such that the reader may progress as far through the text, or through the various sections, as is appropriate to give the required level of detail. The text includes the circuit analysis and design techniques needed by students destined for more advanced circuit design courses. However, it is presented after a more general introduction to the topic such that non-specialists may easily skip the detail without harming their understanding of the basics.

Throughout the text parallels between analogue and digital techniques are highlighted and a systems approach to design is encouraged. Numerous examples are given to illustrate the techniques being discussed and design case studies are included at the end of each chapter to reinforce the material and to promote a systematic approach.

Who should read this book

This text is intended for undergraduate students in all fields of engineering and science. For students of electronics and related disciplines it provides a coherent set of material suitable for first level courses in analogue and digital design. For other students it includes most of the electronics material that will be required throughout their course.

New material within this edition

This second edition has given an opportunity to update the text and to take account of developments in a very rapidly changing field. It has also permitted several major revisions of the text. The major changes to the book are:

- The text now includes a large number of computer simulation exercises as an aid to comprehension. The exercises make use of a set of demonstration files that may be downloaded free of charge over the Internet and a software simulation package that can be obtained free of charge from its supplier.
- A short section on laboratory instruments has been included which describes the operation of oscilloscopes and digital multimeters.
- A new section has been added on electromagnetic compatibility (EMC) to reflect the growing recognition of the importance of this topic to all engineers.
- A substantial new section has been added on programmable logic devices (PLDs) to replace the limited treatment of array logic in the first edition.
- The number of worked examples within the text has been increased.
- Additional self assessment exercises have been added to those at the end of the chapters.
- A new appendix describes the IEC 617 symbols for logic elements.
- An appendix now gives numerical answers to many of the end of chapter exercises. If the number of the exercise is set in italic type, for example, see Exercise ***1.11***, on page 14, this indicates that the answer can be found in this appendix.

Circuit simulation

Throughout the book there are numerous *Computer simulation exercises* that support the material in the text. These are marked by a computer icon in the margin as shown on the left. The exercises may be performed using any suitable circuit simulation package, although the use of **PSpice** is recommended. The PSpice simulation package is produced

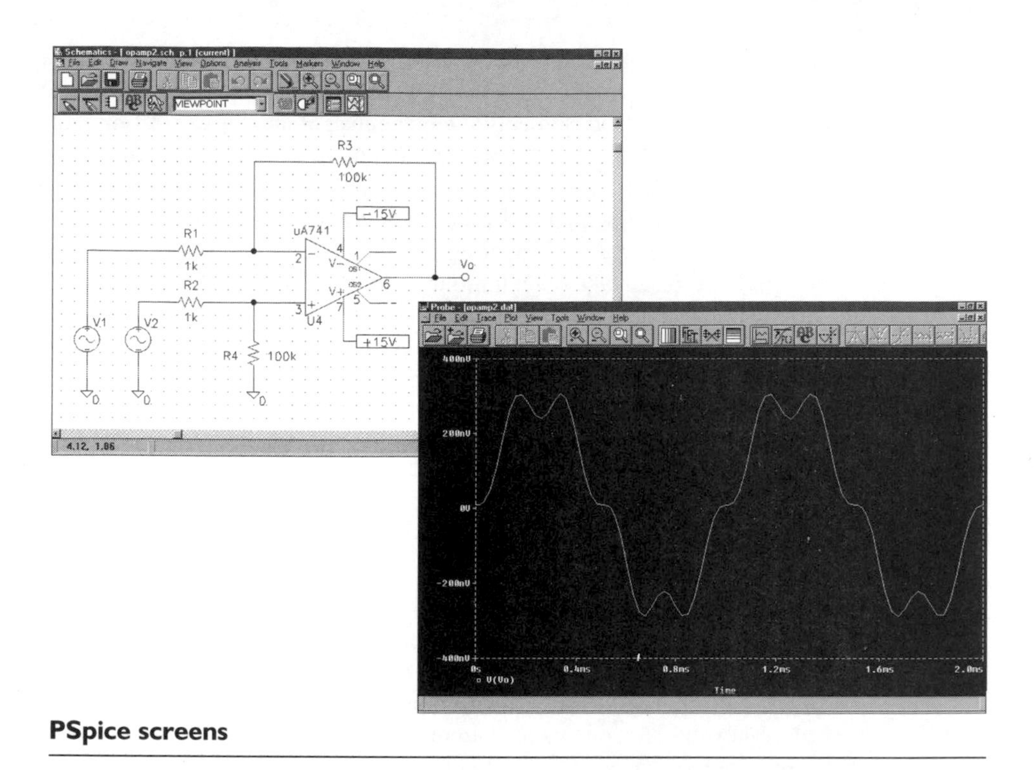

PSpice screens

and marketed by the MicroSim Corporation and is one of the most widely used packages both within industry and within Universities and Colleges. PSpice is based on the industry standard SPICE simulation package and comes with a suite of programmes allowing the schematic capture of circuits and the graphical display of simulation results.

MicroSim produce a free evaluation version of the software which incorporates most of the important features of the package but is restricted in the complexity of the circuits that may be simulated. This evaluation version is completely adequate for the needs of this text. Readers with their own computer may obtain a free copy of the software by contacting MicroSim at their website on the Internet. Their URL is:

http://www.microsim.com

After completing an on-line application form the free software will be dispatched by post within a few days. Those making use of a computer within a University or College should check with the system manager to see if PSpice is already available. If not, they should request that it be loaded – institutions can also make use of this free software.

Since the main purpose of this text is to teach electronics rather than simulation, the text does not attempt to teach the use of PSpice or any other simulation package. Most electronics programmes now include a course on computer simulation that looks at the modelling methods used and considers the limitations and restrictions of these tools. Such courses are not normally taken by non-electronics students. I believe that *all* students can benefit from the use of simulation when studying electronics since it gives a very graphic insight into the operation and characteristics of circuits. Fortunately, the skills required to perform simple simulations can be learnt in a few minutes and I would suggest that all readers of this text should take advantage of these powerful packages. Indeed, it may be beneficial for electronics students to gain an insight into the use of simulation by using it at a basic level with this text *before* taking a more detailed simulation course.

One of the barriers to the use of simulation in the study of electronics is the time taken to prepare the circuit files required by the simulator. Even though packages such as PSpice permit circuits to be entered in a straightforward way through the use of a schematics capture package, it can take several minutes for even a simple circuit to be entered (particularly for the inexperienced). In order to speed up the use of simulation, and to encourage readers to make maximum use of these powerful tools, a suite of demonstration files is available to complement this book. The presence of the demonstration files means that it is possible to simulate the various circuits without a detailed knowledge of the simulation package being used. All that is required is the ability to load a file, run the simulation and make minor modifications to the circuit. Such skills could be gained from a short demonstration within a laboratory session, or could be obtained in a few minutes by referring to one of the many excellent texts available on this topic (those using PSpice can find a list of books on this package on the MicroSim website).

Each demonstration file is provided in two formats: firstly as a *net list* and secondly as a PSpice *schematics file*. The first of these formats is compatible with all versions of SPICE and allows the demonstration files to be used with a wide variety of simulation packages. Unfortunately, while this format is compatible with a range of simulators it describes only the components of a circuit and their connectivity. This information is all that is required to simulate a circuit, but does not allow the circuit diagram to be drawn. This makes the files difficult to interpret and does not aid comprehension. For this reason the demonstration circuits are also provided in the form of PSpice schematics files. These can be read by the PSpice schematics editor which displays the circuits and allows them to be easily modified. The schematics package provides a user friendly interface to the PSpice simulator and greatly aids comprehension of the circuits being studied. For this reason the reader is encouraged to take advantage of the free PSpice software described above, to gain maximum benefit from the various demonstration files.

The names of the demonstration files associated with the computer simulation exercises are given under the computer icons in the margin. An example is shown on the left. Icons will also be found alongside many of the diagrams within the text indicating that simulation files are also available for these circuits to simplify their study using simulation. Associated with the demonstration files is a README file that contains configuration information for each simulation. This document, which assumes the use of PSpice, explains how to configure the simulation package for each circuit. All the demonstration files, and the associated README file may be obtained over the Internet from the following website:

FILE 1A

http://www.awl-he.com/engineering

Full details of how to download and install the files are given on the website.

Problems using simulation have also been included within the exercises given at the end of each chapter. These exercises do *not* have demonstration files and are set to test the reader's understanding of the use of simulation as well as the circuits concerned.

Assumed knowledge

Throughout the book it is assumed that the reader has an understanding of basic passive components such as resistors and capacitors and their parallel and series combinations. A knowledge of the AC behaviour of these components is also assumed.

To the instructor

A comprehensive **instructor's guide** is available for this text. This gives guidance on course preparation and suggests how material might be selected to meet the needs of students with different backgrounds and interests. The guide also gives fully worked solutions to all the numerical exercises within the text and sample answers for the non-numerical exercises. Instructors adopting this book as a course text should contact their local Addison Wesley Longman representative to obtain a complimentary copy of this guide.

Acknowledgements

I would like to express my gratitude to several people who have provided help and support during the production of this text. In particular I would like to thank my colleagues at the University of Warwick who have provided useful feedback on the first edition and David Dyer for his comments on the first draft of the new material on PLDs.

I would also like to thank the companies who have given permission to reproduce their material. These are: RS Components Limited for the photographs of Figures 2.1, 2.2, 2.4, 2.6(b) and 2.12; Farnell Electronic Components Limited for the photographs of Figures 2.3, 2.14(a) and 2.15; R. D. P. Electronics Limited for the photograph of Figure 2.7(b); Tektronix U.K. Limited for the photograph of Figure 2.18(a); Fluke Corporation for the photograph of Figure 2.19(a); Texas Instruments for the diagram at the end of the Design Study in Chapter 8 and Figures 11.20 and 11.32; and Advanced Micro Devices Inc for the diagrams of Figures 11.47, 11.48 and 11.49.

Finally, I wish to give special thanks to my family for their help and support during the writing of this book. In particular I wish to thank my wife Jill for her constant encouragement and understanding.

Electronic Systems

Objectives

When you have studied the material in this chapter you should:

- have an understanding of the concepts of analogue and digital quantities, and their representation by electrical signals;
- be aware of the need for electronic systems in a wide range of applications, and be able to identify the principal components of such systems;
- be familiar with the effects and characteristics of distortion and noise in electronic systems;
- be able to describe the principal stages of system design using good design methodology;
- appreciate the need for an appropriate choice of technology in the design of electronic systems;
- be aware of the existence of powerful electronic computer aided design and development tools (ECAD) to simplify the designer's task.

Contents

1.1 Introduction

The world in which we live is constantly changing.

Our survival amidst this change relies, in part, on our ability to sense our environment and to respond appropriately. This may involve the simple response of moving our hand from a flame, or the more complex response of buying gold when the value of a particular currency goes down. A characteristic of such a response is that a changing quantity is sensed (the **input** quantity) and that a quantity is changed in response (the **output** quantity).

Over the last few thousand years man has developed a great many machines to help him in his interaction with his environment. These machines perform work, and therefore generate a change in some physical quantity. Often these machines also sense physical quantities and use this information to control their operation. The physical quantities that may be sensed or controlled in this way are numerous. Examples include temperature, displacement, force, humidity, light intensity, time and mass.

Most physical quantities vary in a continuous manner. That is, they change smoothly from one value to another without any sudden steps or **discontinuities**. Such quantities can take an infinite number of values and are referred to as **continuous**.

Some quantities, however, do not change smoothly but change abruptly between a number of fixed values. Such quantities are referred to as **discrete**.

Some of the physical quantities given as examples above are discrete, some are continuous and some may be either. Often one's decision on whether a quantity is discrete or continuous will depend on whether one is considering a microscopic, macroscopic or perhaps a relativistic model of the world. Mass, for example, might be considered a continuous variable when viewed at a macroscopic level, but discrete when considered at the atomic level. When considering the mass of a fast moving object one may need to consider the relativistic change in mass as the object approaches the speed of light; this change would appear to be continuous.

In electronic systems a varying physical quantity is usually represented by an electrical **signal** which varies in a manner which describes that quantity. Signals may also be continuous or discrete but it is not obligatory to describe a continuous quantity by a continuous signal or to describe a discrete quantity by a discrete signal. Sometimes it may be advantageous to approximate a continuous quantity by a discrete signal or vice versa.

Continuous signals are generally referred to as **analogue** signals (or sometimes by the American variant **analog**), while discrete signals are often called **digital** signals.

In fact the terms **analogue** and **digital** are not ideal. Analogue comes from the word 'analogy' meaning that the signal resembles the physical quantity being represented. However, a discrete signal can also be an analogy of a changing quantity. Similarly 'digital' implies the use of numbers but many discrete signals are not direct representations of numbers. It would be more precise to use the terms 'continuous' and 'discrete' but since the terms 'analogue' and 'digital' are used universally they will be used throughout this text.

Digital signals may take a number of forms dependent on the number of discrete levels the signal may take. Some digital signals take only two levels; these are referred to as **binary signals**. Such signals are of great importance as they are widely used in electronic logic circuitry and computers. Other digital signals may take three, four or any

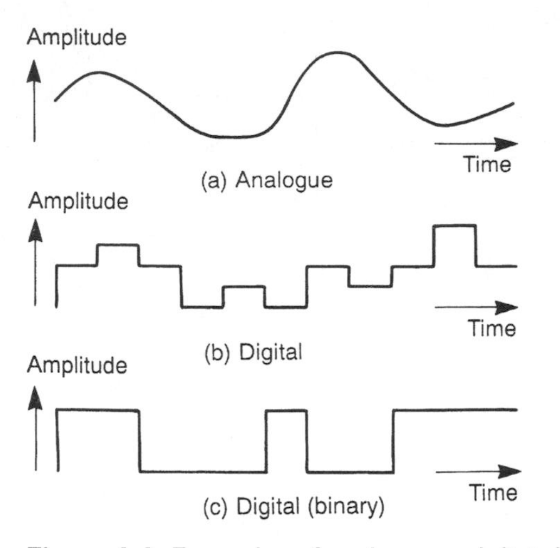

Figure 1.1 Examples of analogue and digital signals.

number of levels; such signals are often known by the general term **m-ary signals**. Figure 1.1 shows examples of analogue and digital signals.

There are many types of analogue signal but a common form is that the **voltage** of the signal at any time is proportional to the magnitude of the quantity being represented. It is also possible to produce signals in which the magnitude of the **current** at any instant represents the physical quantity, or perhaps, using a sinusoidal signal one could arrange that the **frequency** or **phase** of the signal represented the physical quantity. All of these types of signal are used in different applications but the first is probably the most common and we will initially concentrate on this form of signal.

The reasons for representing physical quantities by **electrical signals** are many fold. Electrical signals are very easy to process using electronic circuitry which is both inexpensive and reliable. Electronic signals may be easily transmitted over large distances and may also be stored and reproduced later. However, there would be no point in generating and processing these signals if they were not ultimately used for something!

1.2 Electronic systems

A **system**, in general terms, is any closed volume for which all the inputs and outputs are known. An **electronic system** is any arrangement of electronic components with a defined set of inputs and outputs. Useful electronic systems have the property that they take in information in the form of **input signals** (or simply inputs), perform some operations on it and then produce some useful **output signals** (or outputs).

In recent years electronic systems have found their way into almost all aspects of our lives. Such systems wake us in the morning; control the operation of our cars as we drive to work; maintain a comfortable working environment in our offices and homes; allow us to communicate world-wide; provide access to information at the touch of a button; manage the provision of power to maintain our high technology life

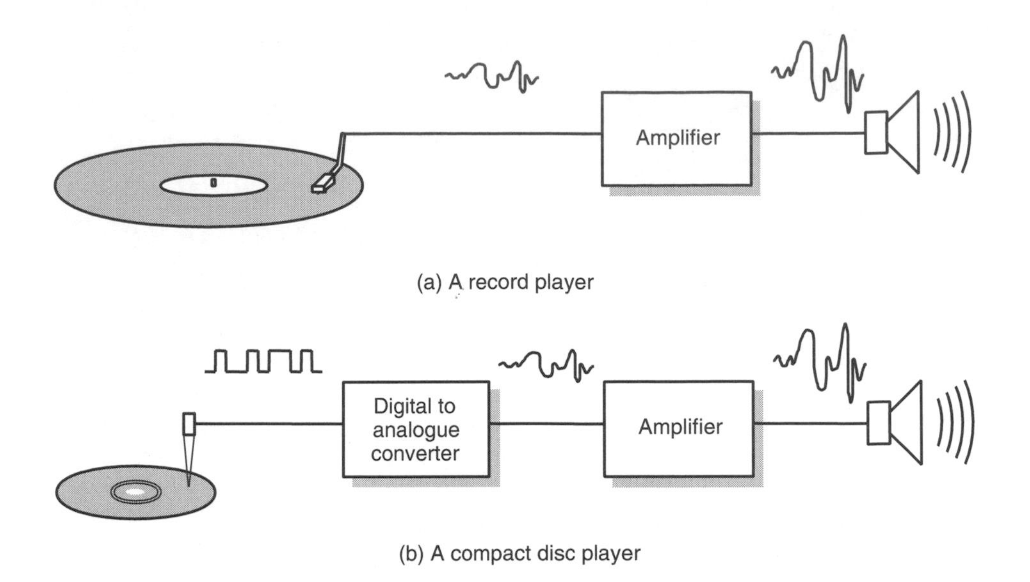

(a) A record player

(b) A compact disc player

Figure 1.2 Typical electronic systems.

styles; and provide restful entertainment after a day of 'electronically controlled' excitement.

In many cases electronic systems are used in these applications because they provide a more cost-effective solution than other available techniques. However, in many cases electronics provide the **only** solution, and the application would be impossible without its use. Our way of life increasingly depends on an ability to monitor or control our environment and to communicate efficiently. In these areas electronic systems are supreme and seem certain to remain so for the foreseeable future.

Figure 1.2 shows two common electronic systems: a record player and a compact disc (CD) player. In the former a sensor is used to detect the motion of a stylus as it tracks the displacement of a groove in a vinyl disc (record). The signal produced by this sensor is analogue in form and corresponds to the sound pressure of the original recording. To reproduce this sound the sensor's output is amplified to produce a signal of sufficient power to drive a loudspeaker. Sound is stored on a compact disc in a digital format. This is achieved by repeatedly measuring the magnitude of the original analogue sound signal and representing these measurements by numbers. This information is converted into a binary form and is written onto the CD as a series of 'pits'. The data stored on the disc is retrieved by a sensing arrangement that uses a laser to detect the presence of these pits as the disc spins at high speed. The result is a train of binary information that is a coded form of the original measurements. In order to reproduce the recorded sound this data is first decoded and then converted back into an analogue form using a digital to analogue converter. This signal is then amplified to drive a loudspeaker as in a record player.

At first sight it might seem strange that we choose to convert an analogue sound signal into a digital form to store it on a CD, when we then have to convert it back into an analogue form in order to listen to it. The reason for this apparently odd behaviour is that it enables us to recover the stored information much more accurately than is possible using analogue techniques. The price we pay for this improvement in performance is an increase in the complexity of the system. In practice, both record players and CD players are more complex than is suggested by the diagrams of Figure 1.2. For example, both will normally have circuitry for changing the relative magnitudes of signals of different frequency ranges. However, CD players also require complex mechanical and electronic subsystems to control the scanning of the laser across the disc and to accurately control the disc's speed of rotation. It is worth noting at this point that while a CD player is based on the use of a digital recording technique, it also contains a substantial amount of analogue circuitry. Many electronic systems make use of a combination of analogue and digital elements.

In general the inputs to an electronic system come from the measurement of one or more **physical quantities**. For example, in the case of a record player the physical quantity being measured is the displacement of the groove on the record, and in a CD player the physical quantity is the presence or absence of pits on the disc. Physical quantities are measured using appropriate **sensors** which, in the case of electronic systems, generally produce electrical signals related to the quantities being detected. Input signals are **processed** in order to produce appropriate output signals. These are used to drive one or more **actuators** that then affect other physical quantities. In both the examples in Figure 1.2 the actuator is a loudspeaker and the physical quantity being affected is air pressure (or if you prefer, sound).

The nature of the processing performed within an electronic system is determined by its required function, the nature of the signals produced by the sensors, and the form of the signals required by the actuators. However, there are certain elements in this processing which are common to a large number of systems. These include: **amplification**; the **addition** and **subtraction** of signals; **integration**; **differentiation**; and **filtering** (that is, changing the relative magnitudes of components of different frequencies). In more complex systems a sequence of operations may be required which in turn requires functions such as the **counting** and **timing** of events. In such systems quite complex decision making may be needed to modify the action taken depending on the inputs received. In some applications it may also be necessary to generate sinusoidal or other signals within the system. These individual processing operations are performed by different **electronic circuits.** We will be looking at such elements in later chapters.

1.3 Distortion and noise

No electronic circuits are ideal; all impose limits on the amplitude and frequency of the signals that pass through them. This may result in signals being **distorted** as they pass through the system. Distortion may take many forms and may, for example, result in alteration of the amplitude, frequency or phase of a signal.

Figure 1.3 shows some examples of distortion of a sinusoidal signal. The causes of these forms of distortion, and methods of overcoming them, will be discussed in later chapters.

The effects of distortion will clearly vary between applications. In an audio amplifier large amounts of distortion would be audible and would degrade the quality of the

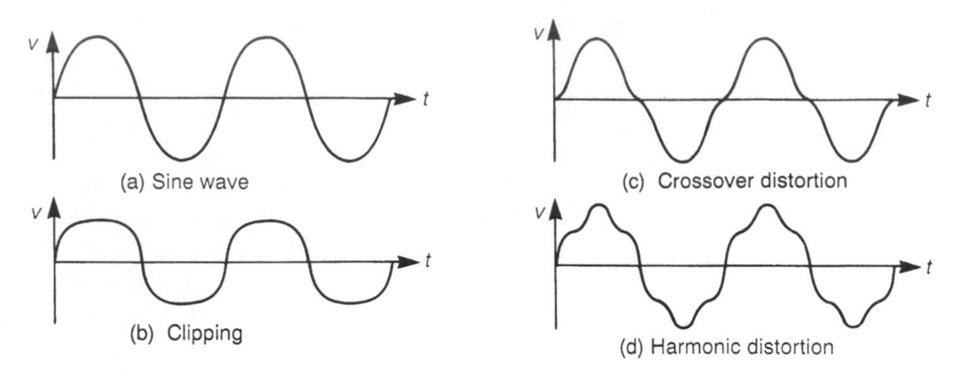

Figure 1.3 Examples of the effects of distortion on a sine wave.

sound produced. In a temperature monitoring system distortion might result in a reduction in the accuracy of the measurements.

Although some distortion is always present, its magnitude will depend on the circuits used. Good design must ensure that distortion is at an acceptable level for a given application.

Signals are also affected by **noise**. This is a random fluctuation of a signal which is produced either by variations within the system or by external effects of the environment. It has a number of causes and is always present within electronic systems.

Since the amount of noise produced within a system is not related to the input signal, its effects will be most marked when the input signal is small. This effect is readily experienced when listening to a radio or record player. Noise, experienced as a background 'hiss', is much more apparent during quiet, rather than loud, passages.

Figure 1.4 illustrates the effects of noise on both analogue and digital (binary) signals. It can be seen that the addition of noise to the signals changes the waveforms in both cases. However, in the digital case it is still apparent which parts of the signal represent the higher voltage and which the lower, and it is therefore possible to extract the original information from the signal. This is not the case with the analogue signal, where separating the noise from the original signal is not usually possible. This illustrates an important distinction between analogue and digital signals which will be developed further in later chapters.

In electronic circuits the presence of distortion and noise often limits the ultimate performance of a system. One of the major tasks of electronic design is to reduce the magnitude of these effects.

1.4 System design

The task of designing an electronic system has many facets and can be greatly simplified by adopting a methodical, rational approach.

Most experienced engineers agree that a **top-down approach** to system design offers many advantages as it focuses on the major aspects of a system first and fills in the details later.

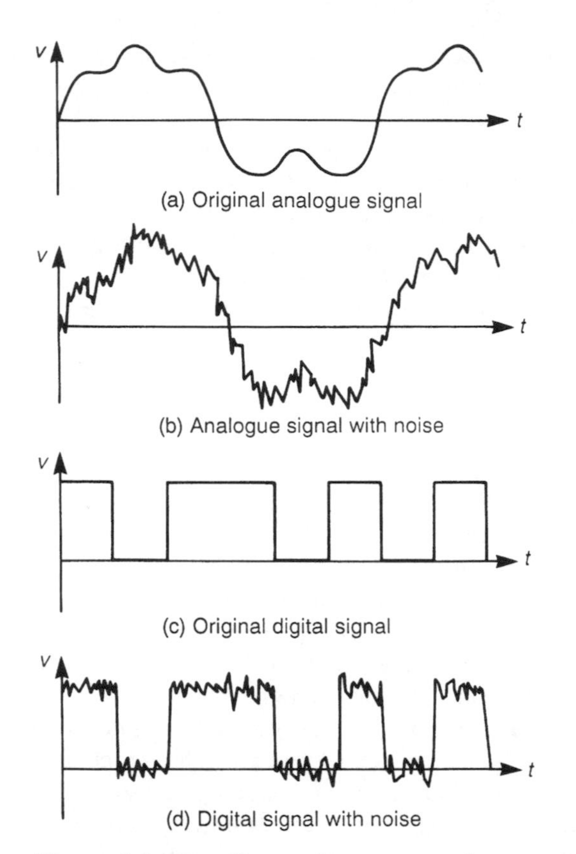

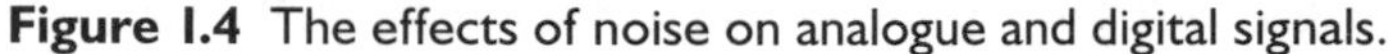

Figure 1.4 The effects of noise on analogue and digital signals.

Customer requirements

Most engineering projects begin with a **customer requirement** for a system to perform a particular task. Often the customer, or the final user of the system, will not be an electronic engineer and indeed may have no engineering or scientific background. Therefore the requirements of the system will often be expressed in non-technical and sometimes imprecise terms.

Top-level specification

The first task is to produce, from the customer requirements, a **top-level specification** for the system. This specification must define precisely what the system is to do in response to all possible inputs, and must also set out any restrictions on the design of the system. These restrictions might include requirements concerning the physical size of the unit, its weight, power consumption or any operating temperature limitations.

Often, a major task in the production of the specification is to determine how the system will appear to someone using it, and how this user will interact with the system.

This involves defining the **man–machine interface** of the system. The characteristics of this interface may play an important part in determining the success or failure of the project.

The production of a system block diagram, identifying all the inputs and outputs of the system, is often of great help in clarifying the operation of the system. A single diagram showing the complete system is much simpler to comprehend than many pages of written explanation of circuit configurations. **It should be noted that the specification sets out what the system is to do but in no way defines how it is to do it.**

When completed the specification must be accepted by the customer as a correct expression of his requirements.

Technology choice

The top-level specification defines the functions to be performed by the system but does not dictate the methods to be used to achieve this goal. It does not, for example, stipulate whether analogue or digital techniques should be used, or even whether the system should be electronic rather than mechanical.

One of the early stages of design is to make an appropriate **technology choice** for the project. This involves deciding whether an electronic solution is suitable for the task and, if so, whether analogue, digital or a mixture of techniques should be used. Assuming that an electronic system is selected for the application in hand, the designer must also decide on an appropriate device technology for the system. In a largely digital system this might involve a choice between building the system using discrete components, standard integrated circuits, some form of logic array or a microprocessor.

The choice of component technology will be affected by a multitude of factors. It requires a detailed knowledge not only of the characteristics of the various devices, but also of the requirements of an individual application. Some component families are faster in operation than others, but component costs tend to increase with speed. Power consumption is another consideration, as are operating voltages, noise levels, physical size, temperature range and development cost.

Choosing the correct method of implementation for an electronic system can make the difference between a successful project and a dismal failure. When you have studied the material in the other chapters of this book you will have a great deal of information to help you in making such decisions.

Top-level design

The process of **top-level design** involves deciding on the techniques to be used to produce the functions described in the specification.

In large projects it will also be necessary to split the work into a number of modules to produce tasks of a manageable size. This operation is sometimes called **system partitioning**. A specification is then produced for each module, allowing an engineer, or group of engineers, to work on an individual module without reference to the other members of the team. This process is essential for projects above a certain size since no engineer can keep in his mind the details of the whole system. Often modules will themselves be subdivided to produce subsystems which will fit on to individual printed circuit boards. This allows boards to be designed and tested separately, thereby simplifying development.

In systems that involve microprocessors it is also necessary to decide which parts of the system will be implemented in hardware and which in software – the so-called **hardware/software trade-off**.

When the system has been divided into suitable sections, the next task is to determine the techniques to be used in each section to achieve its required function, be they hardware or software. This does not involve detailed design of the circuits or generation of programs, but simply determining their overall method of operation.

Detailed design

Once the function and principles of operation of each section are known, the detailed circuitry, or software, can be designed. It is at this stage that the various features and characteristics of the specification are implemented. The ease with which this can be done is greatly affected by the quality of the top-level design.

Module construction and testing

Following the detailed design stage the modules must be constructed and tested individually to ensure that they conform to their specification. If they do not, the design must be changed and the process repeated until its operation is correct. Modules are tested individually rather than as a complete system since this greatly simplifies the testing process, allowing each module to be investigated thoroughly. It also reduces the likelihood of serious problems arising as the result of incorrect operation of the complete system.

System testing

When each of the modules has been proved correct they can be assembled to produce a complete system. The entire system can then be assessed to ensure that it meets the top-level specification.

If faults become apparent at this stage they must be the result either of faults in one of the modules which were not found because of incomplete module testing, or of a mistake in the top-level or module specifications. In any event this will require modification to one or more modules which will then have to be retested individually before returning to testing the complete system.

The final stage of testing is often the process of proving to the customer that the system fulfils the original requirements.

Electronic design aids

In recent years a great many electronic aids have appeared to simplify system design. These often go under the general term of **Electronic Computer Aided Design** or **ECAD** tools.

It should be noted that the emphasis here is on an aid to design, not a tool that does the design for you.

Early tools in this area were little more than electronic drawing boards. These greatly simplified the task of producing circuit diagrams but did little else. Later, printed circuit board (PCB) layout packages appeared which, given an electronic circuit, could enable an engineer to construct a photographic mask to produce a corresponding circuit board. The early PCB layout packages required considerable input from the designer. Although many claimed to perform automatic placement of components and automatic routing of connections, they could not match the skill of a competent engineer. Such tools could greatly reduce the time taken for board design but still required a highly skilled engineer to achieve good results. Modern PCB layout packages are extremely sophisticated and can often be delegated to perform the complete task of placing components and automatically connecting them together. They also achieve considerable time savings, performing complex layouts in minutes rather than the hours taken by a human designer.

While saving considerable amounts of time in laying out PCBs, these packages provide no assistance with the task of designing the original electronic circuit. However, a wide range of tools is available to simplify the design of circuits and to improve their reliability. One of the major classes of tools in this area is the circuit simulator. This package allows the designer to investigate the behaviour of a circuit without constructing it, and can predict its susceptibility to noise and its sensitivity to variations in its components' values. Again it is up to the designer to produce the original circuit, but the simulator can give him a great insight into its characteristics.

In the area of integrated circuit design, ECAD packages are available to perform the tasks of component layout and circuit simulation on the scale required for the production of Very Large Scale Integration (VLSI) circuits. The complexity of these circuits is such that the design of modern VLSI components, such as microprocessors, would be impractical (or even impossible) without such tools.

The development of microprocessor systems relies heavily on the use of electronic tools to aid the writing and simulation of software, the designing and simulation of hardware, software and hardware testing, and many other aspects of system development.

We will be looking in more detail at the use of electronic design tools in the chapter dedicated to system design. **It is important to note, however, that the existence of these tools does not remove the requirement to understand the details of circuit design or the operation of electronic components. The tools complement the designer's skills – they do not replace them.**

The use of circuit simulation within this text

Circuit simulation is an invaluable tool in the design of electronic systems. It is also becoming widely used as an aid to the comprehension of electronics within universities and colleges. As part of their undergraduate programme, most electronics students will take a course which teaches the use of a **Spice circuit simulator** and looks at the mathematical models used to represent the behaviour of various components. Unfortunately, students from other disciplines are less likely to be exposed to this material.

Fortunately, modern packages are very easy to use and powerful circuit simulation software is available free of charge to anyone with access to a personal computer (details of how to obtain such software are given in the preface). This text does not set out to teach the use of any particular package but does provide guidance on how to use simulation as an aid to understanding the material covered. If you are familiar with circuit simulation you should have no difficulty with the various simulation exercises within the text. If you have not used a simulator before you are strongly advised to gain access to a suitable package (such as the free software mentioned above) and to spend a short time experimenting with its features. You are advised to get familiar with these tools *now*, so that you can use them efficiently while you are studying the remainder of the book. There are a number of excellent texts available that provide a basic introduction to these tools. You do not need a detailed understanding of simulation in order to use it with this book. You simply need to be able to load and edit circuits, and to be able to set up the appropriate simulation parameters.

To simplify the use of simulation to investigate the various circuits within this book, **a series of demonstration files is available**. These files remove the need to enter circuits manually and therefore save a great deal of time. Information is also given on the appropriate simulation parameters for each circuit. Details of how to obtain these files are given in the preface. Simulation exercises are marked by a computer icon within the left-hand margin. Where a demonstration file is available, the filename is given below the icon.

FILE 1A

Computer simulation exercise 1.1

To test your ability to use a simulation package, use it to determine the voltage across *R6* in the following circuit.

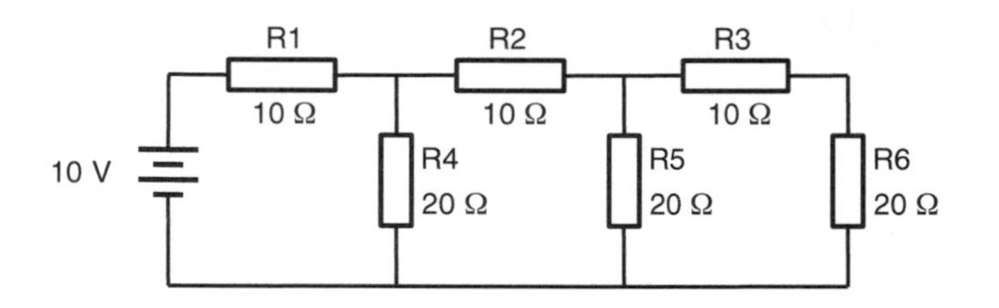

Social and environmental implications

All engineers must take responsibility for the work they perform. Often engineering projects have implications for society and for the environment which must be considered at an early stage in the design. Not the least amongst these considerations are those of safety. It is not within the scope of this text to cover these matters in detail, but it is hoped that having studied the material in this book, an engineer will be better equipped to tackle these problems responsibly.

Key points

- Physical quantities may be either continuous, that is they change smoothly from one value to another, or discrete, meaning that they change abruptly.

- In electronic systems, physical quantities are normally represented by signals.
- Continuous signals are normally referred to as analogue signals, and discrete signals are described as digital signals.
- Some digital signals take one of two levels. Such signals are known as binary signals.
- Useful electronic systems take input signals, process this information and produce appropriate outputs.
- The processing required depends on the application. Common forms of processing include amplification, the adding and subtraction of signals, and filtering.
- Distortion and noise impose major constraints on system performance.
- Both analogue and digital signals are affected by distortion and noise. However, it may be possible to remove their effects from digital signals.
- System design normally follows a 'top-down' methodology.
- Electronic design aids are invaluable. Amongst the most important of these are circuit simulation packages.
- Automated tools complement the designer's skill – they do not replace it.

Design study

An electronic controller is to be designed for a domestic automatic washing machine. This is to have similar characteristics to existing electronic washing machines and should use similar input and output devices.

As a first step in the generation of a detailed specification, produce a block diagram indicating all the inputs and outputs of the controller.

Approach

The task here is simply to identify the input and output devices used in electronic washing machines and to construct a block diagram showing their connections to the controller.

At this stage, the characteristics of the sensors and actuators to be used are not required. This is detailed information which can be added at a later stage of design.

The block diagram below shows a possible representation of the system. There are many acceptable ways of representing the system. It would, for example, be possible to consider the display to be internal to the controller and therefore not show it separately. Similarly, clock circuitry used to time the operation of the machine is considered here to be contained within the controller. It could equally well be considered as an external component.

The block diagram is a good starting point for the generation of the specification, since it shows very clearly the structure of the complete system.

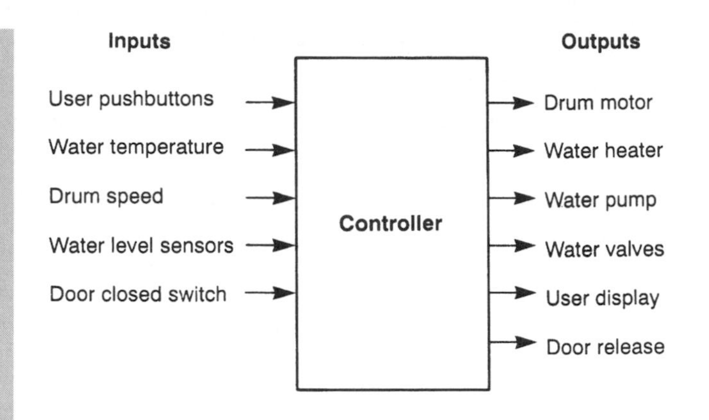

It is interesting to note that the block diagram makes no assumptions concerning the form of the controller. It could be implemented using an electromechanical timer, a microcomputer or a range of other technologies. The top-level specification defines what the system is to do, not how these functions are to be achieved.

Further reading

French M. (1994) *Invention and Evolution, Design in Nature and Engineering*. Cambridge: Cambridge University Press

Friedrichs G. and Schaff A. (1982) *Microelectronics and Society, For Better or For Worse*. Oxford: Elsevier

Herniter M. E. (1996) *Schematic Capture with MicroSim PSpice*, 2nd edn. Englewood Cliffs, NJ: Prentice-Hall

Schumacher E. F. (1993) *Small is Beautiful*. London: Vintage

Exercises

1.1 At the beginning of this chapter several examples were given of physical quantities, namely: temperature, displacement, force, humidity, light intensity, time and mass. Which of these physical quantities are continuous and which are discrete?

1.2 Give three examples, not mentioned in the text, of quantities that are continuous. Similarly give three examples of quantities that are discrete, and three examples of quantities that could be considered to be either.

1.3 What properties of an electrical signal other than its voltage could be used to represent an analogue signal?

1.4 List seven common processing operations which are regularly carried out by electronic circuits.

1.5 In a television, what is the sensor and what are the actuators? What physical quantity is being measured? Is it continuous or discrete?

1.6 In a personal computer, what are the sensors and what are the actuators? What physical quantity is being measured ? Is it continuous or discrete?

1.7 In a digital watch, what is the sensor and what is the actuator? What physical quantity is being measured? Is it continuous or discrete?

1.8 Why is electronic noise more often apparent at times when the input signal is small?

1.9 Give three everyday examples of the effects of electronic noise.

1.10 We have seen that system design is normally performed in a 'top-down' manner. When testing a system during development we normally test individual modules and then progress to testing the complete system. Is this approach 'top-down' or 'bottom-up'?

1.11 Use circuit simulation to determine the voltage across *R3* in the following circuit.

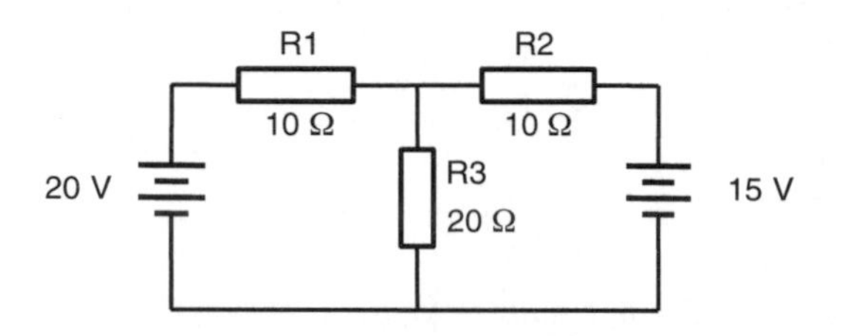

Check your answer by computing the value 'by hand' using nodal analysis, mesh analysis or Ohm's law.

1.12 Draw a block diagram showing the inputs and outputs of an electronic unit to control the operation of a heating system for a large building. The system must switch the heating on and off at appropriate times each day and must provide separate control of temperature in a number of areas. It must also sense the outside temperature to enable it to determine when to switch on the heating in the morning.

Measurement, Sensors and Actuators

Objectives

When you have studied the material in this chapter you should:

- have an understanding of the basic principles of measurement;
- be aware of the primary causes of errors and inaccuracies in measurement;
- understand the limitations on measurement imposed by noise;
- be familiar with the operation and characteristics of a wide variety of sensors and actuators, both analogue and digital, that are frequently used in electronic systems;
- appreciate the requirement for a range of sensors and actuators of different types to meet the needs of varied applications;
- be aware of the forms of the input and output signals associated with various transducers and appreciate the need for electronic processing to allow their interconnection.

Contents

2.1 Introduction

To perform any useful task an electronic system must communicate with the outside world. To do this it uses **sensors** to sense external quantities and **actuators** to control external quantities. Sensors and actuators are often called **transducers** since they usually convert one physical quantity into another. Different transducers convert between a wide range of physical quantities. A mercury-in-glass thermometer, for example, is a transducer that converts temperature into displacement, while a loudspeaker is a transducer that converts electrical signals into sound.

The mercury-in-glass thermometer and the loudspeaker are both examples of transducers that convert one form of analogue quantity into another. Other transducers can be used with digital quantities, producing a digital output from a digital input. An example of such a device is the stepping motor, which converts electrical pulses into discrete movements. The operation of the stepping motor will be described later in this chapter.

Transducers of a third class take an analogue input quantity and generate from it a digital output. In some instances the output is a simple binary representation of the input, as in the thermostat which produces one of two output values depending on whether a temperature is above or below a certain threshold. In other devices the analogue quantity being sensed is represented by a multi-valued output, as in the case of a digital voltmeter where an analogue input signal is represented by a numeric (and therefore discrete) output. Representing an analogue quantity by a digital quantity is, by necessity, an approximation. However, if the number of allowable discrete states is sufficient, the representation may be adequate for a given application. Indeed, in many cases the error caused by this approximation is small compared with the noise or other errors within the system and can therefore be ignored.

Usually the task of defining the nature of the sensors and actuators required for a project is performed at a fairly late stage in the development process, as part of the detailed design. We are considering them at this early stage in the text since an understanding of the nature of the signals generated and required by various transducers is essential for a clear comprehension of the characteristics required of the electronic systems that use them.

Often the objective in sensing a physical quantity is to **measure** it, that is, to obtain a quantitative comparison between it and some standard. Therefore, before moving on to look at sensors and actuators in more detail we will look briefly at the fundamentals of measurement.

2.2 Measurement

The process of measurement involves the comparison of some quantity (the measurand) and a standard.

Sometimes this comparison can be made directly. When measuring distance, for example, it may be possible to compare directly the distance to be measured with a standard metre rule. However, it is often convenient (or necessary) to use a transducer to represent a measurand by a quantity of another form. Very often this alternative form is an electrical quantity.

It is not always possible to have a standard present when making a measurement. This problem is overcome by using a standard to calibrate a measuring system. The standard may then be removed and, provided the characteristics of the system remain unchanged, measurements can be made by comparing an input quantity with the system's memory of the standard. Often calibration itself is achieved not with a standard, but by a model of the standard that is available locally. For example, an engineer may calibrate his measuring system by comparing its readings with those of a laboratory precision voltmeter. The voltmeter was in turn calibrated by its manufacturer who probably adjusted it with the aid of a calibration 'standard' which in turn is an approximation to the actual international standard. The accuracy of the final system depends on the accuracy of the calibration 'standard' used and the care taken when the instruments were set up down the line. In some critical applications it may be necessary to be able to demonstrate the calibration route back to the international standard and to determine the effects of each step on the accuracy of the final system. This leads to the concept of **traceability** which allows the performance of a measuring system to be certified.

No measurement made by a real system is perfect. A small **error** is always present which means that the **result** obtained from the measuring system is always slightly different from the **true** or **actual value** of the quantity being measured. In fact the actual value is never truly known but is only approximated by measurements.

Although all measuring systems have errors associated with them, some have smaller errors than others. We may therefore determine, experimentally or otherwise, the **accuracy** of a given system, this being the maximum expected error. Clearly the more accurate the system the smaller the expected error. Sometimes accuracy is expressed as an absolute value for the maximum expected error. Alternatively it may be expressed as a percentage of the actual value or as a percentage of the full-scale value for the measuring system.

Another useful measure of the performance of a measuring system is its **linearity**. This is best understood by imagining a graph of the measured value of some quantity plotted against its actual value, using equal scales for the two axes. For a perfect measuring system we would expect a straight line at 45 degrees to the horizontal throughout its range. Any real system will deviate from this ideal line to some extent. Its maximum deviation from the ideal line is a measure of its accuracy, but in some applications it is more important to know how linear the system is, rather than how accurate. The linearity of the system is represented by the deviation of the line from the 'best-fit' line through the origin.

2.2.1 Errors

The errors associated with measurement are caused by a number of factors but may be grouped into two main categories.

Random errors

These errors generate different results in repeated readings and may be caused by noise or by environmental variations. Since they are random, such errors are often susceptible to statistical techniques to reduce their effect, for example, by taking the average of a number of readings.

Systematic errors

These errors are constant and are caused by characteristics of the system, such as incorrect calibration, limits of resolution or loading effects where the measuring system changes the value of the quantity it is measuring. Errors in the techniques being used may also generate this kind of error. Systematic errors are not susceptible to statistical techniques. For example, if a system is incorrectly calibrated each reading will be affected and taking a large number of readings will not overcome the problem. A major problem associated with this class of errors is that they cannot be detected by a disparity in the obtained readings. Systematic errors may be reduced by improved calibration or more attention to technique.

2.2.2 Noise

It was noted in Chapter 1 that **noise** is present in all electronic systems. The presence of noise increases the amount of random error associated with any measurement and ultimately places a restriction on the smallest values that may be measured.

When measuring large quantities, the magnitude of the signal representing these quantities will normally be large compared with the noise in the system and the noise will therefore produce a relatively small error in measurement. We say in these cases that the ratio of the magnitude of the signal to that of the noise (the **signal-to-noise ratio**) is high. As smaller and smaller quantities are measured, the relative size of the signal with respect to the noise becomes smaller and thus the percentage error caused by the noise increases. At some point the signal-to-noise ratio drops to a value at which meaningful measurements can no longer be made. This places a limit on the **resolution** or **discrimination** of the measuring system (the resolution or discrimination of a system is the smallest change in the input signal which can be detected).

Thermal noise

All electronic components that possess resistance (in practice this means all real components) generate what is called **thermal noise** or **Johnson noise** as a result of the random, thermally induced motion of their atoms.

Thermal noise has components at all frequencies with equal noise power in all parts of the spectrum. For this reason it is often described as **white noise** by analogy with white light. Although theoretically the noise has an infinite bandwidth (that is, an infinite frequency range), within a given application only the noise that is within the bandwidth of the operation of the system will have any effect. It can be shown that the **noise power** that results from this form of noise is constant for any resistance, and is related to its absolute temperature T and the bandwidth B of the measuring system by the expression

$$P_n = 4kTB \qquad (2.1)$$

where k is Boltzmann's constant ($k \approx 1.3805 \times 10^{-23}$J/K).

For a given resistance of value R, the power dissipated in it is related to the root mean square (r.m.s.) value of the voltage V across it by the expression

$$P = \frac{V^2}{R}$$

which may be rearranged to give

$$V = (PR)^{\frac{1}{2}}$$

Combining this result with that of Equation 2.1 gives an expression for the r.m.s. voltage produced by a resistance of value R as a result of thermal noise. This is

$$V_n\,(\text{r.m.s.}) = (4kTBR)^{\frac{1}{2}} \qquad (2.2)$$

Thus, although the noise power generated by all resistors is equal, the noise voltage increases with the value of the resistance.

The noise obeys a Gaussian amplitude distribution, but since it is random in nature it is not possible to predict its instantaneous value. However, the expression above may be used to determine the r.m.s. noise voltage for a given resistance.

Since all sensors have resistance, all will produce thermal noise which is added to their output signal. This represents a fundamental limit to the resolution of a measuring system.

Clearly noise is a major consideration in any electronic system and we will return to this topic in later chapters.

2.3 Sensors

Almost any physical property of a material that varies in response to some excitation can be used to produce a **sensor**

Commonly used devices include those whose operation is:

- resistive
- inductive
- capacitive
- piezoelectric
- photoelectric
- elastic
- thermal.

Sensors may produce their output in a number of forms. A mercury-in-glass thermometer, for example, produces an output in terms of the length of its column. In this text we are primarily interested in sensors that are used within electronic systems and so we will confine our attention to those sensors that produce an electrical signal.

2.3.1 Temperature

The measurement of temperature is a fundamental part of a vast number of control and monitoring systems ranging from simple temperature regulating systems for buildings to complex industrial process control plants.

Temperature sensors may be divided between those that give a simple binary output to indicate that the temperature is above or below some threshold value, and those that allow temperature measurements to be made.

Binary output devices are effectively temperature operated switches, an example being the **thermostat** which is often based on a **bimetallic strip**.

A large number of different techniques are used for temperature measurement but four main techniques are of interest. These are described below.

Resistive thermometers

The electrical resistance of all conducting materials changes with temperature. Metals generally have low resistivities that vary with their absolute temperature. This allows temperature to be measured by determining the resistance of a sample of the metal and comparing it with its resistance at a known temperature. Typical devices use platinum wire; such devices are known as **platinum resistance thermometers** or **PRT**s.

PRTs can produce very accurate measurements at temperatures from less than −150 °C to nearly 1000 °C to an accuracy of about 0.1 °C or 0.1%. However, they have poor **sensitivity** (that is, a given change in the input temperature produces only a small change in the output signal) because platinum has a low thermal coefficient of resistivity. An advantage of these devices is that since they are extremely linear and have a resistance that is directly proportional to their absolute temperature (to a good approximation), they may be calibrated using only a single set point. A typical PRT might have a resistance of 100 Ω at 0 °C which increases to about 140 Ω at 100 °C.

Figure 2.1(a) shows a typical PRT element. The device shown is 30 mm long, 4 mm wide and 0.82 mm thick. Such a sensor would normally be bonded to a flat surface and would have wires soldered to its contacts. PRTs are also available in a sheathed form as shown in Figure 2.1(b). This has wires attached to simplify installation.

Thermistors

Like PRTs, these devices also change their resistance with temperature. However, they use materials with high thermal coefficients of resistance to give much improved sensi-

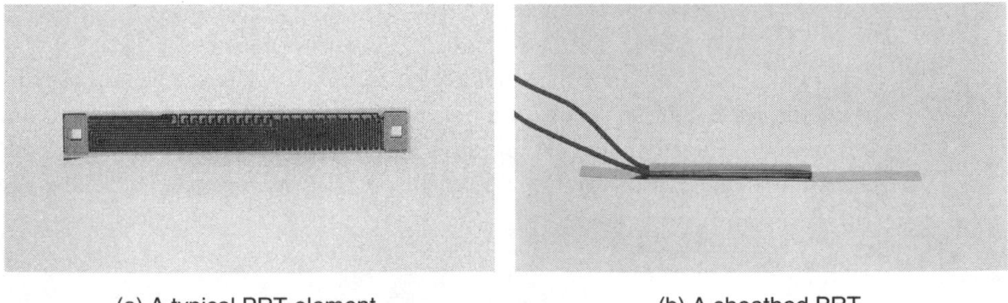

(a) A typical PRT element (b) A sheathed PRT

Figure 2.1 Platinum resistance thermometers (PRTs).

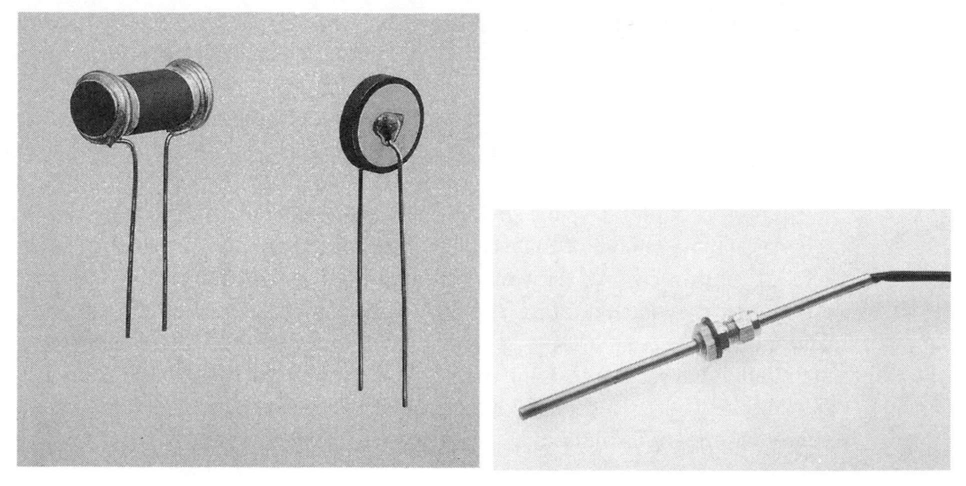

(a) Typical thermistors

Figure 2.2 Thermistors.

tivity. The materials chosen are ceramic-like semiconductors, usually oxides of cobalt, manganese or nickel. Most thermistors have a **negative temperature coefficient** of resistance (**NTC**), that is, their resistance goes down with increasing temperature, although devices with a **positive temperature coefficient** (**PTC**) are also available. A typical NTC device might have a resistance of 5 kΩ at 0 °C and 100 Ω at 100 °C. Thermistors are inexpensive and robust but are very non-linear and often suffer from great variability in their nominal value between devices. It is worth noting, however, that since the non-linearities are a result of the physics of the materials used, they are repeatable. For applications where high accuracy is required it is possible individually to select or calibrate the devices and to process their outputs to 'linearize' the readings.

Figure 2.2(a) shows two examples of the many diverse forms of thermistors that are available. Figure 2.2(b) shows a probe containing a thermistor.

Thermocouples

When two electrical conductors of different materials are joined together an electrical potential is created between them. Normally these **thermoelectric potentials** cancel out around the circuit and are thus not observed. However, since these potentials vary with temperature, if one junction is heated with respect to another the effects no longer cancel. This effect is utilized in the thermocouple in which two dissimilar materials (such as copper and constantan) are joined by twisting or welding to form two junctions. One junction is kept at a constant temperature (the reference junction) and the other (the measuring junction) is used as a temperature sensor. The temperature difference between the two junctions may now be determined from the difference in the thermoelectric potentials. Knowing the reference temperature, the temperature of the measuring junction can be calculated. In many applications it is inconvenient to keep the reference junction at a fixed temperature. An alternative, and more usual, approach is simply to measure the temperature of the reference junction and use this value when calculating the temperature of the measuring junction.

It is worth noting that this effect can be produced unintentionally within electronic circuits which contain conductors of more than one material. This can cause unexpected errors in otherwise high precision measurements.

Thermocouples have the advantages that the junctions can be made small and robust. Also a suitable choice of materials enables temperatures from about –200 °C to about 1800 °C to be measured. However, the thermoelectric voltages are very small requiring very sensitive, and therefore expensive, amplifiers to make accurate measurements. Thermocouples are therefore generally used only where their wide temperature range is essential or where a large number of sensors is required. In the latter case, a single amplifier may be switched between a number of inexpensive sensing junctions, producing a very cost-effective solution.

pn junction

Semiconductor materials may be doped with small amounts of specific impurities to produce an excess of positive charge carriers (a *p-type* semiconductor) or an excess of negative charge carriers (an *n-type* semiconductor). A junction between materials of these two types has very interesting properties – as in a semiconductor diode. The properties and uses of semiconductor devices will be discussed in more detail in Chapter 5.

At a fixed current the voltage across a typical semiconductor junction changes by about 2 mV per °C. Devices based on this property use additional circuitry to produce an output voltage or current that is directly proportional to the junction temperature. Typical devices might produce an output voltage of 1 mV per °C, or an output current of 1 μA per °C, above 0 °C. These devices are inexpensive, linear and easy to use, but are limited to a temperature range from about –50 °C to about 150 °C by the semiconductor materials used.

2.3.2 Light

Sensors for measuring light intensity fall into two main categories: those which generate electricity when illuminated, the magnitude of which may be used as a measure of the light intensity, and those whose properties (for example, their resistance) change under the influence of light. We will consider examples of both these classes of device.

Photovoltaic

Light falling on a *pn* junction produces a voltage and can therefore be used to generate power from light energy. This principle is used in solar cells. On a smaller scale, **photodiodes** can be used to measure light intensity since they produce an output voltage which depends on the amount of light falling on them. A disadvantage of this method of measurement is that the voltage produced is not linearly related to the incident light intensity. Figure 2.3 shows examples of typical photodiode light sensors.

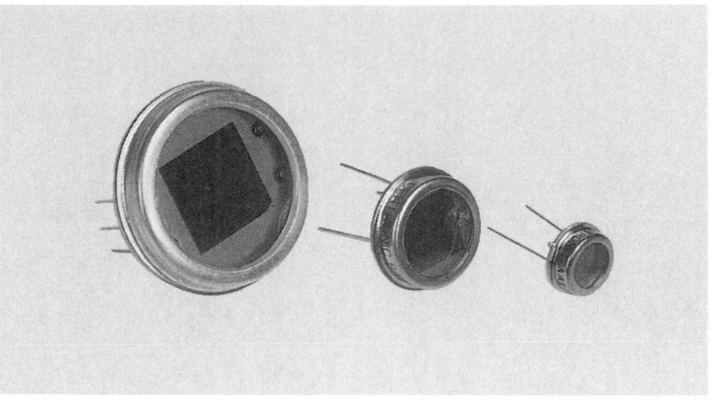

Figure 2.3 Photodiode light sensors.

Photoconductive

Photoconductive sensors do not generate electricity, but their conduction of electricity changes with illumination. The photodiode described above as a photovoltaic device may also be used as a photoconductive device. If a photodiode is reverse-biased by an external voltage source, in the absence of light it will behave like any other diode and conduct only a negligible leakage current. However, if light is allowed to fall on the device, charge carriers will be formed in the junction region and a current will flow. The magnitude of this current is proportional to the intensity of the incident light making it more suitable for measurement than the photovoltaic arrangement described earlier.

The currents produced by photodiodes in their photoconductive mode are very small. An alternative is to use a **phototransistor** which combines the photo-conductive properties of the photodiode with the current amplification of a transistor to form a device with much greater sensitivity. The operation of transistors will be discussed in later chapters.

A third class of photoconductive device is the **light dependent resistor** or **LDR**. As its name implies, this is a resistive device which changes its resistance when illuminated. Typical devices are made from materials such as cadmium sulphide (CdS) which have a much lower resistance when illuminated. One advantage of these devices in some applications is that they respond to different wavelengths of light in a manner similar to the human eye. Unfortunately their response is very slow, taking perhaps a hundred milliseconds to respond to a change in illumination, compared with a few microseconds, or less, for the semiconductor junction devices.

Opto-switches

In addition to sensors that measure light intensity there are also a large number of sensors that use light to measure other quantities, such as position, motion and temperature. One of the most common of these devices is the **opto-switch** which, as its name suggests, is a light operated switch.

The opto-switch consists of a light sensor, usually a phototransistor, and a light source,

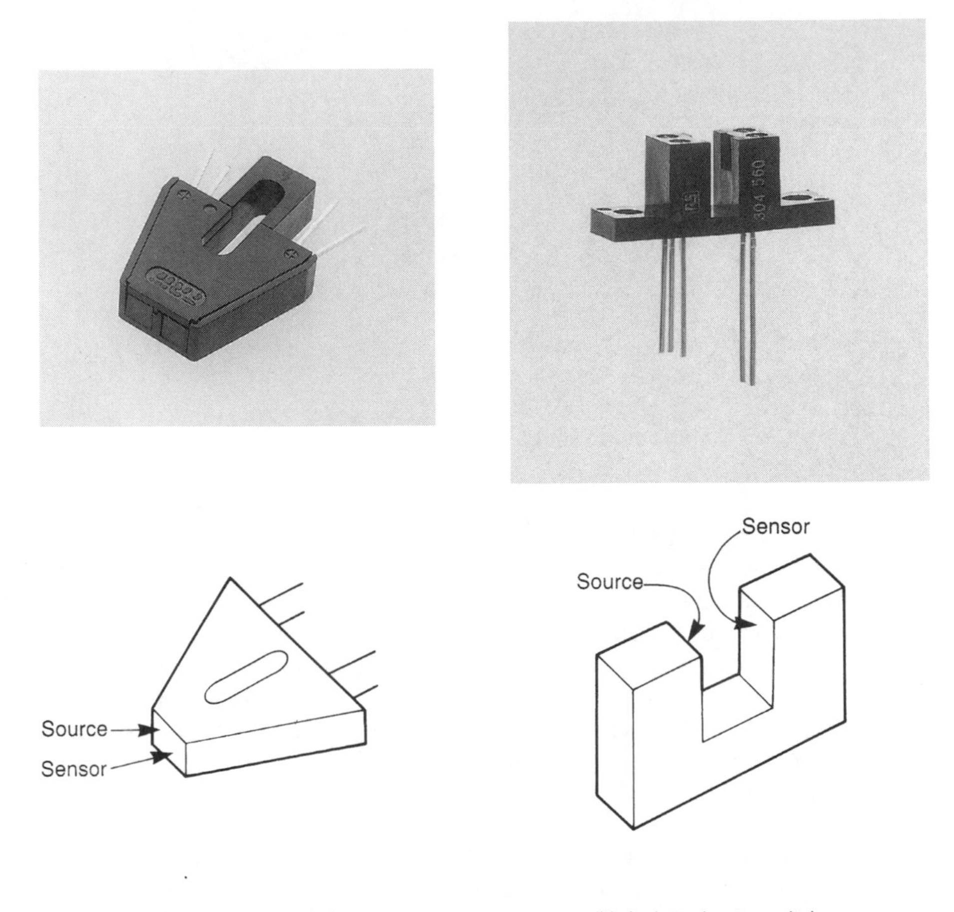

(a) A reflective opto-switch (b) A slotted opto-switch

Figure 2.4 Reflective and slotted opto-switches.

usually a **light emitting diode** (**LED**s are described in Section 2.4.2), housed within a single package. Two physical arrangements are widely used, as illustrated in Figure 2.4.

Figure 2.4(a) shows a reflective device in which the light source and sensor are mounted adjacent to each other on one face of the unit. The presence of a reflective object close to this face will cause light from the source to reach the sensor causing current to flow in the output circuit. Figure 2.4(b) shows a slotted opto-switch in which the source and sensor are arranged to face each other on either side of a slot in the device. In the absence of any object within the slot, light from the source will reach the sensor and this will produce a current in the output circuit. If the slot is obstructed, the light path will be broken and the output current will be reduced.

Although opto-switches may be used with external circuitry to measure the current flowing and thus to determine the magnitude of the light reaching the sensor, it is more common to use them in a binary mode. In this arrangement the current is compared with some threshold value to decide whether the opto-switch is 'ON' or 'OFF'. In this way the switch detects the presence or absence of objects, the threshold value being adjusted to vary the sensitivity of the arrangement.

We will consider some applications of the opto-switch later in this section.

2.3.3 Force

Strain gauge

The resistance between opposite faces of a rectangular piece of uniform electrically conducting material is proportional to the distance between the faces and inversely proportional to its cross-sectional area. The shape of such an object may be changed by applying an external force to it. The term **stress** is used to define the force per unit area applied to the object, and the term **strain** refers to the deformation produced. In a strain gauge, an applied force deforms the sensor, increasing or decreasing its length (and its cross-section) and therefore changing its resistance. Figure 2.5 shows the construction of a typical device.

The gauge is in the form of a thin layer of resistive material arranged to be sensitive to deformation in only one direction. The long thin lines of the sensor are largely responsible for the overall resistance of the device. Stretching or compressing the gauge in the direction shown will extend or contract these lines and will have a marked effect on the total resistance. The comparatively thick sections joining these lines contribute little to the overall resistance of the unit. Consequently, deforming the gauge perpendicular to the direction shown will have little effect on the total resistance of the device.

In use, the gauge is bonded to the surface in which strain is to be measured. The fractional change in resistance is linearly related to the applied strain. Figure 2.5 also shows how a strain gauge may be used to measure the strain induced in a beam when a force is applied. If it is bonded to a structure with a known stress-to-strain characteristic, the gauge can be used to measure force. Thus it is often found at the heart of many force transducers or **load cells**. Similarly, strain gauges may be connected to diaphragms to produce **pressure sensors**.

The above is a simplified description of the operation of a strain gauge. In fact much of the change in resistance is due to piezo-resistive effects which will not be discussed here. The mode of operation of the device is less important than its overall characteristics, which have been described.

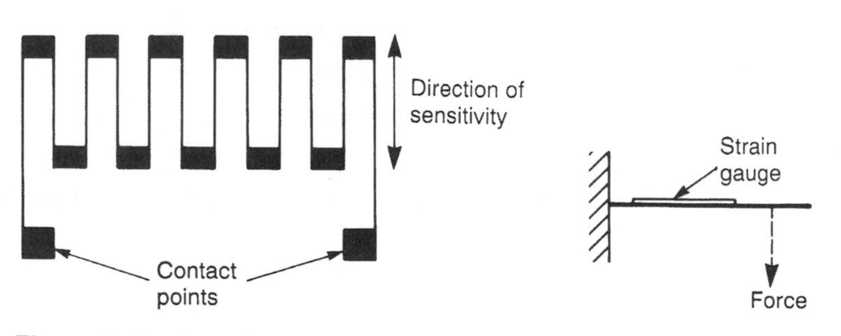

Figure 2.5 A strain gauge.

Piezoelectric

Piezoelectric materials have the characteristic that they generate an electrical output when subjected to mechanical stress. Unfortunately, the output is not a simple voltage proportional to the applied force, but an amount of electrical charge which is related to the applied stress. For most applications this requires some electronic circuitry to convert the signal into a more convenient voltage signal. This is not in itself difficult, but such circuits tend to suffer from 'drift', that is, a gradual increase or decrease in the output. For this reason, piezoelectric transducers are more commonly used for measuring variations in forces, rather than their steady value.

2.3.4 Displacement

As with many classes of sensors, both analogue and digital types are used.

Potentiometers

Resistive potentiometers are amongst the most common position transducers and most people will have encountered them as the controls used in radios and other electronic equipment. Potentiometers may be angular or linear, consisting of a length of resistive material with an electrical terminal at either end and a third terminal connected to a sliding contact on to the resistive *track*. When used as a position transducer, a potential is placed across the two end terminals and the output is taken from the terminal connected to the sliding contact. As the sliding contact moves the output voltage changes between the potentials on either end of the track. Generally there is a linear relationship between the position of the slider and the output voltage.

Older types of potentiometer use a film of carbon for the resistive track, but these suffer from electrical noise, poor linearity and short life. Other devices use resistive wire for the track, either in the form of a straight wire (a slide-wire potentiometer) or wound around an insulating former (a wire-wound potentiometer). Resistive wire devices have good linearity but have limited life and are generally electrically noisy. Slide-wire techniques are generally only suitable for measuring quite large distances while wire-wound devices lack resolution as the slider jumps from one winding to the next. In recent years conducting plastics have been used for potentiometers. These produce a smooth, low-friction track with low noise, excellent resolution and long life. Early conducting plastic devices suffered from poor linearity but more modern devices combine conducting plastic and wire-wound techniques to produce potentiometers which combine the best properties of both methods.

Inductive sensors

The inductance of a coil is affected by the proximity of ferromagnetic materials, an effect that is used in a number of position sensors. One of the simplest of these is the inductive **proximity sensor** in which the proximity of a ferromagnetic plate is determined by measuring the inductance of a coil. Figure 2.6(a) illustrates the form of such a device, while Figure 2.6(b) shows typical sensors.

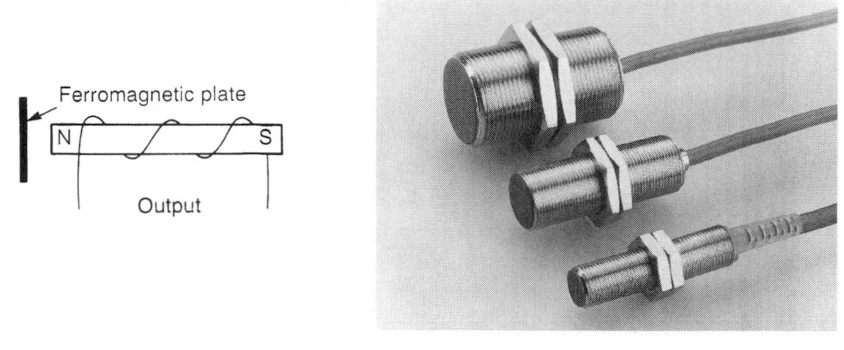

(a) Principle of operation

(b) Typical sensors

Figure 2.6 Inductive displacement or proximity sensors.

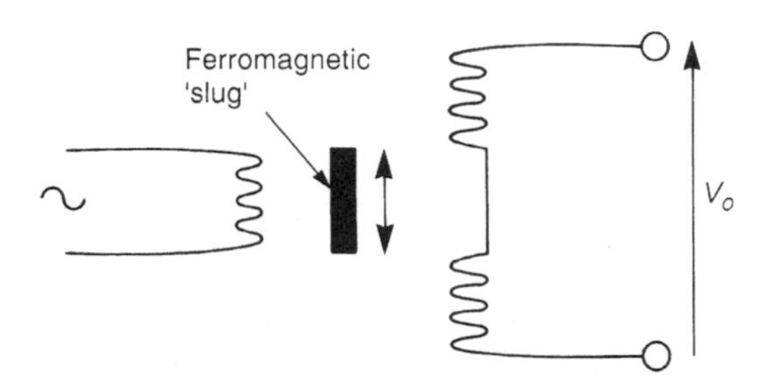

(a) Principle of operation

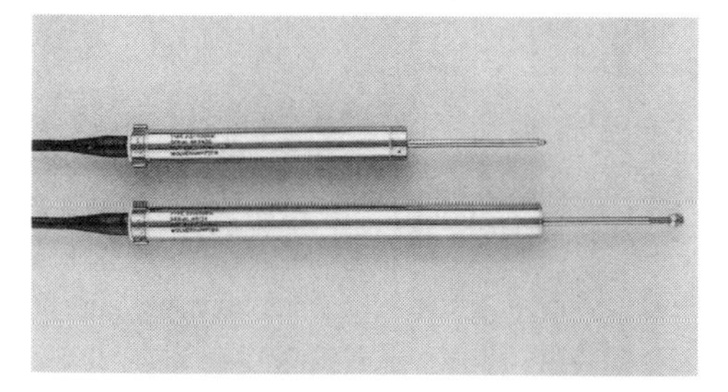

(b) Typical devices

Figure 2.7 Linear variable differential transformers.

Another inductive position transducer is the **linear variable differential transformer (LVDT)** which uses three coils, as shown in Figure 2.7.

The three coils are wound on a hollow tube and are arranged to produce a transformer, one coil forming the primary and the other two coils forming identical secondaries positioned symmetrically either side of the primary. The two secondary coils are connected in series in such a way that their output voltages are out of phase and so cancel. If a sinusoidal voltage is applied to the primary, the voltages induced in the secondary coils will cancel each other and the resultant output will be zero. This arrangement is turned into a useful transducer by the addition of a movable 'slug' of a ferromagnetic material, such as soft iron, within the tube. The presence of the ferromagnetic material increases the mutual inductance between the primary and the secondary coils and thus increases the magnitude of the voltages induced in the secondary coils. If the slug is positioned centrally with respect to the coils it will affect the two secondary coils equally and the output voltages will still cancel. If, however, the slug is moved slightly to one side or the other it will increase the coupling to one secondary more than to the other. The arrangement will now be out of balance and an output voltage will be produced. The greater the movement of the slug from its central position, the greater the resulting output signal. The output signal is in the form of an alternating signal the magnitude of which represents the offset from the central position and the phase of which indicates the direction in which the slug is displaced. A simple circuit may be used to convert this alternating signal into a more convenient DC signal.

LVDTs can be constructed with ranges of a few metres down to a fraction of a millimetre. They typically have a resolution of about 0.1% of their full range (or better if multiple ranges are used) and good linearity. Unlike resistive potentiometers they do not require a frictional contact and so can have a very low operating force and a long life.

Switches

The simplest digital displacement sensors are switches. These are used in many forms and may be manually operated or connected to a mechanism of some form. Manually operated switches include toggle switches, which are often used as power on/off switches on electrical equipment, and momentary action push-button switches as used in computer keyboards. It may not be immediately apparent that switches of this type are position sensors, but clearly they output a value dependent on the position of the input lever or surface, and are therefore binary sensors.

When a switch is connected to some form of mechanism its action as a position sensor becomes more obvious. A common form of such a device is the **microswitch** which consists of a small switch mechanism attached to a lever or push-rod allowing it to be operated by some external force. Microswitches are often used as **limit switches** which signal that a mechanism has reached the end of its safe travel. Such an arrangement is shown in Figure 2.8(a). Switches are also used in a number of specialized position measuring applications, such as liquid level sensors. One form of this sensor is shown in Figure 2.8(b). Here the switch is operated by a float which rises with the level of a liquid until it reaches some specific level.

Most switches have two contacts which are electrically connected when the switch is in one state (the closed state) and are disconnected (or open circuit) when the switch is in the other state (the open state). This arrangement can be used to generate binary electrical signals simply by adding a voltage source and a resistance, as shown in Figure 2.9. When the switch is closed, the output is connected to the zero volts line and therefore the output voltage V_0 is zero. When the switch is open the output is no longer connected to the zero volts line, but is connected through the resistance R to the voltage supply V. The output voltage will therefore be equal to the supply voltage minus any voltage drop across the resistance. This voltage drop will be determined by the value of the resistance

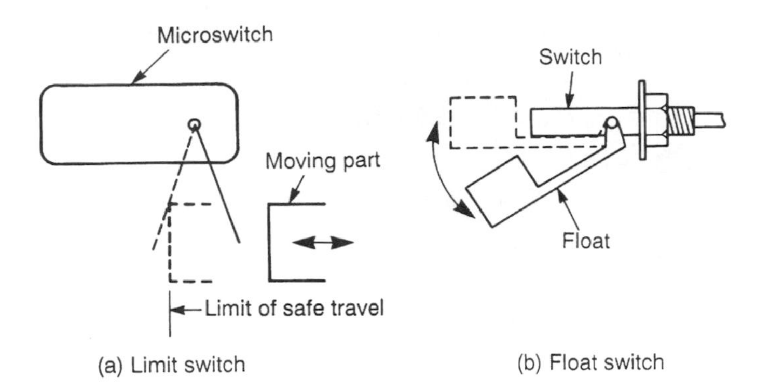

Figure 2.8 Switch position sensors.

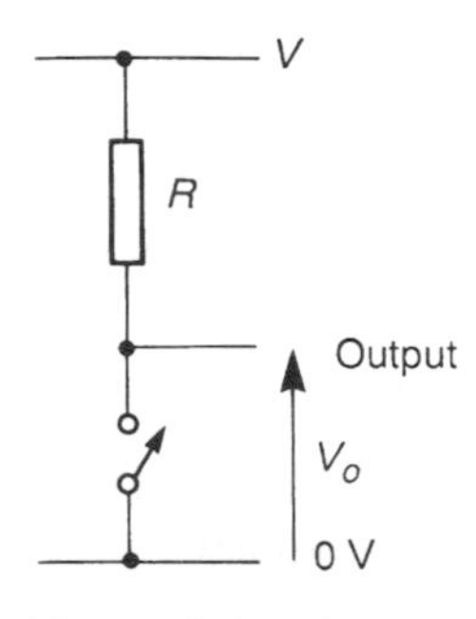

Figure 2.9 Generating a binary electrical signal using a switch.

R and the current flowing into the output circuit. If the value of *R* is chosen such that this voltage drop is small compared with *V*, we can use the approximation that the output voltage is zero when the switch is closed and *V* when it is open. The value chosen for *R* clearly affects the accuracy of this approximation; we will be looking at this in the next chapter when we look at equivalent circuits

One problem experienced by all mechanical switches is that of **switch bounce**. When the moving contacts within the switch come together they have a tendency to bounce, rather than to meet cleanly. Consequently the circuit is first made, then broken, then made again, sometimes several times. This can cause severe problems, particularly if contact closures are being counted. Although good mechanical design can reduce this problem it cannot be eliminated, making it necessary to overcome this problem in other ways. Several electronic solutions are possible, one of which will be described in Section 10.2.1. It is also possible to tackle this problem using computer software techniques in systems that incorporate microcomputers.

Absolute position encoders

Figure 2.10 illustrates the principle of a simple linear absolute position encoder.

A pattern of light and dark areas is printed on to a strip and is detected by a sensor which

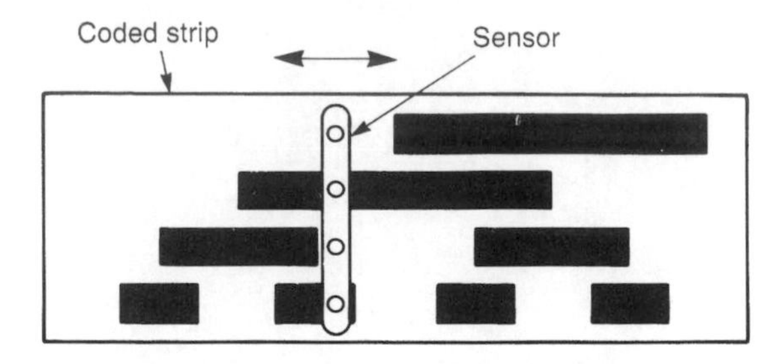

Figure 2.10 An absolute position encoder.

moves along it. The pattern takes the form of a series of lines which each alternate between light and dark. It is arranged so that the combination of light and dark areas on the various lines is unique at each point along the strip. The sensor, which may be a linear array of phototransistors or photodiodes, one per line, picks up the pattern and produces an appropriate electrical signal which can be decoded to determine the sensor's position. The combination of light and dark lines at each point represents a **code** for that position. The choice of codes, and their use, will be discussed in more detail in Section 9.5.4.

Since each point on the strip must have a unique code, the number of distinct positions along the strip which can be detected is determined by the number of lines in the pattern. For a sensor of a given length, increasing the number of lines in the pattern increases the resolution of the device but also increases the complexity of the detecting array and the accuracy with which the lines must be printed.

Although linear absolute encoders are available, the technique is more commonly applied to angular devices. These often resemble rotary potentiometers, but have a coded pattern in a series of concentric rings in place of the conducting track, and an array of optical sensors in place of the wiper. Position encoders have excellent linearity and a long life, but generally have poorer resolution than potentiometers and are usually more expensive.

Incremental position encoders

The incremental encoder differs from the absolute encoder in that it has only a single detector which scans a pattern consisting of a regular series of stripes, perpendicular to the direction of travel. As the sensor moves over the pattern, the sensor will detect a series of light and dark regions. The distance moved can be determined by counting the number of transitions. One problem with this arrangement is that the direction of motion cannot be ascertained as motion in either direction generates similar transitions between light and dark. This problem is overcome by the use of a second sensor, slightly offset from the first. The direction of motion may now be determined by noting which sensor is first to detect a particular transition. This arrangement is shown in Figure 2.11 which also illustrates the signals produced by the two sensors for motion in each direction.

In comparison with the absolute encoder, the incremental encoder has the disadvantage that external circuitry is required to count the transitions, and that some method of resetting this must be provided to give a reference point or datum. However, the device

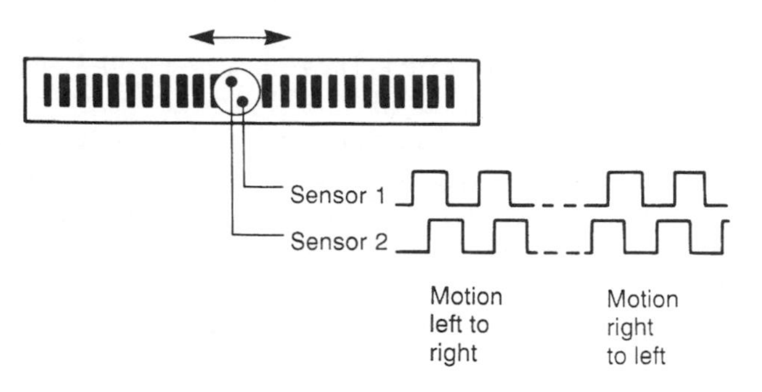

Figure 2.11 An incremental position encoder.

Figure 2.12 An angular incremental position encoder.

is simple in construction and can provide high resolution. Again, both linear and angular devices are available. Figure 2.12 shows a small angular incremental position encoder.

Optical gratings

The incremental encoder described above relies on counting individual lines in a pattern of stripes. To measure very small displacements these lines must be very close together, making them difficult to detect. One approach to this problem is to use optical gratings to simplify the task.

Optical gratings can take many forms, a simple version being formed by printing a pattern of opaque stripes on a transparent film. The stripes are parallel, and have lines and spaces of equal widths. If one piece of this film is placed on top of another, the pattern produced will depend on the relative positions and orientations of the stripes on the two films. If the lines of each sheet are parallel, the overall effect will depend on the relative positions of the stripes. If the lines of each coincide precisely, half the sheet will be transparent and half opaque. If, however, the lines of each sheet are side by side, the combination will be completely opaque.

Imagine now that one grating is placed on a white background and that a second grating is placed on top of the first such that the lines of each are parallel. Movement of one sheet perpendicular to the direction of the stripes will result in an overall pattern which alternates from dark to light as the stripes pass over each other. A light sensor placed above the gratings will detect these variations. By counting the number of transitions, the distance moved can be determined. Since one bright 'pulse' is produced each time the lines coincide, the distance travelled is the product of the number of pulses counted and the spacing of the lines. These bands of light and dark regions are known as **moiré fringes**. The closer together the lines, the greater will be the resolution of the distance measurement.

The measurement technique thus far described suffers from the same problem as the incremental encoder described above, in that if a single sensor is used, motion in either direction generates similar signals at the sensor. This problem can be tackled in two ways. One of the gratings can be rotated slightly so that the lines of each are no longer parallel. Relative motion will now produce diagonal bands of light which move up, for motion in one direction, and down, for motion in the other direction. Alternatively, the line spacing of one of the sheets can be changed so that it is slightly different from the other. When placed together, as before, this will produce a wave of light and dark bands called **Vernier fringes** in a direction perpendicular to the stripes. As the films move with respect to each other, these bands move in a direction determined by the direction of motion. In both methods a second sensor is used to allow the direction of motion to be detected, and in each case signals are produced which are similar to those produced by the incremental encoder, as shown in Figure 2.11.

Practical displacement measuring systems use gratings which are produced photographically allowing a very high resolution. A typical application might use a linear array fixed to a static component and a small moving sensor containing a grating assembly with integrated light sensors. Line spacings down to 1 μm are readily obtainable, although line spacings of 10 μm to 20 μm are typical, with interpolation being used to obtain a measurement resolution of about 1 μm. Gratings of this type are produced in lengths of up to about 1 m, but may be joined (with some loss of accuracy) to produce greater lengths. The ability to measure distances of the order of a metre to a resolution of the order of a micron makes the use of gratings extremely attractive in some demanding applications. However, the high cost of the gratings and sensors limits their use.

Other counting techniques

Incremental encoders and optical gratings employ event counting to determine displacement. Several other techniques use this method. Figure 2.13 shows two examples.

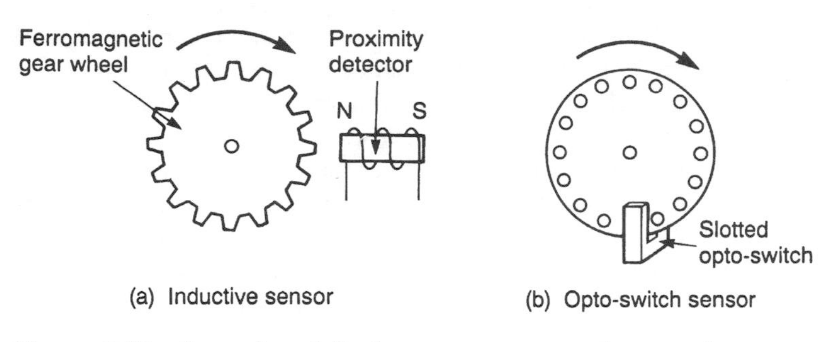

Figure 2.13 Examples of displacement sensors using counting.

Figure 2.13(a) shows a technique which uses an inductive proximity sensor as described earlier in this section. Here a ferromagnetic gear wheel is placed near the sensor; as the wheel rotates the teeth pass close to the sensor, increasing its inductance. The sensor can therefore detect the passage of each tooth and thus determine the distance travelled. A great advantage of this sensor is its tolerance to dirty environments.

Figure 2.13(b) shows a sensor which uses the slotted opto-switch discussed in Section 2.3.2. This method uses a disc which has a number of holes or slots equally spaced around its perimeter. The disc and opto-switch are mounted such that the edge of the disc is within the slot of the switch. As the disc rotates the holes or slots cause the opto-switch to be periodically opened and closed producing a train of pulses with a frequency determined by the speed of rotation. The angle of rotation can be measured by counting the number of pulses. A similar method uses an inductive proximity sensor in place of the opto-switch, and a ferromagnetic disc.

Rangefinders

Measurement of large distances usually requires a non-contact method. Both passive systems (which simply observe their environment) and active systems (which send signals out into the environment) are available. Passive techniques include optical triangulation methods in which two slightly displaced sights are aligned on a common target. The angular difference between the two sights can then be measured using one of the angular sensors described above. Geometry is then used to calculate the distance between the sights and the target. This method is employed in rangefinding equipment used for surveying. Active systems transmit either sound or electromagnetic energy and detect the energy reflected from a distant object. By measuring the time taken for the energy to travel to the object and back to the transmitter, the distance between them may be determined. Because the speed of light is so great, some optical systems use the phase difference between the transmitted and received signals, rather than time-of-travel, to determine the distance.

2.3.5 Motion

In addition to the measurement of displacement, it is often necessary to determine information concerning the motion of an object, such as its velocity or acceleration. These quantities may be obtained by differentiation of a position signal with respect to time, although such techniques often suffer from noise since differentiation tends to amplify high frequency noise present in the signal. Alternatively, velocity and acceleration can be measured directly using a number of sensors.

The counting techniques described earlier for the measurement of displacement can also be used for velocity measurement. This is achieved by measuring the frequency of the waveforms produced instead of counting the number of pulses. This gives a direct indication of speed. In fact many of the counting techniques outlined earlier are more commonly used for speed measurement than for measurement of position. In many applications the direction of motion is either known or is unimportant and these techniques often provide a simple and inexpensive solution.

A range of other velocity sensors exist for different applications. A **tacho-generator** can be used to measure rotational speed. This is a small DC generator which produces a voltage proportional to its speed of rotation. Linear motion can be measured by converting it to a rotational movement (for example, by using a friction wheel running along a flat surface) and applying this to a tacho-generator. Alternatively, there are several methods for measuring linear motion directly, such as those employing the **Doppler effect** as used in 'radar' speed detectors. Here a beam of sound or electromagnetic radiation is directed at the moving object and the reflected radiation is detected and compared with the original transmission. The difference between the frequencies of the outgoing and reflected waveforms gives a measure of the relative speed of the object and the transducer. The velocity of fluids may be measured in many ways including pressure probe, turbine, magnetic, sonic and laser methods. These techniques are highly specialized and will not be discussed here.

Direct measurement of acceleration is made using an **accelerometer**. Most accelerometers make use of the relationship between force, mass and acceleration:

$$Force = mass \times acceleration$$

A mass is enclosed within the accelerometer. When the device is subjected to an acceleration the mass experiences a force which can be detected in a number of ways. In some devices a force transducer, such as a piezoelectric sensor or a strain gauge, is incorporated to measure the force directly. In others, springs are used to convert the force into a corresponding displacement which is then measured with a displacement transducer. Because of the different modes of operation of the devices, the form of the output signal also varies.

2.3.6 Sound

A number of techniques are used to detect sound.

Carbon microphones

Carbon microphones are one of the oldest and simplest forms of sound detector. Sound waves are detected by a *diaphragm* which forms one side of an enclosure containing carbon particles. Sound waves striking the diaphragm cause it to move, compressing the carbon particles to a greater or lesser degree and thus affecting their resistance. Electrodes apply a voltage across the particles and the resulting current thus relates to the sound striking the device.

Capacitive microphones

Capacitive microphones are similar in operation to the carbon microphone described above except that movement of the diaphragm causes a variation in capacitance rather than resistance. This is achieved by arranging that motion of the diaphragm changes the separation of two plates of a capacitor, thereby varying its capacitance.

Moving coil microphones

A moving coil microphone consists of a permanent magnet and a coil connected to a diaphragm. Sound waves move the diaphragm which causes the coil to move with respect to the magnet, thus generating an electrical signal.

Piezoelectric microphones

The piezoelectric force sensor described earlier can also be used as a microphone. The diaphragm is made of piezoelectric material which is distorted by sound waves producing a corresponding electrical signal. This technique is often used for **ultrasonic sensors** which are used over a wide range of frequencies, sometimes up to many megahertz.

2.3.7 Sensors – a summary

It is not the purpose of this section to provide an exhaustive list of all possible sensors. Rather it sets out to illustrate some of the important classes of sensor that are available and to show the ways in which they provide information. It will be seen that some sensors generate output currents or voltages related to changes in the quantity being measured. In doing so they extract power from the environment and can deliver power to external circuitry (although usually the power available is small). Examples of such sensors are thermocouples, photovoltaic sensors, piezoelectric sensors, tacho-generators and moving coil microphones.

Other devices do not deliver power to external circuits but simply change their physical attributes, such as resistance, capacitance or inductance, in response to variations in the quantity being measured. Examples include resistive thermometers, photoconductive sensors, potentiometers, inductive position transducers, strain gauges and carbon microphones. When using this form of sensor, external circuitry must be provided to convert the variation in the sensor into a useful signal. Often this circuitry is very simple. For example, if a constant voltage is placed across the outer terminals of a linear potentiometer, the voltage produced on the centre wiper contact is directly proportional to the displacement of the wiper and hence to the input displacement. Similarly, since the resistance of a platinum resistance thermometer (PRT) varies linearly with its absolute temperature, if an external circuit passes a constant current through the device, the voltage across it will also be linearly related to its absolute temperature.

Unfortunately, some sensors do not produce an output that is linearly related to the quantity being measured (for example, a thermistor). In these cases it may be necessary to overcome the problem by using electronic circuitry or processing to compensate for any non-linearity. This process is called **linearization**. The ease or difficulty of linearization depends on the characteristics of the sensor and the accuracy required.

Example 2.1 Selecting appropriate position sensors for a computer mouse

In this section we have looked at a number of displacement and motion sensors. Armed with this information, we will select a suitable method of determining the displacement of a mouse for use as a computer pointing device. The resolution of the sensing arrangement should be such that the user can select an individual pixel (the smallest definable point within the display). A typical screen might have a 1024 by 768 pixel displays, although more sophisticated displays may have a resolution several times greater than this. Movement of the cursor from one side of the screen to the other should require a movement of the mouse of a few centimetres (the sensitivity of the mouse is often selectable using software within the computer).

A mouse senses motion using a small rubber ball that projects from its base. As the mouse is moved over a horizontal surface the ball rotates about two perpendicular axes, and this motion is used to determine the position of a cursor on a computer screen.

We have looked at several sensors that may be used to measure angular position. These include simple potentiometers and position encoders. Sensing the *absolute* position of the ball (and hence the mouse) could represent a problem since for high performance displays this could require a resolution of better that one part in 2000. Sensors with

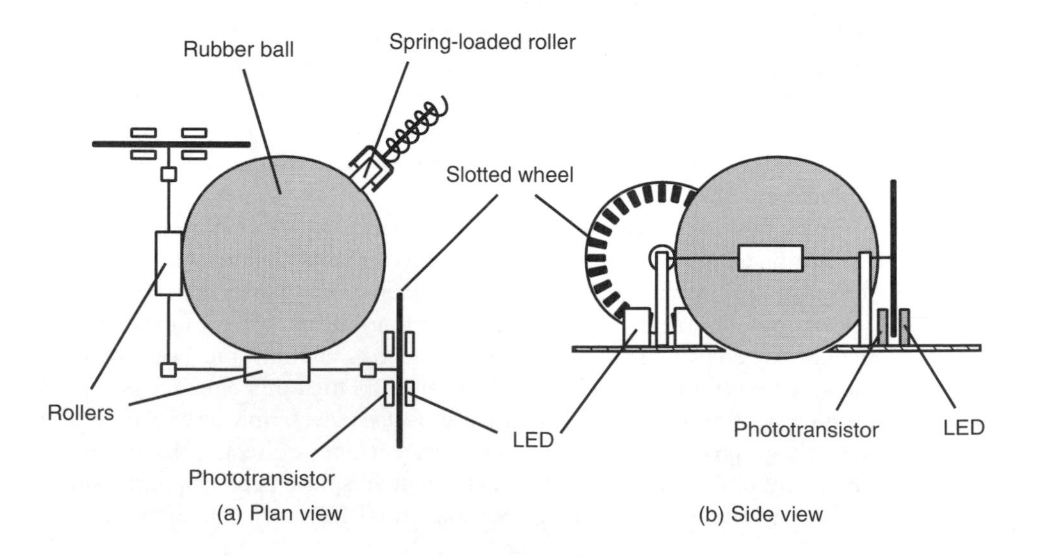

(a) Plan view

(b) Side view

such a high resolution are often expensive and physically large. In this application it is probably more appropriate to sense *relative* motion of the mouse. This reduces the complexity of the sensing mechanism and also means that the mouse is not tied to a fixed absolute position.

Measurement of the relative motion of the rubber ball suggests the use of some form of incremental sensor. This could use a proprietary incremental encoder, but because this is a very high volume application, it is likely that a more cost-effective solution would be found. The diagram above shows a possible arrangement based on the use of slotted wheels and optical sensors (as described in Section 2.3.4).

In order to resolve rotation of the ball into two perpendicular components, the ball is pressed against two perpendicular rollers by a third, spring-loaded roller. Rotation of the ball in a particular direction causes one or both of the sensing rollers to turn. Each of these rollers is connected to a slotted wheel which is placed between two slotted optical switches. The switches are positioned to allow the direction of rotation to be detected in a manner similar to that shown in Figure 2.11. The signals from the sensors are fed to the computer which keeps track of the movement of the ball and hence determines the appropriate cursor position. This arrangement has a range limited only by the method used to count the moving slots. The sensitivity is determined by the relative sizes of the ball and the pulleys, and by the number of slots within the wheels.

2.4 Actuators

Having looked at a range of sensors, which produce an electrical representation of some physical quantity, we will now turn our attention to **actuators** which take an electrical input signal and produce a corresponding variation in a physical quantity.

2.4.1 Heat

Most heating elements may be considered as simple **resistive heaters** which output the dissipated power as heat. For applications requiring only a few watts of heat output, ordinary resistors of the appropriate power rating may be used. Special heating cables and elements are available for larger applications.

Since the devices are effectively resistive they may be driven like any other resistive load using voltage or current to control the output power. However, when using high power heaters it is more common to use switching techniques to reduce the power dissipated in the controlling device. These techniques will be discussed in Chapter 8.

2.4.2 Light

Most lighting for general illumination is generated using conventional incandescent or fluorescent lamps. As with high power heaters, these lamps are usually controlled by switching techniques to reduce the power dissipated in the controlling device. Such techniques are used in conventional domestic **light dimmers**.

For signalling and communication applications, the relatively low speed of response of conventional lamps makes them unsuitable and other techniques are required.

Light emitting diodes

One of the most common light sources used in electronic circuits is the **light emitting diode** or **LED**. This is a semiconductor diode constructed in such a way that it produces light when a current passes through it. A range of semiconductor materials can be used to produce infra-red or visible light of various colours. Typical devices use materials such as gallium arsenide, gallium phosphide or gallium arsenide phosphide.

The characteristics of these devices are similar to those of other semiconductor diodes (these will be discussed in Chapter 5) but with different operating voltages. The light output from an LED is approximately proportional to the current passing through it; a typical small device might have an operating voltage of 2.0 V and a maximum current of 30 mA.

LEDs can be used individually or within multiple element devices. One example of the latter is the LED **seven-segment display** as shown in Figure 2.14(a). This consists of seven LEDs which can be switched ON or OFF individually to display a range of patterns. Figure 2.14(b) shows how such a device can be used to represent the digits 0 to 9.

(a) A typical device

(b) Representations of the digits 0–9

Figure 2.14 An LED seven-segment display.

Infra-red LEDs are widely used with photodiodes or phototransistors to enable short range, wire-less communication. Variations in the current applied to the LED are converted into light with a fluctuating intensity which is then converted back into a corresponding electrical signal by the receiving device. This technique is widely used in **remote control** applications for televisions and video recorders. In these cases the information transmitted is generally in a digital form. Because there is no electrical connection between the transmitter and the receiver, this technique can also be used to couple digital signals between two circuits which must be electrically isolated. This is called **opto-isolation**. Small self-contained **opto-isolators** are available which combine the light source and sensor in a single package. The input and output sections of these devices are linked only by light, enabling them to produce isolation between the two circuits. This is particularly useful when the two circuits are operating at very different voltage levels (see, for example, Section 7.8.1). Typical devices will provide isolation of up to 4 kV.

Fibre-optic communication

For long distance communication, the simple techniques used in television remote control units are not suitable as they are greatly affected by ambient light, that is, light present in the environment. This problem can be overcome by the use of a **fibre-optic cable** which captures the light from the transmitter and passes it along the cable to the receiver without interference from external light sources. Fibres are usually made of either an optical polymer or glass. The former are inexpensive and robust but their high attenuation makes them suitable for only short range communications of up to about 20 metres.

Glass fibres have a much lower attenuation and can be used over several hundred kilometres, but are more expensive that polymer fibres. For long range communications the power available from a conventional infra-red LED is insufficient. In such applications **laser diodes** may be used. These combine the light emitting properties of an LED with the light amplification of a laser to produce a high power, coherent light source.

2.4.3 Force, displacement and motion

In practice, actuators for producing force, displacement and motion are often closely related. A simple DC permanent magnet motor, for example, if opposed by an immovable object will apply a force to that object determined by the current in the motor. Alternatively, if resisted by a spring, the motor will produce a displacement which is determined by its current, and if able to move freely, it will produce a motion related to the current. We will therefore look at several actuators which can be used to produce each of these outputs, as well as some which are designed for more specific applications.

Solenoids

A solenoid consists of an electrical coil and a ferromagnetic slug which can move into, or out of, the coil. When a current is passed through the solenoid the slug is attracted towards the centre of the coil with a force determined by the current in the coil. The motion of the slug may be opposed by a spring to produce a displacement output, or the slug may simply be free to move. Most solenoids are linear devices, the electric current producing a linear force/displacement/motion. However, rotational solenoids are also available that produce an angular output. Both forms may be used with a continuous analogue input, or with a simple on/off (digital) input. In the latter case the device is generally arranged so that when the device is energized (that is, turned ON) it moves in one direction until it reaches an end stop. When de-energized (turned OFF) a return spring forces it to the other end of its range of travel where it again reaches an end stop. This produces a binary position output in response to a binary input. Figure 2.15 shows examples of small linear solenoids.

Meters

Panel meters are important output devices in many electronic systems providing a visual indication of physical quantities. Although there are various forms of panel meter, one of the simplest is the **moving-iron meter** which is an example of the rotary solenoid described above. Here a solenoid produces a rotary motion which is opposed by a spring. This produces an output displacement which is proportional to the current flowing through the coil. A needle attached to the moving rotor moves over a fixed scale to indicate the magnitude of the displacement. Moving-iron meters can be used for measuring AC or DC quantities. They produce a displacement which is proportional to the r.m.s. value of the current and independent of its polarity.

Although moving-iron meters are used in some applications, a more common arrangement is the **moving-coil meter**. Here, as the name implies, it is the coil that moves with respect to a fixed magnet, producing a meter that can be used to determine the polarity of a signal as well as its magnitude. The deflection of a moving-coil meter is proportional to the average value of the current rather than the r.m.s. value as in the moving-iron type. AC quantities can be measured by incorporating a rectifier and applying suitable calibration. However, it should be noted that the calibration usually assumes that the quantity being measured is sinusoidal, and incorrect readings will result if other waveforms are used.

Typical panel meters will produce a full-scale deflection for currents of 50 μA to 1 mA. Using suitable series and shunt resistances it is possible to produce meters that will display either voltages or currents with almost any desired range.

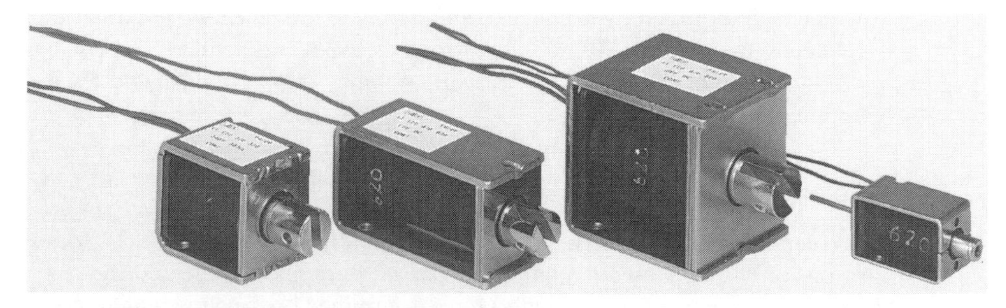

Figure 2.15 Small linear solenoids.

Motors

Electric motors fall into three broad types: AC motors, DC motors and stepper motors. **AC motors** are primarily used in high power applications and situations where great precision is not required. Control of these motors is often by simple on/off techniques, although variable power drives are also used.

DC motors are extensively used in precision position control systems and other electronic systems, particularly in low power applications. These motors have very straightforward characteristics with their speed being determined by the applied voltage and their torque being related to their current. The speed range of DC motors can be very wide with some devices being capable of speeds from tens of thousands of revolutions per minute down to a few revolutions per day. Some motors, in particular DC permanent magnet motors, have an almost linear relationship between speed and voltage, and between torque and current. This makes them particularly easy to use.

Stepper motors, as their name implies, move in discrete steps. The motor consists of a central rotor surrounded by a number of coils (or windings). The form of a simple stepper motor is shown in Figure 2.16.

Diametrically opposite pairs of coils are connected together so that energizing any one pair of coils will cause the rotor to align itself with that pair. By applying power to each set of windings in turn the rotor is made to 'step' from one position to another and thus generate rotary motion. In order to reduce the number of external connections to the motor, groups of coils are connected together in sequence. In the example shown, every third coil is joined to give three coil sets which have been labelled A, B and C. If initially winding A is energized the rotor will take up a position aligned with the nearest winding in the A set. If now A is de-energized and B is activated, the motor will 'step' around to align itself with the next coil. If now B is de-energized and C is activated, the rotor will again step to the adjacent coil. If the activated coil now reverts to A the rotor will move on in the same direction to the next coil, since this is the closest coil in the A set. In this way the rotor can be made to rotate by activating the coils in the sequence 'ABCABCA...'. If the sequence in which the windings are activated is reversed (CBACBAC...) the direction of rotation will also reverse. Each element in the sequence produces a single step which results in an incremental movement of the rotor.

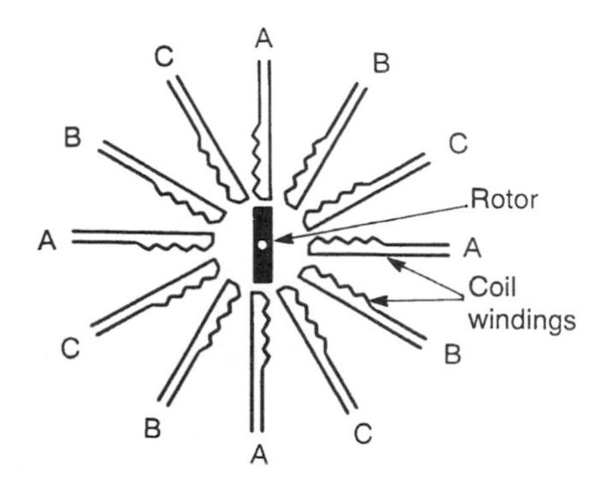

Figure 2.16 A simple stepper motor.

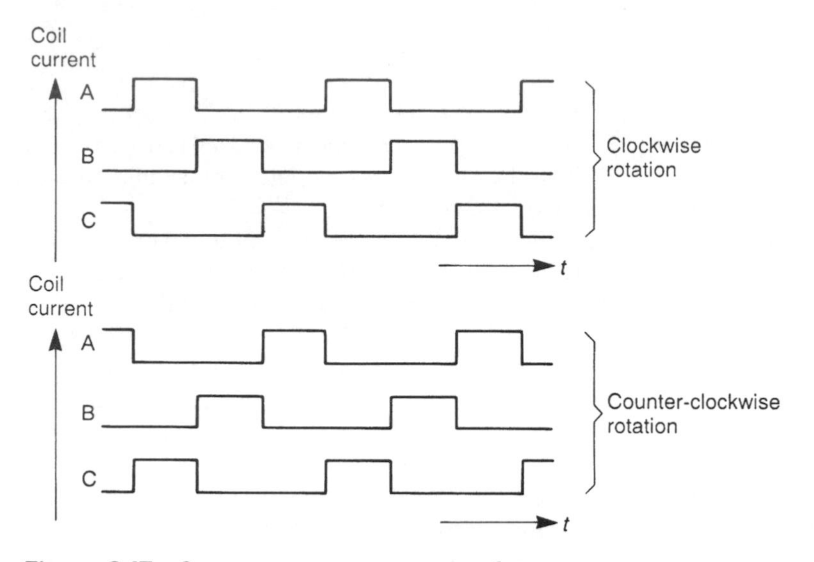

Figure 2.17 Stepper motor current waveforms.

The waveforms used to activate the stepper motor are binary in nature, as shown in Figure 2.17.

The motor shown in Figure 2.16 has 12 coils. and consequently 12 steps would be required to produce a complete rotation of the rotor. Typical small stepper motors have more than 12 coils and might require 48 or 200 steps to perform one complete revolution. The voltage and current requirements of the coils will vary with the size and nature of the motor.

The speed of rotation of the motor is directly controlled by the frequency of the waveforms used. Some stepper motors will operate at speeds of several tens of thousands of revolutions per minute, but all have a limited rate of acceleration determined by the inertia of the rotor. All motors have a 'maximum pull-in speed' which is the maximum speed at which they can be started from rest without losing steps. To operate at speeds above this rate they must be accelerated by gradually increasing the frequency of the applied waveforms. Since the movement of the rotor is directly controlled by the waveforms applied to the coils, the motor can be made to move through a prescribed angle by applying an appropriate number of transitions to the coils. This is not possible using a DC motor since the speed of rotation is greatly affected by the applied load.

2.4.4 Sound

Speakers

Most speakers (or loudspeakers) have a fixed permanent magnet and a movable coil connected to a diaphragm. Input to the speaker generates a current in the coil which causes it to move with respect to the magnet, thereby moving the diaphragm and generating sound. The nominal impedance of the coil in the speaker is typically in the range 4 to 15 Ω and the power handling capacity may vary from a few watts for a small domestic speaker to several hundreds of watts for speakers used in public address systems.

Ultrasonic transducers

At very high frequencies, the permanent magnet speakers described earlier are often replaced by **piezoelectric actuators**. These are similar in construction to piezoelectric ultrasonic sensors, but instead of using sound waves to generate electrical signals, electrical signals are applied to the piezoelectric diaphragm causing it to deform, thus generating sound waves. Such transducers are usually designed to operate over a narrow range of frequencies.

2.4.5 Actuators – a summary

All the actuators we have discussed take an electrical input signal and from it generate a non-electrical output. In each case power is taken from the input in order to apply power at the output. The power requirements are quite small in some cases, such as an LED or a panel meter which consume only a fraction of a watt. In other cases the power required may be considerable. Heaters and motors, for example, may consume hundreds or even thousands of watts.

The efficiency of conversion also varies from device to device. In a heater, effectively all the power supplied by the input is converted to heat. We could say that the conversion efficiency is 100%. LEDs, however, despite being one of the more efficient methods of converting electrical power into light, have an efficiency of only a few percent, the remaining power being dissipated as heat.

Some actuators can be considered as simple resistive loads in which the current will vary in direct proportion to the applied voltage. Most heaters and panel meters would come into this category. Other devices, such as motors and solenoids, have a large amount of inductance as well as resistance while others possess a large capacitance. Such devices behave very differently from simple resistive loads, particularly when a rapidly changing signal is applied. A third group of devices are non-linear and cannot be represented by simple combinations of passive components. LEDs and semiconductor laser diodes come into this third group. When designing electronic systems it is essential to know the characteristics of the various devices to be used so that appropriate circuitry can be produced.

2.5 Laboratory measuring instruments

A study of measurement would not be complete without a brief look at two of the most widely used electronic measuring instruments, the oscilloscope and the digital multimeter.

2.5.1 The oscilloscope

An oscilloscope is an instrument that allows a voltage to be measured in terms of the deflection that it produces in a spot of light on a **cathode ray tube**. Usually a **timebase** circuit is used to repeatedly scan the spot from left to right across the screen at a constant speed, by applying a 'saw tooth' waveform to the horizontal deflection circuitry. An input signal is then used to generate a vertical deflection proportional to the magnitude of the input voltage. In this way the oscilloscope effectively acts as an automated 'graph plotter' that plots the input voltage against time. Most oscilloscopes can display two input quantities by switching the vertical deflection circuitry between two input signals. This can be done by displaying one complete 'trace' of one waveform, and then displaying one trace of the other (**ALT** mode), or by rapidly switching between the two waveforms during each trace (**CHOP** mode). The choice between these modes is governed by the timebase frequency, but in either case the goal is to switch between the two waveforms so quickly that both are displayed steadily and with no noticeable flicker or distortion. In order to produce a stable trace the timebase circuitry includes a **trigger** circuit which attempts to synchronize the beginning of the timebase sweep so that it always starts at the same point in a repetitive waveform – thus producing a stationary trace.

Simple oscilloscopes use analogue circuitry to implement the various functions. However, in recent years there has been a move towards the use of digital techniques to store and manipulate the input data. Digital oscilloscopes are particularly useful when looking at very slow waveforms or short transients, where their ability to store information enables them to display a steady trace. At present digital instruments are generally more expensive than their analogue counterparts and so analogue oscilloscopes are widely used in teaching laboratories. Figure 2.18(a) shows a typical analogue laboratory oscilloscope and Figure 2.18(b) shows a simplified block diagram of such an instrument.

2.5.2 The digital multimeter

A standard measuring instrument in any electronics laboratory is a **digital multimeter** (**DMM**). This combines high accuracy and stability in a device that is very easy to use. While these instruments are normally capable of measuring voltage, current and resistance, they are often (inaccurately) referred to as **digital voltmeters** or simply **DVMs**. At the heart of the meter is a **digital to analogue converter** (**DAC**) that takes as its input a voltage signal, and produces as its output a digital measurement that is used to drive a numeric display. We shall look at the operation of such DACs in Chapter 13.

Measurements of voltage, current and resistance are achieved by using appropriate circuits to generate a voltage proportional to the quantity to be measured. When measuring voltages the input signal is connected directly to either a DC or an AC attenuator depending on the nature of the quantity to be measured. When measuring currents the input signal is connected across an appropriate shunt resistor which generates a voltage proportional to the input current. The value of the shunt resistance is switched to select different input ranges. In order to measure resistance the inputs are connected to an 'ohms converter' which passes a small current between the two input connections. The resultant voltage is a measure of the resistance between these terminals.

In simple DMMs, an AC voltage is simply rectified and the resultant signal smoothed to give its average value. For a sinusoidal waveform the relationship between this quantity and its r.m.s. voltage is known, and so the smoothed signal is measured and the resultant value scaled appropriately. This approach works satisfactorily for sinus-

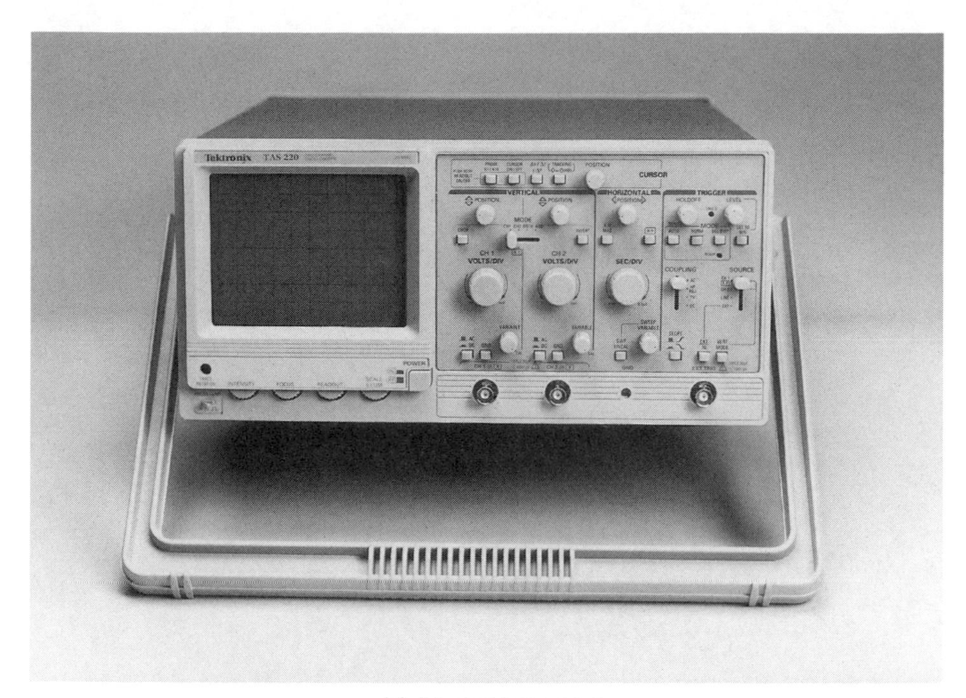

(a) A typical instrument

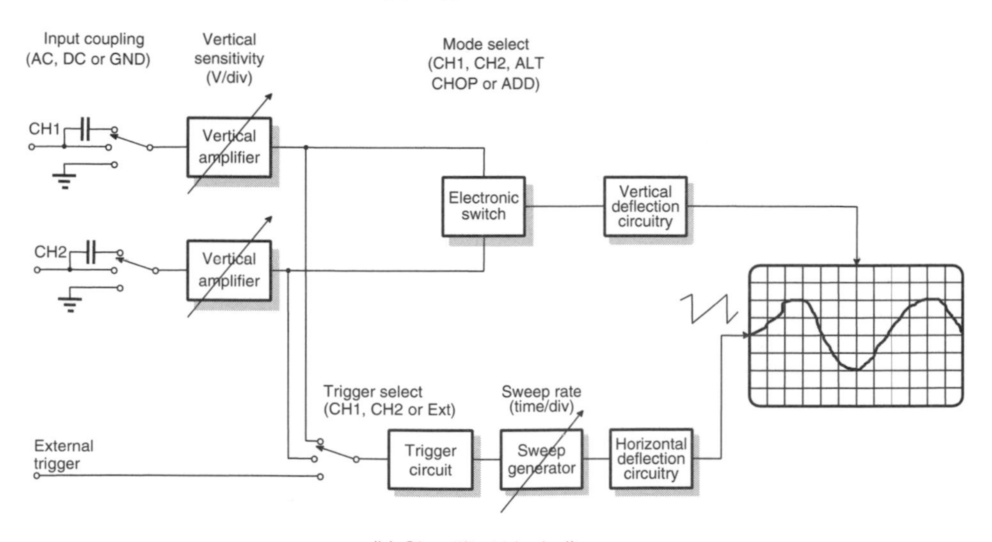

(b) Simplified block diagram

Figure 2.18 An analogue laboratory oscilloscope.

oidal voltages but produces incorrect values for other input waveforms, since the relationship between the r.m.s. value and the rectified and smoothed value is different. For this reason more sophisticated DMMs use a **true r.m.s. converter** which accurately produces a voltage proportional to the r.m.s. value of an input waveform.

(a) A typical digital multimeter

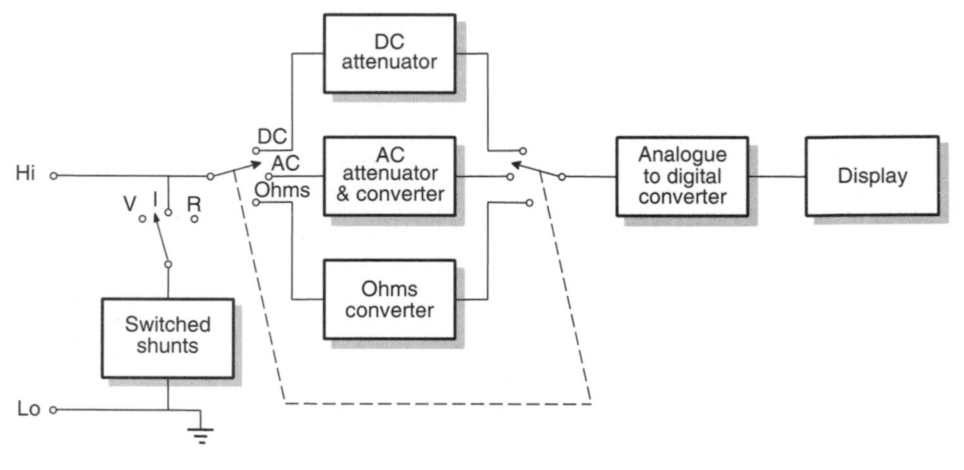

(b) A simplified block diagram of a DMM

Figure 2.19 A hand-held digital multimeter.

Such instruments can be used to make measurements of alternating quantities even when they are not sinusoidal. However, all DMMs are accurate over only a limited range of frequencies.

Figure 2.19(a) shows a typical hand-held digital multimeter and Figure 2.19(b) is a simplified block diagram of a such a device.

Key points

- Measurement involves a comparison between the quantity to be measured and some standard.
- All measurements are subject to errors.
- Errors may be divided into random errors and systematic errors.
- A wide range of sensors is available to meet the needs of a spectrum of possible applications.
- Some sensors produce an output voltage or current that is related to the quantity being measured. They therefore supply power (albeit in small quantities).
- Other devices simply change their physical attributes, such as resistance, capacitance or inductance, in response to changes in the measured quantity.
- Most actuators take power from their inputs in order to deliver power at their outputs. The power required varies tremendously between devices.
- In most cases the energy conversion efficiency of an actuator is less than 100%, and sometimes it is much less.
- Common laboratory measurement instruments include the oscilloscope and the digital multimeter. The former allows an input waveform to be displayed so that the form of the signal can be seen. The latter provides a very simple method of measurement but may be affected by the frequency or shape of the signal.

We noted in the last chapter that the processing requirements of an electronic system are determined by the nature of the sensors and actuators, as well as by the functional requirements of the application. However, from our study of sensors and actuators we can already predict one function of the circuitry that will generally be required. That is, that it must take signals from sensors that provide little or no power, and produce from them signals of sufficient power to drive the required actuators. This function is called **amplification** and is the subject of the next chapter.

Design study

An automatic drilling machine is to be designed for drilling holes in printed circuit boards during manufacture. The essential components of the system are shown below. The system includes a high speed drill mounted on an electrically controlled mechanism which must raise and lower the drill to form holes in the board below. The amount of movement required from this mechanism is approximately 5 cm, and is constant for each drilling operation. The lowering operation must be performed quickly, but must be at a controllable rate to suit the drill size and speed. Raising the drill must also be performed quickly, but in this case the rate is not critical.

The printed circuit board is moved under the drill by an X–Y table (that is, a table which can be moved in two perpendicular directions) which positions the board in an appropriate

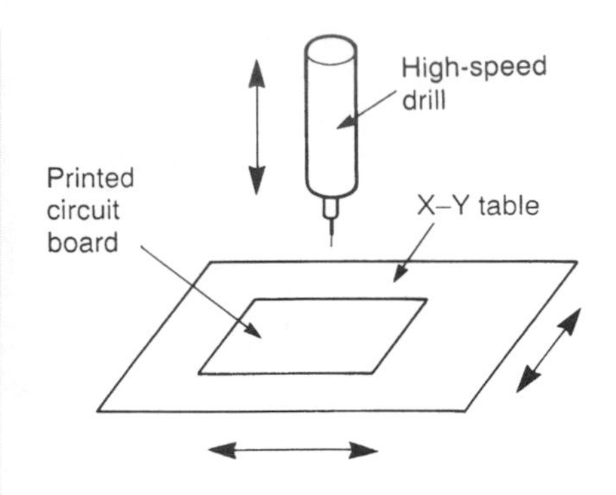

location for each hole in turn. The table must be moved as quickly as possible, as the time taken to move between hole positions will greatly affect the performance of the system. The board must be positioned to an accuracy of at least ± 0.1 mm and preferably to within ± 0.05 mm. The maximum board size is 350 mm × 350 mm.

Ignoring the nature of the electronic control system itself and the man–machine interface, consider the sensors and actuators that could be used to implement the above system.

Approach

The problem may be divided into two parts: the sensors and actuators associated with the drill raising and lowering mechanism, and the sensors and actuators associated with the X–Y table.

The drill mechanism

The requirements of this section of the system are that the mechanism must be able to raise and lower the drill at high speed and at a controllable rate. It also seems sensible that the system should have some way of knowing when the drill has reached the bottom of its travel and when it has returned to the top. This latter point is essential since attempting to move the table while the drill is down could cause damage to the drill and possibly to the board and table.

There are many displacement actuators which could be used to produce the required motion. A linear solenoid with an appropriate length of travel could be used to produce a very simple arrangement. The speed of the downward travel of the drill would be controlled by the current passed through the solenoid allowing the rate to be varied as required. A strong return spring could be used to produce the rapid upward motion since the speed of this motion is not critical. Having set the value of the solenoid current accordingly, control of the drill could then be by a simple binary signal to lower or raise the mechanism. Detection of the drill's position could be achieved using two microswitches, positioned to sense the two extremes of the drill's travel.

While producing a low-cost solution, the use of a solenoid has some disadvantages. The operation of a solenoid is such that the application of a constant current tends to generate a constant force at the output. This will tend to accelerate the output in a manner which is greatly affected by the friction and inertia of the system. This will tend to produce an arrangement in which the drill is accelerated towards the board, rather than driven towards it at a constant rate.

An alternative arrangement might use a rotary actuator, such as a DC permanent magnet motor, and some rotary-to-linear translation mechanism such as a 'lead-screw'. This consists of a rotating threaded shaft (the screw) and a threaded 'nut'. The nut is prevented from turning with the shaft, and so it runs up and down the shaft as it rotates. If the gear ratio of this arrangement is chosen appropriately, the friction and inertia of the drill mechanism will have little effect on the motor speed, which will be fairly accurately controlled by the applied voltage. The drill can thus be raised and lowered at a speed determined by the voltage applied to the motor.

While the DC motor arrangement gives more precise control of the rate of raising and lowering the drill, it is almost certainly a more expensive solution. Before deciding between these options (and several other possible methods) one would need to know more about the requirements of the system and to compare the cost of each method as well as their performance.

The X–Y table

We are concerned with two aspects of the design of the X–Y table: how to move the table about and how to determine its position to allow the controller to drive it to the appropriate location.

While there are several possible solutions to the first of these problems the most obvious methods are to use either DC or stepper motors. In either case it is likely that one of the motors must be mounted on a moving part of the table. This is not obligatory, but most arrangements that avoid this requirement need correspondingly more complex arrangements of pulleys, gears or other linkages. We will not consider the mechanics of the table further, but simply look at the sensors and actuators to be used.

Small DC motors have good acceleration and can produce very rapid movement. However, although the speed and torque produced by these motors are related to the applied voltage and the current, the position of the output of the motor is greatly affected by the applied load and the inertia in the system. To produce a position control system using such motors it is necessary to use some form of displacement transducer to measure the output position. One solution to this problem is to use a rotary transducer to measure the movement of the motor (or some rotating part of the mechanism which is used to drive the table) and to deduce from this the position of the table. An incremental encoder might be a suitable device for this purpose. This produces a low-cost solution but relies on the accuracy of the mechanics of the system to ensure the accuracy of the conversion from rotation to linear motion. Play in a lead screw, for example, would not be detected by such a method. This may necessitate the use of very accurate and therefore very expensive mechanical parts to overcome these problems.

An alternative solution is to measure the position of the table directly, overcoming any problems of play or non-linearities in the drive mechanism. The resolution required for this application (to measure to an accuracy of ±0.1 mm over a range of 350 mm) is outside the capabilities of almost all the displacement transducers described in this chapter. The only techniques discussed which can match this requirement are the optical gratings methods. These are capable of resolutions considerably greater than those required and can easily operate over the required range. The main disadvantage of these techniques is their very high cost. If optical gratings were used in this application it is likely that the transducers would cost as much as all the other components of the system added together.

The need for very expensive position sensors may be overcome by the use of stepper motors. Since these motors move in discrete steps determined by the waveforms applied to their coils it is possible to keep track of the output position by counting the number of steps in each direction. In order to produce high-speed motion the motors must be controlled by sophisticated circuitry which accelerates and decelerates the motors appropriately to achieve maximum speed without losing steps. This control circuitry is more expensive than that required for DC motors. As with an earlier method, this solution will be affected by any play or non-linearities in the drive mechanism which links the motor to the table.

Determining position from either optical gratings or step counting produces an indication of *relative* movement. A reference position must be provided to determine the *absolute* position of the table. This can be achieved manually by requiring the user to press a button when the table is in an appropriate position, or automatically using two microswitches which detect when the table has moved to a specific X and Y location. In either case the counters' used to determine the X and Y positions are zeroed at this time and are then used to measure movements relative to this point.

Although both DC and stepper motors can provide very fast movements in position control systems, DC motors are generally considered to be superior if ultimate speed performance is required As we have seen, in this application the use of DC motors requires the use of very expensive sensors or the reliance on accurate and therefore possibly expensive mechanics to form an accurate positioning system. The designer must therefore consider the financial and performance implications of his choice of actuators.

Further reading

Beckwith T. G., Marangoni R. D. and Lienhard J. H. (1993) *Mechanical Measurements*, 5th edn. Reading, MA: Addison-Wesley

Bell D. A. (1994) *Electronic Instrumentation and Measurement*, 2nd edn. Englewood Cliffs, NJ: Prentice-Hall

Bolton W. (1996) *Measurement and Instrumentation Systems*. Oxford: Newnes

Hauptmann P. (1991) *Sensors Principles and Applications*. Hemel Hempstead: Prentice-Hall

Usher M. J. (1985) *Sensors and Transducers*. Basingstoke: Macmillan

Exercises

2.1 Explain briefly the meanings of the terms transducer, sensor, actuator, measurand, standard, accuracy and errors.

2.2 Identify likely sources of random and systematic errors in the process of measuring the dimensions of a room using a tape measure.

2.3 Draw a graph that illustrates the concept of linearity as applied to a measuring system.

2.4 What is meant by the term 'traceability' when it is applied to a measuring system?

2.5 Estimate the r.m.s. noise voltage produced by thermal noise in a 47 kΩ resistor at normal room temperature (≈300 K) when measured over a bandwidth of 20 kHz (the typical audio spectrum).

2.6 Explain the meanings of the terms 'stress' and 'strain'.

2.7 Suggest a suitable method of employing two strain gauges to measure the vertical force applied to the end of a beam which is supported at one end.

2.8 The arrangement of Figure 2.9 produces an output of 0 volts if the switch is closed, and V if the switch is open. Devise a similar circuit which reverses these two voltages.

2.9 A PRT has a resistance of 100 Ω ± 0.1 Ω at 0 °C and a temperature coefficient of + 0.385 Ω per °C. What would be its resistance at 100 °C?

The PRT is connected to an external circuit which measures the resistance of the sensor by passing a constant current of 10 mA through it and measuring the voltage across it. What would this voltage be at 100 °C?

The PRT is heated by this current and its temperature rises by 0.2 °C per mW. Estimate the error caused by this 'self-heating' (a simple treatment is all that is required).

2.10 The PRT described in the last exercise is connected as shown in the diagram to form an arrangement where the output voltage V_o is determined by the temperature of the PRT.

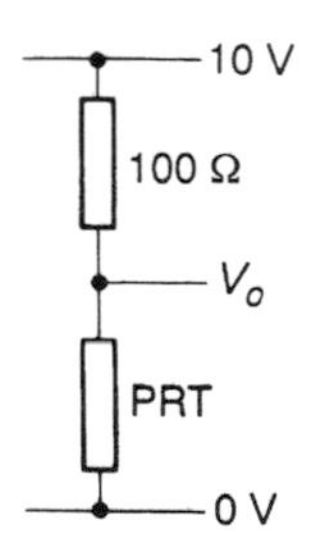

Derive an expression for V_o in terms of the temperature of the PRT.

The resistance of the PRT is linearly related to its absolute temperature. Is V_o linearly related to temperature?

2.11 Optical grating techniques permit displacements of up to a metre or more to be measured to an accuracy of better than 1 micron (1 μm). What is the principal disadvantage of this sensing method?

2.12 Describe two methods of measurement which would be suitable for a non-contact, automatic rangefinder for distances up to 10 m.

2.13 Suggest ten physical quantities, not discussed in this section, which are regularly measured, giving in each case an application where this measurement is required.

2.14 A moving-coil meter has a coil resistance of 75 Ω and produces a full-scale deflection for a current of 1 mA. Show how, with the aid of suitable resistors, this meter may be used to measure currents in the range 0–1 A and voltages in the range 0–1 V.

2.15 Use circuit simulation to verify your solution to Exercise 2.14. Represent the meter by an appropriate resistance and the input by a suitable voltage or current source. Vary the input and monitor the current in the meter to confirm that it is of the correct magnitude.

2.16 Many modern wrist watches use electric motors, driven from a quartz crystal oscillator, to turn second, minute and hour hands. What form of motor would seem suitable for this application?

2.17 In addition to displaying the digits 0–9, the seven-segment display of Figure 2.14 can be used to indicate some alphabetic characters. List the upper and lower case letters which can be shown in this way and give examples of simple status messages (such as 'start' and 'stop') which could be used with an array of these devices.

Amplification

Objectives

When you have studied the material in this chapter you should:

- have a clear understanding of the concept of amplification;
- be aware of examples of both active and passive amplifiers;
- be familiar with the use of 'equivalent circuits' to represent voltage sources, resistive inputs and loads;
- be able to explain the meanings of terms such as output power, power gain, voltage gain and frequency response;
- understand the effects of connecting several amplifiers in series;
- be familiar with several common forms of amplifier, including differential and operational amplifiers.

Contents

3.1 Introduction

In the last chapter we looked at sensors and actuators and observed the need for some form of electronic processing to make the signals produced by the former suitable for use with the latter. Many forms of processing are possible and we will look at several useful examples in later chapters. However, in this chapter we will concentrate on one particular processing operation, that of **amplification**.

Simplistically, amplification means making things bigger.

One can think of many examples of non-electronic amplification, two of which are shown in Figure 3.1. In the first of these examples the unequal lengths of the two arms of the lever mean that the output end will move further than the input end. We therefore have amplified the input movement. Note that the direction of movement is reversed. When the input goes down the output goes up. We could call such an amplifier an **inverting amplifier** since it inverts the direction of the input. Note also that because of the mechanical disadvantage of the arrangement, the force applied at the output is less than that applied at the input. Thus, although the movement has been amplified, the force has been reduced or **attenuated**.

If the position of the fulcrum were moved to be closer to the output end of the lever, the movement of the output would be less than that at the input but the force would be greater. We would now have a force amplifier and a movement attenuator.

The second example in Figure 3.1 shows a pulley arrangement in which a force applied to the input produces a larger force at the load. We therefore have a force amplifier and again, in this case, it is an inverting amplifier since the direction of the force is inverted (clearly other pulley arrangements may be non-inverting). It should be noted that the movement of the load is less than that of the input and therefore the arrangement attenuates movement.

Both these examples are **passive systems**. That is, they have no external energy source other than the inputs. For such systems the **output power** (that is, the power delivered at the output) can never be greater than the **input power** (that is, the power absorbed by the input) and in general it will be smaller because of losses. In our examples, losses would be caused by friction at the fulcrum and the pulleys.

Some amplifiers are not passive but are **active**. This means that they have some form of external energy source which can be harnessed to produce an output which has more power than the input. Figure 3.2 shows an example of such an amplifier called a **torque amplifier**.

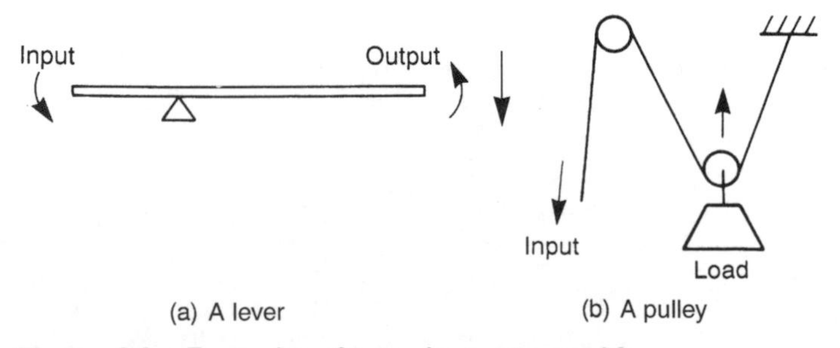

Figure 3.1 Examples of non-electronic amplifiers.

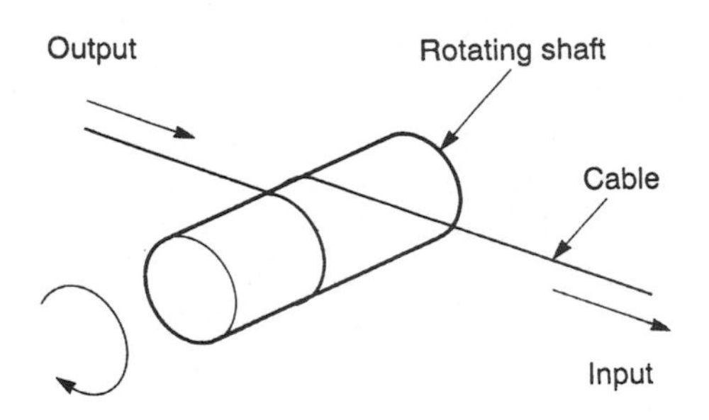

Figure 3.2 A torque amplifier.

This form of torque amplifier consists of a rotating shaft with a rope or cable wound around it. The amplifier can be used as a power winch and is often found in boats and ships. One end of the rope (the output) is attached to a load and a control force is applied to the other end (the input). If no force is applied to the input the rope will hang loosely around the rotating shaft and little force will be applied at the output. The application of a force to the input tightens the rope around the drum and increases the friction between them. This frictional force is applied to the rope and results in a force being exerted at the output. The greater the force applied to the input, the greater the frictional force experienced by the rope and the greater the force exerted at the output. We therefore have an amplifier where a small force applied at the input generates a larger force at the output. The magnitude of the amplification may be increased or decreased by changing the number of turns of the rope around the drum.

It should be noted that since the rope is continuous the distance moved by the load at the output will be equal to the distance moved by the rope at the input. However the force applied at the output is greater than that at the input and we therefore have an increase in power at the output. The extra power available at the output is supplied by the rotating drum and will result in an increased drag being experienced by whatever force is causing it to rotate.

3.2 Electronic amplifiers

In electronics there are also examples of both passive and active amplifiers. Examples of the former include a step up transformer, where an alternating voltage signal applied to the input will generate a larger voltage signal at the output. Although the voltage at the output is increased, the effective impedance of the output circuit is higher than the input circuit, and thus the ability of the output to provide current to an external load is reduced. The power supplied to a load will always be less than the power absorbed at the input. Thus the transformer may be a voltage amplifier but it is *not* a power amplifier.

Although there are several examples of passive electronic amplifiers, the most important and useful electronic amplifiers are active circuits. These take power from an external energy source, usually some form of **power supply**, and use it to boost the input signal. Unless the text indicates differently, for the remainder of this book when we use the term amplifier, we will be referring to an active electronic amplifier.

We saw earlier when looking at mechanical amplifiers that several different forms of amplification are possible. Such devices can, for example, be movement amplifiers or force amplifiers, and can provide power amplification or attenuation.

Electronic amplifiers may also be of different types. One of the most common is the **voltage amplifier**, the main function of which is to take an input voltage signal and to produce a corresponding amplified voltage signal. Also of importance is the **current amplifier** which takes an input current signal and produces an amplified current signal. Usually both these types of amplifier, as a result of the amplification, also increase the power of the signal. However, the term **power amplifier** is usually reserved for circuits that have the primary function of supplying large amounts of power to a load. In such cases the efficiency of the circuit is important, in terms of the ratio of the power delivered to the load to that taken from the power supply. The efficiency is of importance since it determines the power that is dissipated as heat by the circuit itself. This in turn affects the required power rating, cost and size of the components. Clearly, power amplifiers must also provide either voltage or current amplification, or both.

The amplification produced by a circuit is described by its **gain**. From the above we can define three quantities, namely **voltage gain**, **current gain** and **power gain**. These quantities are given by the expressions

$$\text{Voltage gain } (A_v) = \frac{V_o}{V_i}$$

$$\text{Current gain } (A_i) = \frac{I_o}{I_i}$$

$$\text{Power gain } (A_p) = \frac{P_o}{P_i}$$

where V_i, I_i and P_i represent the input voltage, input current and input power, respectively, and V_o, I_o and P_o represent the output voltage, output current and output power, respectively. Initially we will look at voltage amplification and leave consideration of current and power amplification until later.

A widely used symbol for an amplifier is shown in Figure 3.3. This device has a single input and produces an amplification determined by the circuitry used. In this case the input and output quantities are voltages and the circuit is described by its voltage gain.

Clearly the input and output voltages must be measured with respect to some reference voltage or reference point. This point is often called the **earth** of the circuit and is given the symbol shown in the diagram.

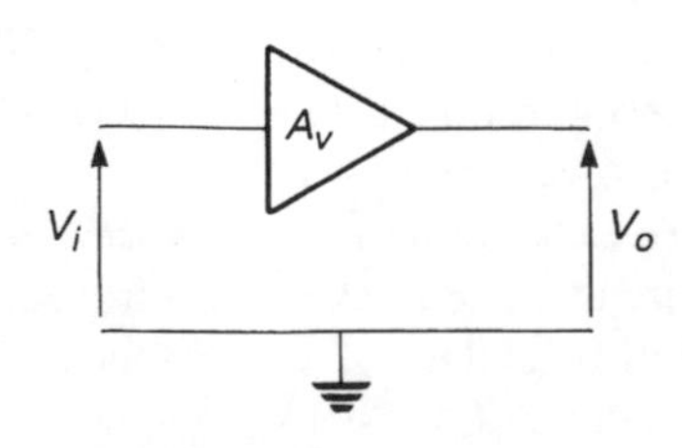

Figure 3.3 An amplifier.

In order for the amplifier to perform some useful function, something must be connected to the input to provide an **input signal** and something must be connected to the output to make use of the **output signal.** In a simple application the input signal could come directly from one of the sensors described in the last chapter and the output could drive an actuator. Alternatively, the input and output could be connected to other electronic circuits. The transducer or circuitry providing an input to the amplifier is sometimes called the **source** while the transducer or circuitry connected to the output is called the **load** of the amplifier.

An **ideal** voltage amplifier would always give an output voltage which was determined only by the input voltage and the gain, irrespective of what was connected to the output (the load). Also an ideal amplifier would not affect the signal produced by the source, implying that no current is taken from it.

In fact, **real** amplifiers cannot fulfil these requirements. To understand why this is so we need to know more about the nature of sources and loads.

3.3 Sources and loads

If one takes a voltage source, such as a battery, and connects it across a resistance R, the current which flows I is related to the battery voltage V by Ohm's law:

$$I = \frac{V}{R}$$

If we connect the terminals of the source together, R is zero, so the current should be infinite. Of course, in practice the current is not infinite because any real voltage source has some resistance associated with it.

We can represent a voltage source by an ideal **voltage generator** (that is, a voltage generator that has no internal resistance) in series with a resistance. This arrangement is shown in Figure 3.4(a). An ideal voltage generator is normally represented by the circular symbol shown in Figure 3.4(b).

It is also possible to represent a voltage source by a **current generator** in parallel with a resistance. This arrangement is shown in Figure 3.5 which illustrates the symbol used for an ideal current generator.

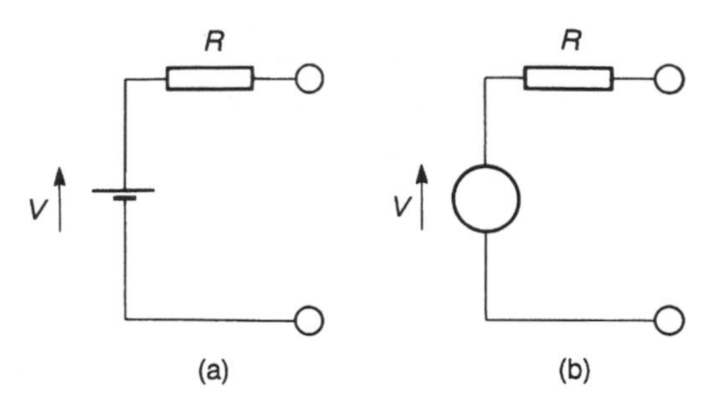

Figure 3.4 Voltage sources.

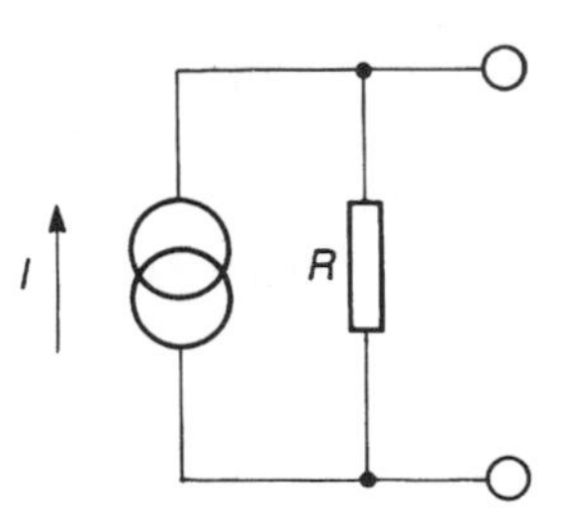

Figure 3.5 Representing a voltage source using a current generator.

Any linear circuit that produces an output voltage may be represented by *either* of these arrangements. They are referred to as **equivalent circuits** for the output.

The first of these representations is described by **Thévenin's theorem**, which may be paraphrased as:

> *As far as its appearance from outside is concerned, any two terminal network of resistors and energy sources can be replaced by a series combination of an ideal voltage source V and a resistor R, where V is the open-circuit voltage of the network and R is the ratio of the open-circuit voltage and the short-circuit current.*

The voltage source–resistor combination described by the theorem is known as the **Thévenin equivalent circuit** of the network. It is important to note that the equivalence is only valid 'as far as its appearance from outside' is concerned. The internal characteristics of the network, such as its power consumption, are not represented by the equivalent circuit.

The second of the representations is described by **Norton's theorem**, which may be paraphrased as:

> *As far as its appearance from outside is concerned, any two terminal network of resistors and energy sources can be replaced by a parallel combination of an ideal current source I and a resistor R, where I is the short-circuit current of the network and R is the ratio of the open-circuit voltage and the short-circuit current.*

This current source–resistor combination is the **Norton equivalent circuit**; again the same restrictions are placed on its applicability. Notice incidentally that the values of R are the same in each case.

It is possible to prove the validity of these two theorems but these proofs will not be repeated here.

Equivalent circuits of these types are not restricted to representing sources with constant voltages and currents. The ideal voltage and current sources within these circuits can represent varying quantities, if appropriate, allowing the sinusoidal output of an oscillator to be represented or the less predictable output of a sensor.

The use of an equivalent circuit greatly simplifies analysis since it allows a complex circuit to be replaced by a much simpler representation. In some cases the component values for the equivalent circuit may be calculated from a knowledge of the construction of the network. In other cases it is necessary to determine their values by making appropriate measurements.

Determination of equivalent circuit values

PROBLEM: A box contains an unknown circuit which may consist of many independent voltage and current sources and many resistors. The circuit has two external connections. Our task is to produce an equivalent circuit for this arrangement.

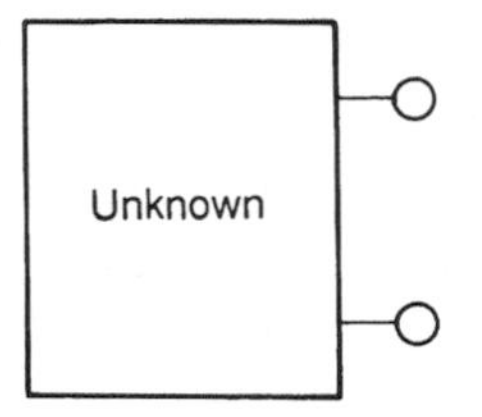

METHOD:

(1) Determine the open-circuit voltage V_{oc} of the circuit.
(2) Determine the short-circuit current I_{sc} of the circuit.
(3) Divide these two quantities to give the internal resistance R

$$R = \frac{V_{oc}}{I_{sc}}$$

THEN:

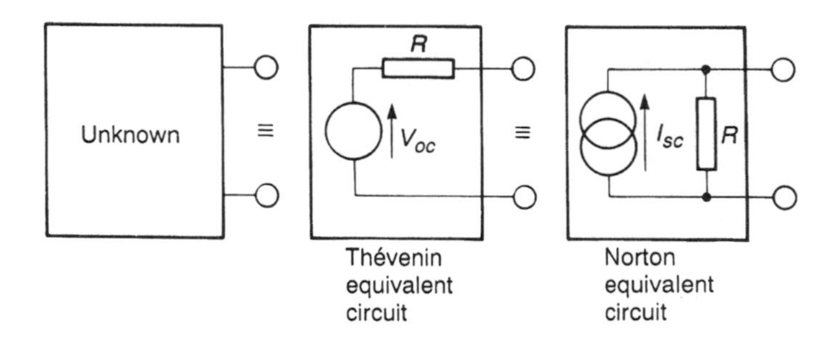

Measurement of the short-circuit current

PROBLEM: Although it is usually easy to measure V_{oc} directly using a high resistance voltmeter (provided the resistance of the meter is high compared with the internal resistance, the error will be small), it is often inappropriate to measure I_{sc} by short-circuiting the source, as this may cause damage. We therefore require a method of determining the short-circuit current indirectly. Fortunately there are several methods available.

METHOD 1: Connect a variable resistance across the source and measure the output voltage and current for different values of external resistance. Plot the voltage against the current to give a linear graph which can be extrapolated to give the current for zero voltage which is the short-circuit current I_{sc}.

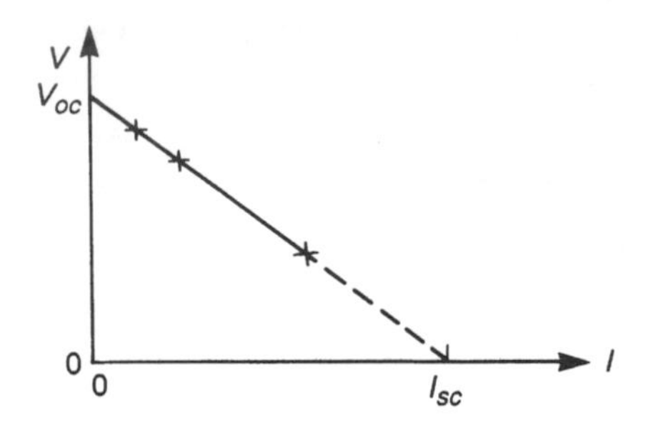

METHOD 2: Measure the open-circuit output voltage V_{oc} using a high resistance meter. Now connect an external resistance r across the output, and measure the output voltage V across r.

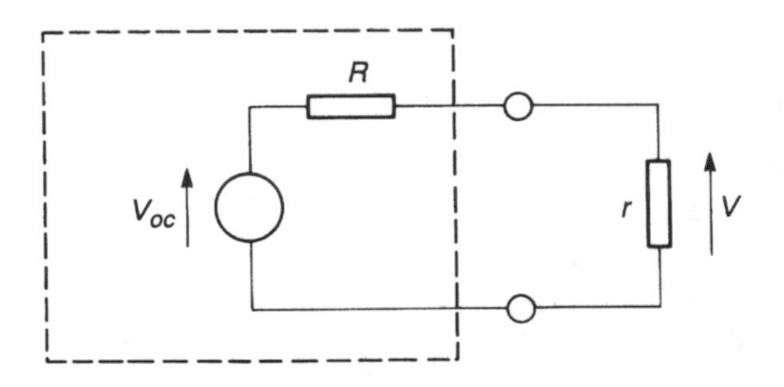

We can now determine the internal resistance R of the source from the relationship

$$\frac{V}{V_{oc}} = \frac{r}{r+R}$$

and hence determine I_{sc} from the relationship

$$I_{sc} = \frac{V_{oc}}{R}$$

METHOD 3: This method is a combination of methods 1 and 2.

Measure the open-circuit voltage as before and connect a variable resistance across the output. Measure the output voltage V and adjust the resistance until V is half of the open-circuit voltage V_{oc}. The external resistance is now equal to the internal resistance R, and I_{sc} can again be found from the ratio of V_{oc} to R, as shown earlier.

When considering an arrangement of unknown internal form it is necessary to determine values for V_{oc} and I_{sc} by making measurements. However, if the construction of a circuit is known, these quantities may be calculated from the component values.

Example 3.1 Determination of Thévenin and Norton equivalent circuits

Derive Thévenin and Norton equivalent circuits for the following arrangement.

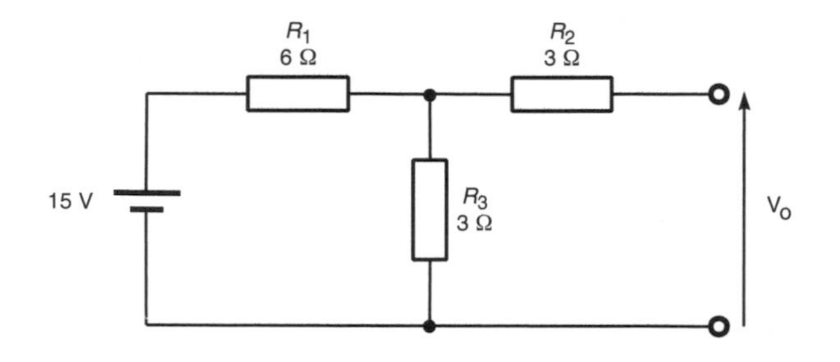

In order to determine values for the components of the equivalent circuits we need to determine the open-circuit voltage V_{oc} and the short-circuit current I_{sc}.

Since no current flows through R_2 when the output is open-circuit, there is no voltage across it and the output voltage is determined by the potential divider formed by R_1 and R_3. Therefore

$$V_{oc} = 15\text{ V} \times \frac{R_3}{R_1 + R_3} = 15\text{ V} \times \frac{3\ \Omega}{6\ \Omega + 3\ \Omega} = 5\text{ V}$$

The short-circuit current can be obtained by considering the equivalent resistance of the circuit when the output is shorted to ground. Under these circumstances R_2 is connected directly across R_3 giving an effective resistance of 1.5 Ω in series with R_1. The total resistance across the voltage source is therefore 7.5 Ω giving a current of 2 A. Since R_2 and R_3 are equal, half of this current will flow in each resistor and hence the short-circuit current (the current through R_2) is 1 A.

The resistance R of the equivalent circuits can then be calculated as

$$R = \frac{V_{oc}}{I_{sc}} = \frac{5\text{ V}}{1\text{ A}} = 5\ \Omega$$

Hence the equivalent circuits are

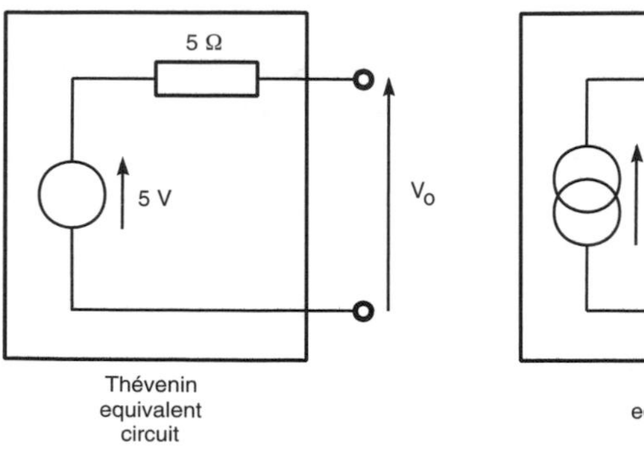

Thévenin equivalent circuit

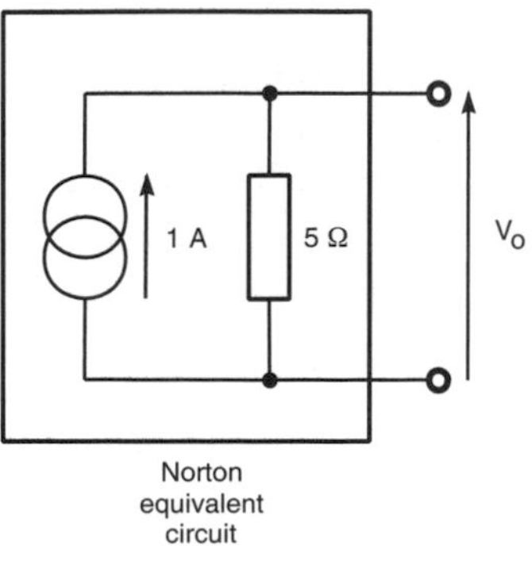

Norton equivalent circuit

The resistance R in these equivalent circuits may be thought of as the resistance seen between the terminals when 'looking into' the network with the effects of all the *independent* energy sources removed. All the energy sources considered so far are independent of currents and voltages at other points in the circuit. We shall see later that some energy sources are *dependent* on (or *controlled* by) quantities elsewhere in the circuit. To remove the effects of the independent energy sources within the network, the voltages or currents which they represent must be set to zero. This can be achieved by replacing voltage sources by short circuits (where the voltage is zero) and by replacing current scurces with open circuits (where the current is zero). The validity of this approach may be illustrated by considering the Thévenin and Norton equivalent circuits given earlier.

3.4 Input resistance

When a voltage is applied to the input terminal of any real network a current will flow. The magnitude of this current will depend on the **input impedance** of the network. We can represent such an arrangement by an equivalent circuit of the input, as shown in Figure 3.6, where Z_i represents the input impedance.

In many cases we can adequately represent an input circuit by a simple resistance, in which case we refer to this as the **input resistance** R_i of the circuit. This is shown in Figure 3.7.

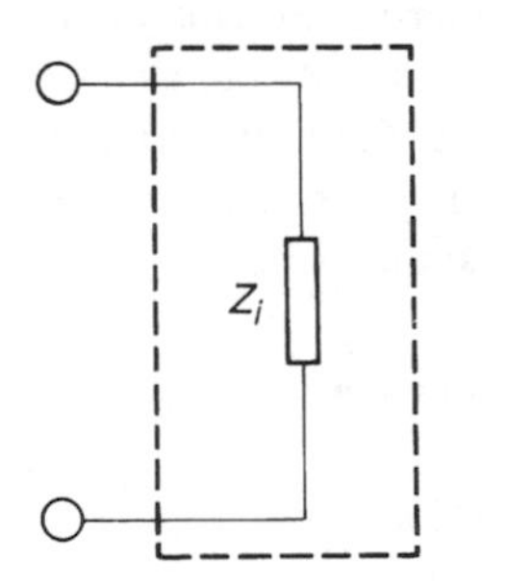

Figure 3.6 An equivalent circuit for an input.

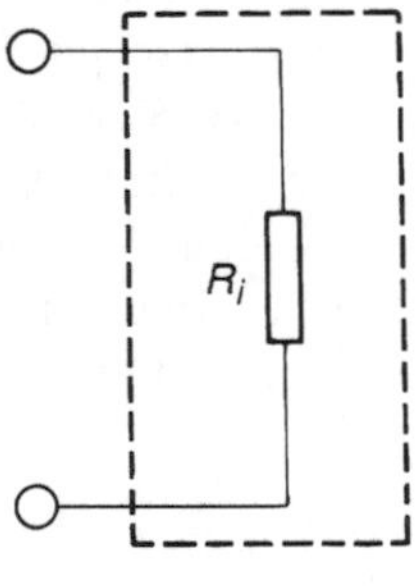

Figure 3.7 A simplified equivalent circuit for an input.

Often the value of the input resistance of a network can be determined from a knowledge of its circuit. In other instances it may be necessary to find this value by performing measurements on the network.

Measurement of input resistance

PROBLEM: To determine the input resistance of an unknown circuit.

METHOD: Connect the circuit under test in series with a variable resistance and a voltage source.

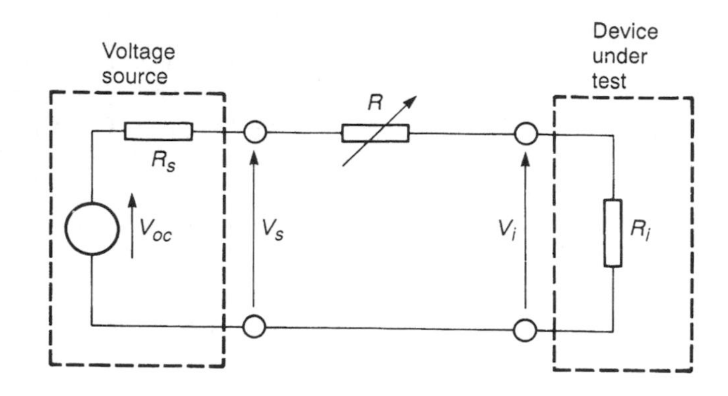

Measure the output voltage of the source V_s and the voltage across the terminals of the circuit under investigation V_i. Adjust the external resistance R until V_i is equal to half of V_s. The external resistance R is now equal to the input resistance R_i. Measure R to determine R_i.

This method is not affected by the internal resistance of the source R_s since it is the source output voltage V_s that is used, not the open-circuit source voltage V_{oc}.

3.5 Equivalent circuit of an amplifier

Since we now have equivalent circuits for both voltage sources and inputs, we are now in a position to draw an equivalent circuit for the amplifier of Figure 3.3.

Figure 3.8 shows an equivalent circuit for the amplifier. The input is represented by a simple input resistance R_i and the voltage which is applied across this input is the **input voltage** V_i. The output is represented by a Thévenin equivalent circuit. The voltage generator used is an example of a **dependent** or *controlled* voltage source, as mentioned earlier. Its value is determined by the input voltage and the voltage gain of the amplifier A_v. The series resistor is the **output resistance** of the amplifier R_o. You will notice that

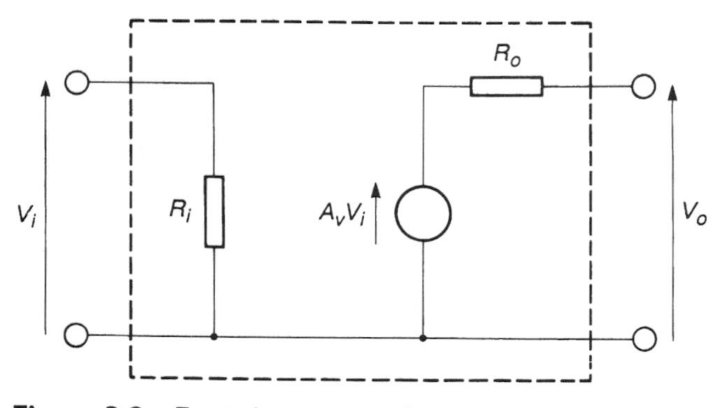

Figure 3.8 Equivalent circuit of an amplifier.

the equivalent circuit has two input terminals and two output terminals as in the amplifier shown in Figure 3.3. The lower terminal in each case forms a reference point, since all voltages must be measured with respect to some reference voltage. In this circuit the input and the output references are joined together. This common reference point is often joined to the **earth** or **chassis** of the system (as in Figure 3.3), and its potential is then taken as the 0 V reference.

Note that the equivalent circuit does not show any connection to a power source. The voltage (or current) source within the equivalent circuit would in practice take power from an external **power supply** of some form, but in this circuit we are only interested in its functional properties – not its source of power.

The advantage of an equivalent circuit is that it allows us easily to calculate the effects that external circuits (which also have input and output resistances) will have on the amplifier. This is illustrated in the following example.

Example 3.2

An amplifier has a voltage gain A_v of 10, an input resistance of 1 kΩ and an output resistance of 10 Ω.

The amplifier is connected to a sinusoidal voltage source of 2 V r.m.s., which has an output resistance of 100 Ω, and to a load resistance of 50 Ω.

What will be the r.m.s. value of the output voltage?

Solution

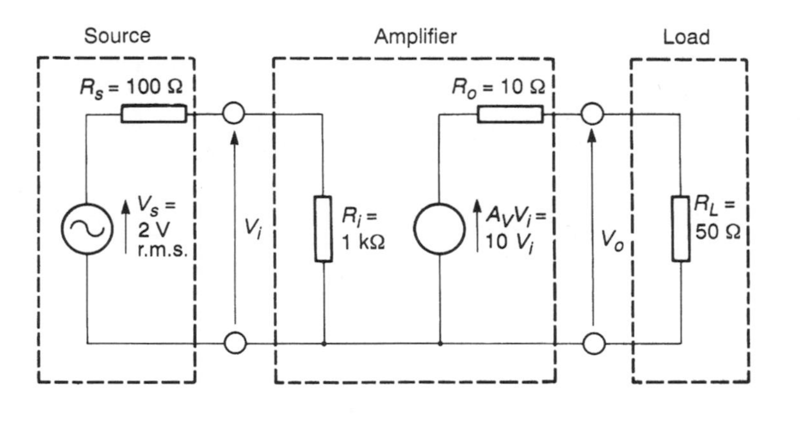

$$V_i = \frac{R_i}{R_s + R_i} V_s$$

$$= \frac{1\ \text{k}\Omega}{100\ \Omega + 1\ \text{k}\Omega} 2\ \text{V r.m.s.}$$

$$V_o = A_v V_i \frac{R_L}{R_o + R_L}$$

$$= 10\, V_i \frac{50\ \Omega}{10\ \Omega + 50\ \Omega}\ \text{V r.m.s.}$$

$$= 10 \frac{1\ \text{k}\Omega}{100\ \Omega + 1\ \text{k}\Omega}\, 2 \frac{50\ \Omega}{10\ \Omega + 50\ \Omega}\ \text{V r.m.s.}$$

$$= 15.2\ \text{V r.m.s.}$$

Computer simulation exercise 3.1

FILE 3A

Simulate the circuit of Example 3.2 replacing the AC voltage source with a DC source chosen to give the same output power. Experiment with different values of the various resistors to investigate their effects on the voltage gain of the circuit. How should the resistors be chosen to maximize the voltage gain?

Example 3.2 shows that the output voltage is considerably less than the product of the source voltage and the voltage gain of the amplifier. The gain of the complete circuit is affected by the values of source and load resistances as well as the characteristics of the amplifier. This effect is known as **loading**. For this reason we should perhaps refer to the gain of the amplifier in isolation as the **open-circuit gain**, since it is the ratio of the output voltage to the input voltage when nothing is connected to the output.

In our discussions of ideal voltage amplifiers we noted that an ideal device would draw no current from the source, and would give an output which was independent of the load. It can be seen from the analysis of the foregoing example that this requires the input resistance R_i to be infinite and the output resistance R_o to be zero. Under these circumstances the voltage gain of the complete circuit would be equal to the open-circuit voltage gain of the amplifier, irrespective of the source and load resistances.

No real amplifier can have an infinite input resistance or zero output resistance. However, if the input resistance is large compared with the source resistance and the output resistance is small compared with the load resistance, the effects of these resistances will be small and may often be neglected. This will produce the maximum voltage gain from the circuit. For these reasons, a good voltage amplifier is characterised by a high input resistance and a low output resistance.

Often the designer does not have a totally free choice for the values of the input and output resistances of the amplifier for a given application. This may be because a particular amplifier has already been chosen or because there are other constraints on the design. In this situation, an alternative is to change the values of the source and load resistances. If the source resistance could be made small compared to the input resistance of the amplifier and the load resistance made large compared with the output resistance of the amplifier, this would again mean that these resistances could be neglected and the maximum voltage gain would be achieved. Thus, when used with voltage amplifiers, it is advantageous for a signal source to have a low resistance (that is, the output resistance of the source should be low) and for a load to have a high resistance (that is, the input resistance of the load should be high). Thus, in circuits concerned with voltage amplification, the input resistances of each stage should be high and the output resistances of each stage should be low to maximize the overall voltage gain.

It should be noted that these characteristics are not beneficial in all forms of amplifier. **Current amplifiers**, for example, should have a low input resistance so that they do not affect the current flowing into the input from a current source, and a high output resistance so that the output current is not affected by an external load resistance.

In many applications the voltage source supplying an input to an amplifier will be a sensor of some form and often the load will be some form of actuator. In these circumstances it may not be possible to change the input or output resistances of these devices to suit the application in question. Under these circumstances it will be necessary to tailor the input and output resistances of the amplifier to suit the sensors and actuators being used. It may also be necessary to allow for the effects of input and output resistances, as illustrated in Example 3.2.

3.6 Output power and power gain

In the last section we looked at the performance of an amplifier in terms of its voltage gain and how this performance is affected by internal and external resistances. We will now consider the performance of the amplifier in terms of the power that it can deliver to an external load.

3.6.1 Output power

The power dissipated in the load resistor (the **output power** P_o) of the circuit given in Example 3.2 may be calculated using Ohm's law. It is simply

$$P_o = \frac{V_o^2}{R_L} = \frac{(15.2)^2}{50} \approx 4.6 \text{ W}$$

We have seen in the previous section that changing the load resistance affects the voltage gain of the circuit and it is not unreasonable to assume that it will also affect the output power. Table 3.1 shows the output voltage and output power produced by the circuit of Example 3.2 for different values of load resistance.

Table 3.1 Variation of output voltage and output power with load resistance in Example 3.2.

Load resistance R_L Ω	Output voltage V_o V r.m.s.	Output power P_o watts
1	1.65	2.7
2	3.03	4.6
3	4.20	5.9
10	9.10	8.3
33	14.0	5.9
50	15.2	4.6
100	16.5	2.7

You will notice that the output voltage increases steadily as the resistance of the load is increased. This is because the output rises as the amplifier is less heavily loaded.

However, the power output initially rises as the resistance of the load is increased from 1 Ω until a maximum is reached. It then drops as the load is increased further. To determine the position of this maximum value we need to look in more detail at the characteristics of the amplifier.

From the analysis of Example 3.2 we have seen that

$$V_o = A_v V_i \frac{R_L}{R_o + R_L} \tag{3.1}$$

Since the power dissipated in a resistance is given by V^2/R, the power dissipated in the load resistance (the output power P_o) is given by

$$P_o = \frac{V_o^2}{R_L} = \frac{\left(A_v V_i \frac{R_L}{R_L + R_o}\right)^2}{R_L} = A_v^2 V_i^2 \frac{R_L}{(R_L + R_o)^2} \tag{3.2}$$

Differentiating this expression for P_o with respect to R_L gives:

$$\frac{dP_o}{dR_L} = \frac{(R_L + R_o)^2 A_v^2 V_i^2 - 2A_v^2 V_i^2 (R_L + R_o) R_L}{(R_L + R_o)^4} \tag{3.3}$$

which must equal zero for a maximum or minimum. This condition is given when the numerator is zero, that is, when

$$(R_L + R_o) - 2R_L = 0$$

giving

$$R_L = R_o$$

Further differentiation of Equation 3.3 will confirm that this is indeed a maximum rather that a minimum value.

Substituting for the component values used in Example 3.2 shows that maximum power should be dissipated in the load when its resistance is equal to R_o which is 10 Ω. The tabulated data in Table 3.1 confirms this result.

Thus in circuits in which the output impedance can be adequately represented by a simple resistance, maximum power is transferred to the load when the load resistance is equal to the output resistance. This result holds for transfers between any two circuits and is a simplified statement of the **maximum power theorem**.

A similar analysis may be performed to investigate the transfer of power between circuits that have complex impedances rather than simple resistances. This produces the more general result that for maximum power transfer the impedance of the load must be equal to the complex conjugate of the impedance of the output. Thus if the output impedance of a network has the value $R + jX$, for maximum power transfer the load should have an impedance of $R - jX$. This implies that if the output impedance has a capacitive component, the load must have an inductive component to obtain maximum output power. It can be seen that the simpler statement given above is a special case of this result in which the reactive component of the output impedance is zero.

The process of choosing a load impedance to maximize **power transfer** is called **impedance matching** and is a very important aspect of circuit design in certain areas. It should be remembered, however, that since maximum power transfer occurs when the load and output impedances are equal, the voltage gain is far from its maximum value under these conditions. In voltage amplifiers it is more common to attempt to maximize input impedances and minimize output impedances to maximize voltage transfer. Similarly, in current amplifiers it may be advisable to have a high output impedance and a low input impedance to produce a high current transfer.

Impedance matching is of importance in circuits where the efficiency of power transfer is paramount. It should be noted, however, that when perfectly matched the power dissipated in the output stage of the source is equal to the power dissipated in the load. Also, when $R_L = R_o$ Equation 3.2 produces the result that

$$P_{o(max)} = \frac{A_v{}^2 V_i{}^2}{4R_L}$$

In other words, the maximum output power is only one-quarter of the power that would be dissipated in the load if the output impedance were zero.

For this reason, impedance matching is seldom used in high power amplifiers as **power efficiency** is of more importance than power transfer. In such cases it is common to attempt to make the output resistance as small as possible to maximize the power delivered to the load for a given output voltage. We will return to this topic in Chapter 8 when we look in more detail at power amplifiers.

Impedance matching finds its main application in low-power **radio frequency (r.f.) amplifiers** where very small signals must be amplified with maximum power transfer.

Computer simulation exercise 3.2

FILE 3A

Use the circuit of Computer simulation exercise 3.1 to investigate the way in which the power output of the circuit of Example 3.2 changes for different values of the various resistors. Use the 'parameter sweep' function of the simulator to determine the value of R_L that produces the greatest power output. Repeat this process using different values of the output resistor R_o.

3.6.2 Power gain

The **power gain** of an amplifier is the ratio of the power supplied by the amplifier to a load to the power absorbed by the amplifier from its source. The input power can be calculated from the input voltage and the input current or, by applying Ohm's law, from a knowledge of the input impedance and either the input voltage or current. Similarly, the output power can be determined from the output voltage and output current, or from one of these and a knowledge of the load impedance.

The power gain of a modern electronic amplifier may be very high, gains of 10^6 or 10^7 being common. With these large numbers it is often convenient to use a logarithmic expression of gain using **decibels** rather than a simple ratio.

The decibel (dB) is a dimensionless figure for **power gain** and is defined by

$$\text{Power gain (dB)} = 10 \log_{10} \frac{P_2}{P_1}$$

where P_2 is the output power and P_1 is the input power of the amplifier or other network.

The use of decibels has several advantages. Firstly, the combined effects of many stages of amplification or attenuation can be found by simply adding the individual gains of each stage when they are expressed in decibels. Secondly, the variations of gain with frequency of many circuits have a simple form when plotted in decibels against frequency on a logarithmic scale. These points will be discussed further in later sections.

Decibels may be used to represent both amplification and attenuation. Although not always trivial to evaluate, they are easily remembered for certain values such as powers of 10. By remembering that $\log_{10} n$ is simply the number to which 10 must be raised to equal n, it is obvious that gains of 10, 100 and 1000 are simply 10 dB, 20 dB and 30 dB, respectively. Similarly, attenuations of 1/10, 1/100 and 1/1000 are simply −10 dB, −20 dB and −30 dB. Clearly a circuit that leaves the power unchanged (a power gain of 1) has a gain of 0 dB.

Other values of importance are gains of +3 dB and −3 dB. These correspond to a doubling and halving of the power, respectively.

From Ohm's law we know that the power dissipated in a resistance R as a result of the application of a voltage V is given by V^2/R. Therefore the gain of an amplifier expressed in decibels may be written as:

$$\text{Power gain (dB)} = 10 \log_{10} \frac{P_2}{P_1} = 10 \log_{10} \frac{V_2^2/R_2}{V_1^2/R_1}$$

where V_1 and V_2 are the input and output voltages, respectively, and R_1 and R_2 are the input and load resistances, respectively.

If, **and only if**, R_1 and R_2 are equal, the power gain of the amplifier is given by:

$$\begin{aligned}\text{Power gain (dB)} &= 10 \log_{10} \frac{V_2^2}{V_1^2} \\ &= 20 \log_{10} \frac{V_2}{V_1} \\ &= 20 \log_{10}(\text{voltage gain})\end{aligned}$$

Some networks do have uniform input and load impedances. In these cases it is often useful to express the gain in decibels rather than as a simple ratio. Note that it is not strictly correct to say, for example, that a circuit has a voltage gain of 10 dB, even though you will often hear such statements. Decibels represent power gain, and what is meant is that the circuit has a voltage gain that corresponds to a power gain of 10 dB. However, it is very common to describe the voltage gain of a circuit in dB as

$$\text{Voltage gain (dB)} = 20 \log_{10} \frac{V_2}{V_1}$$

even when R_1 and R_2 are not equal.

Example 3.3 Calculation of gains in decibels

Express the following in decibels:

(a) a power gain of 100

(b) a voltage gain of 100

(c) a power gain of 1/100

(d) a voltage gain of 1/100

Calculations:

(a) Gain (dB) = 10 $\log_{10}$(power gain) = 10 $\log_{10}$(100) = 20 dB

(b) Gain (dB) = 20 $\log_{10}$(voltage gain) = 20 $\log_{10}$(100) = 40 dB

(c) Gain (dB) = 10 $\log_{10}$(power gain) = 10 $\log_{10}$(0.01) = −20 dB

(d) Gain (dB) = 20 $\log_{10}$(voltage gain) = 20 $\log_{10}$(0.01) = −40 dB

Example 3.4 Interpreting gains expressed in decibels

Determine the voltage and power gains of:

(a) a circuit with a gain of 10 dB

(b) a circuit with a gain of 3 dB

Calculations:

(a) From definition — Gain (dB) = 10 $\log_{10}$ (power gain)
Therefore — 10 = 10 $\log_{10}$ (power gain)
Rearranging — power gain = antilog(1) = 10

From definition — Gain (dB) = 20 $\log_{10}$ (voltage gain)
Therefore — 10 = 20 $\log_{10}$ (voltage gain)
Rearranging — voltage gain = antilog(0.5) = 3.16

(b) From definition — Gain (dB) = 10 $\log_{10}$ (power gain)
Therefore — 3 = 10 $\log_{10}$ (power gain)
Rearranging — power gain = antilog(0.3) = 2

From definition	Gain (dB) = 20 $\log_{10}$ (voltage gain)
Therefore	3 = 20 $\log_{10}$ (voltage gain)
Rearranging	voltage gain = antilog(0.15) = 1.41

3.7 Voltage gain and frequency response

As defined earlier, the **voltage gain** of an amplifier is the ratio of the output voltage to the input voltage. This ratio may also be used to describe **attenuators** which is the name given to networks that reduce the size of a signal. The magnitude of the voltage gain of a voltage attenuator will be less than unity since the output voltage will be less than the input voltage.

No real amplifier has equal gain at all frequencies. Normally when describing the gain of an amplifier we use the **mid-band gain**, that is, the gain of the amplifier in the middle of its normal operating frequency range. The variation of gain with frequency is called the **frequency response** of the amplifier.

The gain of all amplifiers falls at high frequencies for reasons that will be explained in Section 3.7.2. This effect is quantified for a particular amplifier by giving its **high frequency cut-off** (or **upper cut-off**), that is, the frequency at which the gain falls by 3 dB compared with the mid-band gain. This corresponds to the power gain falling to half of its mid-band value. Since the power output of an amplifier is proportional to the square of the output voltage, it follows that the power gain is proportional to the square of the voltage gain. This implies that the voltage gain is proportional to the square root of the power gain. Thus, at the cut-off point, since the power gain is half of its mid-band value, the voltage gain has fallen to $1/\sqrt{2}$, or about 0.707, times its mid-band value.

Some amplifiers have a frequency response that remains constant at frequencies down to DC. Such amplifiers are referred to as **DC coupled amplifiers** since the DC component of any input signal is not removed by the circuitry which couples the various sections of the amplifier together. In other circuits, **coupling capacitors** are used to pass the signal from one stage to another. If appropriately chosen, these have little effect on alternating signals in the frequency range for which the amplifier is designed but block any direct component of the signal. Circuits of this type are referred to as **AC coupled amplifiers**. We shall see in later chapters that this blocking effect is of great importance in many circuits. However, the presence of coupling capacitors also reduces the gain of the amplifier at low frequencies and produces a **low frequency cut-off** (or **lower cut-off**). As with the high frequency cut-off, this is the frequency at which the gain falls by 3 dB from its mid-band value.

Figure 3.9 shows a typical amplifier frequency response. Because of the wide frequency range of many amplifiers it is common to plot frequency on a logarithmic scale. We shall see shortly that this has other advantages since it produces graphs of a simpler form.

3.7.1 Low frequency cut-off

A low frequency cut-off is often caused by the use of a coupling capacitor. Figure 3.10 illustrates a typical arrangement in which the output of one stage is fed into the input of

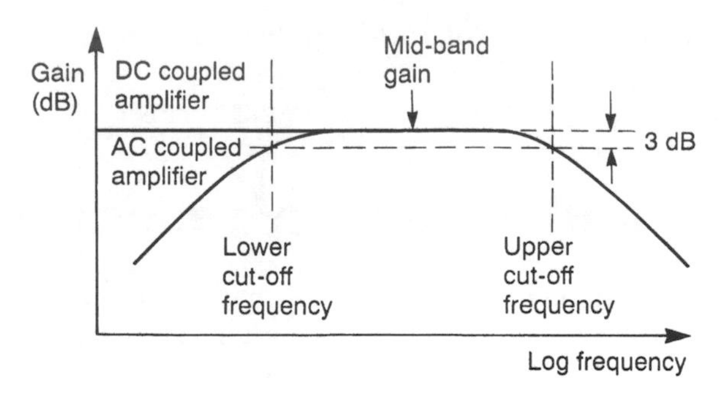

Figure 3.9 A typical amplifier frequency response.

the next through a coupling capacitor. Here the source is shown as an ideal voltage generator; we shall see the effects of the output resistance of the source later in this section.

The arrangement shown in Figure 3.10 has an input voltage V_i and an output voltage V_o and we may therefore refer to the gain of the network as:

$$\text{Voltage gain } A_v = \frac{V_o}{V_i} = \frac{R}{R + \dfrac{1}{\mathrm{j}\omega C}} = \frac{1}{1 + \dfrac{1}{\mathrm{j}\omega CR}} \tag{3.4}$$

Clearly at high frequencies ω is large and the value of $1/\mathrm{j}\omega CR$ is small compared with 1. Therefore the voltage gain is approximately equal to 1 and the coupling capacitor has negligible effects.

However, at lower frequencies the magnitude of $1/\mathrm{j}\omega CR$ becomes more significant and the gain of the network decreases. The magnitude of the voltage gain is given by:

$$|\text{Voltage gain}| = \frac{1}{\sqrt{1 + \left(\dfrac{1}{\omega CR}\right)^2}} \tag{3.5}$$

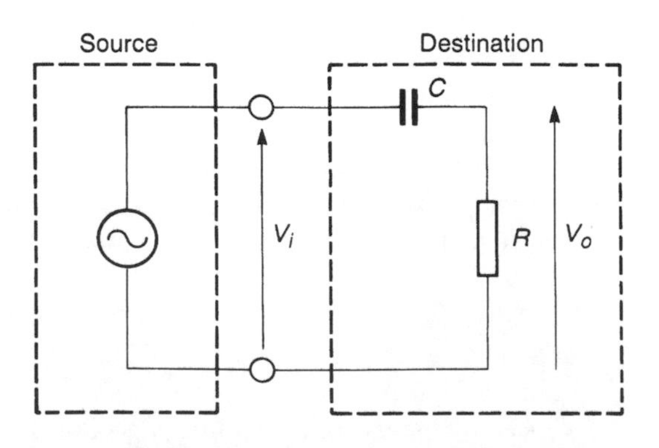

Figure 3.10 A capacitively coupled network.

When the value of $\dfrac{1}{\omega CR}$ is equal to 1,

$$|\text{Voltage gain}| = \frac{1}{\sqrt{1+1}} = \frac{1}{\sqrt{2}} = 0.707$$

This is equivalent to a power gain of −3 dB and this condition therefore corresponds to the **lower cut-off frequency**. If the angular frequency corresponding to the lower cut-off frequency is given the symbol ω_L, then

$$\omega_L = \frac{1}{CR} = \frac{1}{\mathrm{T}_L} \text{ rad/s}$$

where $\mathrm{T}_L = CR$ is the **time constant** of the capacitor–resistor combination which produces the lower cut-off frequency.

Since it is more convenient to deal with frequencies in hertz than in radians per second, we may use the relationship $\omega = 2\pi f$ to calculate the corresponding lower cut-off frequency f_L expressed in hertz.

$$f_L = \frac{\omega_L}{2\pi} = \frac{1}{2\pi CR} \text{ Hz}$$

If we substitute for ω ($\omega = 2\pi f$) and CR ($CR = 1/2\pi f_L$) in Equation 3.4, we obtain an expression for the gain of the network in terms of the signal frequency f and the cut-off frequency f_L:

$$\text{Voltage gain } A_v = \frac{1}{1+\dfrac{1}{\mathrm{j}\omega CR}} = \frac{1}{1+\dfrac{1}{\mathrm{j}(2\pi f)\left(\dfrac{1}{2\pi f_L}\right)}} = \frac{1}{1+\dfrac{f_L}{\mathrm{j}f}} = \frac{1}{1-\mathrm{j}\dfrac{f_L}{f}} \tag{3.6}$$

It can be seen that the inclusion of a capacitor has added an imaginary component into the expression for the gain of the arrangement. This will have the effect of producing a **phase shift** between the output and the input signals. In practice all amplifiers produce phase shifts at some frequencies. This means that to describe the gain of an amplifier we need to specify not only the *magnitude* of the gain but also its *phase*. A common way of representing this is to give the magnitude followed by the phase, such as $20\angle 90$, meaning a gain with a magnitude of 20 and a phase of 90°, or $10\angle 180$, meaning a gain with a magnitude of 10 and a phase of 180°. The phase angles given represent the phase of the output with respect to the input; positive values indicate a phase *lead,* while negative values indicate a phase *lag*. If a gain is given without a phase angle this normally means that the gain phase angle is zero – that is, the output is in phase with the input. Another 'special case' is the condition that the phase angle is 180°. For a sinusoidal signal this means that the output is shifted by 180° with respect to the input (this is often described as the output being 'out-of-phase' with the input) while for DC signals it means that a positive input produces a negative output. This kind of amplifier is called an **inverting amplifier** and these will be discussed in more detail in the next chapter.

It is common to give the gain of an inverting amplifier as a negative number. Thus giving the gain of an amplifier as −10 indicates that the magnitude of the gain is 10 and the phase of the gain is 180°.

Returning to our discussion of the capacitively coupled network, it is interesting to look at the characteristics of this relationship in different regions of the frequency spectrum.

When $f \gg f_L$

The first area of interest is the region well above the cut-off frequency, when f is much greater than f_L. From Equation 3.6 it is clear that under these conditions f_L/f is much less than unity so the magnitude of the voltage gain is approximately 1. Note that here the imaginary part of the gain is negligible and so the gain is effectively real and there is negligible phase shift. Thus, at high frequencies the network passes the signal with little effect on its magnitude or phase; the gain is therefore approximately $1 \angle 0$, or simply 1.

When $f = f_L$

When the frequency f is equal to the cut-off frequency f_L, the voltage gain is given by

$$\text{Voltage gain } A_v = \frac{1}{1 - j\dfrac{f_L}{f}} = \frac{1}{1 - j}$$

Multiplying the numerator and the denominator by $(1 + j)$ gives

$$\text{Voltage gain } A_v = \frac{(1 + j)}{(1 - j)(1 + j)} = \frac{(1 + j)}{2} = 0.5 + j0.5$$

This can be represented by a phasor diagram, as shown in Figure 3.11, which shows that the magnitude of the gain at the cut-off frequency is 0.707. This is consistent with our earlier analysis which predicted that the magnitude of the gain at the cut-off point should be 0.707 times the mid-band gain. In this case the mid-band gain is the gain at a frequency some way above the lower cut-off frequency, which we have just shown to be 1. The phasor diagram also shows that the phase angle at the cut-off frequency is +45°. This shows that the output *leads* the input by 45°. The gain of the arrangement at the cut-off frequency is therefore $0.707 \angle 45$.

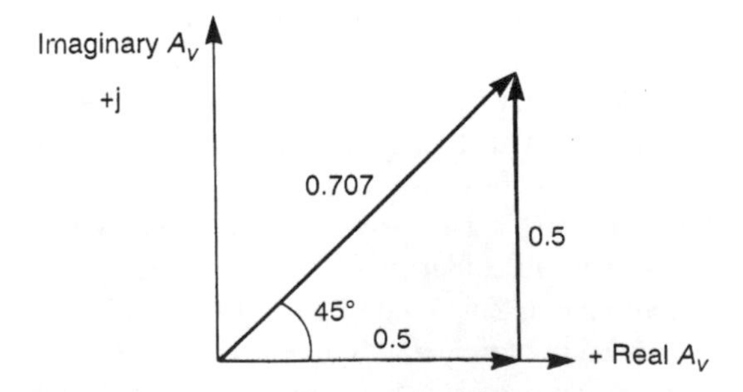

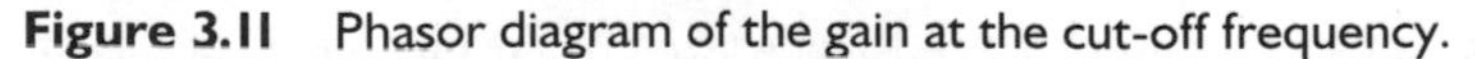

Figure 3.11 Phasor diagram of the gain at the cut-off frequency.

When $f \ll f_L$

The third region of interest is where the frequency is considerably below the cut-off frequency. Here f is much less than f_L and so f_L/f is much greater than 1. Therefore the voltage gain is given by

$$\text{Voltage gain } A_v = \frac{1}{1 - j\dfrac{f_L}{f}} \approx \frac{1}{-j\dfrac{f_L}{f}} = j\frac{f}{f_L}$$

The 'j' signifies that the gain is imaginary, as shown in the phasor diagram of Figure 3.12. This shows that the magnitude of the gain is simply f/f_L and that there is a phase shift of 90°. The polarity of the shift indicates that the output leads the input by 90°.

Since f_L is constant, in this region the voltage gain is linearly related to frequency. If the frequency is halved the voltage gain will be halved. Therefore the gain falls by a factor of 0.5 for every octave drop in frequency (an **octave** is a doubling or halving of frequency and is equivalent to an octave jump on a piano or other musical instrument). A fall in voltage gain by a factor of 0.5 is equivalent to a change in gain of −6 dB. Therefore the rate of fall of gain can be expressed as a fall of 6 dB per octave. An alternative way of expressing the rate of change of gain is to specify the change of gain for a decade change in frequency (a **decade**, as its name suggests, is a change in frequency of a factor of 10). If the frequency falls to 0.1 of its previous value, the voltage gain will also drop to 0.1 of its previous value. This represents a change in gain of −20 dB. Thus the rate of change of gain is 20 dB per decade.

Figure 3.13 shows the gain and phase response of the network of Figure 3.10 for frequencies on either side of the cut-off frequency.

At frequencies much greater than the cut-off frequency, the magnitude of the gain tends to a straight line corresponding to a gain of 0 dB (that is, a gain of 1). Therefore this line (shown dotted in Figure 3.13) forms an asymptote to the response. At frequencies much less than the cut-off frequency, the response tends to a straight line drawn at a slope of 6 dB per octave (20 dB per decade) change in frequency. This line forms a second asymptote to the response and is also shown dotted on Figure 3.13. This line intersects the 0 dB line at the cut-off frequency. At frequencies considerably above or below the cut-off frequency, the gain response tends towards these two asymptotes. Near the

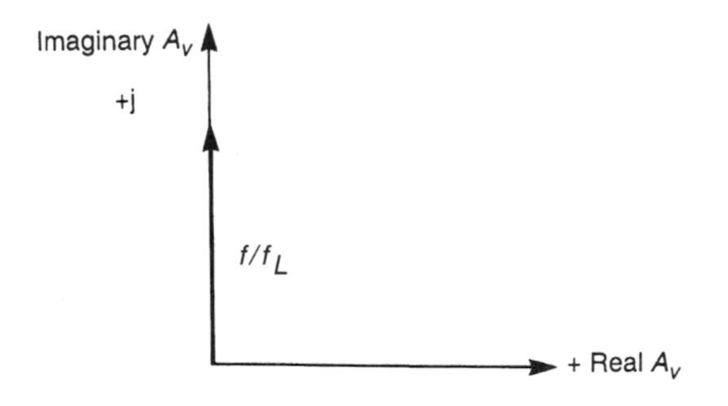

Figure 3.12 Phasor diagram of the gain when $f \ll f_L$.

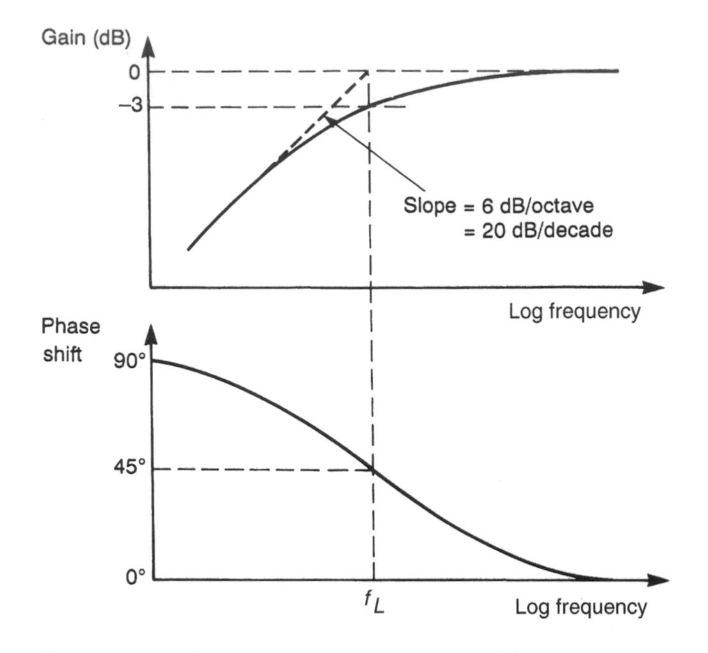

Figure 3.13 Gain and phase responses of the capacitor coupled network.

cut-off frequency the gain deviates from the two straight lines and is 3 dB below their intersection at the cut-off frequency.

Figure 3.13 also shows the variation of phase with frequency of the capacitively coupled network. At frequencies well above the cut-off frequency the network produces very little phase shift and its effects may generally be ignored. However, as the frequency decreases the phase shift produced by the arrangement increases, reaching 45° at the cut-off frequency and increasing to 90° at very low frequencies.

It can be seen that the capacitively coupled network passes signals of some frequencies with little effect but that signals of other frequencies are attenuated and are subjected to a phase shift. Thus signals that have a frequency near or below the cut-off frequency (or which have components that are near or below the cut-off frequency) are subjected to **amplitude** and **phase distortion** by this arrangement. The network has the characteristics of a **high-pass filter** since it allows high frequency signals to pass but filters out low frequency signals. We shall look at filters in more detail in Chapter 8.

Example 3.5 Octaves and decades

Determine the frequencies corresponding to:

(a) an octave above 100 Hz

(b) three octaves above 1 kHz

(c) an octave below 10 kHz

(d) a decade above 50 Hz

(e) three decades below 1 MHz

(f) two decades above 100 kHz

Calculations:

(a) an octave above 100 Hz = 100 × 2 = 200 Hz

(b) three octaves above 1 kHz = 1000 × 2 × 2 × 2 = 8 kHz

(c) an octave below 10 kHz = 10 000 ÷ 2 = 5 kHz

(d) a decade above 50 Hz = 50 × 10 = 500 Hz

(e) three decades below 1 MHz = 1 000 000 ÷ 10 ÷ 10 ÷ 10 = 1 kHz

(f) two decades above 100 kHz = 100 000 × 10 × 10 = 10 MHz

The effects of source resistance

So far in our discussions of the effects of coupling capacitors we have assumed, as in Figure 3.10, that the source resistance is zero. Let us now consider the arrangement shown in Figure 3.14 which is similar to the earlier network with the addition of a source resistance R_s.

If we derive an expression for the voltage gain of this network, equivalent to that given for the previous network in Equation 3.4, we obtain:

$$\text{Voltage gain } A_v = \frac{V_o}{V_i} = \frac{R}{R_s + \dfrac{1}{j\omega C} + R} \tag{3.7}$$

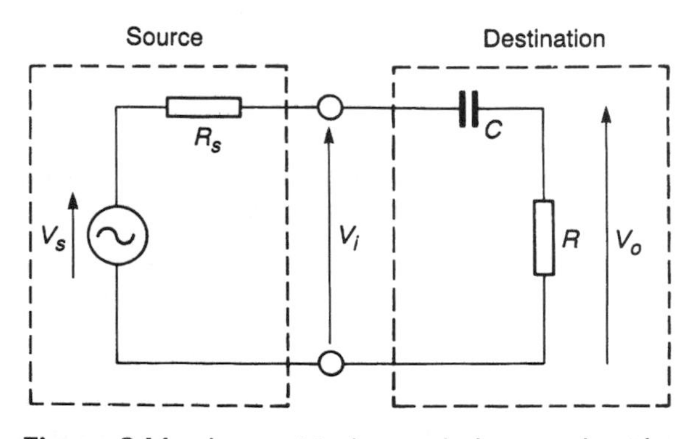

Figure 3.14 A capacitively coupled network with source resistance.

$$= \frac{R}{R+Rs} \times \frac{1}{1+\frac{1}{j\omega C(R+R_s)}}$$

or, by rearranging

$$\text{Voltage gain } A_v = \frac{A'}{1+\frac{\omega_L}{j\omega}} = \frac{A'}{1-j\frac{\omega_L}{\omega}} = \frac{A'}{1-j\frac{f_L}{f}} \tag{3.8}$$

where $A' = R/(R+R_s)$, $\omega_L = \frac{1}{C(R+R_s)} = \frac{1}{T_L}$ and $f_L = \frac{\omega_L}{2\pi}$.

A' represents the gain which the circuit would have if the coupling capacitor were a short circuit. This is simply the attenuation produced by the potential divider formed by R_s and R. By similarity with Equation 3.6 it is clear that f_L represents the cut-off frequency of the new network. Note that this value for f_L is not the same as before, since the time constant T_L is different. The time constant is given by the product of the capacitance and the total resistance in series with it. Hence the presence of the source resistance changes not only the gain of the network but also its cut-off frequency. However, the similarity between Equations 3.6 and 3.8 indicates that the overall form of the responses will be the same. Thus the rate of fall of the gain and the variation in the phase angle will be similar but with a different cut-off frequency.

A fall in gain at low frequencies is not always caused by the presence of coupling capacitors. Other circuit configurations may also cause this effect, but in general the relationship between the magnitude and phase of the gain, and frequency, is of a similar form to that described above. We shall be discussing some of these circuits in later chapters. We will also consider the effects of combining circuits that each have a low frequency cut-off when we come to look at the cascading of amplifiers later in this chapter.

FILE 3B

Computer simulation exercise 3.3

Pick sample values for the components of Figure 3.10 and calculate the cut-off frequency. Simulate the circuit using the component values chosen and plot the gain and phase of the output as the input frequency is swept from several decades below to several decades above the calculated frequency. Estimate the cut-off frequency from these plots and compare this with the predicted value.

Add a source resistor to the simulated circuit to produce the arrangement of Figure 3.14. Calculate the effect of this modification on the gain and the cut-off frequency and compare the new values with those obtained by simulation.

3.7.2 High frequency cut-off

We have seen that the presence of capacitance in *series* with the input of a network (as in a coupling capacitor) produces a reduction of gain at low frequencies. It is therefore not perhaps totally surprising to discover that the presence of capacitance in *parallel*

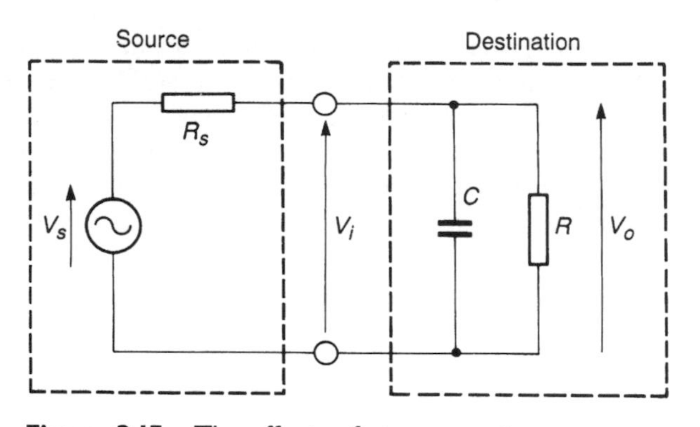

Figure 3.15 The effects of stray capacitance.

with the input of a network produces a reduction of gain at *high* frequencies. Unlike coupling capacitors, which are introduced into the circuit by the designer, parallel capacitance is often present as an undesired side effect of the method of construction of the circuit. Any two adjacent conductors exhibit capacitance; when this is unintentional it is called **stray capacitance**. Since signals are fed from point to point within circuits by conductors, there is inevitably a certain amount of stray capacitance between the signal paths and ground.

Figure 3.15 represents the effects of stray capacitance on the transfer of a signal from one stage to another. The capacitor C corresponds to the stray capacitance between the signal path and ground. The arrangement is a potential divider network in which the applied voltage is divided between the voltage across the source resistance and that across the parallel combination of R and C. The impedance Z of the parallel combination of R and C is given by

$$\frac{1}{Z} = \frac{1}{R} + j\omega C$$

and thus

$$Z = \frac{1}{\frac{1}{R} + j\omega C} = \frac{R}{1 + j\omega CR}$$

Therefore the voltage gain of the network may be expressed as

$$\text{Voltage gain } A_v = \frac{\frac{R}{1 + j\omega CR}}{R_s + \frac{R}{1 + j\omega CR}} = \frac{\frac{R}{R + R_s}}{1 + j\omega C\left(\frac{RR_s}{R + R_s}\right)}$$

$$= \frac{A'}{1 + j\omega CR_{eff}} = \frac{A'}{1 + j\frac{\omega}{\omega_U}}$$

where $A' = R/(R+R_s)$ is the gain which the network would have in the absence of the stray capacitance, $R_{eff} = RR_s/(R+R_s)$ is the effective resistance of the parallel combination of R and R_s and $\omega_U = 1/CR_{eff}$ is the upper cut-off frequency. As with the lower cut-off frequency, the product of the resistance and the capacitance gives the **time constant** of the arrangement. In this case the product of R_{eff} and C gives the time constant T_U which produces the upper cut-off frequency. Hence

$$\omega_U = \frac{1}{CR_{eff}} = \frac{1}{T_U} \text{ rad/s}$$

or

$$f_U = \frac{\omega_U}{2\pi} = \frac{1}{2\pi CR_{eff}} \text{ Hz}$$

Hence the voltage gain is given by

$$\text{Voltage gain } A_v = \frac{A'}{1 + j\dfrac{\omega}{\omega_U}} = \frac{A'}{1 + j\dfrac{f}{f_U}} \tag{3.9}$$

Since the voltage across the voltage generator is constant, irrespective of the current through it, it behaves as though it has zero internal resistance. In determining the resistance across the capacitor C_s one can therefore consider that the source resistance R_s is effectively connected directly across the capacitor and so R_{eff} represents the effective resistance in parallel with the capacitance C. Thus the time constant T_U is given by the product of the capacitance C and the effective resistance in parallel with it.

Comparing Equations 3.8 and 3.9 it is apparent that the form of the expressions for the voltage gain is similar in each case but that the positions of ω and ω_U are interchanged. This causes the voltage gain of the network with stray capacitance to fall at high frequencies rather than at low frequencies, as in the earlier case. For this reason the cut-off frequency produced by stray capacitance is called an **upper cut-off frequency**. The polarities of the imaginary parts of the denominators of the two expressions are also different. The network with stray capacitance thus generates a *phase lag* rather than a *phase lead*, as in the capacitively coupled arrangement.

It is again interesting to look at the response of the network in three frequency ranges.

When $f \ll f_U$

When $f \ll f_U$ then f/f_U is considerably less than unity and the voltage gain is approximately equal to A'. Under these conditions the imaginary part of the voltage gain is negligible, and thus the circuit produces no significant phase shift. Thus, at frequencies considerably below the cut-off frequency the stray capacitance has negligible effect on the gain or phase of signals passing through it.

When $f = f_U$

When $f = f_U$, the ratio f/f_U is equal to 1, and the voltage gain is given by

$$\text{Voltage gain } A_v = \frac{A'}{1 + j\dfrac{f}{f_U}}$$

$$= \frac{A'}{1 + j}$$

$$= A' \times \frac{1 - j}{(1 + j)(1 - j)}$$

$$= \frac{A'}{2} \times (1 - j)$$

$$= 0.5A' - j0.5A'$$

This may be represented by the phasor diagram shown in Figure 3.16. This shows that the magnitude of the gain of the network has dropped to 0.707 of its low frequency value, corresponding to −3 dB, confirming our assumption that this condition represents the upper cut-off frequency. The phase angle of the gain is −45°, that is, a phase *lag* of 45°. The gain of the network is thus 0.707 ∠−45. You may care to compare these results with those obtained for the lower cut-off frequency.

When $f \gg f_U$

When $f \gg f_U$, the ratio f/f_U is considerably greater than unity and the voltage gain of the network is approximately given by

$$\text{Voltage gain } A_v = \frac{A'}{1 + j\dfrac{f}{f_U}}$$

$$\approx \frac{A'}{j\dfrac{f}{f_U}}$$

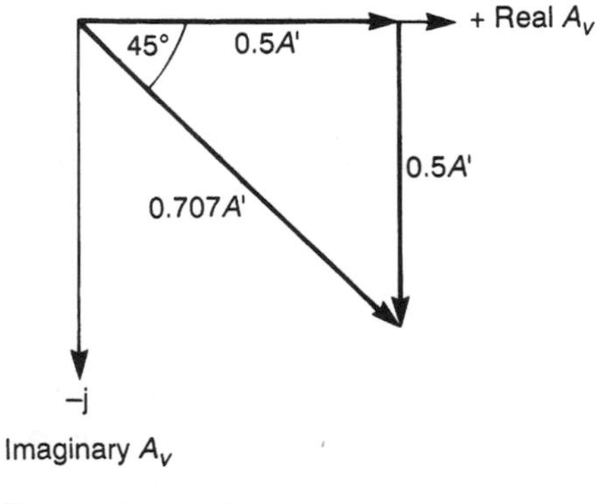

Figure 3.16 Phasor diagram of the gain at the upper cut-off frequency.

$$= A' \times -j\frac{f_U}{f}$$

This represents a gain which is imaginary and has a phase shift of –90°, that is, a phase *lag* of 90°. The magnitude of the gain in this region falls with increasing frequency such that a doubling of the frequency results in a halving of the gain, and a ten fold increase in frequency results in a fall in gain of 90%. Thus the gain changes by –6 dB/octave or –20 dB/decade. Figure 3.17 shows the gain and phase response of the network of Figure 3.15 for frequencies on either side of the cut-off frequency.

A fall in gain at high frequencies is not always associated with the presence of stray capacitances, although their presence always limits the ultimate high frequency performance which a designer can obtain from a circuit. Sometimes a designer may intentionally introduce a high frequency cut-off to determine the high frequency performance, rather than letting it be set by unknown stray capacitance. However, the variation of the magnitude and phase of the gain with frequency is usually of the form outlined above and may therefore be understood in the same way.

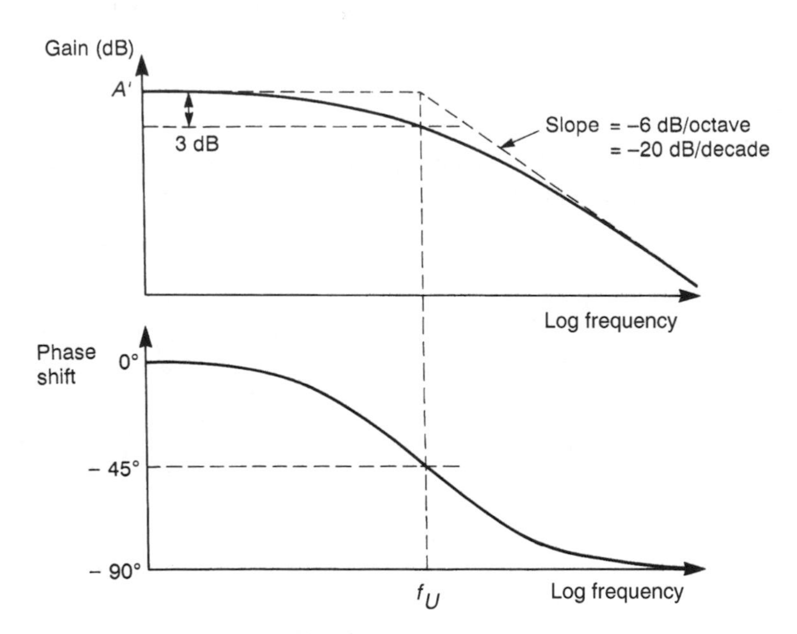

Figure 3.17 Gain and phase responses of the network with stray capacitance.

FILE 3C

Computer simulation exercise 3.4

Select sample values for the circuit of Figure 3.15 and calculate the low frequency gain and the cut-off frequency. Simulate the circuit and plot the gain and phase responses. Determine the gain and the cut-off frequency from these plots and compare these with the calculated values.

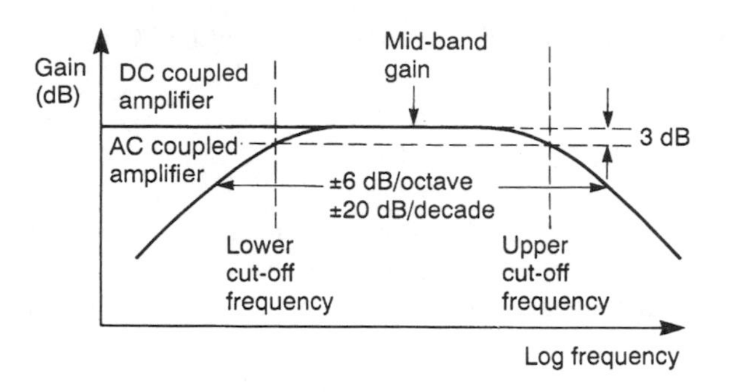

Figure 3.18 Typical Bode plots for AC and DC coupled amplifiers.

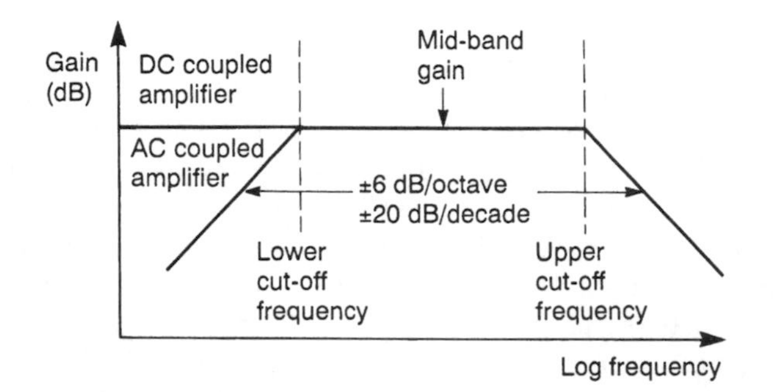

Figure 3.19 Straight-line approximation to the Bode plot of Figure 3.18.

3.7.3 The combined effects of low and high frequency cut-offs

The characteristics of amplifiers at low and high frequencies outlined above may be represented in a single diagram, as shown in Figure 3. 18. This method of describing the magnitude and phase of the gain of an amplifier, using logarithmic gain and frequency scales, is called the **Bode diagram** or **Bode plot** for the circuit.

It is often useful to represent the magnitude response of a Bode plot by a **straight-line approximation** using the asymptotes discussed earlier. This is shown in Figure 3.19. Such a diagram, sometimes called the **asymptotic response** of the network, has lines which intersect at points known as **break** or **corner frequencies**. In the simple amplifier described here these occur at the upper and lower cut-off frequencies described earlier.

3.8 Noise

All real electronic circuits add noise to signals passing through them. In most cases efforts are made to reduce the magnitude of this noise as it ultimately limits the performance of the system. Noise is generated by a number of different mechanisms. In the last chapter we discussed **thermal noise** and we will now consider this further. Other noise sources will be covered in later chapters.

3.8.1 Thermal noise

We noted in Chapter 2 that all electronic components generate a **thermal noise** voltage as a result of the motion of their atoms. We also noted that the noise voltage increases with the resistance of the component, its temperature and the bandwidth over which the noise is measured.

Having derived equivalent circuits for the output of a circuit we can now represent the presence of noise by adding a voltage generator V_n or current generator I_n as shown in Figure 3.20. These circuits may be used to represent the noise produced by a sensor and to allow its effects to be investigated.

A typical amplifier circuit will contain many resistors, each contributing thermal noise to the signal. Clearly noise introduced near the input of the amplifier is going to have more effect than that introduced close to the output since the former will be amplified by the gain of the amplifier. For this reason, in applications where a low noise level is required, great emphasis is placed on producing a low-noise input stage.

Rather than consider a large number of separate noise sources within an amplifier, it is common to combine their effects and to represent them by a single equivalent noise source. Since the noise produced at the output of an amplifier is affected by its gain, it is common to represent noise by an equivalent noise source at the input, as shown in Figure 3.21.

The applied input voltage V_i is applied across the series combination of the input resistance R_i and the noise source V_n. The voltage across the input resistance (which in turn determines the output voltage) is thus $V_i - V_n$. Since the noise is random, the polarity of the noise is unimportant.

Here we have considered only thermal noise. The equivalent circuit shown in Figure 3.21 may also be used to represent the combined effects of noise from a number of sources by using an appropriate noise voltage V_n. We will return to the discussion of noise sources in Chapter 8.

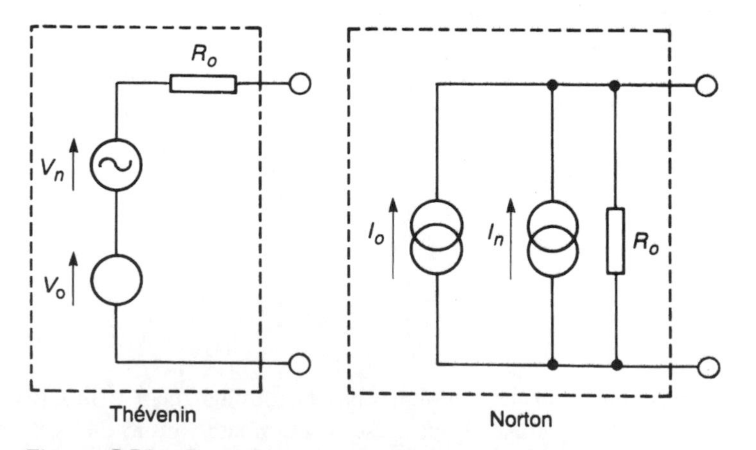

Figure 3.20 Equivalent circuits of output networks with noise.

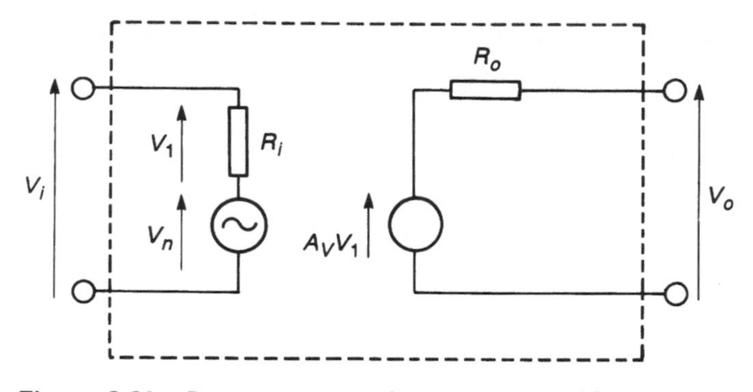

Figure 3.21 Representation of noise in an amplifier.

3.8.2 Signal-to-noise ratio

It is often useful to have a quantitative method of describing the quality of a signal in terms of its corruption by noise. This may be done by measuring its **signal-to-noise ratio S/N**, that is, the ratio of the magnitude of the signal to that of the noise. This is most commonly expressed as the ratio of the signal power P_s to the noise power P_n. Since this is a power ratio it may be expressed in decibels as discussed earlier.

The signal-to-noise ratio may therefore be defined as

$$\text{S/N ratio} = 10 \log_{10}\left(\frac{P_s}{P_n}\right) \text{dB}$$

Since both the signal and the noise are present at the same point within a circuit, they are applied to the same impedance. Thus the power ratio may be expressed as a ratio of the root mean square (r.m.s.) signal voltage V_s to the r.m.s. noise voltage V_n.

$$\text{S/N ratio} = 10 \log_{10}\left(\frac{V_s}{V_n}\right)^2 \text{dB}$$

$$= 20 \log_{10}\left(\frac{V_s}{V_n}\right) \text{dB}$$

The signal-to-noise ratio within a given system will vary with the magnitude of the signal. If the signal becomes very small, the relative size of the noise will increase and the signal-to-noise ratio will become smaller. For this reason it is advantageous to represent quantities by the largest possible signals to produce the highest signal-to-noise ratio. It is often useful to define the maximum achievable signal-to-noise ratio for a given system. This is given by the ratio of the largest possible signal to the noise present. When expressed as a ratio of r.m.s. voltages this is

$$\text{max S/N ratio} = 20 \log_{10}\left(\frac{V_{s(\max)}}{V_n}\right) \text{dB}$$

Example 3.6 Calculation of signal-to-noise ratio

At a particular point in a circuit a signal of 2.5 volts r.m.s. is corrupted by 10 mV r.m.s. of noise. What is the S/N ratio at this point?

From the above

$$\text{S/N ratio} = 20\log_{10}\left(\frac{V_s}{V_n}\right) = 20\log_{10}\left(\frac{2.5}{0.01}\right) = 48\text{ dB}$$

3.9 Cascaded amplifiers

So far we have considered systems that use only single amplifiers. In many cases a single stage of amplification will not produce all the characteristics we require and it is necessary to combine many stages by connecting them in series, that is, by **cascading** them.

Having derived methods of describing the characteristics of an amplifier in terms of its gain, input impedance, output impedance, frequency response and noise performance, we must now consider how we may deduce the overall performance of a combination of stages from their individual characteristics.

Gain

If an amplifier with a gain of A is connected in series with an amplifier of gain B, the gain of the resulting arrangement will be simply the product of their gains AB. This will be true whether A and B are both voltage gains, current gains or power gains, provided that they are gains of the same type. It should be remembered, however, that the gain of an amplifier is affected by the impedance of the source and the load applied to it. Therefore, it is important that the gains A and B in our example are the gains of the amplifiers when connected in this configuration. Some systems arrange that the input and output impedances of all stages are equal. This simplifies gain calculations since the source and load impedances are known and the gain of each stage will not be affected by the order in which they are connected. This arrangement is common in communication systems where the input and output impedances are often chosen to match the **characteristic impedance** of the cable being used. This minimises reflection of the signals at the inputs and outputs of the amplifiers.

The overall gain of a large number of stages of amplification and attenuation may be found simply by determining the product of the gains of the individual stages. This may be calculated more easily if the gain of each stage is expressed in decibels, as shown in Figure 3.22.

Input and output impedances

In most cases the input impedance of an amplifier is not affected by what is connected to its output, and the output impedance is not affected by what is connected to its input. This is shown in the simple equivalent circuits given in Section 3.5. Thus the input impedance of a series of cascaded amplifiers is generally simply the input impedance of the first amplifier in the chain while the output impedance is the output impedance of the last.

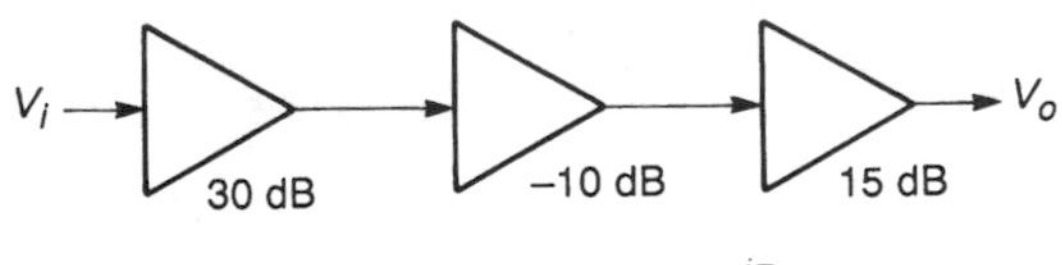

Figure 3.22 Gain calculation for cascaded amplifiers.

In some cases, the operation of the amplifier is such that the load does influence the input impedance. In such cases a more sophisticated equivalent circuit is required to incorporate this characteristic; this must be taken into account when determining the input impedance of a combination of amplifiers.

Frequency response

The frequency response of a cascaded series of amplifiers may be determined by combining the effects of each stage at each frequency. The resultant gain is determined, as above, by multiplying the gains of each stage at each frequency, while the resultant phase shift is determined by summing the phase shifts produced by each stage. As before, determination of the gain response is much easier if the gains are expressed in decibels, since the gain values may then be simply added together. This is illustrated in Figure 3.23 which shows the result of combining two AC coupled amplifiers with different cut-off frequencies.

In the example given in Figure 3.23, the lower cut-off frequency of the combination of amplifiers is equal to the higher of the cut-off frequencies of the constituent stages (in this case f_2). This is because f_1 is sufficiently far below f_2 that the former has a negligible effect on the overall gain at frequencies close to f_2. Had f_1 and f_2 coincided, the asymptote representing the gain would have fallen at 12 dB/octave from this frequency, but the cut-off frequency would be at a slightly higher frequency since both stages produce a fall in gain of 3 dB at this frequency (compared to their mid-band gain). Thus their combined effect at this frequency would be a fall of 6 dB. The 3 dB point, which determines the cut-off frequency, would therefore be at a slightly higher frequency.

The upper cut-off frequency of the combination is determined by the lower of the upper cut-off frequencies of the various stages. As before, if the upper cut-off frequencies are close together the overall cut-off frequency of the combination will be slightly lower than the lowest of the amplifiers in the chain.

Combinations of more than two amplifiers may be dealt with in precisely the same manner by summing the effects of each stage.

Noise

Since noise is a **random phenomenon**, it is not possible simply to add the r.m.s. noise voltages of the various stages to obtain the r.m.s. noise voltage of the combination.

As was noted earlier, it is noise generated near the input of an amplifier that is most important since this is amplified by the gain of the amplifier. Similarly, for a chain of cascaded amplifiers, noise generated by the first amplifier will tend to dominate the noise

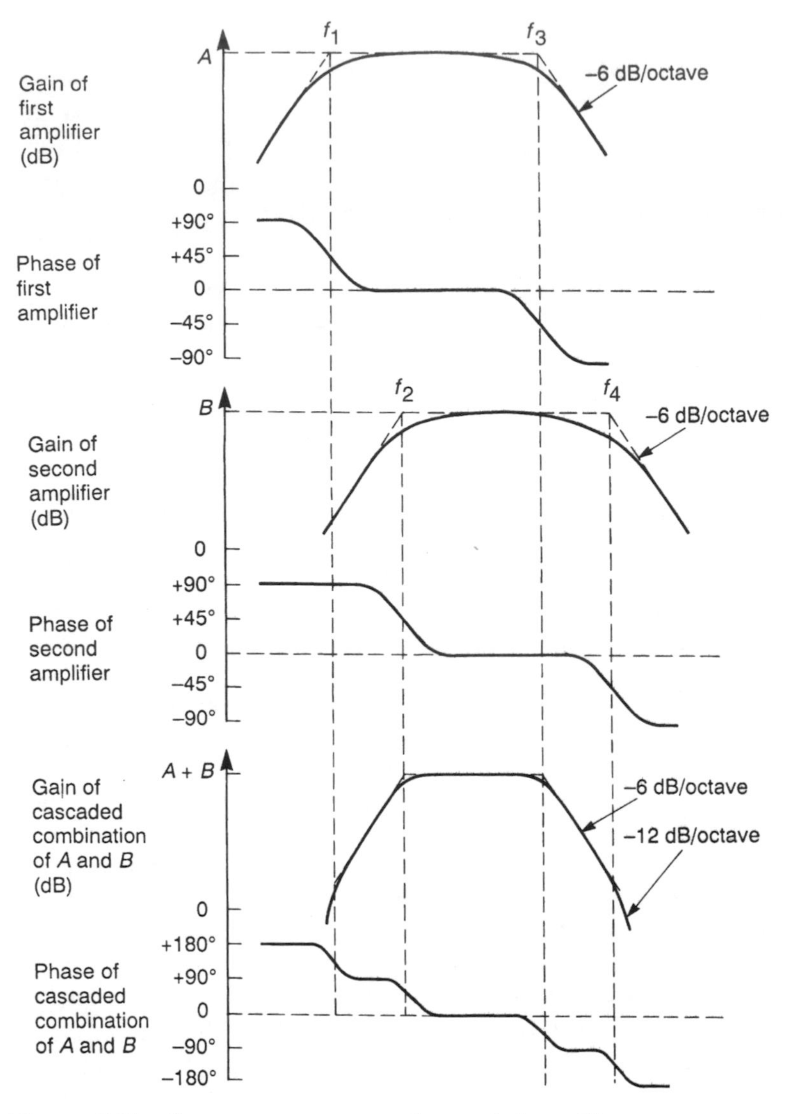

Figure 3.23 Frequency response of cascaded amplifiers.

performance of the system, particularly if the gain of this stage is high. Thus, to a first approximation, the signal-to-noise ratio of the series of amplifiers will be equal to that of the first amplifier in the chain.

3.10 Differential amplifiers

The amplifiers considered so far take as their input a single voltage which is measured with respect to some reference voltage, which is usually earth (0 V). Some amplifiers have

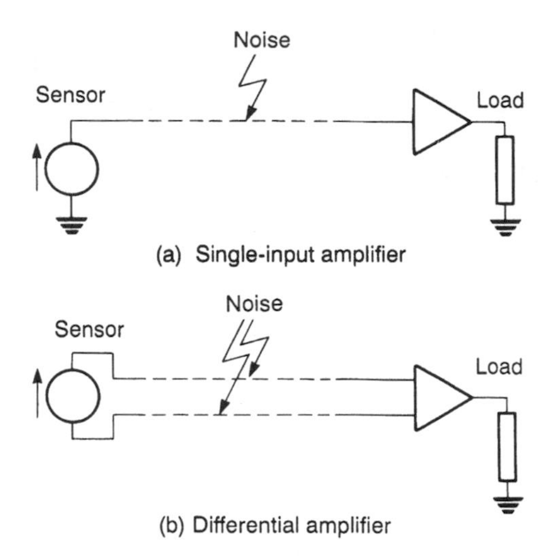

Figure 3.24 A comparison of single and differential input methods.

not one but two inputs and produce an output proportional to the difference between the voltages on these inputs. Such amplifiers are called **differential amplifiers**.

Since a differential amplifier produces an output that is proportional to the difference between two input signals, it goes without saying that if the same voltage is applied to both inputs, no output will be produced. Voltages that are common to both inputs are called **common-mode signals**, and the ability of an amplifier to reject such input signals is termed its **common-mode rejection**.

The ability of an amplifier to reject common-mode signals and to respond only to **differential-mode signals** is often extremely useful. One such example is in the transmission of signals over great distances. Consider the situations illustrated in Figure 3.24.

Figure 3.24(a) shows a sensor connected to a single-input amplifier by a long cable. Any long cable is influenced by **electromagnetic interference** (EMI) and inevitably some noise will be added to the signal from the sensor. This noise will be amplified along with the wanted signal and will therefore appear at the load.

Figure 3.24(b) shows a similar sensor connected by a twin conductor cable to the inputs of a differential amplifier. Again the cable will be affected by noise, but in this case, because of the close proximity of the two cables (which are kept as close as possible to each other), the noise picked up by each cable will be almost identical. Therefore, at the amplifier this noise appears as a common-mode signal and is ignored. The signal from the sensor is a differential-mode signal and is amplified and applied to the load.

The ability of differential amplifiers to reject common-mode signals makes them invaluable in a large number of applications.

A common form of differential amplifier is the **operational amplifier** (or **op-amp**) shown in Figure 3.25. The circuit symbol for the device is shown in Figure 3.25(a); Figure 3.25(b) shows its equivalent circuit. Note that the lower input terminal is not common to the output.

The two inputs are labelled '+' and '−'. The former is called the **non-inverting**

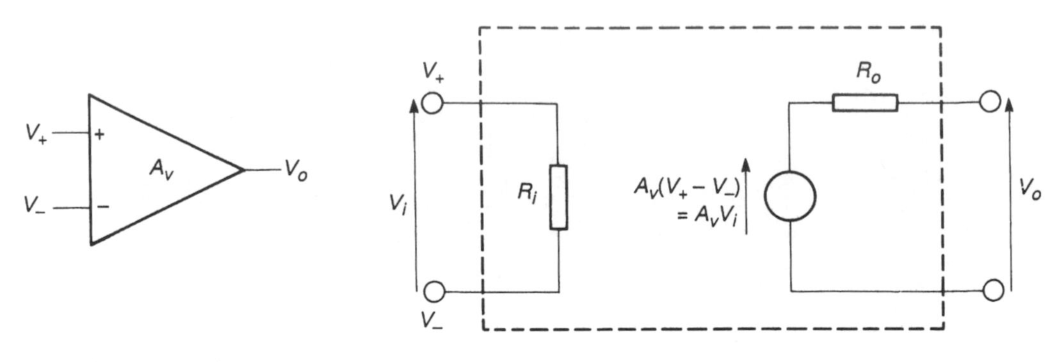

(a) Circuit symbol (b) Equivalent circuit

Figure 3.25 An operational amplifier.

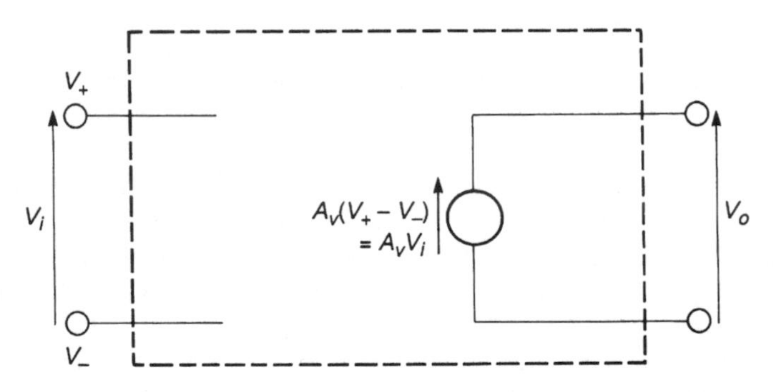

Figure 3.26 Equivalent circuit of an ideal operational amplifier.

input because a positive voltage on this input with respect to the other input will cause the output to become positive. The latter is called the **inverting input** because a positive voltage on this input with respect to the other input will cause the output to become negative.

If the difference between the voltages on the two input terminals is taken as the input voltage V_i, the operation and analysis of the differential amplifier is identical to that of the single-input amplifier discussed earlier. Indeed, if the inverting input is connected to ground (zero volts) and the non-inverting input is taken as the input, the device acts exactly as a single-input amplifier. However, differential amplifiers have several useful features which will be discussed in later chapters.

Earlier, when looking at voltage amplifiers, we concluded that a good voltage amplifier is characterised by a high input impedance and a low output impedance. Operational amplifiers are generally considered as voltage amplifiers. Thus it is useful at this point to imagine what would characterize an **ideal operational amplifier**.

From the above discussion we may deduce that an ideal operational amplifier would have an infinite input resistance ($R_i = \infty$) and zero output resistance ($R_o = 0$). It is perhaps less obvious that an ideal device should have an infinite voltage gain ($A_v = \infty$). This last point will be simply stated here and assumed in the remainder of this chapter. The reason for this condition will become clear in the next chapter when we look at feedback systems.

Having defined the characteristics of our ideal operational amplifier, we are now in a position to draw its equivalent circuit. This is shown in Figure 3.26.

Clearly, no **real operational amplifier** can achieve these idealized characteristics, but in many cases their performance is such that their deviation from ideal devices may be neglected. Before looking at applications of operational amplifiers, let us consider the characteristics of some typical devices.

3.11 Operational amplifiers

Most operational amplifiers are constructed by fabricating all the necessary components on a single chip of silicon forming a **monolithic integrated circuit**.

Operational amplifiers are constructed using a number of **device technologies**, each technology having its own set of characteristics. Part of the designer's task is to choose an appropriate device family for a given application. The operation of the various devices will be considered in later chapters, but at this stage we will simply consider the characteristics of a broad range of operational amplifiers.

One of the most widely used and best known general purpose operational amplifiers is the **741**. Its internal circuitry consists of about twenty bipolar transistors and a handful of resistors and capacitors; we will look at the circuitry of this device at a later stage. Although still popular in undemanding applications, the 741 is a fairly elderly device and more modern general purpose designs are superior in many areas. Also various device technologies have particular characteristics which make them more suitable for certain classes of applications. Such characteristics might include a high input impedance, low power consumption or the ability to operate from very low supply voltages. We will now look at several aspects of operational amplifier performance and see how these real devices differ from the 'ideal' device described above.

3.11.1 Voltage gain

No real amplifier has an infinite gain.

The 741 has a gain of about 106 dB, which corresponds to a voltage gain of about 2×10^5. Most modern devices have gains in the range 90 to 135 dB (voltage gains of about 3.2×10^4 to 5.6×10^6), but devices are available with gains of 150 dB (a voltage gain of about 3.2×10^7) or more.

In many cases the voltage gains of the amplifiers are sufficiently high that they may be considered to be 'high enough' and are thus treated as if they were infinite. Unfortunately one problem suffered by all operational amplifiers is that the gain, though high, is extremely variable between devices. This means that even devices from the same batch may have gains which differ by a factor of two or more. The gain also varies with temperature and, as we shall see later, with frequency.

The gain of the amplifier in the absence of any external circuitry is often called the

open-loop gain of the device, for reasons which will become apparent when you study the next chapter.

3.11.2 Input resistance

The typical input resistance of a 741 is 2 MΩ, but again this quantity varies considerably between devices and may be as low as 300 kΩ. This value is low for modern op-amps and it is not uncommon for devices which (like the 741) use bipolar transistors to have input resistances of 80 MΩ or more. In many applications this value will be very large compared with the source resistance and may be considered to be high enough for loading effects to be ignored. In applications where higher input resistances are required it is common to use devices that use field-effect transistors (FETs) in their input stages. These have input resistances of typically 10^{12} Ω. When using these devices, loading effects can almost always be ignored.

3.11.3 Output resistance

The 741 has a typical output resistance of 75 Ω, this being a typical figure for bipolar transistor op-amps. Some low power designs have much higher output resistances, perhaps up to several thousand ohms. Often of more importance than the output resistance of a device is the maximum current that it will supply. The 741 will supply 20 mA, values in the range 10 to 20 mA being typical for general purpose op-amps. Special high power devices may supply output currents of an amp or more.

3.11.4 Output voltage range

With voltage gains of several hundred thousand times, it would seem that if 1 volt were applied to the input of an operational amplifier one would have to keep well clear of the output! However, in practice the output voltage is limited by the supply voltage. Most op-amps based on bipolar transistors (like the 741) produce a maximum output voltage swing which is slightly less than the supply voltage. An amplifier connected to a positive supply of +15 V and a negative supply of −15 V, for example (a typical arrangement), might produce an output voltage range of about ±13 V. Op-amps based on field-effect transistors can often produce output voltage swings which go very close to both supply voltages.

3.11.5 Supply voltage range

A typical arrangement for an operational amplifier is to use supply voltages of +15 V and −15 V, although a wide range of supply voltages is usually possible. The 741, for example, may be used with supply voltages in the range ±5 V to ±18 V, this being fairly typical. Some devices allow higher voltages to be used, perhaps up to ±30 V, while others are designed for low voltage operation, perhaps down to ±1.5 V.

Many amplifiers allow operation from a single voltage supply which may be more convenient in some applications. Typical voltage ranges for a single supply might be 4 to 30 V, though devices are available which will operate down to 1 V or less.

3.11.6 Common-mode rejection ratio

An ideal operational amplifier would not respond to common-mode signals. In practice all amplifiers are slightly affected by common-mode voltages, though in good amplifiers the effects are very small. A measure of the ability of a device to ignore common-mode signals is its **common-mode rejection ratio** or **CMRR**. This is the ratio of the response produced by a differential-mode signal to the response produced by a common-mode signal of the same size. The ratio is normally expressed in decibels.

Typical values for CMRR for general purpose operational amplifiers are between 80 and 120 dB. High performance devices may have ratios of up to 160 dB or more. The 741 has a typical CMRR of 90 dB.

3.11.7 Input bias current

From the simplified equivalent circuit of an operational amplifier given in Figure 3.25(b) we would expect that if the two input terminals were joined together no current would flow into the inputs. In fact this is not the case. For the amplifier to work correctly, a small input current is required into each input terminal. This current is termed the **input bias current** and must be provided by external circuitry. The polarity of this current will depend on the input circuitry used in the amplifier. The cause of the input bias current will become apparent in later chapters when we look at the form of the input circuitry used in operational amplifiers.

The current taken by the two terminals of the amplifier will be approximately equal, the input bias current being defined as their average value. Typical values for this current, in bipolar op-amps, range from a few microamps down to a few nanoamps or less. For the 741 this value is typically 80 nA. Operational amplifiers based on FETs have much smaller input bias currents, with values of a few picoamps being common and with values down to less than a femtoamp (10^{-15} A) being possible.

The presence of the input bias currents causes voltage drops across resistors connected to the input terminals of the amplifier, and can lead to inaccuracies if these are not taken into account. This is made more complicated by the fact that in practice the bias currents into each terminal are not identical. The difference between these two currents is termed the **input offset current** and is generally an order of magnitude smaller than the input bias current. For the 741 it is typically about 10 nA.

3.11.8 Input offset voltage

One would expect that if the input voltage of the amplifier was zero, the output would also be zero.

In general this is not the case. The transistors and other components in the circuit are not precisely matched and a slight error is usually present which acts like a voltage source added to the input. This is the **input offset voltage** V_{ios}. The input offset voltage is defined as the small voltage required at the input to make the output zero.

The input offset voltage of most op-amps is generally in the range of a few hundred microvolts up to a few millivolts. For the 741 a typical value is 2 mV. This may not seem

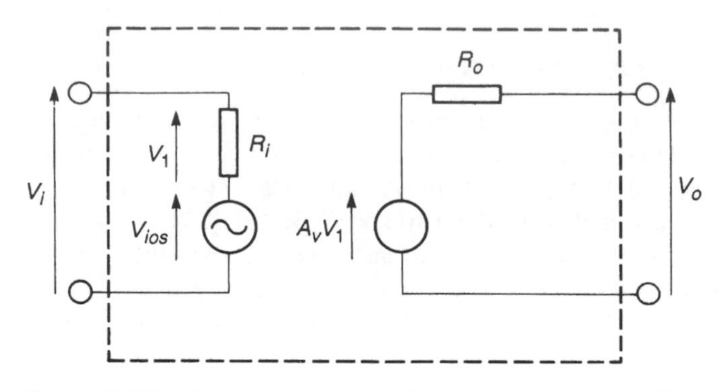

Figure 3.27 Equivalent circuit of an op-amp with an offset voltage.

very significant, but remember that this is a voltage added to the input and it is therefore multiplied by the gain of the amplifier. Fortunately, the offset voltage is approximately constant and so its effects can be reduced by subtracting an appropriate voltage from the input. The 741, in common with many operational amplifiers, provides connections to allow an external potentiometer to 'trim' the offset to zero. Some op-amps are **laser trimmed** during manufacture to produce a very low offset voltage without the need for manual adjustment. Unfortunately, the input offset voltage varies with temperature by a few μV/°C making it generally not possible completely to remove the effects of the offset voltage by trimming alone.

Figure 3.27 shows a modified equivalent circuit for an operational amplifier which includes the effects of the input offset voltage V_{ios}.

3.11.9 Frequency response

Operational amplifiers are DC coupled amplifiers and therefore have no lower cut-off frequency. The **open-loop gain** mentioned earlier is therefore the gain of the amplifier at DC.

We noted in Section 3.7.2 that all amplifiers have an **upper cut-off frequency**, and one would perhaps imagine that to be generally useful, operational amplifiers would require very high upper cut-off frequencies. In fact this is not the case and in many devices the gain begins to roll off above only a few hertz. Figure 3.28 shows a typical frequency response for the 741 op-amp.

The magnitude of the gain of the amplifier is constant from DC up to only a few hertz. Above this frequency it falls steadily at 6 dB/octave until it reaches unity at about 1 MHz. Above this frequency the gain falls more rapidly. The upper cut-off frequency is introduced intentionally by the designer to ensure that the fall in gain is dominated by a single time constant, as described in Section 3.7.2. This ensures that the phase angle between the input and the output is kept to well below 180° until the gain has dropped below unity, a very important consideration in the stability of the system. We will return to the question of **stability** in Chapter 4.

Amplifiers which include this dominant, high frequency cut-off to ensure stability,

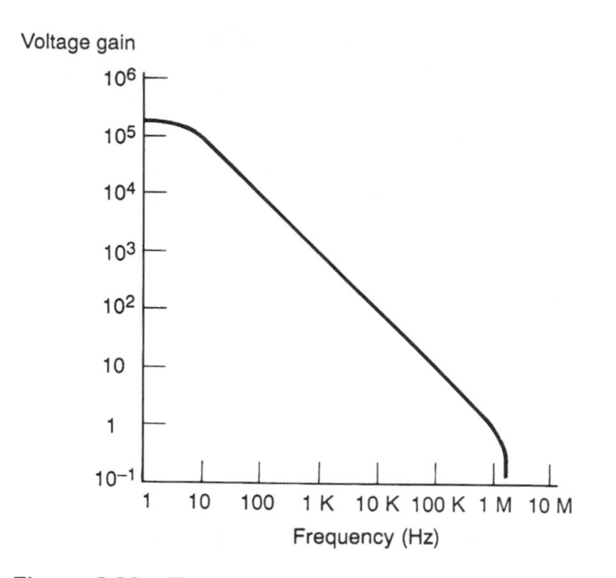

Figure 3.28 Typical gain magnitude vs frequency characteristic for a 741.

are described as **internally compensated**. Other devices do not include this feature and leave it to the circuit designer to add external components to tailor the frequency response to suit the application.

The frequency range of an operational amplifier is usually described by the frequency at which the gain drops to unity f_T, or by its **unity-gain bandwidth**. The latter is the bandwidth over which the gain is greater than unity, and it is clear that for an operational amplifier these two measures are equal. From Figure 3.28 it can be seen that the 741 has an f_T of about 1 MHz. Typical values for f_T for other operational amplifiers vary from a few hundred kilohertz up to a few tens of megahertz. However, a high-speed device may have an f_T of several gigahertz.

3.11.10 Noise

All operational amplifiers add noise to the signals that pass through them. Noise is generated by a number of mechanisms. Some is the product of **thermal noise** (also called **Johnson noise**) as described in Section 2.2.2. Additional noise is generated by the semiconductor devices within the circuit, by mechanisms which will be described in later chapters.

The noise components generated by these various mechanisms have different frequency characteristics. Some produce essentially **white noise** in that it has equal power density at all frequencies (that is, the noise power within a given bandwidth is equal at all frequencies). Others produce more power in some parts of the frequency spectrum than others. For this reason it is difficult accurately to describe the noise performance of a given device without being specific about the frequency range over which it is being used. However, for many applications a less stringent measure is adequate and most manufacturers give a single figure to characterize the noise generated by the device over its normal operating range. Clearly, since noise is present at all frequencies, the amount of noise detected will depend on the bandwidth over which measurements are made. For this reason the overall noise characteristics may be described by specifying the noise power present in a given bandwidth. In fact, when generating equivalent circuits and performing analysis, it is often more convenient to know the **noise voltage** rather than the noise power, this being proportional to the square root of the noise power. Manufacturers normally give a figure indicating the r.m.s. noise voltage divided by the square root of the bandwidth of measurement.

Low-noise op-amps are likely to have noise voltages of about 3 nV/√ Hz. General purpose devices may have noise voltages several orders of magnitude greater. The noise voltage represents the equivalent noise at the input of the amplifier, and may be used within an equivalent circuit of the device, as shown in Figure 3.21, to predict the noise at the output for a given configuration.

3.11.11 Integrated circuit operational amplifiers

Most operational amplifiers are in the form of **integrated circuits** which typically contain one, two or four amplifiers. Figure 3.29 shows examples of device pin-outs. The pins of integrated circuits are numbered anticlockwise when viewed from above (the side away

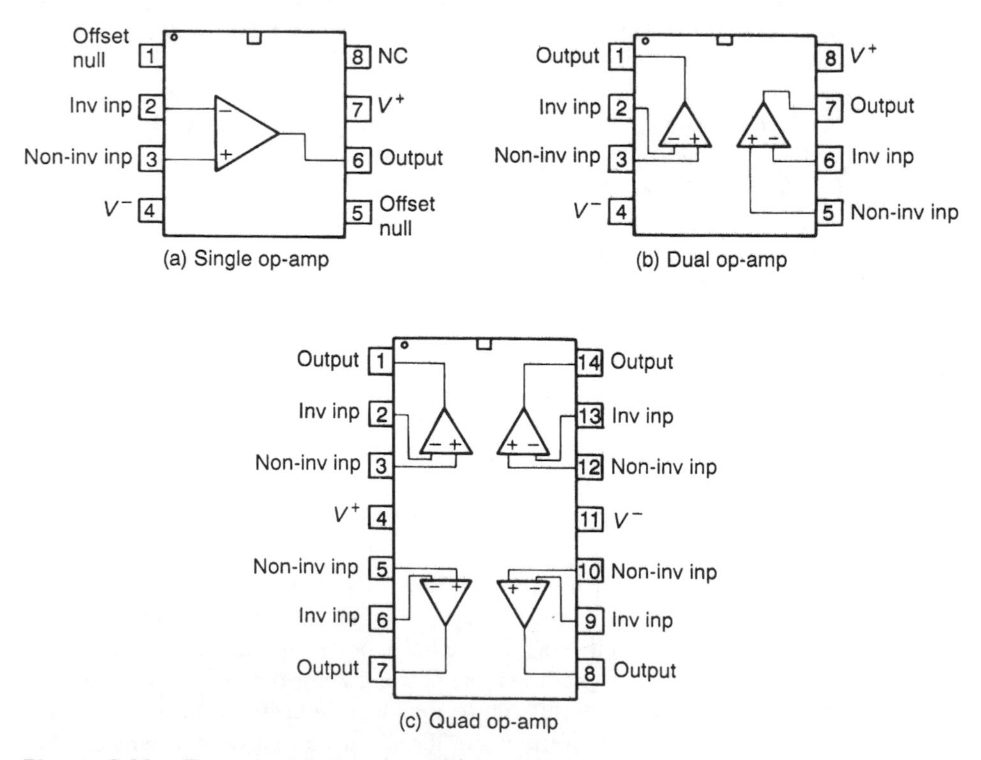

Figure 3.29 Typical operational amplifier packages.

from the pins), as shown in Figure 3.29. The orientation of the device is indicated by a notch at one end of the package, or by a dot against pin number one, or both. The label 'NC' against a pin represents 'no connection', V^- indicates the negative supply connection and V^+ the positive supply. Some devices have 'offset null' inputs which may be used to remove the effects of an offset voltage. The circuitry required for this null function differs from one device to another.

3.12 Simple amplifiers

Operational amplifiers are fairly complex circuits that contain a number of semiconductor devices. Amplifiers may also be formed using single transistors or other active devices. Before we look at the operation of transistors, it is perhaps worth seeing how a single control device may be used to form an amplifier.

Consider the circuit of Figure 3.30. This shows a pair of resistors arranged as a potential divider. The output voltage V_o is related to the circuit parameters by the expression

$$V_o = \frac{R_2}{R_1 + R_2} V$$

If the variable resistance R_2 is adjusted to equal R_1, the output voltage V_o will clearly be half the supply voltage V. If R_2 is reduced, V_o will also reduce. If R_2 is increased, V_o will increase. If we replace R_2 by some, as yet undefined, control device which has an input voltage V_i which controls its resistance, varying V_i will vary the output voltage V_o. Figure 3.31 shows such an arrangement.

In fact the control device does not have to be a voltage controlled resistance. Any device in which the current is determined by a control input may be used in such an arrangement and if the gain of the control device is suitable it can be used as an amplifier.

The output voltage V_o from the amplifier may be used to drive external circuitry connected to its output and may generate currents into or out from the load. This is illustrated in Figure 3.32.

Figure 3.32(a) shows a resistive load R_L connected between the output of the amplifier and 0 V (earth). In this arrangement, current flows from the positive supply, through

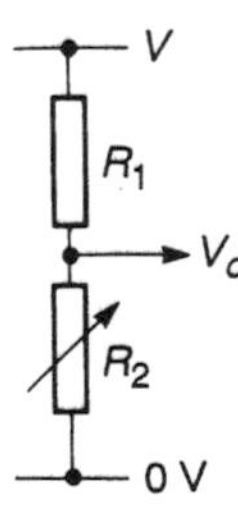

Figure 3.30 A simple potential divider.

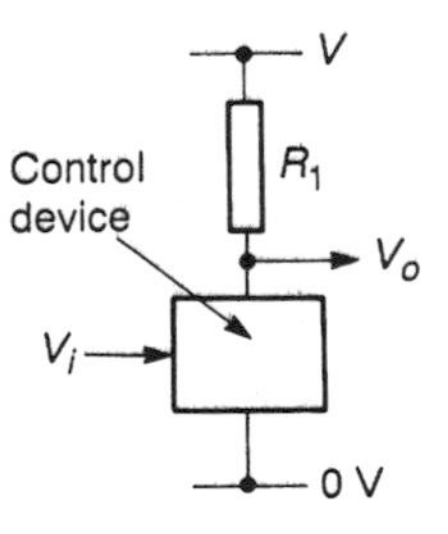

Figure 3.31 The use of a control device.

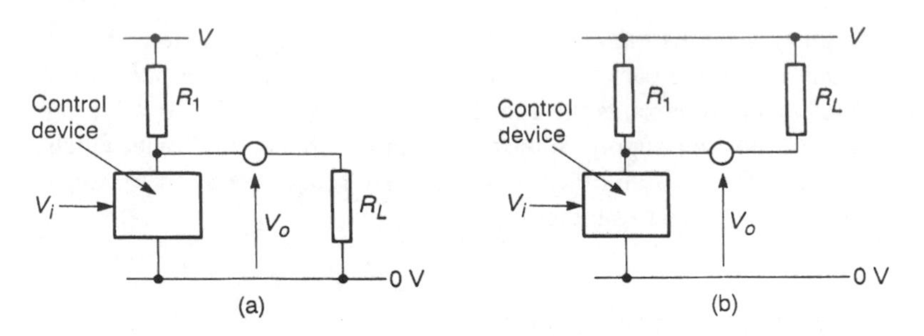

Figure 3.32 Load connections to a simple amplifier.

R_1 and *into* the load. The amplifier is therefore acting as a **current source**. If the value of R_L is much greater than R_1 then its presence will have little effect on the operation of the circuit. However, the maximum output voltage of this arrangement is limited by the ratio of R_L to R_1, since even if the control device passes no current the output voltage will not go above $V \times R_L/(R_L + R_1)$.

Figure 3.32(b) shows a resistive load connected between the output of the amplifier and the positive supply V. Here current flows from the positive supply, into the load and then *from* the load into the amplifier. The amplifier is therefore acting as a **current sink**. In this arrangement the output voltage swing is not restricted by the presence of the load. Current flowing through the control device is split between the load and R_1.

In many cases it may be more efficient to remove R_1 altogether, as shown in Figure 3.33(a). This prevents power being dissipated unnecessarily in R_1, and is particularly useful in high power circuits. Since it is sometimes inconvenient to have the load connected to a supply line, this circuit may be rearranged, as shown in Figure 3.33(b), to produce a current source.

Power dissipation in the control device

Consider the circuit shown in Figure 3.33(a). It is apparent that when V_o is half the supply voltage, the power dissipated in the control device is equal to that dissipated in the load R_L, since the current and voltage of both are identical.

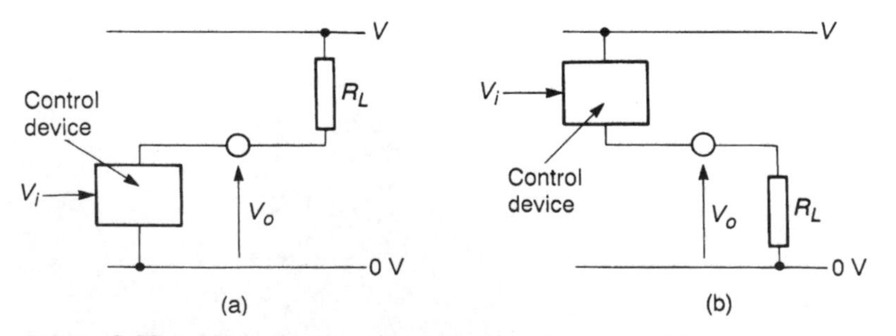

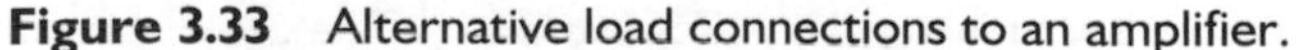

Figure 3.33 Alternative load connections to an amplifier.

As V_o varies, the power dissipated in the control device also varies but, unless V_o is 0 or V, power is always dissipated in the control device.

The situations where V_o is 0 or V are special cases. When V_o is zero there is clearly no voltage across the control device and so no power is dissipated in it. When V_o is equal to V there is no voltage across R_L which implies that there is no current flowing through it. This in turn implies that there is also no current flowing through the control device and again no power is dissipated in it.

It is useful to look at the power dissipated in the control device P_C as V_o varies. The power dissipated is calculated from the product of the current passing through it I and the voltage across it. The former is also the current flowing through the load R_L while the latter is simply the output voltage V_o. Thus

$$P_C = V_o I$$

I may be found from the ratio of the voltage across R_L to its resistance.

$$I = \frac{V - V_o}{R_L}$$

and therefore

$$P_C = V_o \frac{V - V_o}{R_L} \tag{3.10}$$

Differentiating this expression with respect to the output voltage allows us to determine when the power dissipated in the control device is at a maximum and to calculate this maximum value. This gives

$$\frac{dP_C}{dV_o} = \frac{(V - 2V_o)}{R_L}$$

The maximum value of P_C occurs when the above derivative is equal to zero. This occurs when $V = 2V_o$, or rearranging, when $V_o = V/2$. In other words, the power dissipated in the control device is at a maximum when the output voltage is half the supply voltage. Clearly at this point the effective resistance of the control device must be equal to that of the load. It is interesting to compare this result with that obtained in Section 3.6.1 when we considered impedance matching.

Substituting this condition into Equation 3.10 gives the maximum value for the power dissipated in the control device.

$$P_{C(\max)} = \frac{V}{2} \frac{\left(V - \frac{V}{2}\right)}{R_L} = \frac{V^2}{4R_L} \tag{3.11}$$

Note that since the output voltage is half the supply voltage, this is also the power dissipated in the load R_L at this time.

Simple amplifiers of this type are used extensively within all forms of electronic circuits and we will be returning to them when we have considered the operation of some active components that can be used as control devices in such arrangements.

Key points

- Amplification is a fundamental part of most electronic systems.
- Amplifiers may be active or passive. The power delivered at the output of a passive amplifier cannot be greater than that absorbed at its input. Active amplifiers take power from some external energy source and so can produce power amplification.
- When designing and analysing amplifiers, equivalent circuits are invaluable. They allow the interaction of the circuit with other components to be investigated without a detailed knowledge or understanding of the internal construction of the amplifier.
- Amplifier gains are often measured in decibels (dB).
- The gain of all amplifiers falls at high frequencies. In some cases the gain also falls at low frequencies. The gain normally falls at some multiple of 6 dB/octave, above or below some cut-off frequency. This cut-off frequency is normally simply related to component values.
- The quality of a signal, in terms of its corruption by noise, may be described by its signal-to-noise ratio S/N.
- It is often necessary to use several stages of amplification. The characteristics of a series of cascaded amplifiers may usually be derived from the characteristics of the individual stages.
- Differential amplifiers take as their input the difference between the voltages on their two inputs.
- Operational amplifiers (op-amps) are a common form of differential amplifier. An ideal operational amplifier would have an infinite voltage gain, an infinite input resistance and zero output resistance.
- Real operational amplifiers do not have these ideal characteristics.
- In many applications simple amplifiers, perhaps based on single transistors, may be more appropriate than operational amplifiers.

The concept of an ideal operational amplifier has been introduced without any real justification at this stage. The reason behind this apparent diversion will become clear in later chapters. The features and characteristics of real operational amplifiers, such as the 741, differ widely from those of an ideal device. In particular the gain, while high at low frequencies, varies with both frequency and temperature and is different even in devices in the same batch. The input and output resistance are also not of their ideal values and the devices have other non-ideal characteristics such as offset voltages and noise.

Operational amplifiers are not alone in having these problems. All real active amplifiers suffer from them to a greater or lesser extent. To make these devices useful we need to overcome these deficiencies. Methods of achieving this are discussed in the next chapter.

Design study

A temperature display system consists of a semiconductor temperature sensor, an amplifier and a voltmeter which has a scale calibrated in degrees centigrade. This arrangement is shown below.

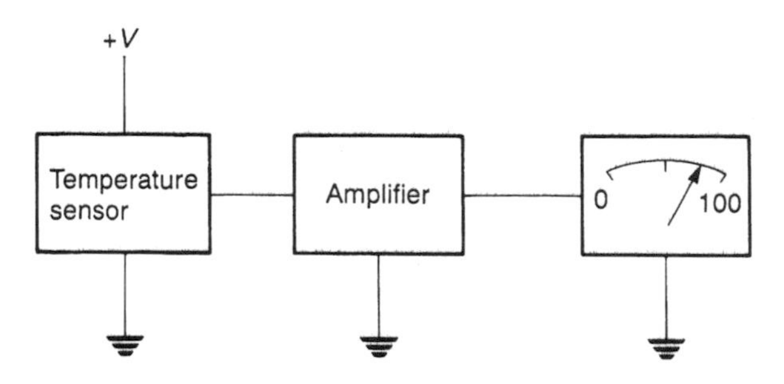

The temperature sensor produces an output voltage of 0 V at 0 °C and an output voltage of 1 mV/°C at temperatures above 0 °C. The output resistance of the sensor is 300 Ω.

The meter has an input resistance of 10 kΩ and produces a full-scale output for an input voltage of 1 V. The meter scale is marked linearly such that full-scale deflection corresponds to 100 °C.

The amplifier is required to take the signal from the temperature sensor and convert it into a form suitable for driving the meter. The device to be used has an input resistance of 2 kΩ and an output resistance of 400 Ω, but has a gain which may be adjusted to suit the application.

Calculate the open-circuit gain required for the amplifier to produce a correct temperature reading on the meter.

Approach

The first step in solving the problem is to derive an equivalent circuit for the arrangement. In this circuit the open-circuit voltage gain of the amplifier is represented by A_v.

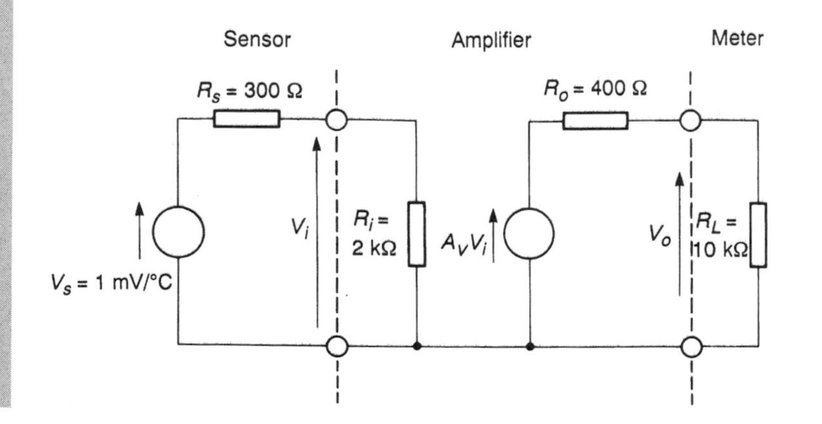

From the equivalent circuit we can see that the input voltage to the amplifier V_i is given by

$$V_i = V_s\left(\frac{R_i}{R_s + R_i}\right)$$

and that the output voltage V_o is given by

$$V_o = A_v V_i\left(\frac{R_L}{R_o + R_L}\right) = A_v V_s\left(\frac{R_i}{R_s + R_i}\right)\left(\frac{R_L}{R_o + R_L}\right)$$

Therefore the ratio of V_o to V_s is given by

$$\frac{V_o}{V_s} = A_v\left(\frac{R_i}{R_s+R_i}\right)\left(\frac{R_L}{R_o+R_L}\right)$$

Substituting for the actual component values we have

$$\frac{V_o}{V_s} = A_v\left(\frac{2\ \text{k}\Omega}{300\ \Omega+2\ \text{k}\Omega}\right)\left(\frac{10\ \text{k}\Omega}{400\ \Omega+10\ \text{k}\Omega}\right)$$

$$= A_v \times 0.836$$

The sensor produces an output voltage of 1 mV/°C, so for a temperature of 100 °C it will produce an output voltage of 0.1 V. The meter is calibrated so that an input of 1 V produces an indication of 100 °C. Therefore an amplification of 10 is required to produce the correct reading. The amplification produced by the circuit arrangement in use is 0.836 A_v. Therefore the required value for A_v may be obtained from the expression:

$$A_v \text{ required} = \left(\frac{V_o}{V_s}\right) \text{required} \times \frac{1}{0.836} = 10 \times \frac{1}{0.836} \approx 12.0$$

Therefore the open-circuit voltage gain required for the amplifier is 12.0, which is 20% higher than the overall gain to be produced. This increase in gain is due to the effects of **loading**.

Further reading

Clayton G. B. and Newby B. (1992) *Operational Amplifiers*, 3rd edn. Oxford: Butterworth-Heinemann

Irvine R. G. (1994) *Operational Amplifier Characteristics and Applications*, 3rd edn. Englewood Cliffs, NJ: Prentice-Hall

Sanderson M. (1987) *Electronic Devices: A Top-down Systems Approach*, Englewood Cliffs, NJ: Prentice-Hall

Exercises

3.1 Conventional automotive hydraulic braking systems are an example of passive amplifiers. What physical quantity is being amplified?

Such systems may also be regarded as attenuators. What physical quantity is attenuated?

Power assisted automotive brakes are active amplifiers. What is the source of power?

3.2 Identify examples (other than those given in the text and in Exercise 3.1) of both passive and active amplifiers for which the operation is mechanical, hydraulic, pneumatic, electrical and physiological.

In each case identify the physical quantity that is amplified and, for the active examples, the source of power.

3.3 Derive Thévenin and Norton equivalent circuits for the following circuit.

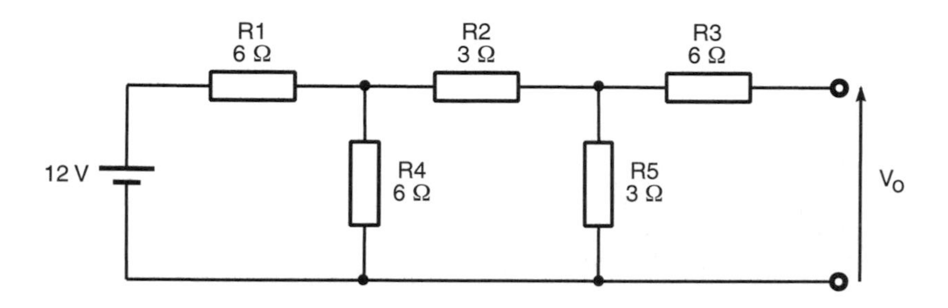

3.4 A voltage source of unknown form is investigated by connecting resistors across its output and measuring the output voltage.

When a resistance of 300 Ω is connected the output voltage is 3 V. When the resistance is replaced by a 100 Ω resistance the output drops to 2 V.

Deduce both the Thévenin and Norton equivalent circuits of the voltage source.

3.5 Estimate the input resistance of the following arrangement.

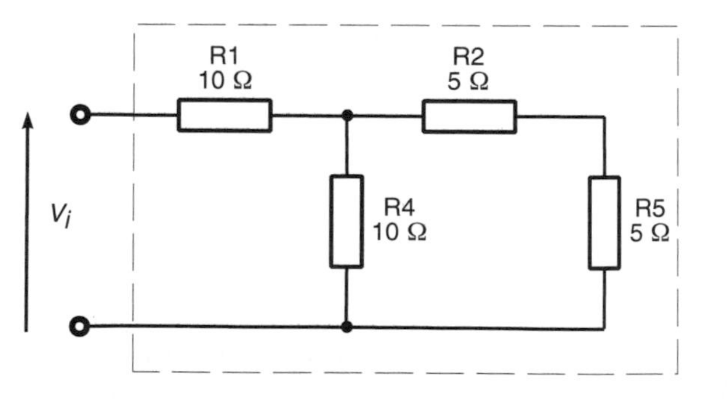

3.6 An amplifier has an open-circuit voltage gain of 20, an input resistance of 10 kΩ and an output resistance of 75 Ω.

The amplifier is connected to an AC source of 1 V r.m.s. which has an output resistance of 200 Ω, and to a load resistor of 1 kΩ.

Assuming that noise and the input offset voltage may be neglected, what will be the r.m.s. value of the output voltage?

3.7 A displacement sensor produces an output of 10 mV per centimetre of movement and has an output resistance of 200 Ω. It is connected to an amplifier which has an open-circuit voltage gain of 15, an input resistance of 10 kΩ and an output resistance of 75 Ω. If the output of the amplifier is connected to a voltmeter with an input resistance of 1 kΩ, what voltage will be displayed on the meter for a displacement of the sensor of 1 metre?

3.8 Convert the following into decibels.

(a) A power gain of 10
(b) A voltage gain of 1
(c) A power gain of 0.5
(d) A voltage gain of 1 000 000.

3.9 Determine the following:

(a) The power gain of a circuit with a gain of 20 dB
(b) The voltage gain of a circuit with a gain of 20 dB
(c) The power gain of a circuit with a gain of −15 dB
(d) The voltage gain of a circuit with a gain of −15 dB.

3.10 What frequency is

(a) An octave above 10 Hz
(b) Three octaves above 1 Hz
(c) Two octaves below 1 kHz
(d) A decade above 100 Hz
(e) Two decades below 10 kHz?

3.11 An amplifier has a mid-band voltage gain of 10, a single time-constant lower cut-off frequency at 100 Hz and a single time-constant upper cut-off frequency of 10 kHz.

Sketch the gain and phase responses of the amplifier indicating the gain and phase at the cut-off frequencies, and the asymptotes of the gain response.

3.12 The amplifier of Exercise 3.11 is connected in cascade with an amplifier with a mid-band voltage gain of 1, a single time-constant lower cut-off frequency of 10 Hz and a single time-constant upper cut-off frequency of 100 kHz.

Sketch the gain and phase responses of the combination indicating, as before, the gain and phase at each cut-off frequency, and the asymptotes of the gain response.

3.13 An amplifier with an input resistance of 10 kΩ is connected to a transducer by a coupling capacitor of 100 nF. If the transducer has an output resistance of 1 kΩ, what will be the cut-off frequency produced by this arrangement?

Is this an upper or lower cut-off frequency?

3.14 Check your answer to Exercise 3.13 by simulating the circuit and plotting the gain and phase responses.

3.15 A transducer may be represented by an ideal, noiseless voltage source of 1 mV r.m.s., in series with a resistor of 1 kΩ. If the resistor is at a temperature of 300 K, and the bandwidth of the measuring system is 100 kHz, what is the signal-to-noise ratio of the output signal?

Feedback

Objectives

When you have studied the material in this chapter you should:

- have an understanding of the principles of 'open-loop' and 'closed-loop' systems and be able to cite electronic, mechanical and biological examples of each;
- be familiar with the major components of feedback systems;
- be capable of analysing the operation of simple amplifier circuits;
- be aware of the uses of negative feedback in overcoming problems of variability in active devices such as transistors and integrated circuits;
- be able to design circuits based on operational amplifiers to perform basic functions such as amplification, addition and subtraction;
- recognize the importance of negative feedback in improving input resistance, output resistance, bandwidth and distortion;
- be familiar with the use of positive feedback in the production of oscillators.

Contents

4.1 Introduction

We noted in the last chapter that all active amplifiers suffer from variability in their gain and other characteristics. To overcome these problems we often use **feedback** to improve the performance of the system.

Feedback systems look at the output of the system and use this information to modify the input signal to achieve the desired result. Such techniques are very widely used and form the basis of most forms of automatic control system. Almost all biological systems incorporate feedback, as do many man-made systems, whether utilizing electrical, mechanical, hydraulic, pneumatic or chemical processes.

A good understanding of the concepts of feedback is vital for all engineers and scientists. It is also encouraging to note that the use of feedback often *simplifies* system design and reduces the importance of linearity and accuracy in many of the key components. Before looking in detail at the design of feedback systems we will take a brief look at the overall principles of feedback.

4.2 Open-loop and closed-loop systems

Let us consider two approaches to controlling the temperature of a room.

The first is to use a heater which has a control that varies the heat output. The user sets the control to give a certain amount of heat output and hopes that this will achieve the desired temperature. He may learn an appropriate setting by experience. If the setting is too low the room will not reach the desired value, while if the setting is too high the temperature will rise above the desired value. If an appropriate setting is chosen the room should stabilize at the right temperature but will become too hot or too cold if external factors, such as the outside temperature or the level of ventilation, are changed. Such a system is called an **open-loop system**.

An alternative approach is to use a heater equipped with a thermostat. The user then sets the thermostat to the temperature required and it then increases or decreases the heat output to achieve and then maintain this value. It does this by comparing the desired and actual temperatures and using the difference between them to determine the appropriate heat output. Such a system should maintain the temperature of the room even if external factors change. Such a system is called a **closed-loop system**.

These two approaches are illustrated in Figure 4.1. The **user** is simply the person using the system, the **goal** is the desired result and the **output** of the system is the achieved result. In our example, the goal would be the required room temperature and the output would be the actual room temperature.

The **forward path** is the part of the system that converts the inputs into the outputs. For example, the elements that convert electricity or gas into heat. Closed-loop systems also have a **feedback path** through which the output is fed back for comparison with the goal. The difference between the output and the goal is represented by an **error signal** which is used as an input to the forward path.

In open-loop systems the goal is not an input to the system but simply guides the user in his input to the system. In closed-loop systems the goal is the primary input to

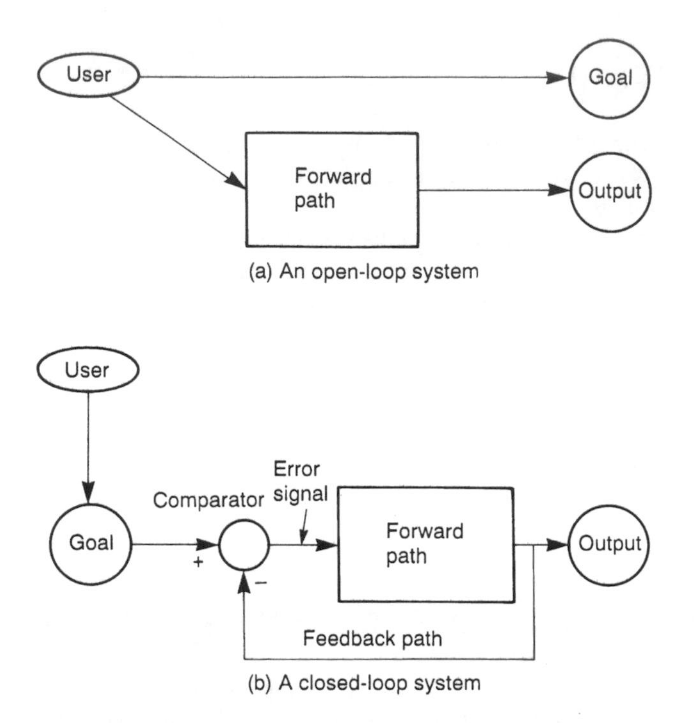

Figure 4.1 Open-loop and closed-loop systems.

the system. In both cases the forward path may have secondary inputs, for example, electrical energy. As in the last section on amplifiers, such inputs are not shown but are simply assumed to be present.

Closed-loop systems form the basis of most automatic control systems. As with amplifiers there are both electrical and non-electrical examples. Figure 4.2 shows an example of a mechanical automatic control system in the form of an engine speed controller. This particular example shows a steam engine, although exactly the same principle could be applied to other types of engine or motor.

The governor uses a series of weights which are made to rotate by the motion of the engine. The weights are hinged so that they may move outwards against a spring under the influence of the centrifugal force as the assembly rotates. A mechanical arrangement closes a steam valve, reducing the power to the engine, when the weights move by a certain amount. As the speed of the engine increases, the centrifugal force on the weights forces them to move outwards and gradually to close the steam valve. The speed of the engine therefore stabilizes at a particular value at which the steam valve is partially closed. The speed at which the engine stabilizes may be determined by adjusting the tension in the return springs attached to the weights.

The speed governor described above is one of the earliest examples of man-made control systems. However, automatic control systems have been around for a great deal longer, since they form the heart of many biological systems. The temperature control system of the human body is a clear example of an automatic, closed-loop control system.

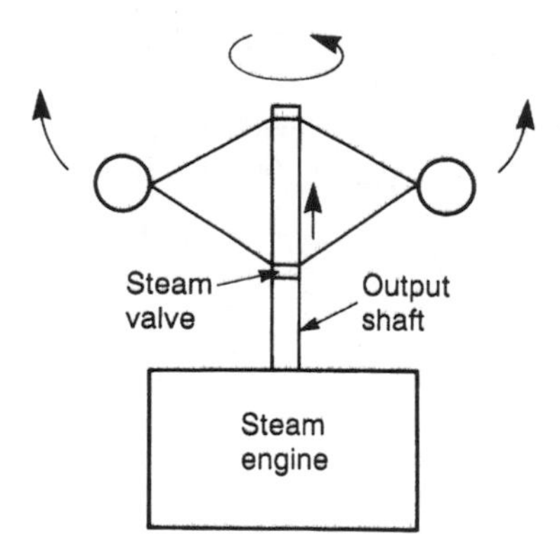

Figure 4.2 A mechanical speed governor.

4.3 Feedback systems

In the previous section we noted the use of feedback in the production of automatic control systems. In fact this is only one example of a range of applications of **feedback systems**. A feedback system is one in which part of the output is combined with the input.

Feedback systems may be of two types. **Negative feedback** systems have the characteristic that the feedback tends to reduce the input to the forward path. This form of feedback is also called **degenerative feedback**. In **positive feedback** systems the feedback tends to increase the input to the forward path. This form of feedback is also called **regenerative feedback**.

Figure 4.3 shows a generalized feedback system which need not be an electrical system – but could be. The input and output of the system are X_i and X_o respectively. Box A represents the **forward path** discussed earlier. In this case A is the gain of the forward path (also called the **forward gain**). Since the output is X_o we can deduce that the input to the forward path is X_o/A.

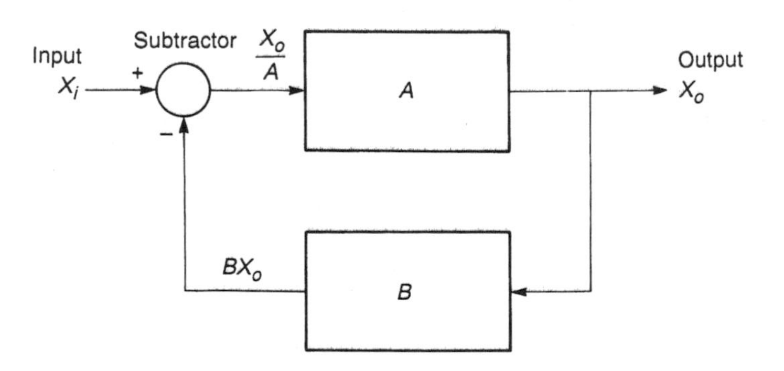

Figure 4.3 A feedback system.

Box B represents the **feedback path**. Again B is the gain of this path and thus the signal reaching the subtractor via the feedback path is BX_o.

From the diagram we can see that the difference between the input X_i and the feedback signal BX_o is equal to the input to the forward path X_o/A. Therefore

$$\frac{X_o}{A} = X_i - BX_o$$

or, by rearranging

$$\frac{X_o}{X_i} = \frac{A}{1+AB} = \frac{\text{Forward gain}}{1+\text{Loop gain}}$$

$$= G$$

The term **loop gain** refers to the gain through the forward path and then back through the feedback path. In other words, the gain right round the loop.

G represents the ratio of the output to the input. In other words it is the **overall gain** of the system. It follows that

$$G = \frac{A}{1+AB} \tag{4.1}$$

It is common to refer to the forward gain A as the **open-loop gain** (as it is the gain from the input to the output with the feedback disconnected) and to refer to the overall gain G as the **closed-loop gain** (as this is the gain from the input to the output with the feedback present). It is important not to confuse these terms with the **loop gain** AB discussed above which is the gain through the forward and feedback paths of the system.

The overall characteristics of the system depend on the values of A and B, or more directly on the term $1+AB$. Let us consider some possible values for A, B and the term $1+AB$ to see how they affect the overall gain G.

No feedback B = 0

Substituting in Equation 4.1 gives

$$G = \frac{A}{1+0} = A$$

This is clearly what we would expect from the block diagram. If the feedback path is not present we simply see the effect of the forward path which has a gain of A.

The loop gain *AB* is negative and <1

If either A or B are negative (but not both), the product AB will be negative. If now the term $(1+AB)$ is less than unity, G is greater than A. The physical significance of this is that the signal fed back increases the size of the error signal and thus makes the overall gain greater than the forward path gain. This is **positive feedback**.

The loop gain AB is equal to -1

A special case of positive feedback occurs when $AB = -1$. Under these circumstances

$$G = \frac{A}{1 + AB} = \frac{A}{1 - 1} = \text{Infinity!}$$

Since the gain of the circuit is infinite it cannot be used as a conventional amplifier, but this arrangement does have some uses. Generally the circuit will produce an output even in the absence of an input and this characteristic makes it useful as an **oscillator**. The use of positive feedback in oscillator circuits will be discussed later in this chapter.

The loop gain AB is positive

If A and B are either both positive or both negative, the loop gain AB will be positive. This implies that the signal fed back to the input is of the same polarity as the input itself, and as it is subtracted from the input signal it has the effect of reducing the size of the error signal and thus the output. This is therefore **negative feedback**.

If the loop gain AB is large compared with unity, the expression

$$G = \frac{A}{1 + AB}$$

may be simplified to

$$G \approx \frac{A}{AB} = \frac{1}{B}$$

This special case of negative feedback is of great importance as we now have a system in which the overall gain is independent of the gain of the forward path, being determined solely by the characteristics of the feedback path. The reasons for the importance of this arrangement will become clear later when we look at applications of negative feedback circuits.

Note: Since the feedback signal is being subtracted from the input signal, a positive loop gain will produce negative feedback and a negative loop gain will produce positive feedback!

In some texts the subtractor in Figure 4.3 is replaced by an adder. This is an equally valid feedback arrangement, and a similar analysis to that given above will generate an expression for the overall gain of the form

$$G = \frac{A}{1 - AB}$$

This equation clearly places different requirements on A and B to obtain positive and negative feedback.

Both arrangements have advantages and disadvantages. The arrangement using a subtractor has the advantage that it more closely relates to the real circuits which we will consider later. You should, however, be aware that other representations exist.

4.4 Negative feedback

We saw in the last chapter that one characteristic of real operational amplifiers is that their gain, while being very large, is also very variable between devices. We also noted that their gain varies with temperature. These characteristics are common to almost all **active devices**.

In contrast, passive components, such as resistors and capacitors, can be made to very high precision and can be very stable as their temperature varies.

In the last section, when looking at negative feedback, it was noted that a feedback circuit with a loop gain which was positive and much greater than unity had an overall gain that was independent of the forward gain, being determined entirely by the feedback gain. If we construct a negative feedback system, as described above, using an operational amplifier as the forward path A and a passive network as the feedback path B, we can produce an amplifier with a stable overall gain independent of the actual value of A.

Example 4.1 illustrates this principle. Here an operational amplifier of indeterminate gain is combined with a feedback network, which has a stable gain, to form an amplifier with a gain of 100.

Example 4.1 An amplifier with a notional gain of 100

Consider the overall gain of the following feedback circuit when the forward path gain A varies from 100 000 to 200 000. The feedback path gain B is constant at $1/100$.

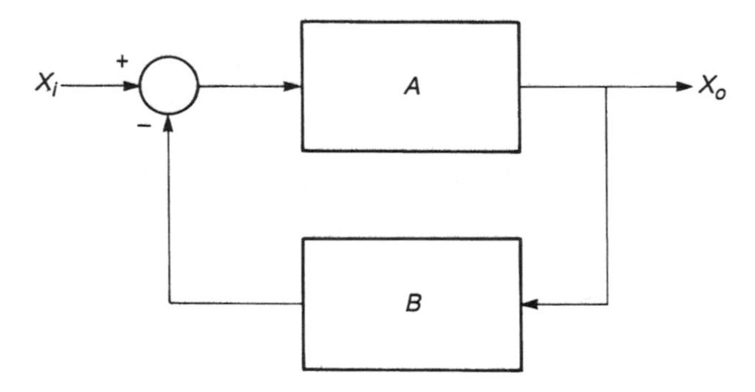

Firstly, let

$$A = 100000$$

$$B = \frac{1}{100}$$

$$G = \frac{A}{1 + AB} = \frac{100000}{1 + 1000} = 99.90$$

$$\approx \frac{1}{B}$$

Alternatively, let

$$A = 200\,000$$

$$B = \frac{1}{100}$$

$$G = \frac{A}{1 + AB} = \frac{200\,000}{1 + 2000} = 99.95$$

$$\approx \frac{1}{B}$$

Notice that a change of 100% in the value of the forward gain A produces a change of only 0.05% in the value of the overall gain G.

Example 4.1 shows that the large variation in gain associated with active circuits, such as operational amplifiers, can be overcome by the use of negative feedback, provided a stable feedback arrangement can be produced. In order to make the feedback path stable it must be constructed using only passive components. Fortunately, this is a simple task.

We have seen that the overall gain of the feedback circuit is $1/B$. Therefore to have an overall gain of greater than unity, we require B to be less than 1. In other words our feedback path may be a **passive attenuator**.

Construction of such a feedback arrangement using passive components is simple. If we take as an example the value used in Example 4.1 we require a passive attenuator with a gain of 1/100. This can be achieved as shown in Figure 4.4. The circuit is a simple potential divider with a ratio of 99:1. The output voltage V_o is related to the input voltage V_i by the expression

$$V_o = V_i \frac{R}{R + 99R}$$

$$= V_i \frac{1}{100}$$

The resistor values of R and $99R$ are shown simply

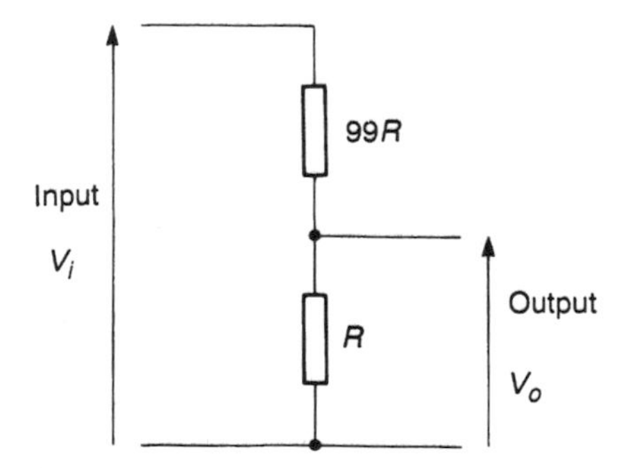

Figure 4.4 A passive attenuator with a gain of 1/100.

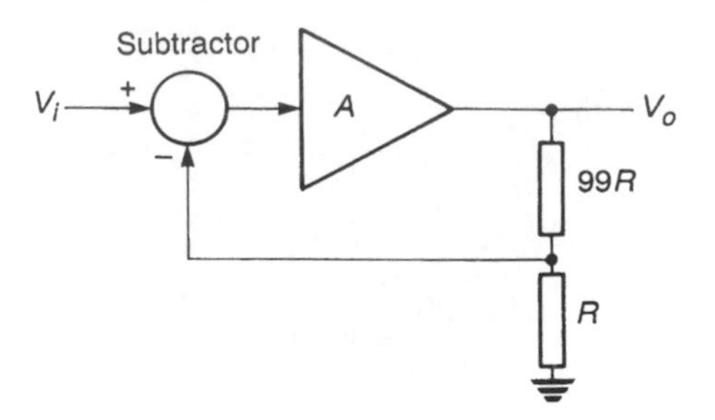

Figure 4.5 An amplifier with a gain of 100.

to indicate their relative magnitudes. In practice R might be 1 kΩ and $99R$ would then be 99 kΩ. The actual values used would depend on the circuit configuration.

Having decided that the forward path of our feedback circuit will be an operational amplifier and that the feedback path will be a resistive attenuator, we are now in a position to complete the circuit. Continuing with the values given in Example 4.1 we may now draw our circuit diagram as shown in Figure 4.5. The figure shows an amplifier with a forward gain of A and a feedback gain B of $1/100$. This produces an amplifier with an overall gain G of 100 (that is $1/B$).

Operational amplifiers, as discussed earlier, have **differential inputs**. That is, they amplify the difference between two input signals. We could visualize such an amplifier as a single-input amplifier with a subtractor connected to its input. Therefore an operational amplifier may be used to replace both the amplifier and the subtractor in the circuit of Figure 4.5. This arrangement is shown in Figure 4.6. The input and output signals of this circuit are of the same polarity (that is, a positive input voltage generates a positive output voltage). For this reason this circuit is called a **non-inverting amplifier.**

Inherent in the design of our simple amplifier is the assumption that the overall gain is equal to $1/B$. From our earlier discussions we know that this assumption is only valid provided that the loop gain AB is much greater than unity. In this case the forward path is to be implemented using an operational amplifier and from the last chapter we

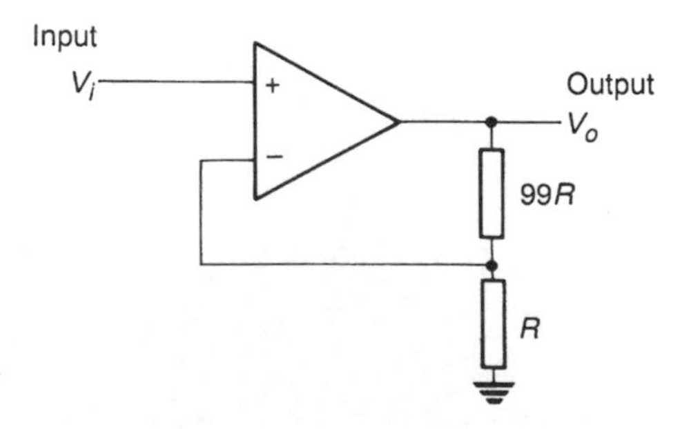

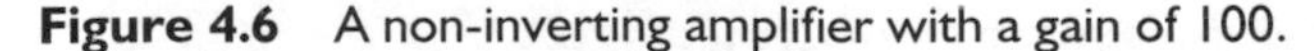

Figure 4.6 A non-inverting amplifier with a gain of 100.

know that this is likely to have an open-loop gain (A) of perhaps 10^5 or 10^6. Since B is $1/100$ in our example, this means that the loop gain AB will have a value of about 10^3 to 10^4. Since this is much greater than unity it would seem that our assumption that the gain is equal to $1/B$ is valid. However, let us consider another example.

Example 4.2 An amplifier with a notional gain of 10 000

Consider the overall gain of the following feedback circuit when the forward path gain A varies from 100 000 to 200 000. The feedback path gain B is constant at 1/10 000.
Firstly, let

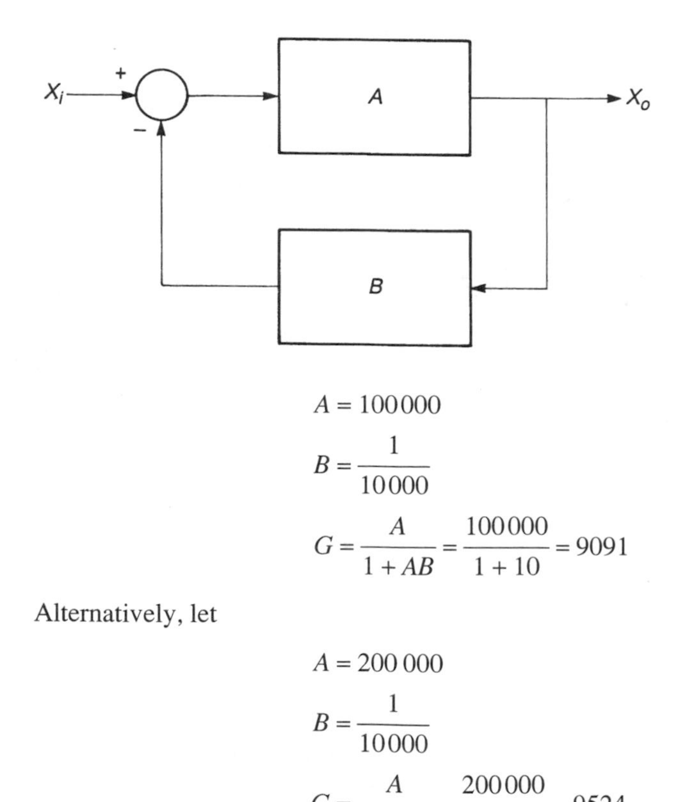

$$A = 100000$$

$$B = \frac{1}{10000}$$

$$G = \frac{A}{1 + AB} = \frac{100000}{1 + 10} = 9091$$

Alternatively, let

$$A = 200\,000$$

$$B = \frac{1}{10000}$$

$$G = \frac{A}{1 + AB} = \frac{200000}{1 + 20} = 9524$$

It can be seen that in this case the resultant gain is not very close to $1/B$ (10 000) and that variations in the gain of the forward path (A) have significant effects on the overall gain. This example shows that in order for the gain to be stabilized by the effects of negative feedback, the open-loop gain (A) must be much greater than the closed-loop gain ($1/B$).

FILE 4A

Computer simulation exercise 4.1

Simulate the arrangement of Example 4.2 using gain blocks and a subtractor element (GAIN and DIFF components within PSpice). Apply a 1 volt DC input to this circuit and investigate the output voltage for different values of forward and feedback gain.

We have seen that negative feedback allows us to generate amplifiers with overall characteristics that are constant despite variations in the gain of the active components. In fact, negative feedback can produce many other desirable effects and is widely used in many forms of electronic circuitry. At this point it is perhaps useful to note some of the major characteristics of negative feedback systems.

Properties of negative feedback systems

(1) They tend to maintain their output despite variations in the forward path or in the environment.

(2) They require a forward path gain which is greater than that which would be necessary to achieve the required output in the absence of feedback.

(3) The overall behaviour of the system is determined by the nature of the feedback path.

4.5 Feedback circuits

In the last section we saw how the circuit of a non-inverting amplifier can be derived from first principles.

In many applications it is not necessary to design from first principles. Instead, standard **cook book circuits** can be used as a starting point, their component values being derived by very simple calculations based on a few simplifying assumptions. These assumptions effectively correspond with the use of an **ideal op-amp**, as outlined in the last chapter.

These assumptions include:

(1) that the open-circuit gain of the amplifier is so high that it may be considered infinite;

(2) that the input impedance of the op-amp is so high that input currents may be neglected;

(3) that the output impedance of the op-amp is so low that it may be considered to be zero.

In some demanding applications it may be necessary to take into account the non-ideal nature of the components being used. This will involve considering the effects of the input offset voltage, input bias current, input offset current, input noise voltages and other non-ideal characteristics. However, a simplistic approach still offers a useful starting point from which to consider these effects.

We will look at a few examples, all of which use negative feedback to overcome the effects of variations in the characteristics of the operational amplifier.

Example 4.3 A non-inverting amplifier

Since the gain A is infinite, if V_o is finite, the input voltage to the op-amp $V_+ - V_-$ must be zero. Therefore

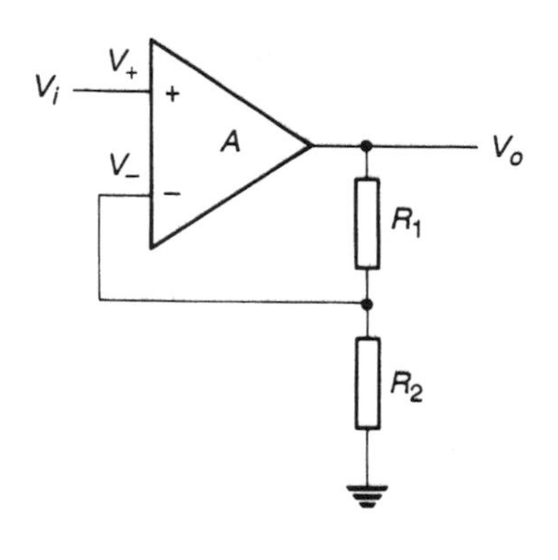

$$V_- = V_+ = V_i$$

Now since negligible current flows into the inputs

$$V_- = V_o \frac{R_2}{R_1 + R_2}$$

and therefore

$$V_i = V_o \frac{R_2}{R_1 + R_2}$$

Therefore the overall gain G is given by

$$G = \frac{V_o}{V_i} = \frac{R_1 + R_2}{R_2}$$

This result is consistent with the design obtained in the last section for a non-inverting amplifier with a gain of 100, as shown in Figure 4.6.

Example 4.4 An inverting amplifier

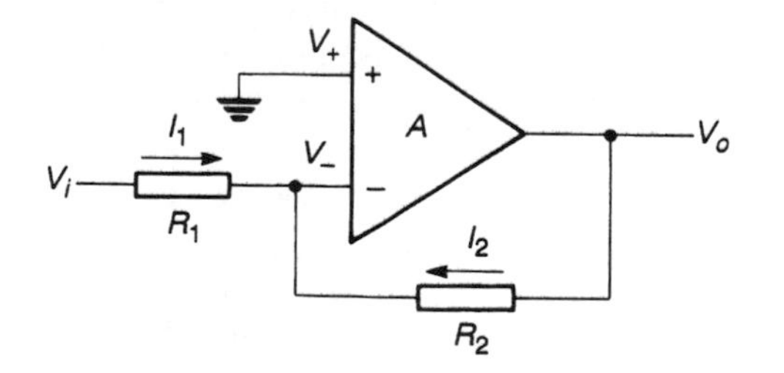

The negative feedback of this circuit tends to maintain the voltage at the inverting input V_- at a voltage equal to that on the non-inverting input V_+, which is earth potential (0 V). This may be understood by noting that if V_- becomes more positive than earth the output will be driven negative which, through R_2, will tend to make V_- more negative. Similarly, if V_- becomes more negative than earth, the output will become positive which will tend to make V_- more positive. For an ideal amplifier where A is infinite, this stabilization is perfect and V_- is maintained at 0 V. Therefore, the op-amp input voltage $V_+ - V_-$ is zero, and

$$V_- = V_+ = 0$$

Since the input current of the op-amp is negligible, the currents I_1 and I_2 must be equal and opposite.

Now

$$I_1 = \frac{V_i - V_-}{R_1} = \frac{V_i - 0}{R_1} = \frac{V_i}{R_1}$$

and

$$I_2 = \frac{V_o - V_-}{R_2} = \frac{V_o - 0}{R_2} = \frac{V_o}{R_2}$$

Therefore, since

$$I_1 = -I_2$$

clearly

$$\frac{V_i}{R_1} = -\frac{V_o}{R_2}$$

and thus the gain G is given by

$$G = \frac{V_o}{V_i} = -\frac{R_2}{R_1}$$

The circuit of Example 4.4 has a gain that is negative. This means that the polarity of the output is opposite to that of the input. A positive input voltage will generate a negative output and a sinusoidal input will produce a sinusoidal output which is 180° out of phase with the input. This kind of circuit is called an **inverting amplifier**.

In this circuit the inverting input (–) is maintained at zero volts even though it is not directly connected to ground ('ground' or 'earth' are names often given to the circuit's common reference point, often the zero volt line from the power supply). For this reason, this point in the circuit is often called a **virtual earth** and this kind of amplifier is sometimes called a **virtual earth amplifier**.

In the circuits of both Example 4.3 and Example 4.4, the gain is determined by the ratios of the resistor values and there is therefore no uniquely suitable value for each resistor. However, we have made simplifying assumptions in our design which place some constraints on our choice of resistor values when using **real** amplifiers.

One such assumption is that the output impedance of the op-amp is zero. This is clearly not the case for a real amplifier, but is a reasonable approximation if the output resistance of the op-amp is small compared with the resistance of the external circuit connected to its output. We have also assumed that the input impedance is so high that input currents can be neglected. This again is a reasonable approximation provided that the resistance of the external circuit connected to the input is small compared with the actual input impedance.

Therefore, our approximations will not cause too many problems if the chosen external resistor values are large compared with the output resistance, and small compared with the input resistance of the operational amplifier being used.

As we saw in the last chapter, typical values for the input resistances of bipolar operational amplifiers are in the 1 to 100 MΩ range, and a typical value for output resistance might be 75 Ω. Therefore, for circuits using such devices, resistors in the 10 to 100 kΩ range would be appropriate.

Operational amplifiers based on FETs have much higher input resistances, of the order of 10^{12} Ω or more. Circuits using these devices may therefore use higher value resistors, of the order of 1 MΩ or more, if desired. However, resistors in the 10 to 100 kΩ range will generally produce satisfactory results with all forms of op-amps.

Using this rule of thumb and the standard circuits of Examples 4.3 and 4.4, it is possible to design inverting and non-inverting amplifiers with specific gains without a detailed analysis of circuit operation. Example 4.5 shows this procedure for a non-inverting amplifier with a gain of 10.

Example 4.5 A non-inverting amplifier with a gain of 10

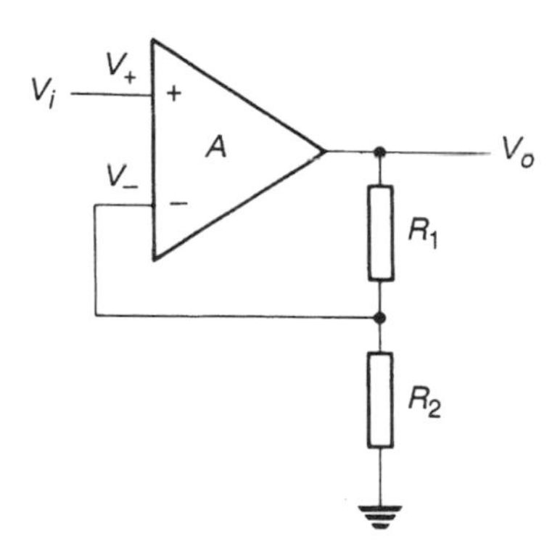

From Example 4.3 we have an expression for the gain G in terms of R_1 and R_2.

$$G = \frac{R_1 + R_2}{R_2} = 10$$

$$R_1 = 9R_2$$

Therefore choose

$$R_2 = 10\ k\Omega$$

and

$$R_1 = 90\ k\Omega$$

In circuits where high accuracy is required, the choice of resistor values must also take into account the input bias and offset currents, since high value resistors will tend to increase the effects that these currents have on the circuit. It is also worth noting at this point that although we are making the simplifying assumption that the input resistance is infinite, all real amplifiers require an input bias current into each input. This requires that there must be DC paths to both inputs.

In addition to simple amplifiers, there are several other op-amp circuits which can be designed using standard cook book techniques. Examples 4.6 to 4.11 show circuits for subtracting and adding signals, and for performing several other processing functions. Suitable combinations of these standard building blocks can perform a large proportion of the electronic manipulation required in electronic systems. PSpice files are available for each of these examples to allow you to investigate their characteristics using simulation.

Example 4.6 A differential amplifier (subtractor)

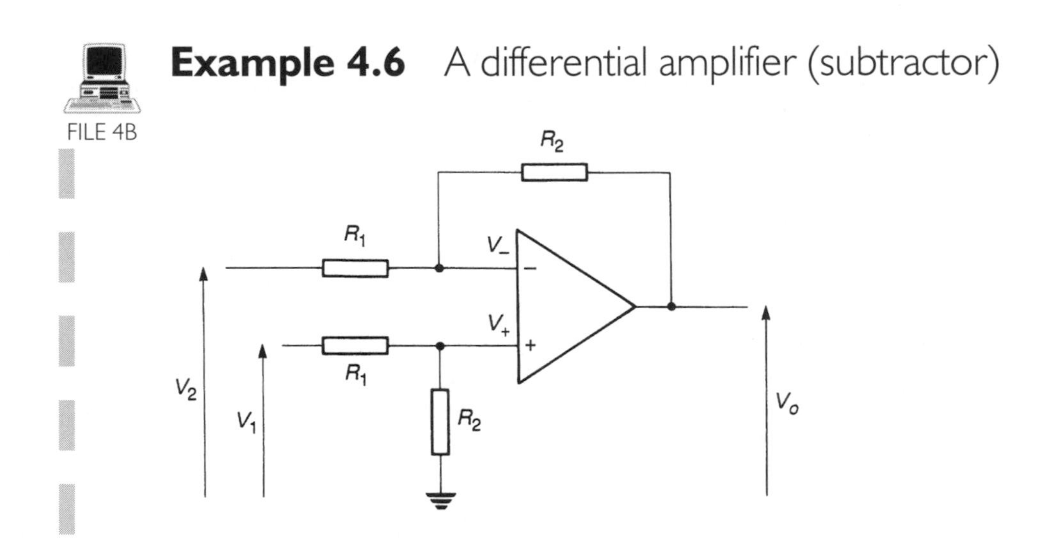

Since negligible current flows into the inputs of the op-amp, the voltages on the two inputs are determined simply by the potential dividers formed by the external resistors.

Thus

$$V_+ = V_1 \frac{R_2}{R_1 + R_2}$$

$$V_- = V_2 + (V_o - V_2) \frac{R_1}{R_1 + R_2}$$

As before, the negative feedback forces V_- to equal V_+, and therefore

$$V_1 \frac{R_2}{R_1 + R_2} = V_2 + (V_o - V_2) \frac{R_1}{R_1 + R_2}$$

Multiplying through by $(R_1 + R_2)$ gives

$$V_1 R_2 = V_2 R_1 + V_2 R_2 + V_o R_1 - V_2 R_1$$

which may be arranged to give

$$V_o = \frac{V_1 R_2 - V_2 R_2}{R_1}$$

and hence the ouput voltage V_o is given by

$$V_o = (V_1 - V_2) \frac{R_2}{R_1}$$

Thus the output voltage is simply the differential input voltage $(V_1 - V_2)$ times the ratio of R_2 to R_1. Note that if $R_1 = R_2$, the output is simply $V_1 - V_2$.

FILE 4C

Example 4.7 An inverting summing amplifier (adder)

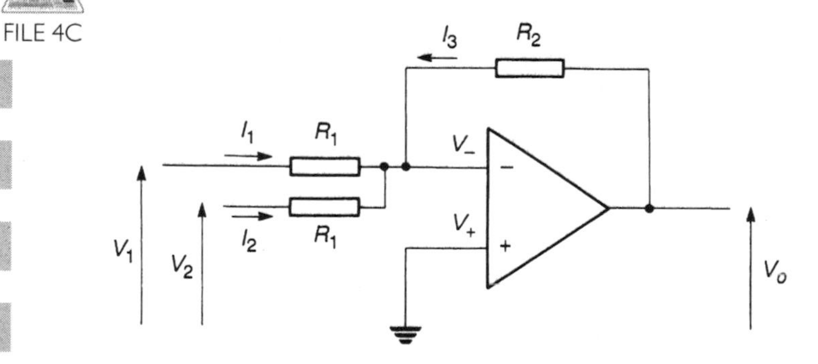

As in Example 4.4, V_- is a virtual earth point. Therefore it is at ground potential (zero volts) and the currents into this point are simple to calculate.

$$I_1 = \frac{V_1}{R_1}$$

$$I_2 = \frac{V_2}{R_1}$$

$$I_3 = \frac{V_o}{R_2}$$

Since the op-amp has a negligible input current, the currents flowing into the inverting input must sum to zero. Therefore

$$I_3 = -(I_1 + I_2)$$

and

$$\frac{V_o}{R_2} = -\left(\frac{V_1}{R_1} + \frac{V_2}{R_1}\right)$$

Therefore the output voltage V_o is given by

$$V_o = -(V_1 + V_2)\frac{R_2}{R_1}$$

The output voltage is determined by the sum of the input voltages $(V_1 + V_2)$ and the ratio of the resistors R_2 and R_1. The minus sign in the expression for the gain indicates that this is an inverting adder. Note that if $R_1 = R_2$ the output is simply $-(V_1 + V_2)$.

This circuit can be easily modified to add more than two input signals. Any number of input resistors may be connected in parallel, and provided they are all of value R_1 the output will become

$$V_o = -(V_1 + V_2 + V_3 + \cdots)\frac{R_2}{R_1}$$

FILE 4D

Example 4.8 A unity gain buffer amplifier (voltage follower)

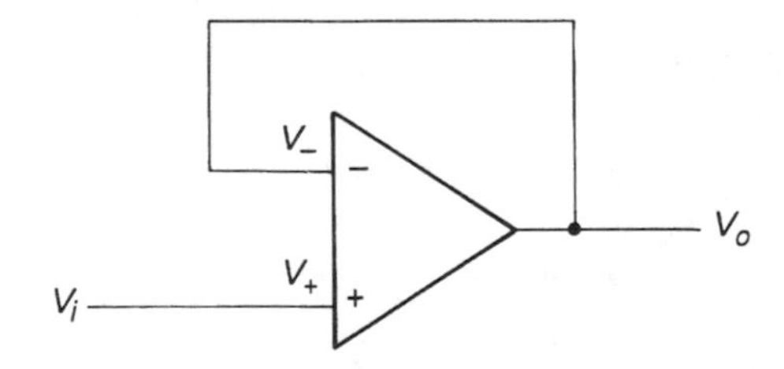

This amplifier is a special case of the non-inverting amplifier of Example 4.3 with R_1 equal to zero and R_2 equal to infinity. In the earlier example we derived an expression for the output voltage, namely

$$G = \frac{R_1 + R_2}{R_2}$$

This may be arranged to give

$$G = \frac{R_1}{R_2} + 1$$

If we substitute appropriate values for R_1 and R_2 we get

$$G = \frac{0}{\infty} + 1 = 1$$

and we therefore have an amplifier with a gain of unity.

At first sight this may not seem a very useful circuit, but one must remember that magnitude is not the only important attribute of a signal. The importance of this circuit is that it has a very high input resistance and a very low output resistance, making it ideal as a buffer. Input and output resistances will be discussed in more detail in the next section.

Example 4.9 A current to voltage converter

FILE 4E

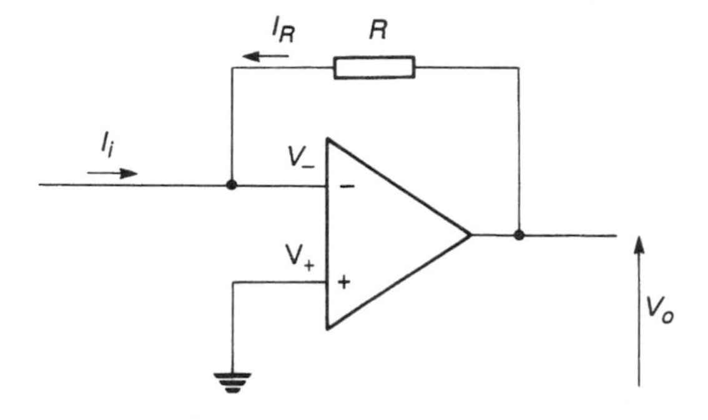

As in earlier circuits, V_- is a virtual earth point and thus the voltage at this point is zero. Since the current going into the inverting input of the op-amp is negligible, the external currents into the virtual earth point must sum to zero. Therefore the input current I_i and the feedback current I_r must sum to zero.

$$I_i + I_R = 0$$

or

$$I_i = -I_R$$

Now since V_- is zero, I_R is given by

$$I_R = \frac{V_o}{R}$$

and therefore

$$I_i = -I_R = -\frac{V_o}{R}$$

or rearranging

$$V_o = -I_i R$$

Thus the output voltage is directly proportional to the input current. The minus sign indicates that the output is inverted.

This circuit differs from a simple voltage or current amplifier in that the output *voltage* is proportional to the input *current*. Unlike a simple gain which is dimensionless, this ratio has the units of voltage/current, which is resistance. For this reason this form of amplifier is sometimes called a **trans-resistive** or **trans-impedance** amplifier.

From the above analysis it is clear that the ratio is given by

$$\frac{V_o}{I_i} = -R$$

Circuits that produce an output *current* which is determined by the input *voltage* are termed **trans-conductance** amplifiers. We will discuss examples of such amplifiers in later chapters.

Example 4.10 An integrator

FILE 4F

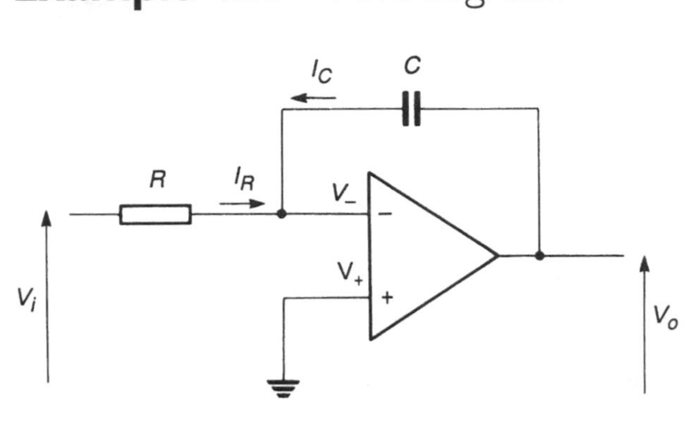

Replacing R_2 in the inverting amplifier of Example 4.4 by a capacitor produces a circuit which acts as an integrator.

Again V_- is a virtual earth point and again the currents into this point must sum to zero. Thus

$$I_C + I_R = 0$$

$$I_C = -I_R = -\frac{V_i}{R}$$

Since V_- is zero, the output voltage V_o is simply the voltage across the capacitor. The voltage across any capacitor is proportional to its charge and inversely proportional to its capacitance. The charge is in turn equal to the integral of the current into the capacitor. Thus

$$V_o = \frac{q}{C} = \frac{1}{C}\int_0^t I_C \,\mathrm{d}t + \text{constant}$$

where the constant represents the initial charge on the capacitor at $t = 0$. If we assume that initially there is no charge on the capacitor, substituting for I_C gives

$$V_o = -\frac{1}{C}\int_0^t \frac{V_i}{R}\,\mathrm{d}t$$

or

$$V_o = -\frac{1}{RC}\int_0^t V_i \,\mathrm{d}t$$

Therefore the output voltage is proportional to the integral of the input voltage, the constant of proportionality being determined by a **time constant** equal to the product of R and C.

One problem associated with the use of integrators is that any DC component in the input (for example, an input offset voltage) is integrated to produce a continuously increasing output which eventually results in the output saturating at one of the supply rails. To overcome this problem the circuit is usually modified to reduce its DC gain.

Example 4.11 A differentiator

FILE 4G

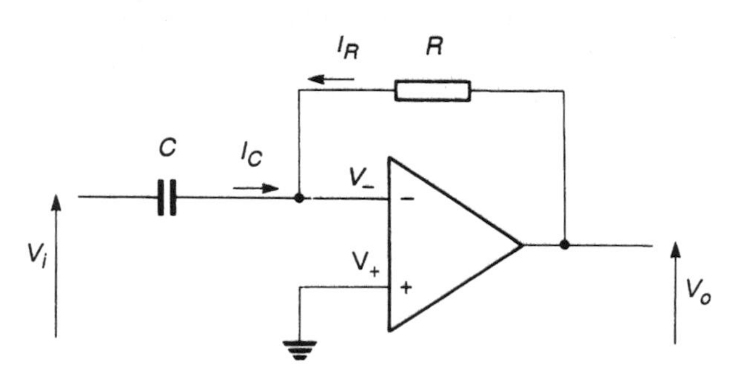

Exchanging the position of the resistor and the capacitor in the integrator produces a differentiating circuit. As before, V_- is a virtual earth point and the currents into this point must sum to zero. Therefore

$$I_C + I_R = 0$$

and thus

$$I_C = -I_R = -\frac{V_o}{R}$$

Since V_- is zero, the voltage across, the capacitor is simply the input voltage V_i, and therefore

$$\begin{aligned} V_i &= \text{voltage across capacitor} \\ &= \frac{1}{C}\int_o^t I_c \, \mathrm{d}t + \text{constant} \end{aligned}$$

and differentiating both sides with respect to t

$$\frac{\mathrm{d}V_i}{\mathrm{d}t} = \frac{I_C}{C}$$

Substituting for I_c gives

$$\frac{\mathrm{d}V_i}{\mathrm{d}t} = -\frac{V_o}{RC}$$

and rearranging gives

$$V_o = -RC\frac{\mathrm{d}V_i}{\mathrm{d}t}$$

Therefore the output voltage is proportional to the derivative of the input voltage with respect to time. In fact, the circuit given above is rarely used in this form since it greatly amplifies high frequency noise and unwanted spikes in the signal and is inherently unstable. The addition of a resistor in series with the capacitor reduces the undesirable amplification of noise at the expense of a slightly less precise differentiation.

From these examples it can be seen that a large number of functions may be produced using an operational amplifier and a few passive components. Simple rules have also been derived for determining the values of the external components required.

It is also worthy of note that the rules have been obtained without a detailed knowledge of the circuitry of the operational amplifier itself, and without any complicated analysis or mathematics. It is one of the great virtues of negative feedback that it makes circuit design much easier!

Computer simulation exercise 4.2

FILE 4H

Use circuit simulation to study the behaviour of the non-inverting amplifier circuit of Example 4.3, using a 741 operational amplifier. Pick appropriate values to produce a gain of 100 and confirm the correct operation of the circuit using transient analysis.

Look at the effects of changing the ratio of the resistor values and note the performance of the circuit when using a ratio that should produce a gain comparable with the open-loop gain of the operational amplifier.

While keeping the ratio of the resistor values constant, look at the effects of using very large and very small resistor values.

Computer simulation exercise 4.3

FILE 4J

Repeat the investigations of Computer simulation exercise 4.2 using the inverting amplifier circuit of Example 4.4. Again use a 741 operational amplifier and pick component values to produce a gain of −100.

4.6 The effects of negative feedback

4.6.1 Gain

We have seen that negative feedback reduces the gain of an amplifier.

In the absence of feedback the gain of an amplifier G is simply its **open-loop gain** A. We know from Equation 4.1 that with feedback the gain becomes

$$G = \frac{A}{1 + AB}$$

and thus the effect of the feedback is to reduce the gain by a factor of $1 + AB$.

In the absence of negative feedback, the gain of most active amplifiers is extremely variable. The manufacturing process generates devices of very different gains even within the same batch of components, and during operation the gain may vary considerably with temperature.

The addition of negative feedback reduces the overall gain of the circuit but produces an arrangement where the gain is determined predominantly by the feedback network rather than the forward path. The overall gain of the resultant feedback system (the **closed-loop gain**) will be numerically equal to the reciprocal of the gain of the feedback network. Therefore, to produce an overall gain of greater than unity, the feedback arrangement must have a gain of less than one. This allows passive, rather than active, components to be used in the feedback network, with a resulting improvement in both tolerance and stability. This principle is illustrated in Example 4.1.

In order for the closed-loop gain to be constant despite variations in the open-loop gain, it is necessary that the **loop gain** of the system (that is, the product of the open-loop gain A and the feedback path gain B) is much greater than unity. This condition will be satisfied provided the open-loop gain is much greater than the closed-loop gain. That is, provided that the gain of the amplifying device (e.g. the op-amp) is much greater than the gain of the complete circuit. This requirement explains why it is advantageous for an op-amp to have an extremely high gain.

4.6.2 Frequency response

In Figures 3.18 and 3.23 we looked at the frequency responses of single and cascaded amplifiers and noted that the gain of all amplifiers falls at high frequencies and that, in many cases, it also drops at low frequencies. The reasons for these effects were discussed in Section 3.7.

From the previous discussion of gain we know that the closed-loop gain of a feedback amplifier is largely independent of the open-loop gain of the amplifier *provided* that the latter is considerably greater than the former. Since the open-loop gain of all amplifiers falls at high frequencies (and often at low frequencies) it is clear that the closed-loop gain will also fall in these regions. However, if the open-loop gain is considerably greater than the closed-loop gain, the former will be able to fall by a considerable amount before this has an appreciable effect on the latter. Thus the closed-loop gain will be stable over a wider frequency range than that of the amplifier without feedback. This is illustrated in Figure 4.7.

The solid line in Figure 4.7 shows the variation of gain with frequency of an amplifier without feedback, that is, its open-loop frequency response. The addition of negative feedback (shown by the broken line) reduces the gain of the arrangement. The resultant closed-loop gain is constant over the range of frequencies where it is considerably less than the amplifier's open-loop gain. The addition of negative feedback thus results in an increase in the bandwidth of the amplifier.

If a wide bandwidth amplifier is required, two stages of amplification with negative feedback, one after the other (in **cascade**), will have a higher bandwidth than a single stage of the same gain. Thus wide bandwidth amplifiers often use many stages of amplification with large amounts of negative feedback.

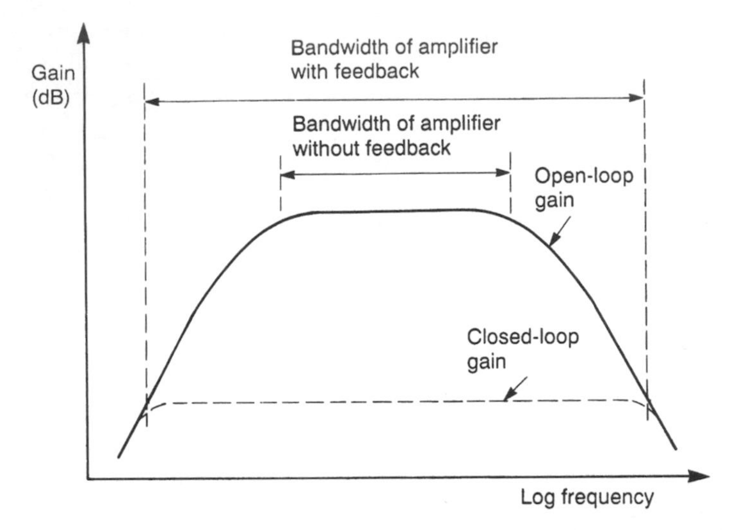

Figure 4.7 The effects of negative feedback on frequency response.

Frequency response of operational amplifiers with negative feedback

Figure 3.28 showed the gain magnitude versus frequency characteristic of a typical operational amplifier, the 741, without feedback, that is, its open-loop frequency response. We observed in Section 3.11.9 that the response is characterized by a steady fall in gain of 6 dB/octave at frequencies above a few hertz. It was also noted that this is due to the intentional introduction of a single dominant upper cut-off frequency.

This effect may be quantified, as described in Section 3.7.2, by considering the gain of an amplifier with a single upper cut-off frequency caused by a single resistor–capacitor combination. From Equation 3.9 we know that the gain of such an arrangement is of the form

$$G = \frac{A}{1 + \mathrm{j}\dfrac{f}{f_U}} \tag{4.2}$$

Therefore, with feedback the gain of such an arrangement becomes

$$G = \frac{A}{1 + AB} = \frac{\dfrac{A}{1 + \mathrm{j}\dfrac{f}{f_U}}}{1 + \dfrac{A}{1 + \mathrm{j}\dfrac{f}{f_U}} B} = \frac{A}{1 + \mathrm{j}\dfrac{f}{f_U} + AB}$$

$$= \frac{\dfrac{A}{1+AB}}{1+j\dfrac{\dfrac{f}{f_U}}{1+AB}} \tag{4.3}$$

Looking back at Equation 4.2 (no feedback) we see that at low frequencies (when $f/f_U \ll 1$), the denominator is approximately unity and the overall gain is simply A. In Equation 4.3 (with feedback) we see that the corresponding term is now $A/(1+AB)$ and thus, as discussed in the last section, the closed-loop gain is reduced by a factor of $1+AB$.

The upper cut-off frequencies in these two cases are determined by the points at which the real and imaginary parts of the denominators become equal, as shown in Figure 3.16. Without feedback (Equation 4.2) this occurs when $f/f_U = 1$, that is, when the frequency f is equal to the upper cut-off frequency of the amplifier f_U. With feedback (Equation 4.3) this occurs when

$$\frac{\dfrac{f}{f_U}}{1+AB} = 1$$

when the frequency f is given by

$$f = f_U(1+AB)$$

Thus negative feedback increases the upper cut-off frequency (and the **bandwidth**) of the system by a factor of 1 + AB.

Since the gain of the system has been *decreased* by a factor of $1+AB$ and the bandwidth has been *increased* by the same factor, clearly the product of these two remains constant. This leads to the conclusion that

Gain × bandwidth = constant

The value of this constant is given by the bandwidth of the system when the gain is unity, the **unity-gain bandwidth** as outlined in Section 3.11.9. Thus

Gain × bandwidth = unity-gain bandwidth

From Figure 3.28 it can be seen that the unity-gain bandwidth of a 741 is typically about 1 MHz. Thus using such a device an amplifier with a closed-loop gain of 100 will have a bandwidth of approximately 10 kHz and an amplifier with a gain of 10 will have a bandwidth of about 100 kHz.

Figure 4.8 shows the frequency response of an amplifier using a 741 with negative feedback, for different values of closed-loop gain G, corresponding to different values of the feedback gain B.

High-speed op-amps may have unity-gain bandwidths of perhaps a gigahertz or more. With such devices it is possible to construct wide bandwidth amplifiers which also provide very high gain.

It should be noted that this relationship between closed-loop gain and bandwidth is produced by the way in which we are using the operational amplifier and is not true for

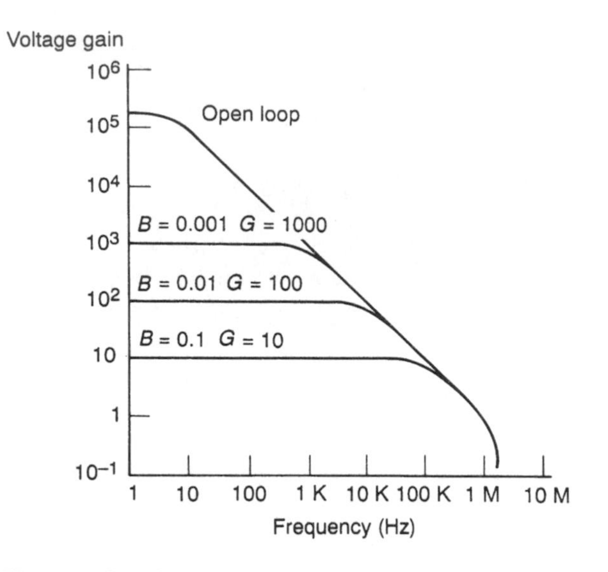

Figure 4.8 Gain versus frequency characteristics of a 741 amplifier with feedback.

all forms of amplifier. However, it is a useful characteristic of operational amplifiers when used as voltage amplifiers.

A similar analysis to the above may be carried out on the effects of feedback on lower cut-off frequencies, when they are present. This shows that, for arrangements with a single dominant time-constant, these are lowered by the same factor. Therefore, negative feedback has the effect of increasing the upper cut-off frequency and lowering the lower cut-off frequency (if present) and thus widening the overall bandwidth of the system, as shown in Figure 4.7.

FILE 4K

Computer simulation exercise 4.4

Use circuit simulation to investigate the relationship between gain and bandwidth for an inverting amplifier constructed using a 741 op-amp.

Apply an alternating voltage to the input of the amplifier and use an AC sweep to measure its frequency response. Plot the gain of the amplifier in dBs against log frequency and hence determine the circuits bandwidth and its gain–bandwidth product.

Repeat this process for a range of amplifiers with resistor values chosen to give a selection of low-frequency gains. Include within your selection an amplifier with a gain of unity. Compare the gain–bandwidth products obtained with the various amplifiers with the bandwidth of the unity gain circuit, and with the form of Figure 4.8.

4.6.3 Input and output resistances

In the general representation of a feedback system shown in Figure 4.3 you will notice that the input and output variables are given the symbols X_i and X_o rather than V_i and V_o. This notation was used since there is nothing in the following theory that limits it to voltage amplifiers. Indeed, X_i could represent an input current or another input quantity (for example, an input force into a mechanical amplifier). If we restrict ourselves to electrical systems, the input and output quantities will usually represent voltages or currents, but again the block diagram does not define the nature of the feedback signal or the way in which it is combined with the input to produce the error signal.

The fact that feedback may be applied in a variety of ways is illustrated by the various circuits of Section 4.5. In Example 4.4, for instance, a *current* I_2 proportional to the output voltage is fed back and subtracted from the input current I_1. However, in Example 4.5 a *voltage* proportional to the output voltage is fed back and subtracted from the input voltage. In order to understand the effects of negative feedback on the input and output resistances of circuits, we need first to consider the various ways in which feedback can be applied.

Sensing the output quantity

The most common methods of generating a feedback signal are to sense the output voltage or the output current. These two techniques are illustrated in Figure 4.9.

Figure 4.9(a) shows the use of **voltage feedback**, where the quantity fed back is proportional to the output voltage V_o. This form of feedback will tend to maintain the output voltage constant, *reducing the apparent output resistance* of the circuit. To understand this effect consider the equivalent circuit of Figure 4.10.

The amplifier may be represented by a dependent voltage source with a magnitude that is A times the error signal V_i', in series with an output resistance of R_o. The magnitude of this voltage source may be thought of as the open-circuit output voltage V_o'. The output current I_o is conventionally drawn going *into* the amplifier. If we assume that the current taken by the feedback network is negligible, this is the current flowing through the output resistance R_o.

Therefore, from the equivalent circuit it can be seen that

$$\begin{aligned} V_o &= AV_i' + R_o I_o \\ &= A(V_i - BV_o) + R_o I_o \end{aligned}$$

To allow us to investigate this condition, let us consider the situation where the input voltage is zero. In this case

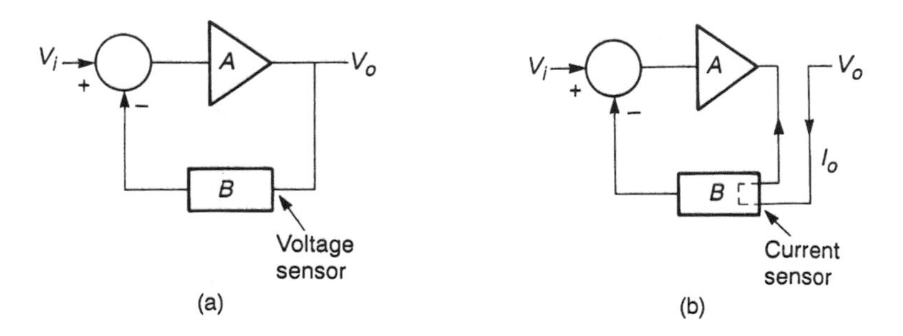

Figure 4.9 Sensing the output quantity.

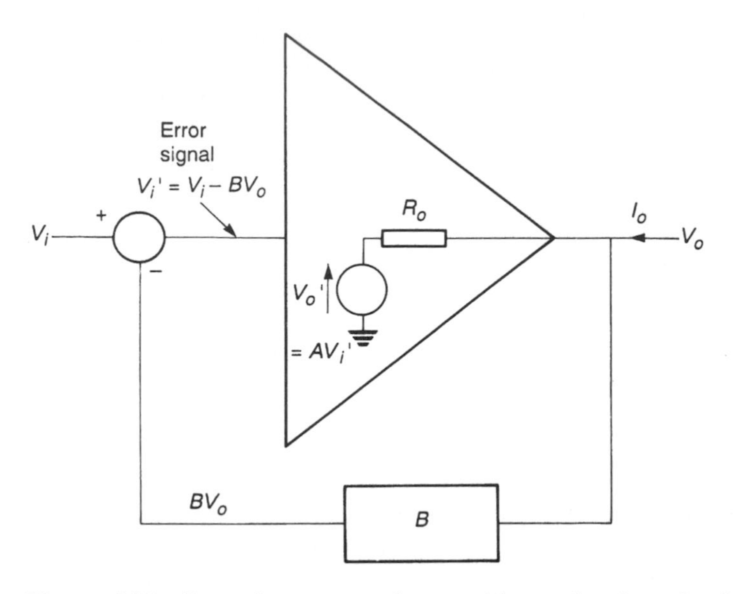

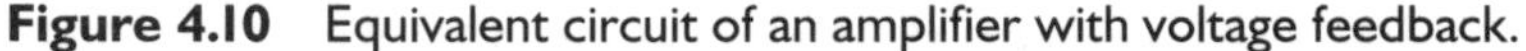

Figure 4.10 Equivalent circuit of an amplifier with voltage feedback.

$$V_o = -ABV_o + R_o I_o$$

and rearranging gives

$$V_o(1 + AB) = R_o I_o$$

Therefore the effective output resistance of the amplifier is given by

$$\text{Output resistance} = \frac{V_o}{I_o} = \frac{R_o}{1 + AB}$$

Thus the effective output resistance is reduced by a factor of $(1 + AB)$.

Figure 4.9(b) shows the use of **current feedback**, where the quantity fed back is proportional to the output current I_o. This form of feedback will tend to maintain the output current constant, *increasing the apparent output resistance* of the circuit. Consider the equivalent circuit of Figure 4.11.

In this case it is more convenient to represent the amplifier by a dependent current source in parallel with an output resistance R_o. In this case, the output voltage V_o is given by

$$V_o = R_o(I_o - I_o')$$
$$= R_o(I_o - A(I_i - BI_o))$$

Again to investigate this expression it is convenient to put the input to zero, so in this case $I_i = 0$, and hence

$$V_o = R_o I_o(1 + AB)$$

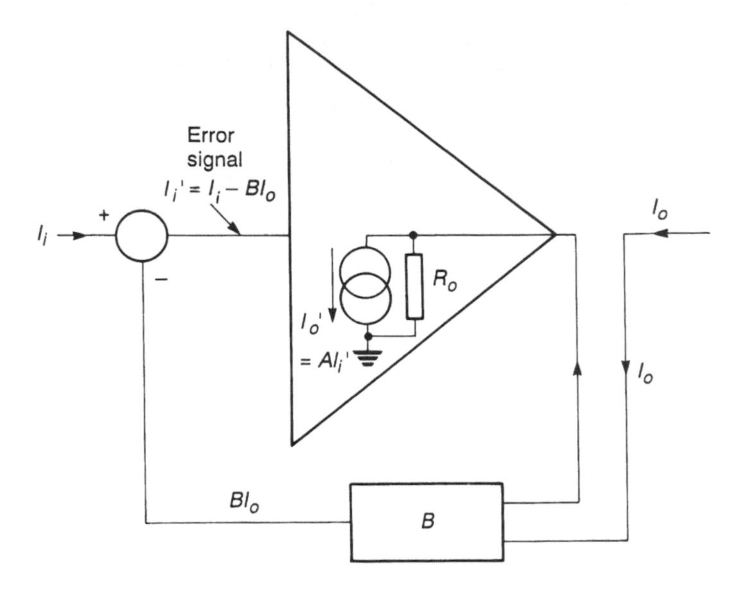

Figure 4.11 Equivalent circuit of an amplifier with current feedback.

and thus

$$\text{Output resistance} = \frac{V_o}{I_o} = R_o(1 + AB)$$

Thus the effective output resistance is increased by a factor of $(1 + AB)$.

Applying the feedback signal

The most common methods of applying the feedback signal are to subtract a voltage related to the sensed output quantity from the input voltage, or to subtract a current related to the sensed output quantity from the input current. These two techniques are illustrated in Figure 4.12.

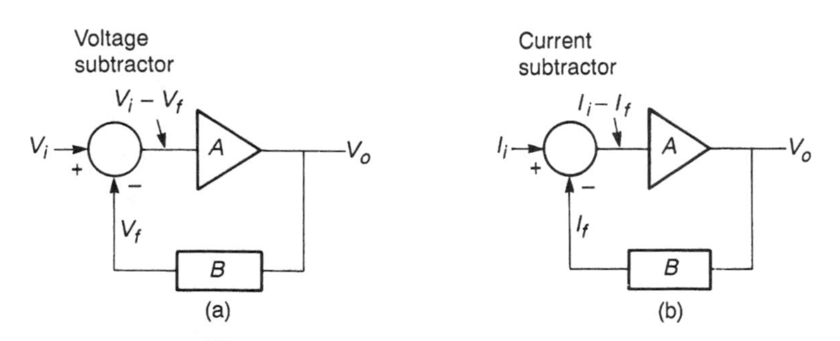

Figure 4.12 Applying the feedback signal.

In Figure 4.12(a) a feedback voltage is subtracted from the input. This has the effect of improving the performance of the arrangement as a voltage amplifier by *increasing the effective input resistance* of the circuit. To see why this is so, consider the equivalent circuit of Figure 4.13.

From the equivalent circuit it can be seen that

$$V_i' = V_i - V_f$$
$$= V_i - ABV_i'$$

and therefore

$$V_i'(1 + AB) = V_i$$

Since

$$R_i = \frac{V_i'}{I_i}$$

then clearly

$$\text{Input resistance} = \frac{V_i}{I_i} = \frac{V_i'(1 + AB)}{I_i} = R_i(1 + AB)$$

Therefore the effective input resistance of the amplifier is increased by a factor of $(1 + AB)$. This method of applying feedback is often referred to as a **series input** arrangement, because the feedback appears in series with the input signal as shown in Figure 4.13.

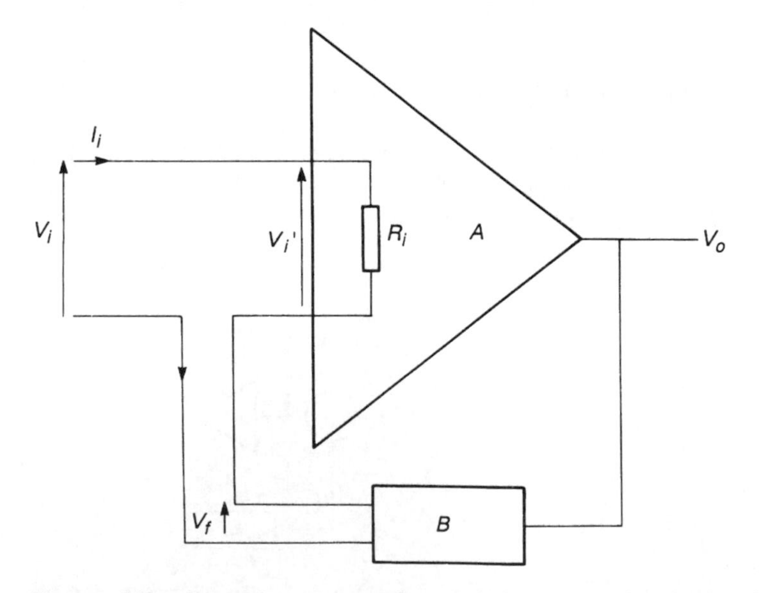

Figure 4.13 Equivalent circuit of an amplifier using a feedback voltage (series input).

In Figure 4.12(b) a feedback current is subtracted from the input. This has the effect of improving the performance of the arrangement as a current amplifier by *decreasing the effective input resistance* of the circuit. Consider the equivalent circuit of Figure 4.14.

From the equivalent circuit it can be seen that

$$I_i' = I_i - I_f$$
$$= I_t - ABI_i'$$

and thus

$$I_i'(1 + AB) = I_i$$

Since

$$R_i = \frac{V_i}{I_i'}$$

then clearly

$$\text{Input resistance} = \frac{V_i}{I_i} = \frac{V_i}{I_i'(1 + AB)} = \frac{R_i}{(1 + AB)}$$

Therefore the effective input resistance of the amplifier is decreased by a factor of $(1 + AB)$. This method of applying feedback is often referred to as a **shunt input** arrangement, because the feedback appears in parallel with the input signal as shown in Figure 4.14.

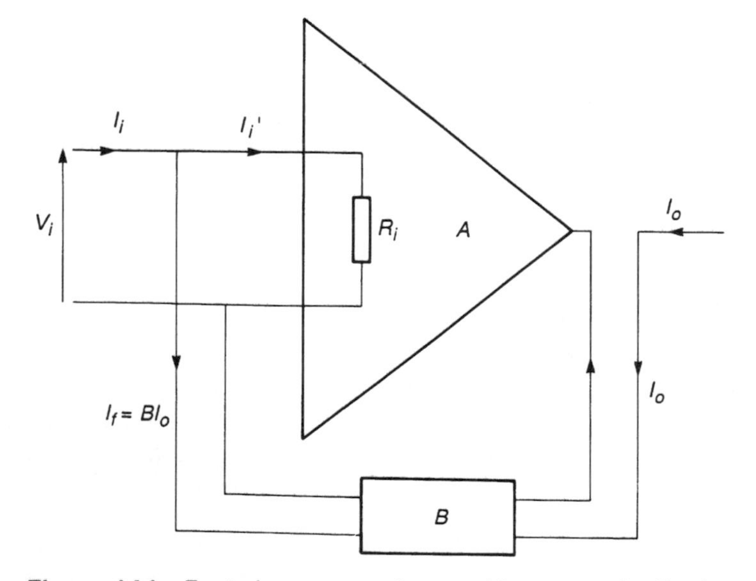

Figure 4.14 Equivalent circuit of an amplifier using a feedback current (shunt input).

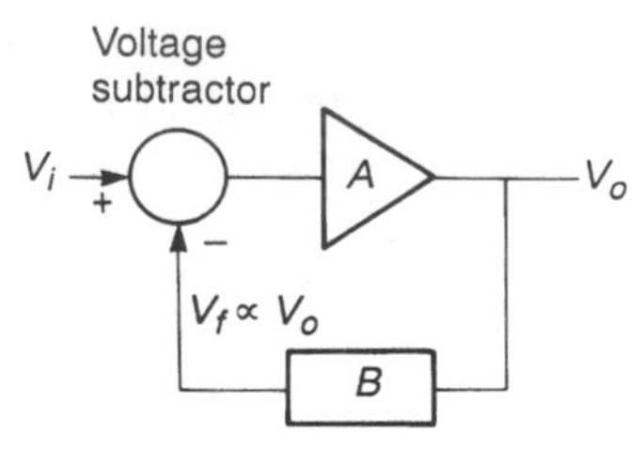

(a) Series input, voltage feedback

(b) Series input, current feedback

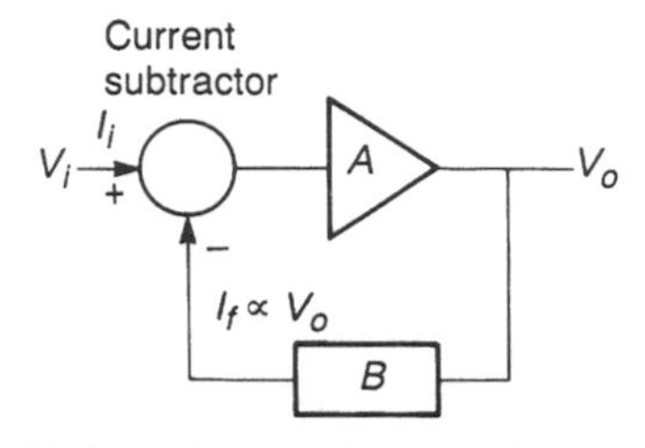

(c) Shunt input, voltage feedback

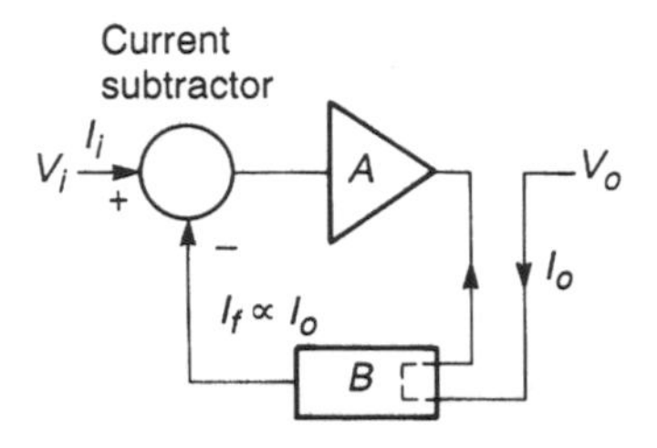

(d) Shunt input, current feedback

Figure 4.15 Negative feedback configurations.

Input and output resistances – a summary

Negative feedback can be achieved in a number of ways. Circuitry can be arranged to measure the output voltage or the output current of a system and to feed back a voltage or a current related to the measured quantity. This gives rise to four primary configurations, as shown in Figure 4.15.

Generally, negative feedback tends to **improve** whatever characteristic is being fed back. For example:

- If the output **voltage** is sensed and fed back to the input this will tend to improve the constancy of the output voltage, effectively **reducing** the **output resistance** (or impedance).
- If the output **current** is sensed and fed back to the input this will tend to improve the constancy of the output current, effectively **increasing** the output resistance (or impedance).
- If the signal subtracted from the input is a current related to some characteristic of the output, this will tend to improve the current handling properties of the input, effectively **lowering** the **input resistance** (or impedance).
- If the signal subtracted from the input is a voltage related to some characteristic of the output, this will tend to improve the voltage handling properties of the input, effectively **increasing** the **input resistance** (or impedance).

The factor by which the appropriate characteristic is improved is generally $(1 + AB)$. That is, 1 plus the loop gain.

Example 4.12 The effects of feedback on the input and output resistances of a non-inverting amplifier

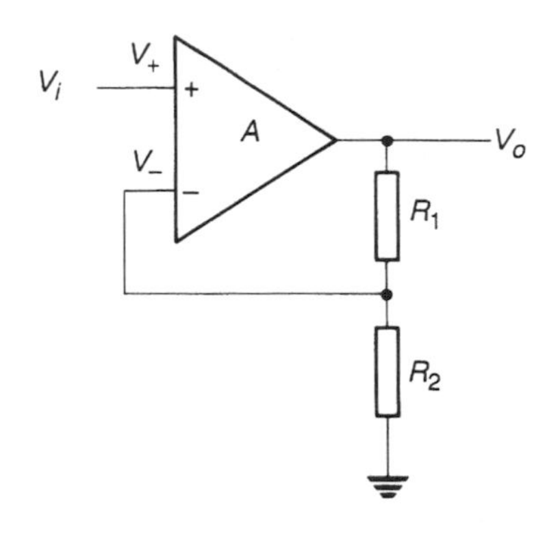

Here the output **voltage** is sensed so the feedback will tend to improve the consistency of the output voltage by **reducing** the output resistance by a factor of $(1+AB)$, where

$$B = \frac{R_2}{R_1 + R_2}$$

The feedback is applied in the form of a **voltage** which is subtracted from the input. Therefore the feedback will tend to improve the circuit as a voltage amplifier by **increasing** the input resistance by a factor of $(1+AB)$.

Note: Since A varies with frequency, so do the input and output resistances.

Example 4.13 The effects of feedback on the input and output resistances of an inverting amplifier

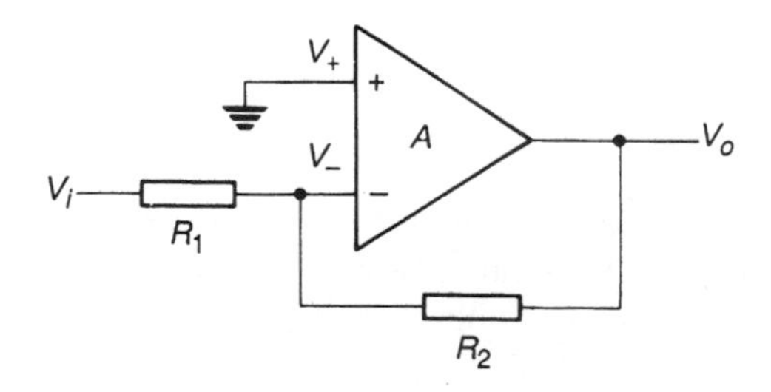

Again it is the output **voltage** that is sensed, so again the output resistance is **reduced** by a factor of $(1+AB)$, thus improving the consistency of the output voltage.

However, here a **current** is being subtracted from the input, so the feedback tends to make the circuit a better current amplifier by **reducing** the input resistance.

Note: Here the input resistance R_1 is in series with the input of the amplifier and this will normally dominate the input resistance of the circuit. Since the inverting input is a virtual earth point, the input sees an input resistance of R_1 to ground. Therefore the input resistance is simply R_1.

It can be seen that the effect of negative feedback is to *improve* the input resistance and the output resistance of a circuit by a factor of $(1+AB)$. To get some feel for the significance of this improvement let us look back at one of the circuits we have already considered.

In Example 4.5 we looked at a non-inverting amplifier with a gain of 10. If this circuit were to be implemented with a 741 we might expect the op-amp to have an open-circuit gain of about 200 000, an input resistance of about 2 MΩ and an output resistance of about 75 Ω. To obtain a gain of 10 we require a value of B of 1/10 or 0.1. Thus the improvement factor $(1+AB)$ is equal to $(1+200\,000\times0.1)$ or about 20 000. This would suggest that the amplifier with feedback would have an input resistance of about 40 GΩ (20 000 × 2 MΩ), and an output resistance of about 4 mΩ (75 Ω ÷ 20 000). Thus the negative feedback produces a very marked improvement in these two characteristics.

4.6.4 Distortion

In Figure 1.3 we looked at examples of the effects of distortion on a sinusoidal signal. The examples shown represent typical forms of distortion caused by a **non-linear amplitude response** in an amplifier. In the case of clipping, the gain of the amplifier falls as the magnitude of the signal increases, producing a non-sinusoidal output. Crossover distortion is produced when the gain falls for very small amplitude signals. Non-linearities may also change the waveform by adding components at frequencies which are a multiple of the input signal. This gives rise to harmonic distortion.

It is clear that these forms of distortion are produced because the gain of the amplifier is not consistent for signals or different amplitudes. These problems are particularly acute when the amplifier is required to operate with large voltage excursions, as in power amplifiers.

Since negative feedback has the effect of making the overall system gain stable, regardless of variations in the open-loop gain of the amplifying device, it is not surprising that it also reduces the amount of distortion produced. It is perhaps also not unexpected, in the light of earlier sections, that the improvement is by a factor of $1+AB$. To illustrate this point let us consider an amplifier with distortion as shown in Figure 4.16.

Figure 4.16(a) shows a representation of an amplifier with distortion consisting of an idealized distortionless amplifier followed by a voltage source V_d representing the added distortion. The output voltage of this combination is given simply by the expression

$$V_o = AV_i + V_d \tag{4.4}$$

Figure 4.16(b) shows this amplifier with added negative feedback. Since the amount of distortion produced by the amplifier is determined by the magnitude of the output signal V_o this has been kept constant and the input voltage has been increased to V_{if} to compensate

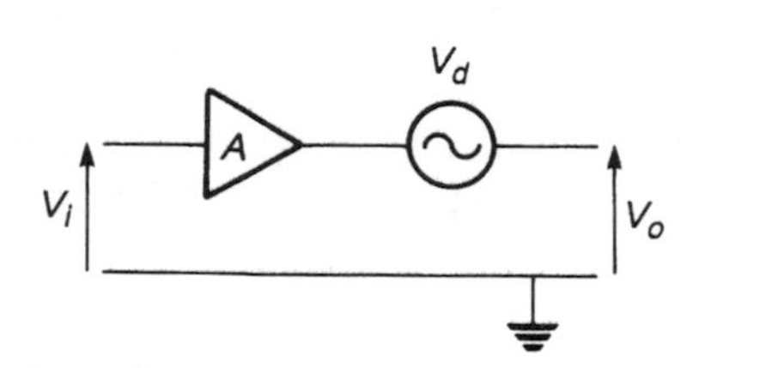

(a) Representation of an amplifier with distortion

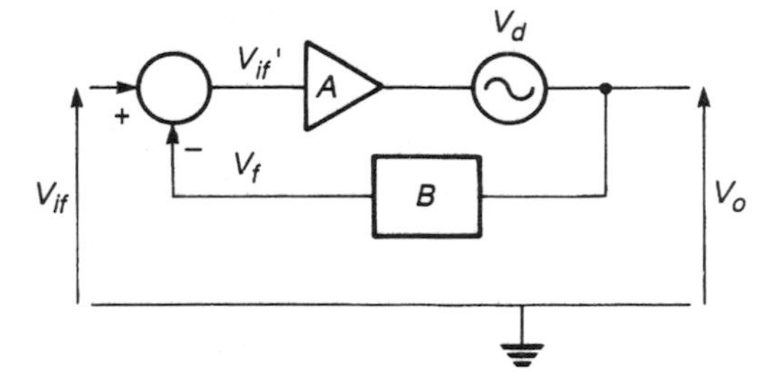

(b) Negative feedback applied to an amplifier with distortion

Figure 4.16 The effects of distortion in a feedback amplifier.

for the reduction in gain caused by the feedback. Thus

$$V_{if} = (1 + AB)V_i \tag{4.5}$$

From Figure 4.16(b) it is clear that

$$V_{if}' = V_{if} - V_f$$

and

$$\begin{aligned} V_o &= AV_{if}' + V_d \\ &= A(V_{if} - V_f) + V_d \\ &= A(V_{if} - BV_o) + V_d \end{aligned}$$

Rearranging gives

$$V_o(1 + AB) = AV_{if} + V_d$$

and substituting for V_{if} from Equation 4.5 gives

$$V_o(1 + AB) = AV_i(1 + AB) + V_d$$

thus

$$V_o = AV_i + \frac{V_d}{(1 + AB)} \tag{4.6}$$

Comparing the result of Equation 4.6 (with feedback) with that of Equation 4.4 (without feedback) shows that the amount of distortion produced for a given magnitude of undistorted output AV_i is reduced by a factor of $(1 + AB)$ by the use of feedback.

4.6.5 Noise

Noise produced *within* an amplifier may be represented by an equivalent circuit similar to that shown in Figure 4.16, simply by changing the voltage generator from V_d to V_n. A similar analysis to that given above can then be used to show that the noise generated within the amplifier will also be reduced by the effects of negative feedback, producing the result

$$V_o = AV_i + \frac{V_n}{(1 + AB)} \tag{4.7}$$

However, noise introduced along with the input signal is indistinguishable from the input and so is amplified in a similar manner. The relationship of Equation 4.7 holds only for noise generated *within* the amplifier.

4.6.6 Stability

From Equation 4.1 we know that

$$G = \frac{A}{1 + AB}$$

and this implies that provided $|1 + AB|$ is greater than 1, the gain with feedback G will be less than the open-loop gain of the amplifier A.

So far in this section we have assumed that both A and B can be described by simple real gains, such that their product AB is a positive real number. Under these circumstances $|1 + AB|$ is always greater than unity. However, we noted in Section 3.7 that all amplifiers produce a phase shift in the signals passing through them, the magnitude of which varies with frequency. We also saw in Figure 3.23 that these phase shifts add if a number of stages of amplification are cascaded. The result of this phase shift is that at some frequencies $|1 + AB|$ may be less than 1. Under these circumstances the feedback *increases* the gain of the amplifier and is now positive, rather than negative feedback. This can result in the amplifier becoming unstable and the output oscillating, independent of the input signal. We shall see in the next section that positive feedback is often used intentionally to produce oscillation. However, its presence can prevent the normal operation of an amplifier.

Gain and phase margins

From Figure 3.23 it is clear that the maximum phase shift produced by a multi-stage amplifier is related to the number of stages or, more correctly, to the number of **time constants**, in the system. Here the term 'time constant' refers to a circuit characteristic which produces a variation of gain with frequency of 6 dB/octave. A single resistor–capacitor network, for example, represents a single time constant, and each time constant will add 90° of phase shift at very high, or very low, frequencies. In practice it is impossible to build high performance amplifiers without using many stages. Hence most amplifiers produce large phase shifts at certain frequencies.

The stability of an amplifier is determined by the term $(1 + AB)$. If this term is positive we have negative feedback and the stability of the circuit is assured. If, as a result of a phase shift, the term $(1 + AB)$ becomes less than 1 (because AB becomes negative), the feedback becomes positive and all the advantages of negative feedback are lost. In the extreme case, if $(1 + AB)$ becomes equal to 0 the closed-loop gain of the arrangement is infinite and the system becomes unstable and will oscillate.

The condition that $(1 + AB) = 0$ represents the case where $AB = -1$, or, in other words, where the loop gain has a magnitude of 1 and a phase of 180°. Indeed, the ampli-

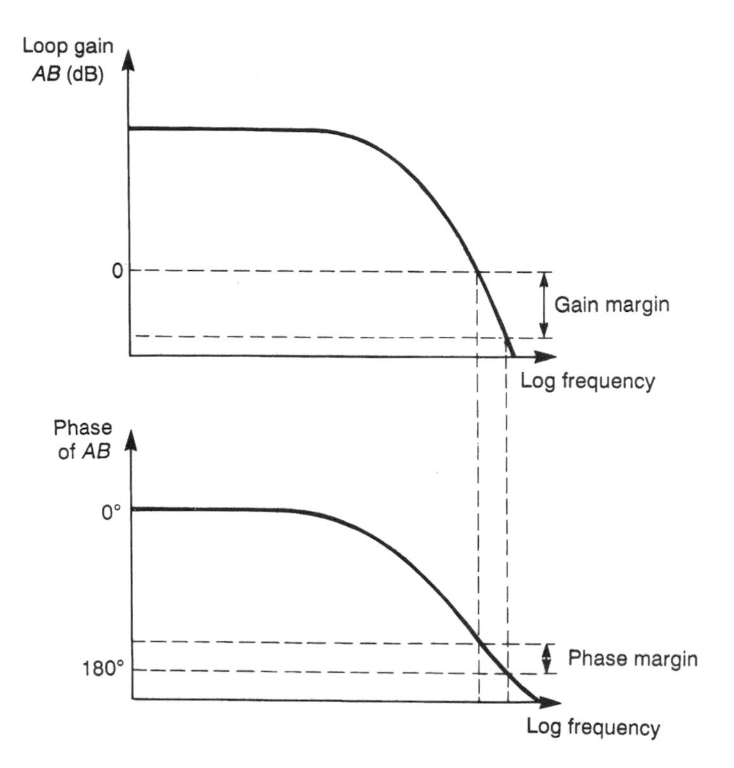

Figure 4.17 Gain and phase margins.

fier will remain stable even if the phase shift is equal to 180° provided that the magnitude of the loop gain is less than unity. The task of the designer is thus to ensure that the loop gain of the amplifier falls below unity *before* the phase shift reaches 180°. In practice it is advisable to allow some margin for variability within the phase and gain values. This leads to the concept of the **phase margin**, which is the angle by which the phase is less than 180° when the loop gain falls to unity, and the **gain margin**, which is the amount (in dBs) by which the loop gain is less than 0 dB (that is unity gain) when the phase reaches 180°. These quantities may be illustrated using a Bode diagram, as shown in Figure 4.17.

The above discussion makes clear why designers of operational amplifiers such as the 741 described in Chapter 3, choose to add a single dominant time constant to the amplifier to roll off the gain as shown in Figure 3.28. This ensures that the gain falls to less that 0 dB well before the phase shift reaches 180°. This produces large gain and phase margins and ensures good stability.

Nyquist diagrams

An alternative method of investigating the stability of a circuit is the use of a **Nyquist diagram** which illustrates the relationship between gain and phase in a single plot. The diagram is essentially a plot of the real and imaginary parts of the loop gain *AB* for all

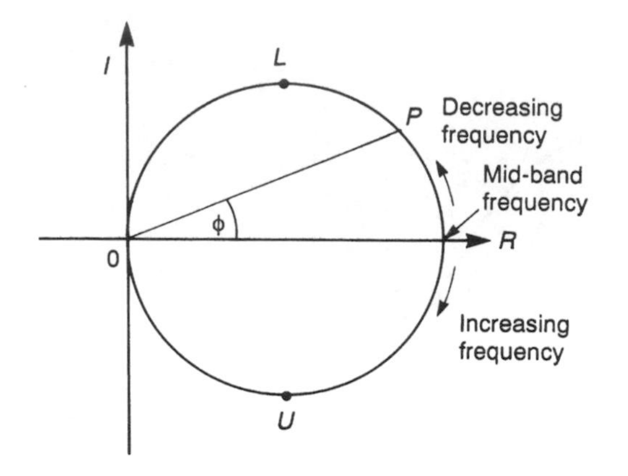

Figure 4.18 Nyquist diagram for an amplifier with single upper and lower cut-offs.

frequencies. An example of a Nyquist diagram for an amplifier with a single low frequency cut-off and a single high frequency cut-off is shown in Figure 4.18.

The diagram is formed by the locus of P, where the distance of P from the origin (the magnitude of OP) represents the magnitude of AB, and the angle ϕ represents its phase. At mid-band frequencies the phase of the output is zero and the gain is simply real. As the frequency is reduced the lower cut-off frequency causes the magnitude of the gain to fall and produces a positive phase angle. When the frequency is equal to the lower cut-off frequency the gain will have dropped to 0.707 of its mid-band value and its phase will be $+45°$. Thus the lower cut-off frequency corresponds to point L. As the frequency goes to very low values the magnitude of the gain tends to zero and the phase tends to $+90°$. The locus of P thus approaches the origin along the positive imaginary axis. At high frequencies the upper cut-off frequency reduces the magnitude of the gain and produces a phase lag. At the upper cut-off frequency the magnitude of the gain has again dropped to 0.707 of its mid-band value and the phase angle is $-45°$. This corresponds to point U. As the frequency increases the magnitude of the gain falls towards zero and the phase angle tends to $-90°$. The locus therefore approaches the origin along the negative imaginary axis. For this idealized case (an amplifier having one upper and one lower cut-off frequency) the Nyquist diagram is a circle.

Figure 4.19 shows some examples of Nyquist diagrams for a range of amplifiers. Figure 4.19(a) represents an amplifier with no low frequency cut-off (a DC coupled amplifier). The gain therefore stays constant at low frequencies rather than falling to zero. This example is for an amplifier having a single high frequency cut-off. The maximum phase shift is therefore $-90°$, and the locus approaches the origin along the negative imaginary axis. Figure 4.13(b) represents a DC amplifier with two high frequency cut-offs. This has a maximum phase shift of 180° and thus approaches the origin along the negative real axis. Figure 4.19(c) shows a system with three high frequency cut-offs and the locus therefore approaches the origin along the positive imaginary axis. Figure 4.13(d) shows the response of an amplifier with two low frequency cut offs and three high frequency cut-offs. This has a maximum phase shift of $+180°$ at low frequencies and $-270°$ at high frequencies.

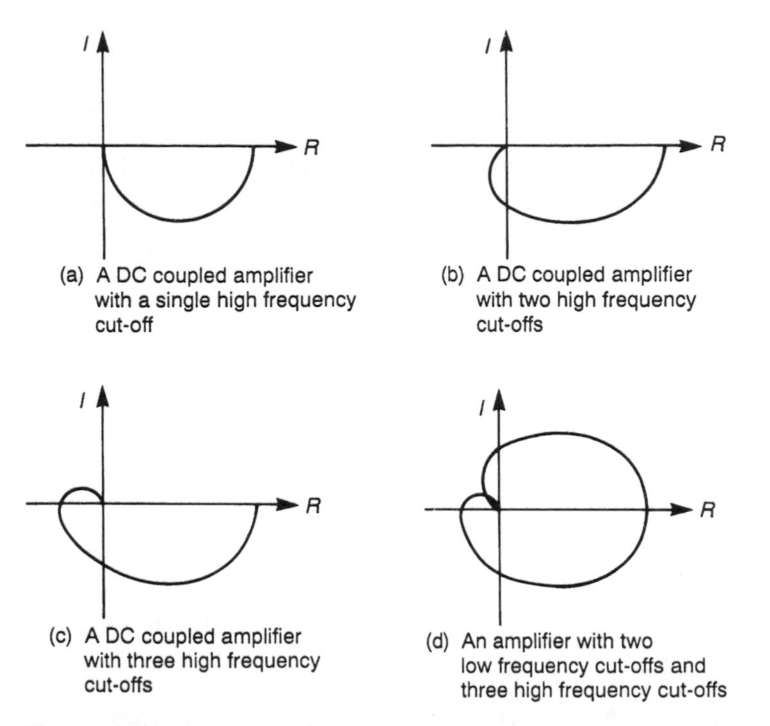

Figure 4.19 Examples of Nyquist diagrams.

The stability of the amplifier is determined by the magnitude of the term $1+AB$ which may be represented on the Nyquist diagram by a line drawn from the point $(-1,0)$ to the locus P. This is shown in Figure 4.20.

If a circle of unit radius is drawn with its centre at $(-1,0)$, whenever P lies within this circle it represents a point at which $1+AB$ is less than unity. Under these circumstances

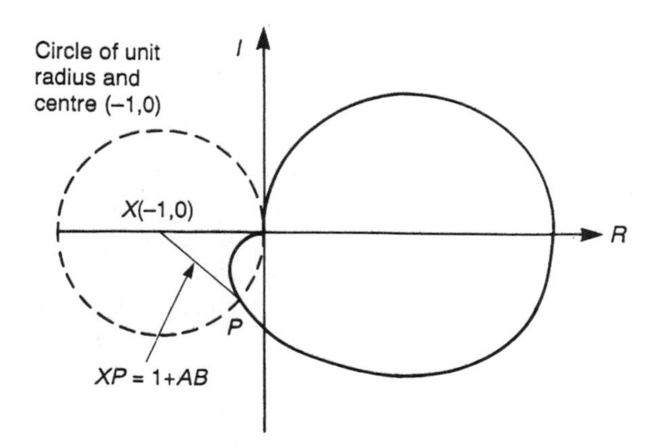

Figure 4.20 Investigations of stability using a Nyquist diagram.

the feedback is positive rather than negative. This implies that the gain of the amplifier is greater than A and all the advantages of negative feedback are lost. If the locus passes through the point $(-1, 0)$ this represents the condition that $1 + AB$ is equal to 0. This means that the gain of the amplifier is infinite and it is thus unstable and will oscillate.

The Nyquist diagram can therefore be used to investigate the stability of a system, the general principles of which are summed up in the **Nyquist stability criterion**. This may be paraphrased as:

(1) If the locus of P does not enter the unit circle centred on $(-1, 0)$ the circuit is stable and has negative feedback.

(2) If the locus enters the unit circle the feedback is positive within that region.

(3) If the locus encircles the point $(-1, 0)$ the amplifier will oscillate.

It can be seen that amplifiers with no more than a single upper and lower cut-off will always be stable since the locus of P is always to the right of the origin. Amplifiers with two upper or two lower cut-offs can enter a region where the feedback is positive but cannot encircle the point $(-1, 0)$ and are thus always stable. Systems with more than two upper or lower cut-offs may be unstable if the locus of P encircles the point $(-1, 0)$.

Note. We saw at the end of Section 4.3 that some texts adopt an alternative form of description for feedback circuits which uses an adder rather than a subtractor. We also noted that this notation produces an expression for the overall gain of the form

$$G = \frac{A}{1 - AB}$$

One of the effects of this change is that the condition for stability is that AB must be less than $+1$. This results in Nyquist diagrams that are mirror images of those given above.

4.6.7 Negative feedback – a summary

We have seen that negative feedback has many beneficial effects on the performance of an amplifier. It stabilizes the gain against variations in the open-loop gain of the amplifying device. It increases the bandwidth of the amplifier. It can be used to increase or decrease input and output resistances as required. It reduces distortion caused by non-linearities in the amplifier. It reduces the effects of noise produced within the amplifier.

In exchange for these benefits, negative feedback reduces the gain of the amplifier. In most cases this is a small price to pay, since the majority of modern amplifying devices have high gains and are inexpensive, allowing many stages to be cascaded. However, in applying negative feedback attention must be paid to the stability of the amplifier because of the effects of phase shifts at high and low frequencies. The designer can ensure stability by ensuring that the gain drops to below unity before the phase shift reaches 180°.

We shall see that negative feedback plays a vital role in electronic circuits for a wide range of applications. These include not only circuits based on operational amplifiers, but also those based on discrete amplifying devices such as transistors.

4.7 Positive feedback

In the last section we looked at the many advantages of the use of negative feedback in amplifiers. We will now consider the use of **positive feedback** for the production of **oscillators**.

4.7.1 Oscillators

In earlier sections we looked at feedback systems and considered the characteristics of such systems for different values of the loop gain AB. We discovered that the overall or closed-loop gain G is given by

$$G = \frac{A}{(1 + AB)}$$

and that if $AB = -1$ the closed-loop gain is infinite. Under these circumstances the circuit will generally produce an output even in the absence of any input.

The requirements for oscillation are expressed by the **Barkhausen criterion** which, using our notation, may be represented by the condition that for oscillation to occur:

(1) The magnitude of the loop gain AB must be equal to 1.

(2) The phase shift of the loop gain AB must be 180°, or 180° plus an integer multiple of 360°.

The second condition comes about because our representation of the feedback system, as shown in Figure 4.3, assumes that the feedback signal is subtracted from the input. Subtracting a signal which is 180° out of phase with the input signal is equivalent to adding a signal which is in phase with it. Hence the feedback increases the effective magnitude of the input and this condition represents positive feedback. A circuit that satisfies this condition is shown in Figure 4.21.

Consider the operation of the circuit of Figure 4.21. A small positive signal at the inverting input δV will be amplified and produce an output voltage of $-A \times \delta V$. This will be attenuated by the feedback arrangement which has a gain of $B = -1/A$, and thus the

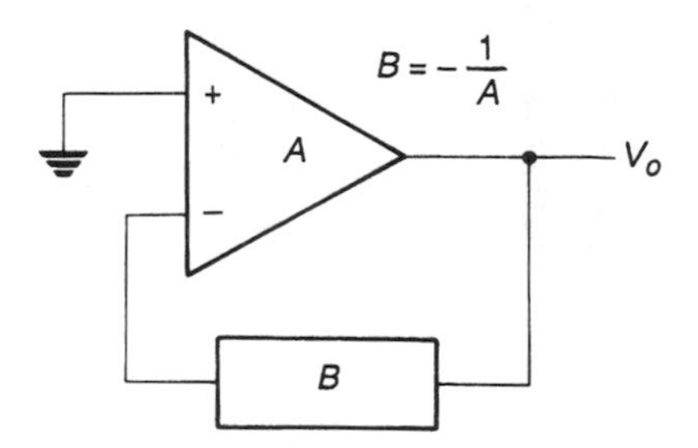

Figure 4.21 A feedback circuit with $AB = -1$.

signal fed back to the input will be $-A \times \delta V \times -1/A = \delta V$. This condition is stable and so the voltages at the input and the output remain constant. A similar situation occurs if a small negative signal is applied to the inverting input. Thus the circuit simply stays in whatever state it finds itself when it is turned on, being affected only by the small amount of noise that is always present in such circuits.

In practice such an arrangement is not only of little use but also impractical. It relies on the fact that the gain of the feedback path *exactly* matches the open-loop gain of the amplifier. This is not achievable, especially since the open-loop gain of the amplifier will certainly vary with temperature and other factors. In reality, the gain of the feedback path will either be a little too low or a little too high. If the feedback gain is too low, $1 + AB$ will be positive, producing negative feedback, and the output signal will decay to zero. If the feedback gain is too high, $1 + AB$ will be negative thus producing positive feedback. Whatever small voltage is present at the output of the amplifier will be amplified until the device saturates at one of the supply rails.

A more useful arrangement is produced if we add a frequency selective element to the circuit. If we could arrange that the loop gain was equal to -1 at a single frequency, this would cause the circuit to oscillate continuously at that frequency.

The *RC* or phase-shift oscillator

A simple way of producing a phase shift of 180° at a single frequency is to use an *RC* ladder network, as shown in Figure 4.22. Here a number of *RC* stages are cascaded, each producing an additional high frequency cut-off. From the discussion at the end of Section 4.6.6 it is clear that at least three stages are required to produce a phase shift of 180° (since two stages only produce a 180° phase shift at infinite frequency).

If we adopt a ladder with three identical *RC* stages, then standard circuit analysis reveals that the ratio of the output voltage to the input voltage is given by the expression

$$\frac{v_0}{v_i} = \frac{1}{1 - \dfrac{5}{(\omega CR)^2} - \mathrm{j}\left(\dfrac{6}{\omega CR} - \dfrac{1}{(\omega CR)^3}\right)} \tag{4.8}$$

The magnitude and phase angle of this ratio is clearly dependent on the angular frequency ω. We are interested in the condition where the phase shift is 180°. This implies

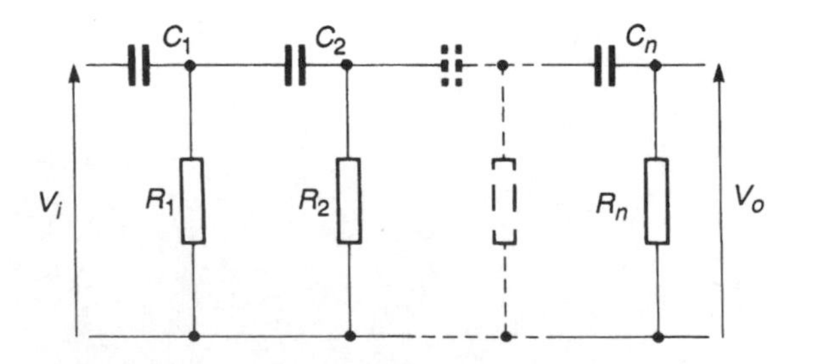

Figure 4.22 An *RC* ladder network.

that the gain is negative and real, and that the imaginary part of the ratio is zero. This condition is met when

$$\frac{6}{\omega CR} = \frac{1}{(\omega CR)^3}$$

or

$$6 = \frac{1}{(\omega CR)^2}$$

This may be rearranged to give

$$\omega = \frac{1}{CR\sqrt{6}}$$

and therefore

$$f = \frac{1}{2\pi CR\sqrt{6}}$$

Substituting for $(\omega CR)^2$ in Equation 4.8 gives

$$\frac{v_o}{v_i} = \frac{1}{1 - 5 \times 6} = -\frac{1}{29}$$

If we use the *RC* ladder network as our feedback path it is clear that $B = -1/29$. In order for the loop gain *AB* to be equal to -1, we therefore require the forward gain of the arrangement *A* to be +29. Oscillators based on this principle are called ***RC* oscillators** or sometimes **phase-shift oscillators**. Figure 4.23 shows their basic form.

It can be seen from Figure 4.23 that the phase-shift oscillator consists of an inverting amplifier (the input is applied to the inverting input) and a feedback network with a phase shift of 180°. The same result may be achieved using a non-inverting amplifier and a feedback network with a phase shift of 0°. This approach is used in the Wien-bridge oscillator.

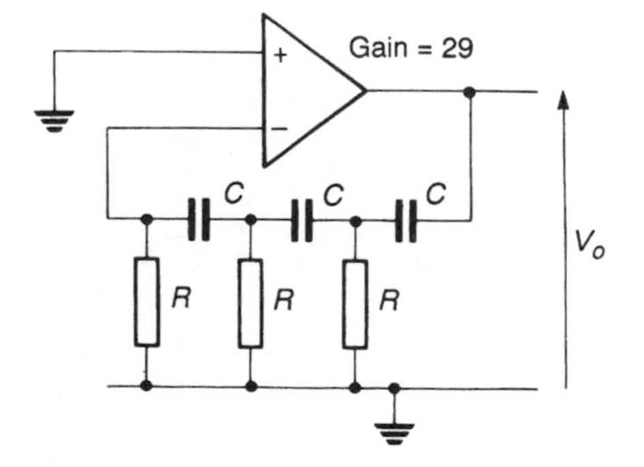

Figure 4.23 An *RC* or phase shift oscillator.

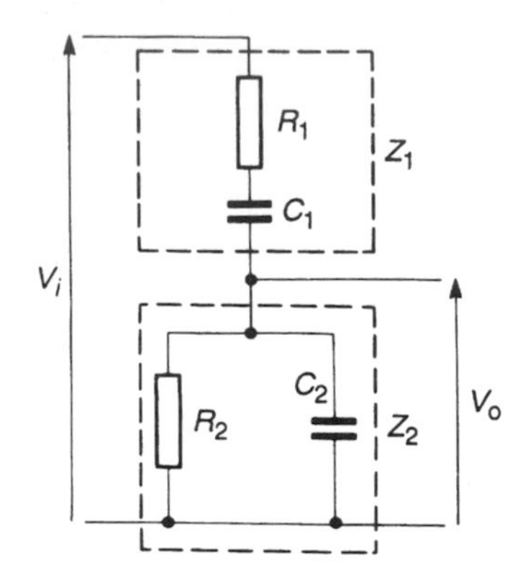

Figure 4.24 The Wien-bridge network.

Wien-bridge oscillator

The Wien-bridge oscillator uses a series/parallel combination of resistors and capacitors for the feedback network, as shown in Figure 4.24.

If we consider that R_1 and C_1 together constitute an impedance Z_1, and that R_2 and C_2 are represented by an impedance Z_2, it is clear that the output of the network is related to the input by the expression

$$\frac{v_o}{v_i} = \frac{Z_2}{Z_1 + Z_2}$$

Since

$$Z_1 = R_1 + \frac{1}{j\omega C_1}$$

and

$$Z_2 = \frac{1}{\dfrac{1}{R_2} + j\omega C_2}$$

if we make the resistors and capacitors equal, it is relatively straightforward to show that

$$\frac{v_o}{v_i} = \frac{1}{3 - j\left(\dfrac{1 - \omega^2 R^2 C^2}{\omega RC}\right)} \tag{4.9}$$

In order for the phase shift of this network to be zero the imaginary part must also be zero. This is true when

$$\omega^2 R^2 C^2 = 1$$

that is, when

$$\omega = \frac{1}{RC}$$

Substituting for ω in Equation 4.9 gives

$$\frac{v_o}{v_i} = \frac{1}{3}$$

Thus at the selected frequency the network has a phase shift of zero and a gain of 1/3. Further investigation of Equation 4.9 will show that the gain is a maximum at this point and that this is therefore the **resonant frequency** of the circuit. To form an oscillator this network must be combined with a non-inverting amplifier with a gain of 3, making the magnitude of the loop gain unity. Figure 4.25 shows a possible arrangement using the non-inverting amplifier of Example 4.3.

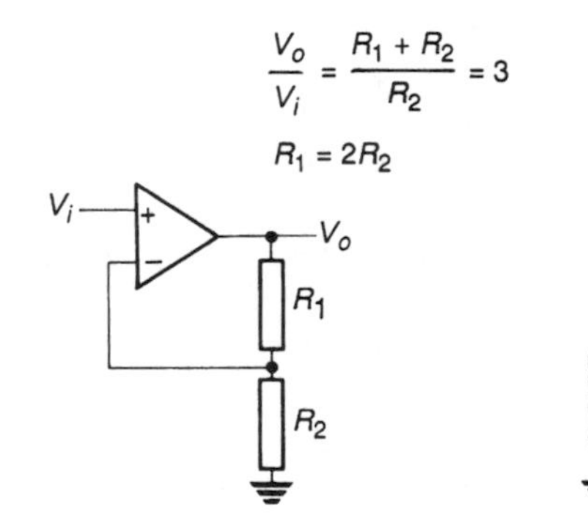

(a) A non-inverting amplifier

(b) Redrawn circuit

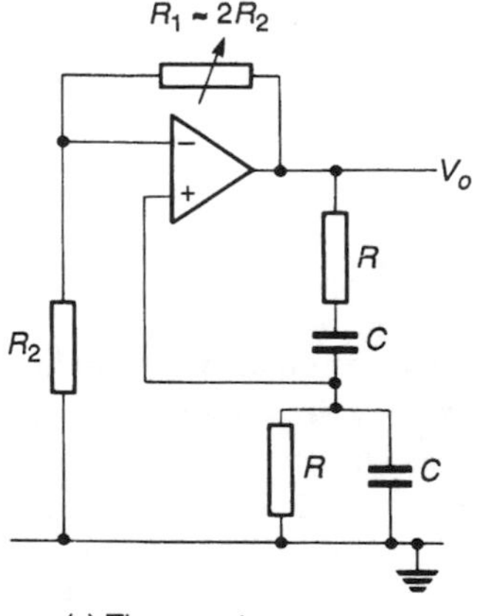

(c) The complete oscillator

Figure 4.25 The Wien-bridge oscillator.

Figure 4.25(a) shows the non-inverting amplifier of Example 4.3 with the resistors chosen to give a gain of 3. Figure 4.25(b) shows the same circuit redrawn in a more convenient form. Figure 4.25(c) shows the oscillator formed by adding the feedback network.

Amplitude stabilization

In the phase-shift and Wien-bridge circuits discussed above, the loop gain of the circuit is determined by component values within the oscillator. If the gain thus set is too low, the oscillations will die; if it is too high the oscillations will grow until limited by circuit constraints.

In Figure 4.25(c), R_2 has been shown as a variable resistor to allow it to be adjusted to the correct value. In practice the gain must be set such that the magnitude of the loop gain is slightly greater than unity, to ensure that any oscillation grows rather than decays and to allow for any downward fluctuation in the gain of the amplifier.

Several methods exist for limiting the magnitude of the oscillation. In the circuit shown in Figure 4.25(c) the amplitude is restricted simply by the limitations on the output swing of the amplifier. Fortunately for this application, operational amplifiers have non-linear gain characteristics and the gain tends to drop as the amplitude approaches the supply rails. Thus if the gain is set to slightly greater than that required to maintain the oscillation for small signals, as the signal amplitude increases it will enter a region where the gain falls and the magnitude will stabilize at that value. While this is a simple method, it does produce some distortion (resembling the clipping shown in Figure 1.3) because the amplifier is being used in its non-linear region.

A possible solution is to replace the variable resistor R_1 in the circuit of Figure 4.25(c) with a suitable negative-temperature-coefficient (NTC) thermistor (see Section 2.3.1). The resistor values are chosen such that when the thermistor is at normal room temperature, the gain is slightly greater than that required for oscillation, and thus the amplitude of the output increases. This increases the power dissipated in the thermistor, causing it to heat up. The increase in temperature causes the resistance of the thermistor to fall, reducing the gain of the circuit. The amplitude of the oscillation therefore stabilizes at a point where the magnitude of the loop gain is exactly unity. This limits the amplitude of the output signal without causing distortion.

Although the use of a thermistor is a possible solution to this problem there are several more elegant solutions, including some utilizing the properties of FETs. We will return to this topic in Section 6.6.2 after we have looked at the characteristics of these devices.

Crystal oscillators

The **frequency stability** of an oscillator is largely determined by the ability of the feedback network to select a particular operating frequency. In a **resonant circuit** this ability is described by its **quality factor** or **Q**, which is the ratio between its resonant frequency and its bandwidth (we will return to this topic in Section 8.2 when we look at filters). A circuit with a very high Q will be very frequency selective and will therefore tend to have a stable frequency. Networks based on resistors and capacitors have relatively low values of Q. Those based on combinations of inductors and capacitors are better in this respect, with Q values up to several hundred. These are suitable for most purposes but are not adequate for some demanding applications, such as the measurement of time. In such cases it is normal to use a frequency selective network based on a crystal.

We noted that some materials have a **piezoelectric** property in that deformation of the substance causes them to produce an electrical signal. The converse is also true; an applied electric field will cause the material to deform. A result of these properties is that if an alternating voltage is applied to a crystal of one of these materials, it will vibrate. The mechanical resonance of the crystal, caused by its size and shape, produces an electrical resonance with a very high Q. Resonant frequencies from a few kilohertz to many megahertz are possible with a Q as high as 100 000.

These piezoelectric resonators are commonly referred to simply as **crystals** and are most commonly made from **quartz** or some form of **ceramic** material. The devices have a pair of resonant frequencies: at one (the parallel resonant frequency) the impedance approaches infinity while at the other (the series resonant frequency) it drops almost to zero. Over the remainder of the frequency range the device looks like a capacitor. The parallel resonance occurs at a slightly higher frequency than the series resonance, but the frequency difference is normally so small that it may be ignored. The presence of these two forms of resonance allows the device to be used in a number of different circuit configurations.

Crystal oscillators are widely used in a range of analogue and digital applications. They form the basis of the time measurement in digital watches and clocks and are used to generate the timing reference (clock) in most computers.

Oscillators – a summary

Oscillators are one of the most important (though by no means the only) applications of positive feedback. They consist of a forward path and a feedback path, with the gain and phase arranged to produce the required conditions for oscillation at a single frequency. A range of circuit techniques may be used to produce appropriate feedback and to limit the amplitude of the output signal. We have looked at only a few examples of these circuits. However, the general principles of operation of all these arrangements are similar to those described.

Key points

- Feedback systems form an essential part of almost all automatic control systems, be they electronic, mechanical or biological.
- Feedback systems may be divided into two types. In negative (or degenerative) feedback systems the feedback tends to *reduce* the input to the forward path. In positive (or regenerative) feedback systems the feedback tends to *increase* the input to the forward path.
- If the gain of the forward path is A, the gain of the feedback path is B and the feedback signal is subtracted from the input, then the overall gain G of the system is given by

$$G = \frac{A}{1 + AB} = \frac{\text{Forward gain}}{1 + \text{Loop gain}}$$

- If the loop gain AB is positive we have negative feedback. If the loop gain is also large compared with unity the expression for the gain simplifies to $1/B$. In these circumstances the overall gain is independent of the gain of the forward path.
- Electronic circuits are often constructed using standard 'cookbook circuits' rather than being designed from first principles.
- Analysis of operational amplifier circuits is often greatly simplified if we assume the use of 'ideal' op-amps.
- Negative feedback tends to increase the bandwidth of an amplifier at the expense of a loss of gain. The factor by which the bandwidth is increased and the gain is reduced is $(1 + AB)$. Thus the gain–bandwidth product remains constant.
- Negative feedback also tends to improve the input resistance, output resistance, distortion and noise of an amplifier. In each case the improvement is generally by a factor of $(1 + AB)$.
- If the loop gain AB is negative and less than unity we have positive feedback. For the special case where AB is equal to unity the gain is infinite. This condition is used in the production of oscillators.

Design study

In the design study at the end of the last chapter we looked at an application in which a temperature sensor was required to drive a meter. The sensor had a relatively high output resistance of 300 Ω and the meter had an input impedance of 10 kΩ. A gain of 10 was required to produce the correct reading on the meter.

In that design study the amplification was produced using an unspecified amplifier with an input resistance of 2 kΩ and an output resistance of 400 Ω. Because the input resistance of the amplifier was of the same order as the output resistance of the sensor, and the output resistance of the amplifier was of the same order as the input resistance of the meter, the gain required from the amplifier was somewhat greater than 10 to allow for the effects of loading.

The requirement for more gain from the amplifier is not in itself a problem. High gain amplifiers are relatively easy to produce. What is significant is that the gain had to be carefully calculated to allow for the effects of loading and a detailed knowledge of the input and output resistances was required. This is inconvenient and unnecessary. In this case study we will consider using an operational amplifier to fulfil the same requirement.

Approach

In Example 4.5 we designed a non-inverting amplifier with a gain of 10.

This is clearly an undemanding application for an operational amplifier and almost any device could be used. Since we have outlined many of the characteristics of the 741, let us assume that this device is used.

The input resistance of the 741 without feedback is typically about 2 MΩ. In this circuit the presence of negative feedback *increases* this value, as shown in Example 4.12. This increase is by a factor of $1 + AB$, where A is the open-loop gain (about 10^5) and B is the feedback gain

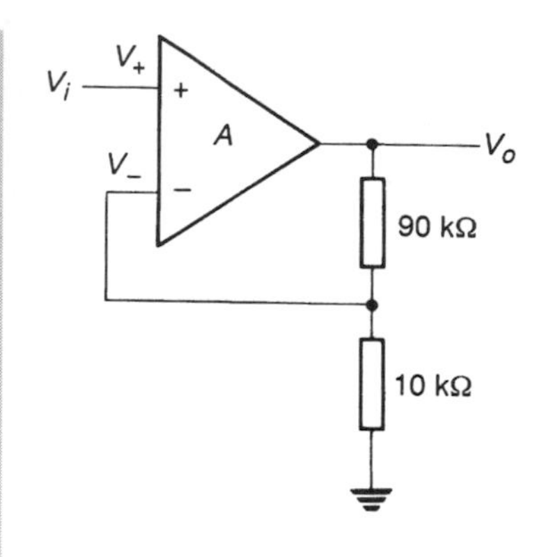

which is 1/10 in this case. Thus the input resistance is increased by about 10^4 giving a value of about $2 \times 10^{10}\Omega$. This is so much greater than the output resistance of the sensor that loading effects can be ignored.

The output resistance of the 741 is typically about 75 Ω in the absence of negative feedback. In this circuit this is *decreased* by a factor of $1 + AB$ which, as before, is about 10^4 in this case. This gives an effective output resistance of about 8 mΩ. This is so much smaller than the input resistance of the meter that again the effects of loading may be neglected.

Thus in this arrangement the effects of the input resistance and the output resistance are completely negligible and need not be considered. This greatly simplifies the design process, and makes the circuit's operation insensitive to variability in the op-amp used.

Further reading

Ahmed H. and Spreadbury P. J. (1984) *Analogue and Digital Electronics for Engineers*, 2nd edn. Cambridge: Cambridge University Press

Horowitz P. and Hill W. (1989) *The Art of Electronics*, 2nd edn. Cambridge: Cambridge University Press

Nelson J. C. C. (1995) *Operational Amplifier Circuits: Analysis and Design*, Oxford: Butterworth-Heinemann

Exercises

4.1 Identify two examples in each case of both open-loop and closed-loop systems, the operations of which are electronic, mechanical, hydraulic, pneumatic and biological.

4.2 Sketch a block diagram of a generalized feedback system and derive an expression for the output in terms of the input and the gains of the forward and feedback paths.

4.3 Draw the circuit diagram of a non-inverting amplifier with a gain of 100 using a 741 operational amplifier. Estimate the input and output resistances of your circuit at very low frequencies using the data on the 741 given in the last chapter.

4.4 Use circuit simulation to confirm your estimates of the input and output resistance in the last exercise. *Hint*: to measure the input resistance add a resistor in series with the input and measure the ratio of the voltage across this resistor to the voltage at the input. To measure the output resistance investigate the effect on the output voltage of changes in the load resistance.

4.5 Draw the circuit diagram of an inverting amplifier with a gain of −50 using a 741 operational amplifier. Estimate the input and output resistances of your circuit at very low frequencies.

4.6 Use circuit simulation to confirm your estimates of the input and output resistance in the last exercise.

4.7 Draw the circuit of a subtractor which produces an output signal that is ten times the magnitude of the difference between two input signals. That is

$$V_o = 10(V_1 - V_2)$$

4.8 Draw the circuit of an inverting adder circuit which produces an output equal to the sum of its two inputs.

4.9 Consider the following circuit

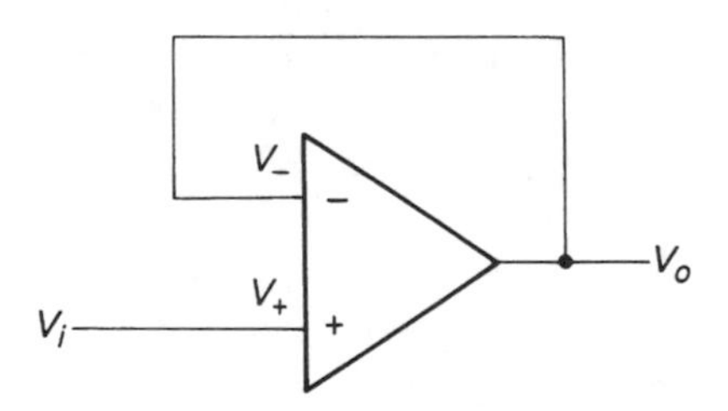

Calculate the gain and estimate the input and output resistances of the circuit at low frequencies, assuming that the amplifier is a 741. Can you suggest a possible use for such a circuit?

4.10 Deduce an expression for the output voltage V_o of the following circuit, in terms of the input voltages V_1 and V_2, and the component values.

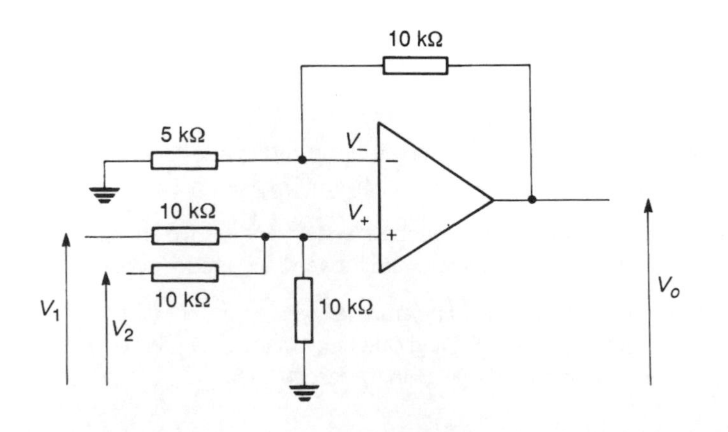

4.11 An amplifier has a gain of 1000 and produces 15% of harmonic distortion at its output. It is decided to improve the performance of the circuit by applying negative feedback to reduce the distortion to 1%. Calculate the value of feedback gain *B* required to produce this effect and the increase in input signal required to maintain the same output level.

4.12 Explain how a Bode diagram may be used to predict the stability of an amplifier.

4.13 Explain how a Nyquist diagram may be used to investigate the stability of an amplifier. What is the significance of the point $(-1, 0)$?

4.14 Derive Equation 4.8.

4.15 Calculate the frequency of oscillation of a phase-shift oscillator which uses a three-stage ladder network, each with $R = 1\ \text{k}\Omega$ and $C = 1\ \mu\text{F}$.

4.16 Calculate the frequency of oscillation of a phase-shift oscillator constructed using four identical *RC* ladder stages. What would be the required gain of the forward path of such an oscillator?

4.17 Assuming that the operational amplifier has an open-circuit voltage gain *A* of 200 000, an input resistance of 1 MΩ and an output resistance of 75 Ω, calculate the voltage gain, input resistance and output resistance of the following circuits at low frequencies.

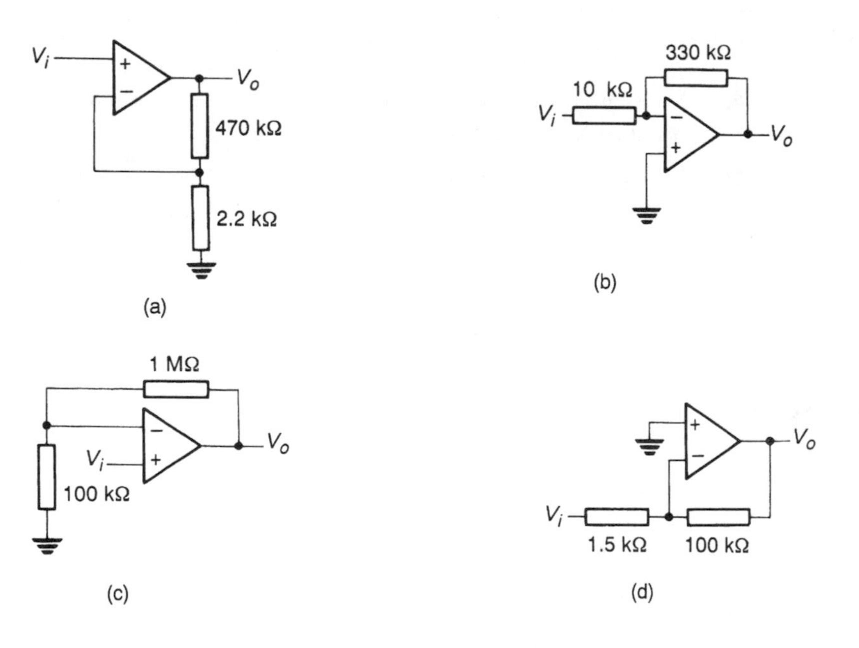

4.18 Derive Equation 4.9.

4.19 A Wien-bridge oscillator of the form shown in Figure 4.25(c) has $R = 100\ \text{k}\Omega$ and $C = 10$ nF. Calculate the frequency of oscillation.

4.20 Explain how a positive-temperature-coefficient (PTC) thermistor could be used to stabilize the output level of the Wien-bridge oscillator of Figure 4.25(c).

4.21 Calculate the percentage accuracy required in the frequency of oscillation of a clock which must keep time to within 1 second per month.

Semiconductors and Diodes

Objectives

When you have studied the material in this chapter you should:

- be familiar with the basic electrical properties of conductors, insulators and semiconductors;
- have an understanding of the mechanics of electrical conduction in pure and doped semiconductor materials;
- be aware of the operation and characteristics of *pn* junction diodes for both forward and reverse bias conditions;
- be able to describe the uses of various forms of diode including Zener diodes, Schottky diodes, tunnel diodes and varactor diodes;
- understand the operation of a range of circuits based on the use of diodes, including power rectifiers, demodulators, 'wave-shapers' and voltage references.

Contents

5.1 Introduction

So far we have considered 'black-box' amplifiers and operational amplifiers but have not yet looked in detail at the operation of the devices at the heart of these systems. In many applications we may ignore the internal operation of these components and look simply at their external characteristics. However, it is sometimes necessary, and very interesting, to look in more detail at the construction of the active components of our system to gain more insight into their characteristics and their operation.

Most modern electronic systems are based on **semiconductor devices** of one form or another. In this chapter we will look at the nature and characteristics of semiconductors and discover why they are so useful.

The material in this chapter includes the minimum amount of physics and mathematics since these are not the primary interest of this text. Those whose interest is fired by such subjects can find many books that cover semiconductor physics from these viewpoints in any good technical library. A primary aim of this chapter is to explain the rudimentary concepts necessary to understand the following chapters, which look at various forms of transistor. An appreciation of these devices is essential for all engineers and scientists.

5.2 Electrical properties of solids

Solid materials may be divided with respect to their electrical properties into three categories: conductors, insulators and semiconductors. The different characteristics of these groups are produced by the atomic structure of the materials and in particular by the distribution of electrons in the outer orbits of the atoms. These outermost electrons are termed **valence electrons** and they play a major part in determining many of the properties of the material.

Conductors

Conductors such as copper or aluminium have a cloud of free electrons at all temperatures above absolute zero. This is formed by the weakly bound 'valence' electrons in the outermost orbits of the atoms. If an electric field is applied across such a material, electrons will flow causing an electric current.

Insulators

In insulating materials, such as polythene, the valence electrons are tightly bound to the nuclei of the atoms and very few are able to break free to conduct electricity. The application of an electric field does not cause a current to flow since there are no mobile charge carriers.

Semiconductors

At very low temperatures semiconductors have the properties of an insulator. However, at higher temperatures some electrons are free to move and the materials take on the properties of a conductor – albeit a poor one. Nevertheless semiconductors have some useful characteristics which make them distinct from both insulators and conductors.

5.3 Semiconductors

Semiconductor materials have very interesting electrical properties that make them extremely useful in the production of electronic devices. The most commonly used semiconductor material for such applications is **silicon**, but **germanium** is also used, as are several more exotic materials such as **gallium arsenide**. Many metal oxides have semiconducting properties (for example, the oxides of manganese, nickel and cobalt). We have already come across applications of these materials when discussing **thermistors** in Chapter 2.

At temperatures near absolute zero the valence electrons in a semiconductor are tightly bound to their nuclei and the material has the characteristics of an insulator. The reasons for this effect may be understood by considering the structure of a typical semiconductor.

Figure 5.1 shows a two-dimensional representation of a crystal of silicon. Silicon is a tetravalent material, that is, it has four valence electrons. The outermost electron shell of each atom can accommodate up to eight electrons, and the atom is most stable when the shell is fully populated. In a crystal of pure silicon each atom shares its valence electrons with its four neighbouring atoms so that each atom has a part-share in eight valence electrons rather than sole ownership of four. This is a very stable arrangement which is

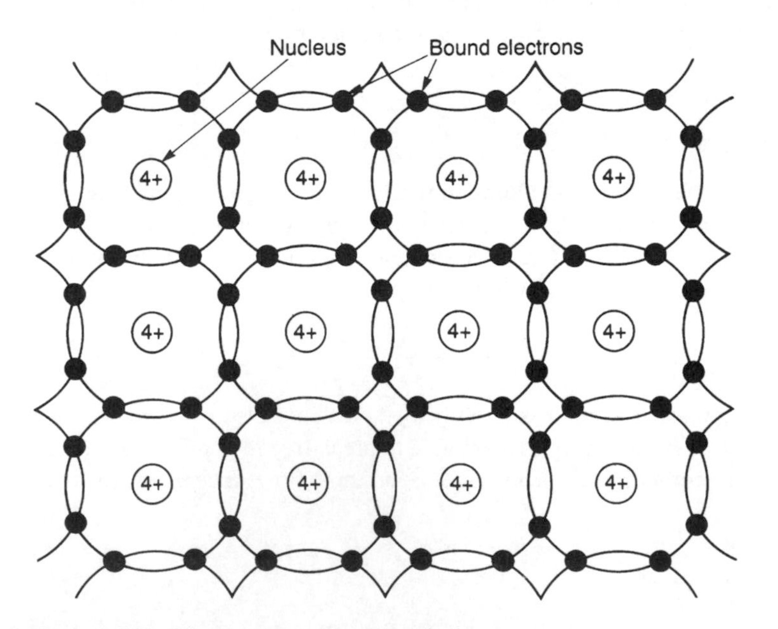

Figure 5.1 The atomic structure of silicon.

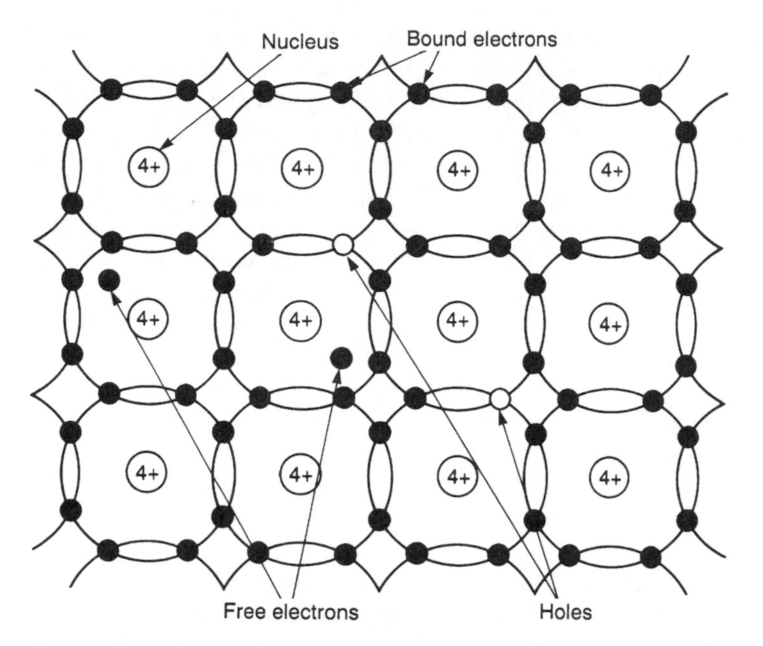

Figure 5.2 The effects of thermal vibration on the structure of silicon.

also found in materials such as diamond. This method of atomic bonding is called **covalent bonding**.

At low temperatures the tight bonding of the valence electrons in semiconductor materials leaves no electrons free to conduct electricity, resulting in the insulating properties described above. However, as the temperature rises, thermal vibration of the crystal lattice results in some of the bonds being broken, generating a few **free electrons** which are able to move throughout the crystal. This also leaves behind **holes** which accept electrons from adjacent atoms and therefore also move about. Electrons are negative charge carriers and will move against an applied electric field generating an electrical current. Holes, being the absence of an electron, act like positive charge carriers and will move in the direction of an applied electric field and will also contribute to current flow. This process is illustrated in Figure 5.2.

At normal room temperatures the number of charge carriers present in pure silicon is small and consequently it is a poor conductor. This form of conduction is called **intrinsic conduction**.

5.4 Doping

The addition of small amounts of impurities to a semiconductor can drastically affect its properties. Of particular interest are impurities of materials that can fit within the crystal lattice of the semiconductor, but which have a different number of valence electrons. An example of such an impurity is the presence of **phosphorus** in silicon. Phosphorus is a pentavalent material, that is, it has five valence electrons in its outer electron shell. When a phosphorus atom is present within the lattice of a piece of silicon, four of its valence electron are tightly bound by the covalent bonding described earlier. However, the fifth electron is only weakly bound and is therefore free to move within the lattice and contribute to an electric current. Materials such as phosphorus are known as **donor impurities** since they produce an excess of free electrons. Semiconductors containing such impurities are called ***n*-type semiconductors** since they have free *negative* charge carriers.

Boron has three valence electrons and is thus a trivalent material. When a boron atom is present within a silicon crystal the absence of an electron in the outer shell leaves a space (a hole) which can accept an electron from an adjacent atom to complete its covalent bonds. This hole moves from atom to atom and acts as a mobile positive charge carrier in exactly the same manner as the holes generated in the intrinsic material by thermal vibration. Materials such as boron are known as **acceptor impurities** since they accept electrons to produce holes. Semiconductors containing such impurities are called ***p*-type semiconductors** since they have free *positive* charge carriers.

The intentional inclusion of these impurities into semiconductor materials is called **doping**.

Both *n*-type and *p*-type semiconductors have much greater conductivities than that in the intrinsic material, the magnitude depending on the doping level. This is called **extrinsic conductivity**. The dominant charge carriers in a doped semiconductor (that is, electrons in an *n*-type material and holes in a *p*-type material) are called the **majority charge carriers**. The other charge carriers are called the **minority charge carriers**.

5.5 *pn* Junctions

Although *p*-type and *n*-type semiconductor materials have some useful characteristics individually, they are of greater interest when they are used together.

When *p*-type and *n*-type materials are joined, the charge carriers in each interact in the region of the junction. Although each material is electrically neutral, each has a much higher concentration of majority charge carriers than of minority charge carriers. Thus on the *n*-type side of the junction there are far more free electrons than on the *p*-type side. Consequently, electrons diffuse across the junction from the *n*-type side to the *p*-type side where they are absorbed into the lattice by recombination with free holes which are plentiful in the *p*-type region. Similarly holes diffuse from the *p*-type side to the *n*-type side and combine with free electrons.

This process of diffusion and recombination of charge carriers produces a region close to the junction which has very few mobile charge carriers, but in which the bound charge carriers are still present within the lattice. This region is referred to as a **depletion layer** or sometimes as a **space-charge layer**. The net diffusion of negative charge carriers in one direction and positive charge carriers in the other generates a net charge imbalance across the junction. This process is illustrated in Figure 5.3

Once the deletion layer has become established, it limits any further diffusion of holes or electrons. The positive charge on the *n*-type side repels holes in the *p*-type side of the junction while the negative charge on the *p*-type side similarly repels electrons in the *n*-type side.

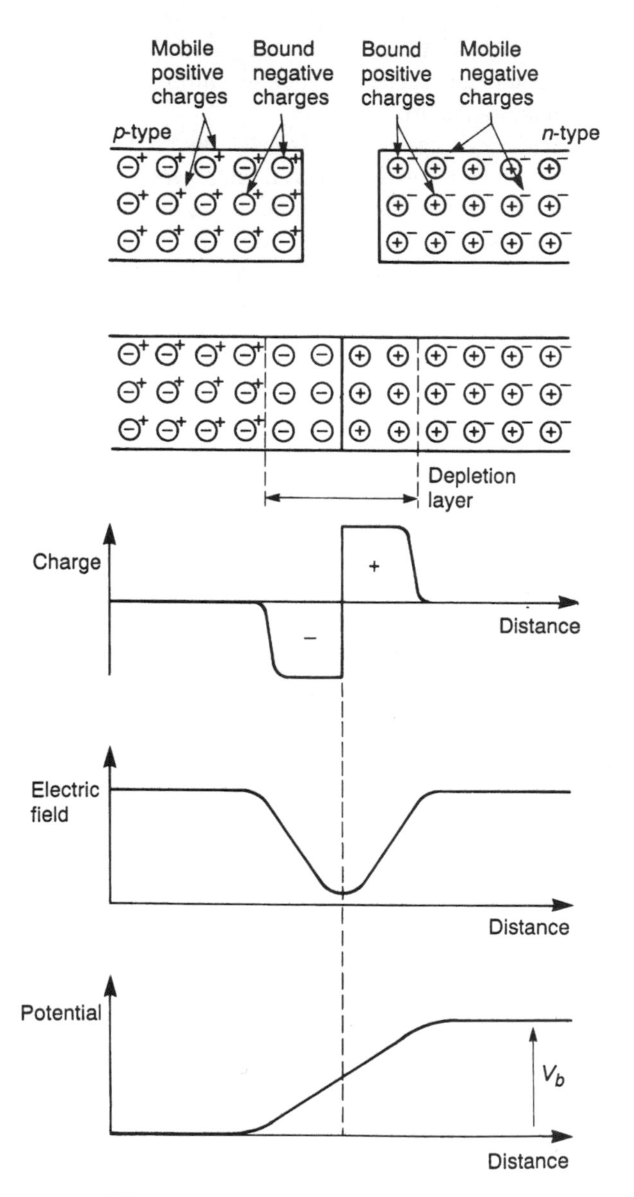

Figure 5.3 A *pn* junction.

The existence of positive and negative charges on either side of the junction produces an electric field across it, as shown in Figure 5.3. This produces a **potential barrier** which charge carriers must overcome to cross the junction. Only a small number of *majority* charge carriers have sufficient energy to surmount this barrier and these generate a small **diffusion current** across the junction. However, the field produced by the space-charge region does not oppose the movement of *minority* charge carriers across the junction; rather, it assists it. Any such charge carriers that stray into the depletion layer, or which are formed there by thermal vibration, are accelerated across the junction forming a small **drift current**. In an isolated junction a state of dynamic equilibrium exists in which the diffusion current exactly matches the drift current.

The necessity for dynamic equilibrium in an isolated junction determines the magnitude of the potential barrier V_b. For a given material the potential will be such that the diffusion and drift currents cancel to produce zero net current. For silicon, V_b is about 0.7 V and for germanium it is about 0.3 V. This potential, also called the **contact potential**, cannot be used to generate current flow in an external circuit. Connecting external components to the device would produce additional 'contacts' with contact potentials that cancel that of the junction.

The application of an external potential across the device will affect the height of the potential barrier and change the state of dynamic equilibrium.

Forward bias

If the *p*-type of the device is made positive with respect to the *n*-type side, the applied potential neutralizes some of the space-charge and the width of the depletion layer decreases. The height of the barrier is reduced and a larger proportion of the majority carriers in the region of the junction now have sufficient energy to surmount it. The diffusion current produced is therefore much larger than the drift current and a net current flows across the junction. This situation is shown in Figure 5.4.

Reverse bias

If the *p*-type side of the device is made negative with respect to the *n*-type side, the space-charge increases and the width of the depletion layer is increased. This produces a larger

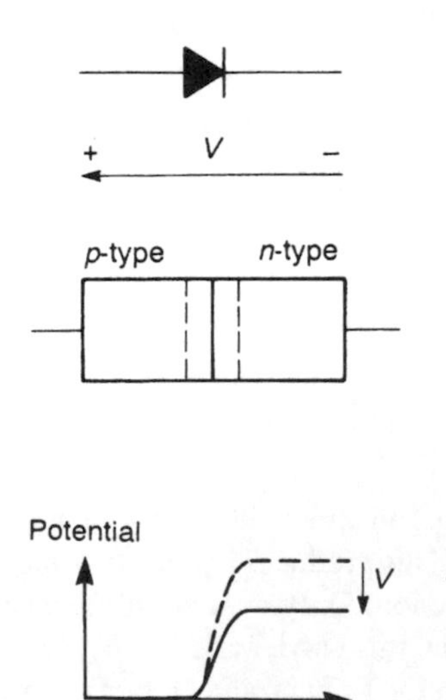

Figure 5.4 A forward biased *pn* junction.

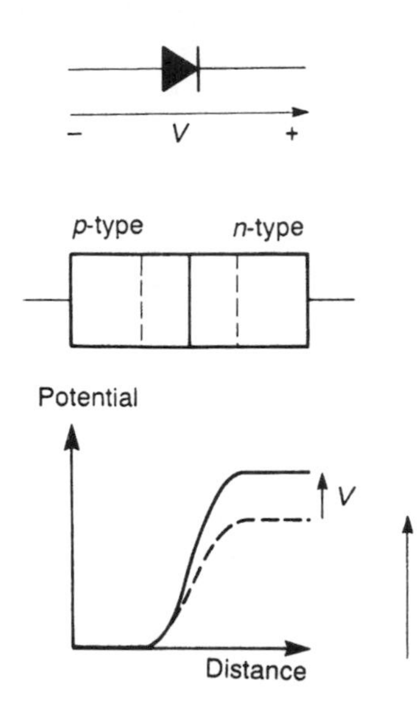

Figure 5.5 A reverse biased *pn* junction.

potential barrier and reduces the number of majority carriers that have sufficient energy to surmount it, reducing the diffusion current across the junction. This situation is shown in Figure 5.5.

Even a small negative bias, of perhaps 0.1 V, is sufficient to reduce the diffusion current to a negligible value. This leaves a net imbalance in the currents flowing across the junction which are now dominated by the drift current. Since the magnitude of this current is determined by the rate of thermal generation of minority carriers in the region of the junction it is not related to the applied voltage. At normal room temperatures this reverse current is very small, typically a few nanoamps for silicon devices and a few microamps for germanium devices. It is however, exponentially related to temperature and doubles for a temperature rise of about 10 °C.

Forward and reverse currents

The current flowing through a *pn* junction can be approximately related to the applied voltage by the expression

$$I = I_s\left(\exp\frac{eV}{\eta kT} - 1\right)$$

where I is the current through the junction, e is the electronic charge, V is the applied voltage, k is Boltzmann's constant, T is the absolute temperature and η is a constant in the range 1 to 2 determined by the junction material. Here a positive applied voltage represents a forward bias voltage and a positive current a forward current.

The constant η is approximately 1 for germanium and about 1.3 for silicon. However,

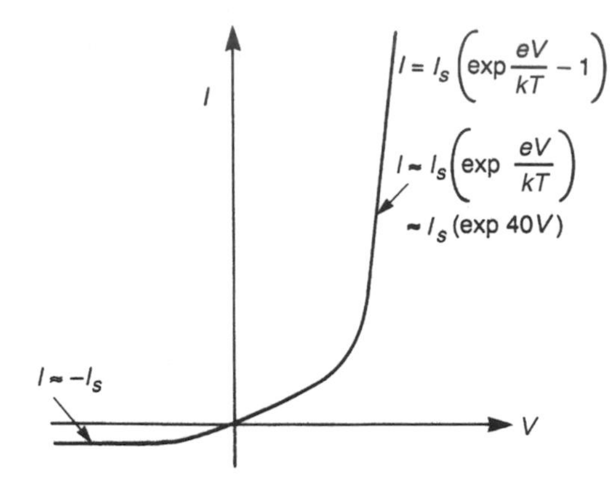

Figure 5.6 Current–voltage characteristics of a *pn* junction.

for our purposes it is reasonable to use the approximation that

$$I \approx I_s\left(\exp\frac{eV}{kT} - 1\right) \tag{5.1}$$

and we will make this assumption for the remainder of this text.

At normal room temperatures e/kT has a value of about 40 V^{-1}. If V is less than about −0.1 V the exponential term within the brackets in Equation 5.1 is small compared to 1, and I is given by

$$I \approx I_s(0 - 1) = -I_s \tag{5.2}$$

Similarly, if V is greater than about +0.1 V the exponential term is much greater than 1, and I is given by

$$I \approx I_s\left(\exp\frac{eV}{kT}\right) = I_s \exp(40V) \tag{5.3}$$

We therefore have a characteristic for which the reverse bias current is approximately constant at $-I_s$ (the **reverse saturation current**), and the forward bias current rises exponentially with the applied voltage.

In fact the expressions of Equations 5.1 to 5.3 are only approximations of the junction current in a real device, as effects such as **junction resistance** and **minority carrier injection** tend to reduce the current flowing. However, this analysis gives values that indicate the form of the relationship and are adequate for our purposes. Figure 5.6 shows the current–voltage characteristic of a *pn* junction.

5.6 Semiconductor diodes

One could characterize an **ideal diode** as a component that would conduct no current when a voltage was applied across it in one direction, but would appear as a short circuit when a voltage was applied in the opposite direction. The characteristic of such a device is shown in Figure 5.7(a).

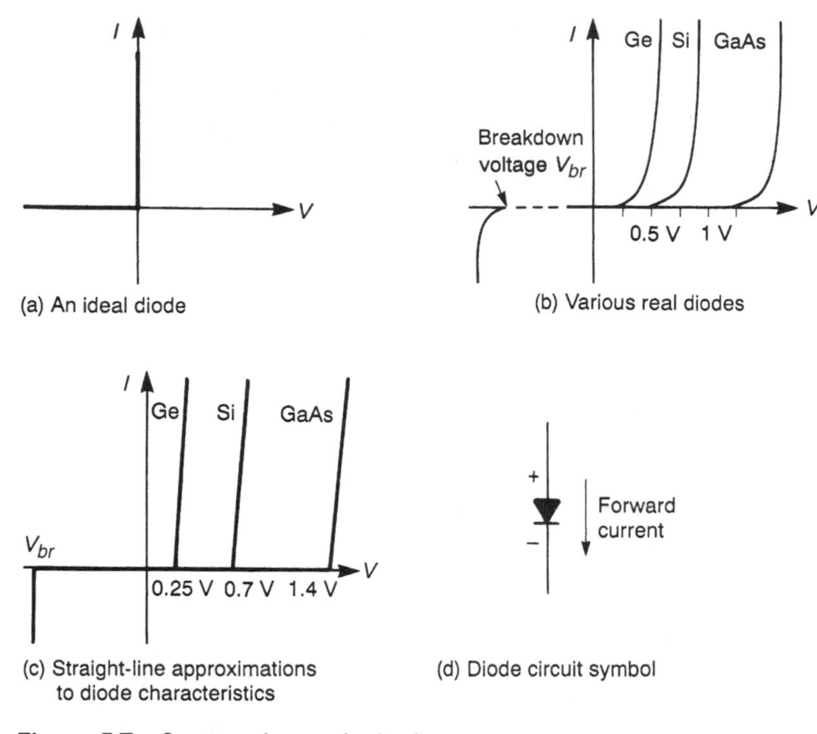

Figure 5.7 Semiconductor diode characteristics.

A *pn* junction is not an ideal diode, but it does have a characteristic that approximates to such a device. The exponential current–voltage relationship shown in Figure 5.6, when viewed on a larger scale, has the form shown in Figure 5.7(b).

As we have seen, if the *p*-type side of a junction is made negative with respect to the *n*-type side (**reverse bias**) only a very small current will flow – the **reverse saturation current**. For a silicon junction this saturation current is typically 1 nA and is negligible in almost all applications. The reverse current is approximately constant as the reverse voltage is increased to a critical voltage called the **reverse breakdown voltage** V_{br}. If the negative voltage is increased beyond this point the junction breaks down and begins to conduct. This limits the useful voltage range of the diode.

If the *p*-type side is made positive with respect to the *n*-type side (**forward bias**), a negligible current will flow for a small applied voltage but this increases exponentially as the voltage is increased. *When viewed on a large scale* it appears that the current is zero until the voltage reaches a so-called **turn-on voltage**, and that as the voltage is increased beyond this point the junction begins to conduct and the current increases rapidly. This turn-on voltage is clearly related to the magnitude of the potential barrier shown in Figure 5.3. It is about 0.2 V for a germanium junction, about 0.5 V for a silicon junction and about 1.3 V for a gallium arsenide junction. A further increase in the applied voltage causes the junction current to increase rapidly. This results in the current–voltage characteristic being almost vertical, showing that the voltage across the diode is approximately constant, irrespective of the junction current. In many applications it is reasonable to approximate the characteristic by a straight-line response as shown in Figure 5.7(c). This simplified form represents the diode by a forward voltage drop (of about 0.25 V for germanium, 0.7 V for silicon and 1.4 V for gallium arsenide devices) combined with a forward resistance. The latter results in the slope of the characteristic above the turn-on voltage. In many cases the forward resistance of the diode may be ignored, and the diode considered simply as a near ideal diode with a small forward voltage drop. This is termed the **conduction voltage** of the diode. As the current through the diode increases, the voltage across the junction also increases. At 1 amp the conduction voltage might be about 1 V for a silicon diode, rising to perhaps 2 V at 100 amps. In practice most diodes would be destroyed long before the current reached such large values.

Figure 5.7(d) shows the circuit symbol for a diode. The arrow in this symbol shows the direction of forward current flow.

Diodes are used for a number of purposes within electronic circuits. In many cases relatively low voltages and currents are present and devices for such applications are usually called **signal diodes**. A typical device might have a maximum forward current of 100 mA and reverse breakdown voltage of 75 V. Other common applications for diodes include their use within power supplies to convert alternating currents into direct currents. Such diodes will usually have a greater current handling capacity (usually measured in amps or tens of amps) and are generally called **rectifiers** rather than diodes. Reverse breakdown voltages for such devices will vary with the application but are typically hundreds of volts.

Diodes and rectifiers can be made using a variety of semiconductor materials and may use other techniques in place of simple *pn* junctions. This allows devices to be constructed with a wide range of characteristics in terms of current handling capability, breakdown voltage and speed of operation.

FILE 5A

Computer simulation exercise 5.1

Use simulation to investigate the relationship between the current and the applied voltage in a small signal diode (such as a 1N4002). Measure the current while the applied voltage is swept from 0 to 0.8 volts and plot the resulting curve.

Look at the behaviour of the device over different voltage ranges including both forward and reverse bias conditions. Estimate from these experiments the reverse breakdown voltage of the diode.

Effects of temperature

From Equation 5.1 we have

$$I \approx I_s\left(\exp\frac{eV}{kT} - 1\right)$$

Clearly, for a given value of diode current I, the voltage across the junction V is inversely proportional to the absolute junction temperature T. For silicon devices the junction voltage *decreases* by about 2 mV/°C rise in temperature.

The diode current is also affected by the reverse saturation current I_s. We noted in the last section that this current is related to the production of minority carriers as a result of thermal vibration. As the temperature rises, the number of minority carriers produced increases and the reverse saturation current goes up; I_s approximately doubles for an increase in temperature of about 10 °C. This corresponds to an increase of about 7%/°C.

Reverse breakdown

The reverse breakdown of a diode may be brought about by one of two phenomena.

In devices with heavily doped *p*- and *n*-type regions, the transition from one to the other is very abrupt and the depletion region is often only a few nanometres thick. Under these circumstances, junction voltages of only a few volts produce fields across the junction of several hundreds of megavolts per metre. Such high field strengths result in electrons being pulled from their covalent bonds producing additional charge carriers and a large reverse current. The current produced by this **Zener breakdown** must be limited by external circuitry to prevent damage to the diode. The voltage at which Zener breakdown occurs is determined by the energy gap of the semiconductor used and is thus largely independent of temperature. However, it *decreases* very slightly with increasing temperature. Zener breakdown normally occurs at voltages below 5 V.

Diodes in which one, or both, of the semiconductor regions are lightly doped, have a less abrupt transition and consequently a wider depletion layer. In such devices the field generated by an applied voltage is usually insufficient to produce Zener breakdown but does accelerate current carriers within the depletion layer. As these carriers are accelerated they gain energy which they may lose by colliding with atoms within the lattice. If the carriers gain sufficient energy they may ionize these atoms by freeing their electrons. This will generate additional carriers which are themselves accelerated by the applied field. At some point the applied field is large enough to produce an 'avalanche' effect in which the current increases dramatically. This gives rise to **avalanche breakdown**. For high voltage operation it is possible to construct devices with breakdown voltages of several thousands of volts. Alternatively, it is possible to arrange that breakdown occurs at only a few volts. The voltage at which avalanche breakdown occurs *increases* with junction temperature.

Generally, if a diode suffers reverse breakdown at a voltage of less than 5 V it is likely to be caused by Zener breakdown. If it occurs at a voltage of greater than 5 V, it is likely to be the result of avalanche breakdown.

5.7 Zener diodes

When the reverse breakdown voltage of a diode is exceeded, the current that flows is generally limited only by external circuitry. If steps are not taken to limit this current, the power dissipated in the diode may destroy it. However, if the current is limited by the circuitry connected to the diode, the breakdown of the junction need not cause any damage to the device. This effect is utilized in special purpose devices called **Zener diodes**, although it should be noted that the name is largely historical and is used to describe devices the operation of which may depend on either Zener or avalanche breakdown. From Figure 5.7 it is clear that when the junction is in the breakdown region the junction voltage

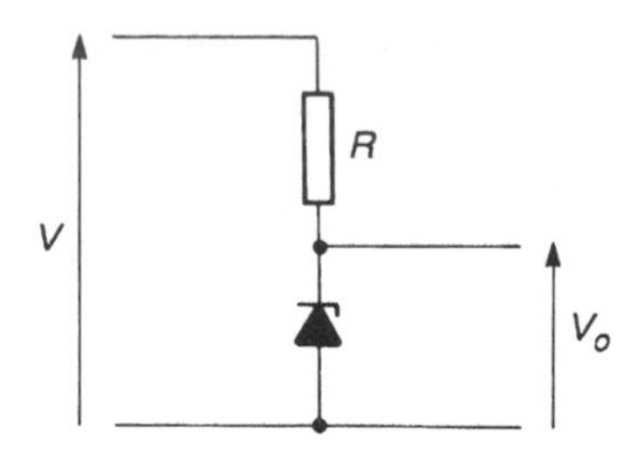

Figure 5.8 A simple voltage reference using a Zener diode.

is approximately constant irrespective of the reverse current flowing. This allows the device to be used as a **voltage reference**. In such devices the breakdown voltage is often given the symbol V_Z. Zener diodes are available with a variety of breakdown voltages to allow a wide range of reference voltages to be produced.

A typical circuit using a Zener diode is shown in Figure 5.8. Here a poorly regulated voltage V is applied to a series combination of a resistor and a Zener diode. The diode is connected so that it is reserved biased by the positive applied voltage V. If V is greater that V_Z the diode junction will break down and will conduct current drawn from the resistance R. The diode prevents the output voltage going above its breakdown voltage V_Z and thus generates an approximately constant output voltage irrespective of the value of the input voltage, provided it remains greater than V_Z. If V is less that V_Z the diode will conduct negligible current and the output V_o will be approximately equal to V. In this situation the Zener diode has no effect on the circuit.

If circuitry is connected to the output of the arrangement shown in Figure 5.8, this will also draw current through the resistance R. The value of R must be chosen so that the voltage drop across R caused by this current is not great enough to reduce the voltage across the Zener diode to below its breakdown voltage. This requirement must be balanced against the fact that the power dissipated in the diode and in the resistor increases as R is reduced. The choice of resistor value is illustrated in the Design Study at the end of this chapter.

It should be noted that although the voltage across a Zener diode is approximately constant in its breakdown region, irrespective of the current passing through it, it is not completely constant. You will observe from Figure 5.7 that the characteristic is not vertical above the breakdown voltage but has a finite slope. This slope represents an effective output resistance which is typically a few ohms up to a few hundred ohms, causing the output voltage to vary slightly with current. For a high precision voltage reference it is necessary to pass a fairly constant current through the Zener diode to produce a more constant output voltage. We also noted earlier that the breakdown voltages of these devices vary slightly with temperature. Typical devices have **temperature coefficients** for their breakdown voltages of between 0.001%/°C and 0.1%/°C.

FILE 5B

Computer simulation exercise 5.2

The D1N750 is a 4.7 V Zener diode. Simulate the circuit of Figure 5.8 using this diode and a suitable resistor.

Apply a swept DC input voltage to the circuit and plot the output voltage against the input voltage for a range of values of R. Investigate the effect of connecting a load resistor to the circuit.

5.8 Other forms of diode

We have already encountered several forms of semiconductor diode. These include the *signal diode* and *Zener diode* described in this chapter, and the *light emitting diode* (*LED*), *laser diode* and *photodiode* discussed in Chapter 2.

There are several other kinds of diode which are widely used, each having its own unique characteristics and applications. We will look briefly at some of the more popular forms.

5.8.1 Schottky diodes

Unlike conventional *pn* junction diodes which are formed at the junction of two layers of doped semiconductor material, **Schottky diodes** are formed by a junction between a layer of metal (such as aluminium) and a semiconductor. The rectifying contact formed relies only on **majority charge carriers** and is consequently much faster in operation than *pn* junction devices which are limited in speed by the relatively slow recombination of **minority charge carriers**.

Schottky diodes also have a low forward voltage drop of about 0.25 V. This characteristic is used to great effect in the design of high-speed logic gates, as described in Chapter 11.

5.8.2 Tunnel diodes

The **tunnel diode** was first described by **Esaki** (1958). It uses high doping levels to produce a device with a very narrow depletion region. This region is so thin that a quantum mechanical effect known as *tunnelling* can take place. This results in charge carriers being able to cross the depletion layer, even though they do not have sufficient energy to surmount it. The combination of the tunnelling effect and conventional diode action produces a characteristic as shown in Figure 5.9

This rather strange characteristic finds application in a number of areas. Of particular

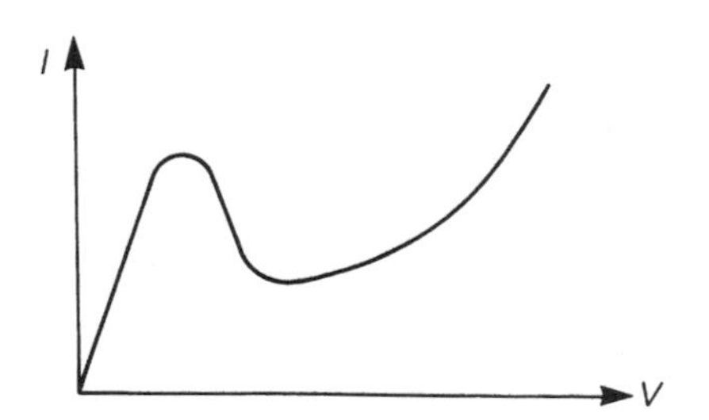

Figure 5.9 Characteristic of a tunnel diode.

interest is the fact that for part of its operating range the voltage across the device falls for an increasing current. This corresponds to a region where the incremental resistance of the device is *negative*. This property is utilized in high frequency oscillator circuits in which the negative resistance of the tunnel diode is used to cancel losses within passive components.

5.8.3 Varactor diodes

A reverse biased diode has two conducting regions of p- and n-type semiconductor separated by a depletion region. This structure resembles a capacitor with the depletion region forming the insulating dielectric. Small silicon signal diodes have a capacitance of a few picofarads which changes with the reverse bias voltage since this varies the thickness of the depletion region.

This effect is used by **varactor diodes** which act as voltage dependent capacitors. A typical device might have a capacitance of 160 pF at 1 V, falling to about 9 pF at 10 V. Such devices are used at the heart of many automatic tuning arrangements, where the varactor is used within an L–C or R–C tuned circuit. The capacitance of the device, and therefore the frequency characteristics of the circuit, may then be varied by the applied reverse bias voltage.

5.9 Diode circuits

Diodes are used in a wide range of circuits in both analogue and digital applications. In this section we will look at several analogue examples, leaving digital circuits until Chapter 11.

Example 5.1 A half-wave rectifier

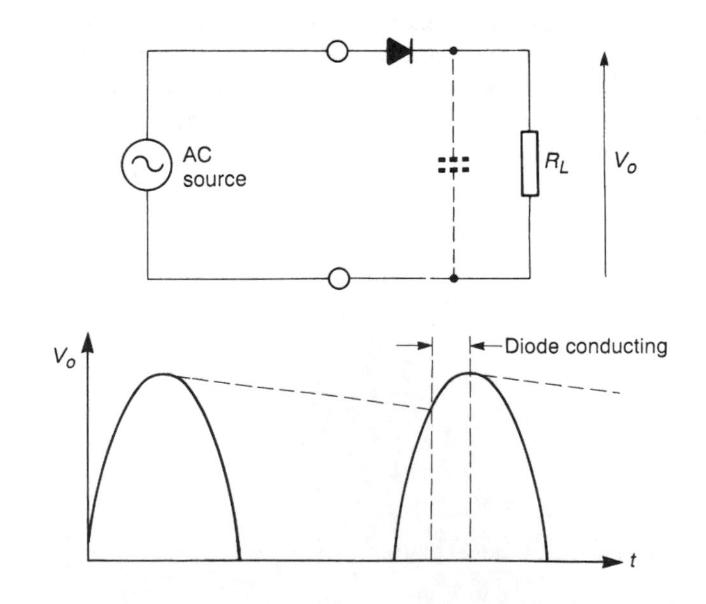

One of the most common uses of diodes is as a rectifier within a power supply to generate a direct voltage from an alternating supply.

A simple arrangement to achieve this is the half-wave rectifier shown above. While the input voltage is greater than the turn-on voltage of the diode, the diode conducts and the input voltage (minus the small voltage drop across the diode) appears across the load. During the part of the cycle in which the diode is reverse biased, no current flows in the load.

To produce a steadier output voltage a **reservoir capacitor** is normally added to the circuit as shown. This is charged while the diode is conducting and maintains the output voltage when the diode is turned off by supplying current to the load. This current gradually discharges the capacitor, causing the output voltage to decay. One effect of adding a reservoir capacitor is that the diode conducts for only short periods of time. During these periods the diode currents are thus very high. The magnitude of the *ripple* in the output voltage is affected by the current taken by the load, the size of the capacitor and the frequency of the incoming signal. Clearly, as the supply frequency is increased the time for which the capacitor must maintain the output is reduced.

FILE 5C

Computer simulation exercise 5.3

Simulate the circuit of Example 5.1 and investigate the behaviour of the circuit.

While typical half-wave rectifier arrangements might have input voltages of several hundred volts, the operation of the circuit is more apparent if smaller voltages are used so that the turn-on voltage of the diode is more easily observed.

Simulate the circuit with and without a reservoir capacitor and use transient analysis to study the circuit's behaviour. Plot the output voltage and the current through the diode, and see how these relate to the input voltage. Vary the input frequency and see how this affects the output.

Note the peak voltage at the output and the ripple voltage for a given set of circuit parameters.

Example 5.2 A full-wave rectifier

One simple method of effectively increasing the frequency of the waveform applied to the capacitor in the last example is to use a full-wave rectifier arrangement as shown

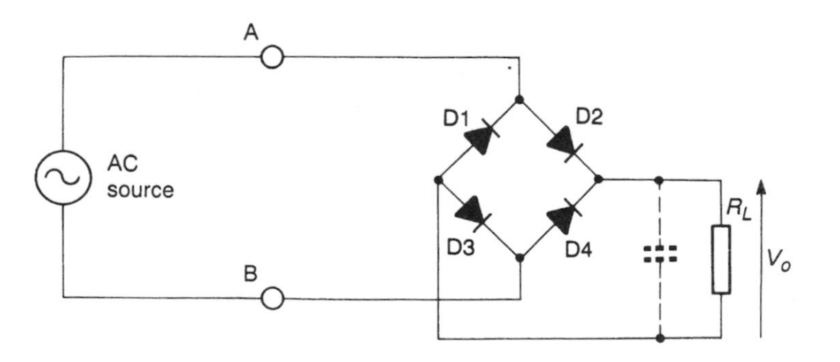

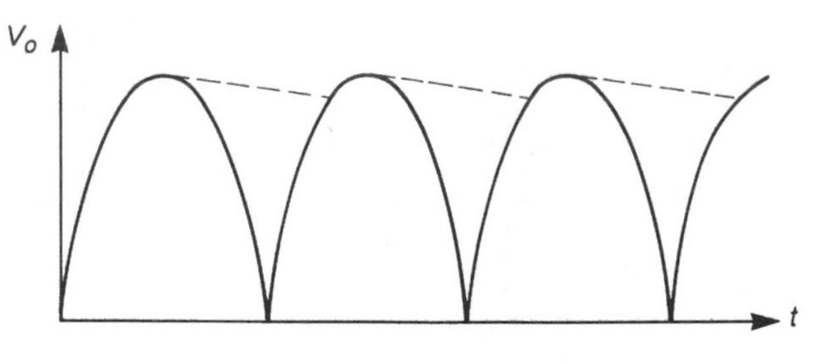

above. When terminal A of the supply is positive with respect to B, diodes D2 and D3 are forward biased and diodes D1 and D4 are reversed biased. Current therefore passes from terminal A, through D2, through the load R_L, and returns to terminal B through D3. This makes the output voltage V_o positive. When terminal B is positive with respect to terminal A, diodes D1 and D4 are forward biased and D2 and D3 are reverse biased. Current now flows from terminal B through D4, through the load R_L and returns to terminal A through D1. Since the direction of the current in the output resistor is the same, the polarity of the output voltage is unchanged. Thus both positive and negative half-cycles of the supply produce positive output peaks and the time during which the capacitor must maintain the output voltage is reduced.

Note that for a short period of time between each half-cycle, the supply voltage is less than the turn-on voltage of the diodes and all the diodes are turned off.

FILE 5D

Computer simulation exercise 5.4

Repeat the investigations of Computer simulation exercise 5.3 for the full-wave rectifier circuit of Example 5.2. Compare the peak voltage at the output and the ripple voltage with those obtained using the earlier circuit for similar circuit parameters.

Example 5.3 A voltage doubler

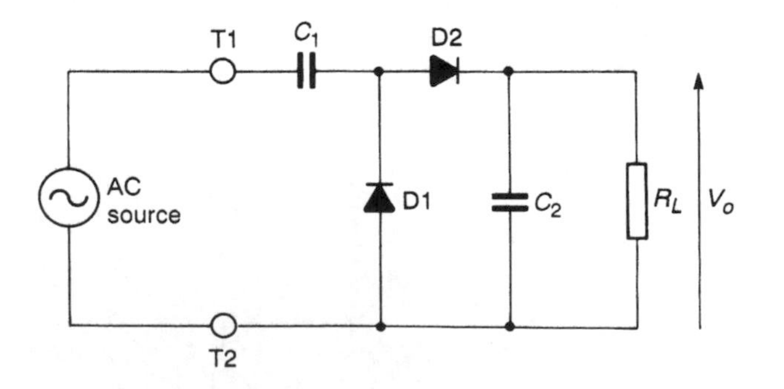

The voltage doubler shown above produces an output voltage considerably greater than the peak voltage of the input.

To understand its operation, consider initially the half-cycle when the input terminal T1 is negative with respect to terminal T2. Diode D1 is forward biased by the input voltage and so conducts, charging the capacitor C_1 to close to the peak voltage of the input waveform. During the next half-cycle, input terminal T1 is positive with respect to terminal T2 and diode D1 is reverse biased and therefore does not conduct. As T1 becomes more positive with respect to T2, the voltage across D1 increases, being equal to the input voltage plus the voltage across C_1. When the input reaches its peak value the voltage across D1 is nearly twice the input voltage. This forward biases diode D2, charging C_2 to close to twice the peak input voltage. This forms the output voltage of the arrangement.

If greater output voltages are required, several stages of voltage doubling can be cascaded to produce progressively higher voltages. These **voltage multiplier** circuits are ideal for applications that require high voltages at relatively low currents. Common applications include the extra high tension (EHT) supplies of cathode ray tubes (CRTs) and photomultiplier tubes.

Example 5.4 A signal rectifier

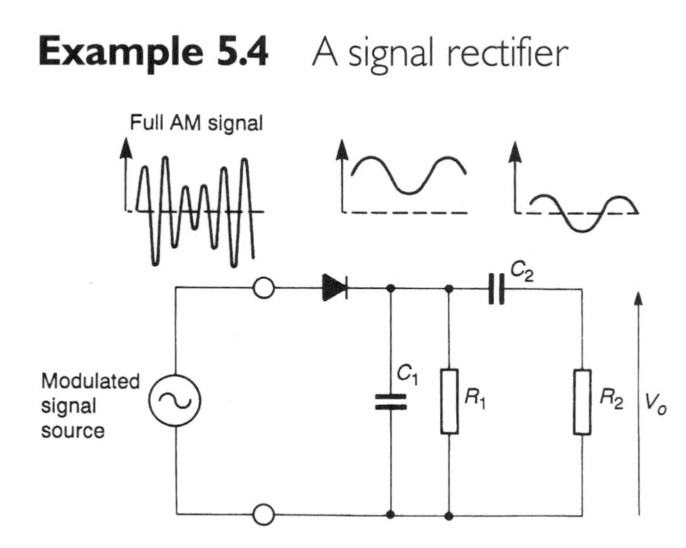

A common use of signal diodes is in the rectification (also called **demodulation** or **detection**) of modulated signals, such as those used for radio frequency broadcasting. Such signals often use **full amplitude modulation (full AM)** which produces a waveform of the type shown above. The signal consists of a high frequency carrier component, the amplitude of which is modulated by a lower frequency signal. It is this low frequency signal which conveys the useful information, and which must be recovered by **demodulating** the signal.

The demodulator works in a similar manner to the half-wave rectifier described earlier. The modulated signal is passed through a diode which applies only the positive half of each cycle to a parallel R–C network formed by R_1 and C_1. This behaves in a similar manner to the circuit described in Section 3.7.2 when we looked at amplifiers with high frequency cut-offs. The values of R_1 and C_1 are chosen such that they produce a high frequency cut-off which is above the signal frequency but below that of the radio frequency carrier. Thus the carrier is removed leaving only the required signal plus a DC component. This direct component is removed by a second capacitor C_2 which applies the demodulated signal to R_2. The behaviour of this second R–C network is similar to that described in Section 3.7.1.

The output voltage developed across R_2 represents the envelope of the original signal. For this reason the circuit is often called an **envelope detector**. This arrangement forms the basis of most radio receivers from simple **crystal sets**, which consist largely of the detector and a simple frequency selective network, to complex **superheterodyne receivers**, which use sophisticated circuitry to select and amplify the required signal.

Example 5.5 Signal clamping

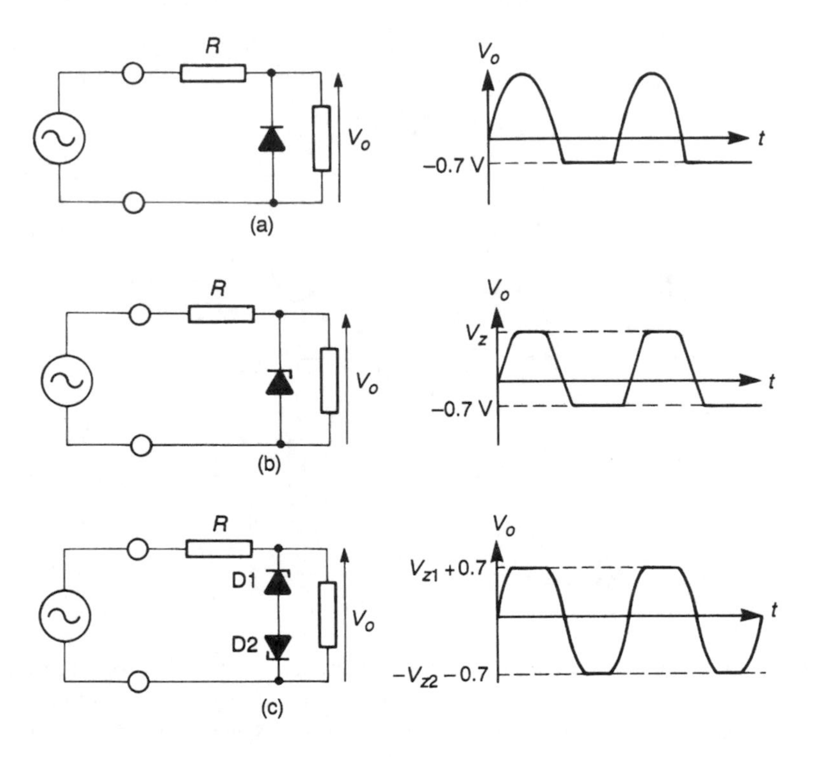

Diodes may be used in a number of ways to change the form of a signal. Such arrangements come under the general heading of **wave-shaping circuits**.

Circuit (a) shows a simple arrangement for limiting the negative excursion of a signal. When the input signal is positive the diode is reverse biased and so has no effect. However, when the input is negative and larger than the turn-on voltage of the diode, the diode conducts, clamping the output signal. This prevents the output from going more negative than the turn-on voltage of the diode (about 0.7 V for a silicon device). If a second diode is added in parallel with the first but connected in the opposite sense, the output will be clamped to ±0.7 V.

If the diode of (a) is replaced with a Zener diode, as shown in (b), the waveform is clamped for both positive and negative excursions of the input. If the input goes more positive than the breakdown voltage of the Zener diode V_Z, breakdown will occur preventing the output from rising further. If the input goes negative by more than the forward turn-on voltage of the Zener, it will conduct, again clamping the output. The output will therefore be restricted to the range $+V_Z > V_o > -0.7\ V$.

Two Zener diodes may be used, as shown in (c), to clamp the output voltage to any chosen positive and negative voltages. Note that the voltages at which the output signal is clamped are the sums of the breakdown voltage of one of the Zener diodes V_Z and the turn-on voltage of the other Zener diode.

Computer simulation exercise 5.5

FILE 5E

Use simulation to investigate the behaviour of the various circuits of Example 5.5. Apply a sinusoidal input voltage of 10 volts peak and use various combinations of simple diodes and Zener diodes. Suitable components might include 1N4002 signal diodes and D1N750 4.7 V Zener diodes.

Use transient analysis to look at the relationship between the input and the output waveforms.

Example 5.6 Catch diodes

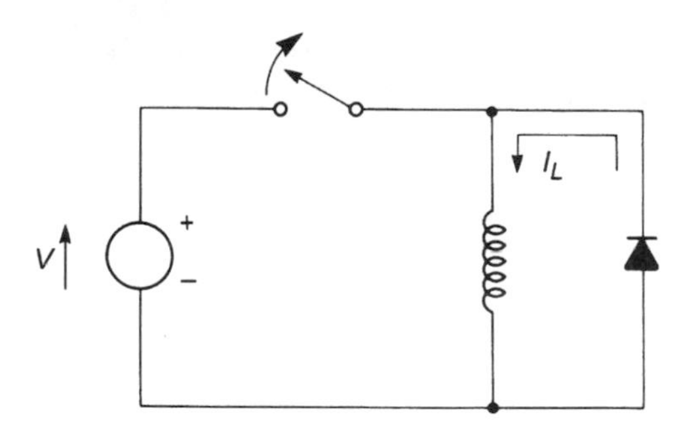

We saw in Chapter 2 that many actuators are inductive in nature. Examples include relays and solenoids. One problem with such actuators is that large back EMFs are produced if they are turned off rapidly. This effect is used to advantage in most automotive ignition systems in which a circuit breaker (the 'points') is used to interrupt the current in a high voltage coil. The large back EMF produced is used to generate the spark required to ignite the fuel in the engine. In other electronic systems these reverse voltages can do serious damage to delicate equipment if they are not removed. Fortunately in many cases the solution is very simple. This involves placing a **catch diode** across the inductive component to reduce the magnitude of this reverse voltage using the arrangement shown above.

The diode is connected so that it is normally reverse biased by the applied voltage and is therefore normally nonconducting. However, when the supply voltage is removed any back EMF produced by the inductor will forward bias the diode which therefore conducts and dissipates the stored energy. The diode must be able to handle a current equal to the forward current flowing before the supply was removed.

Key points

- Semiconductor materials are used at the heart of a multitude of electronic devices.
- The electrical properties of materials are brought about by their atomic structure.
- At very low temperatures semiconductors have the properties of an insulator. At higher temperatures thermal vibration of the atomic lattice leads to the generation of mobile charge carriers.
- Pure semiconductors are poor conductors even at high temperatures. However, the introduction of small amounts of impurities dramatically changes their properties.
- Doping of semiconductors with appropriate materials can lead to the production of *n*-type or *p*-type materials.
- A junction between *n*-type and *p*-type semiconductors (a *pn* junction) has the properties of a diode.
- Semiconductor diodes approximate ideal diodes but have a conduction voltage. Silicon diodes have a conduction voltage of about 0.7 V.
- In addition to conventional *pn* junction diodes there are a wide variety of more specialized diodes such as Zener diodes, Schottky diodes, tunnel diodes and varactor diodes.
- Diodes are used in a range of applications in both analogue and digital systems. Analogue uses include rectification, demodulation and signal clamping.

Design study

A reference voltage of 3.6 V is required to drive a circuit which has an effective input resistance of 200 Ω. The reference voltage is to be produced from a supply voltage *V* which can vary between 4.5 and 5.5 V.

Approach

A suitable circuit might be that shown on the next page.

Clearly a Zener diode with a breakdown voltage V_Z of 3.6 V is used, but calculations must be performed to determine the value required for the resistor *R* and the power ratings required for the resistor and the diode.

For power dissipation considerations we would like *R* to be as high as possible. The maximum value of *R* is determined by the requirement that the voltage drop across *R* must not take the voltage at the output (and across the diode) below the required output voltage. This condition is most critical when the input voltage *V* is at its lowest, so the value of *R* is

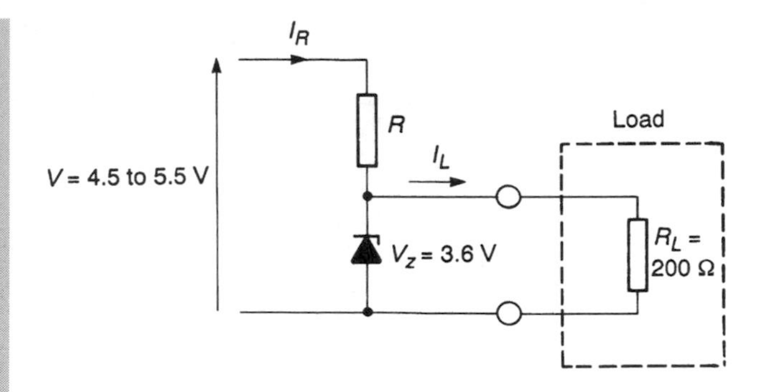

determined by calculating the value of R for which the output would be equal to the required value (3.6 V) when the input voltage V is at its lowest permissible value (4.5 V). we can ignore the effects of the Zener diode at this point since all the current flowing through R will be flowing into the load.

The current flowing in the load I_L can be calculated from

$$I_L = \frac{V_Z}{R_L}$$

$$= \frac{3.6}{200} = 18 \text{ mA}$$

The voltage drop across the resistance R caused by I_L must be less than the difference between the minimum supply voltage (4.5 V) and the Zener voltage (3.6 V), therefore

$$I_L R < 4.5 - 3.6\ V$$

and

$$R < \frac{4.5 - 3.6\ V}{I_L}$$

$$< \frac{0.9\ V}{18 \text{ mA}}$$

$$< 50\ \Omega$$

So a standard value of 47 ohms would probably be chosen.

Maximum power dissipation in the various components is generated when the input voltage is at its maximum value (5.5 V).

Since the output voltage is fixed the voltage across the resistance is easy to calculate; it is simply $V - 3.6$ V. Therefore the maximum power dissipation in the resistance is simply

$$P_{R(max)} = \frac{V^2}{R} = \frac{(5.5 - 3.6)^2}{47} W$$

$$= 77 \text{ mW}$$

The power dissipated in the Zener diode P_Z is also simple to calculate. The voltage across it is fixed (3.6 V) and the current through it is simply the current through the resistance R (I_R)

minus the current taken by the load (I_L). This is also at a maximum when the supply voltage has its maximum value.

Thus

$$P_{Z(max)} = V_Z I_{Z(max)} = V_Z(I_{R(max)} - I_L)$$
$$= 3.6\left(\frac{5.5-3.6}{47} - 0.018\right) = 81 \text{ mW}$$

Thus the Zener diode must have a power rating of greater than 81 mW.

References

Esaki L. (1958) New phenomenon in narrow germanium p-n junctions. *Phys. Rev.* **109**, 603

Further reading

Streetman B. G. (1995) *Solid State Electronic Devices*, 4th edn. Englewood Cliffs, NJ: Prentice-Hall

Sze S. M. (1985) *Semiconductor Devices: Physics and Technology*. New York: John Wiley

Tocci R. J. and Oliver M. E. (1991) *Fundamentals of Electronic Devices*, 4th edn. New York: Merrill

Williams G. E. (1996) *Analogue Electronics: Devices, Circuits and Techniques*. St Paul, MN: West

Exercises

5.1 Explain, with the aid of diagrams, where appropriate, the meaning of the terms tetravalent material, covalent bonding, semiconductor, intrinsic conduction and doping.

5.2 Name three semiconductor materials commonly used for semiconductor devices. Which material is most widely used for this purpose?

5.3 Why cannot the potential barrier formed at a *pn* junction be measured externally with a voltmeter?

5.4 Design a circuit to apply a constant voltage of 5.6 V to a network with an input resistance of 1 kΩ. The circuit must derive its power from a supply which may vary from 10.5 to 12.5 V. Calculate the power rating required for the components of your circuit.

5.5 Repeat the last exercise assuming the input resistance of the network may vary from 500 Ω to 2 kΩ.

5.6 A half-wave rectifier is connected to a 50 Hz supply and generates a peak output voltage of 10 V across a 10 mF (10 000 μF) reservoir capacitor.

Estimate the peak ripple voltage produced if this arrangement is connected to a load which takes a constant current of 200 mA.

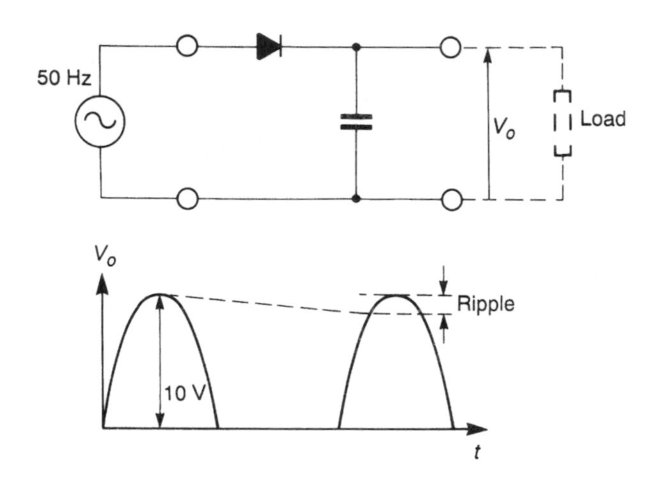

5.7 What would be the effect on the ripple voltage calculated in the last exercise of replacing the half-wave rectifier with a full-wave rectifier of similar peak output voltage?

5.8 Sketch the output waveforms of the following circuits. In each case the input signal is a sine wave of ±5 V pk.

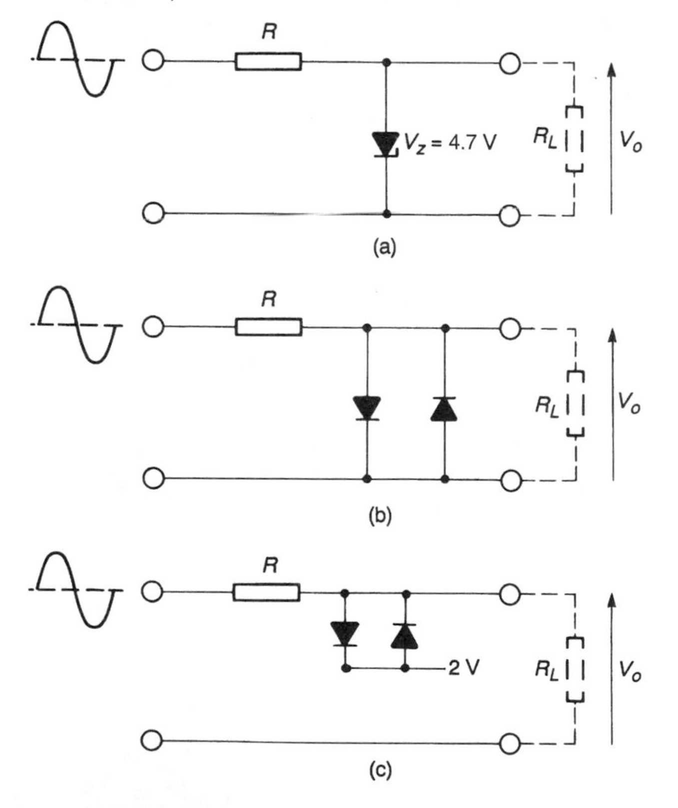

5.9 Use circuit simulation to verify your answers to the last exercise. How does the value of R_L affect the operation of the circuits?

5.10 Design a circuit which will pass a signal unaffected, except that it limits its excursion to the range $+10.4\text{ V} > V > -0.4\text{ V}$.

5.11 Use circuit simulation to verify your design for the last exercise.

Field Effect Transistors

Objectives

When you have studied the material in this chapter you should:

- be familiar with the construction and operation of common forms of field effect transistor (FET);
- be able to derive and use equivalent circuits of such devices to describe their behaviour and characteristics;
- be familiar with the use of FETs in the formation of amplifiers and be able to design and analyse simple circuits;
- be able to explain the distinction between 'small signal' and 'large signal' aspects of a circuit's design;
- have an understanding of the limitations imposed on the frequency response of FET circuits by device capacitances;
- be aware of a wide range of suitable applications for FETs in both analogue and digital systems.

Contents

6.1 Introduction

In Section 3.12 we saw how a three terminal 'control device' could be used to form an amplifier. The control devices described in that section had the characteristic that a signal on one terminal (the input terminal) controlled the current flowing between the other two. Several devices of this type can be made using the properties of semiconductor materials outlined in the last chapter.

In some of these devices the output current is determined by the *current* flowing into the control input. In others it is the *voltage* applied to the input terminal that determines the output current. In this chapter we will look at devices that fall into this latter category. These devices are called **field effect transistors**, or simply **FET**s, since their operation relies on the electric field generated by the input voltage.

Field effect transistors were first proposed by **Shockley** (1952), but it was more than ten years before the fabrication process was perfected to enable the construction of dependable devices. FETs are probably the simplest forms of transistor to understand, and are widely used in both analogue and digital applications. They are characterised by very high input resistances, low power requirements and small physical dimensions. These characteristics combine to make them ideal for the construction of very high density circuitry such as that used in **very large-scale integration** (**VLSI**) circuits. We will deal primarily with their uses in analogue circuits in this chapter and leave consideration of digital circuits until Chapter 11.

There are two main forms of field effect transistor, namely the *junction-gate FET* and the *insulated-gate FET*. We will start by looking at the latter, which is known by a variety of names including the *MOSFET*.

6.2 The MOSFET

The term **MOSFET** refers to the method of construction of this device and its mode of operation. It stands for metal oxide semiconductor field effect transistor. The first part of its name indicates that its construction is a sandwich of layers of a metal, an oxide (usually silicon oxide) and a semiconductor. The second part indicates that the device is operated by an electric field rather than an electric current. A MOSFET may also be called a **MOST** (metal oxide semiconductor transistor) or an **IGFET** (insulated gate field effect transistor).

6.2.1 Construction of a MOSFET

The construction of a typical MOSFET is illustrated in Figure 6.1(a) and its circuit symbol is shown in Figure 6.1(b).

Two n-type regions (called the **drain** and **source**) are formed in a p-type silicon bar (called a **substrate**); these two regions are joined by a thin **channel** of n-type material at the surface of the bar. The channel is covered by a conducting electrode (the **gate**) which is insulated from it by a layer of oxide. The gate electrode is normally formed by a layer of metal but may alternatively be a layer of polycrystalline silicon. Electrical contacts

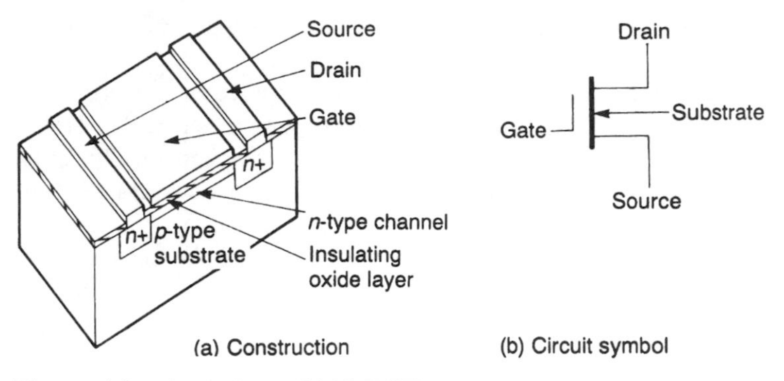

Figure 6.1 An n-channel MOSFET.

are made to the drain, source, gate and substrate. The substrate is usually connected to the source either internally or externally.

6.2.2 MOSFET operation

Let us consider initially the situation in which a positive voltage V_{DS} is applied between the drain and the source, as shown in Figure 6.2. The drain–channel–source regions being all of n-type material form a pn junction with the p-type substrate. The source end of this junction is at the same potential as the substrate and so no current flows between the source and the substrate. The drain and the channel, being positive with respect to the source, are also positive with respect to the substrate. Thus the junction between them and the substrate is reverse biased and again no current flows between these regions and the substrate.

The presence of the thin n-type channel provides a conduction path between the drain and the source, and the voltage applied between them will cause a current to flow. The

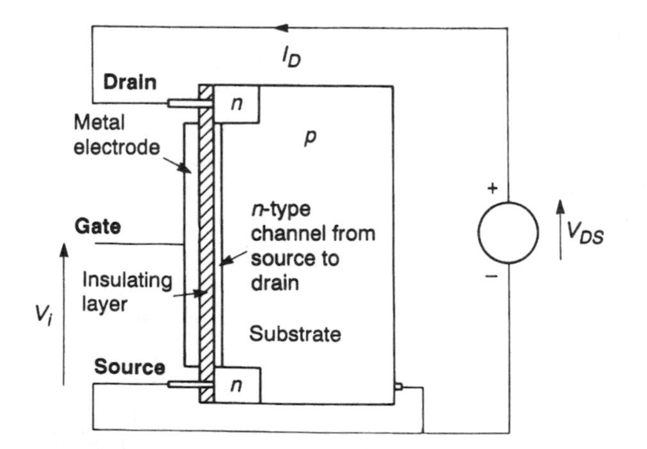

Figure 6.2 Operation of an n-channel MOSFET.

magnitude of this current will depend on the thickness of the channel and on the applied voltage.

If the gate is made positive with respect to the source (and, therefore, with respect to the substrate), the positive charge on the gate repels holes in the substrate close to the gate and thus increases the effective depth of the *n*-type channel. This has the effect of increasing the current from the drain to the source. This process is called **enhancement** of the channel.

Conversely, if the gate is made negative with respect to the source, electrons are repelled from the region of the gate and the effective thickness of the channel is reduced, decreasing the current from the drain to the source. This is known as **depletion** of the channel.

It is clear that the voltage on the gate of the MOSFET controls the current flowing from the drain to the source. We therefore have a 'control device', as described in Section 3.12, which can be used as an amplifier.

At this point it is worth explaining the **labelling convention** used for the voltages and currents associated with electronic components. In this section we have used the term V_{DS} to refer to a voltage applied across the MOSFET. The subscript of this label defines the nodes between which the voltage is measured and the order of the letters shows the polarity. Thus V_{DS} is the voltage on the drain with respect to the source, while V_{GS} would be the voltage on the gate with respect to the source. Currents can also be labelled in this way and, for example, I_{DS} would be the current flowing from the drain to the source.

6.2.3 Types of MOSFET

In the MOSFET described, the voltage on the gate can be used to produce depletion or enhancement of the channel. It is therefore called a **depletion/enhancement MOSFET** or **DE MOSFET** (or sometimes simply a **depletion MOSFET**).

Other devices are constructed so that no channel exists between the drain and the source in the absence of a gate voltage. In such devices a voltage must be applied to the gate to produce a conducting channel. If, as before, we consider a device constructed by forming *n*-type drain and source regions in a *p*-type substrate, a positive voltage applied to the gate attracts electrons and repels holes from the region around the gate to form a conducting *n*-type channel between the drain and source. This region is called an inversion layer since it is an *n*-type layer in a *p*-type material. A device of this type can only be used in enhancement mode and is therefore called an **enhancement MOSFET**.

The devices described so far have an *n*-type channel in a *p*-type substrate and are therefore called ***n*-channel MOSFET**s. Devices can also be constructed by forming a *p*-type channel in an *n*-type substrate. Logically enough these devices are called ***p*-channel MOSFET**s. The operation of *p*-channel devices is similar to that of *n*-channel devices except that the polarities of all the voltages and currents are reversed. For simplicity we will use *n*-channel devices for most of the examples in this chapter.

In all MOSFETs the current is carried by the **majority charge carriers** in the channel. That is, by electrons in an *n*-channel device and by holes in a *p*-channel device. Since conduction is performed by charge carriers of only one polarity, MOSFETs are called **unipolar transistors**.

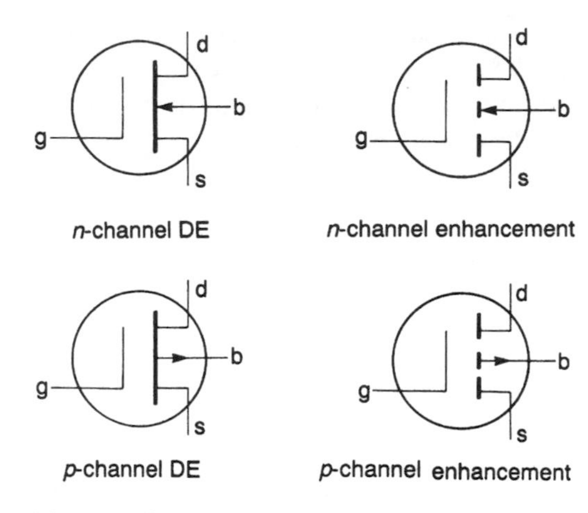

Figure 6.3 MOSFET circuit symbols.

Figure 6.3 shows the circuit symbols for the four basic types of MOSFET. Notice that the substrate is marked 'b' for bulk to distinguish it from the 's' for source. Sometimes the substrate is joined to the source internally to give a device with three rather than four terminals. The line connecting the drain and source represents the channel. It is shown solid in a DE MOSFET to represent a channel which is present even when no gate voltage is applied, and broken in an enhancement MOSFET to indicate that the channel is not established with zero gate voltage.

The arrow between the substrate and the channel indicates the polarity of the device. The arrow points from the *p*-type material towards the *n*-type material, as in a diode, and thus points towards an *n*-type channel and away from a *p*-type channel.

The relationship between the gate voltage and the drain current will obviously be affected by the geometry of the device. Typical characteristics for a DE MOSFET and an enhancement MOSFET are shown in Figure 6.4.

Clearly there is not a linear relationship between I_D and V_{GS}. However, if we operate the device over a small range of values for V_{GS}, the characteristic could reasonably be approximated by a straight line. The gradient of this line can be used to predict the

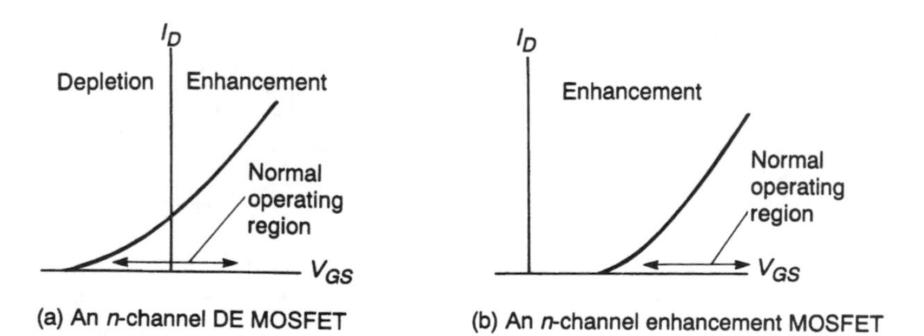

Figure 6.4 Typical DE MOSFET characteristics.

relationship between a small change in the gate voltage ΔV_{GS} and the corresponding change in the drain current ΔI_D. For the DE MOSFET it is convenient to operate the device over a small range on either side of the point where V_{GS} is equal to zero volts. For the enhancement MOSFET one must operate in a region where V_{GS} has some positive value (for an *n*-channel device).

6.2.4 A simple MOSFET amplifier

In a MOSFET an input voltage V_{GS} controls an output current I_D and the device can therefore be used to produce an amplifier as discussed in Section 3.12.

The circuit of such an amplifier using an *n*-channel DE MOSFET is shown in Figure 6.5(a), and a similar amplifier using an *n*-channel enhancement device is shown in Figure 6.5(b). Compare these circuits with that given in Figure 3.31.

The capacitor C in each case is a **coupling capacitor**, which is also known as a **blocking capacitor**. This *couples* an AC input signal to the active device, while it *blocks* the DC component of any input signal, preventing it from upsetting the mean voltage applied to the gate of the MOSFET. The value of this capacitor would be selected so that it presented a negligible impedance at the frequencies of the AC signals to be amplified. Because of the presence of the capacitor, this circuit cannot be used to amplify DC signals and is thus referred to as an **AC coupled amplifier**.

V_{DD} and V_{SS} are the supply voltages (sometimes called the **supply rails**) for the circuit. Following the labelling convention discussed earlier it would seem that V_{DD} should represent the voltage on the drain with respect to the drain. Since this is a rather pointless quantity it is conventional to use a repeated subscript to represent a constant supply voltage within the circuit. It is common to give these lines names which reflect the terminals of the active device to which they are connected. Thus V_{DD} is the *drain supply*, which in this case is positive, and V_{SS} is the *source supply*, which in this case is the reference or zero volt line. In a circuit using *p*-channel MOSFETs V_{DD} would be negative with respect to V_{SS}.

The resistor(s) connected to the gate of each circuit are used to maintain the mean value of the gate voltage in the centre of its normal operating region. This is referred to as **biasing** the transistor. In the circuit of Figure 6.5(a), a single resistor is connected from the gate to zero volts since the normal operating region of a DE MOSFET is centred at

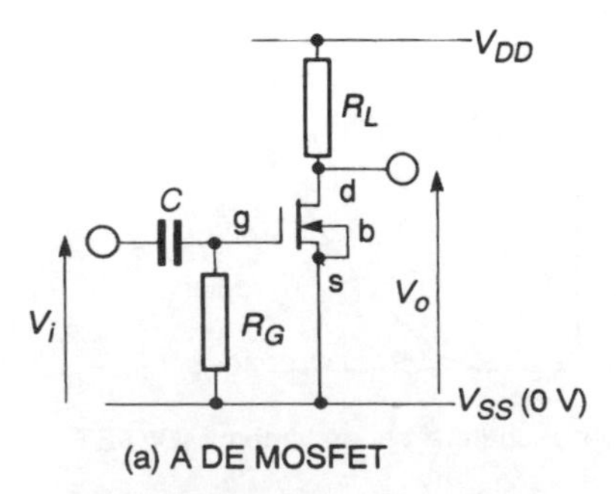

(a) A DE MOSFET

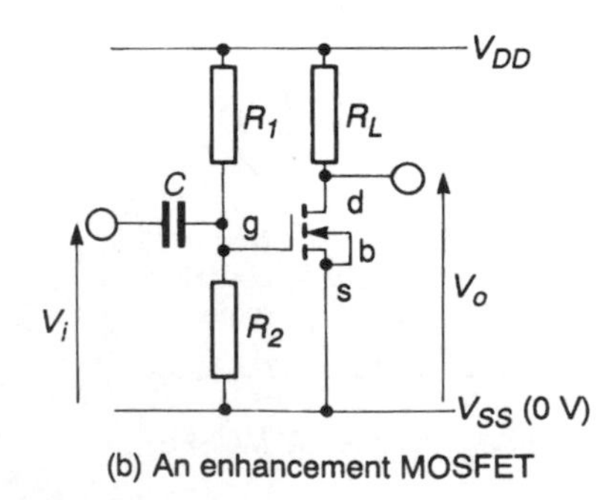

(b) An enhancement MOSFET

Figure 6.5 AC coupled MOSFET amplifiers.

the point where $V_{GS} = 0$ V. In the circuit of Figure 6.5(b), this resistor is replaced by a pair of resistors forming a potential divider across the supply voltage. This biases the gate to a positive voltage to place the MOSFET in its normal operating region. The ratio of these resistors would be chosen to generate the appropriate bias voltage for a given MOSFET and supply voltage.

Having determined the mean voltage on the gate, a fluctuating (small signal) input voltage will cause the gate voltage to vary, which will, in turn, change the drain current of the FET. This will produce a varying voltage drop across the load resistor and will therefore produce a small signal output voltage. If the component values are chosen appropriately, this will produce voltage amplification. We will look at amplifier circuits in more detail in Section 6.5.

Computer simulation exercise 6.1

FILE 6A

Simulate the simple amplifier of Figure 6.5(b). If you are using PSpice you may use the circuit within the demonstration file. If you do not have access to the demonstration file, or are using another simulation package, you may need to experiment to find appropriate values for the various components depending on the FET used. Suitable values to start with might be: $V_{DD} = 12$ V; $R_1 = 300$ kΩ; $R_2 = 100$ kΩ; $R_L = 130$ Ω and $C = 1$ μF. PSpice provides a model of the IRF150 FET. If you are using another package you will need to pick a suitable device from the models available.

Apply a 1 kHz sinusoidal input voltage of about 50 mV peak and use transient analysis to investigate the relationship between the input and output waveforms. Determine the magnitude and phase of the gain of your circuit. What happens to the output if the input magnitude is progressively increased?

6.3 The JFET

The term JFET stands for **junction-gate field effect transistor**. The latter portion of its name indicates that, as with the MOSFET, the device's operation is controlled by an *electric field* rather than an *electric current*. The former part of the name indicates that the gate of the device forms a *pn junction* with the substrate. Such devices are also called **JUGFET**s.

6.3.1 Construction of a JFET

The construction of a typical *n*-channel JFET is shown in Figure 6.6(a) and its circuit symbol is given in Figure 6.6(b).

A narrow **channel** of *n*-type material has electrical connections at either end, called the **drain** and **source** as in the MOSFET. Within the channel is a region of *p*-type material which forms the **gate** of the device; a layer of conductor allows electrical contact to be made to this region. Unlike in the MOSFET, there is no layer of insulation to separate the gate from the channel. Instead the gate is electrically connected to the channel, the two dissimilar materials forming a *pn* junction.

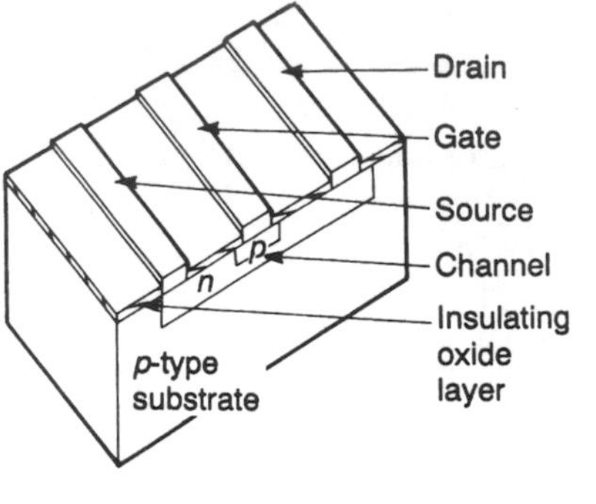

(a) Construction

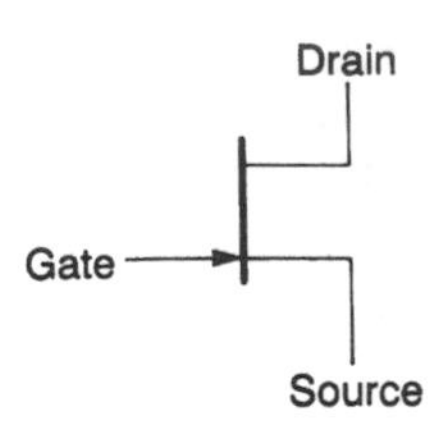

(b) Circuit symbol

Figure 6.6 An *n*-channel JFET.

6.3.2 JFET operation

The operation of the JFET relies on the junction formed between the gate and the channel being reverse biased. This forms a **depletion layer** around the gate as shown in Figure 6.7.

If the drain is made positive with respect to the source by the application of a voltage V_{DS}, current will flow from the drain to the source through the *n*-type channel. If now a negative voltage is applied to the gate with respect to the source, this reverse biases the gate-channel junction and prevents any gate current from flowing. The negative voltage on the gate repels electrons in the region of the gate and thus increases the size of the depletion layer. This reduces the effective area of cross-section of the channel and thus reduces its conductivity.

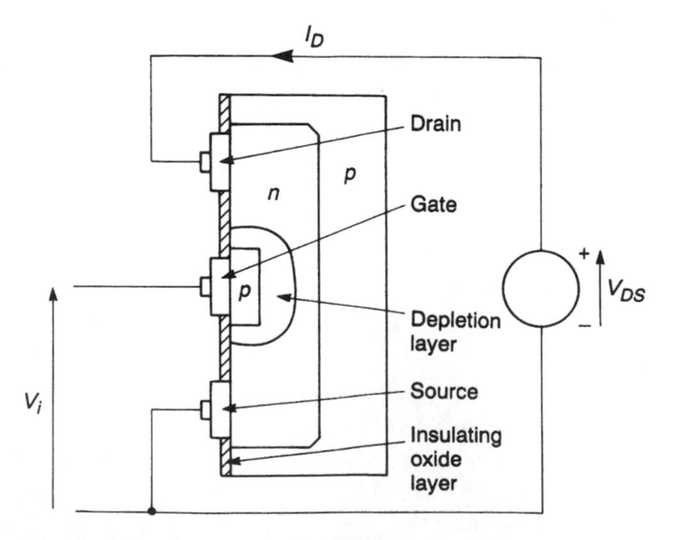

Figure 6.7 Operation of a JFET.

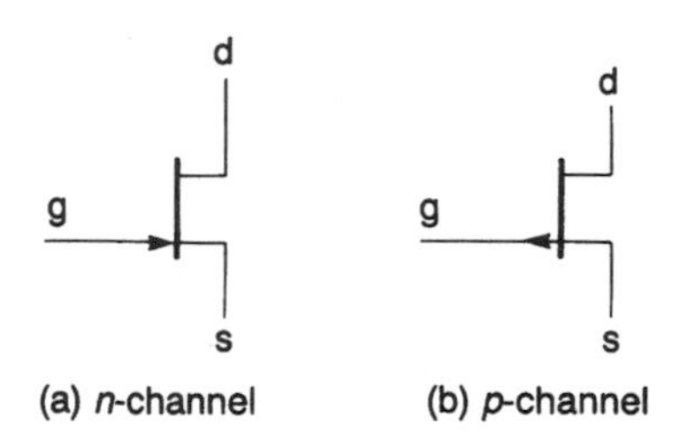

Figure 6.8 Circuit symbols for *n*-channel and *p*-channel JFETs.

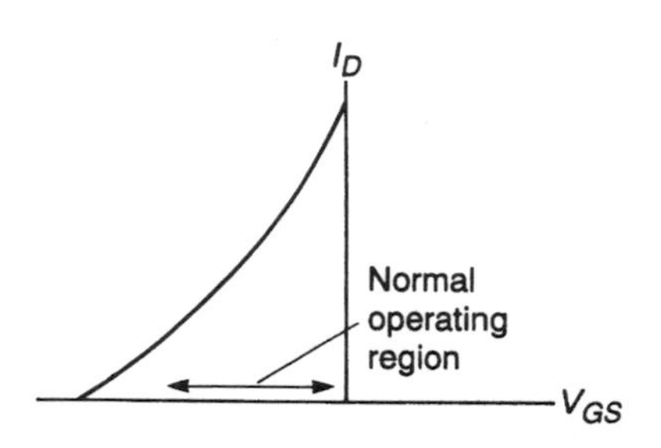

Figure 6.9 Typical *n*-channel JFET characteristics.

As the gate is made more and more negative, the effective width of the channel decreases until the drain current is stopped completely. This mode of operation is analogous to the use of the MOSFET in its **depletion** mode.

6.3.3 Types of JFET

The JFET can be used in a *depletion* mode but cannot be used to *enhance* the channel since this would imply forward biasing the gate junction which would cause current to flow into the gate. JFETs can, however, be formed with either *n*-type or *p*-type channels, these being called *n*-channel and *p*-channel devices. The circuit symbols used for these two types of device are shown in Figure 6.8. As with the MOSFET, the arrow points towards an *n*-type channel and away from a *p*-type channel.

The relationship between gate voltage and drain current for a typical *n*-channel JFET is shown in Figure 6.9.

As with the characteristics of the MOSFET shown in Figure 6.4, it is clear that the relationship between I_D and V_{GS} is not linear. However, if we restrict ourselves to fairly small changes of V_{GS} we may approximate the relationship by a straight line. Unlike the MOSFET, the normal region of operation of a JFET corresponds to a negative range of values of V_{GS} (for an *n*-channel device).

6.3.4 A simple JFET amplifier

As with the MOSFET, it is possible to utilize the characteristics of the JFET to produce

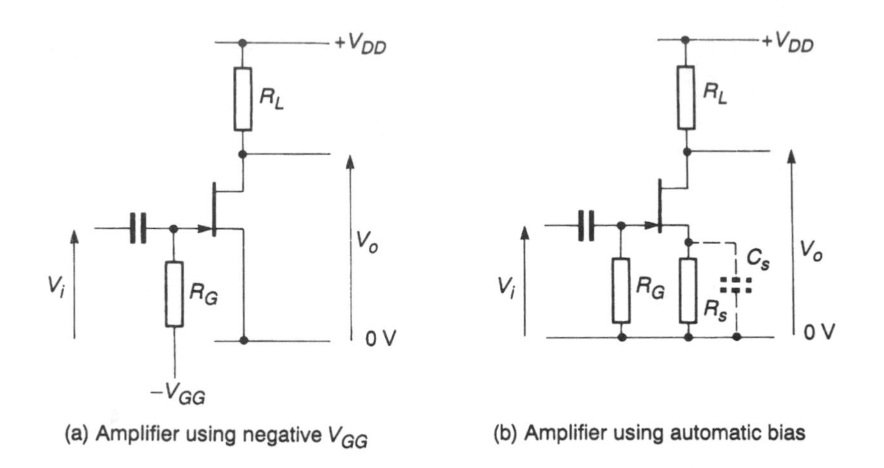

Figure 6.10 AC coupled JFET amplifiers.

an **AC coupled amplifier**. The circuits of two such amplifiers are shown in Figure 6.10.

Figure 6.10(a) shows an amplifier which uses a negative supply voltage to provide a **bias voltage** V_{GG} to place the JFET in an appropriate position on its characteristic. If a negative supply voltage is not available it may be more convenient to use the circuit arrangement shown in Figure 6.10(b). This uses a source resistor R_S to provide the bias voltage. Current flowing from the source of the transistor generates a voltage drop across this resistor making the source more positive than ground (that is, zero volts). The mean voltage on the gate is then set at zero volts by connecting a gate resistor R_G from the gate to ground. Since the gate is now at ground potential and the source is positive with respect to ground, the source is negative with respect to the source. By choosing an appropriate value for the source resistor, the gate-to-source voltage can be set to the desired value. One disadvantage of this approach is that although the source resistor correctly sets the bias voltage of the circuit, it also reduces the small signal gain of the amplifier. This is because any input signal applied to the gate appears across the combination of the gate-to-source junction and the source resistor. Thus only a fraction of the input voltage is applied across the gate junction of the transistor. In small signal amplifiers this problem is normally overcome by placing a capacitor C_S across the source resistor. If this capacitor is sufficiently large it will act as a short circuit for AC signals and thus the source will be effectively joined to ground for small signals. Any small signal input voltage will now appear directly across the gate-to-source junction giving maximum gain. C_S is referred to as a **decoupling capacitor** (or sometimes as a **bypass capacitor**) and its value should be such that *at the frequencies of interest* its impedance is small compared with the impedances within the circuit between the source and ground. The impedance from the source to ground is given by the parallel combination of R_S and the resistance seen looking into the source of the FET (determination of this resistance will be discussed in Section 6.5.6). Clearly, since the impedance of a capacitor goes up as the frequency goes down, it is the lowest frequency at which the circuit is to operate that will determine the appropriate value of C_S. We will return to the biasing of FETs in Section 6.5.

As with the MOSFET amplifiers considered earlier, having set the mean value of the gate voltage, a small signal input will vary the gate voltage producing a changing drain current and hence a fluctuating output voltage.

Computer simulation exercise 6.2

FILE 6B

Simulate the simple amplifier of Figure 6.10(a). If you are using PSpice you may use the circuit within the demonstration file. If you do not have access to the demonstration file, or are using another simulation package, you may need to experiment to find appropriate values for the various components depending on the FET used. Suitable values to start with might be: $V_{DD} = 12$ V; $V_{GG} = -1$ V; $R_G = 100$ kΩ; $R_L = 1$ kΩ and $C = 1$ μF. PSpice provides a model of the 2N3819 JFET. If you are using another package you will need to pick a suitable device from the models available.

Apply a 1 kHz sinusoidal input voltage of about 50 mV peak and use transient analysis to investigate the relationship between the input and output waveforms. Determine the magnitude and phase of the gain of your circuit. What happens to the output if the input magnitude is progressively increased?

6.4 FET characteristics

In order to get a clear view of the characteristics of field effect transistors we need to look in slightly more detail at their operation. In Sections 6.2.2 and 6.3.2 we took a fairly simple view of the operation of MOSFETs and JFETs which ignored, for example, the effects of the drain-to-source voltage V_{DS} on current flow. We will now look again at these devices to get a deeper understanding of their mechanics.

6.4.1 The MOSFET

We saw in Figure 6.2 that conduction between the drain and source of a MOSFET is provided by the channel, and noted that the thickness of this channel is controlled by the voltage on the gate V_{GS}. With zero voltage between the drain and the source the channel has a fairly uniform thickness along its length, this being determined by the voltage on the gate. If now a small voltage is applied between the drain and the source, the channel behaves like a simple resistor with the drain current I_D increasing linearly with the applied voltage V_{DS}. The effective resistance of the channel R_{DS} is controlled by the gate voltage and thus the device functions as a **voltage dependent resistance**. We shall look at some applications that use this effect in Section 6.6.

As the voltage between the drain and the source is increased, the thickness of the channel soon ceases to be of uniform thickness. The drain-to-source voltage is distributed along the length of the channel such that the voltage between the gate electrode and the channel is very different at one end of the gate from that at the other. The effect of this is shown in Figure 6.11.

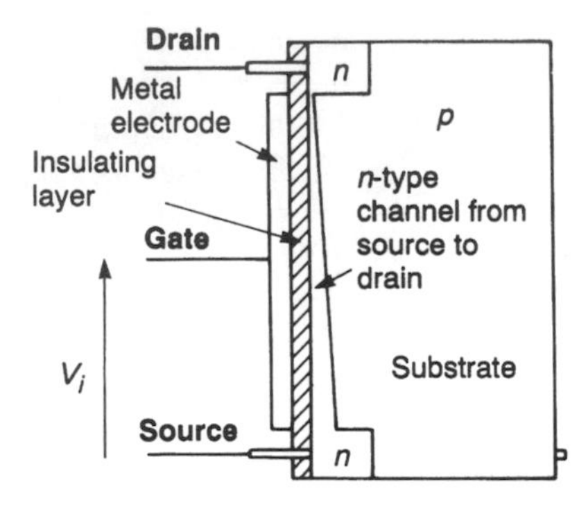

Figure 6.11 The effect of drain-to-source voltage on channel thickness in a MOSFET.

With an n-channel device, as shown in Figure 6.11, the drain would normally be positive with respect to the source. Therefore, near the drain the channel is much more positive with respect to the gate than at the source end. Consequently the channel is much thinner near the drain than it is near the source. This reduces the conduction of the channel as the drain voltage increases. When the drain voltage reaches a sufficiently large value the channel becomes **pinched-off** and any further increase in the drain voltage has very little effect on the drain current. When in this state the device is said to be in **saturation**.

The MOSFET therefore has two distinct regions of operation with a transition area between them. For low values of V_{DS} the device is in its **ohmic** region, in which the drain current is directly proportional to the drain voltage. For higher values of V_{DS} the device is saturated, in which case the drain current is largely independent of the drain voltage. In this latter region the drain current is determined by the gate voltage. Typical characteristics of both depletion and enhancement MOSFETs are shown in Figure 6.12. These graphs are often called **output characteristics** since they relate the output current to the output voltage of the device. They are also called **drain characteristics**.

When used as an amplifier, the MOSFET is normally operated in its saturated region where the output current is determined primarily by the input voltage and is not greatly affected by the output voltage (that is, the voltage on the drain). The characteristics given in Figure 6.4 show the behaviour of the device in this region. These curves show the effects of the input voltage on the output current and are generally called the **transfer characteristics** of the device. Figure 6.13 indicates relevant points on these transfer characteristics.

The characteristics of a DE MOSFET are normally described by defining V_T and I_{DSS}. V_T represents the gate-to-source **threshold voltage** and is the voltage below which the channel is effectively cut off. I_{DSS} is the drain-to-source **saturation current**; it is the drain-to-source current with the gate connected to the source. In an enhancement device the characteristic is normally described by giving V_T and some particular value of the drain current $I_{D(ON)}$ corresponding to a specified value of the gate voltage $V_{GS(ON)}$. In both cases the relationship between I_D and V_{GS} in the saturation region is approximately parabolic (a square law) and is described by the simple expression

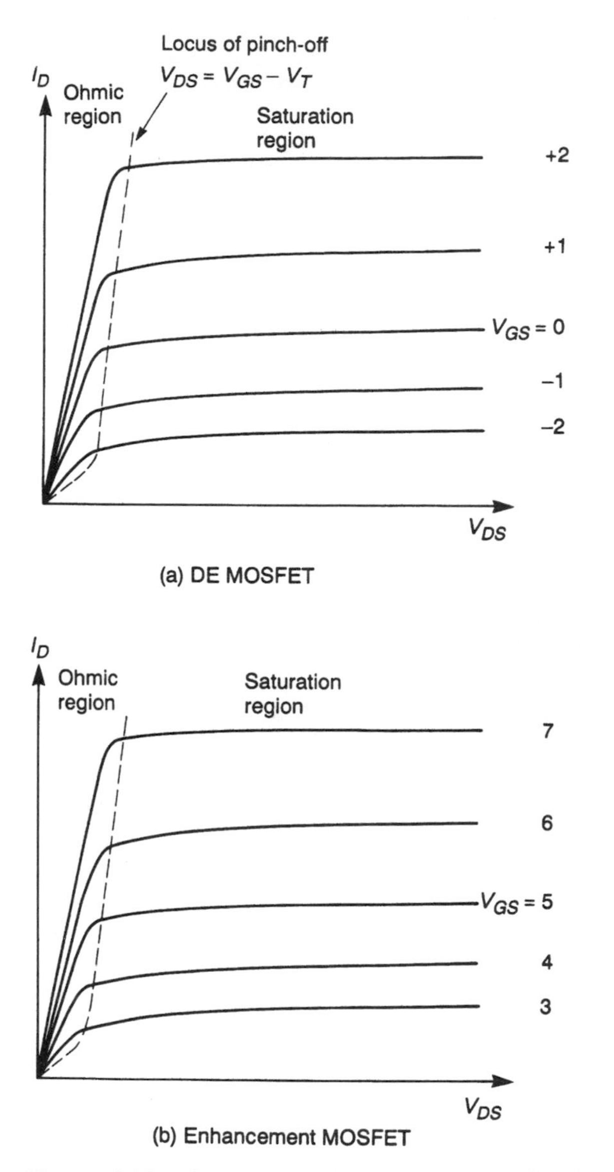

Figure 6.12 Output (drain) characteristics of *n*-channel MOSFETs.

$$I_D = K(V_{GS} - V_T)^2 \tag{6.1}$$

where K is a constant that depends on the physical parameters and the geometry of the device.

The locus of pinch-off shown in Figure 6.12 (that is, the boundary that separates the ohmic and the saturation regions) is related to V_T by the expression

$$V_{DS}(\text{at pinch-off}) = V_{GS} - V_T$$

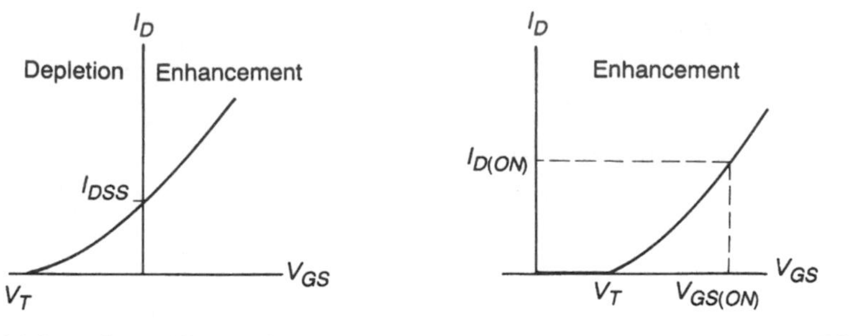

(a) An n-channel DE MOSFET (b) An n-channel enhancement MOSFET

Figure 6.13 Transfer characteristics of n-channel MOSFETs.

6.4.2 The JFET

The effects of saturation described above for the MOSFET occur in a very similar manner in the JFET. The simple diagram of Figure 6.7 becomes that of Figure 6.14 when a voltage is applied between the drain and source.

At low drain voltages the channel behaves linearly, the drain current being directly proportional to the drain voltage. As the drain voltage is increased the size of the depletion layer increases, reducing the width of the channel and reducing its conductance. Eventually the channel becomes **pinched off** and any further rise in the drain voltage does not produce an increase in the drain current. This is the **saturation** region.

The output characteristic for the JFET is similar in appearance to that of the DE MOSFET with the exception that V_{GS} is restricted to being negative to prevent forward biasing the gate–channel junction. A typical JFET output characteristic is shown in Figure 6.15.

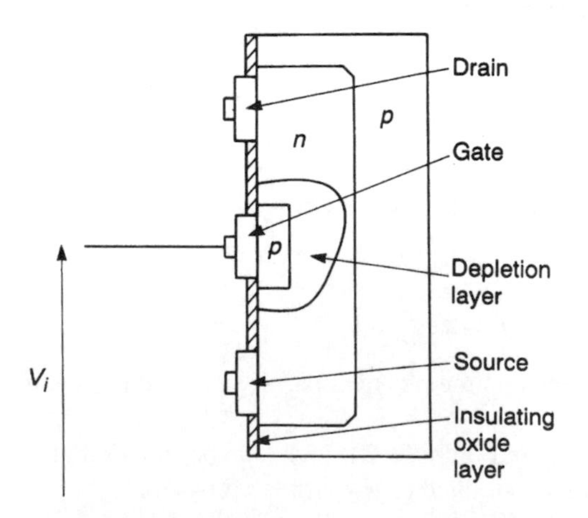

Figure 6.14 The effect of drain-to-source voltage on channel thickness in a JFET.

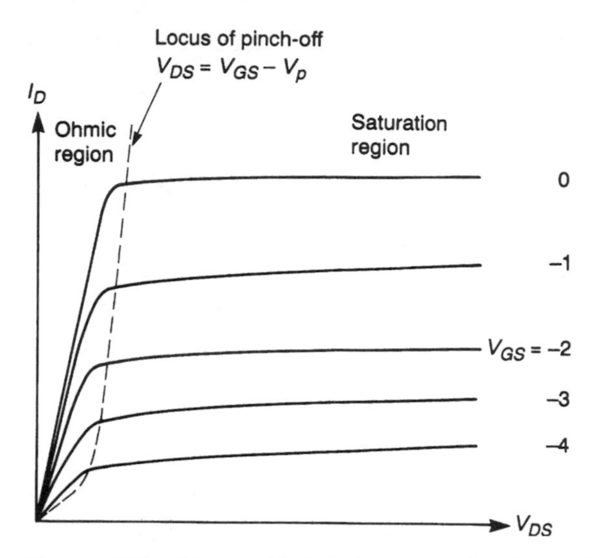

Figure 6.15 Output (drain) characteristic of an *n*-channel JFET.

Figure 6.16 shows a typical transfer characteristic for an *n*-channel JFET in its saturation region.

As before, I_{DSS} is the drain current with the gate connected to the source and V_p is the **pinch-off voltage**. This characteristic is clearly similar to those of the MOSFETs discussed earlier. Again, in the saturation region, the relationship between the drain current and the gate voltage obeys a square law. The relationship is

$$I_D = I_{DSS}\left(1 - \frac{V_{GS}}{V_p}\right)^2 \tag{6.2}$$

which may be rearranged into the form

$$I_D - K'(V_{GS} - V_p)^2$$

which is clearly of the same form as Equation 6.1. As in the earlier equation, K' is a constant related to the physical parameters and the geometry of the device.

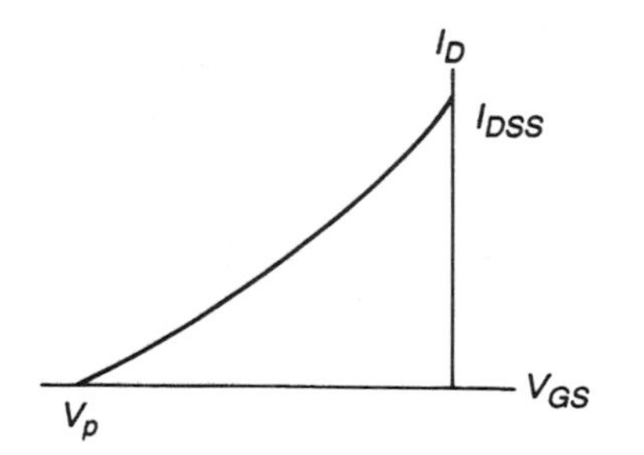

Figure 6.16 Transfer characteristic for an *n*-channel JFET.

As in the case of the MOSFET, the locus of pinch-off is defined by a simple expression. In this case

$$V_{DS}(\text{at pinch-off}) = V_{GS} - V_p$$

as shown in Figure 6.15.

6.4.3 Equivalent circuit of a FET

In Section 3.5 we looked at equivalent circuits for amplifiers and deduced such a circuit for an operational amplifier. We can also produce an equivalent circuit for a FET which can be used to simplify the design of circuits using these components.

The gate input of a MOSFET is insulated from the remainder of the device by the oxide layer. It is therefore effectively isolated, with the exception of a small amount of capacitance which can usually be ignored at low frequencies. The gate input of a JFET forms a *pn* junction with the channel region. This junction can pass an appreciable current if forward biased but, under normal operating conditions, this junction is kept reverse biased and so only **leakage currents** are present. These are normally of the order of a nanoamp and can normally be neglected. Therefore, the gate terminal is effectively insulated from the remainder of the device in both MOSFETs and JFETs.

Because the output of a FET is in the form of a current that is determined by the gate voltage, it is common to represent the output of such devices by a Norton equivalent circuit rather than a Thévenin form (though both are possible). We therefore represent the output by a current generator in parallel with a resistance.

Figure 6.17 shows an equivalent circuit for a FET. The gate input is shown open circuit (that is, not connected to anything). The output is represented by a current source, the magnitude of which is some (as yet unknown) function of the input voltage V_{GS}, in parallel with an output resistor R_o. In order for this equivalent circuit to be useful we need to be able to define the relationship between the current produced by the current generator and the input voltage, as well as the value of the output resistance.

We looked at the relationship between drain current and gate voltage in both MOSFETs and JFETs in Figures 6.13 and 6.16, which, for simplicity, are duplicated in Figure 6.18.

When we looked at an equivalent circuit for a 'black box' amplifier in Section 3.5, we represented the output by a voltage source with a magnitude that was simply the gain of the device times the input voltage. This approach relies on the fact that the amplifier may be reasonably represented as a linear system, which implies that the output is directly proportional to the input signal.

In FETs there is not a linear relationship between I_D and V_{GS}, and in Figure 6.17, $f(V_{GS})$

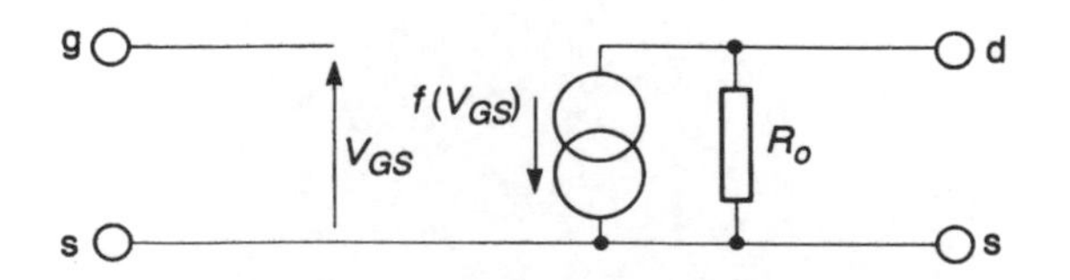

Figure 6.17 An equivalent circuit for a FET.

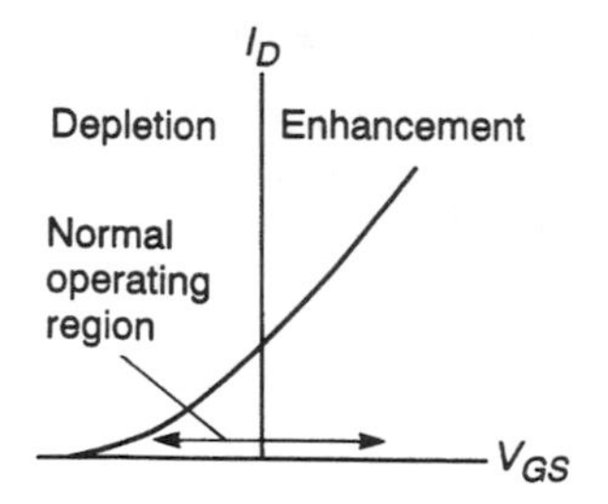

(a) An n-channel DE MOSFET

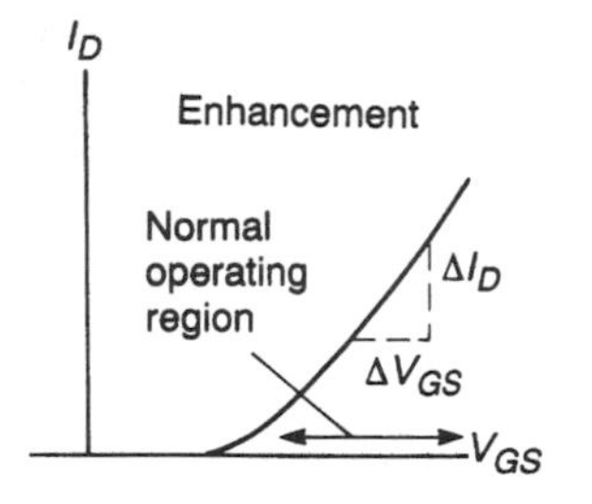

(b) An n-channel enhancement MOSFET

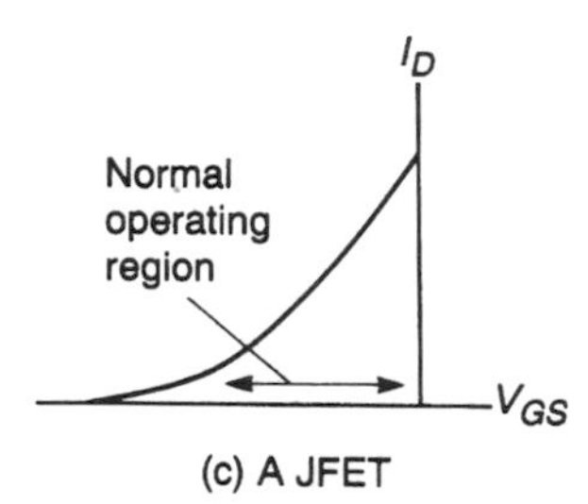

(c) A JFET

Figure 6.18 Transfer characteristics of n-channel FETs.

clearly represents a non-linear function. We cannot, therefore, represent the output by a voltage or current generator with a magnitude that is a simple multiple of the input voltage. However, it is reasonable to represent the output by a linear function over a small range of values of V_{GS}.

The normal method of representing this situation is to use a model which looks not at the absolute values of the input and output but at the effects of small changes in the input. This model, which is described as a **small signal** model of the system, allows us to produce an equivalent circuit for the device which can be used to describe its behaviour for small changes in the input. This is a **small signal equivalent circuit** of a FET; such a circuit is shown in Figure 6.19.

In Figure 6.19 the value of g_m represents the relationship between small changes in the input voltage ΔV_{GS} and the resultant small changes in the drain current ΔI_D. This relationship corresponds to the gradient of the graphs given in Figure 6.18 within the operating region. Thus g_m is given by the ratio $\Delta I_D/\Delta V_{GS}$, as illustrated in Figure 6.18(b). The units of this ratio are current divided by voltage, which corresponds to those of

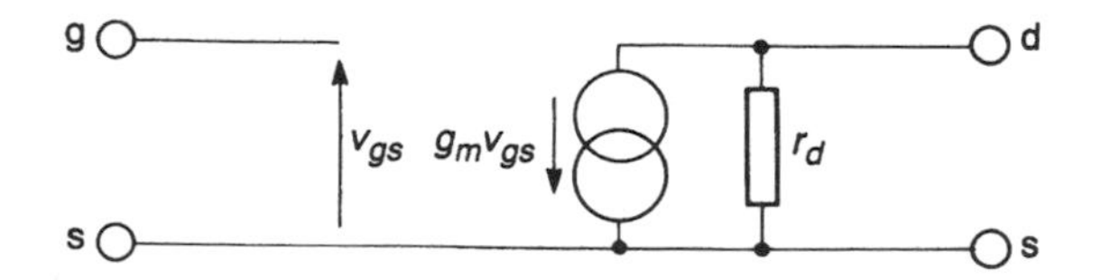

Figure 6.19 A small signal equivalent circuit for a FET.

conductance. The term therefore takes the name **transconductance** to reflect the fact that it represents the effects of the input on the output.

Note that g_m is $\frac{\Delta I_D}{\Delta V_{GS}}$ **not** $\frac{I_D}{V_{GS}}$. g_m has the units of Siemens (S) which is equivalent to Ω^{-1}.

Clearly, in the limit g_m is given by

$$g_m = \frac{dI_D}{dV_{GS}}$$

From Equation 6.2 we know that for a JFET

$$I_D = I_{DSS}\left(1 - \frac{V_{GS}}{V_p}\right)^2$$

and we can therefore find g_m by differentiation. This gives

$$g_m = -\frac{2I_{DSS}}{V_p}\left(1 - \frac{V_{GS}}{V_p}\right) \tag{6.3}$$

$$= -2\frac{\sqrt{I_{DSS}}}{V_p} \times \sqrt{I_D}$$

Thus in a JFET, g_m is proportional to the square root of the drain current. A similar analysis can be performed to obtain a similar result for the MOSFET.

You will notice that in Figure 6.19 the various voltages and component values have been given in lower case letters rather than in upper case as used earlier. For example, v_{gs} rather than V_{GS}. This is done to emphasize the fact that these are small signal values and are only valid for small input changes. In the figure, r_d represents the **drain resistance**, that is, the small signal resistance from drain to source. The presence of r_d means that the drain-to-source voltage increases with drain current and it gives rise to the slope of the lines in the saturation region of the output characteristic of the device, as shown in Figures 6.12 and 6.15. The value of r_d is given by the slope of these lines and for this reason it is sometimes called the output **slope resistance** of the device.

The small signal equivalent circuit is a very useful model to represent the behaviour of a device in response to small changes in its input signal. It must, however, be used in conjunction with data on the DC characteristics of the device, that is, the behaviour of the device in response to steady, DC voltages. As we have seen, the DC characteristics of MOSFETs and JFETs are not the same since they require different bias voltages to place them in their normal operating regions. However, their small signal characteristics and small signal equivalent circuits are similar. The design of circuits using FETs must take both of these considerations into account.

6.4.4 FETs at high frequencies

In Figure 6.19 we looked at a small signal equivalent circuit for a FET. The circuit given at that time is sufficient for most purposes but does not adequately describe the behaviour of these devices at high frequencies.

The MOSFET consists of two conducting regions, the gate and the channel, separated by an insulator. This construction forms a capacitor with the insulating layer forming the dielectric. In the JFET the insulator is replaced by a depletion layer which has the same effect. In both cases capacitance is present between the gate and the channel, and since the channel is joined to the drain at one end and to the source at the other, capacitance is present between the gate and each of the other terminals. The reverse biased junction between the channel and the substrate also acts as a capacitor, the depletion layer separating the two conducting regions. This produces capacitance between the drain and the substrate and thus between the drain and the source (since the latter is normally joined to the substrate). There is, therefore, capacitance between each pair of terminals of the device.

At low frequencies the effects of these capacitances are small and they can normally be neglected (as in Figure 6.19). However, at high frequencies their effects become more significant and it is necessary to include them in the small signal equivalent circuit as shown in Figure 6.20(a). Each of the capacitors shown in this figure has a magnitude of the order of 1 pF.

The presence of C_{gd} makes analysis of this circuit much more complicated. Fortunately it is possible to represent the effects of this capacitance by increasing the magnitude of the capacitance between the gate and the source. In fact the capacitance required between the gate and the source to give the same effect as C_{gd} is $(A+1)C_{gd}$ where A is the voltage gain between the drain and the gate. This apparent increase in capacitance is brought about by a phenomenon known as the **Miller effect**. Thus although C_{gd} and C_{gs} are of the same order of magnitude, it is C_{gd} which tends to dominate the high frequency performance of the device.

It is therefore possible to represent the FET by the equivalent circuit shown in Figure 6.20(b) where the effects of both C_{gs} and C_{gd} are combined into a single capacitance C_T

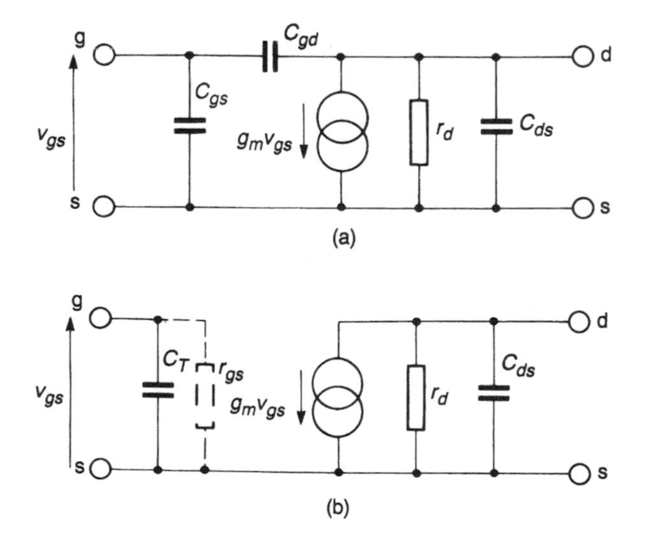

Figure 6.20 Small signal equivalent circuits for a FET at high frequencies.

which represents the total input capacitance. From the discussion of Section 3.7.2 it is clear that the presence of this capacitance will produce a fall in gain at high frequencies and will give rise to a breakpoint, at a frequency determined by the value of the capacitance and the impedance of the input to ground. This impedance will almost always be dominated by the source resistance. However, in some cases it may be appropriate to include in the equivalent circuit a resistance r_{gs} representing the **small signal gate resistance** of the device.

The effects of capacitance greatly reduce the performance of FETs at high frequencies. The presence of capacitance across the input reduces the input impedance from several hundreds of megohms at low frequencies, to perhaps a few tens of kilohms at frequencies of the order of 100 MHz. There is also a reduction in g_m at high frequencies.

6.5 FET amplifiers

FETs are widely used in applications that require low noise and a high input resistance. Both n-channel and p-channel types are used but, as before, for simplicity we will concentrate on circuits using n-channel devices.

The design of amplifiers based on FETs must satisfy both DC and small signal conditions. In Figures 6.5 and 6.10 we looked at simple amplifiers based on MOSFETs and JFETs. In essence these circuits are simply a transistor, a load resistor and a biasing arrangement. The differences between these circuits arise from the differing requirements of each type of transistor in respect of its **biasing**.

All types of FET can be accommodated by the circuit given in Figure 6.21 by choosing suitable values for the gate supply voltage V_{GG}. When using n-channel devices this voltage would be positive for enhancement MOSFETs, negative for JFETs and normally zero for DE MOSFETs. For p-channel devices these polarities would be reversed.

In the circuit of Figure 6.21, the input signal is applied between the gate and the source of the FET, and the output is measured between the drain and the source. The source is

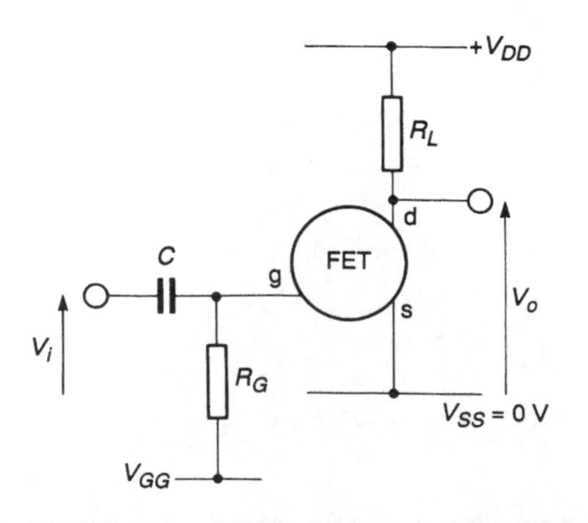

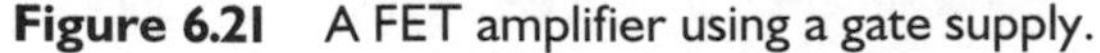

Figure 6.21 A FET amplifier using a gate supply.

therefore common to both the input circuit and the output circuit. For this reason, amplifiers of this general form are called **common-source amplifiers**. The circuits of Figures 6.5 and 6.10 are also common-source amplifiers.

Although the arrangement of Figure 6.21 is feasible, it is often inconvenient to use a separate gate supply and in general this bias voltage is obtained from the other supply rails. For the DE MOSFET the bias voltage is normally zero volts which can be obtained simply by connecting R_G to ground, as in Figure 6.5(a). The situation is slightly more complicated with the enhancement MOSFET since the required bias voltage is non-zero. However, since the required bias voltage is between the drain supply voltage V_{DD} and the source supply voltage V_{SS}, the bias voltage may be easily obtained using a pair of resistors in a potential divider, as shown in Figure 6.5(b). For the JFET the bias voltage lies outside the range of the drain and source supply rails. In this case it is normal to use a source resistor, as shown in Figure 6.10(b). Source current flowing through this resistor produces a voltage drop which causes the source voltage to be above V_{SS}. If now the gate resistor is connected to V_{SS}, the gate will be biased correctly with respect to the source. This technique is called **automatic bias**.

6.5.1 Equivalent circuit of a FET amplifier

We saw in Section 3.5 that it is often useful to represent an amplifier by an equivalent circuit. In Figures 6.17 and 6.19 we looked at equivalent circuits for the FET itself and it is quite simple to extend these to encompass the complete amplifier circuit. As discussed earlier, the non-linear current generator of Figure 6.17 makes this representation of the overall characteristic of the device of only academic interest. Of more use is the circuit of Figure 6.19 which, although limited to only the small signal behaviour of the device, incorporates a current generator which is linearly related to the input signal. A **small signal equivalent circuit** of the amplifier of Figure 6.21 is shown in Figure 6.22.

It is important to remember that since the supply lines are constant voltages, they do not fluctuate with respect to each other and thus *there is no small signal voltage between the supply lines*. Therefore, as far as small signal (AC) signals are concerned, the rails V_{DD}, V_{SS} and V_{GG} may be considered to be joined together. If this concept seems strange, remember that it is normal to place a large **reservoir capacitor** across the output of the power supply (see Section 5.9) to reduce ripple. At all but the lowest frequencies this capacitor resembles a short circuit between the supply rails.

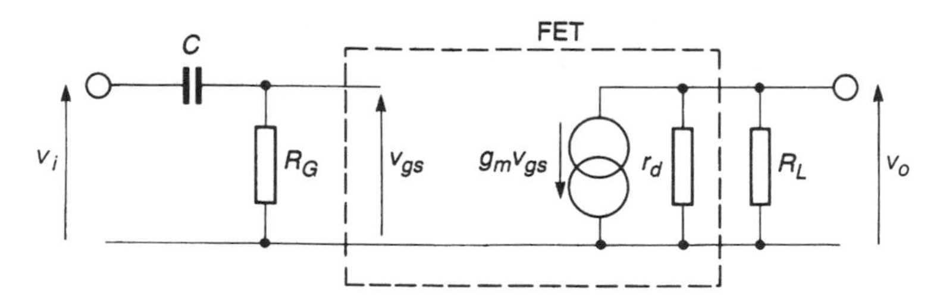

Figure 6.22 A small signal equivalent circuit of a FET amplifier.

Since V_{DD} and V_{SS} are effectively joined as far as AC signals are concerned, the drain resistance r_d and the load resistance R_L appear in parallel. Similarly, since V_{GG} and V_{SS} are effectively joined for small signals, the gate resistor R_G appears from the gate to ground, the common reference point. If a single resistor is used to generate the gate bias this resistor simply appears as R_G, as shown. If the gate is in fact biased by two resistors forming a potential divider, both resistors will appear in parallel. It should be noted that any biasing arrangement consisting of a combination of resistors and voltage sources can be modelled by its **Thévenin equivalent circuit**, as described in Section 3.3. This allows the arrangement to be replaced by a single resistor and a voltage source, as shown in Figure 6.21, which can then be represented by a single resistor in the small signal equivalent circuit since constant voltages are not shown. Normally C would be chosen to have a negligible effect at the frequencies of interest. If this is the case, its effects may be ignored.

It should be remembered that the small signal model of Figure 6.22 is appropriate for all types of FET. The differences between the different types of device affect their biasing arrangements rather than their small signal behaviour.

6.5.2 Small signal voltage gain

Having derived a small signal equivalent circuit for the amplifier we are now in a position to determine its small signal voltage gain.

From Figure 6.22 it is clear that if we ignore the effects of the input capacitor C, the voltage on the gate of the FET is the same as that at the input v_i. The output voltage is determined by the current generator and the effective resistance of the parallel combination of the small signal drain resistance r_d and the load resistance R_L. We often use a shorthand notation for 'the parallel combination of', which is simply two parallel lines. Thus 'the parallel combination of R_1 and R_2' would be written as $R_1//R_2$. Therefore the output voltage is given by

$$\begin{aligned} v_o &= -g_m v_{gs}(r_d//R_L) \\ &= -g_m v_i(r_d//R_L) \end{aligned}$$

and thus

$$\frac{v_o}{v_i} = -g_m(r_d//R_L)$$

The minus sign in the expression for the output voltage reflects the fact that the output voltage falls as the output current increases, hence the output voltage is inverted with respect to the input. We therefore have an inverting amplifier.

The voltage gain is thus given simply by the product of the transconductance of the FET g_m, and the effective resistance of the parallel combination of r_d and R_L. This can be expanded to give

$$\text{Voltage gain} = \frac{v_o}{v_i} = -g_m\frac{r_d R_L}{r_d + R_L} \qquad (6.4)$$

It is also straightforward to calculate the **small signal input resistance** and the **small signal output resistance** of the amplifier from the equivalent circuit. The input resistance is simply equal to the gate resistance R_G. Because of the very high input resistance of the FET, the gate resistance can normally be chosen to be as high as necessary to suit a particular application. The output resistance is given by the parallel combination of r_d and R_L. The input and output resistances calculated from the small signal equivalent circuit are *small signal* resistances. That is, they are the relationship between small signal voltages and small signal currents. They do not relate to the DC voltages and currents within the circuit.

Example 6.1 Determination of the small signal voltage gain, input resistance and output resistance of a FET amplifier

Calculate the small signal voltage gain, input resistance and output resistance of the following amplifier, given that $r_d = 100\ \text{k}\Omega$ and $g_m = 2\ \text{mS}$.

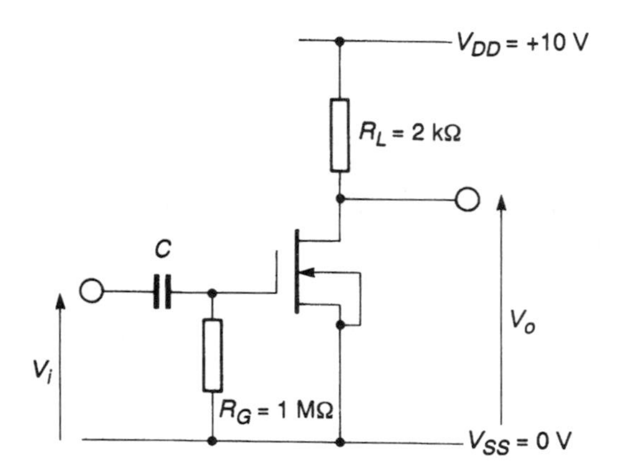

The first step in this problem is to determine the small signal equivalent circuit of the amplifier.

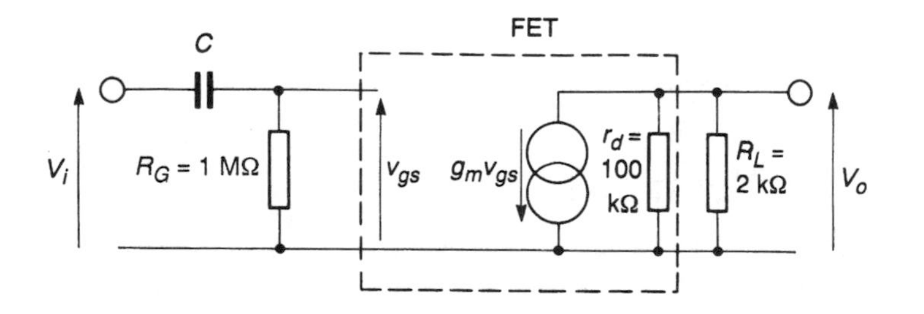

Clearly from the equivalent circuit,

$$\frac{v_o}{v_i} = -g_m(r_d//R_L)$$

$$= -g_m \frac{r_d R_L}{r_d + R_L}$$

$$= -2 \times 10^{-3} \frac{100 \times 10^3 \times 2 \times 10^3}{100 \times 10^3 + 2 \times 10^3}$$

$$= -3.9$$

The minus sign indicates that the amplifier is an inverting amplifier. The small signal input resistance of the amplifier is simply R_G, and thus

$$r_i = R_G$$
$$= 1 \text{ M}\Omega$$

The small signal output resistance is given by

$$r_o = r_d // R_L$$
$$= \frac{r_d R_L}{r_d + R_L}$$
$$= \frac{100 \times 10^3 \times 2 \times 10^3}{100 \times 10^3 + 2 \times 10^3}$$
$$\approx 2.0 \text{ k}\Omega$$

This example considers a circuit containing an *n*-channel DE MOSFET; a similar calculation could be performed for a circuit incorporating another type of FET.

A typical value for the small signal drain resistance r_d would be in the range 50 to 100 kΩ; this will generally be much larger than the load resistance R_L. Under these circumstances the effects of the r_d may usually be neglected, and the gain may be approximated by the expression

$$\frac{v_o}{v_i} \approx -g_m R_L$$

Clearly, by changing the values of R_L we change the small signal voltage gain of the amplifier, but we must remember that this will also affect the DC current, that is, the steady current I_D which flows even in the absence of any input, which in turn affects the value of g_m. We must therefore turn our attention to the DC aspects of the amplifier.

6.5.3 Biasing considerations

The **biasing** arrangement of an amplifier determines the operation of the circuit in the absence of any input signal. This is said to be the **quiescent** state of the circuit. Of central importance in the amplifiers being considered is the **quiescent drain current**, which in turn affects the **quiescent output voltage**.

Returning to the circuit of Figure 6.21, it is clear that the quiescent drain current is affected by the value of the drain resistor R_L and by the voltage–current characteristics of the FET. If the relationship between the drain current and the drain voltage in the FET

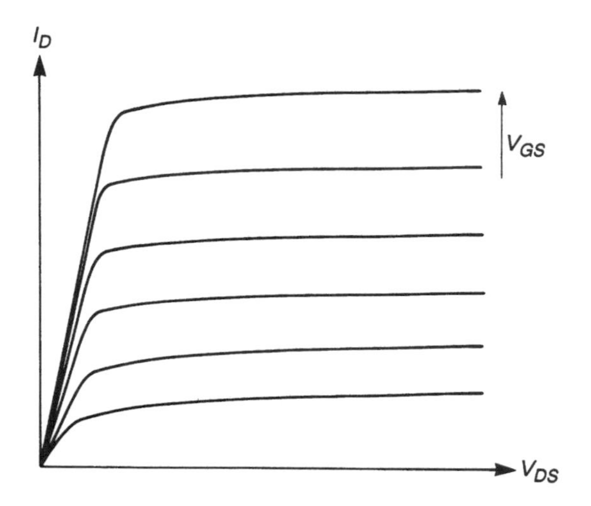

Figure 6.23 Generalized FET output characteristic.

were linear, as in a resistor, it would be simple to calculate the value of the resistor required to give an appropriate current and to arrange the ratio of the resistance of the FET to that of R_L to give an appropriate quiescent output voltage. However, we know from Figures 6.12 and 6.15 that the relationship between drain current and drain voltage is not linear. Indeed, in the section of the characteristic in which we wish to operate (the saturation region) the drain current is largely independent of the drain voltage. This makes determination of the quiescent conditions somewhat more complicated.

One solution to this problem is to use a graphical technique known as a **load line**. It is clear that although the currents flowing through the load resistor and the FET are not easily determined, the voltages across these two devices must sum to the voltage between the supply rails ($V_{DD} - V_{SS}$). The voltage across the FET is determined by its characteristics and by the bias voltage V_{GS}. From Figures 6.12 and 6.15 it can be seen that the basic form of the FET characteristics is the same for all forms of device, the differences being largely the voltages that are applied to the gate V_{GS}. Figure 6.23 shows a generalized FET characteristic which could apply to any *n*-channel FET. This shows the voltage across the device on the horizontal axis against current on the vertical axis for different values of V_{GS}.

When a current flows through the FET it also flows through the load resistor producing a voltage drop across it. The voltage on the drain of the FET is simply the supply voltage V_{DD} minus the voltage drop across the resistor, which is $I_D R_L$. Figure 6.24 shows the voltage on the drain of the FET for different values of drain current. When the drain current is zero, there is no voltage drop across the resistor and the drain voltage is simply the supply voltage V_{DD}. As I_D increases, V_{DS} decreases, the slope of the line being the inverse of the load resistance R_L.

Figures 6.23 and 6.24 both represent the relationship between the drain current and the drain voltage in our simple amplifier. Clearly the actual operating condition must satisfy both these relationships. To determine this condition we simply plot both characteristics on a single graph, as shown in Figure 6.25.

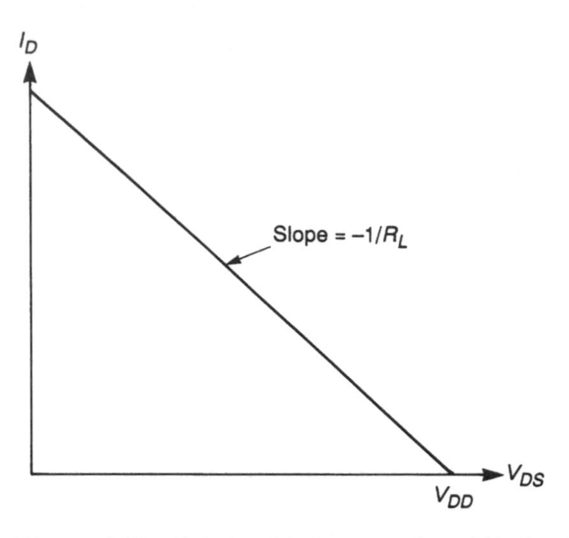

Figure 6.24 Relationship between I_D and V_{DS} for the simple FET amplifier.

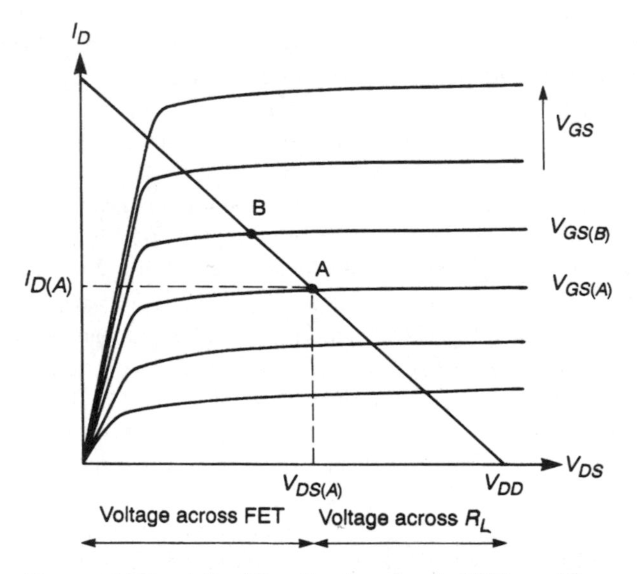

Figure 6.25 A load line for the simple FET amplifier.

The straight line in this graph is called a **load line** since it indicates the effect of the load resistance in reducing the drain voltage. The intersection of this line with one of the output characteristic lines represents a point at which both relationships are satisfied. Consider, for example, point A on this line. The graph indicates that if V_{GS} is set to $V_{GS(A)}$, the drain current will be $I_{D(A)}$ and the drain voltage (which is also the output voltage of the amplifier) will be $V_{DS(A)}$. It may help to visualize the significance of the load line to note that since the voltage across the FET plus the voltage across R_L must equal the supply voltage V_{DD}, the distance from zero to $V_{DS(A)}$ represents the voltage across the FET, and the distance from this point to V_{DD} represents the voltage across the load resistor. If now the gate voltage is increased to $V_{GS(B)}$, the drain current will increase and the drain voltage decrease, as indicated by point B on the characteristic. Thus the load line shows how the drain current and drain voltage vary for different values of the gate voltage. It should be remembered that the various lines of the output characteristic are only a representative few of the infinite number of lines that could be plotted. For simplicity we plot a small number of lines and estimate the behaviour of the device between them.

The graph of Figure 6.25 shows the characteristics of an amplifier with a given value of R_L. If this value were changed the slope of the load line would change thereby affecting the characteristics of the amplifier. In practice, the designer is normally faced with the problem of selecting a value for R_L for optimum performance. In doing this he defines an **operating point** which corresponds to the position on the characteristic under quiescent conditions. This point is also known as the **working point**. The designer therefore starts with the output characteristics of the FET to be used, but without knowing the value of the load resistor. In order to determine the value of this resistor the designer must select the ideal operating point for the system. If we assume that the designer chooses the point corresponding to position A of Figure 6.25, a line would then be drawn through this point to the V_{DD} position on the horizontal axis and this would form the load line. The value of R_L required can then be found by measuring the slope of this line. When the operating point is known the required gate voltage V_{GS} will be known, and the necessary biasing circuitry can be designed, as discussed earlier. The operating point determines the quiescent state of the circuit and thus defines the quiescent drain current and output voltage. When a small signal input is applied to the circuit the variations in the gate voltage cause the circuit to move along the load line on either side of the operating point. If the input signal is sufficiently large, this could cause the circuit to enter the ohmic region or to *limit* as the output reaches the supply rail. Either of these conditions will **distort** the output signal. If a large output swing is required it is important that the operating point is chosen appropriately.

6.5.4 Choice of operating point

In Figures 6.12 and 6.15 we divided the output characteristic of FETs into two regions, the ohmic region and the saturation region, and noted that for normal operation the ohmic region is avoided. In fact other areas of the characteristic are normally avoided when using such a device as a linear amplifier. Figure 6.26 shows these regions.

Region A is the ohmic region discussed earlier. It is not used because the drain current is heavily dependent on the drain voltage in this area. When constructing a linear amplifier we wish the drain current to be controlled by the input signal and not by the voltage across the device. Region B may be caused by one of two factors, depending on the type of FET being used. For all devices there is a maximum allowable drain current before the device is damaged and the designer must ensure that the device is not taken into this region. For JFETs there is also a limit imposed by the fact that the gate voltage V_{GS} must not forward bias the gate junction. One or other of these constraints limits the maximum drain current, or gate voltage, which may be used. Constraints are also imposed by the

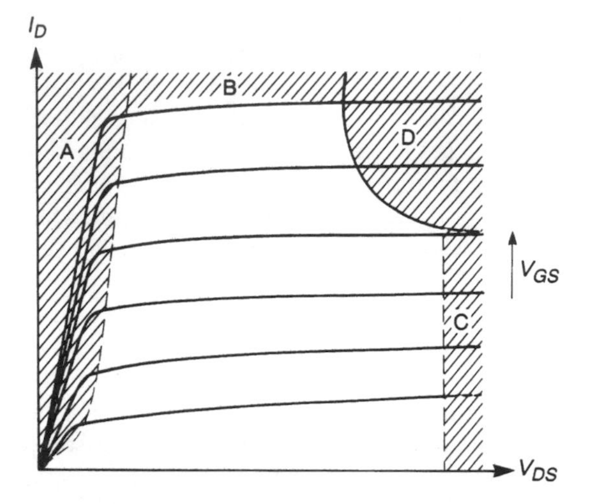

Figure 6.26 The forbidden operating regions of a FET.

breakdown voltage of the device, as shown by region C. If this voltage is exceeded, permanent damage to the device may result. Finally, a fourth prohibited region is imposed by **power dissipation** considerations. The power dissipated by the FET is given by the product of the drain current and the drain voltage (since the gate current is negligible) and results in the generation of heat. This heat causes the temperature of the device to rise and operation is limited by the allowable temperature of the junction. The region of operation which satisfies the power dissipation conditions is bounded by a hyperbola (the locus of the point where current times voltage equals a constant), as shown by region D. These four regions define an allowable area in which the device must operate.

In selecting the operating point for an amplifier, the designer must ensure that the transistor is kept within safe limits and within its normal operating region. This will normally require that the supply voltage is less than the breakdown voltage of the device and that the maximum current and power limits are not infringed. For maximum voltage swing the operating point is normally placed approximately half-way between the supply voltage and the lower end of the saturation region, as shown in Figure 6.27. This allows for maximum transition of the input before the signal distorts.

Once the operating point has been selected the load line is drawn from the supply voltage point on the horizontal axis, through the operating point. The value of V_{GS} corresponding to the operating point determines the biasing circuit that must be used on the gate, and the slope of the load line determines the required value for the load resistor. The position of the operating point shows the **quiescent output voltage** and the **quiescent drain current** for the circuit. A small signal input causes the circuit to move along the load line on either side of the operating point, as illustrated in Figure 6.27, and results in changes to the drain current and the output voltage. By comparing the magnitudes of a small signal input voltage and the resultant small signal output voltage, it is possible to deduce the small signal voltage gain of the arrangement from this diagram.

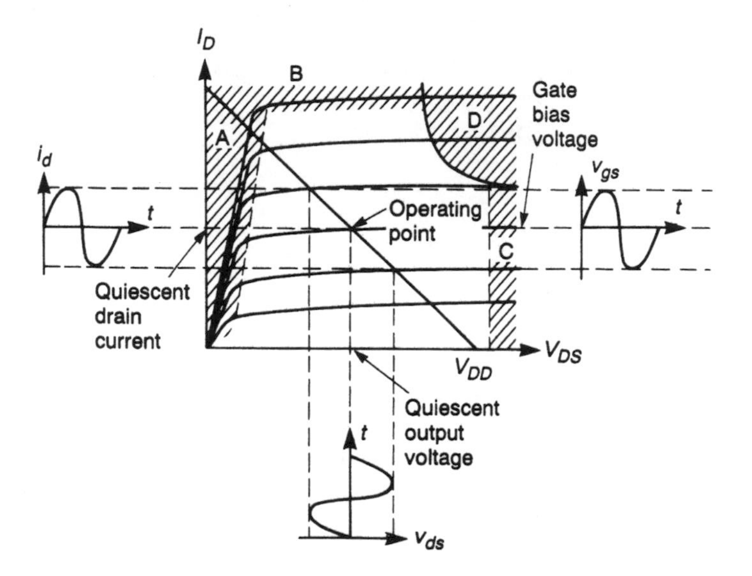

Figure 6.27 Choice of operating point in a FET amplifier.

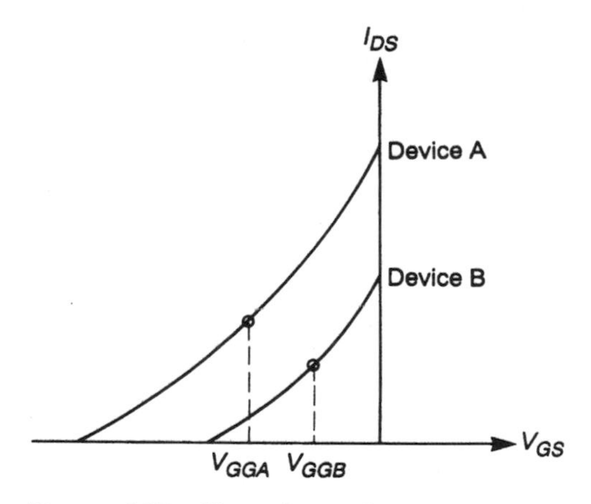

Figure 6.28 Typical transfer characteristics of JFETs.

6.5.5 Device variability

One problem associated with the use of FETs, in common with the use of most active devices, is the large spread in the characteristics of devices that are supposedly of the same type. This variability is illustrated for a JFET in Figure 6.28, which shows the spread of characteristics that might be obtained from devices from a single batch of components of the same type.

When choosing component values for a circuit using device A of Figure 6.28, the

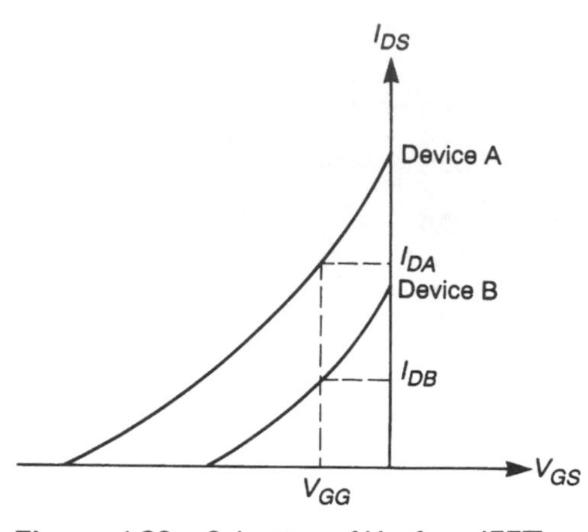

Figure 6.29 Selection of V_{GG} for a JFET amplifier.

designer might choose to set the gate bias voltage to V_{GGA} to obtain the maximum output range. Similarly, when using device B, V_{GGB} might be chosen. However, when choosing component values the designer will not know the characteristics of the device but only that it should lie within the range indicated by these two devices. Clearly, if the gate bias voltage is set to V_{GGA} and the component used has a characteristic similar to device B, the usable range of the amplifier will be very limited. The designer is therefore forced to assume the extreme case and to choose the gate voltage to be V_{GGB} in all cases, as shown in Figure 6.29.

One effect of this is that the **quiescent drain current** $I_{D(quies)}$ is ill defined. In Figure 6.29 it may be anywhere between I_{DA} and I_{DB}.

The variability of the quiescent drain current may be reduced by the use of **negative feedback**. In fact, the **automatic bias** network described earlier, and shown in Figure 6.10, uses negative feedback to stabilize the drain current. The operation of this arrangement is illustrated in Figure 6.30.

As the current flowing through the FET increases, the voltage drop across the source resistor also increases. Since this voltage drop is determining the gate bias voltage, it has the effect of making the gate-to-source voltage more negative as the current increases, which tends to reduce the current. This therefore constitutes *negative* feedback. Using suitable component values, the circuit of Figure 6.30 would operate at point B using a device with characteristics corresponding to the lower line, and at point A with a device corresponding to the upper line. Devices with intermediate characteristics would operate at some point along the line from A to B. If a fixed bias voltage V_{GG} were used, as described earlier, the operating point of such a circuit would move from point B to point C across the spectrum of devices. Thus the variation of I_D is much less with automatic biasing than with a fixed bias voltage. The circuit of Figure 6.30 uses a JFET but this technique may also be used with other forms of FET.

It might seem that an alternative method of solving the problems of device variability would be to **measure** the characteristics of the actual device to be used and to

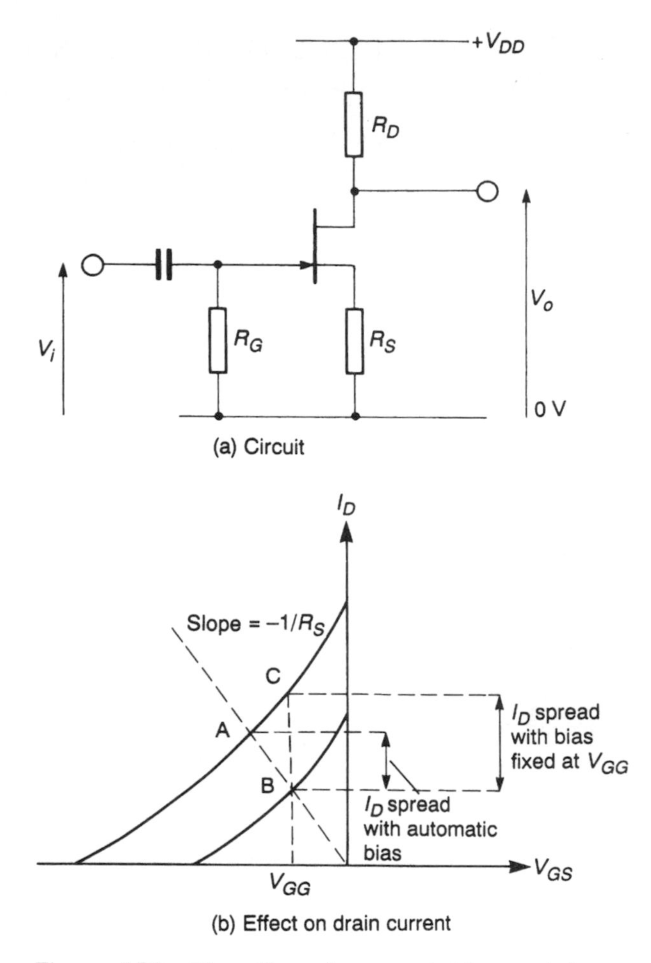

Figure 6.30 The effect of automatic bias on drain current.

tailor the circuit accordingly. This approach has many disadvantages. Firstly, measuring devices is a slow process if performed manually and is therefore expensive. It also implies that component values need to be individually calculated, which may also be time consuming. A second disadvantage is that should the component fail in the field, it would need to be replaced and there is no guarantee that the replacement would exactly match the original. An alternative arrangement is to **select** components that match a particular characteristic. This again implies measuring each device, although this is often performed automatically by the manufacturer who divides his products into distinct ranges of performance. Almost all active components are subject to manufacturing spread, and the more precisely they are characterized into distinct groups, the more expensive they become. In general good designs are tolerant of a broad spread of active component characteristics, rather than needing more expensive selected components.

Example 6.2 Design of the biasing network of a FET amplifier using device transfer characteristics

A 2N5486 *n*-channel JFET has $V_P = -6$ V and $I_{DSS} = 8$ mA. Design a biasing arrangement such that, when used with a supply voltage of 15 V and a load resistance of 2.5 kΩ, the amplifier has a quiescent output voltage of 10 V.

A suitable circuit is given below.

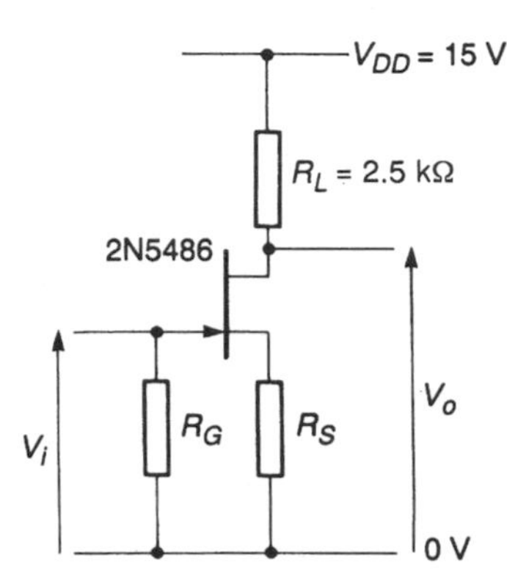

From Equation 6.2 we know that

$$I_D = I_{DSS}\left(1 - \frac{V_{GS}}{V_P}\right)^2$$

which, using the figures given for V_P and I_{DSS} may be plotted to give

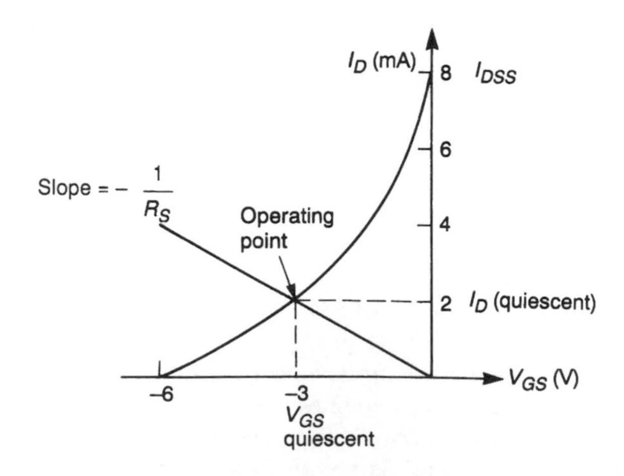

The quiescent output voltage $V_{o(quiescent)}$ is given by

$$V_{o(quiescent)} = V_{DD} - V_L$$

where V_L is the voltage drop across the load resistor R_L. Therefore the required value of V_L is given by

$$V_L = V_{DD} - V_{o(quiescent)} = 15 - 10 = 5 \text{ V}$$

and the required quiescent drain current $I_{D(quiescent)}$ is

$$I_{D(quiescent)} = \frac{V_{o(quiescent)}}{R_L} = \frac{5\text{V}}{2.5 \text{ k}\Omega} = 2 \text{ mA}$$

From the transfer characteristic, this value of drain current corresponds to a gate-to-source voltage of −3 V. Since the gate is at ground potential this gate-to-source voltage must be obtained by a voltage drop across R_S of +3 V. Thus the value of R_S is given by

$$R_S = \frac{V_{GS}}{I_D} = \frac{3 \text{ V}}{2 \text{ mA}} = 1.5 \text{ k}\Omega$$

The value of R_G is not critical as it is simply required to bias the gate to zero volts. It would normally be chosen to give a high input resistance, but must not be so high that the voltage drop caused by the effects of the gate current (a few nanoamps) become significant. A value of 470 kΩ would be suitable.

Example 6.3 Design of the biasing network of a FET amplifier by direct calculation

The design of Example 6.2 can also be performed numerically rather than using graphs.

As before

$$I_{D(quiescent)} = \frac{V_{o(quiescent)}}{R_L} = \frac{5 \text{ V}}{2.5 \text{ k}\Omega} = 2 \text{ mA}$$

From Equation 6.2

$$I_D = I_{DSS}\left(1 - \frac{V_{GS}}{V_p}\right)^2$$

or by rearranging

$$\begin{aligned} V_{GS} &= V_P\left(1 - \sqrt{\frac{I_D}{I_{DSS}}}\right) \\ &= -6\left(1 - \sqrt{\frac{2}{8}}\right) \\ &= -3 \text{ V} \end{aligned}$$

as before. Hence R_S is 1.5 kΩ as above.

6.5.6 Source followers

So far we have considered a series of *common-source* amplifier circuits. Some other widely used FET amplifier configurations are shown in Figure 6.31. In these circuits the drain terminal is common to the input and the output circuit (remember that V_{DD} is effectively joined to ground for small signals). Consequently they are called **common-drain amplifiers**.

From the definition of g_m we know that

$$g_m = \frac{i_d}{v_{gs}}$$

therefore

$$i_d = g_m v_{gs} = g_m(v_g - v_s)$$

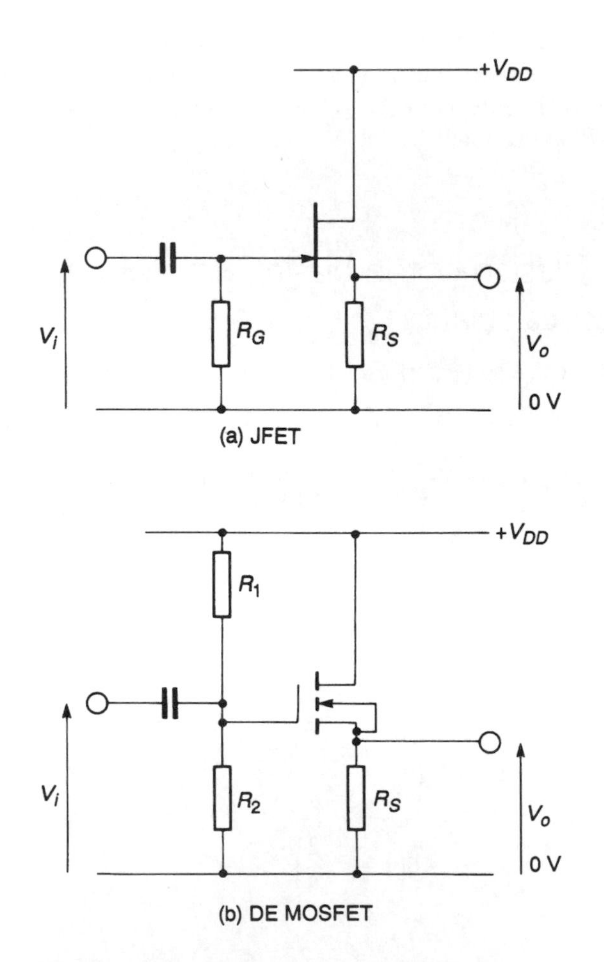

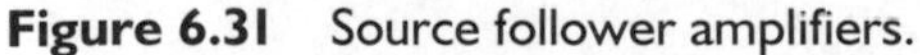
Figure 6.31 Source follower amplifiers.

Since the source voltage v_s is given by

$$v_s = R_S i_d$$

then

$$v_s = \frac{R_S g_m}{1 + R_S g_m} v_g = \frac{1}{\dfrac{1}{R_S g_m} + 1} v_g$$

If $1/R_S g_m \ll 1$, then $v_s \approx v_g$. In other words the source voltage (the output) tends to follow the gate voltage (the input). For this reason these circuits are often called **source followers**; as the output follows the input, circuits of this type are **non-inverting amplifiers**.

Since the output of the source follower is very nearly the same as the input, the gain of the amplifier v_s/v_g is approximately unity. In most cases these circuits are used because of their very high input resistance and comparatively low output resistance. The input resistance is defined by the gate resistance R_G and the output resistance by the characteristics of the FET.

Source follower output resistance

To determine the output resistance of the circuit, we wish to know how the output voltage v_s changes with the output current i_s, in the absence of any change at the input. Thus the output resistance r_o is v_s/i_s with $v_g = 0$

As above

$$i_d = g_m v_{gs} = g_m(v_g - v_s)$$

and substituting $v_g = 0$ gives

$$i_d = g_m v_{gs} = g_m(0 - v_s)$$

Since gate currents are negligible, the magnitude of the source current is equal to that of the drain current. However, currents are conventionally considered to flow *into* the device and thus $i_s = -i_d$. Therefore

$$i_s = -i_d = g_m v_s$$

and

$$r_o = \frac{v_s}{i_s} = \frac{1}{g_m}$$

Since g_m varies with current, so will the output resistance, but it is typically a few hundred ohms for currents of a few milliamps.

Source followers do not have as low an output resistance as similar circuits constructed using bipolar transistors (these will be discussed in the next chapter) but their very high input resistance makes them extremely useful as **unity gain buffer amplifiers**.

6.5.7 Differential amplifiers

In Section 3.10 we discussed the use of differential amplifiers which produce an output proportional to the *difference* between two input signals and which ignore signals that are common to both inputs, the latter property being known as **common-mode rejection**.

A common form of differential amplifier is the **long-tailed pair** which is often used in the input stage of operational amplifiers. Such a circuit is shown in Figure 6.32.

Two FET amplifiers share a common source resistor R_S, and have equal values for their gate and drain resistances. The transistors are chosen to have matched characteristics so that the circuit is as symmetrical as possible. The circuit has two inputs v_1 and v_2, and two output signals v_3 and v_4. A small signal equivalent circuit for this arrangement is shown in Figure 6.33.

The input and output voltages are measured to the common reference point (ground). The gate resistors are normally chosen to be sufficiently high in value to have little effect on the operation of the circuit other than to set the correct DC bias conditions for the FETs.

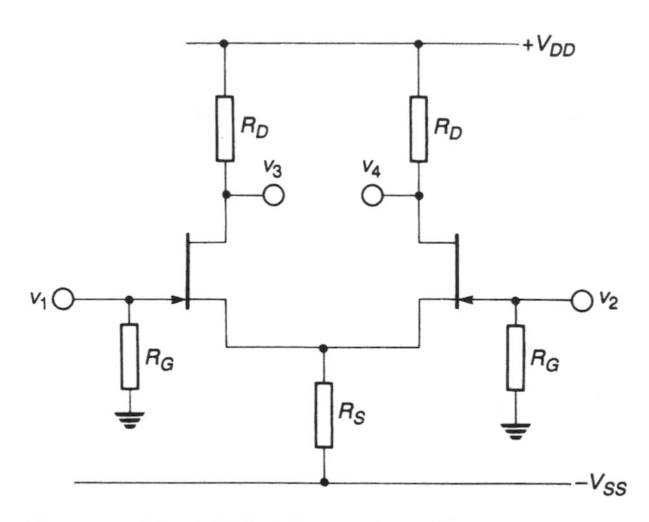

Figure 6.32 A FET differential amplifier.

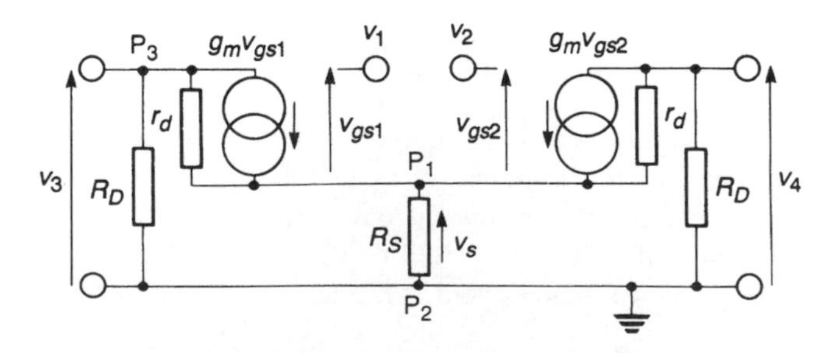

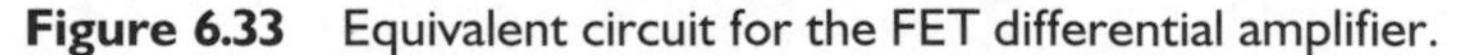

Figure 6.33 Equivalent circuit for the FET differential amplifier.

They are therefore omitted from the small signal equivalent circuit. We will assume that the devices are perfectly matched so that the transconductance g_m and the drain resistance r_d of both are equal.

Since the input voltages v_1 and v_2 are measured with respect to ground, the voltages applied across the gate-to-source junction of each device are simply

$$v_{gs1} = v_1 - v_s$$

and

$$v_{gs2} = v_2 - v_s$$

From **Kirchhoff's Law** we know that the currents flowing into any node of a circuit must sum to zero. We can apply this principle to a number of points in the circuit to produce a series of simultaneous equations.

If we consider point P_1 we see that

$$g_m v_{gs1} + \frac{(v_3 - v_s)}{r_d} + g_m v_{gs2} + \frac{(v_4 - v_s)}{r_d} - \frac{v_s}{R_S} = 0$$

which substituting for v_{gs1} and v_{gs2} gives

$$g_m(v_1 - v_s) + \frac{(v_3 - v_s)}{r_d} + g_m(v_2 - v_s) + \frac{(v_4 - v_s)}{r_d} - \frac{v_s}{R_S} = 0 \qquad (6.5)$$

Applying the same principle to point P_2 yields

$$\frac{v_3}{R_D} + \frac{v_4}{R_D} + \frac{v_S}{R_S} = 0 \qquad (6.6)$$

and to point P_3 gives

$$\frac{v_3}{R_D} + \frac{(v_3 - v_s)}{r_d} + g_m(v_1 - v_s) = 0 \qquad (6.7)$$

From these equations it is possible to deduce an expression for the outputs of the circuit v_3 and v_4 in terms of the inputs, but the analysis is fairly involved. We can simplify the process by making an assumption about the term v_s/R_s in Equation 6.6. This term represents the *small signal* current in the source resistor R_S. That is, the fluctuations in this current as a result of the varying inputs. For reasons that will become apparent later, it is advantageous to assume that this term is very small so that its effects can be neglected. Making this approximation implies that the current through R_S is constant and that it is acting as a **constant current generator**.

If the term v_s/R_s is negligible, Equation 6.6 becomes

$$\frac{v_3}{R_D} + \frac{v_4}{R_D} = 0 \qquad (6.8)$$

which simplifies to give

$$v_3 = -v_4$$

Combining this result with Equations 6.5 and 6.7 we can obtain an expression for the output signals which is

$$v_3 = -v_4 = (v_1 - v_2)\frac{-g_m}{2\left(\frac{1}{r_d}+\frac{1}{R_D}\right)} \tag{6.9}$$

Thus the output signals are equal and opposite and their magnitude is determined by the difference between the two input signals. We therefore have a *differential amplifier*.

The *differential* output voltage of this arrangement v_o is given by $v_3 - v_4$ and since v_3 and v_4 are equal and opposite, the differential voltage gain of the circuit is simply

$$\text{Differential voltage gain} = \frac{v_o}{v_i} = \frac{v_3 - v_4}{v_1 - v_2} = \frac{-g_m}{\left(\frac{1}{r_d}+\frac{1}{R_D}\right)}$$

We noted in Section 6.5.2 that r_d is usually much greater than R_D. If we make this assumption we can simplify the above expression to

$$\text{Differential voltage gain} \approx -g_m R_D$$

which is of a similar form to that derived earlier for a simple FET amplifier.

It should be noted that the input signals are not required to be symmetrical to produce symmetrical outputs. Note also that the magnitude of the output signals is not affected by the actual value of the input signals, but only by the difference between them. Thus common-mode signals are ignored.

FILE 6C

Computer simulation exercise 6.3

Simulate the differential amplifier of Figure 6.32. If you are using PSpice you may use the circuit within the demonstration file. If you do not have access to the demonstration file, or are using another simulation package, you may need to experiment to find appropriate values for the various components depending on the FETs used. Suitable values to start with might be: $V_{DD} = 12$ V; $V_{SS} = -12$ V; $R_D = 2.2$ kΩ; $R_S = 3.3$ kΩ. PSpice provides a model of the 2N3819 JFET. If you are using another package you will need to pick a suitable device from the models available.

Initially apply a 1 kHz sinusoidal input voltage of about 50 mV peak to one input to the amplifier and a constant (DC) voltage to the other. Use transient analysis to look at the form of the outputs and then experiment with different magnitudes and combinations of input signals.

Common-mode rejection ratio of the long-tailed pair amplifier

The common-mode rejection ratio CMRR is the ratio of the gain for difference input signals to that for common-mode input signals. That is

$$\text{CMRR} = \frac{\text{differential-mode gain}}{\text{common-mode gain}}$$

From Equation 6.9 it would seem that the CMRR for the long-tailed pair amplifier is infinite since the output is unaffected by common-mode signals. It should be remembered, however, that this expression was derived by making the simplifying assumption that the current through the source resistor was constant. If the variation of this current is taken into account it can be shown that the common-mode rejection ratio varies directly with the value of R_S such that

$$\text{CMRR} \approx g_m R_S$$

The higher the value of R_S the more closely the circuit represents a constant current source and the higher the CMRR. However, in practice it is not possible to use a very high value for R_S without upsetting the DC conditions of the circuit. For circuits requiring a high CMRR, R_S is normally replaced by a **constant current source** which behaves like a resistor of very high value without affecting the DC operation of the circuit. Using such techniques it is possible to produce amplifiers with CMRRs of more than 100 dB. Such arrangements will be discussed in Section 6.6.

In our analysis we have assumed a perfectly matched pair of transistors. In practice the devices will never be identical which will lead to a reduction in common-mode rejection. When the devices are fabricated within a single integrated circuit, as in an operational amplifier, the manufacturing tolerances will tend to affect both devices equally and a good match will generally be obtained. Because of their close proximity the devices will also generally operate at the same temperature. With circuits constructed using discrete devices the problems are more severe since the devices will generally be less well matched and their operating temperatures will not always be equal. These problems may be reduced by using a matched pair of transistors within a single package. This ensures a close correspondence of device characteristics and a similar operating temperature.

6.6 Other FET applications

6.6.1 A FET as a constant current source

Provided that the drain-to-source voltage is greater than the pinch-off voltage, the drain current of a FET is controlled by the gate-to-source voltage. Therefore a very simple **constant current source** can be formed simply by applying a constant voltage to the gate. For JFETs and DE MOSFETs, the simplest forms of such an arrangement are shown in Figures 6.34(a) and 6.34(b).

In these circuits the gate is simply connected to the source giving a drain current of I_{DSS}. The current produced by these arrangements is determined by the characteristics of the device and is generally in the range 1 mA to 5 mA. Commercially available 'constant current sources' are often simply FETs with their source and gate joined internally to produce two-terminal devices which are then selected for different current ranges.

It is also possible to produce a variable constant current source by using the **automatic bias** technique discussed in Section 6.5. Such an arrangement is shown in Figure 6.34(c). Current through the device produces a voltage drop across the resistor and generates a bias voltage between the gate and the source. The value of this resistor is adjusted to produce the required current.

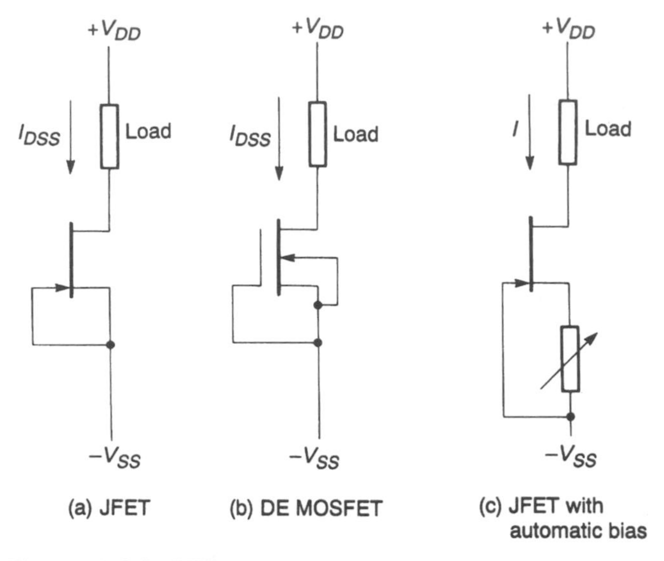

Figure 6.34 FET constant current sources.

FET constant current sources are often used to produce the source current for **long-tailed pair** amplifiers, as described in Section 6.5.7. An example of such an arrangement is shown in Figure 6.35.

FILE 6D

Computer simulation exercise 6.4

Use circuit simulation to investigate the properties of the circuit of Figure 6.34(a).

Use a 15 V supply and a 2N3819 JFET. Look at the current through the load as R_L is varied between 100 Ω and 1 kΩ. What happens if R_L is increased to 2 kΩ? Can you explain this?

6.6.2 A FET as a voltage controlled resistance

From Figures 6.12 and 6.15 it is clear that *for small values of drain-to-source voltage*, FETs have a characteristic that can be described as *ohmic* in that the drain current increases linearly with the drain voltage. The value of the effective resistance (which corresponds to the slope of the characteristics) is controlled by the gate voltage. This allows the device to be used as a **voltage controlled resistance** (**VCR**). The range of resistance that can be produced varies from a few tens of ohms (or less for power devices) up to several gigohms (1 gigohm = 1000 megohms)

A common application of this arrangement is within **automatic gain control** circuits. Here the voltage controlled resistance is used in a potential divider arrangement with a fixed resistor to form a **voltage controlled attenuator**, as shown in Figure 6.36.

The attenuator is used within the feedback path of an amplifier to vary its gain. The

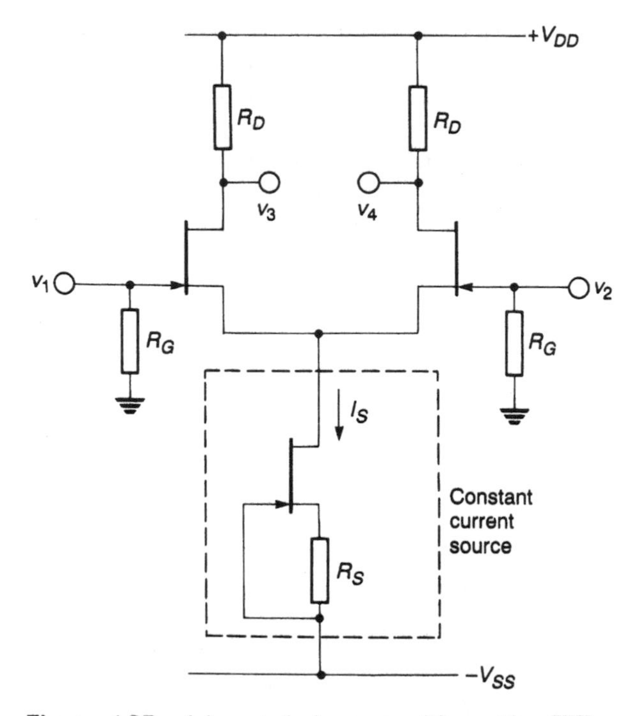

Figure 6.35 A long-tailed pair amplifier with a FET current source.

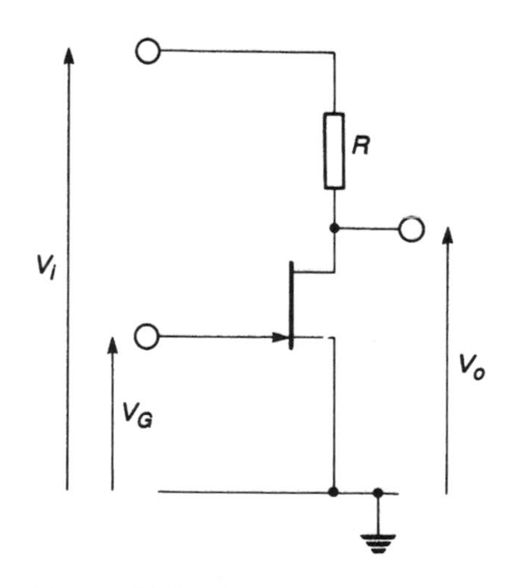

Figure 6.36 A voltage controlled attenuator.

voltage fed to the FET to control its resistance is derived from the output signal of the amplifier and is arranged so that as the magnitude of the output voltage increases, the amount of negative feedback is increased, thereby reducing the amplifier's gain. This allows the output amplitude to be maintained at some fixed value independent of the magnitude of the input signal. This technique is used, for example, to keep the volume of a radio receiver constant, even if the strength of the radio signal changes.

Another use for voltage controlled attenuators is in the design of oscillators. In Section 4.7.1 we noted the problems of adjusting the magnitude of the loop gain of an oscillator circuit to precisely 1 and then maintaining it at that value. The automatic gain control arrangement described above can be used to stabilize the gain of an oscillator circuit without producing distortion of the output.

The voltage controlled attenuator described above may be used with DC or AC input signals since the FET is essentially symmetrical in its operation (although the characteristics of the device for input signals of different polarities are usually very different). However, to avoid excessive distortion the magnitude of the input signals must be restricted to a few tens of millivolts.

6.6.3 A FET as an analogue switch

By applying a suitable voltage to the gate of a FET the effective drain-to-source resistance can be varied from a few tens of ohms or less (effectively a short circuit in many applications) to a value so high that it can almost always be considered to be an open circuit. The resistances of the device in these two states are called the **ON resistance** and the **OFF resistance** of the FET. The ability to turn the device 'ON' and 'OFF' in this way allows it to be used as a switch, as illustrated in Figure 6.37.

Figure 6.37(a) shows a JFET being used as a series switch. MOSFETs can be used in a similar manner. When the device is turned ON, the resistance between the input and output is small. Provided the resistances of the source and destination are large compared with the ON resistance of the FET, the device will resemble a short circuit. When the device is turned OFF, the resistance between the source and destination will be equal to the OFF resistance of the FET. Provided this is large compared with the resistances within the circuit, this will represent an open circuit. Because of the many orders of magnitude difference between the ON and OFF resistances of the FET it is usually easy to satisfy these conditions, allowing the FET to be used as a very efficient switch.

Figure 6.37(b) shows a FET used in a shunt arrangement. Here the series resistor R is chosen to be large compared with R_{ON} and small compared with R_{OFF}. The potential

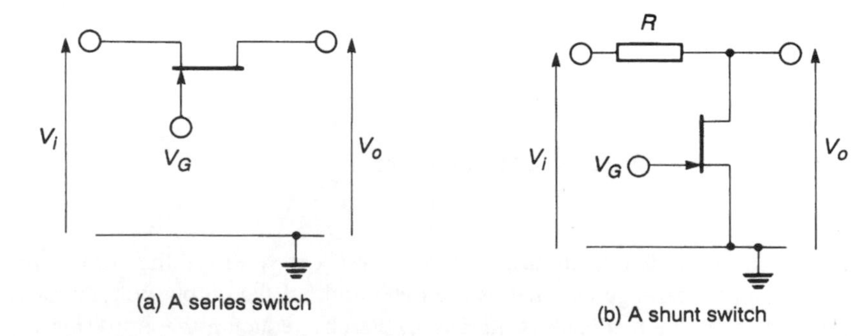

Figure 6.37 The FET as an analogue switch.

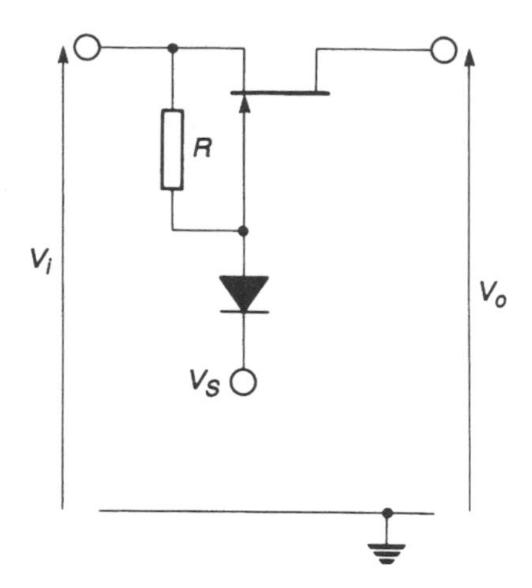

Figure 6.38 A series switch with automatic gate bias.

divider produces an output voltage close to V_i when the device is turned OFF, and close to zero when it is turned ON.

When using FETs as analogue switches, care must be taken to ensure that the operating conditions of the device are correct. It is obviously essential to ensure that the breakdown voltage of the gate is not exceeded, but it is also necessary to guarantee that the gate is taken to an appropriate voltage to turn the device either completely ON or completely OFF. For *n*-channel MOSFETs the gate can be taken to a large positive voltage to turn the device ON and must be made negative with respect to the input voltage by an appropriate amount to turn it OFF. In the case of JFETs the situation is slightly more complicated, particularly when used in the series arrangement, since the gate junction must not be forward biased. A simple circuit that overcomes this problem is shown in Figure 6.38. When the switching voltage V_S is more positive than the input voltage V_i the diode is reverse biased and the gate voltage is set equal to V_i by the resistor R, thus turning the FET ON. If V_S is taken negative, the diode conducts taking the gate negative with respect to the source and turning the FET OFF.

6.6.4 A FET as a logical switch

In addition to their use within analogue circuits, FETs (particularly MOSFETs) are widely used in digital applications. We shall see in Chapter 9 that such circuits usually adopt a two-state, or **binary**, arrangement in which all signals are constrained to be within one of two voltage ranges, one range representing one state (for example, the ON state) and the other representing a second state (for example, the OFF state). These ranges are often referred to as 'logical 1' and 'logical 0'. Within circuits using MOSFETs it is common for voltages close to zero to represent a logical 0 and for voltages close to the positive supply voltage to represent a logical 1.

One of the simplest logic circuits is the **logical inverter** which is required to generate a voltage corresponding to a logical 1 if the input corresponds to a logical 0, and vice versa. A simple circuit to realize this function is shown in Figure 6.39(a). The circuit

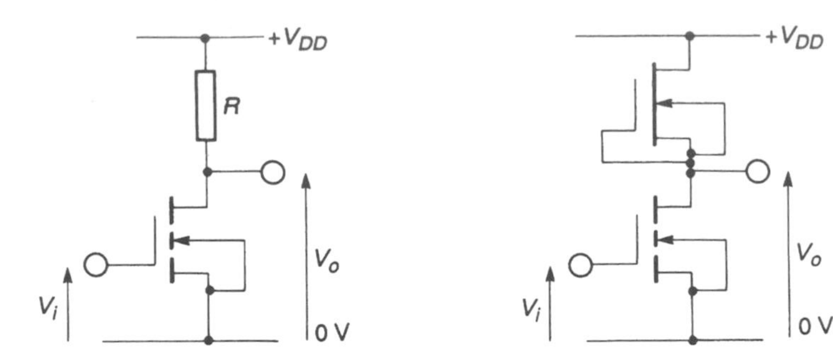

Figure 6.39 Logical inverters using MOSFETs.

uses an n-channel enhancement MOSFET and a resistor. It is similar in appearance, and in operation, to the amplifier described in Section 3.12 and discussed earlier in this chapter.

When used as a logical inverter, the input voltage will be either close to zero (logical 0) or close to the supply voltage V_{DD} (logical 1). When the input voltage is close to zero volts, the enhancement MOSFET is turned OFF since this device needs a positive voltage on the gate to produce a channel between the drain and the source (see Figure 6.13). The drain current is thus negligible and there is no appreciable voltage drop across the resistor R. The output voltage is therefore approximately equal to the supply voltage V_{DD} (logical 1). When the input voltage is close to the supply voltage the MOSFET is turned ON and current flows through R, dropping the output voltage to close to ground (logical 0). Thus, when the input is high the output is low, and when the input is low the output is high. We therefore have the function of an inverter.

The arrangement shown in Figure 6.39(a) is quite acceptable when discrete components are to be used but is unattractive when an integrated circuit is to be produced. One of the reasons why MOSFETs are so widely used in digital integrated circuits is that each transistor requires a very small area of silicon, allowing a very large number of devices to be fabricated on a single chip. Resistors, on the other hand, occupy a proportionately larger area making them components to be avoided wherever possible. Therefore, when producing logic inverters using MOSFETs it is common to use the circuit shown in Figure 6.39(b). Here a second MOSFET is used as an **active load**, greatly reducing the area of silicon required. We will return to the use of active loads in Chapter 8.

The arrangement shown in Figure 6.39(b) uses n-channel MOSFETs. Circuits of this kind are often called **NMOS** circuits, the letters being an abbreviation of **N-channel metal oxide semiconductor**. Similarly, circuits based on p-channel devices are referred to as **PMOS** circuits. We shall return to look at other forms of NMOS and PMOS logic circuitry in Chapter 11.

6.6.5 CMOS circuits

In the NMOS and PMOS circuits described above, the value of the load resistance R (or the effective resistance of the MOSFET used in its place) affects the output resistance

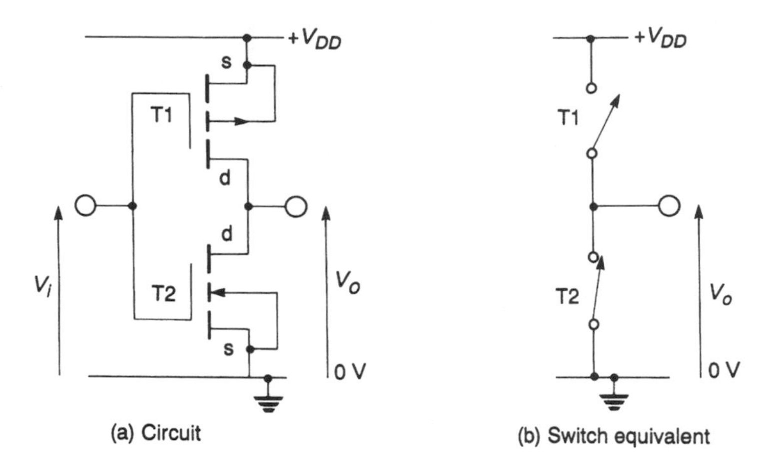

Figure 6.40 A CMOS logic inverter.

of the circuit when the output is high and the power dissipation of the gate when the output is low.

When the input is low, the switching MOSFET is turned OFF and the output is pulled high by the load resistor R. As the output resistance of the device is determined by the value of R, to achieve a low output resistance R should be as small as possible.

When the input is high, the switching MOSFET is turned ON and the output is pulled low. Since the switching MOSFET has a low ON resistance the output resistance is low, enabling the circuit to sink a high current from an external load. In this state almost the entire supply voltage is applied across the load resistor R producing a large current and consequently a high power consumption. To minimize this power consumption, the load resistor should be as large as possible.

Clearly the requirements of a low output resistance and a low power consumption place conflicting requirements on the value of R. This problem can be overcome by using the arrangement shown in Figure 6.40.

Here both NMOS and PMOS devices are combined within a single circuit which is now described as complementary MOS or CMOS logic. When the input voltage is close to zero, the n-channel device T2 is turned OFF but the p-channel device T1 is turned ON. When the input voltage is close to the supply voltage, the position is reversed with T1 OFF and T2 ON. Thus with the input in either state, one of the transistors is ON and the other OFF.

The circuit of Figure 6.40(a) may be represented by the arrangement of Figure 6.40(b). With switch T1 closed and T2 open, the output is pulled high and the output resistance is low, being determined by the ON resistance of T1. With T2 closed and T1 open, the output is pulled low and the output resistance is also low, now being determined by the ON resistance of T2. In both cases, since one of the switches is turned OFF the only supply current flowing is the current drawn by the load. If the load is another circuit of the same form, this will be negligible because of the high input resistance of MOSFETs. Thus, in either state the output resistance is very low and the power consumption is extremely small. In fact, when static (that is, when in one state or the other) the power consumption is generally negligible. In practice the power consumed by a CMOS circuit is determined by the small amount of current that flows as the devices switch from one state to the other. For a short period of time both transistors are conducting, producing a short burst of current from the supply to ground.

Because of their low power consumption, CMOS circuits are widely used for battery operated applications. We will discuss CMOS circuitry in more detail in Chapter 11.

6.7 FET circuit examples

Example 6.4 FET input buffer for an operational amplifier

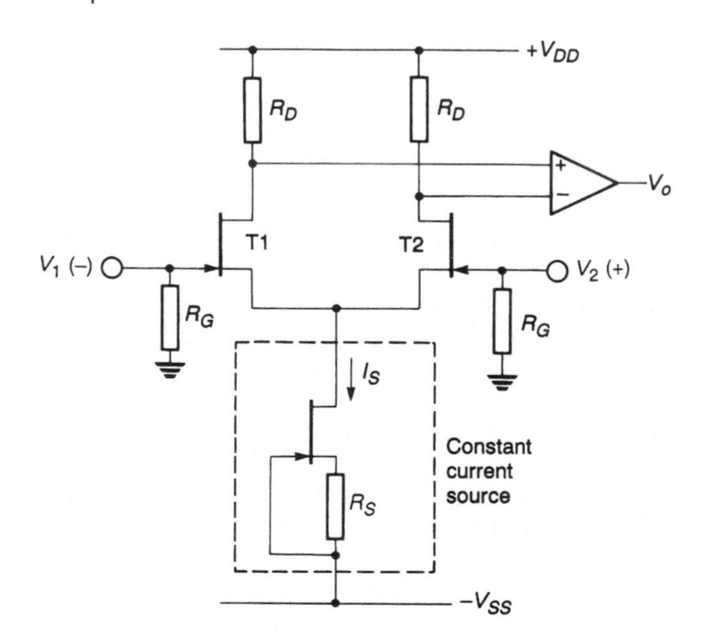

A pair of FETs may be used in a **long-tailed pair amplifier** to improve the performance of an operational amplifier.

If used with a bipolar op-amp, the FET buffer can be used to provide a very high input resistance. If FETs T1 and T2 are matched they may also improve the **common-mode rejection ratio CMRR** of the amplifier. Usually these two transistors will be in the form of a matched pair within a single package. This arrangement would typically produce a CMRR in excess of 120 dB, somewhat more than most general purpose operational amplifiers.

In common with most operational amplifiers, the arrangement above would normally be used with some form of feedback. The circuit may be treated as a single amplifying block with V_1 and V_2 being the inverting and non-inverting inputs and V_o being the output.

Example 6.5 An integrator with reset

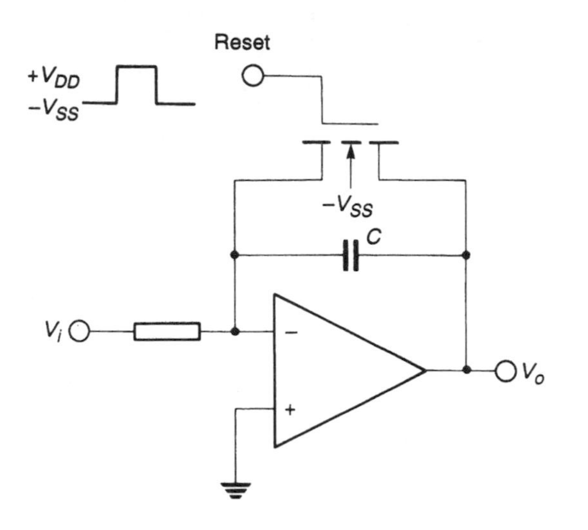

In Example 4.10 we considered the use of an operational amplifier as an integrator. Very often it is necessary to have some method of zeroing the output of the circuit by removing the charge on the capacitor.

One of the simplest ways of achieving this is to place some form of electrically activated switch across the capacitor. FETs are an obvious choice for this application. The above circuit shows a simple arrangement based on the use of an enhancement MOSFET.

The FET switch is controlled by a gate signal which is switched between the two supply voltages. The substrate of the device is connected to the most negative voltage in the system $-V_{SS}$ to ensure that the voltages on the source and drain are never more negative than the substrate.

When the gate input signal is equal to $-V_{SS}$, the gate is at the same potential as the substrate and, therefore, being an enhancement device, the channel is turned OFF. Under these conditions the presence of the FET has no effect and the circuit acts as a simple integrator, as in Example 4.10.

When the gate input voltage is equal to $+V_{DD}$, the channel is turned ON and the capacitor is effectively shorted out. Any charge on the capacitor will quickly be removed and the output voltage of the circuit will be clamped near to ground until the FET is again turned OFF.

If the FET is turned OFF, a fixed DC voltage at the input V_i will produce a steady

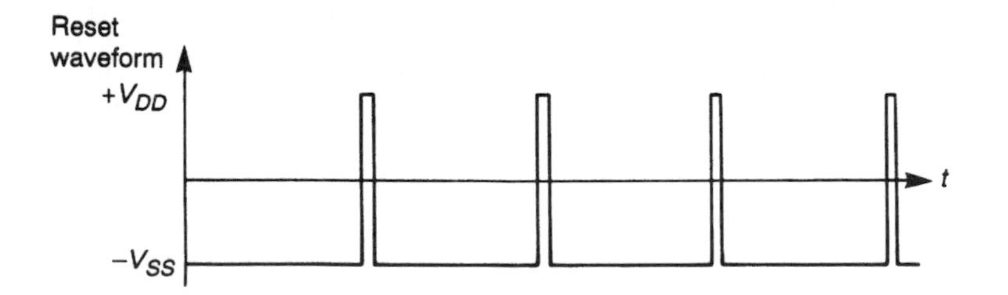

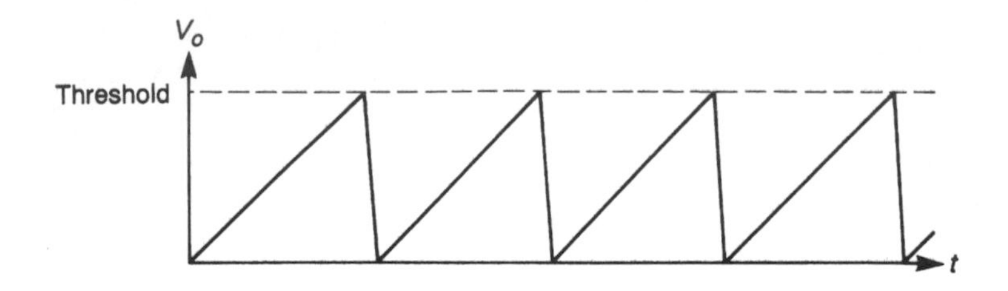

increase in the output voltage. If a voltage level detector is connected to the output of the integrator and is made to generate a reset pulse whenever the output reaches a particular threshold value, this will produce a **sawtooth waveform**, as shown in the above diagram.

Example 6.6 Sample and hold gate

In essence, a sample and hold gate can be simply a capacitor and a switch.

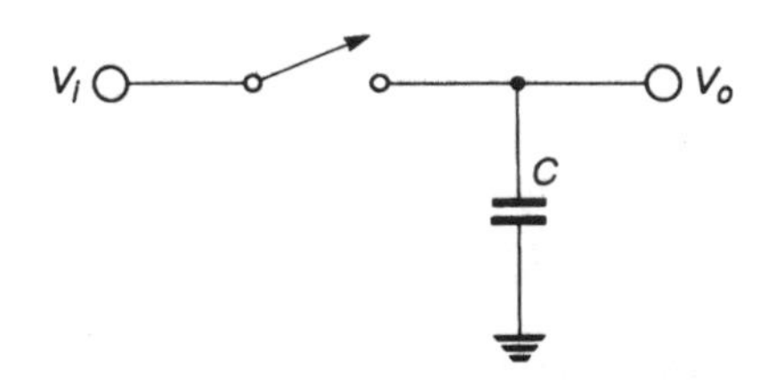

When the switch is closed, the capacitor quickly charges or discharges so that its voltage, and hence the output voltage, equals the input voltage. If the switch is now opened, the capacitor simply holds its current charge and its voltage remains constant. The circuit is used to take a *sample* of a varying voltage by closing the switch, and then to hold that value by opening the switch.

In practice the simple circuit above has a couple of weaknesses. Firstly, when the switch is closed the capacitor represents a very low impedance to the source and thus loads it heavily, possibly distorting the input value. If the source has a fairly high output resistance it may take some time for the capacitor to charge up, reducing the speed at which samples can be taken. Secondly, in practice the capacitor will be connected to a load which will tend to discharge the capacitor because of its finite input resistance.

To overcome these problems, buffer amplifiers are normally used.

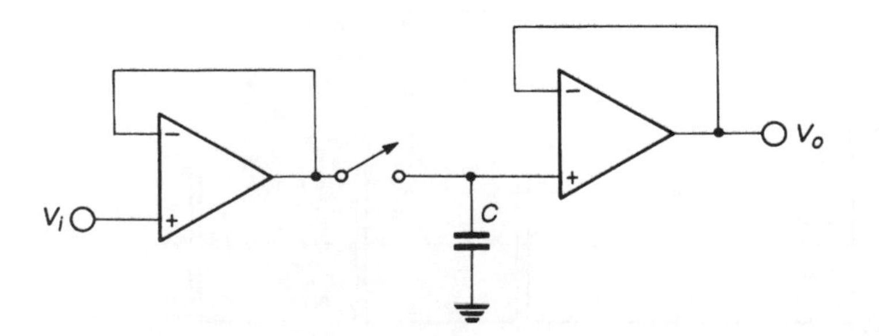

The above arrangement uses two operational amplifiers as unity gain buffers, as described in Example 4.8. The first amplifier provides a high input resistance but a low output resistance to provide rapid charging of the capacitor. The second amplifier would normally be a FET input device which takes very little current from the capacitor, allowing the sampled voltage to be held for a considerable time. The near ideal properties of FET switches make them an obvious choice as the switching device in such circuits.

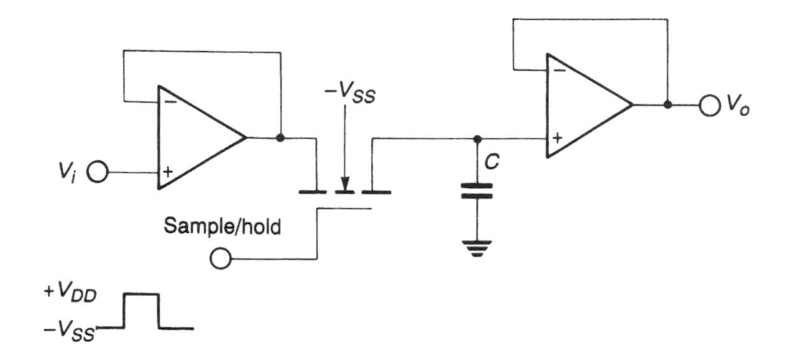

The performance of this circuit is limited by the leakage current of the FET switch and any input bias current of the operational amplifiers. More advanced circuits reduce the effects of these currents.

Key points

- FETs are widely used in both analogue and digital applications.
- The term FET describes a range of components that may be divided into *insulated-gate* devices, also called MOSFETs or MOSTs, and *junction* devices, which are also called JFETs.
- Insulated gate FETs are further divided into *enhancement* and *depletion* types.
- The various forms of FET can be produced in both *n*-channel and *p*-channel versions.
- Although the characteristics of the various classes of FET are slightly different, their principles of operation are similar.
- All are characterized by very high input resistance. In most cases their input currents are so low that they may be considered to be negligible. For this reason they are often used within the input stage of amplifiers to provide a high input impedance.
- FETs can be used as voltage controlled resistances, constant current sources and both analogue (linear) and digital (logical) switches.
- In switching applications their very high OFF resistance and very low ON resistance make them nearly ideal switches.
- Another important characteristic of FETs is their small physical size when implemented in integrated circuit form. This combined with their excellent switching properties has led to

their extensive use in digital very large scale integration (VLSI) circuitry. The majority of microprocessors, memories and associated components are implemented using FETs.

Although the myriad variations of FET devices may be a little daunting, designing circuits using FETs is not as complicated as it might at first appear. In most cases the circuit differences required to cater for the various device types relate simply to the biasing arrangements. The small signal characteristics of all kinds of FET are of the same basic form. Thus when the general principles have been mastered, these may be applied to all types of FET.

Design study

In Figure 6.34 we looked at the design of several fixed and variable constant current sources using FETs. The aim of this design study is to investigate the generation of a precision voltage controlled current source.

The circuit is required to take an input voltage V_i in the range 0 to 1 V, and to sink a current which is 100 mA × (V_i/1.0 V). An accuracy of ±0.1 mA is required.

Approach

In order to implement this system we require a device capable of controlling the current, a means of sensing the current flowing, and a means of comparing the actual and required currents and modifying the current accordingly.

Clearly a FET is capable of controlling the current taken from the load. Both JFETs and MOSFETs are suitable, but the latter have the advantage that it is not necessary to guard against forward biasing the gate. A resistor is a simple current sensor since the voltage across it is linearly related to the current flowing through it. Very stable and high precision resistors are available at low cost. The obvious device to monitor the current and to control the FET is an operational amplifier. This has high gain and is easy to use. A possible circuit is given in the following diagram.

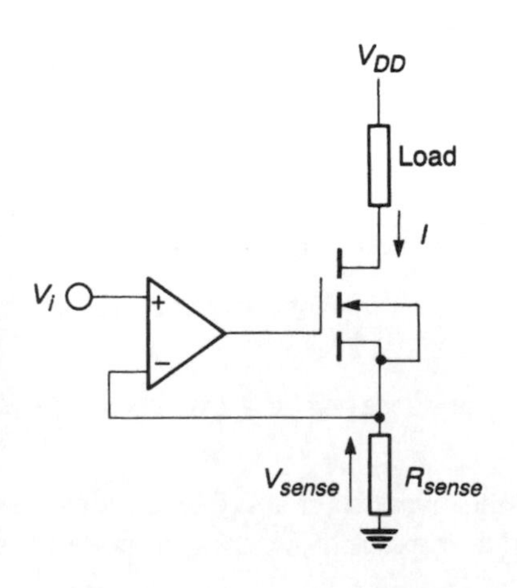

Current is drawn from the positive supply V_{DD}, and through the load, FET and sensing resistor R_{sense}, to ground. The voltage across R_{sense} (V_{sense}) is compared with the input voltage V_i by the op-amp. If V_i is greater than V_{sense} the op-amp will output a positive voltage to the gate of the FET, increasing the current through the load. Conversely, if V_i is less than V_{sense} the current through the load will be decreased. Thus the current through the load will be maintained such that V_{sense} is precisely equal to V_i.

Now since

$$V_i = V_{sense} = IR_{sense}$$

and the required characteristic of the circuit is that

$$I = 100\text{ mA} \times \frac{V_i}{1.0\text{ V}}$$

it follows that

$$V_i = 100\text{ mA}\frac{V_i}{1.0\text{ V}}R_{sense}$$

and hence

$$R_{sense} = \frac{1.0\text{ V}}{100\text{ mA}} = 10\ \Omega$$

The required accuracy of the output is ±0.1 mA in a range of 0 to 100 mA, which requires R_{sense} to have an accuracy of 0.1%.

Either an enhancement or a DE MOSFET can be used, this only affecting the voltage required from the op-amp to produce a given current. A DE MOSFET requires the gate to go both positive and negative whereas an enhancement device requires only positive gate voltages. This is of little importance unless it is desired to run the circuit from a single positive supply, rather than from both positive and negative supplies. If single supply working is required, an enhancement device should be used.

For best results the op-amp should be a FET input device so that its input current can be ignored, making V_{sense} a true reflection of the current through the load (since there is effectively no gate current in a MOSFET). The only remaining source of error is the op-amp's input offset voltage V_{ios}. The desired accuracy requires a current measurement of ±0.1 mA in a 10 Ω resistor. This is a voltage measurement of ±1 mV. Therefore V_{ios} should be small compared with 1 mV to obtain the required accuracy. This can be achieved by specifying a low offset device or by balancing out the offset using an 'offset null' adjustment at set-up time.

FILE 6E

Computer simulation exercise 6.5

Simulate the circuit in the above case study and verify that the proposed design functions as intended.

Use an LF411 FET input op-amp with +15 V and −15 V supplies and use a 100 Ω load resistor connected to the +15 V supply. An IRF150 enhancement FET should be used as the control device. Apply a DC sweep to the input and plot the load current against the input voltage.

References

Shockley W. (1952) A unipolar field-effect transistor. *Proc. I.R.E.* **40**, 1365–76

Further reading

Horowitz P. and Hill W. (1989) *The Art of Electronics*, 2nd edn. Cambridge: Cambridge University Press

Schilling D. L. and Belove C. (1989) *Electronic Circuits: Discrete and Integrated*, 3rd edn. New York: McGraw-Hill

Tocci R. J. and Oliver M. E. (1991) *Fundamentals of Electronic Devices*, 4th edn. New York: Merrill

Exercises

6.1 What type of charge carriers are responsible for conduction in a *p*-channel JFET?

6.2 What is the function of the bias circuitry in a FET amplifier?

6.3 Within a linear amplifier circuit, what would be the polarity of a typical gate bias voltage applied to: an *n*-channel DE MOSFET; a *p*-channel JFET; a *p*-channel enhancement MOSFET; and an *n*-channel JFET?

6.4 Explain the function of the capacitor C in the circuits of Figure 6.5. What effect will this capacitor have on the frequency response of the circuit?

6.5 In this chapter we have considered both common-source and common-drain (source follower) circuits, and have looked at the characteristics of each. Some applications use a common-gate configuration. Sketch a circuit which corresponds to this mode of operation.

6.6 Repeat the calculations of Example 6.1 using a device with a g_m of 2.5 mS, and a load resistor R_L of 3.3 kΩ.

6.7 An *n*-channel JFET has a pinch-off voltage of −4 V and a drain-to-source saturation current of 6 mA. Calculate the transconductance of this device at drain currents of 1, 2 and 4 mA.

6.8 An amplifier is required to operate from a supply of 25 V with a load resistance of 4.7 kΩ and a quiescent output voltage of 15 V. Use a graphical method to design a biasing arrangement to satisfy these conditions, assuming the design uses a JFET as described in Exercise 6.7.

6.9 In Section 6.5.6 we obtained the result that the voltage gain of the circuit of Figure 6.31(a) is approximately unity. Use this result to calculate the voltage gain of the amplifier of Figure 6.30(a) in terms of the component values.

6.10 How would the circuit of Figure 6.38 be modified if a *p*-channel device were used?

6.11 Explain why the circuits of Figure 6.39 use enhancement MOSFETs rather than depletion devices for the switching transistor.

6.12 Use simulation to estimate the common-mode rejection ratio (CMRR) of the circuit of Figure 6.32. Take as your starting point the circuit of Computer simulation exercise 6.3 (FILE 6C).

6.13 Modify your circuit for the last exercise by replacing the source resistor with a constant current source formed using a JFET and a resistor as shown in Figure 6.34(c). Use a 2N3819 JFET and an appropriate value resistor. Hint: before modifying your circuit place a current probe in series with the source resistor to determine the source current. Replace the source resistor with the FET and a resistor of a few hundred ohms and measure the resulting source current. Adjust the value of the resistor until the source current is approximately the same as before – the biasing of the circuit will now be similar to the original circuit.

When the circuit is operating correctly, measure the CMRR of the new circuit and compare this with the value obtained in the previous exercise.

Bipolar Junction Transistors

Objectives

When you have studied the material in this chapter you should:

- have a clear understanding of the construction, operation and characteristics of bipolar transistors;
- be able to perform simple DC and AC analyses of a range of transistor circuits;
- be familiar with a range of small signal equivalent circuits for bipolar transistors including the hybrid-parameter and hybrid-π models;
- be aware of the role of feedback in designing circuits that are less affected by device characteristics and limitations;
- be familiar with the operation and uses of several four-layer devices;
- be able to design a range of transistor circuits using simple design rules.

Contents

7.1 Introduction

The **bipolar transistor** is so called because charge is carried by both majority and minority charge carriers. The device incorporates two *pn* junctions and is sometimes called a **bipolar junction transistor** or **BJT**. In the early days of electronics, bipolar transistors were more widely used than their unipolar counterparts and they were often simply called 'transistors', the term FET being used to distinguish unipolar devices. Bipolar devices still dominate in analogue applications, but FETs are now widely used in digital devices such as microprocessors and memory components. However, it is still common to refer to bipolar transistors simply as 'transistors'.

Simple forms of these devices were first produced in 1947 by Bardeen and Brattain (1948), and then refined in the following year by Shockley (1949). They are more complex than unipolar transistors and often consume more power, but usually have a higher gain and provide greater drive capability. Bipolar transistors are the major 'workhorse' of electronic circuit design and a clear understanding of their characteristics and use is essential for all those who have an interest in electronics.

7.2 Construction

Bipolar transistors are formed from three layers of semiconductor material. Two device polarities are possible. The first is formed by placing a thin layer of *p*-type semiconductor between two layers of *n*-type material to form an ***npn*** **transistor**. The second is formed by placing a thin layer of *n*-type semiconductor between two layers of *p*-type material to give a ***pnp*** **transistor**. Electrical connections are made to each layer of the device, the three electrodes being called the **emitter**, **base** and **collector**. Both kinds of device are widely used and circuits often combine transistors of different polarities. The operation of the two forms is similar, differing mainly in the polarities of the voltages and currents. Figure 7.1 shows the construction and circuit symbols for both an *npn* and a *pnp* transistor. The arrow in the circuit symbol is always shown between the base and the emitter and its direction is from the *p*-type region to the *n*-type region – as in the symbol for a diode.

7.3 Transistor operation

The *npn* or *pnp* structure produces two *pn* junctions connected 'back-to-back', as shown in Figure 7.1. If a voltage is connected across the device between the collector and the emitter, with the base open circuit, one or other of these junctions is reverse biased and so negligible current will flow. If a transistor were nothing more than two 'back-to-back diodes' it would have little practical use. However, the construction of the device, in particular the fact that the base region is very thin, allows the base to act as a control input. Signals applied to this electrode can be used to produce, and control, currents between the other two terminals. To see why this is so, consider the circuit configuration shown in Figure 7.2.

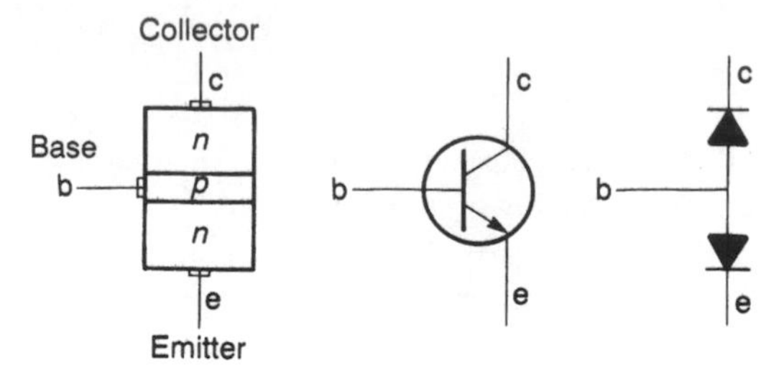

(a) An *npn* transistor

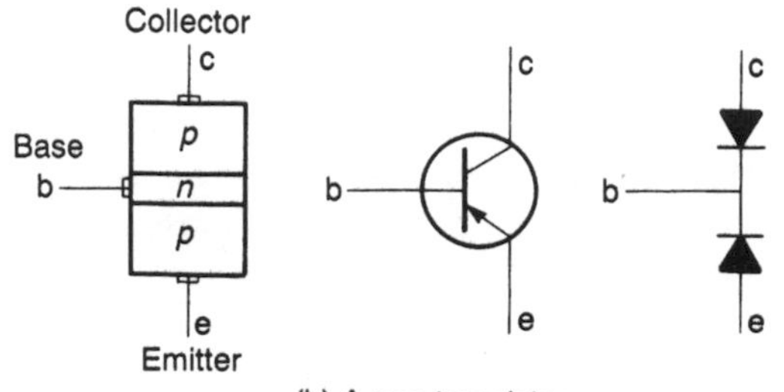

(b) A *pnp* transistor

Figure 7.1 *npn* and *pnp* transistors.

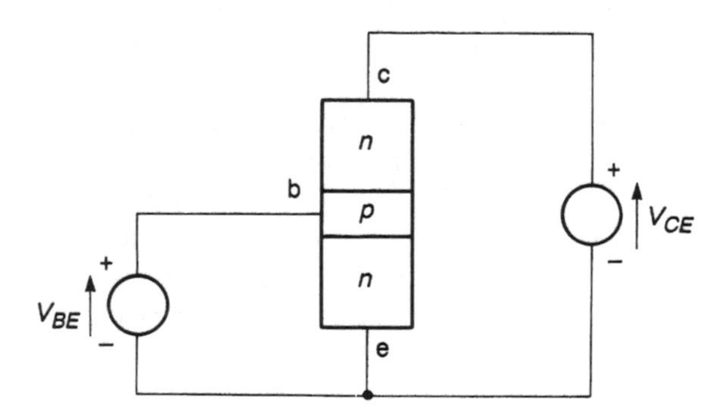

Figure 7.2 Transistor operation.

The normal circuit configuration for an *npn* transistor is to make the collector more positive than the emitter. Typical voltages between the collector and the emitter V_{CE} might be a few volts. With the base open circuit, the only current flowing from the collector to the emitter will be a small leakage current I_{CEO}, the subscript specifying that it is the current from the **C**ollector to the **E**mitter with the base **O**pen circuit. This leakage current is small and can normally be neglected. If the base is made positive with respect to the emitter this will forward bias the base–emitter junction which will behave in a manner similar to a diode (as described in Sections 5.5 and 5.6). For small values of the base-to-emitter voltage V_{BE} very little current will flow, but as V_{BE} is increased beyond about 0.5 V (for a silicon device) the base current begins to rise rapidly.

The fabrication of the device defines that the emitter region is heavily doped while the base is lightly doped. The heavy doping in the emitter region results in a large number of **majority charge carriers** which are electrons in an *npn* transistor. The light doping within the base region generates a smaller number of holes which are the majority carriers within the *p*-type base region. Thus in an *npn* transistor the base current is dominated by electrons flowing from the emitter to the base. In addition to being lightly doped, the base region is very thin. Electrons that pass into the base from the emitter as a result of the base–emitter voltage become **minority charge carriers** within the *p*-type base region. Since the base is very thin, electrons entering the base find themselves close to the space-charge region formed by the reverse bias of the base–collector junction. While the reverse bias voltage acts as a barrier to majority charge carriers near the junction, it actively propels minority charge carriers across it. Thus any electrons entering the junction area are swept across into the collector and give rise to a collector current. Careful design of the device ensures that the majority of the electrons entering the base cross the junction into the collector. Thus the emitter-to-collector current is many times greater than the emitter-to-base current. This allows the transistor to function as a current amplifying device, with a small base current generating a larger collector current. This phenomenon of current amplification is referred to as **transistor action**.

The relationship between the collector current I_C and the base current I_B for a typical silicon bipolar transistor is shown in Figure 7.3(a).

Because of the slight non-linearity of the relationship there are two ways of specifying the **current gain** of the device. The first is the **DC current gain**, h_{FE} or β, which is found simply by dividing the collector current by the base current. This is usually given at a particular value of I_C. Because of the non-linearity of the characteristic, it is slightly different at different values of I_C. The DC current gain h_{FE} (β) is used in large signal calculations.

When considering small signals, we need to know the relationship between a small change in I_B (ΔI_B) and the corresponding change in I_C (ΔI_C) The ratio $\Delta I_C/\Delta I_B$ is called the **small signal current gain** and is given the symbol h_{fe}. It is also called the **AC current gain** of the device. The value of h_{fe} may be obtained from the slope of the characteristic given in Figure 7.3(a).

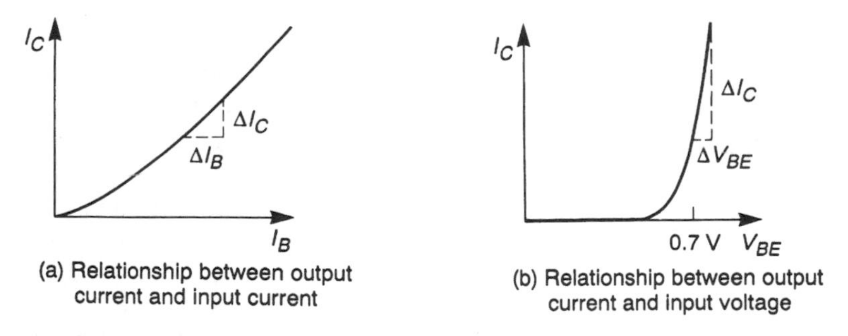

Figure 7.3 Characteristics of a typical silicon bipolar transistor.

For most practical purposes h_{FE} and h_{fe} can be considered to be equal. A typical value for a silicon transistor would be in the range 100 to 300, but the current gain of bipolar transistors varies considerably with temperature and operating conditions. There is also a considerable spread of characteristics between devices of the same nominal type, and even within the same batch.

The characteristics of a bipolar transistor may also be described by the relationship between the output current I_C and the input voltage V_{BE}, as shown in Figure 7.3(b). Since the base–emitter junction resembles a simple *pn* junction, the input current I_B is exponentially related to the input voltage V_{BE}, as illustrated in Figure 5.6. Since the output current I_C is approximately linearly related to the input current (by the current gain h_{FE}), the relationship between I_C and V_{BE} has the same shape (though with correspondingly larger values of current). The slope of this curve at any point is given by the ratio $\Delta I_C/\Delta V_{BE}$ which represents the **transconductance** of the device g_m (you may like to compare this with the similar discussion of the FET given in Section 6.4.3). In the limit

$$g_m = \frac{dI_C}{dV_{BE}}$$

Unlike h_{FE}, which is approximately constant for a given device, g_m varies with the collector current (and hence the emitter current) at which the circuit is operated.

7.4 A simple amplifier

In the field effect transistors discussed in the last chapter, the output current is controlled by the input *voltage* applied to the gate, and in normal operation the input current is negligible. We therefore describe these devices as being *voltage controlled*. In bipolar transistors the output current is related to the input *current* to the base, which is itself related to the input *voltage*. We may therefore consider the bipolar transistor as either a *current controlled* or a *voltage controlled* device, as is most appropriate to the problem in hand. In either case it is clear that the output current is controlled by an input signal, and thus the bipolar transistor may also be used as a control device within an amplifier, as described in Section 3.12.

Figure 7.4 shows a simple amplifier based on a bipolar transistor. The circuit is, in fact, a very poor linear amplifier for reasons that will become apparent shortly. It is used at this stage since it is important to understand the fundamental weaknesses of this simple approach before looking at more appropriate techniques.

You will notice that in this arrangement the input is applied between the base and the emitter while the output is measured between the collector and the emitter. The emitter is thus common to the input circuit and the output circuit. For this reason such an arrangement is called a **common-emitter amplifier**.

The form of the common-emitter amplifier of Figure 7.4 is similar to that of the common-source MOSFET amplifier of Figure 6.5(a). The difference between the two circuits is that in the FET amplifier, a resistor R_G goes from the input (the gate) to ground, whereas in the bipolar amplifier, a resistor R_B goes from the input (the base) to the positive supply. The need for such an arrangement for the bipolar transistor amplifier is clear

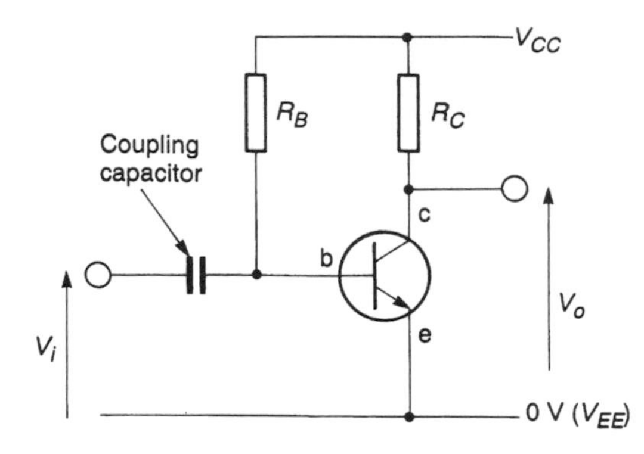

Figure 7.4 A simple amplifier using a bipolar transistor.

from Figure 7.3. You will notice that the collector current I_C falls to zero as the base current I_B goes to zero. If we wish to apply a bipolar signal to our amplifier (that is, one that goes negative as well as positive), we must offset the input from zero in order to amplify the complete signal. This is done by **biasing** the input of the amplifier.

The base resistor R_B applies a positive voltage to the base of the transistor, forward biasing the base–emitter junction and producing a base current. This in turn produces a collector current by the *transistor action* described above. This collector current flows through the collector load resistor R_C producing a voltage drop and making the output voltage V_o less than the collector supply voltage V_{CC}. When no input is applied to the circuit it is said to be in its **quiescent** state. The value of the base resistor R_B will determine the **quiescent base current** which in turn will determine the **quiescent collector current** and the **quiescent output voltage**. The values of R_B and R_C must be chosen carefully to ensure the correct **operating point** for the circuit.

If a positive voltage is applied to the input of the amplifier this will tend to increase the voltage on the base of the transistor and thus increase the base current. This will in turn raise the collector current, increasing the voltage drop across the collector resistor R_C and decreasing the output voltage V_o. If a negative voltage is applied to the input this will decrease the current through the transistor and thus increase the output voltage. We therefore have an **inverting amplifier**. As with the MOSFET amplifiers of Figure 6.5, a **coupling capacitor** is used to prevent input voltages from affecting the mean voltage applied to the base. The circuit therefore cannot be used to amplify DC signals and is an **AC coupled amplifier**.

Example 7.1 Determining the quiescent conditions of a simple amplifier

The quiescent conditions of the amplifier are determined by the DC gain of the amplifier h_{FE}, the base resistor R_B and the collector resistor R_C. Let us assume for this example that $h_{FE} = 100$.

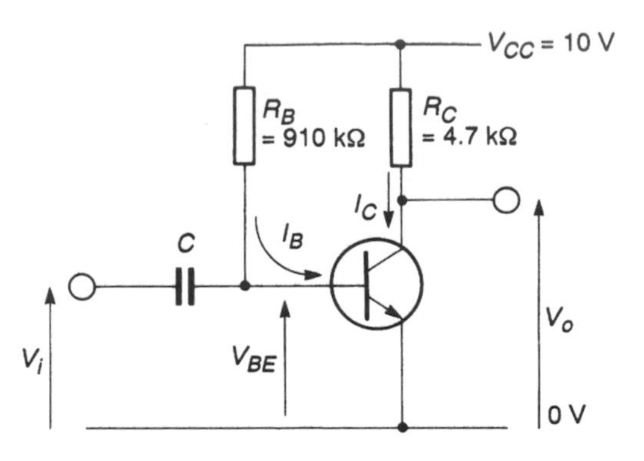

The base-to-emitter junction of the transistor resembles a forward biased *pn* junction. From the discussion of Section 5.6 we know that the voltage across such a junction is approximately constant with a value of about 0.7 V for a silicon device. We will therefore assume that the base-to-emitter voltage V_{BE} is 0.7 V.

From a knowledge of V_{BE} we also know the voltage across R_B since this is simply $V_{CC} - V_{BE}$ which in turn enables us to calculate the base current I_B. In this case

$$I_B = \frac{V_{CC} - V_{BE}}{R_B} = \frac{10 - 0.7\text{ V}}{910\text{ k}\Omega} = 10.2\ \mu\text{A}$$

The collector current I_C is now given by

$$I_C = h_{FE} I_B = 100 \times 10.2\ \mu\text{A} = 1.02\text{ mA}$$

The quiescent output voltage is simply the supply voltage minus the voltage drop across R_C and is therefore

$$V_o = V_{CC} - I_C R_C = 10 - 1.02 \times 10^{-3} \times 4.7 \times 10^3 \approx 5.2\text{ V}$$

Thus the circuit has a quiescent collector current of about 1 mA and a quiescent output voltage of approximately 5.2 V.

Note: In this circuit the quiescent collector current and the quiescent output voltage are both determined by the value of h_{FE}, which varies considerably between devices. For this reason this simple circuit arrangement is rarely used. However, it is useful to look at the disadvantages of this arrangement before progressing to more suitable techniques.

Computer simulation exercise 7.1

FILE 7A

Simulate the circuit of Example 7.1 applying no input signal. Experiment with different values for the various resistors and note their effects on the quiescent voltages and currents within the circuit. Now apply a small alternating voltage to the input and observe the resultant variations at the output.

The circuit of Example 7.1 is based on the use of a transistor with an h_{FE} of 100. Since PSpice does not provide a standard transistor model for such a component, a custom library is provided with the demonstration files. If this library is configured as described in the README file provided with the demonstration files, various additional transistors will become available. One of these devices has the symbol name NPN100. This is an *npn* transistor with a gain of 100 and is suitable for use in this exercise. Later simulation exercises will use other custom components from this library.

If you do not have access to the demonstration files, or are using a different simulation package, you will need to generate your own custom transistor model with a gain of 100. You should look at the documentation for your simulator for details of how to do this.

7.5 Bipolar transistor characteristics

As we shall see later in Section 7.6.8, bipolar transistors can be used in a number of configurations; the device characteristics differ accordingly. We have already met the most widely used of these configurations, the *common-emitter* arrangement, and we will initially concentrate on this mode of operation.

The currents and voltages associated with a transistor in the common-emitter configuration are shown in Figure 7.5.

The arrangement has two sections: the input circuit and the output circuit. The characteristics of the device as a whole are determined by the characteristics of these two sections and by the relationship between them.

7.5.1 Input characteristics

Figure 7.6 shows the relationship between the base current and the base-to-emitter voltage for the transistor. This represents the **input characteristic** of the device in the common-emitter mode.

The characteristic is similar to that of a simple forward biased *pn* junction, as described in Section 5.5 and shown in Figure 5.6. It is clearly non-linear. The variation of the input current with the input voltage of the transistor in this mode may be determined by considering the expression given in Equation 5.1 for the current of a *pn* junction. This is

$$I \approx I_S\left(\exp\frac{eV}{kT} - 1\right)$$

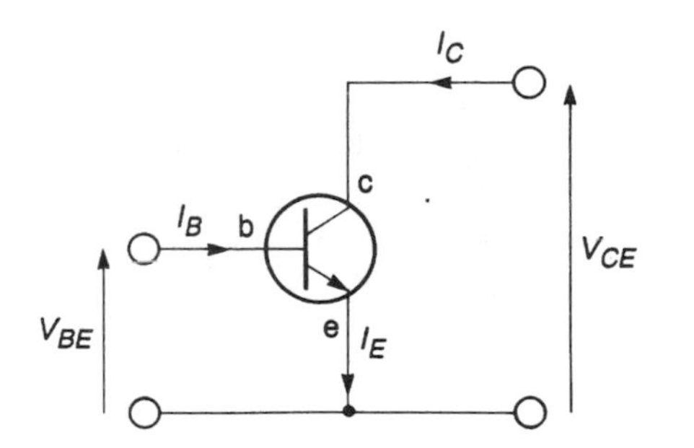

Figure 7.5 Voltages and currents for the common-emitter configuration.

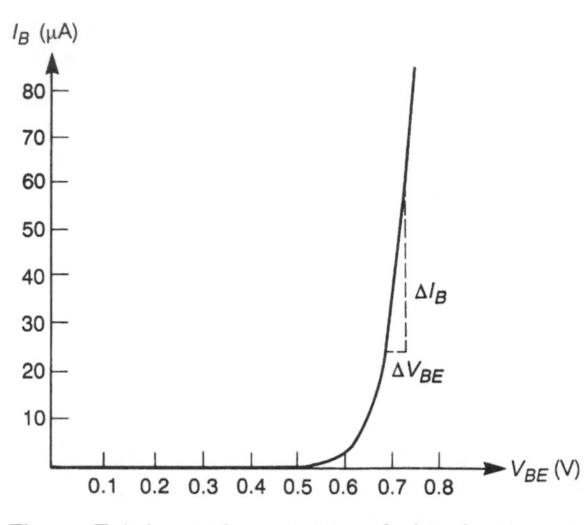

Figure 7.6 Input characteristic of a bipolar transistor in the common-emitter configuration.

where I_S is the **reverse saturation current**. We noted in Equation 5.3 that at room temperature this may be approximated by the expression

$$I \approx I_S \exp(40V)$$

In this case the current I represents the base current I_B and the voltage V represents the junction voltage V_{BE}. We therefore have

$$I_B = I_{BS}\left(\exp\frac{eV_{BE}}{kT} - 1\right) \approx I_{BS}\exp(40V_{BE}) \tag{7.1}$$

where I_{BS} is a constant determined by the base characteristics. This expression represents the **input characteristic** of the common-emitter configuration, as shown in Figure 7.6.

The slope, at any point, of the line in Figure 7.6 represents the relationship between a small change in the base-to-emitter voltage ΔV_{BE} and the corresponding change in the base current ΔI_B. The slope therefore indicates the **small signal input resistance** of the arrangement, which is given the symbol h_{ie}. Clearly the magnitude of the input resistance varies with position along the characteristic. The value at any point may be found by differentiating Equation 7.1 with respect to I_B, which gives

$$1 = 40I_{BS}\exp(40V_{BE})\frac{\mathrm{d}V_{BE}}{\mathrm{d}I_B}$$

$$= 40I_B\frac{\mathrm{d}V_{BE}}{\mathrm{d}I_B}$$

and thus

$$h_{ie} = \frac{\mathrm{d}V_{BE}}{\mathrm{d}I_B} \approx \frac{1}{40I_B}\ \Omega \tag{7.2}$$

The input characteristics of a bipolar transistor in the common-emitter configuration may therefore be described by very simple expressions for the base current and the small signal input resistance. It should be noted, however, that the 'e' subscript in h_{ie} stands for 'emitter' and that h_{ie} is the small signal input resistance in the *common-emitter configuration*. The input resistances in the other modes have different values.

7.5.2 Transfer characteristics

From Figure 7.5 it is clear that the emitter current I_E must be given by the sum of the collector current I_C and the base current I_B. Thus

$$I_E = I_C + I_B$$

and since

$$I_C = h_{FE} I_B$$

it follows that

$$I_E = h_{FE} I_B + I_B = (h_{FE} + 1) I_B$$

Since h_{FE} is usually much greater than unity, we may make the approximation that

$$I_E \approx h_{FE} I_B = I_C \qquad (7.3)$$

Combining this result with that of Equation 7.1 gives us an expression for the emitter current in terms of the input voltage

$$\begin{aligned} I_E &\approx h_{FE} I_B \approx h_{FE} I_{BS} \exp(40V_{BE}) \\ &\approx I_{ES} \exp(40V_{BE}) \end{aligned}$$

where $I_{ES} \approx h_{FE} I_{BS}$ is a constant related to the emitter characteristics. Thus both I_B and I_E are exponentially related to V_{BE}. The difference between these two quantities is reflected in the values of I_{BS} and I_{ES}.

Since $I_C \approx I_E$, we can say that

$$I_C \approx I_E \approx I_{ES} \exp(40V_{BE}) \qquad (7.4)$$

This gives rise to the exponential form of Figure 7.3(b) from which it is clear that g_m may be found by differentiating Equation 7.4 with respect to V_{BE}.

This gives

$$\frac{dI_C}{dV_{BE}} \approx 40 I_{ES} \exp(40V_{BE}) = 40 I_C$$

and therefore

$$g_m \approx 40 I_C \approx 40 I_E \text{ Siemens} \qquad (7.5)$$

It is important to notice that in bipolar transistors g_m is proportional to I_E, whereas in FETs g_m is proportional to the square root of the drain current (see Section 6.4.3). Since g_m is directly controlled by the quiescent collector (emitter) current, the voltage gain of an amplifier formed using the device will also be related to this current. The choice of quiescent circuit conditions thus plays a major part in determining the performance of an amplifier.

The value of g_m represents the ratio of changes in the collector (emitter) current to changes in the base voltage. The inverse of this quantity has the units of resistance and is the ratio of the small signal base-to-emitter voltage to the corresponding change in the emitter current. This is termed the **emitter resistance** r_e. It follows from Equation 7.5 that

$$r_e = \frac{1}{g_m} \approx \frac{1}{40I_C} \approx \frac{1}{40I_E}\ \Omega \tag{7.6}$$

From Equation 7.2

$$h_{ie} \approx \frac{1}{40I_B}$$

it follows that

$$h_{ie} \approx h_{fe} r_e$$

Notice that here we use h_{fe} rather than h_{FE} since both h_{ie} and r_e are small signal quantities.

7.5.3 Output characteristics

Figure 7.7 shows the relationship between the collector current I_C and the collector voltage V_{CE} for various values of the base current I_B. These two quantities represent the output current and the output voltage of the common-emitter configuration, as shown in

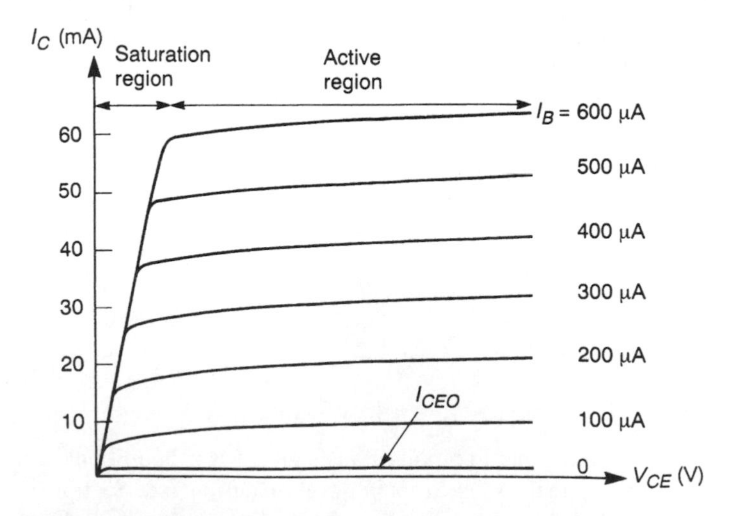

Figure 7.7 Common-emitter output characteristic.

Figure 7.5. The relationship between them is often referred to as the common-emitter **output characteristic**. You might like to compare this characteristic with those given for the various forms of FETs in Figures 6.12 and 6.15.

For a given base current, the collector current initially rises rapidly with the collector voltage as this increases from zero. However, it soon reaches a steady value and any further increase in the collector voltage has little effect on the collector current. The value of collector current at which the characteristic stabilizes is determined by the base current. The ratio between this steady value of collector current and the value of the base current represents the DC current gain of the device h_{FE}.

The section of the characteristic over which the collector current is approximately linearly related to the base current is referred to as the **active region**. Most linear amplifier circuits operate in this region. The section of the characteristic close to the origin where this linear relationship does not hold is called the **saturation region**. The saturation region is generally avoided in linear circuits but is widely used in non-linear arrangements, including digital circuitry. It is important to note that the term 'saturation' has a different meaning when applied to bipolar transistors than when discussing FETs. Saturation occurs in bipolar transistors when V_{CE} is very low, because the efficiency of the transistor action is reduced and many charge carriers pass from the emitter to the base without being swept into the collector region.

In an ideal bipolar transistor the various lines in the output characteristic would be horizontal, indicating that the output current was completely independent of the collector voltage. In practice this is not the case and all real devices have a slight gradient, as shown in Figure 7.7. The slope of these lines indicates the change in output current with output voltage and is therefore a measure of the **output resistance** of the arrangement. It is often more convenient to consider the reciprocal of this quantity, namely the **output admittance** h_{oe}, where again the '*e*' subscript indicates that this value is for a device in the common-emitter configuration.

If the nearly horizontal portions of the output characteristic are extended 'backwards' (to the left in Figure 7.7) they converge at a point on the negative portion of the horizontal axis. This point is referred to as the **Early voltage**, after J. M. Early of Bell Laboratories. The Early voltage is given the symbol V_A and has a typical value of between 50 and 200 volts.

Notice in Figure 7.7 that the collector current is not zero when the base current is zero, as predicted by Equation 7.3, because of the presence of the **leakage current** I_{CEO} discussed in Section 7.3. The effect of I_{CEO} is magnified in the figure to allow it to be visible. In silicon devices its effects are usually negligible.

7.5.4 Equivalent circuits for a bipolar transistor

Having considered the characteristics of both the input and the output sections of the bipolar transistor in the common-emitter configuration, we are now in a position to draw **small signal equivalent circuits** for the device. Figure 7.8 shows two simple small signal equivalent circuits for a bipolar transistor, the first using the device's transconductance and the second using its current gain. The first of these is of the same basic form as those derived in the last chapter for field effect transistors.

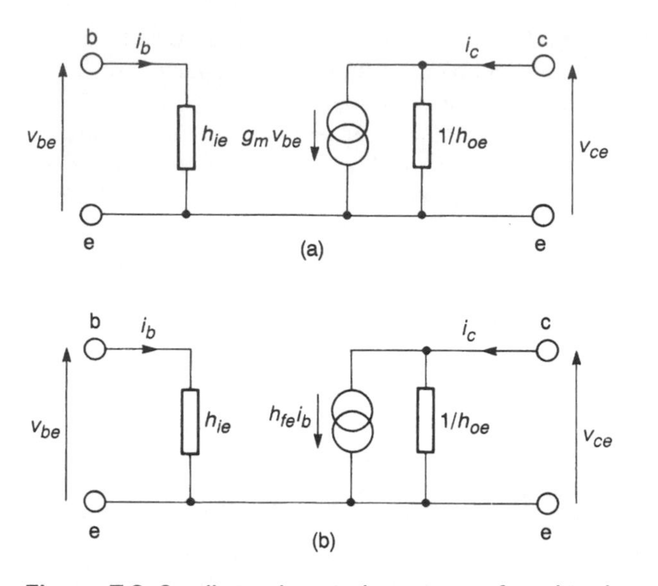

Figure 7.8 Small signal equivalent circuits for a bipolar transistor.

Some of the values of the components in the equivalent circuits may be derived from the analysis given earlier. Others can be found in the manufacturer's data sheet. For example, for a general purpose silicon device h_{fe} might be in the range 80 to 250 and h_{oe} in the range 1 to 300 μS. Remember that h_{oe} is an *admittance* and is therefore equivalent to a resistance of $1/h_{oe}$.

It is worth noting at this point that h_{fe} varies considerably from one device to another, but g_m is defined by the physics of the materials used and can be calculated from the collector (or emitter) current, as shown in Equation 7.5. It must be remembered, however, that we are using an approximation for g_m which ignores, for example, the effects of η in the diode equations (see Section 5.5).

The equivalent circuits given in Figure 7.8 are perfectly acceptable for most applications, but are not the only ways of modelling a device. A number of other representations are used, each using different refinements to provide a better model of the operation of the device.

The hybrid-parameter model

You will have noticed that the symbols used to describe the characteristics of the bipolar transistor in the simple equivalent circuit of Figure 7.8(b), h_{fe}, h_{ie} and h_{oe}, have a similar form. There is, in fact, a fourth commonly used member of the set h_{re} which is the **reverse voltage transfer ratio**. This describes how the output voltage affects the current in the input circuit. To understand the nature of this *feedback* effect we need to look back at the input characteristic of the device. In Figure 7.6 we looked at the variation of base current with base-to-emitter voltage. This figure gives a single characteristic indicating a unique correspondence between one and the other. In practice the position of this characteristic depends on the collector voltage, as shown in Figure 7.9. As V_{CE} increases, the value of V_{BE} required to produce a given base current also increases. The reverse voltage transfer ratio h_{re} is the rate of change of V_{BE} with V_{CE} for a given I_B. This effect is fairly small and h_{re} has a typical value of about 10^{-3} to 10^{-5}. The interaction between V_{BE} and V_{CE} is brought about because a change in the collector voltage alters the width of the base region. This **base width modulation** results in variations of V_{BE} with V_{CE} at constant I_C and is called the **Early effect**. An equivalent circuit which incorporates this additional feature is given in Figure 7.10. You will notice that an additional voltage generator has been added which is controlled by the output voltage.

The parameters used in the equivalent circuit of Figure 7.10 have different forms: h_{ie} is an impedance, h_{oe} is an admittance, h_{fe} is a current ratio and h_{re} is a voltage ratio. For this reason this model is called the **hybrid-parameter model**, often abbreviated to the ***h*-parameter model**, and this accounts for the '*h*' in the symbols used for the parameters. Each parameter is identified by a two-letter suffix. The second letter of each is an '*e*' indicating that these quantities relate to the common-emitter configuration.

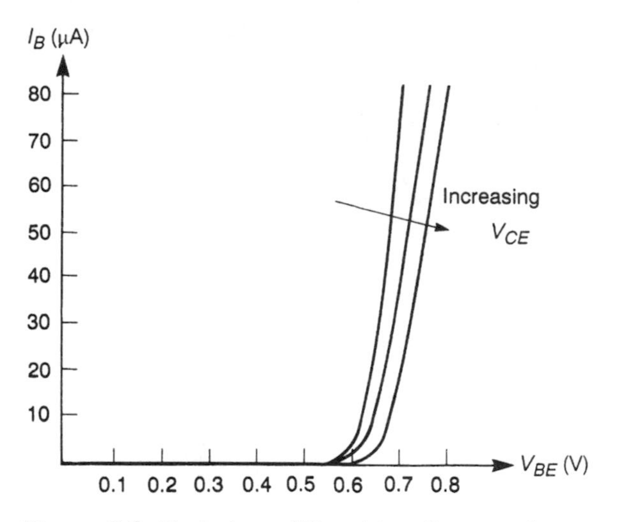

Figure 7.9 Variation of V_{BE} with collector voltage.

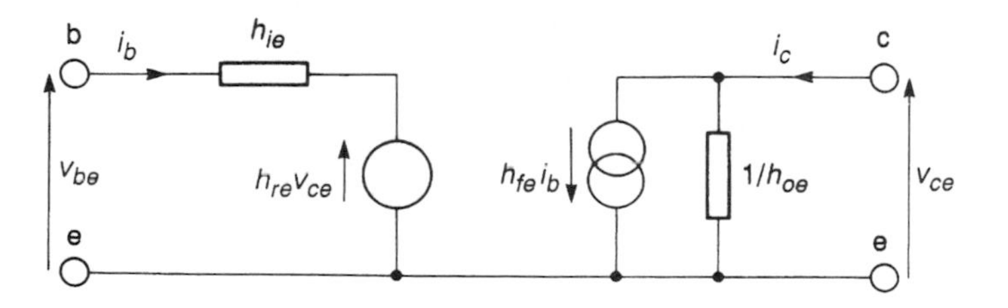

Figure 7.10 The hybrid-parameter equivalent circuit for a common-emitter transistor.

The first suffix indicates the nature of the parameter: h_{ie} describes the characteristics of the *input*, h_{re} refers to the *reverse* effects, h_{fe} gives the *forward* current gain and h_{oe} relates to the *output* of the device.

These h-parameters are frequently given in manufacturers' data sheets and the hybrid model is widely used.

The hybrid-π model

The hybrid-parameter model described above provides a good description of the operation of transistors at low frequencies, but does not take into account the effects of device capacitances which greatly alter a device's performance at high frequencies. A more sophisticated model can be used to take into account these effects and some other additional physical properties of the device.

Figure 7.11 shows the **hybrid-π model** for a bipolar transistor. The circuit still has three nodes e, b and c representing terminals of the device, but now an additional node b′ is added which represents the base *junction*. The ohmic resistance between the base terminal and the base junction is represented by $r_{bb'}$, which is typically in the range 5 to 50 Ω for general purpose devices and a few ohms for high frequency types. The parameter $r_{b'e}$ is the base–emitter junction resistance. A similar analysis to that given in Section 7.5.2 shows that this is given approximately by the expression

$$r_{b'e} \approx h_{fe} r_e \approx \frac{h_{fe}}{40 I_E} \tag{7.7}$$

This is the expression previously derived for h_{ie}, which is the total resistance seen looking into the base. From this analysis it is apparent that a more accurate expression for h_{ie} is

$$h_{ie} = r_{bb'} + r_{b'e} \approx r_{bb'} + h_{fe} r_e \approx r_{bb'} + \frac{h_{fe}}{40 I_E} \tag{7.8}$$

If we take typical values for h_{fe} and I_E of 200 and 10 mA, respectively, this gives a value for $r_{b'e}$ of approximately 500 Ω. As this is large compared with typical values for $r_{bb'}$ (5 to 50 Ω) the approximation that $h_{ie} \approx r_{b'e}$ is normally acceptable.

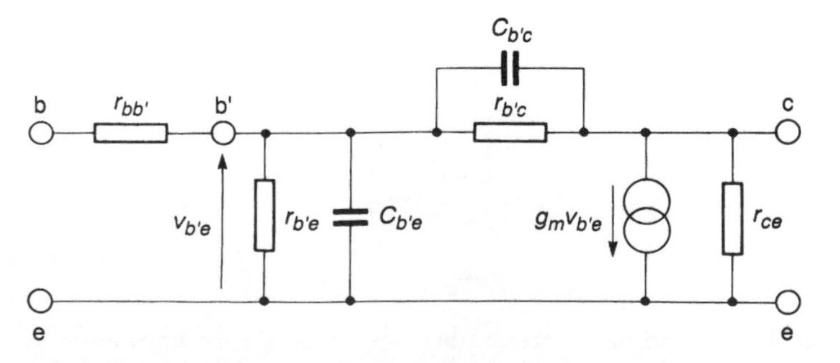

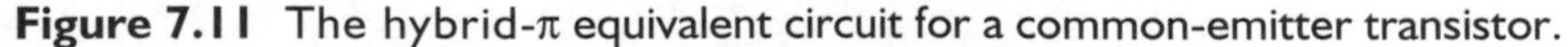

Figure 7.11 The hybrid-π equivalent circuit for a common-emitter transistor.

The effects of variations in the output voltage on the input are represented in the hybrid-π model by a resistor $r_{b'c}$ rather than by a voltage generator as in the hybrid-parameter model. Capacitors $C_{b'e}$ and $C_{b'c}$ are included to represent the capacitance across the base–emitter junction and the capacitance from the base to the collector, respectively. The current produced by the current source in the circuit is proportional to the current that passes through the base–emitter junction resistance $r_{b'e}$. However, because of the presence of $r_{b'c}$ and the two capacitors, this is not equal to the current flowing into the base terminal. The current generator is therefore made equal to g_m times the voltage across the base–emitter junction $v_{b'e}$.

The hybrid-π model agrees more closely with actual device performance than the simpler models described earlier. However, it is much more difficult to analyse. The inclusion of device capacitances allows the high frequency performance of the device to be predicted, although the model breaks down at frequencies considerably below those at which the gain falls to unity.

7.5.5 Bipolar transistors at high frequencies

A detailed analysis of the high frequency performance of bipolar transistors, as predicted by the hybrid-π model, is beyond the scope of this text. However, looking at the model it is clear that the presence of $C_{b'e}$ will produce a fall in gain at high frequencies, as described in Section 3.7.2. $C_{b'c}$ also affects the frequency response of the device as it produces feedback from the output to the input. Since the device inverts the input signal, this is **negative feedback**. As the impedance of the capacitor decreases as the frequency increases, the amount of negative feedback increases with frequency, reducing the gain at high frequencies. The combined effects of the two capacitances produce a frequency response of the general form shown in Figure 7.12, where $h_{fe(0)}$ represents the low-frequency value of h_{fe}.

The frequency at which the gain falls to 0.707 $(1/\sqrt{2})$ of its low-frequency value represents the **bandwidth** of the device and is given the symbol f_β. In Section 3.7.2 we

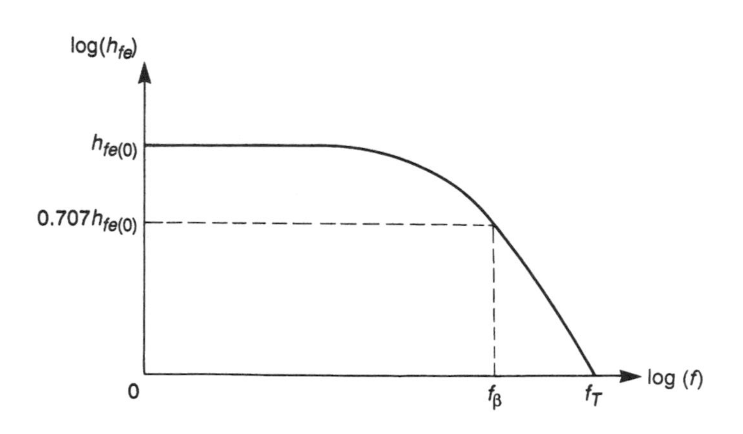

Figure 7.12 Variation of current gain with frequency for a bipolar transistor.

deduced that the upper cut-off frequency of an *RC* network is given by

$$f_{co} = \frac{1}{2\pi CR}\ \text{Hz}$$

It can be shown that f_β is given by the expression

$$f_\beta \approx \frac{1}{2\pi r_{b'e}(C_{b'e} + C_{b'c})}$$

In general the effects of $C_{b'e}$ dominate those of $C_{b'c}$ and so it is usually possible to make the approximation

$$f_\beta \approx \frac{1}{2\pi r_{b'e} C_{b'e}} \tag{7.9}$$

The frequency at which the gain drops to unity is termed the **transition frequency** and is given the symbol f_T. It can be shown that this is simply related to f_β by the expression

$$f_T = h_{fe(0)} f_\beta \tag{7.10}$$

From Equation 7.7 we know that $r_{b'e}$ varies with the emitter current I_E, and from Equations 7.9 and 7.10 it follows that f_β and f_T also vary with current. By substituting from Equation 7.7 we can obtain expressions indicating this relationship. These are

$$f_\beta \approx \frac{40 I_E}{h_{fe(0)} 2\pi(C_{b'e} + C_{b'c})} \approx \frac{40 I_E}{h_{fe(0)} 2\pi C_{b'e}}$$

and

$$f_T \approx \frac{40 I_E}{2\pi(C_{b'e} + C_{b'c})} \approx \frac{40 I_E}{2\pi C_{b'e}}$$

Thus the bandwidth and transition frequency of the transistor are both directly proportional to the emitter current. This result has implications for high bandwidth, low power consumption amplifiers.

7.5.6 Leakage currents

In Section 7.3 we discussed the presence of a leakage current I_{CEO} which flows from the collector to the emitter in the absence of any base current. This current is represented in Figure 7.7.

A leakage current is also present between the collector and the base across the reverse biased collector–base junction. This is given the symbol I_{CBO} since it is the current which flows from the **c**ollector to the **b**ase with the third terminal (the emitter) **o**pen circuit.

An understanding of the relationship between these two currents and their significance within transistor circuits can be gained by remembering that the transistor can be seen as two back-to-back *pn* junctions. One of these, the collector–base junction, is reverse biased in normal operation. From the discussions of Chapter 5 we know that reverse biased junctions have a reverse saturation current which is approximately constant for all but the smallest reverse voltages. This accounts for I_{CBO} which flows across the reverse biased collector–base junction.

When the leakage current I_{CBO} enters the base region it produces a similar effect to current entering from the base terminal. The leakage current is amplified by the current gain of the device producing a much larger current from the collector to the emitter. This is the collector-to-emitter leakage current I_{CEO}. Therefore

$$I_{CEO} \approx h_{FE} I_{CBO}$$

and the collector current is given by

$$I_C = h_{FE} I_B + I_{CEO}$$

In some semiconductors, for example those manufactured using germanium, the effects of leakage currents are very significant, particularly at high temperatures (in the next section we will discuss the effects of temperature on leakage currents). However, in silicon devices the leakage currents are extremely small and can, with good design, normally be neglected.

7.5.7 Temperature effects

In Section 5.6 we looked at the effects of temperature on the behaviour of semiconductor diodes. The junctions within bipolar transistors are also affected by temperature which changes their current gain, base–emitter voltage and leakage currents.

Details of how h_{FE} increases with temperature for a particular device can be found in its data sheet. Typical silicon devices might show an increase in gain of about 15% when taken from 0 to 25 °C, and a doubling in gain when the temperature increases from 25 to 75 °C.

The value of V_{BE} shows a similar variation with temperature to that seen in semiconductor diodes. In silicon devices this corresponds to a *fall* in V_{BE} of about 2 mV per °C rise in temperature.

The transistor leakage currents I_{CBO} and I_{CEO} increase with temperature. I_{CBO} shows a similar rate of increase to that of semiconductor diodes, doubling for an increase of about 10 °C (an increase of about 7%/°C). Since I_{CEO} is related to I_{CBO} by the expression

$$I_{CEO} \approx h_{FE} I_{CBO}$$

this also increases with temperature, but more rapidly since h_{FE} also increases with temperature. With silicon devices, leakage currents are generally of the order of a few nanoamps and their effects are almost always negligible. Consequently the effects of temperature on these currents are not usually of great concern.

7.6 Bipolar amplifier circuits

Having looked at the characteristics of bipolar transistors in some detail let us return to the simple amplifier circuit given in Figure 7.4. Having derived a small signal equivalent circuit for the transistor we are now in a position to produce an equivalent circuit for the complete amplifier.

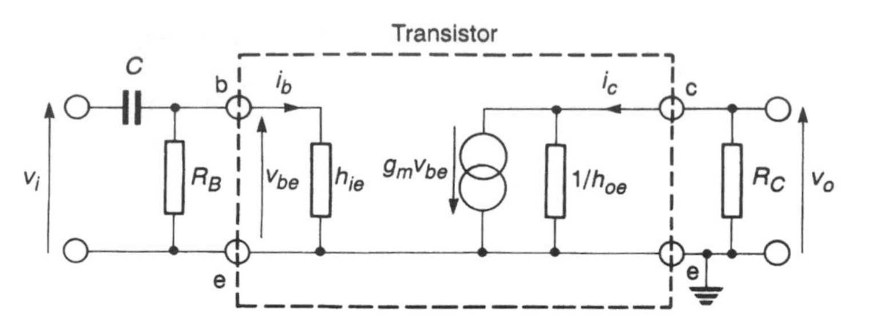

Figure 7.13 Small signal equivalent circuit of a simple transistor amplifier.

7.6.1 Equivalent circuit of a bipolar transistor amplifier

Remembering that the supply line V_{CC} is effectively joined to the earth line (0 V) as far as small signals are concerned (as discussed in Section 6.5.1), we can construct a small signal equivalent circuit for the amplifier of Figure 7.4. This is shown in Figure 7.13. Clearly a more sophisticated model could be produced using a full hybrid-parameter or hybrid-π model, but a simple circuit is sufficient for our present needs.

Both R_B and h_{ie} could be replaced by a single resistor representing the resistance of the parallel combination. This could also be done with $1/h_{oe}$ and R_C, if desired. It should be remembered that since h_{oe} is an admittance it must be inverted to give an equivalent resistance. The input capacitor C is present to prevent any incoming signal from affecting the bias conditions of the transistor. Normally its value would be chosen so that it has a negligible effect at the frequencies of interest. However, the presence of C will introduce a **low-frequency breakpoint** into the frequency characteristic of the circuit, as discussed in Section 3.7.

7.6.2 Small signal voltage gain

From the small signal equivalent circuit for the amplifier we can deduce its small signal voltage gain. If we assume that C can be ignored, it is clear that

$$v_{be} = v_i \tag{7.11}$$

and that

$$v_o = -g_m v_{be}\left(\frac{1}{h_{oe}}//R_C\right)$$

where $(1/h_{oe})//R_C$ simply means the effective resistance of the parallel combination of $1/h_{oe}$ and R_C. Combining these two expressions gives

$$v_o = -g_m v_i\left(\frac{1}{h_{oe}}//R_C\right)$$

and thus the voltage gain is given by

$$\text{Voltage gain} = \frac{v_o}{v_i} = -g_m\left(\frac{1}{h_{oe}}//R_C\right) = -g_m \frac{R_C}{h_{oe}R_C + 1} \tag{7.12}$$

The negative polarity of the gain indicates that this is an **inverting amplifier** and that a positive input voltage will produce a negative output voltage. You might like to compare this expression with that obtained in Equation 6.4 for the voltage gain of an amplifier using a FET.

In many cases the collector resistance R_C will be much smaller than the output resistance of the transistor, and so $(1/h_{oe})//R_C$ is approximately equal to R_C. *In this case* Equation 7.12 may be approximated by

$$\text{Voltage gain} = \frac{v_o}{v_i} \approx -g_m R_C \tag{7.13}$$

Note that since $g_m \approx 40I_E$, the voltage gain may be expressed as

$$\text{Voltage gain} \approx -g_m R_C \approx -40I_E R_C \approx -40I_C R_C \approx -40V_{RC}$$

where V_{RC} is the voltage across R_C. Thus the voltage gain of the amplifier is related to the large signal voltage across R_C. The voltage gain is therefore determined by the quiescent conditions of the circuit as set by the biasing arrangement. We noted in Example 7.1 that the quiescent conditions of this arrangement are affected by the current gain of the device which varies considerably between devices. It is therefore clear that the voltage gain of the circuit will also vary greatly with the current gain of the device used. These characteristics make this a very poor circuit and, in practice, transistors are invariably used with negative feedback to stabilize these parameters.

From the equivalent circuit it is also straightforward to determine the **small signal input resistance** and the **small signal output resistance** of the amplifier. This is illustrated in the following example.

Example 7.2 Determination of the small signal voltage gain, input resistance and output resistance of a bipolar transistor amplifier

Calculate the small signal voltage gain, input resistance and output resistance of the following amplifier, given that $h_{fe} = 100$ and $h_{oe} = 10\ \mu\text{S}$.

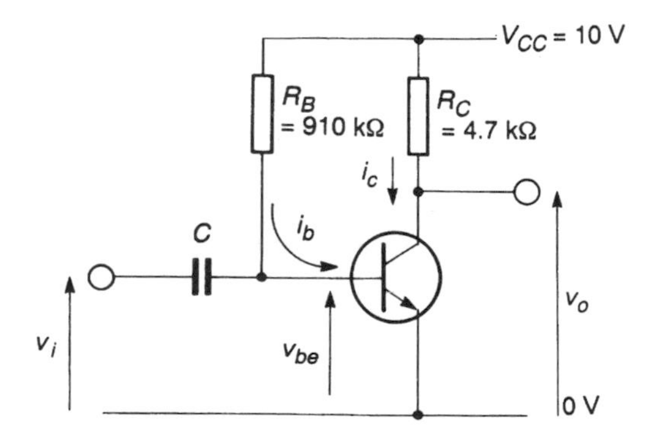

The first step in this problem is to determine the small signal equivalent circuit of the amplifier.

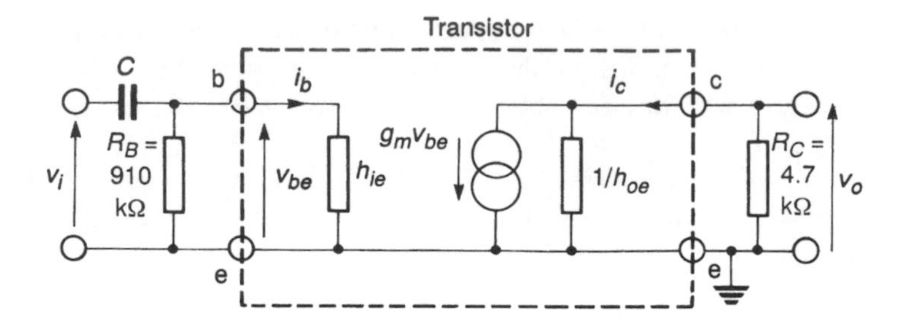

Voltage gain

In order to determine the behaviour of the circuit we need to establish the values of g_m and h_{ie}. To do this we must know the DC operating conditions since both are affected by the quiescent current. Fortunately we have already investigated the DC conditions of the circuit in Example 7.1 from which we know that I_E is 1.02 mA. Therefore

$$g_m \approx 40 I_E \approx 40.8 \text{ mS}$$

and

$$h_{ie} \approx \frac{h_{fe}}{40 I_E} \approx \frac{100}{40 \times 1.02 \times 10^{-3}} \approx 2.45 \text{ k}\Omega$$

From Equation 7.12 we have

$$\text{Voltage gain} = \frac{v_o}{v_i} = -g_m \frac{R_C}{h_{oe} R_C + 1}$$

and substituting for the component values gives

$$\text{Voltage gain} = -40.8 \times 10^{-3} \frac{4700}{10 \times 10^{-6} \times 4700 + 1} \approx -183$$

If we consider that $1/h_{oe}$ is large compared with R_C and assume that the voltage gain is equal to $-g_m R_C$, this gives a value of -192. Given the inaccuracies in our calculations this is probably a reasonable approximation, therefore

$$\text{Voltage gain} = \frac{v_o}{v_i} \approx -g_m R_C$$

Input resistance

From the equivalent circuit it is clear that the **small signal input resistance** is simply $R_B // h_{ie}$. Since $R_B \gg h_{ie}$, it is reasonable to say

$$r_i = R_B // h_{ie} \approx h_{ie} \approx 2.4 \text{ k}\Omega$$

Output resistance

The **small signal output resistance** is the resistance seen 'looking into' the output terminal of the circuit. Since the idealized current generator has an infinite internal resistance, the output resistance is simply the parallel combination of R_C and $1/h_{oe}$. Thus

$$r_o = R_C \bigg/\!\!\bigg/ \frac{1}{h_{oe}} = 4700//100\,000 \approx 4.5\ \text{k}\Omega$$

and again it is reasonable to use the approximation that $r_o \approx R_C$.

Normally the designer's task is to determine the component values required to give particular amplifier characteristics, rather than to predict the characteristics of a given circuit as in this example. However, as has been stated before, this circuit is *not* an ideal amplifier arrangement and we will therefore leave the design process until later in this chapter.

FILE 7A

Computer simulation exercise 7.2

Using the circuit from Computer simulation exercise 7.1, measure the small-signal voltage gain of the amplifier of Example 7.2 and compare this with the value calculated in the text.

Measure the small signal input and output resistance of the circuit and compare these with the predicted values. Hint: measure the input resistance by inserting a resistor in series with the input and comparing the *small-signal* voltage across this resistor with the voltage across the input to the amplifier. Measure the output resistance by connecting different value resistors from the output to ground and measuring the change in the *small-signal* output voltage.

7.6.3 Large signal considerations

The large signal considerations in the design of a transistor amplifier are concerned with the **quiescent operation** of the circuit. In other words, the voltages and currents in the circuit for zero input. These voltages and currents are determined by the biasing arrangements of the circuit and the characteristics of the device itself.

In Section 7.4 and Example 7.1 we considered the determination of the quiescent conditions of a simple amplifier and found them easy to calculate. The quiescent currents and voltages in the simple amplifier investigated were determined by the various resistors in the circuit and by the current gain of the transistor. We have also seen in Section 7.6.2 that calculating the various small signal characteristics of a circuit is straightforward. Unfortunately, although it is simple to calculate the steady state voltages and currents within a given circuit, designing a circuit to give a desired set of characteristics is slightly more complicated. This is because the collector resistor R_C determines not only the quiescent conditions of the circuit, but also the small signal gain. To overcome this problem we can adopt the technique used earlier when dealing with FET amplifiers, that is, the use of a **load line** (see Section 6.5.3).

Figure 7.14 shows a typical **output characteristic** for a bipolar transistor in a common-emitter configuration with a load line superimposed. The function and nature of the load line are identical to those of the load line used for the FET amplifier in Section 6.5. The load line passes through V_{CC}, the collector supply voltage, and has a slope equal

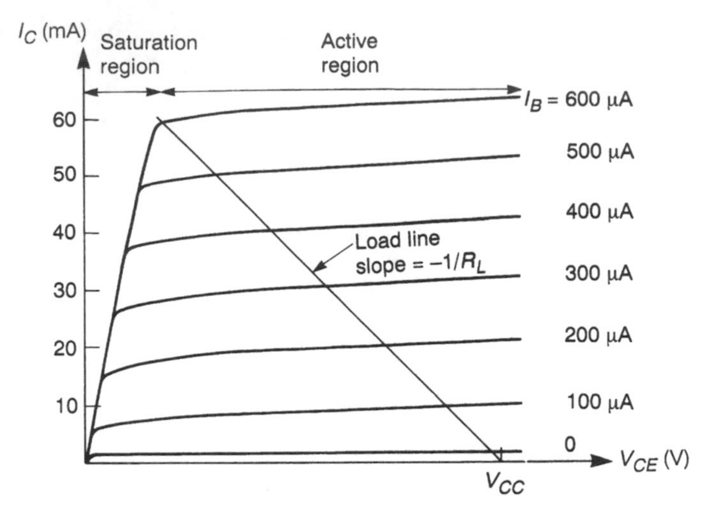

Figure 7.14 A load line.

to the reciprocal of the load resistance. The various lines of the output characteristic represent the relationship between the collector current and the collector emitter voltage for different values of base current. The load line represents the relationship between the collector current and the voltage at the junction of the load resistor R_C and the transistor. The intersection of the load line and one of the lines of the output characteristic represents a possible solution which satisfies both these relationships simultaneously. The point of intersection of the load line and the characteristic line representing the base current to be used is termed the **operating point**. This corresponds to the quiescent state of the circuit.

The load line can be used in a number of ways. If the desired operating point is known, a load line can simply be drawn from this point to the V_{CC} position on the voltage axis. The slope of this line then gives the value of the required load resistor. Alternatively, if the value of R_C is fixed, the load line can be drawn with a slope equal to $-1/R_C$ through the V_{CC} points on the voltage axis, and the operating point chosen appropriately. Choice of the operating point then determines the required quiescent base current. The choice of operating point is of great importance since it directly affects not only the quiescent conditions of the circuit but also the output voltage range.

7.6.4 Choice of operating point

In Section 6.5.4 we considered the choice of operating point for a FET amplifier and noted that there are several forbidden zones within the characteristic of the FET which must be avoided. These forbidden regions also exist for bipolar transistors, as shown in Figure 7.15.

Region A represents the **saturation region** which is normally avoided in linear applications since it is highly non-linear. However, digital circuits often make use of this part of the characteristic and we will discuss the reasons for this in Chapter 11. All devices

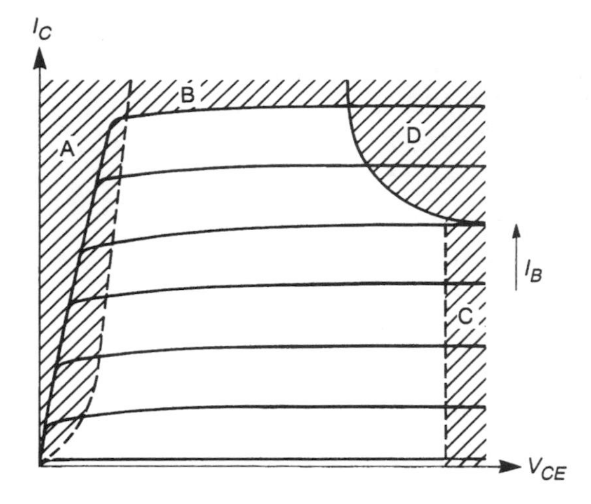

Figure 7.15 The forbidden operating regions for a bipolar transistor.

have a maximum collector current $I_{C(max)}$ which must not be exceeded. This results in forbidden zone B. Devices also have a maximum collector voltage $V_{CE(max)}$, resulting in zone C. If this voltage is exceeded the device will exhibit **avalanche breakdown** at the collector–base junction which can lead to damage (see Section 5.6). The last region which must be avoided, area D, is caused by a need to limit the power dissipated within the device to safe limits. Since the power dissipation is the product of the collector voltage and the collector current (the base current can normally be neglected), this area is bounded by a hyperbola $P = P_{max}$.

To ensure safe operation of a circuit, the operating point and the various signals within the circuit must be chosen to ensure that the device remains within the allowed regions of the characteristic. Clearly the output voltage cannot exceed the supply voltage V_{CC} and to maintain linearity the device should not be operated in the saturation region. Therefore, for a maximum output voltage range the operating point should be positioned half-way between V_{CC} and the saturation region. For maximum collector current swing, the operating point should be chosen at half $I_{C(max)}$. To satisfy both these requirements the operating point would be positioned close to the centre of the allowable region, as shown in Figure 7.16. In many applications neither maximum voltage range nor maximum current swing are required. In such cases the operating point may be chosen to simplify circuit design providing that adequate voltage and current ranges are produced.

The current and voltage corresponding to the operating point are the **quiescent collector current** and the **quiescent output voltage** of the circuit, respectively. Small signal input voltages to the amplifier will cause corresponding small signal changes in the base current, causing the circuit to traverse the load line on either side of the operating point.

In normal operation the input signal should be limited to ensure that the transistor remains within its linear region. If the input is increased so that the device leaves this region, the output waveform will be distorted. This is illustrated in Figure 7.17 which

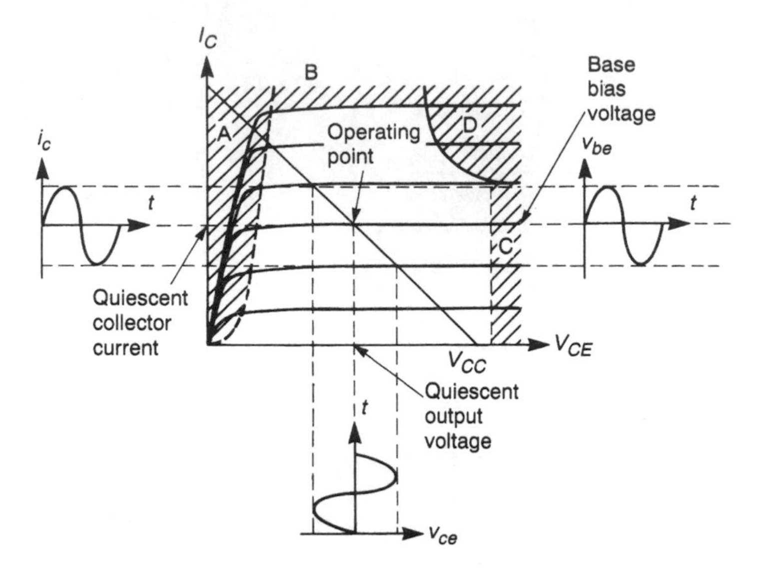

Figure 7.16 Choice of operating point in a bipolar transistor amplifier.

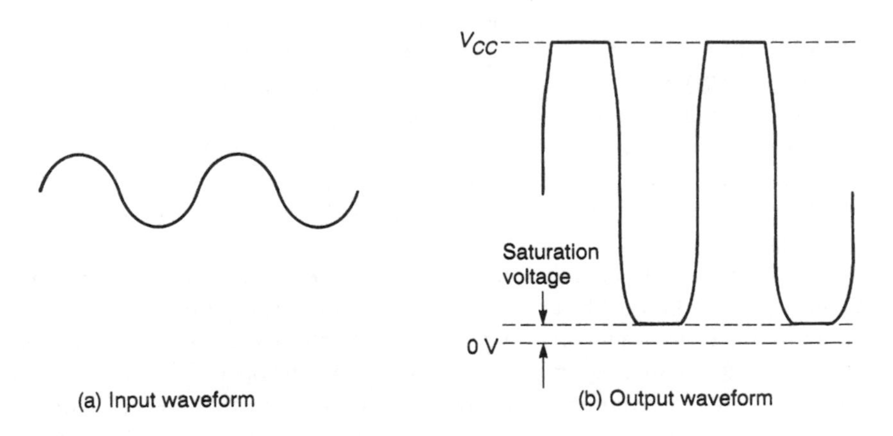

Figure 7.17 Clipping of a sinusoidal signal.

shows the effects of applying an excessive sinusoidal input signal to an amplifier. The output voltage is **clipped** as it attempts to exceed the supply voltage and as it drives the transistor into its saturation region. This clipping is a form of **amplitude distortion**, as discussed in Chapter 1. When driven into saturation, the collector voltage drops to its **saturation voltage** and then remains relatively constant as the base current is increased further. When in this saturated state the collector current is determined predominantly by the external circuit. A typical value for the saturation voltage of a small general purpose silicon transistor would be 0.2 V. Note that this value is less than the voltage between the base and the emitter V_{BE} so the collector is negative with respect to the base and the base-to-collector junction is forward biased by a small amount.

7.6.5 Device variability

Bipolar transistors, in common with almost all active devices, suffer from considerable device variability during their production. Devices from the same batch of components may have parameters that differ by several hundred percent. This leads to considerable problems in the design of circuits using these components and greatly limits the usefulness of the simple amplifier circuits discussed so far.

To illustrate this point let us consider the simple amplifier of Examples 7.1 and 7.2 when used with a typical small signal bipolar transistor. In the examples we assumed that $h_{FE} = 100$. Typical general purpose transistors of a given type might have a spread of current gain in the range 80 to 350. If we repeat the calculations performed in the examples for different values of current gain, we get a range of results, as illustrated in Table 7.1.

From the table of results it is clear that the performance of the amplifier is greatly affected by the current gain of the transistor used. You will notice that the characteristics are only tabulated for values of current gain up to 200, even though the device may have a current gain up to 350. The reason for this will become clear shortly. If we look first at the quiescent collector current, we see that this rises linearly with the current gain. This is because the base current I_B is fixed and the collector current is simply the current gain times the base current. As the collector current rises the quiescent output voltage drops. You will remember that this is given by the expression

$$V_o = V_{CC} - I_C R_C$$

As I_C increases, the voltage drop across the load resistor increases and the output voltage falls. From the previous section we know that the available output swing of the amplifier is limited by the distance of the operating point from the supply voltage and from the saturation region. With a device with a current gain of 100, the operating point is positioned midway between these two limits. The quiescent output voltage is 5.2 V, allowing it to swing by nearly 5 V in either direction. When a device with a current gain of 80 is used the quiescent output voltage rises to 6.2 V. This allows the output to go positive by only a little under 4 V before it is limited by the supply rail. If the transistor has a current gain of 150 the quiescent output voltage is only 2.8 V permitting a maximum negative going excursion of only about 2.5 V before the device enters saturation. The extreme case is

Table 7.1 The effects of device variability on amplifier characteristics.

	Amplifier characteristics		
Current gain	**Quiescent collector current (mA)**	**Quiescent output voltage (V)**	**Voltage gain**
80	0.82	6.2	−147
100	1.02	5.2	−183
150	1.53	2.8	−275
200	2.04	0.4	−366

reached with a device with a current gain of 200. Here the quiescent output voltage is only 0.4 V which is at the edge of the non-linear saturation region allowing no useful output swing in the negative direction. For values of current gain above 200 the transistor is driven hard into saturation and the circuit cannot be used as a linear amplifier.

FILE 7A

Computer simulation exercise 7.3

Using the circuit from Computer simulation exercises 7.1 and 7.2, measure the characteristics of the amplifier for transistors of different gains. Determine the small-signal voltage gain and the quiescent output voltage of the circuit when using a transistor with a gain of: 80; 100; 150; 200; 250; 300 and 350.

The custom component library supplied with the demonstration files includes models for a number of transistors of various gains. These have the designations NPN80, NPN100, NPN150, etc., where the numeric part of the name refers to the current gain.

One apparent solution to these problems would appear to be to measure the current gain of the transistor to be used and to design the circuit appropriately. This is not an attractive solution in a mass production environment in which it would be totally impractical to design each circuit uniquely to match individual components. It would also cause severe problems when a device failed and needed to be replaced. An alternative would be to select devices within a particular close range of parameters. This is possible but is expensive, particularly if a very narrow spread is required.

The only practical solution to this problem is to design circuits that are not greatly affected by changes in the current gain of the transistors within them. Fortunately we have already looked at techniques to achieve this objective, namely the use of feedback.

7.6.6 The use of feedback

We have seen that amplifiers based on bipolar transistors suffer from problems associated with the variability of their gain. In Chapter 4 we saw how negative feedback could be used to overcome the problems associated with the variability of gain in operational amplifier circuits. We also noted that this led to circuits that were easier to design and understand. In Chapter 6 we discussed the use of feedback to overcome variability in the characteristics of FETs and saw how this simplified the design of biasing networks. It will perhaps come as no surprise that feedback can also be used in circuits using bipolar transistors and you will be pleased to hear that here too it produces simpler circuits.

An amplifier using negative feedback

Consider the circuit of Figure 7.18(a).

You will notice that an emitter resistor has been added. The voltage across this resistor is clearly proportional to the emitter current, which is (almost exactly) equal to the collector current. Since the input voltage is applied to the base, the voltage across the emitter resistor is effectively subtracted from the input to produce the voltage that is applied to the transistor. Therefore, we have **negative feedback** with a voltage

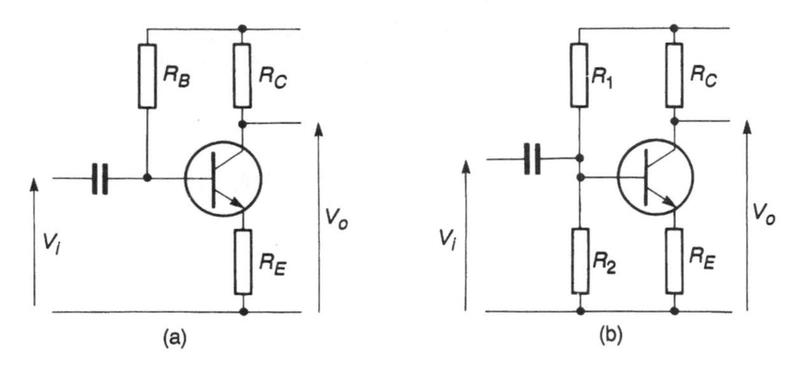

Figure 7.18 Amplifiers with negative feedback.

proportional to the output current being subtracted from the input. From the discussions of Section 4.6.3 it is clear that this will improve the constancy of the output current by **increasing** the **output resistance** and will improve the performance of the circuit as a voltage amplifier by **increasing** the **input resistance**. The voltage gain will also be stabilized, making it less affected by variations in the current gain of the device.

Since the input is no longer applied directly between the base and the emitter it is no longer appropriate to describe this circuit as a common-emitter amplifier. A more accurate description would be a **series feedback amplifier**, although the term does not describe this circuit uniquely.

A further development of this circuit is shown in Figure 7.18(b). Here two resistors are used to provide base bias rather than the single resistor used in the earlier circuit. This arrangement produces good stabilization of the DC operating conditions of the circuit and also gives an arrangement which is very easy to analyse. To illustrate this point let us consider both the large signal (DC) and the small signal (AC) characteristics of the circuit of Figure 7.18(b). As observed earlier, it is normally the designer's task to choose component values to obtain a given set of characteristics, rather than to analyse an existing circuit. However, we will first look at the behaviour of a given circuit and then turn our attention to the problem of designing for a given specification.

DC analysis of a negative feedback amplifier

In performing the DC analysis of the circuit it is useful to make two simplifying assumptions. These are

(1) That the DC current gain h_{FE} is large and therefore, in normal operation, the base current I_B can be neglected.

(2) That the DC base–emitter voltage V_{BE} is constant. Here we will assume that it has a value of 0.7 V.

We will discuss the validity of these assumptions later.

In analysing the DC characteristics of a circuit, the two quantities of greatest interest are the quiescent output voltage and the quiescent collector current. However, it is easier in this case to start by determining values for the quiescent base and emitter voltages since these lead directly to the quantities required.

Example 7.3 DC analysis of a series feedback amplifier

Consider the following circuit.

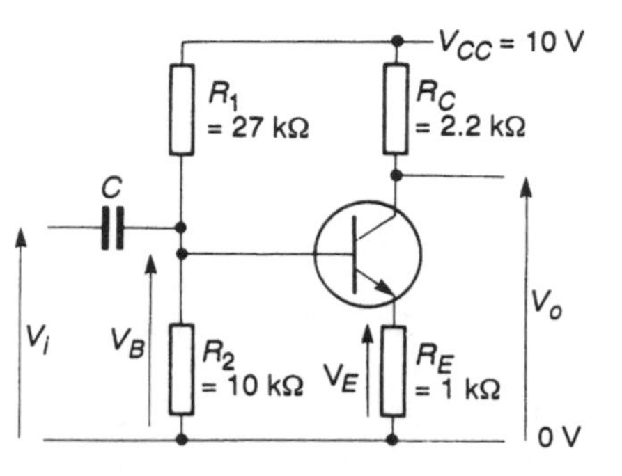

Quiescent base voltage

If we assume that the base current is negligible, since no constant current can flow through the input capacitor, the **quiescent base voltage** is determined simply by the supply voltage V_{CC} and by the potential divider formed by R_1 and R_2. Hence

$$V_B \approx V_{CC} \frac{R_2}{R_1 + R_2}$$

Therefore in our example

$$V_B \approx 10 \frac{10 \text{ k}\Omega}{27 \text{ k}\Omega + 10 \text{ k}\Omega} \approx 2.7 \text{ V}$$

Quiescent emitter voltage

Since the base-to-emitter voltage V_{BE} is assumed to be constant, it is simple to determine the emitter voltage from the base voltage. Thus the **quiescent emitter voltage** is simply

$$V_E = V_B - V_{BE}$$

and in our circuit

$$V_E = 2.7 - 0.7 = 2.0 \text{ V}$$

Quiescent emitter current

Knowing the voltage across the emitter resistor and its value gives us the emitter current

$$I_E = \frac{V_E}{R_E}$$

and therefore

$$I_E = \frac{2.0\ \text{V}}{1\ \text{k}\Omega} = 2\ \text{mA}$$

Quiescent collector current

If the base current is negligible, it follows that the collector current is equal to the emitter current

$$I_C \approx I_E$$

Therefore in our circuit

$$I_C \approx I_E = 2\ \text{mA}$$

Quiescent collector (output) voltage

In this circuit the output voltage is simply the collector voltage. This is determined by the supply voltage V_{CC} and the voltage across the collector resistor R_C. The voltage across R_C is simply the product of its resistance and the collector current and therefore

$$V_{o(quiescent)} = V_C = V_{CC} - I_C R_C$$

In this case

$$V_{o(quiescent)} = 10\ \text{V} - 2\ \text{mA} \times 2.2\ \Omega = 5.6\ \text{V}$$

It is clear from Example 7.3 that the DC conditions of such circuits can be determined quickly and easily. It is interesting to note that at no time in the analysis did we need to know the value of the current gain of the transistor. The fact that we have determined the quiescent conditions of the circuit without a knowledge of the current gain indicates that the gain does not directly affect the DC operation of the circuit. However, we did have to make some assumptions in order to perform the analysis in this way and it is perhaps useful at this point to look back at these assumptions in the light of the results obtained to see if they appear justifiable.

The first assumption made was that the DC current gain h_{FE} is large and that the base current may therefore be neglected. This assumption is required at two stages within the analysis. The first is when determining the base voltage V_B. By neglecting the effects of I_B the determination of V_B becomes trivial and independent of the actual value of I_B. This assumption would seem justifiable provided that the current flowing through the base resistors R_1 and R_2 is large compared with the base current. If we look at the actual circuit values in this example we see that the current I_{bias} flowing through the potential divider formed by R_1 and R_2 is given by

$$I_{bias} = \frac{V_{CC}}{(R_1 + R_2)} = \frac{10\ \text{V}}{(27\ \text{k}\Omega + 10\ \text{k}\Omega)} \approx 270\ \mu\text{A}$$

The base current is given by I_E/h_{FE} and therefore varies depending on the current gain of the transistor used. For a typical general purpose transistor h_{FE} might be in the range 80 to 350, and since I_E was calculated to be 2 mA, this gives a range for I_B of approximately 6 to 25 μA. Thus the base current will be small compared with the current through the potential divider for all values of the current gain within the given range and the base current can safely be ignored. The other stage of the analysis which required the assumption that the base current could be neglected was when it was assumed that the collector current was approximately equal to the emitter current. Again for typical values for h_{FE} of between 80 and 350 this assumption is reasonable.

The second assumption made in the analysis was that the base-to-emitter voltage V_{BE} was constant. This assumption was used when determining the emitter voltage from the base voltage. From Figure 7.6 we know that in practice V_{BE} is not constant but varies with the bias current. However, the figure also shows that the voltage is approximately constant for small fluctuations in the base current. Provided the variation of V_{BE} is small compared with the magnitude of the base voltage, it will have little effect on the value of the emitter voltage. In our example V_B is about 2.7 V, which is large compared with likely fluctuations of V_{BE} and the assumption is therefore reasonable.

We have looked at our two simplifying assumptions with respect to the circuit used in the example and find that they are justified. In other circuits using different component values they may not be and a more complicated analysis would be required. Generally, when designing circuits of this form we aim to ensure that these assumptions are valid, since if they are not, the effects of the negative feedback will not be fully utilized. In this circuit this means that the values of R_1 and R_2 should be chosen to ensure that the current in the potential divider chain is large compared with the base current, and V_B should be chosen to be large compared with likely fluctuations of V_{BE}. In meeting these requirements there may well be other factors to be considered, such as the AC characteristics described below. Inevitably design is a compromise between a number of conflicting requirements.

Computer simulation exercise 7.4

FILE 7B

Simulate the circuit of Example 7.3 using a transistor with a moderate gain (for example an NPN150) and investigate the quiescent voltages and currents. Compare the results with the values predicted in the text.

Replace the transistor with a high gain device (for example an NPN300) and compare the results.

AC analysis of a series feedback amplifier

As before the analysis of the circuit can be simplified if we make a few assumptions. These are

(1) That the small signal current gain h_{fe} is large and therefore the small signal base current i_b may be neglected.

(2) That the large signal base-to-emitter voltage V_{BE} is approximately constant and therefore the small signal base-to-emitter voltage v_{be} is very small.

It can be seen that these assumptions are directly equivalent to those made in the DC analysis earlier.

Clearly, a small signal equivalent circuit can be constructed and used to determine the AC characteristics of the circuit. However, even without the equivalent circuit it is fairly easy to determine the small signal gain of the arrangement, just using our simplifying assumptions. Example 7.4 shows such an analysis while Example 7.5 shows the use of an equivalent circuit to derive additional information about the AC behaviour of the circuit.

Example 7.4 AC analysis of a series feedback amplifier

Let us consider the AC performance of the circuit used in Example 7.3.

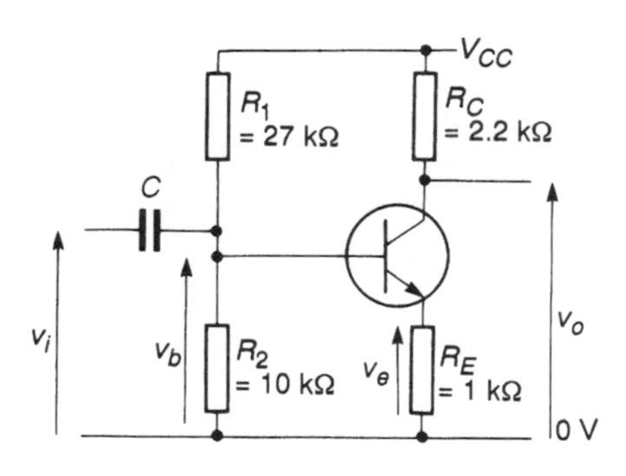

Small signal voltage gain

From the circuit diagram it is clear that the input signal is applied to the base of the transistor through the coupling capacitor C. Normally C would be chosen to have negligible impedance at the frequencies of interest and so it can be ignored. We will discuss later how to determine the effects of C if these are not negligible, but for the moment we will assume that the base voltage v_b is equal to the input voltage v_i.

From the second of our assumptions we have that v_{be} is very small and thus the *small signal* voltage on the emitter is effectively equal to that on the base. That is

$$v_e \approx v_b \approx v_i \tag{7.14}$$

Now, from Ohm's law we know that

$$i_e = \frac{v_e}{R_E}$$

and since

$$i_c \approx i_e$$

it follows that

$$v_o = -i_c R_C \approx -i_e R_C = -\frac{v_e}{R_E} R_C$$

where the '–' sign reflects the fact that the output voltage goes down when the current increases. If you expected to see V_{CC} in this expression, remember that the supply rail has no small signal voltages on it (see Section 6.5.1).

Substituting from Equation 7.14 gives

$$v_o = -v_e \frac{R_C}{R_E} \approx -v_i \frac{R_C}{R_E}$$

and therefore the voltage gain of the circuit is given by

$$\text{Voltage gain} = \frac{v_o}{v_i} \approx -\frac{R_C}{R_E} \tag{7.15}$$

For the component values used this gives

$$\text{Voltage gain} \approx -\frac{2.2\ \text{k}\Omega}{1.0\ \text{k}\Omega} \approx -2.2$$

We therefore have a very simple expression for the voltage gain of the circuit which relies only on the values of the passive components. It is interesting to compare the results of Equation 7.15 with the expression derived for the gain of a simple common-emitter amplifier given in Equation 7.13.

From Equation 7.13 we have

$$\text{Voltage gain (common-emitter)} \approx -g_m R_C = -\frac{R_C}{r_e}$$

Thus with feedback the gain is approximately R_C/R_E and without feedback it is approximately R_C/r_e. An important difference between these two arrangements is that with feedback the gain is determined by the ratio of two stable and well defined passive components. Without feedback the gain is controlled by r_e which varies with the transistor's operating conditions.

Example 7.5 AC analysis of a series feedback amplifier using a small signal equivalent circuit

For a more detailed view of the AC characteristics of the circuit we turn again to a small signal equivalent circuit.

The presence of the emitter resistor makes the equivalent circuit more complicated than that given in Figure 7.13. The emitter is connected to ground through the emitter resistor R_E, but the base resistors R_1 and R_2 and the collector resistor R_C are shown connected to ground since these resistors are connected either to ground or to V_{CC} (which is at ground potential for AC signals). The small signal voltage gain and the input and output resistances can be calculated directly from the equivalent circuit as follows.

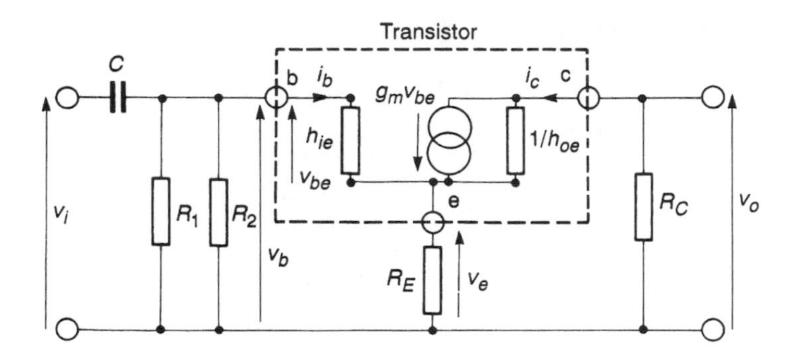

Small signal voltage gain

The analysis of Example 7.4 can be performed with reference to the equivalent circuit, producing an identical result. The equivalent circuit perhaps illustrates more clearly why the voltage gain of the amplifier is negative. If one ignores the effects of i_b and considers a current flowing from the current source, through R_E and then back through R_C, it is clear that the magnitude of the current flowing in each resistor is equal. However, since the current is flowing in opposite directions in the two resistors, the polarity of the voltage across each will be reversed. since the voltage across R_E is approximately equal to the input voltage, the output voltage will be inverted with respect to the input. Also since the same current flows through both R_C and R_E it is logical that the voltage gain will be the ratio of the resistor values. Therefore

$$\text{Voltage gain} = \frac{v_o}{v_i} \approx -\frac{R_C}{R_E}$$

as before.

From the expression for the voltage gain of the amplifier it would seem that using a very small value for R_E would produce a very high gain. Taking this to its logical conclusion we might expect that taking a vanishingly small value for R_E would result in an extremely high gain circuit. A moment's thought should make it clear that the gain of the resulting circuit cannot exceed that of the common-emitter amplifier of Figure 7.4, which is simply $-g_m R_C$, and thus a more complete expression for the gain is required.

If we ignore the effects of the output resistance $1/h_{oe}$ (this is normally large compared with R_E and R_C) and the base current (which is normally small compared with I_E), then the output voltage is given by

$$v_o = -g_m v_{be} R_C$$

since the current in R_C is equal to that produced by the current generator. The negative sign here reflects the fact that a positive current from the current generator produces a negative output voltage.

The input voltage to the amplifier is equal to the sum of the voltages across the base–emitter junction v_{be} and the emitter resistor v_e. Therefore

$$v_i = v_{be} + v_e$$

If we again ignore the effects of h_{oe} and i_b, then v_e is given by

$$v_e = g_m v_{be} R_E$$

and combining this with the previous expression gives

$$v_i = v_{be} + g_m v_{be} R_E$$

This may be rearranged to give

$$v_{be} = \frac{v_i}{1 + g_m R_E}$$

and combining this with the expression above for v_o gives

$$v_o = -\frac{g_m v_i R_C}{1 + g_m R_E}$$

Thus the voltage gain of the amplifier is given by

$$\text{Voltage gain} = \frac{v_o}{v_i} = -\frac{g_m R_C}{1 + g_m R_E}$$

$$= -\frac{R_C}{R_E + 1/g_m}$$

It can now be seen that if R_E is zero the gain becomes equal to $-g_m R_C$ as for the common-emitter amplifier. However, when R_E is much greater than $1/g_m$, the gain tends to $-R_C/R_E$ as before.

Notice that $1/g_m$ is equal to r_e, so the gain may be written as

$$\text{Voltage gain} = -\frac{R_C}{R_E + r_e}$$

and the gain is approximately equal to R_C/R_E when $R_E \gg r_e$.

Small signal input resistance

From the equivalent circuit it is apparent that the input resistance is formed by the parallel combination of R_1, R_2 and the resistance seen looking into the base of the transistor. This last term is not simply the sum of h_{ie} and R_E because of the effect of the current generator. When a current i_b enters the base of the transistor, the current which flows through R_E is the sum of the base current and the collector current $g_m v_{be}$. From Figure 7.8 it is clear that the collector current is equal to $h_{fe} i_b$, and thus the emitter current i_e is given by

$$i_e = i_b + h_{fe} i_b = (h_{fe} + 1) i_b$$

When a current flows through a resistor, a voltage drop is produced and the ratio of the voltage to the current determines the resistance of the component. When i_b flows into the base of the transistor, a much greater current flows in the emitter resistor producing a proportionately larger voltage drop. Therefore, the emitter resistor *appears* much larger when viewed from the base. In fact the emitter resistor appears to be increased by a factor of $(h_{fe} + 1)$. The input resistance h_{ie} is not amplified in this way since the current passing through this is simply i_b. Therefore the **effective input resistance** seen looking into the base of the transistor is

$$r_b = h_{ie} + (h_{fe} + 1)R_E \tag{7.16}$$

and the input resistance of the amplifier is

$$r_i = R_1//R_2//r_b \tag{7.17}$$

It is interesting to look at the relative magnitudes of the three components of the input resistance, as defined in Equation 7.17. R_1 and R_2 are base bias resistors and we noted when considering the DC performance of the circuit that these should be chosen such that the current flowing through them is large compared with the base current. This limits their maximum values, which typically might be a few kilohms or a few tens of kilohms. The magnitude of r_b is given in Equation 7.16. We know from Section 7.5.4 that a typical value for h_{ie} might be 500 Ω to 3 kΩ. Since R_E will typically be a few kilohms and h_{fe} will be perhaps 80 to 350, it is clear that it would be reasonable to use the approximation

$$r_b = h_{ie} + (h_{fe} + 1)R_E \approx h_{fe}R_E$$

From this it can be seen that r_b will generally be several hundred kilohms and will usually be large compared with R_1 and R_2. Therefore, since the three resistors are in parallel the effects of r_b will often be negligible, and it will be the parallel combination of R_1 and R_2 that will determine the input resistance of the amplifier.

If this is so

$$r_i \approx R_1//R_2$$

and in this circuit

$$r_i \approx 27\ \text{k}\Omega//10\ \text{k}\Omega \approx 7.3\ \text{k}\Omega$$

From this it is clear that to achieve a high input resistance it is desirable to make R_1 and R_2 as high as possible. This is in conflict with the biasing considerations discussed earlier, since it is necessary to ensure that the current flowing through R_1 and R_2 is large compared with the base current to ensure that the effects of the latter may be ignored. These opposing requirements are resolved by a compromise, where R_2 is typically chosen to be about ten times the value of R_E.

An advantage of this circuit, in comparison with the simple common-emitter amplifier, is that the input resistance is determined by the passive components within the circuit rather than by the transistor. This makes the circuit much more predictable and less affected by the characteristics of the active device used.

Small signal output resistance

In Example 7.2 we considered the output resistance of a simple common-emitter amplifier and deduced that this was equal to the parallel combination of R_C and $1/h_{oe}$. The addition of the emitter resistor R_E in the series feedback amplifier places an extra resistance in series with $1/h_{oe}$ as shown in the equivalent circuit. The addition of this resistance makes calculation of the output resistance somewhat more complicated. However, since $1/h_{oe}$ is normally much greater than R_C it follows that $1/h_{oe}$ in series with an additional resistance will also be greater than R_C and that the resistance of the parallel combination is dominated by R_C. Therefore in this example

$$r_o \approx R_C = 2.2\ \text{k}\Omega$$

As with the input resistance, the output resistance is determined by the passive components within the circuit rather than the transistor. This improves the predictability of the circuit by reducing its dependence on device characteristics.

In the light of the results of Examples 7.4 and 7.5 we can now re-examine the assumptions made and see if they appear appropriate for the AC analysis of the circuit.

The assumption that h_{fe} is large is used in determining the small signal voltage gain of the arrangement. Here we assumed that the current flowing in R_E is the same as that flowing in R_C. For a general purpose device with a gain of between 80 and 350 this assumption would seem valid.

The second assumption, that v_{be} is very small, was also used in determining the gain where it was assumed that the small signal voltage across the emitter resistor v_e was equal to the input voltage v_i. To investigate this assertion it is useful to consider the derivation of the input resistance of the circuit given in Example 7.5. Here it was observed that the input of the circuit appears as if it were the input resistance of the transistor h_{ie} in series with a resistance that is approximately h_{fe} times the emitter resistance R_E. This is illustrated in Figure 7.19.

The two resistors form a potential divider; the output voltage v_e is given by the expression

$$v_e = v_b \frac{h_{fe}R_E}{h_{ie} + h_{fe}R_E}$$

Thus the assumption that v_{be} may be neglected is valid provided that $h_{fe}R_E \gg h_{ie}$.

In our example R_E is 1.0 kΩ, so $h_{fe} \times R_E$ will be in the range 80 to 350 kΩ. This will always be much greater than h_{ie} which will be of the order of 1 kΩ. Therefore the effects of h_{ie} can be neglected and

$$v_e = v_b \frac{h_{fe}R_E}{h_{ie} + h_{fe}R_E} \approx v_b \frac{h_{fe}R_E}{h_{fe}R_E} \approx v_b$$

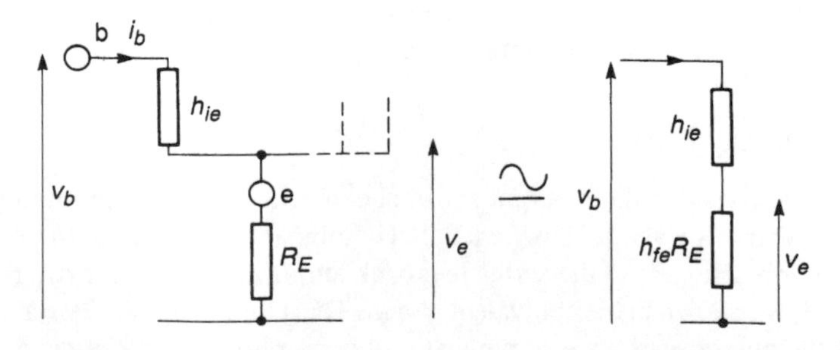

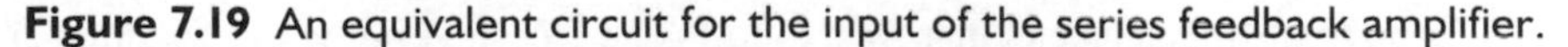

Figure 7.19 An equivalent circuit for the input of the series feedback amplifier.

Having looked at the large signal and small signal characteristics of the series feedback amplifier we have found that the circuit can be analysed very easily provided that a few simplifying assumptions are made. The assumptions required are effectively the same for both the DC and the AC case and may be summarized as follows:

(1) That the current gains h_{FE} and h_{fe} are high, and that therefore the base currents I_B and i_b may be neglected.

(2) That the steady state base–emitter voltage V_{BE} is approximately constant (at say 0.7 V) and that therefore the small signal base–emitter voltage v_{be} is very small.

It should be remembered that these assumptions will not always be justified. It is up to the designer to verify at the end of his design that the approximations made are appropriate.

Computer simulation exercise 7.5

FILE 7C

Simulate the circuit of Example 7.3 using a transistor with a moderate gain (for example an NPN150) and investigate its AC characteristics. Use a coupling capacitor of 1 μF and an input voltage of 50 mV peak at 1 kHz.

Measure the small-signal voltage gain and estimate the input and output resistance (see Computer simulation exercise 7.2 for suggestions on how to measure these quantities).

Replace the transistor with a high gain device (for example an NPN300) and compare the results.

The effects of a coupling capacitor

We noted earlier that if a **coupling capacitor** is used its value is normally chosen so that its effects are negligible at the frequencies of interest. Coupling capacitors are used to prevent the DC component of any input signal from upsetting the biasing of the transistor, and we noted in Section 3.7.1 that their use produces a **low-frequency cut-off**. The frequency of this cut-off was shown to be given by

$$f_{co} = \frac{1}{2\pi CR}$$

where C is the value of the coupling capacitor and R is the input resistance of the amplifier (ignoring the effects of the source resistance).

It is worthy of note that, in this case, the use of negative feedback increases the input resistance of the amplifier, providing a lower cut-off frequency for a given value of C.

Computer simulation exercise 7.6

FILE 7D

Having estimated the input resistance of the circuit used in Computer simulation exercise 7.5, calculate the low frequency cut-off produced by the coupling capacitor.

Use an AC sweep to measure the frequency response of the amplifier and compare the actual cut-off frequency with the calculated value.

Design of an amplifier to meet a given specification

So far we have looked at the analysis of existing circuits and have used this to increase our understanding of the operation of these circuits. We will now turn our attention to the process of designing an amplifier for a given task. The circuit used in Examples 7.3 to 7.5 gives us a good blueprint for a design, but we need to choose component values to suit our purpose. The process involved in this choice will vary from one application to another, since the specification will differ in terms of the parameters that are specified. In some cases the supply voltage and quiescent output voltage may be given, in others the quiescent current may be of importance. An example of the design process is given in Example 7.6.

Example 7.6 Design of a single-stage transistor amplifier

Design a single-stage transistor amplifier with a small signal voltage gain of –4 and a maximum output swing of 10 V peak-to-peak (when used with a high impedance load) which operates from a 15 V supply line. The amplifier should be AC coupled but should have a gain that is approximately constant down to 100 Hz.

We will use an amplifier with an emitter resistor and potential divider biasing, as in earlier examples.

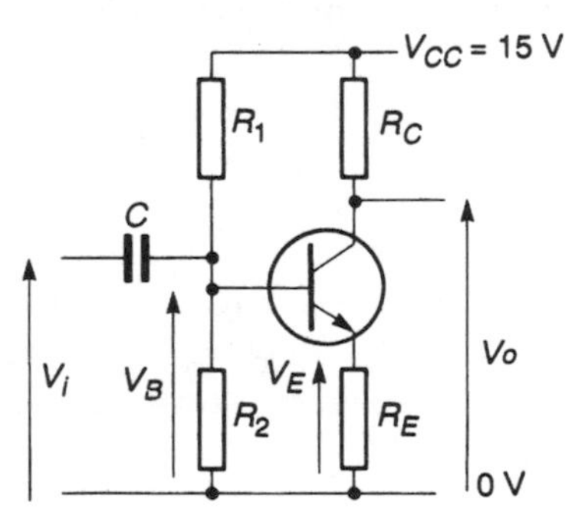

Designing circuits is not a precise art. There is no unique ideal solution and often we use 'rules-of-thumb' to simplify component choice.

Quiescent output voltage and collector current

One of our first tasks is to decide on appropriate values for the quiescent output voltage and the quiescent collector current. The first of these is constrained by the relatively large required output swing. In order for the output to produce 10 V peak-to-peak output, it must be able to go above and below its quiescent value by at least 5 V. In order to leave a reasonable voltage across the emitter resistor (to increase stability) let us choose to make the quiescent output voltage about 5.5 V below V_{CC}, therefore

$$V_{C(quiescent)} \approx V_{CC} - 5.5 = 9.5 \text{ V}$$

The choice of the quiescent collector current is fairly arbitrary, since the load impedance is high. Let us choose a value of 1 mA.

This immediately allows us to calculate an appropriate value for R_C, since

$$V_{C(quiescent)} = V_{CC} - I_{C(quiescent)} R_C$$

and hence

$$R_C = \frac{V_{CC} - V_{C(quiescent)}}{I_{C(quiescent)}} = \frac{15.0\ \text{V} - 9.5\ \text{V}}{1\ \text{mA}} = 5.5\ \text{k}\Omega$$

5.5 kΩ is not a standard resistor value and since the value is not critical we choose the closest standard value which is 5.6 kΩ. Therefore

$$R_C = 5.6\ \text{k}\Omega$$

Small signal voltage gain

From our discussions in Example 7.5 we know that

$$\text{Voltage gain} = -\frac{R_C}{R_E + r_e}$$

Therefore, in order that the gain is determined by the passive components in the circuit we require $R_E \gg r_e$, and in order to obtain a gain of −4 we require $R_C = 4R_E$.

With a quiescent collector current of 1 mA the emitter current is also 1 mA and the emitter resistance is given by

$$r_e \approx \frac{1}{40 I_E} \approx \frac{1}{40 \times 10^{-3}} \approx 25\ \Omega$$

If R_C is equal to 5.6 kΩ, then R_E should be 5.6/4 = 1.4 kΩ, to give a gain of −4. This value satisfies the condition that $R_E \gg r_e$.

Again 1.4 kΩ is not a standard resistor value. If it is vital that the gain is very close to −4, this value could easily be achieved by combining two high tolerance resistors of appropriate values. Alternatively, we choose the closest standard value and accept that the actual gain will be slightly above or below the specified value. In this case let us choose $R_E = 1.3$ kΩ. This gives a value for the gain of

$$\text{Voltage gain} = -\frac{R_C}{R_E + r_e} \approx -\frac{5.6\ \text{k}\Omega}{1.3\ \text{k}\Omega + 25\Omega} \approx -4.2$$

Base bias resistors

If the quiescent emitter current is to be about 1 mA and R_E is 1.3 kΩ, the quiescent emitter voltage must be given by

$$V_{E(quiescent)} = I_{E(quiescent)} \times R_E = 10^{-3} \times 1.3 \times 10^{3} = 1.3\ \text{V}$$

To achieve this emitter voltage the base must be biased to $1.3 + 0.7 = 2.0$ V. The ratio of R_1 to R_2 must therefore be determined by the relationship

$$\frac{R_2}{R_1 + R_2} V_{CC} = 2.0 \text{ V}$$

Choice of the absolute values of the base resistors is a compromise between high values, which give a high input resistance to the circuit, and low values, which make the current through the bias resistors large compared with the base current. The usual 'rule-of-thumb' solution is to choose R_2 to be approximately 10 times R_E. Therefore R_2 becomes 13 kΩ. Rearranging the above expression we have

$$R_1 = \frac{R_2(V_{CC} - 2.0)}{2.0} = \frac{13 \text{ k}\Omega(15.0 - 2.0)}{2.0} = \frac{13 \text{ k}\Omega(15.0 - 2.0)}{2.0} = 84.5 \text{ k}\Omega$$

The nearest standard value for R_1 is 82 kΩ. The use of this value raises the base voltage slightly above 2.0 V, which in turn increases the emitter current and reduces the quiescent collector voltage. Calculation of these values is left as an exercise for the reader.

Input resistance and the choice of *C*

From our earlier discussions we know that the input resistance is given approximately by the parallel combination of R_1 and R_2. Therefore in this case

$$\text{Input resistance} \approx R_1 // R_2 = 13 \text{ k}\Omega // 82 \text{ k}\Omega \approx 11.2 \text{ k}\Omega$$

From Section 3.7.1 we know that, if the effects of the source resistance are neglected, the presence of the coupling capacitor will produce a low-frequency cut-off at a frequency given by

$$f_{co} = \frac{1}{2\pi CR}$$

where R is the input resistance of the amplifier.

In this example we require the gain to be approximately constant down to 100 Hz. We therefore choose a lower cut-off frequency of $100/10 = 10$ Hz, which gives a value for C of

$$C = \frac{1}{2\pi f_{co} R} = \frac{1}{2\pi \times 10 \times 11.2 \times 10^3} = 1.4 \text{ µF}$$

Therefore a non-polarized capacitor of more than 1.4 µF would be used, for example a 2.2 µF polyester type.

FILE 7E

Computer simulation exercise 7.7

Simulate the circuit designed in Example 7.6 and compare its performance with the original specification.

7.6.7 Use of a decoupling capacitor

A comparison of the results of Example 7.2, which looked at a simple common-emitter amplifier, and Example 7.4, which was concerned with a series feedback amplifier, indicates that the use of feedback has several advantages. These include the stabilization of circuit parameters, such as voltage gain and input and output resistances, which become less affected by changes in the transistor's characteristics. However, a comparison between the two examples will show that this is achieved at the expense of a considerable reduction in voltage gain. This fall in gain is a direct result of the negative feedback incorporated. In some applications this reduction in gain is unacceptable and a **decoupling capacitor** is used to reduce the amount of AC negative feedback while maintaining DC feedback. This increases the small signal gain of the circuit but does not affect the DC feedback, which provides stability to the bias conditions of the circuit.

The use of a decoupling capacitor is illustrated in Figure 7.20. The capacitor C_E is placed across the emitter resistor R_E, providing a low impedance path for alternating signals from the emitter to ground but having no effect on the steady bias voltages. Since the emitter is now effectively connected to ground for small signal inputs, we can accurately describe this amplifier as a **common-emitter amplifier** since the emitter is common to both the input and the output circuit.

The decoupling capacitor does not change the DC performance of the circuit since the capacitor is effectively an infinite impedance at DC. Therefore the quiescent conditions of the circuit are unaffected.

For AC signals, the capacitor looks like a low impedance. Its value would normally be chosen so that at the frequencies to be used by the circuit the capacitor looks like a short circuit. A small signal equivalent circuit for the amplifier of Figure 7.20 is shown in Figure 7.21.

The presence of the decoupling capacitor effectively removes R_E from the small signal equivalent circuit, producing an arrangement that is almost identical to that of the simple common-emitter amplifier shown in Figure 7. 13. The only difference between these two circuits is the number of resistors that connect the base to ground. It follows that the AC analysis of the two circuits is similar in respect of their voltage gain and their input and output resistances. The differences in the calculations are simply that R_B in the analysis

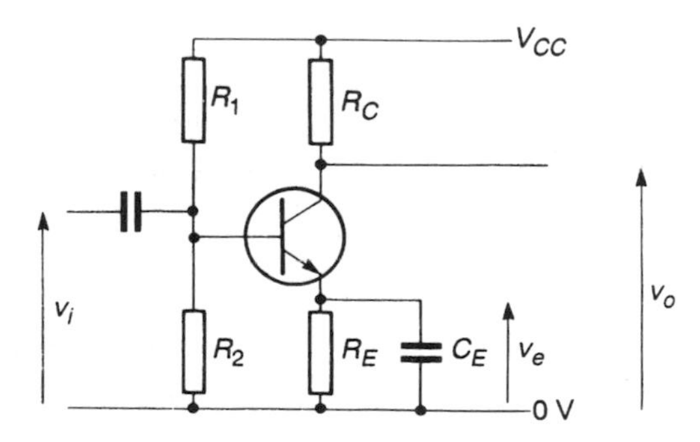

Figure 7.20 Use of a decoupling capacitor.

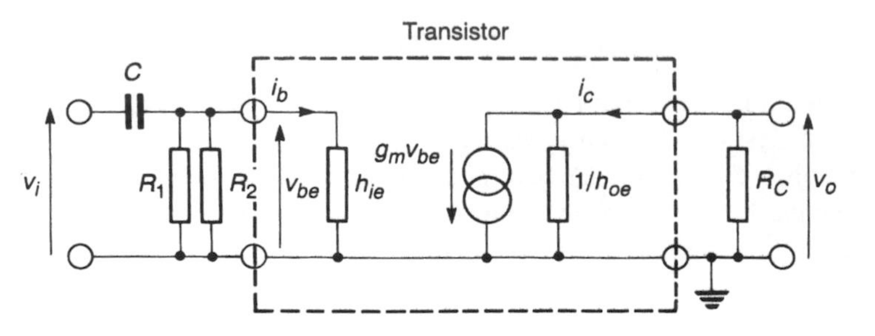

Figure 7.21 Small signal equivalent circuit of an amplifier using a decoupling capcitor.

of the common-emitter amplifier is replaced by the resistance of the parallel combination of R_1 and R_2 when using the capacitively decoupled amplifier.

Although the decoupling capacitor will normally be chosen to provide a low impedance at the frequencies of interest, at low frequencies its effects will diminish and the small signal gain will fall. The point at which this occurs will be determined by the frequency at which the impedance of the decoupling capacitor becomes appreciable compared with the resistance within the circuit across the capacitor, that is, between the emitter and ground. This resistance is given by the parallel combination of the emitter resistor R_E and the resistance seen looking into the emitter r_e. The point at which the impedance of the capacitor matches the effective resistance across it determines the **low-frequency cut-off**. Therefore

$$\omega_{co} = \frac{1}{C_E(R_E//r_e)}$$

and hence

$$f_{co} = \frac{1}{2\pi C_E(R_E//r_e)}$$

From Equation 7.6 we know that

$$r_e \approx \frac{1}{40 I_E}$$

It follows that it will have a typical value of a few ohms and generally will be much smaller than R_E. We can therefore approximate f_{co} by the expression

$$f_{co} \approx \frac{1}{2\pi C_E r_e} \tag{7.18}$$

This expression can be used to calculate the size of decoupling capacitor that will be required to produce a sufficiently low cut-off frequency for a given application. Note that the low value of r_e means that the capacitor must be considerably larger than would be required if the cut-off frequency were determined by the time constant $C_E R_E$. This can prove inconvenient in applications where good low-frequency response is required since large capacitors may be both expensive and bulky.

Below the cut-off frequency the gain drops at 6 dB/octave, as described in Section 3.7.1, until the impedance of C_E becomes comparable with R_E. Below this point the gain levels out as R_E tends to dominate the response. The frequency at which this occurs is given by

$$f_1 = \frac{1}{2\pi C_E R_E}$$

The resultant frequency response of the amplifier is shown in Figure 7.22, which also shows the equivalent responses for a simple common-emitter amplifier and a series feed-

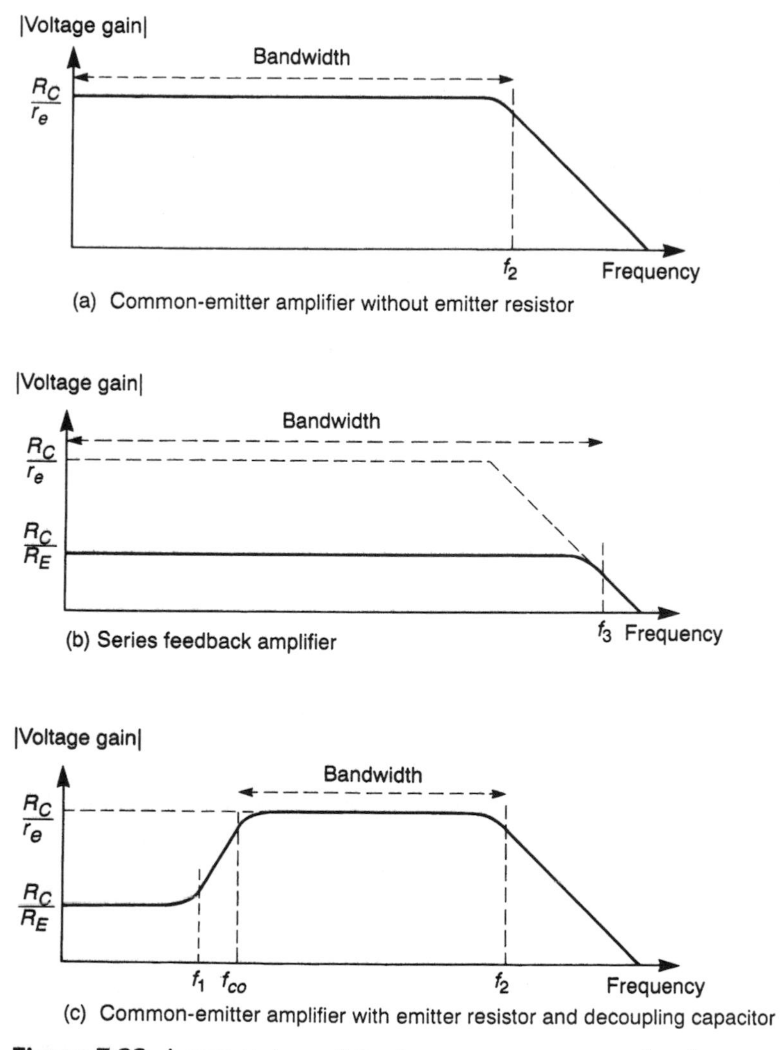

Figure 7.22 A comparison of the frequency responses of various amplifiers.

back amplifier, using the same transistor and component values (with appropriate biasing arrangements). The figure shows the responses of amplifiers that are *not* fitted with coupling capacitors.

From Figure 7.22(a) it is apparent that the usable bandwidth of the simple common-emitter amplifier extends down to DC and is limited at its upper end by the frequency response of the transistor. The **mid-band gain** of the amplifier is approximately R_C/r_e and, since r_e is a function of I_E, will vary considerably with the h_{FE} of the transistor used. Figure 7.22(b) shows the response of the series feedback amplifier with the response of the simple common-emitter amplifier shown as a dotted line for comparison. The response of the series feedback amplifier also extends down to DC but the gain is considerably less than the common-emitter circuit at R_C/R_E. The gain, however, is determined by stable and well defined passive components rather than by the characteristics of the transistor. The upper cut-off frequency of the amplifier is increased from f_2 to f_3 by the negative feedback, and the bandwidth of the circuit is thus increased. The response of the common-emitter amplifier using an emitter resistor and a decoupling capacitor is shown in Figure 7.22(c). This has a gain of R_C/R_E at low frequencies, which then rises to R_C/r_e as the effects of the decoupling capacitor come into play. This produces a low-frequency cut-off frequency at f_{co}. The high frequency response would appear, at first sight, to be unaffected by the addition of the emitter resistor and decoupling capacitor. However, from Equation 7.6 we see that

$$r_e \approx \frac{1}{40I_E}$$

Therefore, since the presence of the feedback resistor stabilizes the emitter current, it also stabilizes the gain of the amplifier.

A variant on the circuit of Figure 7.20 uses two resistors in series in place of R_E, with a decoupling capacitor connected across only the resistor connected to ground. This use of **split emitter resistors** allows the total emitter resistance to be tailored to suit the biasing requirements of the circuit, while permitting only part of this resistance to be decoupled to produce the required small signal performance.

The use of an emitter resistor and a decoupling capacitor produces an amplifier that has a high gain for AC signals but which uses negative feedback to stabilize the DC operating conditions. A disadvantage of this approach is that the arrangement has a frequency response that may be inconvenient in some situations. Circuits that must operate at low frequencies will require very large decoupling capacitors to produce a sufficiently low cut-off frequency. In many applications it is preferable to use several stages of amplification using feedback rather than a single stage which requires a large decoupling capacitor. In discrete circuits, transistors are much less expensive and smaller than large capacitors. In integrated circuits, transistors require much less chip area than small capacitors while large capacitors are completely impractical.

FILE 7F

Computer simulation exercise 7.8

Modify the circuit used in the previous computer simulation exercise by the addition of a 1 µF emitter resistor. Measure the frequency response of this circuit and compare its cut-off points with those predicted in the text.

It is very important to distinguish between the functions of coupling and decoupling capacitors.

(1) **Coupling capacitors** are used to couple one stage of a circuit to the next by passing the AC component of a signal while preventing the DC component from one stage affecting the biasing conditions of the next.

(2) **Decoupling capacitors** remove the AC component of a signal by shorting it to ground, while leaving the DC component unchanged. They therefore decouple a node of the circuit from the AC signal.

7.6.8 Amplifier configurations

So far we have concentrated on circuits in which the input is applied to the base of the transistor and the output is taken from the collector. The common-emitter amplifier is the simplest of these. It is also possible to construct circuits using other transistor configurations in which the input and output are applied to other nodes. Examples of these configurations are shown in Figure 7.23.

Of these arrangements, the common-emitter configuration is by far the most widely used, which is why the text so far has concentrated on such circuits. However, the other configurations are of interest since they have different characteristics that are utilized for particular applications.

Common-collector amplifiers

Figure 7.24 shows the circuit of a simple common-collector amplifier. The circuit is similar to the series feedback circuit used earlier except that the output is taken from the emitter rather than the collector, eliminating the need for a collector resistor. The collector is connected directly to the positive supply which is at earth potential for AC signals since there are no AC voltages between the supply and ground (see Section 6.5.1). Input signals are applied between the base and ground; output signals are measured between the emitter and ground. Since the collector is at ground potential, for small signals the collector is common to both the input and output circuits. Hence the arrangement is a **common-collector** amplifier.

As with the series feedback amplifier considered earlier, this circuit uses **negative feedback**. However, in this case, the voltage that is subtracted from the input is related

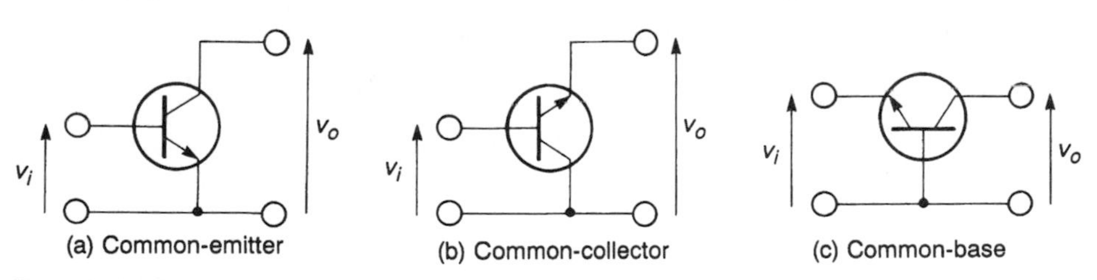

Figure 7.23 Transistor circuit configurations.

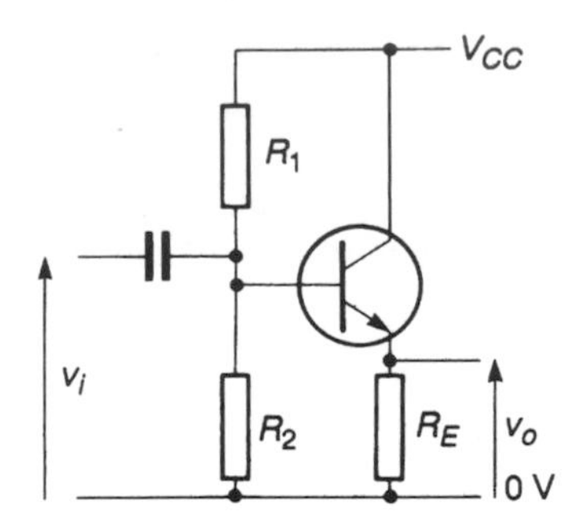

Figure 7.24 A common-collector amplifier.

to the output *voltage* rather than the output current since the emitter voltage *is* the output voltage of the emitter follower. From the discussion of Section 4.6.3 it is clear that this will *increase* the input resistance, as before, but *decrease* the output resistance. Analysis of the circuit is very much simpler than that of the series feedback amplifier, and having done the groundwork, it is easy to write down its characteristics.

The DC analysis of the circuit is similar to that of the series feedback amplifier except that fewer steps are required. Once the quiescent base voltage has been determined, the quiescent emitter voltage is found by subtracting the (constant) base-to-emitter voltage V_{BE} which gives the quiescent output voltage directly. The quiescent emitter current is found by dividing the emitter voltage by the value of the emitter resistor R_E.

The AC analysis is also straightforward. One of the assumptions made when looking at the series feedback amplifier was that the small signal base-to-emitter voltage v_{be} was negligible. This implies that the emitter tracks the base with some constant offset voltage (typically about 0.7 V). Thus the voltage gain of the common-collector amplifier is approximately unity and the emitter simply follows the input signal. For this reason, this form of amplifier is often called an **emitter follower** amplifier. The small signal input resistance is calculated in the same manner as for the series feedback amplifier (see Example 7.5) and thus

$$r_i = R_1 // R_2 // r_b$$

where r_b is the resistance seen looking into the base. Since, as before, r_b is given by

$$r_b = h_{ie} + (h_{fe} + 1)R_E$$

r_b is again likely to be large compared with R_1 and R_2 and its effects may be ignored. Therefore, as for the series feedback amplifier

$$r_i \approx R_1 // R_2$$

The output resistance of the amplifier is given by the parallel combination of the emitter resistor R_E and the resistance seen looking back into the emitter r_e. From Equation 7.6 we know that

$$r_e \approx \frac{1}{40 I_E}$$

which means that r_e will usually be of the order of a few ohms or tens of ohms, and will dominate the output resistance of the circuit. Therefore

$$r_{out} \approx r_e \approx \frac{1}{40I_E}$$

The fact that the emitter follower has a voltage gain of approximately unity might at first sight appear to make it of little use. However, it is its relatively high input impedance and very low output impedance that make it of interest as a **unity gain buffer amplifier**. You might like to compare the characteristics of the emitter follower with those of the source follower described in Section 6.5.6. In doing so, remember that in Section 7.5.2 we deduced that, for the bipolar transistor, $r_e = 1/g_m$.

Common-base amplifiers

The common-base configuration is the least widely used transistor arrangement, but does have characteristics that make it of interest. Figure 7.25 shows a simple common-base arrangement.

The circuit is similar to the series feedback amplifier used earlier except that the base is connected to ground through a capacitor, placing it at earth potential for AC signals. The input is applied between the emitter and ground and the output is measured between the collector and ground. Since the base is at ground potential as far as small signals are concerned, the input and output are effectively measured with respect to the base and the base is thus common to the input and the output circuits. This arrangement is therefore a common-base amplifier.

Since the presence or absence of AC signals makes no difference to the quiescent state of the circuit, the DC analysis is identical to that of the series feedback amplifier performed in Example 7.4.

We found when considering the emitter follower circuit that the output resistance was very low ($r_e//R_E$). It is perhaps no surprise that if we use a similar arrangement with the emitter as the input terminal, this will produce a very low *input* resistance. In fact, the input resistance is also equal to $r_e//R_E$ and, as before, r_e dominates, giving

$$r_i \approx r_e$$

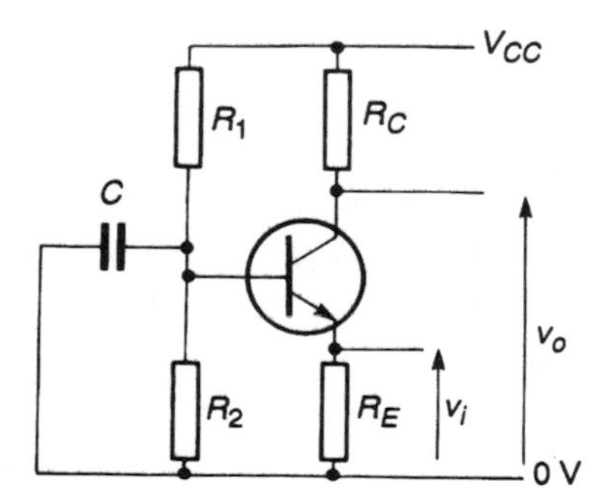

Figure 7.25 A common-base amplifier.

The output arrangement is similar to that used in the common-emitter amplifier and produces a similar (though not identical) output resistance which is dominated by the collector resistor. Therefore

$$r_o \approx R_C$$

Since the base is at earth potential for AC signals, it follows that making the input (the emitter) more positive, *reduces* the voltage across the base–emitter junction and reduces the emitter current. This in turn reduces the collector current which increases the output voltage. Thus this amplifier is a **non-inverting amplifier**. Analysis shows that the gain is given by

$$\text{Voltage gain} \approx g_m R_C \approx \frac{R_C}{r_e}$$

This is of the same magnitude as the gain of the simple common-emitter amplifier (not the series feedback amplifier which it more closely resembles) but with opposite polarity.

Since the emitter and collector currents are almost identical, the current gain of the common-base amplifier is approximately unity.

Common-base amplifiers are characterized by a low input resistance and a high output resistance. These characteristics are not generally associated with good voltage amplifiers but make them useful as **trans-impedance amplifiers** (amplifiers that take an input *current* and produce a related output *voltage*). The common-base configuration is also often used in **cascode amplifiers** (not cascade amplifiers) where it is combined with a common emitter amplifier. The common-base stage then provides voltage gain and the common-emitter stage gives current gain.

Amplifier configurations – a summary

So far we have considered three basic circuit configurations for bipolar transistors, namely common-emitter, common-collector and common-base. These arrangements are shown in Figures 7.18(b), 7.24 and 7.25, respectively.

It is interesting to compare these three configurations in terms of a few key characteristics. Table 7.2 shows approximate maximum values for these characteristics for the circuits given in the figures.

7.6.9 Cascaded amplifiers

If more gain is required than can be obtained from a single amplifier, many stages can be **cascaded** by connecting them in series. Indeed, we have seen that it is often better to use many stages, each with a relatively low gain, than to use one high gain stage.

When connecting the output of one stage to the input of the next, it is important to ensure that the bias conditions of the circuits are not affected. One way of ensuring this is to use **coupling capacitors** between each stage. These pass AC signals, but prevent DC voltages from one stage upsetting the bias conditions of the next. Figure 7.26 shows an example of such an arrangement.

The bias conditions of each stage can be analysed separately since they are effect-

Table 7.2 A comparison of amplifier configurations.

	Common-emitter	Common-collector	Common-base
Input terminal	base	base	emitter
Output terminal	collector	emitter	collector
Voltage gain A_v	$-g_m R_C$ (high)	≈ 1 (unity)	$g_m R_C$ (high)
Current gain A_i	$-h_{fe}$ (high)	h_{fe} (high)	≈ -1 (unity)
Power gain A_p	$A_v A_i$ (very high)	$\approx A_i$ (high)	$\approx A_v$ (high)
Input impedance	$R_1//R_2$ (moderate)	$R_1//R_2$ (moderate)	$\approx r_e$ (very low)
Output impedance	$R_C//\frac{1}{h_{oe}}$ (moderate)	$\approx r_e$ (very low)	$\approx R_C$ (high)
Phase shift (mid-band)	180°	0°	0°

ively isolated as far as DC signals are concerned. The small signal performance of the combination may be determined by calculating the input and output resistances and the gain of each stage, and combining these as described in Section 3.9. If the output resistance of one stage is low compared with the input resistance of the next, then the effects of **loading** can normally be ignored.

The use of capacitive coupling, although simple in concept, does have some disadvantages. Firstly, each coupling capacitor produces a **low-frequency cut-off** which limits the low-frequency response of the amplifier, as described in Section 3.7.1. Secondly, the presence of capacitors makes the circuit more expensive and less suitable for the production of integrated circuits since capacitors require a large amount of 'chip' area.

If care is taken with the design of the circuit it is possible to dispense with the use of coupling capacitors between the stages of cascaded amplifiers by ensuring that the quiescent output voltage of one stage represents the correct biasing voltage for the next. This not only removes the need for coupling capacitors, but also reduces the complexity of

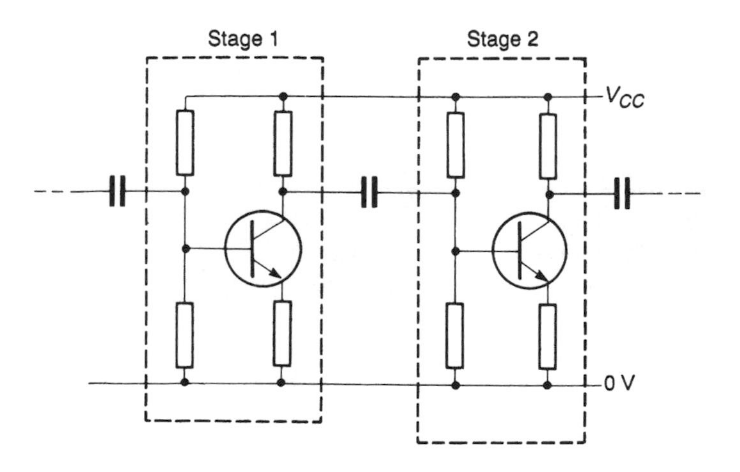

Figure 7.26 Capacitive coupling between amplifier stages.

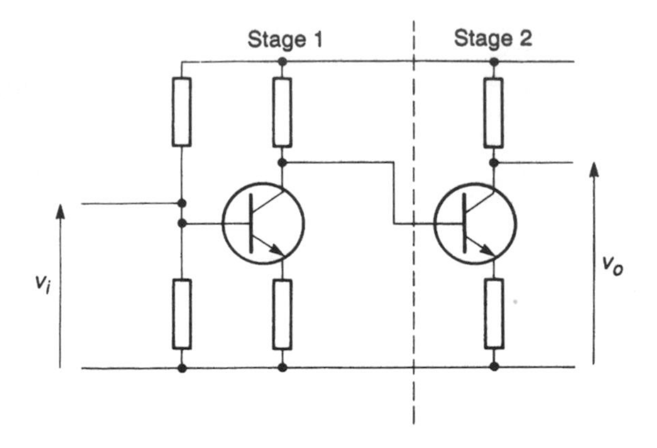

Figure 7.27 A two-stage DC coupled amplifier.

the biasing circuitry required. An example of an amplifier of this form is shown in Figure 7.27.

A major advantage of this technique is that it removes the frequency limitations introduced by capacitive coupling, allowing the production of amplifiers that can be used at frequencies down to DC. Amplifiers that have a low-frequency cut-off as a result of coupling capacitors are often called **AC coupled amplifiers**. Circuits that have no capacitive coupling, and are therefore able to amplify signals down to DC, are referred to as **DC coupled** or **directly coupled** amplifiers. In applications where the signals to be amplified include DC values, directly coupled amplifiers are essential.

Analysis of directly coupled amplifiers is very straightforward, as is illustrated in Example 7.7.

Example 7.7 Analysis of a two-stage directly coupled amplifier

Calculate the quiescent output voltage and the small signal voltage gain of the following circuit.

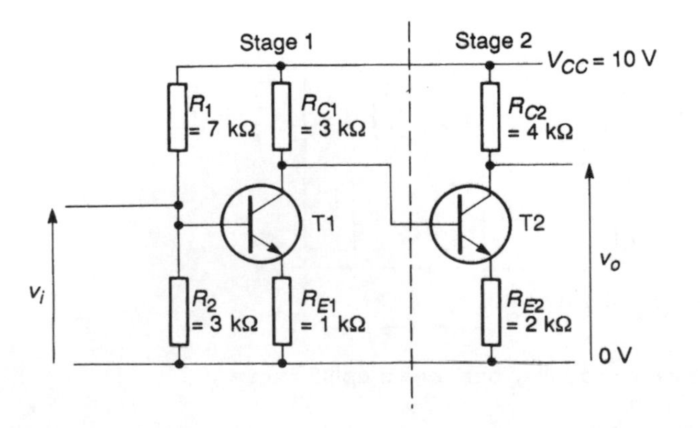

If we adopt the notation that V_{B1} is the voltage on the base of T1, V_{C2} is the voltage on the collector of T2, etc., it follows that

$$V_{B1} = V_{CC}\frac{R_1}{R_1 + R_2} = 10\frac{3\ \text{k}\Omega}{7\ \text{k}\Omega + 3\ \text{k}\Omega} = 3\text{V}$$

$$V_{E1} = V_{B1} - V_{BE} = 3.0 - 0.7 = 2.3\ \text{V}$$

$$I_{C1} \approx I_{E1} = \frac{V_{E1}}{R_{E1}} = \frac{2.3\ \text{V}}{1\ \text{k}\Omega} = 2.3\ \text{mA}$$

and therefore

$$V_{C1} = V_{CC} - I_{C1}R_{C1} = 10\ \text{V} - 2.3\ \text{mA} \times 3\ \text{k}\Omega = 3.1\ \text{V}$$

V_{C1} forms the bias voltage V_{B2} for the second stage, thus

$$\begin{aligned} V_{E2} &= V_{B2} - V_{BE} \\ &= 3.1 - 0.7 \\ &= 2.4\ \text{V} \end{aligned}$$

$$\begin{aligned} I_{C2} \approx I_{E2} &= \frac{V_{E2}}{R_{E2}} \\ &= \frac{2.4\ \text{V}}{2\ \text{k}\Omega} \\ &= 1.2\ \text{mA} \end{aligned}$$

and hence

$$\begin{aligned} \text{Quiescent output voltage} &= V_{C2} = V_{CC} - I_{C2}R_{C2} \\ &= 10\ \text{V} - 1.2\ \text{mA} \times 4\ \text{k}\Omega \\ &= 5.2\ V \end{aligned}$$

Calculation of the voltage gain of the amplifier is also straightforward. In the absence of additional base bias resistors the input resistance of the second stage is at least h_{fe} times R_{E2} (see Example 7.5). This is likely to be greater than 100 kΩ and is certainly large compared with the output resistance of the previous stage, which must be less than R_{C1} (4 kΩ). Therefore loading effects can be ignored and the gain of the combination is simply the product of the gains of the two stages when considered separately. These gains are given simply by the ratios of the collector and emitter resistors (see Example 7.4) and therefore the first stage has a voltage gain of 3 and the second a gain of 2. The combination therefore has a voltage gain of 6.

The process of designing multi-stage amplifiers is similar to that outlined in Example 7.6, except that it must be performed over each section in turn, beginning with the output stage. The specified output swing and output current of the final stage will generally determine its design, and the required biasing conditions of this stage then determine the form of the preceding section. This process is then repeated working back towards the input.

Computer simulation exercise 7.9

Simulate the circuit of Example 7.7 and determine its DC and AC characteristics. Apply an input signal of 0.1 volts peak at 1 kHz.

Compare the measured values with those predicted in the text.

7.6.10 Darlington transistors

An interesting method of combining two or more transistors is the **darlington connection** illustrated in Figure 7.28(a).

The current gain of the first transistor is multiplied by that of the second to produce a combination that acts like a single transistor with an h_{fe} equal to the product of the gains of the two transistors. The two devices are often available within single packages, which are called **darlington transistors**. Looking at the form of the device it is clear that in operation the offset voltage between the input (base) and the output (emitter) of the circuit will be twice that of a single transistor at about 1.4 V. The saturation voltage for the device is also greater than that of a single transistor at slightly more than the conduction voltage of a single junction (about 0.7 V). This is because the emitter of the input transistor must be at least this voltage above the emitter voltage of the output transistor in order for the latter to be turned ON. The collector of the input transistor cannot then be below the voltage on its own emitter.

One disadvantage of the arrangement shown in Figure 7.28(a) is that if T1 is turned OFF rapidly, it stops conducting and there is no path by which the stored charge in the base of T2 can be quickly removed. T2 thus responds relatively slowly. This problem can be resolved by adding a resistor across the base–emitter junction of T2 as shown in Figure 7.28(b). If now T1 is turned OFF, R provides a conduction path to remove the charge stored in T2. R also prevents the small leakage current from T1 being amplified by T2 to produce a significant output current. R is chosen such that the voltage drop across it caused by the leakage current is smaller than the turn-on voltage of T2. Typical values for R might be a few kilohms in a small signal darlington down to a few hundred ohms in a power device. R is normally included within the darlington package.

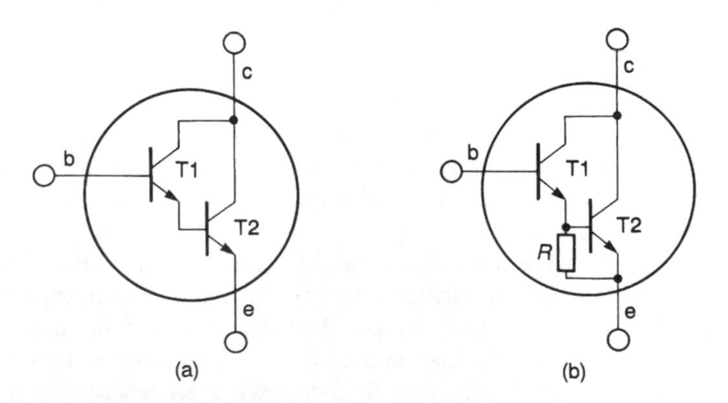

Figure 7.28 The darlington connection.

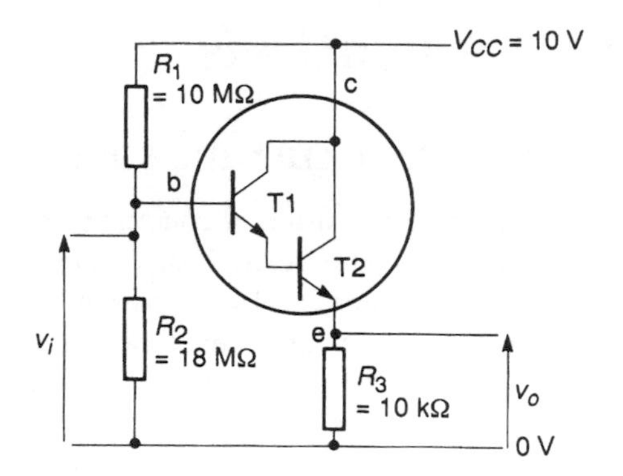

Figure 7.29 A high input resistance buffer amplifier.

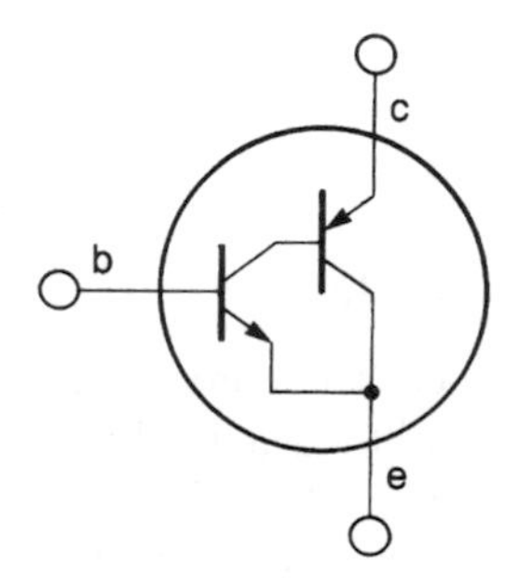

Figure 7.30 The complementary darlington connection.

The very high gain of darlingtons, which may be 10^4 to 10^5 or more, has a number of uses including the production of circuits with very high input resistances. For example, the arrangement of Figure 7.29 has a gain of unity and an input resistance approximately equal to the parallel combination of its base resistors (about 6.4 MΩ). This is because the resistance seen looking into the base of the darlington pair (approximately R_3 times the product of the gains of the two transistors) is so large that its effects may be neglected. The circuit is useful as a **unity gain buffer amplifier**.

Another common form of the darlington configuration uses transistors of opposite polarities, as shown in Figure 7.30. This is known as the **complementary darlington connection**, or sometimes as the **Sziklai connection**.

This arrangement also behaves as a single high gain transistor and has the advantage that its base-to-emitter voltage is equal to that of a single transistor (although its saturation voltage is still greater than the conduction voltage of a single junction).

A common application of the darlington connection is in the production of high gain power transistors. Conventional high power devices have relatively low gains of perhaps 10 to 60. A typical darlington power device might have a minimum gain of 1000 at 10 A.

7.7 Bipolar transistor applications

7.7.1 A bipolar transistor as a constant current source

In Section 6.6.1 we looked at the use of field effect transistors as constant current sources. It is also possible to use bipolar transistors for this purpose, as illustrated in Figure 7.31.

Figure 7.31(a) shows a circuit using an *npn* transistor. Here a pair of base resistors forms a potential divider which applies a constant voltage to the base of the transistor. The constancy of the base-to-emitter voltage results in a fixed emitter voltage which therefore produces a constant emitter current. This, in turn, draws a collector current, equal to the emitter current, through the load. Figure 7.31(b) works in the same manner but uses a *pnp* transistor and produces a current to ground rather than a current from the supply. Strictly speaking, the first of these two circuits is a **current sink** and the second a **current source**. However, it is common to use the general term **current source** to refer to both types of circuit. The circuits may be improved by using a **Zener diode** in place of R_2 to improve the constancy of the emitter voltage. The circuit can be made into a variable current source by using a variable resistance in place of R_E.

7.7.2 A bipolar transistor as a current mirror

A current mirror may be considered as a form of constant current source, where the current produced is equal to some input current. The principle is illustrated in Figure 7.32(a).

Two transistors are joined at the base and at the emitter and thus have identical base-to-emitter voltages. If the transistors are identical, this will produce equal collector currents in each device. The current in T1 represents the input current which, in this example, is supplied by the resistor R. The circuit acts as a current sink, the current through T2 being drawn through the load. A major use of this technique is in creating a number of equal currents within a circuit. This can be achieved by connecting a number of output transistors to a single input transistor, as shown in Figure 7.32(b). Current mirrors are widely used in integrated circuits where it is possible to achieve very good matching of

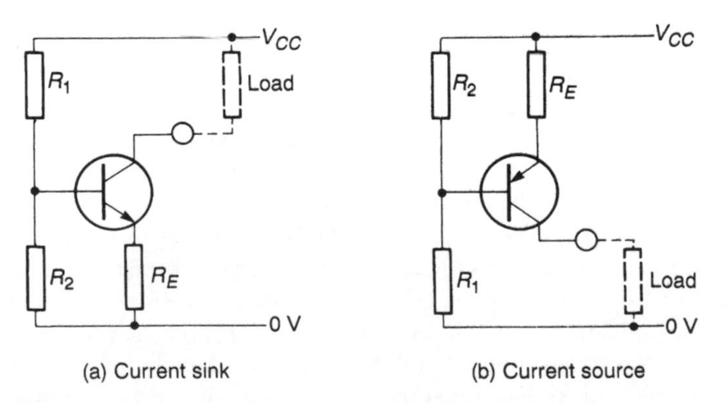

Figure 7.31 Current sources using bipolar transistors.

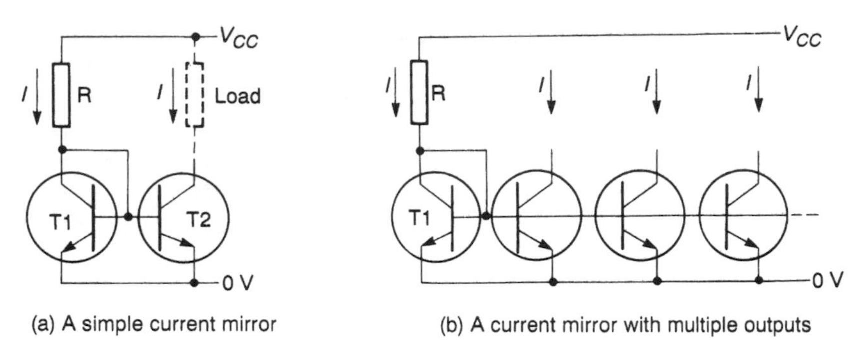

(a) A simple current mirror

(b) A current mirror with multiple outputs

Figure 7.32 Current mirrors.

the transistors. The close proximity of the transistors within the circuit also ensures that they are at approximately the same temperature. This is important because of the variation in the current–voltage characteristics of the transistors with temperature.

7.7.3 Bipolar transistors as differential amplifiers

In Section 6.5.7 we looked at the use of FETs in differential amplifiers. Bipolar transistors are also widely used in such circuits and again a common form is the **long-tailed pair**. Figure 7.33 shows an example of such an arrangement in which, for simplicity, the base bias circuitry has been omitted. You might like to compare this with the FET amplifier shown in Figure 6.32.

The circuit resembles two series feedback amplifiers which share a common emitter resistor. The emitter resistor acts as a current source and the two transistors share the

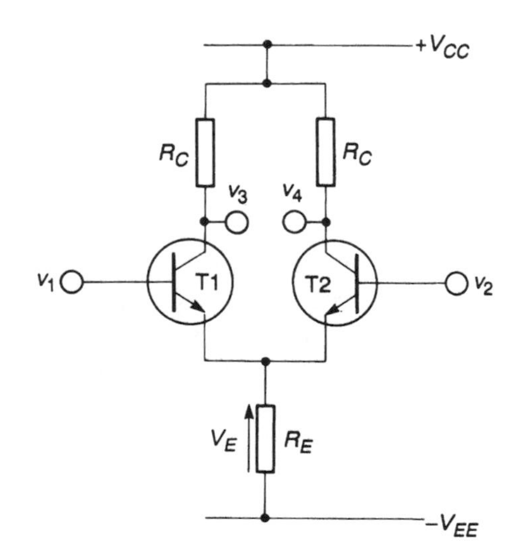

Figure 7.33 A long-tailed pair amplifier.

current that it supplies. With perfectly matched transistors and resistors, the circuit is symmetrical. If the input voltages v_1 and v_2 are identical, the outputs will be equal, with the current from R_E being shared equally between the two devices. The voltage V_E will be simply V_{BE} less than the voltage on the inputs. If now v_1 is made slightly positive with respect to v_2, V_E will rise with v_1, and the base–emitter voltage on T2 will be reduced, tending to turn it OFF. Thus the current supplied by R_E will be split unevenly with more of the current going through T1. This will reduce v_3 and increase v_4. If v_2 is made slightly positive with respect to v_1 the same process occurs in reverse.

Thus if we consider the input signal to be $v_1 - v_2$ and the output signal to be $v_3 - v_4$ we have an inverting differential amplifier.

It can be shown that the voltage gain of this arrangement is given simply by

$$\text{Differential voltage gain} = -g_m R_C = -\frac{R_C}{r_e}$$

You might like to compare this result with that obtained earlier for the common-emitter amplifier, and with that for the FET differential amplifier of Section 6.5.7.

One of the great advantages of this form of amplifier is its great linearity for small differential input voltages (approx ±25 mV), which is considerably better than can be achieved using a single transistor. For this reason, this form of circuit forms the basis of most bipolar operational amplifiers and is used even where a non-differential input is required (simply by earthing the unused input). The temperature stability of the arrangement is also good since the temperature drifts of the two transistors tend to cancel.

A useful measure of the performance of a differential amplifier is its **common-mode rejection ratio CMRR**. That is, the ratio of the differential mode gain to the common mode gain. It can be shown that for this amplifier

$$\text{CMRR} = g_m R_E = \frac{R_E}{r_e}$$

and again you might like to compare this with the performance of the equivalent FET circuit.

To obtain a good CMRR it is necessary to use a high value for R_E. One method of increasing the effective value of R_E, without upsetting the DC conditions of the circuit, is to replace it with a **constant current source**, as described earlier in this section. An ideal current source has an infinite internal resistance, giving the maximum possible CMRR. Figure 7.34 shows a circuit using a current mirror as a constant current source for a long-tailed pair amplifier. Circuits of this type often form the basis of the input stage of bipolar operational amplifiers. Note that, as with the FET long-tailed pair amplifier, the two output signals are symmetrical even if the inputs are not. Thus the two outputs may be used individually as single-ended (that is non-differential) outputs if required.

7.7.4 Bipolar transistors in push-pull amplifiers

We noted in Section 3.12 that a single transistor amplifier can act either as a **current source** or as a **current sink** but cannot perform both tasks. We also noted in that section that power is continuously dissipated in the control device, in this case the transistor, and

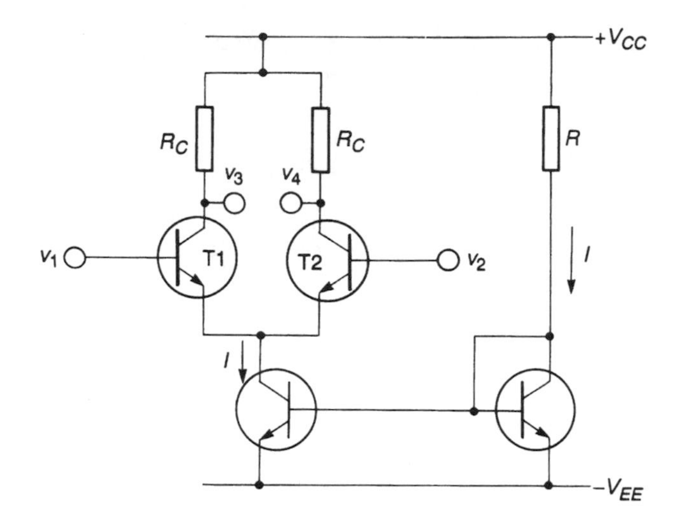

Figure 7.34 A long-tailed pair amplifier using a current mirror.

that the magnitude of this power is related to the size of the load. This arrangement has several limitations when driving high power or capacitive loads, as illustrated in Figure 7.35.

The figure shows an emitter follower amplifier driving a capacitive load. Consider initially the situation shown in Figure 7.35(a) where the input is becoming more positive. The transistor drives the output positive by passing its emitter current into the load, the low output impedance of the transistor allowing the capacitor to be charged quickly. If now the input becomes more negative, as shown in Figure 7.35(b), charge must be removed from the capacitor. This cannot be done by the transistor which can only source current in this configuration. Therefore the charge must be removed by the emitter resistor R_E, which we know from our study of the emitter follower in Section 7.6.8 is considerably greater than the output resistance of the amplifier. Replacing the transistor with

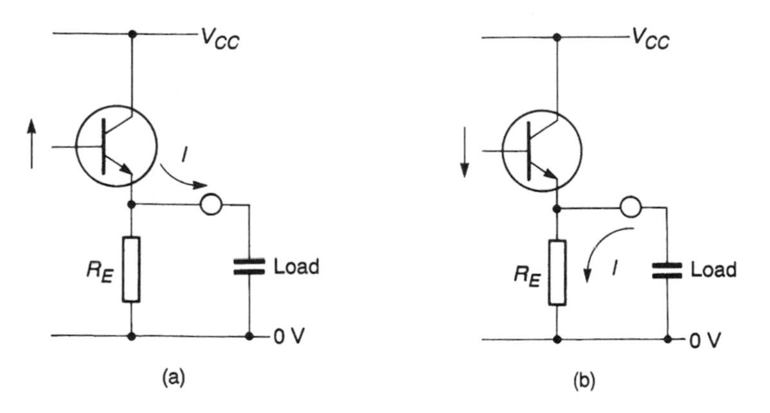

Figure 7.35 An emitter follower with a capacitive load.

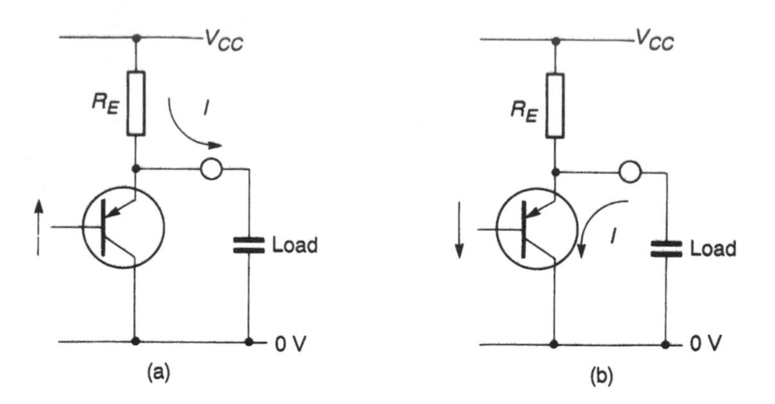

Figure 7.36 An emitter follower using *pnp* transistors.

a *pnp* device, as shown in Figure 7.36, produces an arrangement that can discharge the load quickly but is slow to charge it.

Clearly the rate at which the capacitor can be charged or discharged through the resistor R_E can be increased by decreasing the value of the resistor. However, Equation 3.11 shows that the maximum power dissipated in the transistor is given by

$$P_{max} = \frac{V^2}{4R_E}$$

Therefore, reducing the emitter resistor increases the power dissipated in the transistor. In high power applications this can cause serious problems.

One approach to this problem, often used within the high power output stage of an amplifier, is to use two transistors in a **push-pull** arrangement as shown in Figure 7.37.

Here one transistor is able to source and the other to sink current so that the load can be driven from a low resistance output in either direction. This arrangement is commonly used with a **split power supply**, that is, a supply which provides both positive and negative

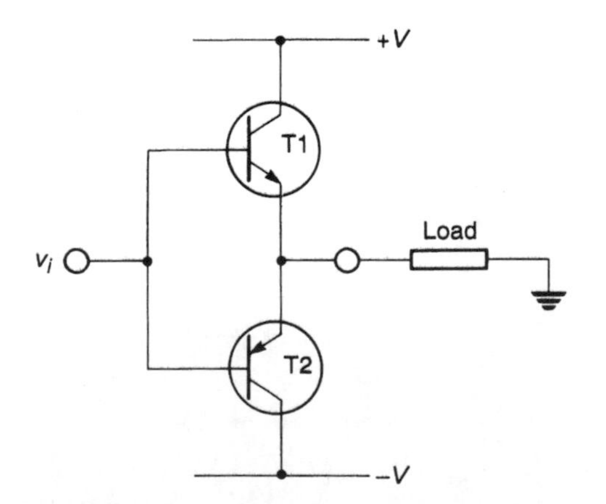

Figure 7.37 A simple push-pull amplifier.

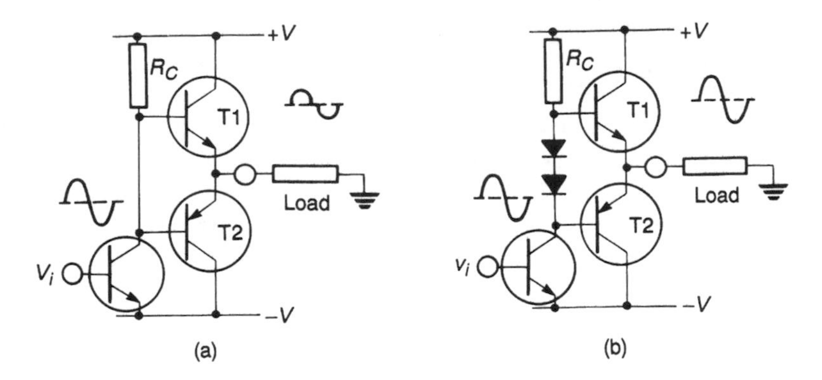

Figure 7.38 Driving a push-pull output stage.

voltages, with the load being connected to earth. For positive input voltages, transistor T1 will be within its active region but T2 will be turned OFF, since its base junction will be reverse biased. Similarly, for negative input voltages T2 will be within its active region and T1 will be turned OFF. Hence at any time only one of the transistors is turned ON, reducing the overall power consumption.

A possible method of driving the push-pull stage is shown in Figure 7.38(a). Here a conventional common-emitter amplifier is used to drive the bases of the two output transistors. A problem with this arrangement is that for small values of the base voltage on either side of zero, both of the output transistors are turned OFF. This gives rise to an effect known as **crossover distortion**. One solution to this problem is shown in Figure 7.38(b). Here two diodes are used to apply different voltages to the two bases. Since the base voltages differ by $2 \times V_{BE}$, one transistor should turn ON precisely where the other turns OFF. A slight problem with this arrangement is that the current passing through the output transistors is considerably greater than that passing through the diodes. Consequently the V_{BE} of the diodes is less than that of the transistors and a small dead band still remains. Crossover distortion is reduced but not completely eliminated. More effective methods of biasing the output transistors are available. These will be discussed later in Section 8.3 when we look at classes of amplifiers.

FILE 7H
FILE 7J

Computer simulation exercise 7.10

Simulate the arrangement of Figure 7.38(a). Suitable transistors would be a 2N2222 for T1 and a 2N2907A for T2. Use +15 V and −15 V supplies, a value of 10 kΩ for R_C and a 10 Ω load. Any *npn* transistor may be used for the drive transistor, although a suitable biasing arrangement must be added to set the quiescent voltage on the bases of the output transistors to about zero volts. Alternatively, the drive transistor can be replaced with a sinusoidal current generator (ISIN in PSpice). If a current generator is used this should be configured to give an offset current of 1.5 mA so that the quiescent voltage on the bases of the transistors is close to zero volts.

Apply a sinusoidal input to the circuit at 1 kHz and observe the form of the output. Set the magnitude of the input to produce an output of about 10 volts peak to peak. You should observe that the output suffers from crossover distortion. Display the fast Fourier transform (FFT) of the output waveform and observe the presence of harmonics of the input signal. Note which harmonics are present within this signal.

Modify the circuit by adding two diodes as in Figure 7.38(b). You may use any conventional small signal diodes – for example 1N914 devices. The drive transistor biasing arrangement, or the offset of the current generator, must now be adjusted to return the quiescent voltage on the bases of the output transistors to zero volts.

Again apply a sinusoidal input and observe the output voltage. Display the FFT of this waveform and notice the effect of the diodes on the crossover distortion.

7.7.5 A bipolar transistor as a voltage regulator

We have seen how a bipolar transistor can be used to produce a constant voltage across a resistor in order to create a constant current source, and we have also seen that, in the emitter follower configuration, it can produce a very low output resistance. If we combine these two functions we can realize a circuit that produces a constant output voltage with a low output resistance. Having a low output resistance implies that the output voltage will not change appreciably with the output current. This is the function of a **voltage regulator**. Such an arrangement is shown in Figure 7.39(a). The resistor and Zener diode form a constant voltage reference V_Z which is applied to the base of the transistor. The output voltage will be equal to this voltage minus the approximately constant base-to-emitter voltage of the transistor.

The circuit of Figure 7.39(a) is shown redrawn as Figure 7.39(b), which illustrates a more common way of representing the circuit. The arrangement as shown provides a considerable amount of regulation, but for large fluctuations in output current suffers from the variation of V_{BE} with current. A more effective regulator circuit is given later in Section 7.9, as Example 7.10.

7.7.6 A bipolar transistor as a switch

We saw in Sections 6.6.3 and 6.6.4 that FETs can be used as both linear and logical switches. Bipolar transistors are not usually used in linear switching applications but can be used as logical switching elements.

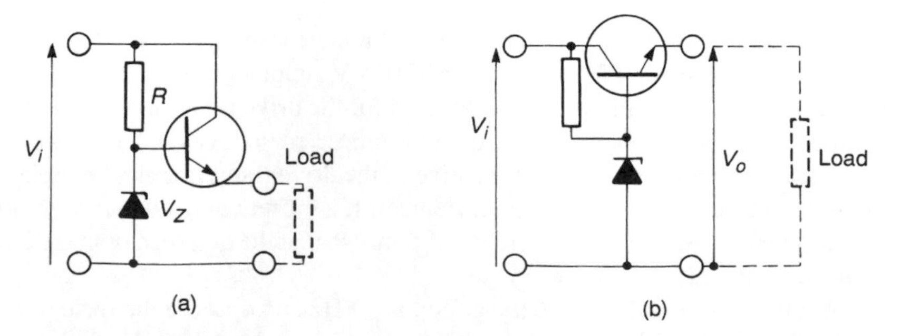

Figure 7.39 The bipolar transistor as a voltage regulator.

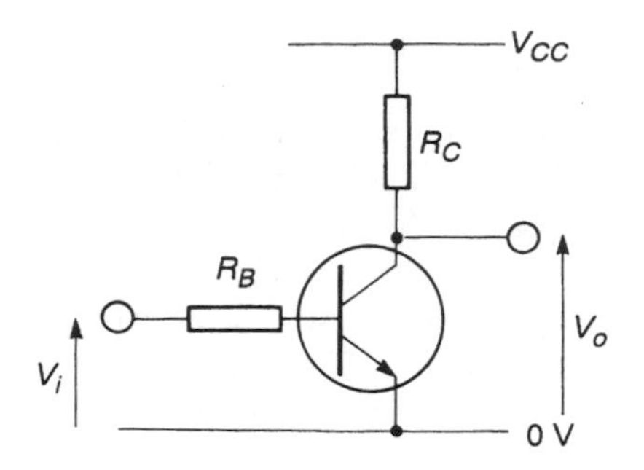

Figure 7.40 A logical inverter based on a bipolar transistor.

Figure 7.40 shows a simple form of **logical inverter** based on a bipolar transistor. The circuit resembles the common-emitter amplifier considered earlier and indeed its operation is similar. The main different between this circuit and a linear amplifier is that the inputs are restricted to two distinct ranges. Input voltages close to zero (representing logical '0') are insufficient to forward bias the base of the transistor and it is therefore cut off. Negligible collector current flows and the output voltage is therefore close to the supply voltage. Input voltages close to the supply voltage (representing a logical '1') forward bias the base junction turning ON the transistor. The resistor R_B is chosen such that the base current is sufficient to saturate the transistor producing an output voltage equal to the transistor's collector saturation voltage (normally about 0.2 V). Thus an input of logical '0' produces an output of logical '1', and vice versa. The circuit is therefore a logical inverter.

If the transistor were an **ideal switch** it would have an infinite resistance when turned OFF, zero resistance when turned ON, and would operate in zero time. In fact, the bipolar transistor is a good, but not an ideal, switch. When turned OFF, the only currents that are passed are small leakage currents, which are normally negligible. When turned ON, the device has a low ON resistance but a small saturation voltage, as described above. The speed of operation of bipolar transistors is very high, with devices able to switch from one state to another in a few nanoseconds (or considerably faster for high-speed devices). We will return to look in more detail at the characteristics of transistors as switches in Chapter 11.

7.8 Four-layer devices

Although transistors make excellent logical switches they have limitations when it comes to switching high currents at high voltages. To make a transistor with a high current gain requires a thin base region which produces a low breakdown voltage. An alternative approach is to use one of a number of devices designed specifically for use in such applications. These components are not transistors, but they are included within this chapter because their construction and mode of operation have a great deal in common with those of bipolar transistors.

7.8.1 The thyristor

The thyristor is a **four-layer device** consisting of a *pnpn* structure as shown in Figure 7.41(a). The two end regions have electrical contacts called the **anode** (*p* region) and the **cathode** (*n* region). The inner *p* region also has an electrical connection called the **gate**. The circuit symbol for the thyristor is shown in Figure 7.41(b).

In the absence of any connection to the gate, the thyristor can be considered as three diodes in series formed by the *pn*, *np* and *pn* junctions. Since two of these diodes are in one direction and one is in the other, any applied voltage must reverse bias at least one of the diodes and no current will flow in either direction. However, when an appropriate signal is applied to the gate, the device becomes considerably more useful.

Thyristor operation

The operation of the thyristor is most readily understood by likening it to two interconnected transistors, as shown in Figure 7.42(a) which can be represented by the circuit of Figure 7.42(b). Let us consider initially the situation in which the anode is positive with respect to the cathode but no current is flowing in either device. Since T1 is turned OFF, no current flows into the base of T2 which is therefore also turned OFF. Since T2 is turned OFF, no current flows from the base of T1 so this transistor remains OFF. This situation is stable and the circuit will stay in this state until external events change the condition of the circuit.

Consider now the effect of a positive pulse applied to the gate. When the gate goes positive, T2 will turn ON causing current to flow from its collector to its emitter. This current will produce a base current in T1 turning it ON. This in turn will cause current to flow through T1 producing a base current in T2. This base current will tend to increase the current in T2 which in turn will increase the current in T1 and the cycle will

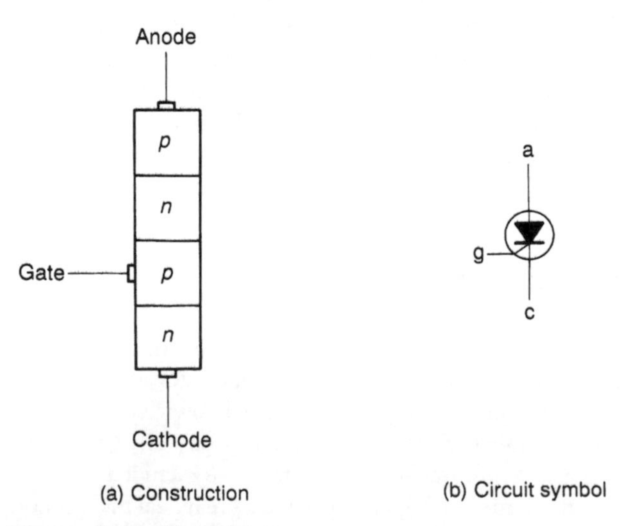

Figure 7.41 The thyristor.

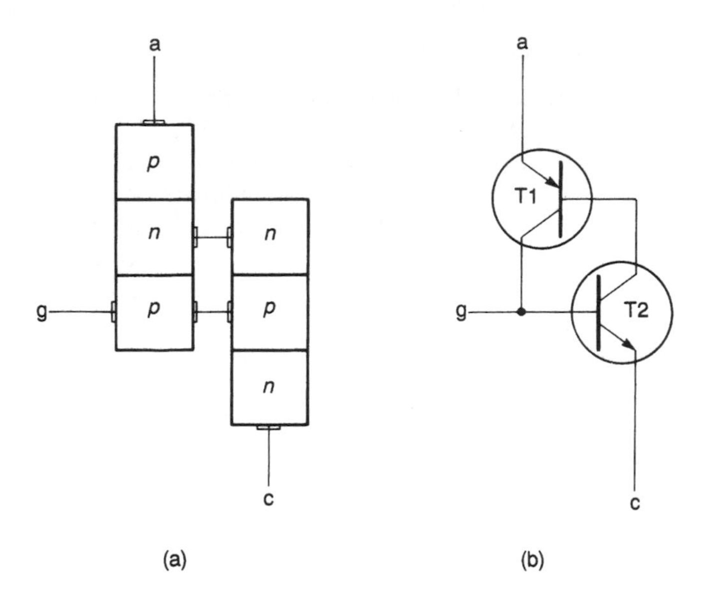

Figure 7.42 Thyristor operation.

continue until both devices are saturated. The current flowing between the anode and the cathode will increase until it is limited by external circuitry. This process is said to be *regenerative* in that the current flow is self-increasing and self-maintaining. Once the thyristor has been 'fired' the gate signal can be removed without affecting the current flow.

The thyristor will only function as a control device in one direction; if the anode is made negative with respect to the cathode, the device simply acts like a reverse biased diode. This is why the thyristor is also called a **silicon controlled rectifier** or **SCR**.

The thyristor may be thought of as a very efficient electrically controlled switch, with the rather unusual characteristic that when it is turned ON, by a short pulse applied to the gate, it will stay ON as long as current continues to flow through the device. If the current stops, or falls below a certain **holding current**, the transistor action stops and the device automatically turns OFF. In the OFF state, only leakage currents flow and breakdown voltages of several hundreds or thousands of volts are common. In the ON state, currents of tens or hundreds of amperes can be passed with an ON voltage of only a volt or so. The current needed to turn ON the device may vary from about 200 μA for a small device to about 200 mA for a device capable of passing 100 A or so. The switching times for small devices are generally of the order of 1 μs and are somewhat longer for larger devices. It should be remembered that even though the thyristor is a very efficient switch, power is dissipated in the device as a result of the current flowing through it and the voltage across it. This power must be removed without allowing the device to exceed its maximum working temperature. For this reason, all but the smallest thyristors are normally mounted on **heat sinks** which are specifically designed to dispel heat.

The thyristor in AC power control

Although the thyristor can be used in DC applications, it is most often found in the control of AC systems. Consider the arrangement shown in Figure 7.43(a).

Here a thyristor is connected in series with a resistive load to an AC supply. External circuitry senses the supply waveform, shown in Figure 7.43(b), and generates a series of gate trigger pulses as shown in Figure 7.43(c). Each pulse is positioned at the same point within the phase of the supply, so the thyristor is turned ON at the same point in each cycle. Once turned ON the device continues to conduct until the supply voltage, and the current through the thyristor, drop to zero, producing the output waveform shown in Figure 7.43(d). In the illustration the thyristor is fired approximately half-way through the positive half-cycle of the supply and therefore the thyristor is ON for approximately one-quarter of the cycle. Therefore, the power dissipated in the load is approximately one-quarter of what it would be if the load were connected directly to the supply. By varying the phase angle at which the thyristor is fired, the power

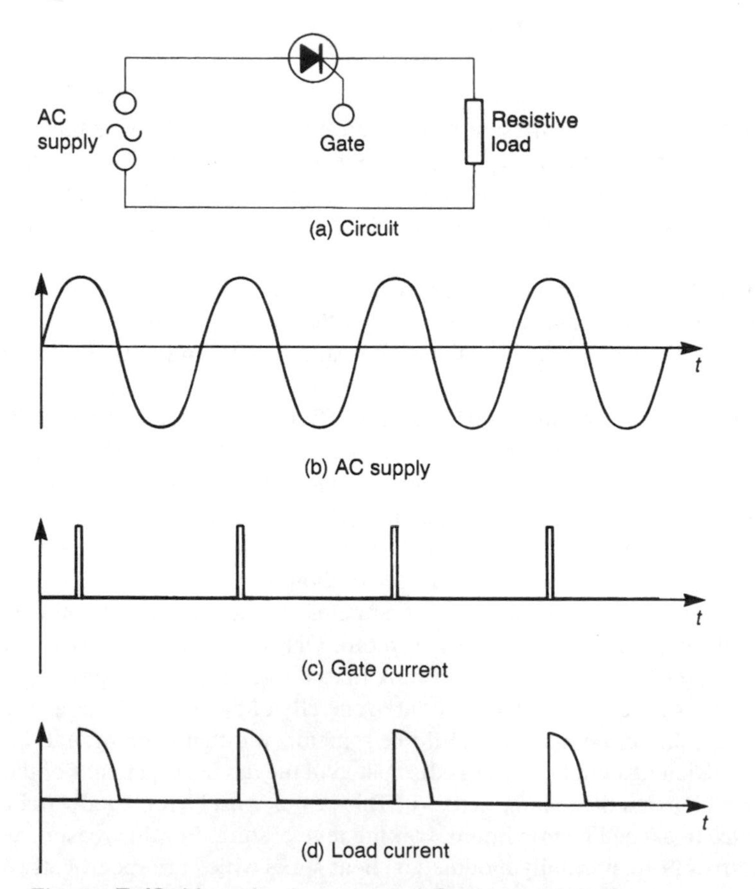

Figure 7.43 Use of a thyristor in AC power control.

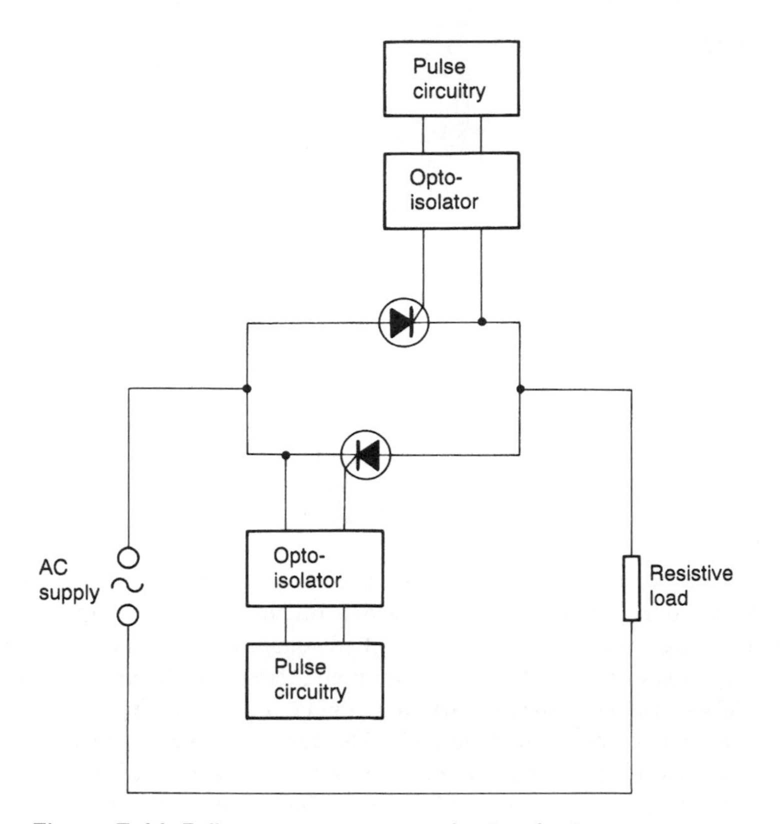

Figure 7.44 Full-wave power control using thyristors.

delivered to the load can be controlled from 0 to 50% of full power. Such control is called half-wave control.

The gate current pulse is generated by applying a voltage of a few volts between the gate electrode and the cathode. Since the cathode is within the supply circuit, it is common to use **opto-isolation** to insulate the electronics used to produce these pulses from the AC supply (as described in Section 2.4.2). To achieve full-wave control using thyristors requires the use of two devices connected in inverse-parallel as shown in Figure 7.44. This allows power to be controlled from 0 to 100% of full power but unfortunately requires duplication of the gate pulse generating circuitry and the isolation network.

7.8.2 The triac

A more elegant solution to full-wave control of AC power is to use a **triac**. This is effectively a bidirectional thyristor which can operate during both halves of the supply cycle. It resembles two thyristors connected in inverse-parallel but has the advantage that gate pulses can be supplied by a single isolated network. Gate pulses of either polarity will trigger the triac into conduction throughout the supply cycle. Since the device is

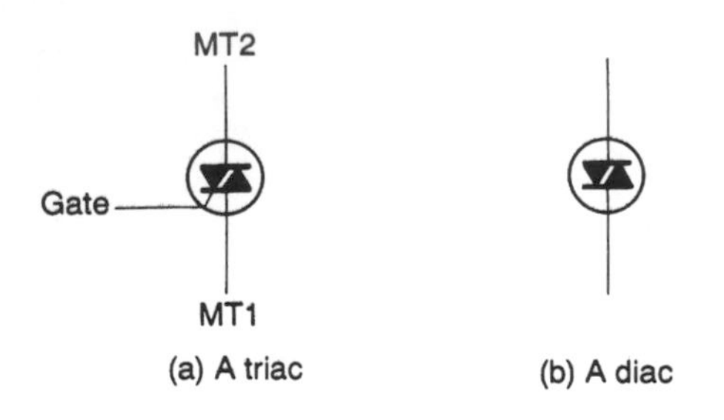

Figure 7.45 A triac and a diac.

effectively symmetrical, the two electrodes of the device are simply given the names MT1 and MT2, where MT simply stands for 'main terminal'. Voltages applied to the gate of the device are applied with respect to MT1. The circuit symbol for a triac is shown in Figure 7.45(a).

Gate trigger pulses in triac circuits are often generated using another four-layer device, the **bidirectional trigger diode** or **diac**. The diac resembles a triac without any gate connection. It has the property that for small applied voltages it passes no current, but if the applied voltage is increased above a certain point, termed the **breakover voltage**, the device exhibits **avalanche breakdown** and begins to conduct. Typical values for the breakover voltage for a diac are 30 to 35 V. The device operates in either direction and is used to produce a burst of current into the gate when a control voltage, derived from. the supply voltage, reaches an appropriate value. The circuit symbol for a diac is shown in Figure 7.45(b).

Triacs are widely used in applications such as **lamp dimmers** and **motor speed controllers**. A circuit for a simple domestic lamp dimmer is given in Figure 7.46.

Operation of the circuit is very straightforward. As the supply voltage increases at the beginning of the cycle, the capacitor is charged through the resistors and its voltage

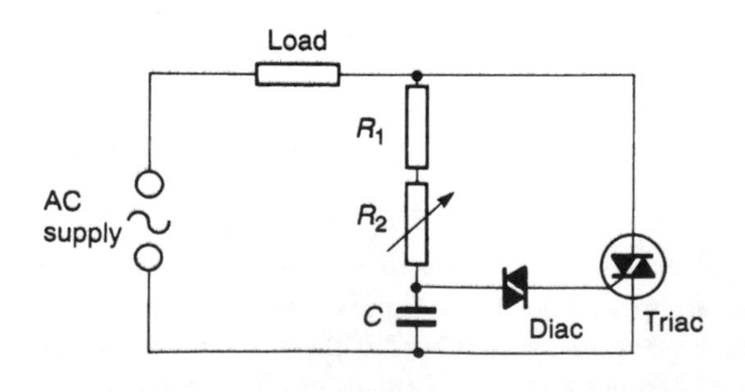

Figure 7.46 A simple lamp dimmer using a triac.

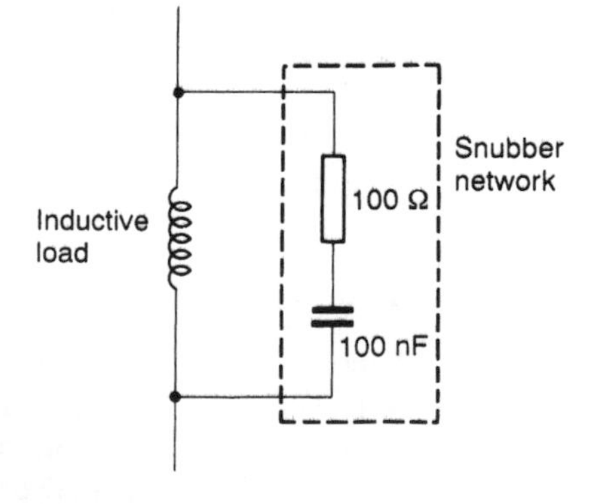

Figure 7.47 A snubber network.

increases. When it reaches the breakover voltage of the diac (about 30 V), the capacitor discharges through the diac producing a pulse of current which fires the triac. The phase angle at which the triac is triggered is varied by changing the value of R_2, which controls the charging rate of the capacitor. R_1 is present to limit the minimum resistance of the combination to prevent excessive dissipation in the variable resistor. Once the triac has been fired it is maintained in its ON state by the load current flowing through it, while the voltage across the resistor–capacitor combination is limited by the ON voltage of the triac which is of the order of a volt. This situation is maintained until the end of the present half-cycle of the supply. At this point the supply voltage goes to zero, reducing the current through the triac below its holding current and turning it OFF. The supply voltage then enters its next half-cycle, the capacitor voltage again begins to rise (this time in the opposite sense), and the cycle repeats. If the component values are chosen appropriately, the output can be varied from zero to nearly full power by adjusting the setting of the variable resistor.

The simple lamp dimmer circuit above controls the power in the lamp by varying the phase angle of the supply at which the triac is fired. Not surprisingly this method of operation is called **phase control**. Using this technique large transients are produced as the triac switches ON part way through the supply cycle. These transients can cause problems of **interference**, either by propagating noise spikes through the supply lines or by producing electromagnetic interference (EMI). An alternative method of control is to turn the triac ON for complete half-cycles of the supply, varying the ratio of ON to OFF cycles to control the power. Switching occurs where the voltage is zero and so interference problems are removed. This technique is called **burst firing** and is useful for controlling processes with a relatively slow speed of response. It is not suitable for use with lamp control since it can give rise to flickering.

In Example 5.6 we looked at the use of **catch diodes** in DC circuits driving inductive loads. Such an arrangement cannot be used in AC circuits since the diode would conduct on alternate half-cycles shorting out the load. In AC applications it is normal to use a **snubber network** to provide this protection. Such a network is illustrated in Figure 7.47 which shows typical component values.

The network consists of a series *RC* combination which has a relatively high impedance at the AC supply frequency but a low impedance at the high frequencies associated with noise spikes and transients. The circuit therefore tends to 'short-out' high voltage spikes, preventing the generation of interference and protecting sensitive equipment. Such an arrangement should normally be connected to all equipment in which inductive loads are switched with an AC supply.

7.8.3 The gate turn-off thyristor

As we have seen, with conventional thyristors, once the device has been turned ON, it will stay ON until the current falls below the holding current. This arrangement has many advantages in AC systems but is less convenient in DC applications. In these cases, a different form of thyristor is often used, this device being the **gate turn-off thyristor** or **GTO**. As the name implies, this can be turned OFF as well as ON by signals applied to the gate. It is turned ON in the same way as a conventional thyristor, and can be turned OFF by applying a negative pulse to the gate.

7.9 Circuit examples

Example 7.8 A phase splitter

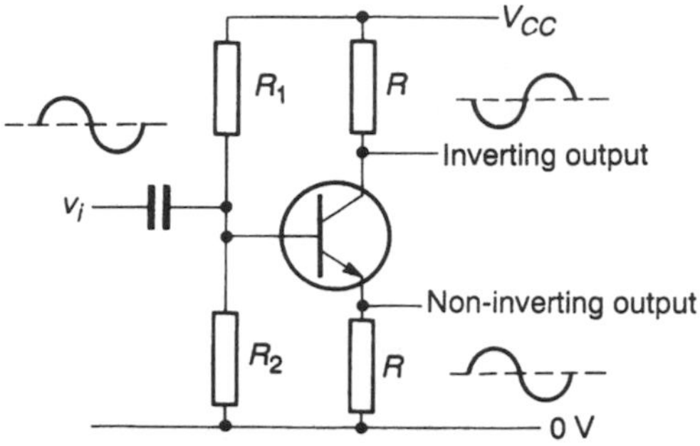

This circuit produces two output signals which are inverted with respect to each other. By taking outputs from both the collector and the emitter we have combined the series negative feedback amplifier and the emitter follower into a single stage. From Example 7.4 we know that the voltage gain of the series negative feedback amplifier is given by the inverse of the ratio of the collector resistor to the emitter resistor. In this circuit the two resistors are equal, giving a gain of minus unity. The emitter follower has a gain of plus unity, so the two signals are identical in size but of opposite polarity.

It should be remembered that the output resistances of the two outputs are very different. Therefore, the signals should be fed into high input resistance circuits if the correct relative signal magnitudes are to be maintained. It is also worth noting that the DC quiescent output voltages are different on the two outputs.

Computer simulation exercise 7.11

FILE 7K

Simulate the circuit of Example 7.8. A suitable choice of components would be: $R = 1$ kΩ; $R_1 = 6.7$ kΩ; $R_2 = 3.3$ kΩ; $C = 1$ μF and any *npn* transistor (for example a 2N2222). Use a supply voltage $V_{CC} = 10$ V, and apply a 1 kHz sinusoidal input.

Display the two output signals and compare their relative magnitude and phase. Then connect load resistors of 1 kΩ between each output and ground, and again compare the outputs.

Example 7.9 An operational amplifier power booster

The power output capabilities of an operational amplifier can be enhanced by attaching a high power transistor to its output in an emitter follower arrangement, as shown in (a) below. This circuit is then used in standard op-amp circuits, taking the output from the emitter of the transistor rather than the output of the op-amp. An example of the production of a high power unity gain buffer is shown in (b). Since the feedback is taken from the output of the circuit, the offset voltage produced by V_{BE} is removed by the effects of the

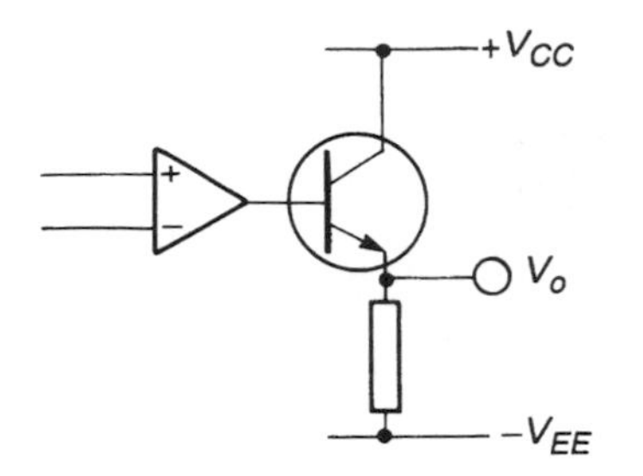

(a) A transistor power booster

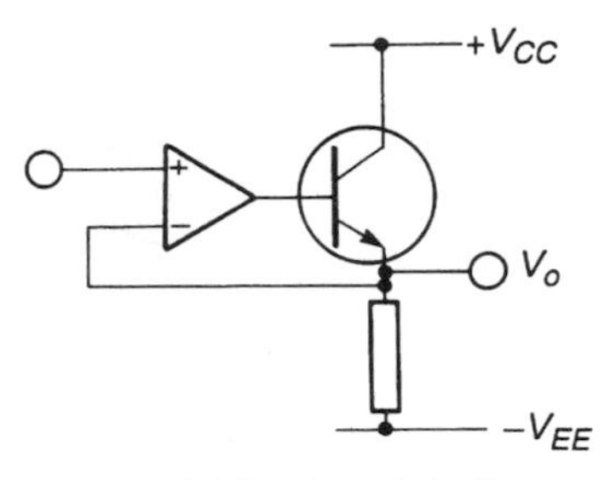

(b) A unity gain buffer

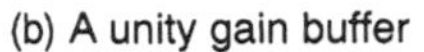

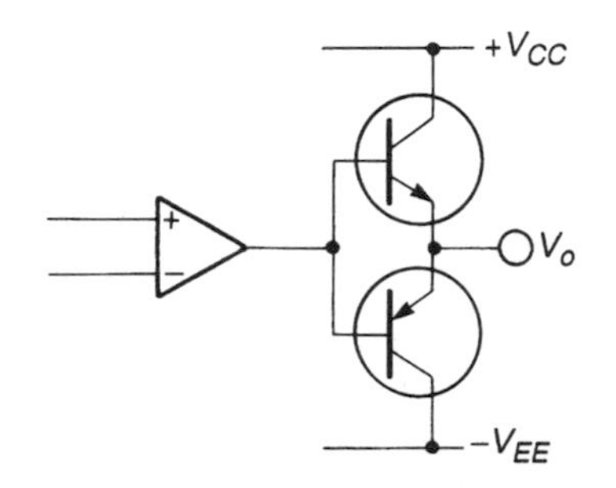

(c) A push-pull arrangement

feedback. Since it uses only a single output transistor, this arrangement can only source current effectively. If appropriate, it can be replaced by a push-pull arrangement, as shown in (c), which performs well as both a current sink and a current source.

Example 7.10 Improved voltage regulator using a bipolar transistor and an op-amp

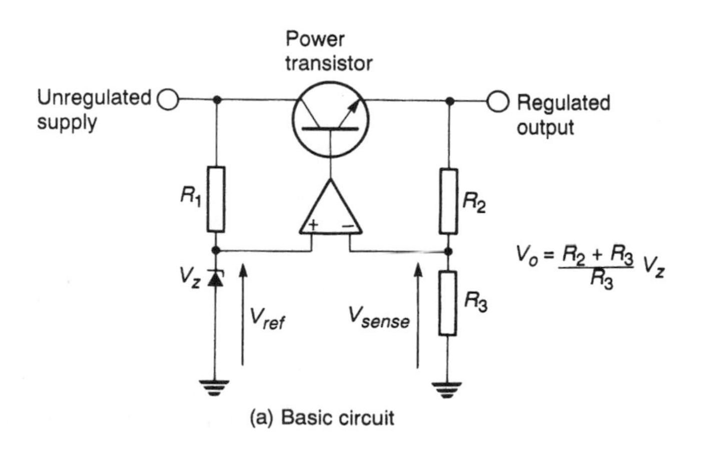

(a) Basic circuit

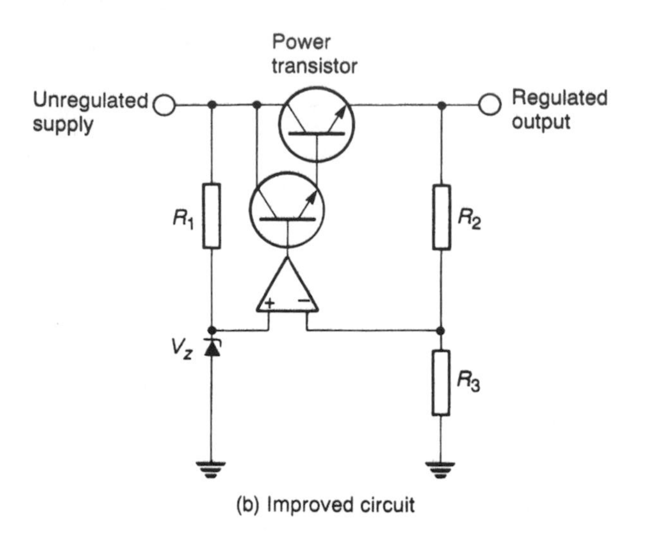

(b) Improved circuit

The performance of the simple voltage regulator described in Section 7.7.5 suffers because the V_{BE} of the transistor varies with the emitter current. The diagrams above show two alternative arrangements.

In circuit (a), an operational amplifier compares the actual output voltage with a voltage reference and drives the transistor to increase or decrease the output voltage until the correct value is achieved. The high gain of the operational amplifier provides very good regulation of the output voltage.

R_1 and the Zener diode form a constant voltage reference V_{ref}, equal to V_Z. The potential divider formed by R_2 and R_3 divides down the output voltage such that the voltage applied to the inverting input of the op-amp is

$$V_{sense} = V_o \frac{R_3}{R_2 + R_3}$$

If V_{sense} is greater than V_{ref}, the output of the op-amp will be negative, decreasing the output voltage. If V_{sense} is less than V_{ref}, the output of the op-amp will be positive, increasing the output voltage. The output will therefore stabilize with

$$V_{sense} = V_{ref}$$

which gives

$$V_o \frac{R_3}{R_2 + R_3} = V_{ref}$$

or

$$V_o = V_{ref} \frac{R_2 + R_3}{R_3}$$

The circuit given in (a) is perfectly satisfactory where small, high gain transistors are used. However, large power transistors often have very low current gains (perhaps only 20), and the op-amp is unlikely to be able to provide sufficient drive current. In such cases it is advisable to use a second transistor to provide additional gain, as shown in (b). Here a smaller, and therefore higher gain, transistor is connected to form a **darlington pair**. Typically these two devices would take the form of a single **darlington transistor**.

Many regulator circuits include some form of **overload protection** to prevent the circuit from being damaged in the event that the load takes an excessive current. A simple **current limiting** arrangement is shown below.

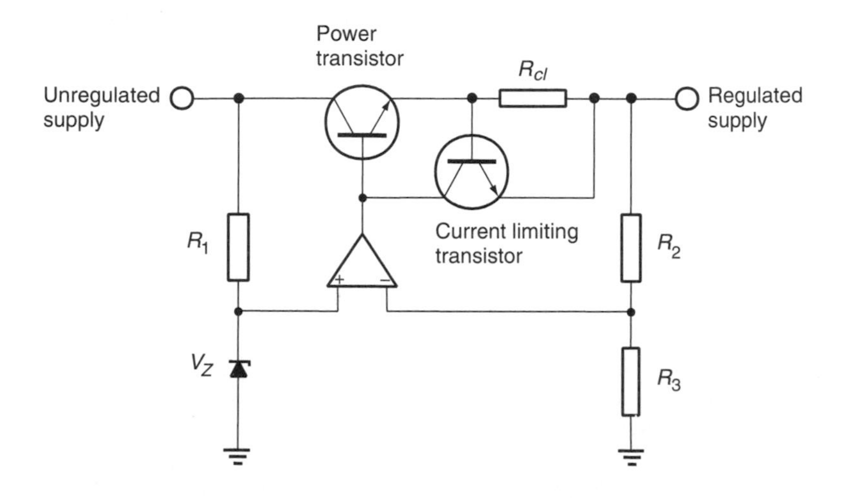

The **protection circuit** uses a resistor R_{cl} in series with the load to measure the output current. As the current increases the voltage across this resistor also increases. If the voltage across R_{cl} exceeds the turn-on voltage of the current limiting transistor, the latter will begin to conduct and will remove the base current from the power transistor, turning it off. R_{cl} is chosen such that the voltage drop across it is just less than the turn-on voltage of the transistor when the output current is at its maximum value.

Key points

- Bipolar transistors are one of the most important forms of electronic component and a clear understanding of their operation and use is essential for anyone working in this area.
- They are used in a wide variety of both analogue and digital circuits.
- Bipolar transistors can be considered as either current controlled or voltage controlled devices.
- If we choose to view them as current controlled devices we depict them as current amplifiers and describe their performance by their current gain.
- If we choose to view them as voltage controlled devices we depict them as transconductance amplifiers and describe their behaviour using their transconductance g_m.

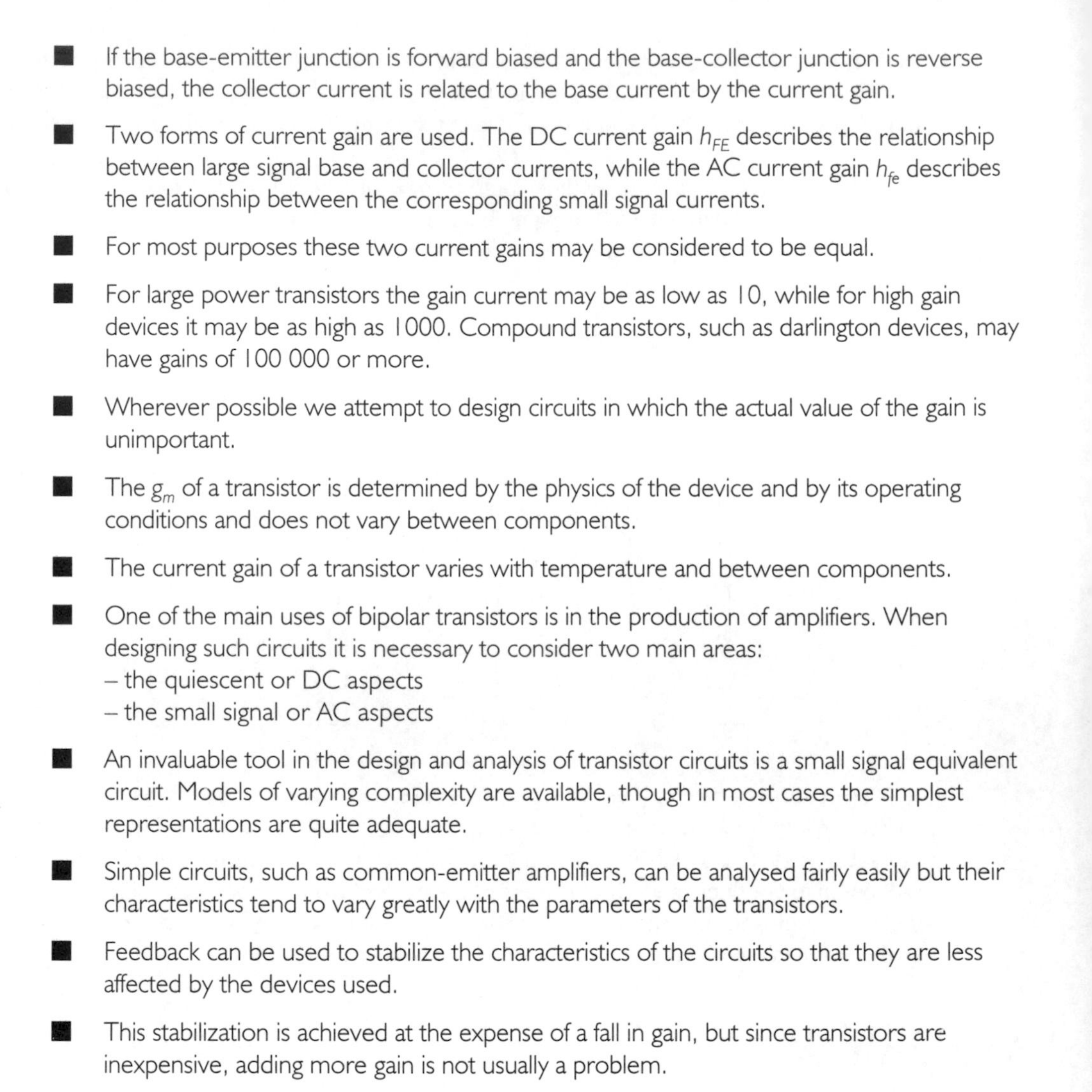

- If the base-emitter junction is forward biased and the base-collector junction is reverse biased, the collector current is related to the base current by the current gain.
- Two forms of current gain are used. The DC current gain h_{FE} describes the relationship between large signal base and collector currents, while the AC current gain h_{fe} describes the relationship between the corresponding small signal currents.
- For most purposes these two current gains may be considered to be equal.
- For large power transistors the gain current may be as low as 10, while for high gain devices it may be as high as 1000. Compound transistors, such as darlington devices, may have gains of 100 000 or more.
- Wherever possible we attempt to design circuits in which the actual value of the gain is unimportant.
- The g_m of a transistor is determined by the physics of the device and by its operating conditions and does not vary between components.
- The current gain of a transistor varies with temperature and between components.
- One of the main uses of bipolar transistors is in the production of amplifiers. When designing such circuits it is necessary to consider two main areas:
  – the quiescent or DC aspects
  – the small signal or AC aspects
- An invaluable tool in the design and analysis of transistor circuits is a small signal equivalent circuit. Models of varying complexity are available, though in most cases the simplest representations are quite adequate.
- Simple circuits, such as common-emitter amplifiers, can be analysed fairly easily but their characteristics tend to vary greatly with the parameters of the transistors.
- Feedback can be used to stabilize the characteristics of the circuits so that they are less affected by the devices used.
- This stabilization is achieved at the expense of a fall in gain, but since transistors are inexpensive, adding more gain is not usually a problem.
- A major advantage of the use of feedback is that by making the circuit less dependent on the characteristics of the transistor, the analysis of the circuit is made much simpler.
- Usually it is possible to simplify the analysis of transistor circuits that use feedback, by making a few simple assumptions. These are:
  – that the gain of the transistor is so high that base currents can be neglected
  – that the base-emitter voltage is approximately constant and therefore the small signal voltage between the base and the emitter is negligible.
- Although the majority of transistor circuits apply the input signal to the base and take the output signal from the collector, other configurations are also used.
- The common-collector (emitter follower) mode produces a unity gain amplifier with a high input resistance and a low output resistance which is thus a good buffer amplifier.

- The common-base configuration produces an amplifier with a low input resistance and a high output resistance which makes a good trans-impedance amplifier.
- Several transistor amplifiers may be cascaded using coupling capacitors to carry the AC signals from one stage to the next while blocking any DC component that would upset the biasing. Unfortunately, the use of capacitors restricts the low-frequency performance of the arrangement and produces a complex and bulky circuit.
- An alternative method is to use direct coupling between stages. This requires fewer components and permits operation down to DC.
- Bipolar transistors are used in a wide range of applications in addition to their use as simple amplifiers. More complex circuits, such as operational amplifiers and other integrated circuits, are often constructed using bipolar transistors.
- In some applications four-layer devices such as thyristors and triacs are more suitable than simple transistors. Their ability to control high voltages and high currents while dissipating relatively little power makes them ideal for controlling AC loads.

Design study

An operational amplifier is to provide differential inputs with good common-mode rejection, a high gain and the ability to both source and sink current at the output.

Approach

Within this chapter we have looked at circuit techniques that can provide the various features required for this design. A long-tailed pair amplifier provides differential inputs and high gain in a simple circuit. To give a high common-mode rejection ratio, the emitter current for the long-tailed pair must be supplied by a constant current source. The ability to both source and

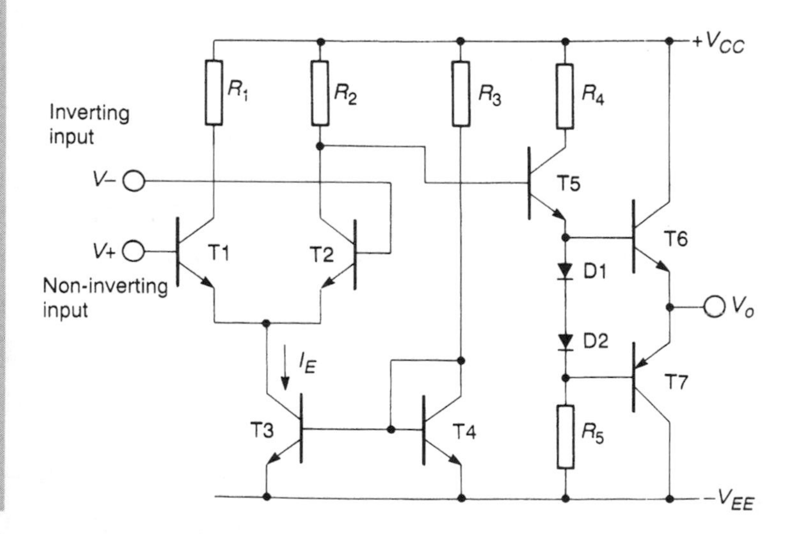

sink current requires the use of a push-pull output stage. This also reduces the quiescent power consumption of the circuit. The diagram above shows a possible circuit.

Transistors T1 and T2 form the long-tailed pair differential amplifier with collector resistors R_1 and R_2. In fact R_1 is unnecessary since the output is taken as a single-ended signal from T2. The constant current generator for the input amplifier is provided by the current mirror of T3 and T4. This produces an output current equal to that flowing in R_3, which is

$$I_E = \frac{V_{CC} + V_{EE} - V_{BE}}{R_3}$$

The output from T2 is amplified by the series negative feedback amplifier formed by T5, which applies its output to the output transistors T6 and T7. Diodes D1 and D2 are present to provide differential bias to the bases of the output transistors in order to reduce crossover distortion.

Most real operational amplifiers are more complicated than the circuit given here, but most use circuit techniques that are similar to those described.

References

Bardeen J. and Brattain W. H. (1948) The transistor: a semiconductor triode. *Phys. Rev.* **74**, 230–1

Shockley W. (1949) The theory of *p-n* junctions in semiconductors and *p-n* junction transistors. *Bell Systems Tech. J.* **28**, 435–89

Further reading

Ahmed H. and Spreadbury P. J. (1984) *Analogue and Digital Electronics for Engineers*, 2nd edn. Cambridge: Cambridge University Press

Horowitz P. and Hill W. (1989) *The Art of Electronics*, 2nd edn. Cambridge: Cambridge University Press

Millman J. and Grabel A. (1989) *Microelectronics*, 2nd edn. New York: McGraw-Hill

Schilling D. L. and Belove C. (1989) *Electronic Circuits: Discrete and Integrated*, 3rd edn. New York: McGraw-Hill

Exercises

7.1 Calculate the quiescent collector current and the quiescent output voltage of the following circuit assuming that $h_{FE} = 100$.

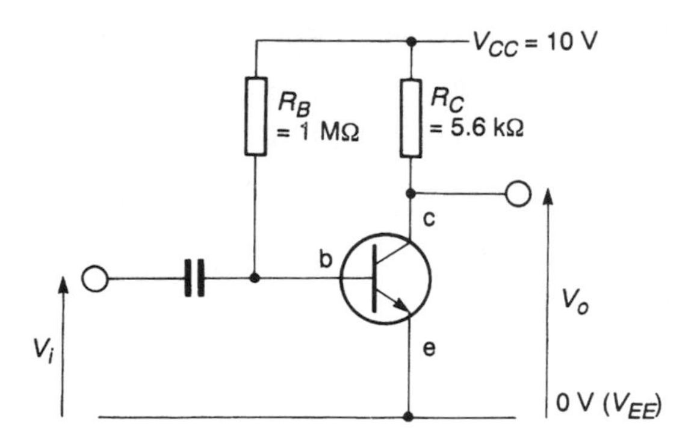

7.2 Repeat Exercise 7.1 with the transistor replaced by one with a current gain of 200.

7.3 Derive a simple small signal equivalent circuit for the following circuit and hence deduce the small signal voltage gain, input resistance and output resistance, given that $h_{FE} \approx h_{fe} = 175$ and $h_{oe} = 15\ \mu S$. How is the small signal voltage gain related to the quiescent voltage across R_C?

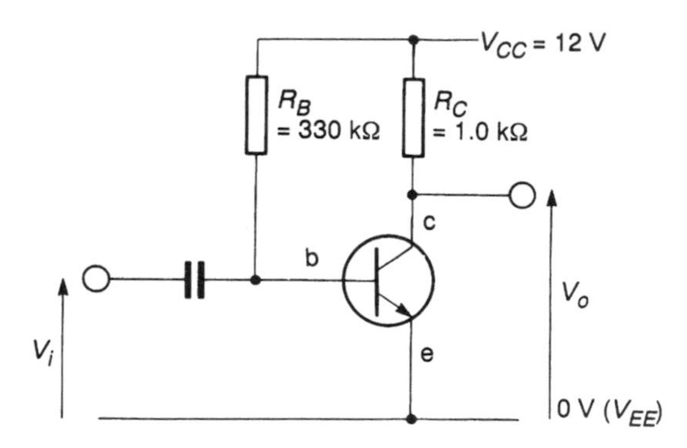

7.4 Repeat the calculations of Exercise 7.3 assuming that the transistor has $h_{oe} = 330\ \mu S$.

7.5 Calculate the quiescent collector current, the quiescent output voltage and the small signal voltage gain of the following circuit.

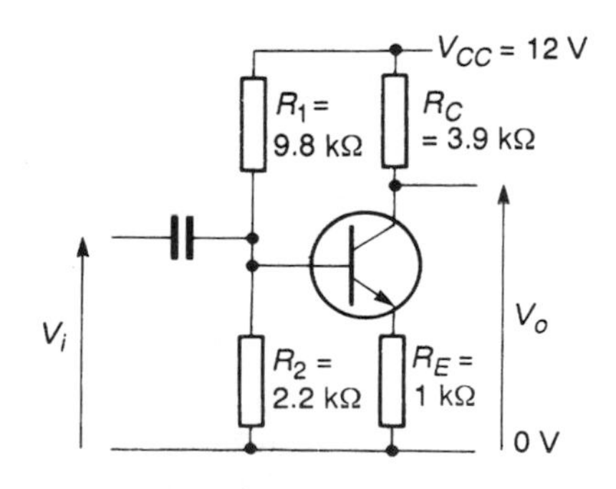

7.6 For the circuit of Exercise 7.5, estimate the small signal input and output resistances.

7.7 For the circuit of Exercise 7.5, estimate the effect on the frequency response of the circuit of using a coupling capacitor C of 1 μF

7.8 If the circuit of Exercise 7.5 were modified by the addition of an emitter decoupling capacitor of 1 μF, estimate the quiescent output voltage, small signal voltage gain, input resistance and low-frequency response of the resulting circuit. What would be a suitable value for the decoupling capacitor if the amplifier were required for use with signals down to 100 Hz?

7.9 Using a circuit of the form shown in Exercise 7.5, design an amplifier with a small signal voltage gain of −3, a quiescent output voltage of 7 V, a supply voltage of 12 V and a collector load resistance of 2.2 kΩ.

7.10 Use simulation to confirm the operation of your solution to the previous exercise.

7.11 For the amplifier designed in Exercise 7.9, calculate the small signal input resistance and hence determine an appropriate value for the input capacitor to allow satisfactory operation at frequencies down to 50 Hz.

7.12 Estimate the quiescent collector current, quiescent output voltage, small signal voltage gain and the input and output resistances of the following circuit.

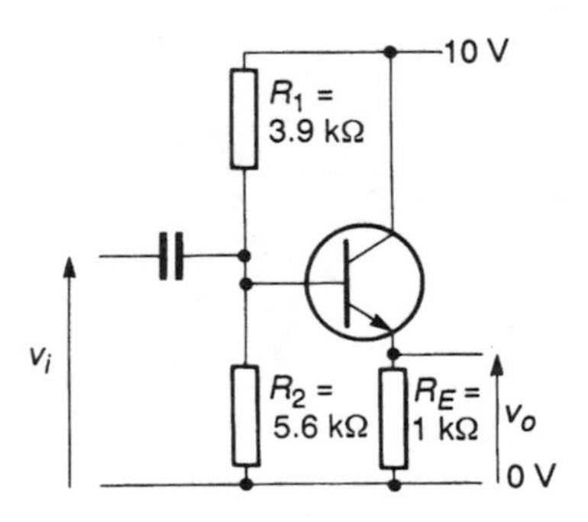

7.13 Determine the quiescent output voltage, the voltage gain and the input and output resistances of the following circuit. You may find it helpful to redraw the circuit in a more familiar form. It may be assumed that the capacitor *C* has a negligible impedance at the frequencies of interest.

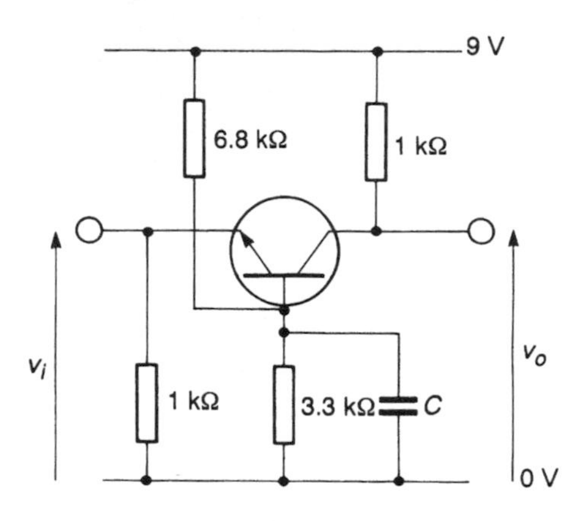

7.14 Design a two-stage, direct coupled amplifier with a voltage gain of 10. The circuit should operate from a 15 V supply and have a maximum output swing of at least 4 volts peak-to-peak.

7.15 Use simulation to investigate the performance of your solution to the previous exercise.

7.16 Design a long-tailed pair amplifier based on the circuit of Figure 7.34 and use simulation to determine the small signal voltage gain of the circuit and its CMRR.

Analogue Signal Processing

Objectives

When you have studied the material in this chapter you should:

- be familiar with the characteristics of single-stage and multiple-stage *RC* filters of both 'high-pass' and 'low-pass' forms;
- understand the distinction between active and passive filters and be aware of the advantages of active techniques in filter design;
- be aware of the different forms of active filters that are used, including the Butterworth, Chebyshev and Bessel forms;
- be able to describe the differences between and characteristics of amplifiers of classes A, B, AB, C and D;
- recognize and understand various common circuit techniques for the design of push-pull power amplifier stages;
- be familiar with the major causes of noise and interference within electronic systems, and with design techniques that can help to reduce their effects.

Contents

8.1 Introduction

Analogue signal processing is a general term used to describe the various operations that are performed on analogue signals as they pass through an electronic system.

One of the most important of these operations is **filtering**. That is, the process of changing the relative magnitudes of components of a signal which are of different frequencies. Applications of filtering are numerous, from the removal of ripple from the output of a power supply to the selection of one broadcast station from another in a radio receiver. Any network or component in which the gain varies with frequency can be considered to be a filter, but here we are concerned with circuits that are specifically designed to produce a particular frequency response. A wide range of circuits is used for this purpose and we will look at both passive and active techniques.

Clearly another operation of great importance is **amplification**, and we have met several aspects of this process in earlier chapters. In this chapter we will consider the various classes of amplifier and say more about power output stages.

One of the major limitations on the performance of electronic systems is the presence of **noise**. In Chapter 2 we discussed thermal noise and noted that this is produced by any component that has resistance. Active devices are also sources of noise and in many cases are the dominant noise source. In some applications a very low level of noise is essential and there are a number of techniques that can be used to help achieve this. In this chapter we will look at various sources of noise in electronic circuits and at the steps that can be taken to reduce their effects.

One aspect of noise relates to interference. The ability of a system to operate correctly in the presence of interference, and to not interfere with other systems, is described by its **electromagnetic compatibility** or **EMC**. We will consider some of the major aspects of this vast topic and discuss its implications for system design.

8.2 Filters

In Section 3.7 we looked at the effects of a coupling capacitor and stray capacitance on an amplifier's frequency response. We noted that the former produces a **low frequency cut-off** and the latter a **high frequency cut-off**, and that in both cases the gain and the phase are affected. The frequency at which the cut-off occurs is determined, in each case, by the **time constant** of the network, which in a simple *RC* network is given by the product *CR*. The time constant is given the symbol T (the Greek letter tau). The results of the analysis performed in Chapter 3 are summarized in Figure 8.1.

You will notice that the circuit responsible for the low-frequency cut-off passes high frequency signals unchanged but attenuates low frequency signals. For obvious reasons this is called a **high-pass filter**. Conversely, the high-frequency cut-off circuit passes low frequency signals and is therefore a **low-pass filter**. These simple circuits, which contain only one time constant, are called **first order** or **single-pole systems**.

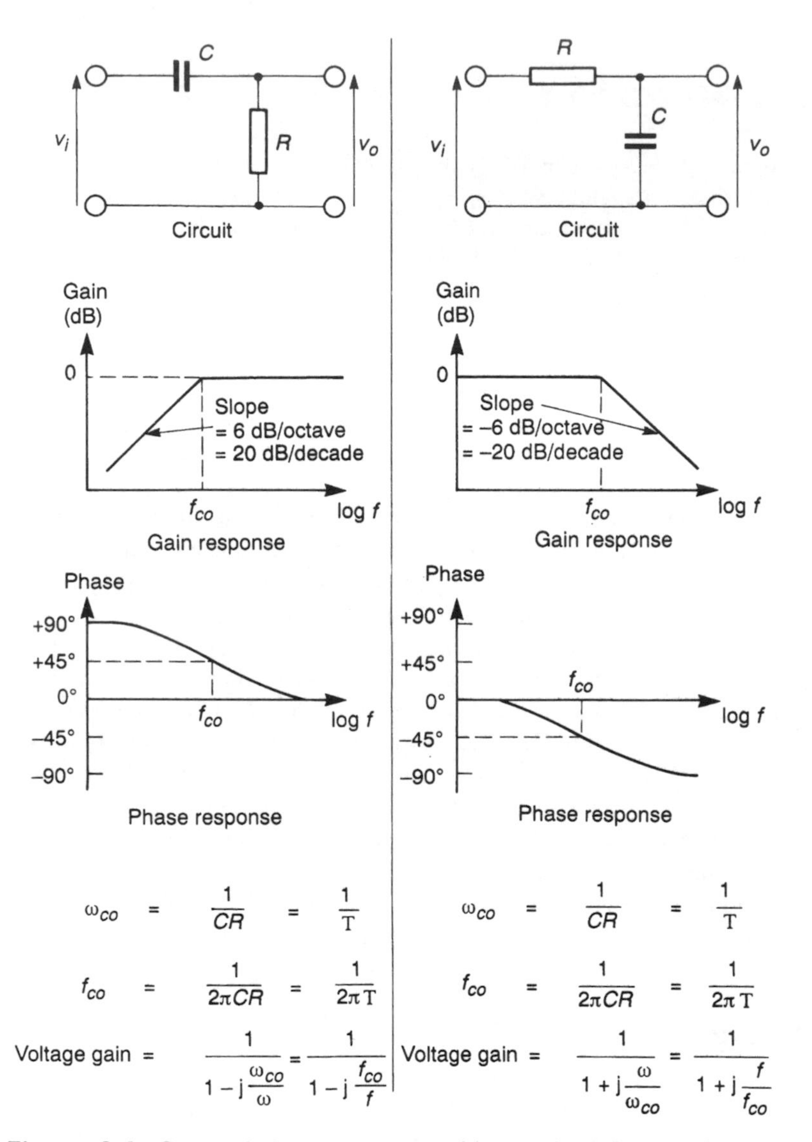

$$\omega_{co} = \frac{1}{CR} = \frac{1}{T} \qquad \omega_{co} = \frac{1}{CR} = \frac{1}{T}$$

$$f_{co} = \frac{1}{2\pi CR} = \frac{1}{2\pi T} \qquad f_{co} = \frac{1}{2\pi CR} = \frac{1}{2\pi T}$$

$$\text{Voltage gain} = \frac{1}{1 - j\dfrac{\omega_{co}}{\omega}} = \frac{1}{1 - j\dfrac{f_{co}}{f}} \qquad \text{Voltage gain} = \frac{1}{1 + j\dfrac{\omega}{\omega_{co}}} = \frac{1}{1 + j\dfrac{f}{f_{co}}}$$

Figure 8.1 Gain and phase responses of first order *RC* networks.

8.2.1 *RC* filters

Simple first order *RC* networks are often used as filters within electronic systems to select or remove components of a signal. However, for many applications the relatively slow roll-off of the gain is inadequate to remove unwanted signals effectively. In such cases, filters with more than one time constant are used to provide a more rapid roll-off of gain.

Figure 8.2(a) shows a second order (2-pole) system using two identical *RC* time constants. This circuit, while being usable as a filter, is difficult to analyse as the impedance

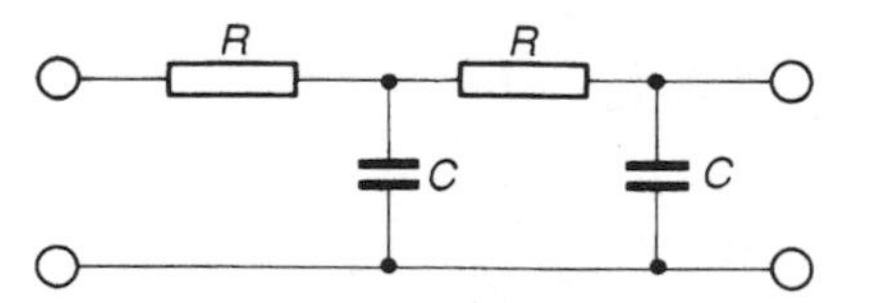

(a) A simple second order filter

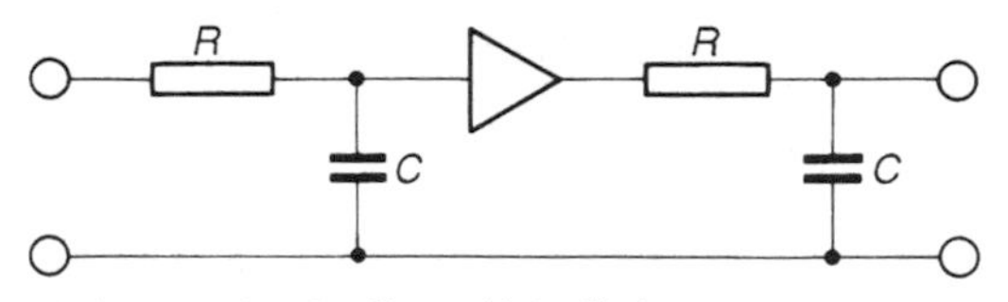

(b) A second order filter with buffering

Figure 8.2 Second order low-pass filters.

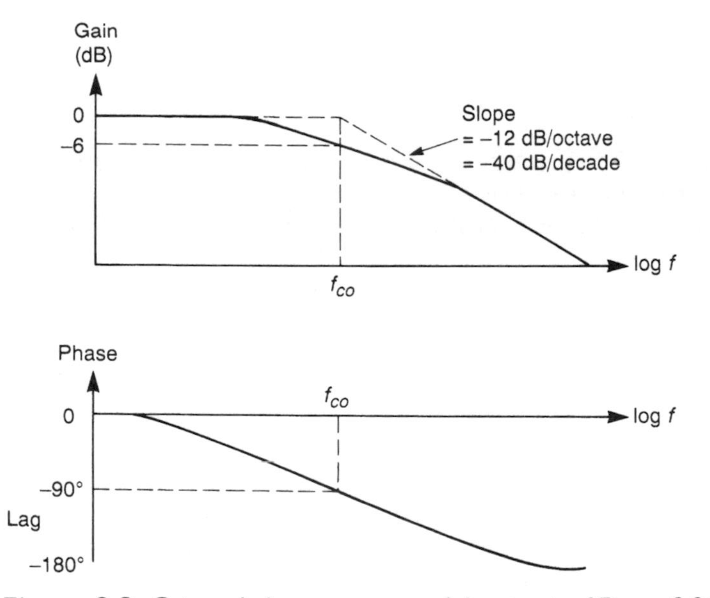

Figure 8.3 Gain and phase response of the circuit of Figure 8.2(b).

of the second stage affects the response of the first. A much simpler analysis is possible if we consider the circuit of Figure 8.2(b). Here a buffer amplifier is used to separate the two stages, preventing one from influencing the other. The gain and phase response of this arrangement are shown in Figure 8.3.

The responses obtained are simply the result of combining the effects of the two individual stages. Consequently, the fall in gain above the cut-off frequency is now −12 dB/octave (−40 dB/decade) and the phase angle is −90° at the cut-off point, increasing to −180° at high frequencies. Note that the fall in gain at the cut-off frequency f_{co} is now 6 dB, since each stage produces a fall in gain of 3 dB at this frequency. Therefore the **bandwidth** of the filter, which is measured to the half-power frequency (where the

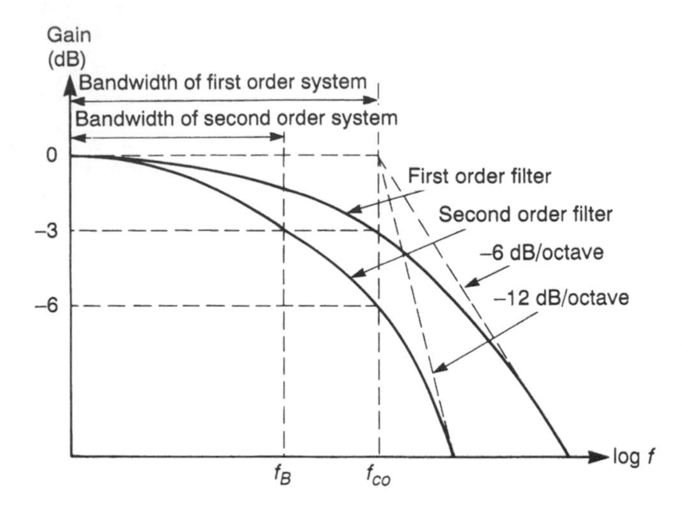

Figure 8.4 Relationship between bandwidth and cut-off frequency.

gain has fallen by 3 dB), is *not* the same as the cut-off frequency of the individual stages. This is illustrated in Figure 8.4.

In principle, any number of stages can be combined in this way to produce an nth order (n-pole) system. This will have a cut-off slope of $-6n$ dB/octave, an attenuation of $3n$ dB at its cut-off frequency and will produce a phase shift of $-45n°$ at the cut-off frequency. It is also possible to combine high-pass and low-pass characteristics into a single **band-pass filter** as illustrated in Figure 8.5.

For many applications, an **ideal filter** would have a constant gain and zero phase shift within its pass-band and zero gain outside. This is illustrated for a low-pass filter in Figure 8.6(a).

Unfortunately, although adding more stages to the *RC* filter increases the *ultimate* rate of fall of gain within the stop-band, the sharpness of the 'knee' of the response is not improved (see Figure 8.6(b)). To produce a circuit that more closely approximates an ideal filter, different techniques are required.

Computer simulation exercise 8.1

FILE 8A

Design a third order low-pass filter using the techniques described in Figure 8.2(b). Pick values for C and R such that the cut-off frequency of each stage is 1 kHz. The demonstration file for this exercise has *inappropriate* values for the capacitors and resistors – you should replace these with your calculated values.

Sketch the expected frequency and phase response of this arrangement, indicating the rate of fall of gain at high frequencies and the gain of the circuit at 1 kHz.

Simulate your design using unity gain buffer amplifiers based on 741 operational amplifiers. Plot the frequency response of the arrangement and compare this with your predictions.

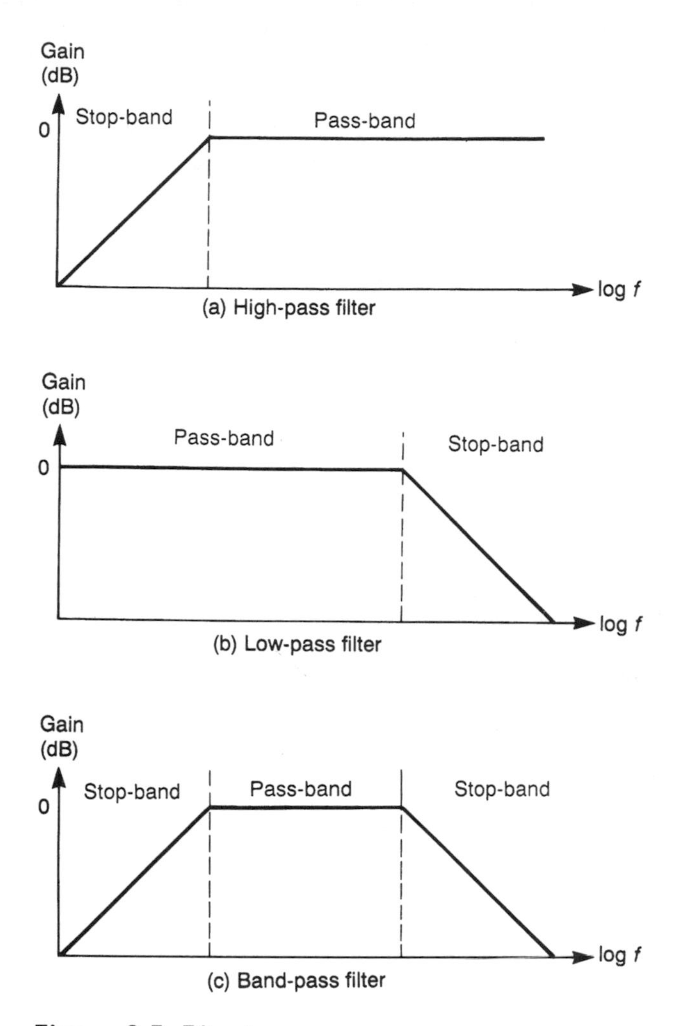

Figure 8.5 Filter types.

8.2.2 *LC* filters

The combination of inductors and capacitors allows the production of filters with a very sharp cut-off rate. Perhaps the best known forms of *LC* filters are the series and parallel **resonant circuits** shown in Figure 8.7, which are also known as **tuned circuits**.

These combinations of inductors and capacitors produce narrow-band filters with centre frequencies given by the expression

$$f_o = \frac{1}{2\pi\sqrt{(LC)}}$$

and bandwidths determined by the source impedance and by losses in the inductor and

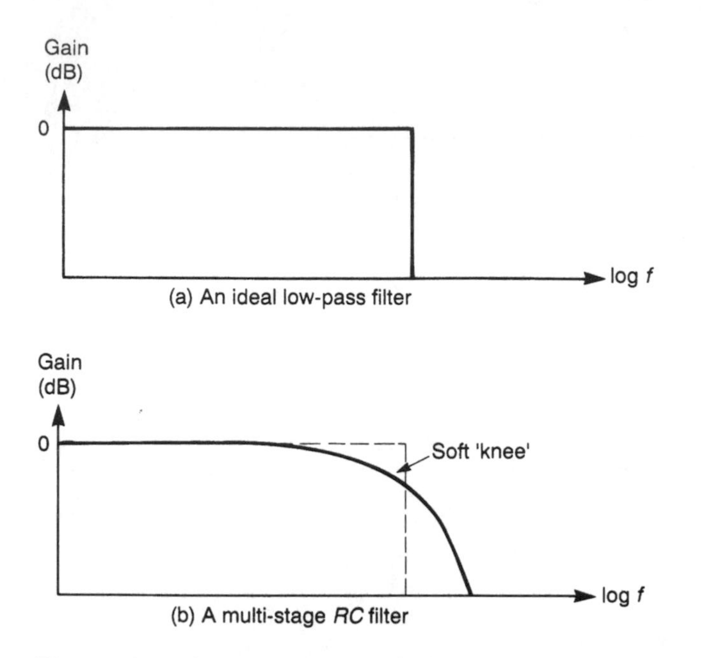

Figure 8.6 Gain responses of ideal and real low-pass filters.

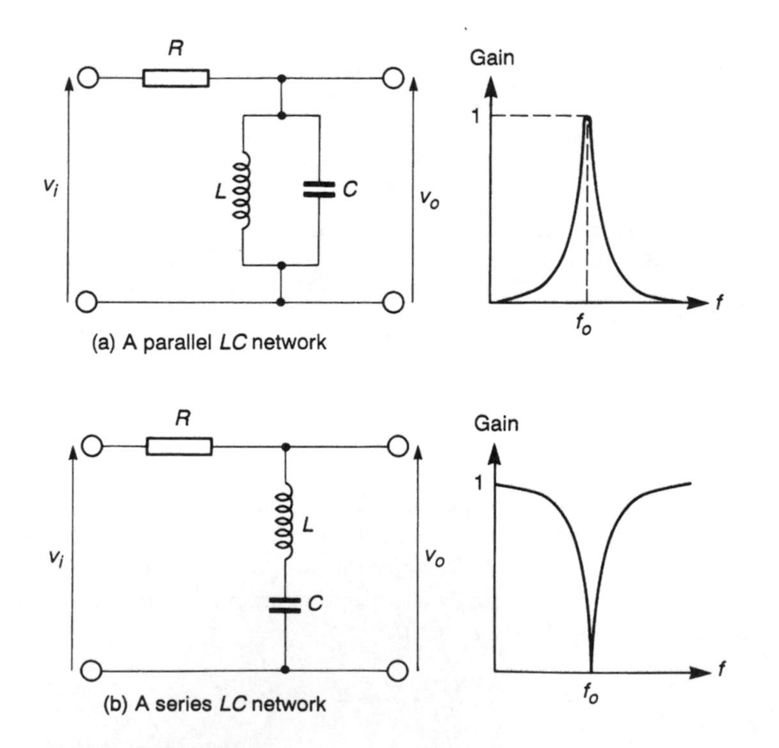

Figure 8.7 *LC* filters.

the capacitor. The narrowness of the response is usually described by the **quality factor** or **Q** of the circuit, which is given by the ratio of the bandwidth (measured between the 3 dB points) and the centre frequency f_o.

Other configurations of inductors, capacitors and resistors can be used to form high-pass, low-pass, band-pass and band-reject filters and can achieve very high cut-off rates.

8.2.3 Active filters

Although combinations of inductors and capacitors can produce very high performance filters, the use of inductors is inconvenient since they are expensive, bulky and suffer from greater losses than other passive components. Fortunately it is possible to **synthesize** any *RLC* network using an operational amplifier and suitable arrangements of resistors and capacitors. Such filters are called **active filters** since they include an active component (the operational amplifier) in contrast to the other filters we have discussed which are purely passive (ignoring any buffering). A detailed study of the operation and analysis of active filters is beyond the scope of this text, but it is worth looking at the characteristics of these circuits and comparing them with those of the *RC* filters discussed earlier.

Filter characteristics

To construct multiple-pole filters it is often necessary to cascade many stages. If the time constants and the gains of each stage are varied in a defined manner, it is possible to create filters with a wide range of characteristics. Using these techniques it is possible to construct filters of a number of different types to suit particular applications.

In simple *RC* filters, the gain starts to fall towards the edge of the pass-band and so is not constant throughout the band. This is also true of active filters, but here the gain may actually rise towards the edge of the pass-band before it begins to fall. In some circuits the gain fluctuates by small amounts right across the band. These characteristics are illustrated in Figure 8.8.

The ultimate rate of fall of gain with frequency for any form of active filter is $6n$ dB/octave where 'n' is the number of poles in the filter, which is normally equal to the number of capacitors in the circuit. Thus the performance of the filter in this respect is related directly to circuit complexity.

Although the ultimate rate of fall of gain of a filter is defined by the number of poles, the sharpness of the 'knee' of the filter varies from one design to another. Filters with a very sharp knee tend to produce more variation of the gain of the filter within the pass-band. This is illustrated in Figure 8.8, where it is apparent that filters B and C have a more rapid roll-off of gain than filter A, but also have greater variation in their gain within the pass-band.

Of great importance in some applications is the **phase response** of the filter. That is, the variation of phase lag or lead with frequency as a signal passes through the filter. We have seen that *RC* filters produce considerable amounts of phase shift within the pass-band. All filters produce a phase shift that varies with frequency. The way in which it is related to frequency varies from one type of filter to another. The phase response of a filter is of particular importance where pulses are to be used.

A wide range of filter designs is available, enabling one to be selected to favour any of the above characteristics. Unfortunately, the requirements of each are often mutually

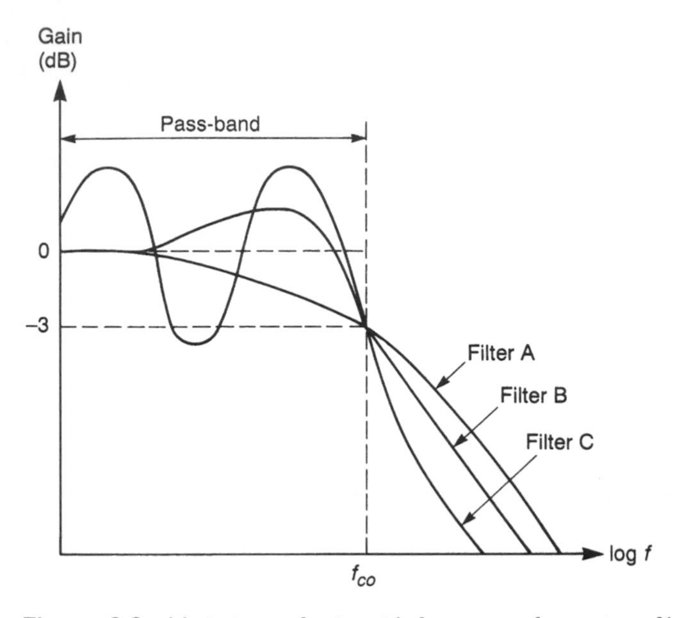

Figure 8.8 Variations of gain with frequency for various filters.

exclusive and so there is no universal optimum design and an appropriate circuit must be chosen for a given application.

Filter types

From the myriad of filter designs, three basic types are of particular importance, firstly because they are widely used, and secondly, because they are each optimized for a particular characteristic.

The **Butterworth filter** is optimized to produce a flat response within its pass-band, which it does at the expense of a less sharp 'knee' and a less than ideal phase performance. This filter is sometimes called a **maximally flat filter** as it produces the flattest response of any filter type.

The **Chebyshev filter** produces the sharpest transition from the pass-band to the stop-band but does this by allowing variations in gain throughout the pass-band. The gain ripples within specified limits, which can be selected according to the application. The phase response of the Chebyshev filter is poor and it creates serious distortion of pulse waveforms.

The **Bessel filter** is optimized for a linear phase response and is sometimes called a **linear phase filter**. The 'knee' is much less sharp than for the Chebyshev or the Butterworth types (though slightly better than a simple *RC* filter), but its superior phase characteristics make it preferable in many applications, particularly where pulse waveforms are being used. The phase shift produced by the filter is approximately linearly related to the input frequency. The resultant phase shift therefore has the appearance of a fixed time delay, with all frequencies being delayed by the same time interval. The result is that complex waveforms that consist of many frequency components (such as pulse

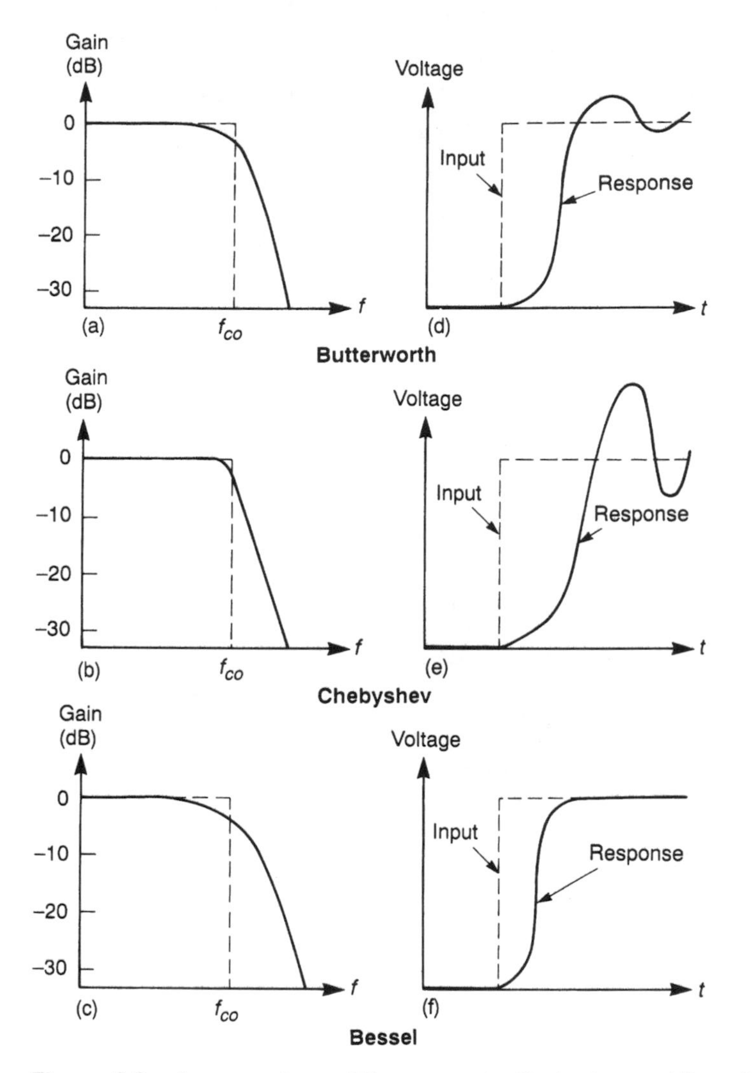

Figure 8.9 A comparison of Butterworth, Chebyshev and Bessel filters.

waveforms) are filtered without distorting the phase relationships between the various components of the signal. Each component is simply delayed by an equal time interval.

Figure 8.9 compares the characteristics of these three types of filter. Parts (a), (b) and (c) show the frequency responses for Butterworth, Chebyshev and Bessel filters, each with six poles (the Chebyshev is designed for 0.5 dB ripple), while (d), (e) and (f) show the responses of the same filters to a step input.

Filter realization

Over the years a number of designs have emerged to implement various forms of filter. The designs have different characteristics and each has advantages and disadvantages.

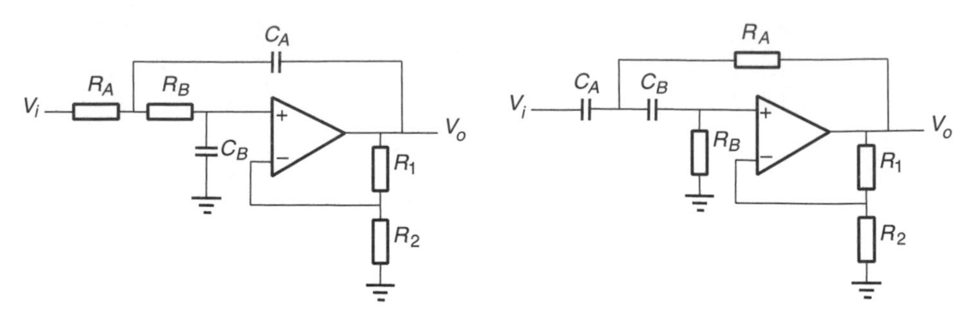

(a) A low-pass filter

(b) A high-pass filter

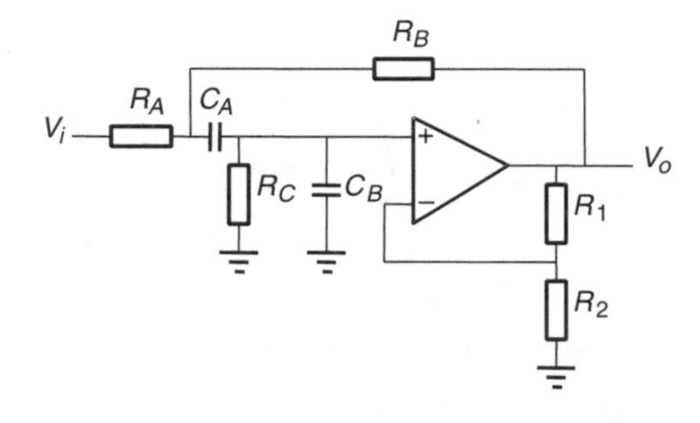

(c) A band-pass filter

Figure 8.10 Operational amplifier filter circuits.

Here we will look briefly at a single family of circuits that can be used to implement a range of filter types.

Figure 8.10 shows three filters each constructed around a non-inverting amplifier. By appropriate choice of the component values these circuits may be designed to produce the characteristics of various forms of filter, such as Bessel, Butterworth or Chebyshev. The circuits shown are 2-pole filters, but several such stages may be cascaded to form higher-order filters. In general the cascaded stages will not be identical, but are designed such that the combination has the required characteristics.

Figure 8.10(a) shows a low-pass filter arrangement. The resistors R_1 and R_2 define the overall gain of the circuit, as in the non-inverting amplifier of Example 4.3. R_B and C_B produce a low-pass arrangement that reduces the input into the amplifier when the frequency exceeds its cut-off frequency. R_A and C_A form a second low-pass network similar to that described in Example 4.10. These combine to form a 2-pole low-pass characteristic. In the circuit of Figure 8.10(b) the positions of the capacitors and resistors are reversed to form a high-pass circuit. If $R_A = R_B = R$ and $C_A = C_B = C$ in these two circuits, the cut-off frequency of the filters will be determined by the time constant RC. However, the actual cut-off frequency is affected by the filter characteristics, which are in turn determined by the gain of the circuit, as set by R_1 and R_2. For a Butterworth filter the cut-off frequency f_o is given by $f_o = 1/2\pi CR$. For other filter types the cut-off frequency may be slightly above or slightly below this value.

The circuit of Figure 8.10(c) has a combination of high-pass and low-pass networks which result in a band-pass characteristic. However, this filter can produce a response with a peak that is much sharper than can be achieved by cascading separate high- and low-pass circuits.

FILE 8B
FILE 8C
FILE 8D

Computer simulation exercise 8.2

Simulate the low-pass filter of Figure 8.10(a) using values corresponding to a Butterworth filter. Suitable component values would be: $R_A = R_B = 16\ \text{k}\Omega$; $C_A = C_B = 10\ \text{nF}$; $R_1 = 5.9\ \text{k}\Omega$ and $R_2 = 10\ \text{k}\Omega$. Plot the frequency response of this arrangement and note the general shape of the response and its cut-off frequency.

Repeat the above using the high-pass filter of Figure 8.10(b) using the same component values.

Look briefly at the frequency response of the band-pass filter of Figure 8.10(c). Use the same component values as before with $R_C = 16\ \text{k}\Omega$.

8.3 Amplifiers

In Chapter 3 we looked in general at the subject of amplification and at the use of operational and other amplifiers. In Chapters 6 and 7 we looked at the use of FETs and bipolar transistors in various forms of amplifier and we considered aspects of biasing and circuit design specific to these devices. In this section we will return to broader issues and consider the various *classes* of amplifier and the design of the output stage of power amplifiers.

An important aspect in the choice of a technique to be used for a power output stage is its efficiency. We may define the **efficiency** of an amplifier as

$$\text{Efficiency} = \frac{\text{power dissipated in the load}}{\text{power absorbed from the supply}} \tag{8.1}$$

Efficiency is of importance since it determines the power dissipated by the amplifier itself. The **power dissipation** of an amplifier is important for a number of reasons. One of the least important, except in battery powered applications, is the actual cost of the electricity used as this is generally negligible. Power dissipated by an amplifier takes the form of **waste heat**, and the production of excess heat requires the use of larger, and more expensive, power transistors to dissipate this heat. It might also require the use of other methods of heat dissipation, such as **heat-sinks** or **cooling fans**. It may also be necessary to increase the size of the power supply to deliver the extra power required by the amplifier. All these factors increase the cost and size of the system.

8.3.1 Classes of amplifier

All amplifiers may be allocated to one of a number of classes depending on the way in which the active device is operated.

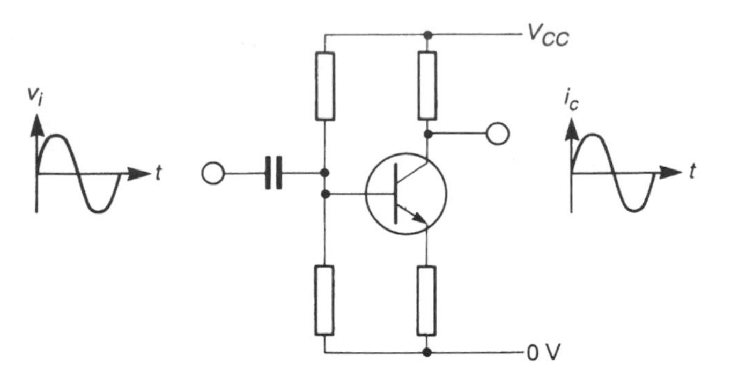

FILE 8E **Figure 8.11** A class A amplifier.

Class A

In class A amplifiers, the active device (for example, a bipolar transistor or a FET) conducts during the complete period of any input signal. An example of such a circuit would be a conventionally biased single-transistor amplifier of the type shown in Figure 8.11.

It can be shown that for conventional class A amplifiers, maximum efficiency is achieved for a sinusoidal input of maximum amplitude, when it reaches only 25%. With more representative inputs, the efficiency is very poor.

The efficiency of class A amplifiers may be improved by coupling the load using a transformer. The primary replaces the load resistor while the load is connected to the secondary to form a **transformer coupled amplifier**. This enables efficiencies approaching 50% to be achieved, but is unattractive because of the disadvantages associated with the use of inductive components, including their cost and bulk.

Class B

In a class B amplifier the output active devices conduct for only half the period of an input signal. These are normally push-pull arrangements in which each transistor is active for half of the input cycle. Such circuits were discussed in Section 7.7.4; an example is given in Figure 8.12.

Class B operation has the advantage that no current flows through the output transistors in the quiescent state and so the overall efficiency of the system is much higher than in class A. If one assumes the use of ideal transistors, it can be shown that the maximum efficiency is about 78%.

Class AB

Class AB describes an amplifier that lies part way between classes A and B. The active device conducts for more than half of the input cycle but less than 100%. A class AB amplifier can be formed from a standard push-pull stage by ensuring that both devices conduct for part of the input waveform, as shown in Figure 8.13.

The efficiency of a class AB amplifier will lie between those of class A and class B designs, and will depend on the bias conditions of the circuit.

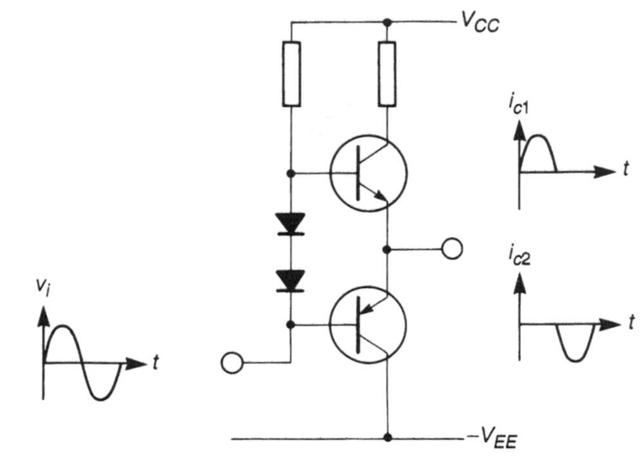

FILE 8F

Figure 8.12 A class B amplifier.

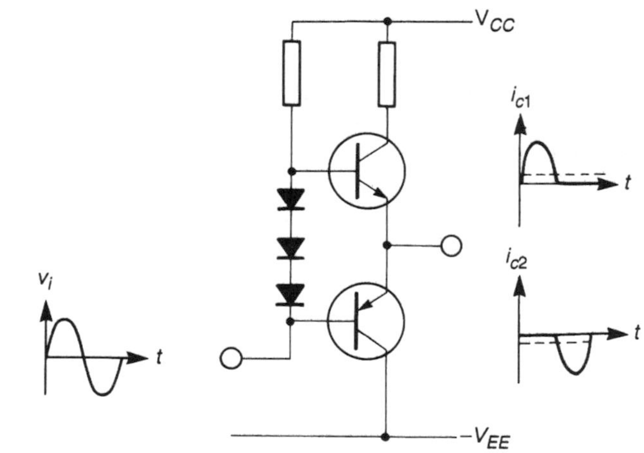

FILE 8G

Figure 8.13 A class AB amplifier.

Class C

Following on from the definitions of classes A and B, it is perhaps not surprising that the definition of class C is that the active device conducts for *less* than half of the input cycle. Class C is used to enable the device to be operated at its peak current limit without exceeding its maximum power rating. The technique can produce efficiencies approaching 100%, but results in gross distortion of the waveform. For these reasons class C is used only in fairly specialized applications. One such use is in the output stage of radio transmitters where inductive filtering is used to remove the distortion. A possible circuit for a class C amplifier is given in Figure 8.14. Often the collector resistor in this figure is replaced by an *RC* tuned circuit.

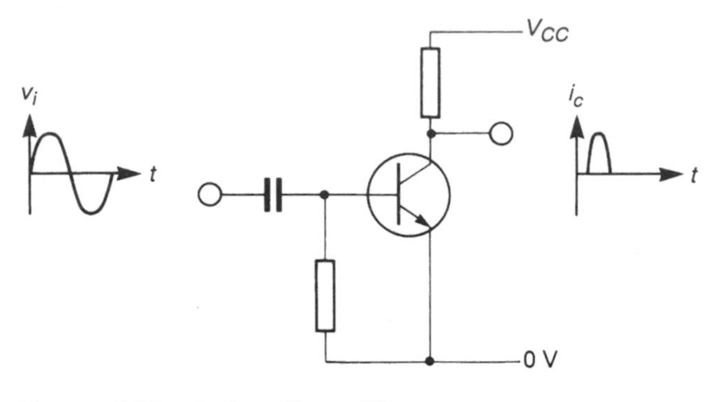

FILE 8H **Figure 8.14** A class C amplifier.

Class D

In class D amplifiers, the active devices are used as switches and are either completely ON or completely OFF. A perfect switch has the characteristics of having infinite resistance when open and zero resistance when closed. If the devices used for the amplifier were perfect switches, this would result in no power being dissipated in the amplifier itself, since when a switch was ON it would have current flowing through it but no voltage across it, and when it was OFF it would have voltage across it but no current flowing through it. Since power is the product of voltage and current, the dissipation in both states would be zero. Although no real device is an ideal switch, bipolar transistors make very good switching devices and amplifiers based on power transistors are both efficient and cost effective. Amplifiers of this type are often called **switching amplifiers** or **switch mode amplifiers**.

Class D amplifiers may use single devices or push-pull pairs. In the latter case, only one of the two devices is ON at any time. An example of such an arrangement is shown in Figure 8.15.

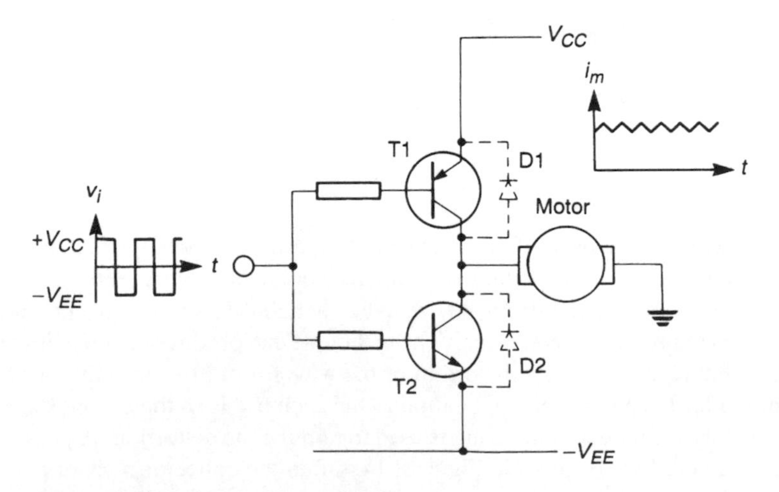

Figure 8.15 A class D amplifier.

Here a push-pull circuit is used to control a DC motor. An input voltage of V_{CC} turns the upper transistor OFF since it has zero volts between the base and the emitter. The lower transistor is driven hard ON and, since it cannot draw current through the other transistor, pulls current from the motor. An input voltage of $-V_{EE}$ has the opposite effect, turning the lower transistor OFF and the upper one ON. When the upper transistor is turned ON the motor is connected to the positive supply which will attempt to push current *into* the motor. When the lower transistor is ON, the motor is connected to the negative supply which will attempt to pull current *from* the motor. Since the motor has a considerable inductance, this will prevent the motor current from changing rapidly. If the input is switched rapidly between its two states, the output current will continuously ramp up and down in an attempt to respond to the changing voltages across it. If the switching rate is fast compared with the mechanical time constant (the rate of response) of the motor, it will respond simply to the average value of the switching waveform. The diodes D1 and D2 are **catch diodes**, as described in Example 5.6, and are normally added to prevent damage to the transistors by the back EMF from the motor. A practical implementation of this system would include a motor current sensing arrangement which would feed back the motor current to a 'motor controller'. The controller would then modify the times spent in each state to bring the average current to the desired value. This arrangement is considerably more complicated than a simple class A or B amplifier but has the advantage of greatly improved efficiency. In the case of ideal switches, the efficiency of the amplifier would be 100%. In practice the transistors do dissipate some power, as do other circuit components required to control and stabilize the circuit. However, this method of power control is one of the most efficient techniques available.

Amplifier classes – a summary

Class A, B, AB and C arrangements are linear amplifiers. Class D circuits are switching amplifiers. The distinction between the linear amplifiers may be seen as the differences between the quiescent currents flowing through the output devices. This in turn is determined by the biasing arrangements of the circuit. These are summarized in Figure 8.16.

8.3.2 Power amplifiers

Class A

In class A circuits the quiescent current in the output device is greater than the maximum output current. This allows the current through the device to decrease by an amount equal to the peak output current without 'bottoming out'. This leads to low distortion but consumes a great deal of power. Since the quiescent current is large, considerable power is dissipated even in the absence of an input.

Class A techniques are occasionally used in power amplifiers but only where the very low distortion produced by these techniques can justify the high cost. Examples of such applications include high performance audio amplifiers.

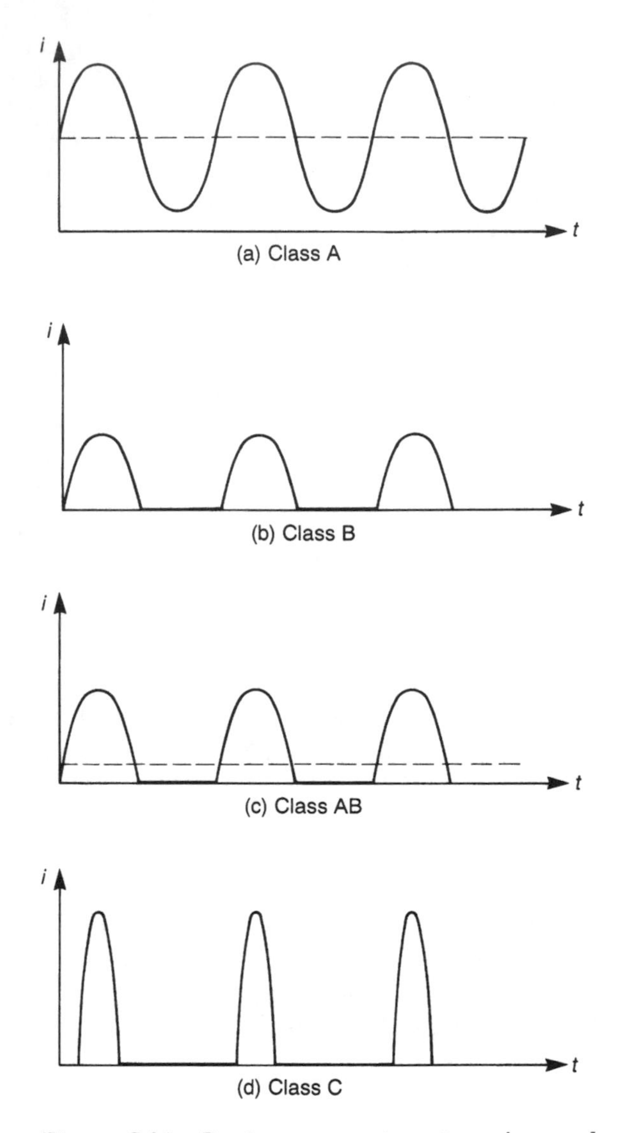

Figure 8.16 Device currents in various classes of amplifier for a sinusoidal input.

Class B

In class B amplifiers the quiescent current is zero, giving a much lower power consumption and greater efficiency. We have already met this form of amplifier in our discussions of **push-pull amplifiers** in Section 7.7.4. Figure 8.17 shows one of the simplest push-pull arrangements.

This amplifier is strictly speaking a class C amplifier since there is a dead-band of $\pm V_{BE}$ for which neither transistor is conducting. Each device therefore conducts for less than half of the input cycle. However, this arrangement is often considered as a poorly

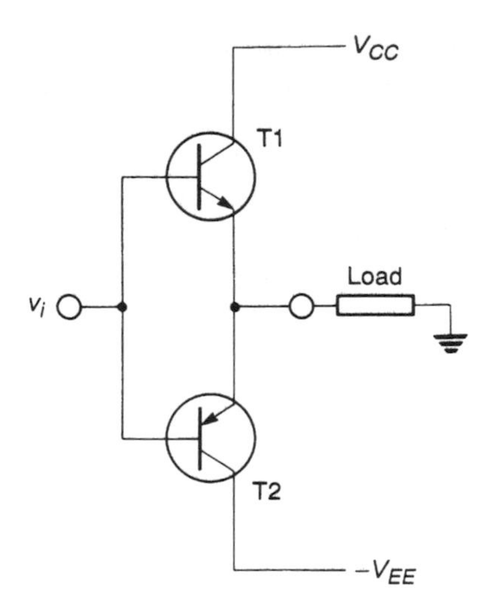

FILE 8J **Figure 8.17** A simple push-pull arrangement.

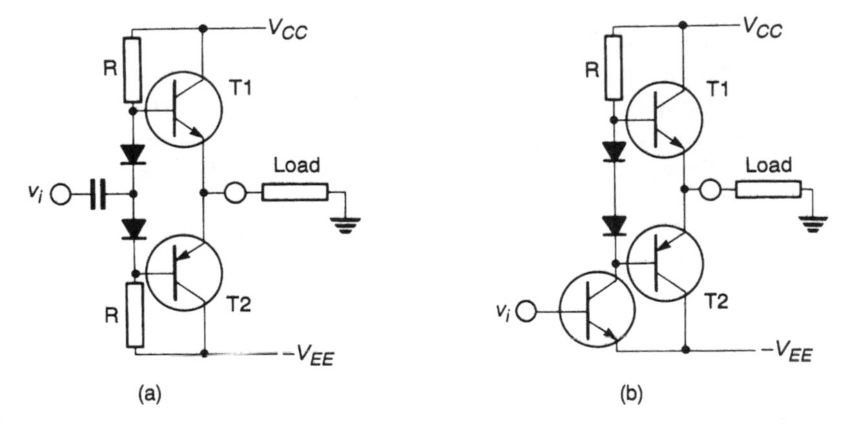

FILE 8K **Figure 8.18** The use of diodes for base biasing.

designed class B amplifier, since for large input signals each device conducts for nearly half the cycle. The limitations of this design are clear. Small fluctuations of the input will produce no output change, while large input swings will be distorted as they traverse the dead-band. This leads to the generation of large amounts of **crossover distortion**.

To convert the amplifier of Figure 8.17 into a true class B amplifier we need to arrange that one transistor turns ON precisely when the other turns OFF. To do this, the bases of the transistors must be biased so that they are separated by twice their normal base-to-emitter voltage. In Section 7.7.4 we looked at a simple method of achieving this result; two variants of this technique are shown in Figure 8.18.

A problem associated with this arrangement is that of **temperature instability** caused by changes in the temperature of the transistors. As the output devices warm up, because of the power that is being dissipated in them, their V_{BE} drops in comparison with the voltages across the relatively cool diodes. If the voltage between the bases of the two transistors becomes greater than the sum of the turned ON voltages of the transistors, a quiescent current will flow, further increasing the power dissipation and hence the temperature of the output transistors. In extreme cases this process can lead to **thermal runaway** and to the ultimate destruction of the circuit. If the diodes of this circuit are placed close to the output transistors they will provide **temperature compensation**, since variations in the V_{BE} of the output transistors, caused by temperature changes, will be matched by similar variations in the voltages across the diodes.

Unfortunately, since the current flowing through the power transistors in the circuits of Figure 8.18 is greater than that flowing through the diodes, the base-to-emitter voltages of the former will be greater than the voltages across the latter and a perfect match will not be achieved. The bias offset must therefore be slightly more than the voltage across the two diodes for optimum results.

Several approaches are used in an attempt to produce the correct offset between the bases of the two output transistors. One of the simplest approaches is to add a small preset variable resistance between the bases, in series with the two diodes, as shown in Figure 8.19.

Current flowing through the biasing network produces a voltage across the preset resistance, increasing the voltage between the bases. If this voltage is increased sufficiently, both transistors will conduct at the same time producing a quiescent current through the output transistors. For true class B operation the resistance should be set to a value which just prevents quiescent current from flowing through the transistors. This technique allows precise adjustment of the offset voltage while maintaining much of the temperature stability produced by the diodes.

Unfortunately, the improvement in efficiency of class B amplifiers as compared with those of class A has a price as there is always some non-linearity associated with the

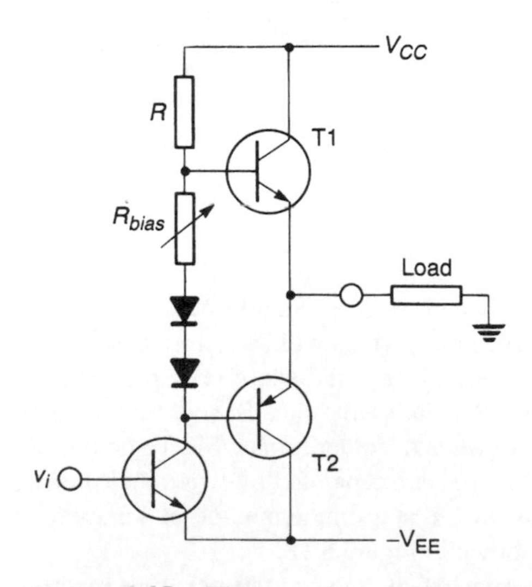

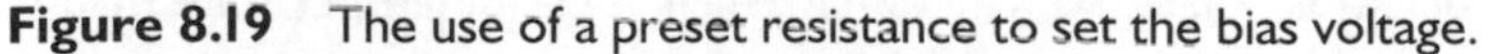
Figure 8.19 The use of a preset resistance to set the bias voltage.

transition from one transistor being turned ON to the other being turned ON. This inevitably leads to a small amount of **crossover distortion**.

Class AB

The distortion produced at the crossover point can be reduced by allowing a small quiescent current to flow through the output devices. This smooths the transition from one transistor to the other and thus reduces crossover distortion. This technique has the effect of increasing the quiescent power consumption of the system and reducing the overall efficiency. However, the reduction in distortion makes this approach extremely attractive for applications such as **audio amplifiers**. Since current flows in both transistors when no input is applied, each output transistor is conducting for more than 50% of the input cycle. This arrangement is therefore class AB rather than class B. The circuit of Figure 8.19 can be used as a class AB amplifier simply by choosing an appropriate value for the preset resistor. This circuit, and those that follow, may therefore be class B or class AB depending on the value set for the quiescent current. If appropriately adjusted, the amplifiers could be used in class A by ensuring that the quiescent current is so large that both transistors remain conducting throughout the input cycle. However, use of class A in this form of power output stage is unusual. In practice, the circuits are normally adjusted to suit the application, the quiescent current being set to zero to minimize power consumption (class B operation) or set for a quiescent current that minimizes crossover distortion (class AB operation).

Output stage techniques

The stability of the quiescent conditions of the circuit of Figure 8.19 can be improved by the addition of small emitter resistors, as shown in Figure 8.20. The voltage drop across

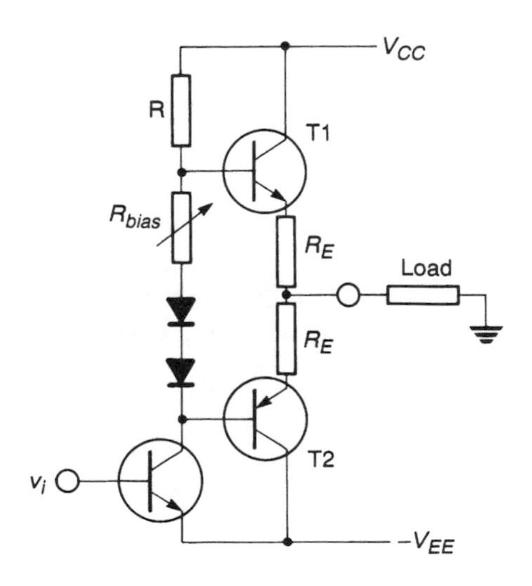

Figure 8.20 The use of emitter resistors.

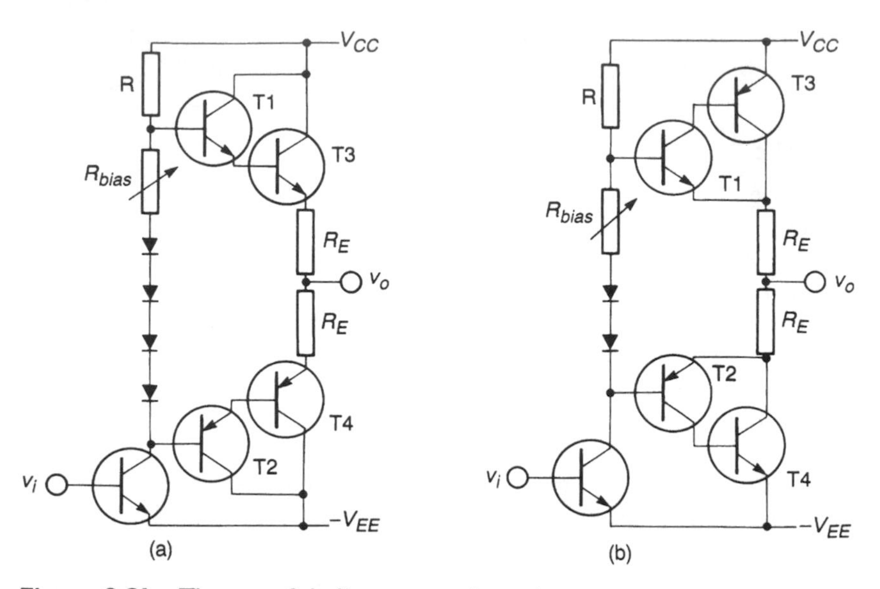

Figure 8.21 The use of darlington transistors in power output stages.

these resistors is subtracted from the base-emitter voltage of the circuit. They therefore provide series negative feedback by reducing the voltage across the base–emitter junction as the current increases, and thus stabilize the quiescent current. Typical values for these resistors might be a few ohms for low power applications and much less for higher power circuits. The power dissipated in these resistors is thus fairly small.

We noted in the last chapter that the current gain of high power transistors is relatively low. For this reason it is often useful in power output stages to use a **darlington** arrangement as shown in Figure 8.21. This figure shows two widely used configurations, the first using darlington pairs of the same type of transistor, the second using **complementary darlington** transistors.

The first circuit of Figure 8.21 uses four diodes. These are required since a voltage of $4V_{BE}$ is needed between the bases of T1 and T2 to turn on the four output transistors. In the second circuit only two diodes are required since a voltage of only $2V_{BE}$ between the bases of T1 and T2 will cause all four output transistors to conduct.

An alternative to the use of a chain of diodes in the bias network is shown in Figure 8.22(a). Here a single transistor is used with a pair of resistors to provide its base bias. since, for a high gain transistor, the base current is generally negligible, the base-to-emitter voltage V_{BE} is determined by the collector-to-emitter voltage V_{CE} and the ratio of the resistors R_1 and R_2. Therefore

$$V_{BE} = V_{CE}\frac{R_1}{R_1 + R_2}$$

and rearranging

$$V_{CE} = V_{BE}\frac{R_1 + R_2}{R_1}$$

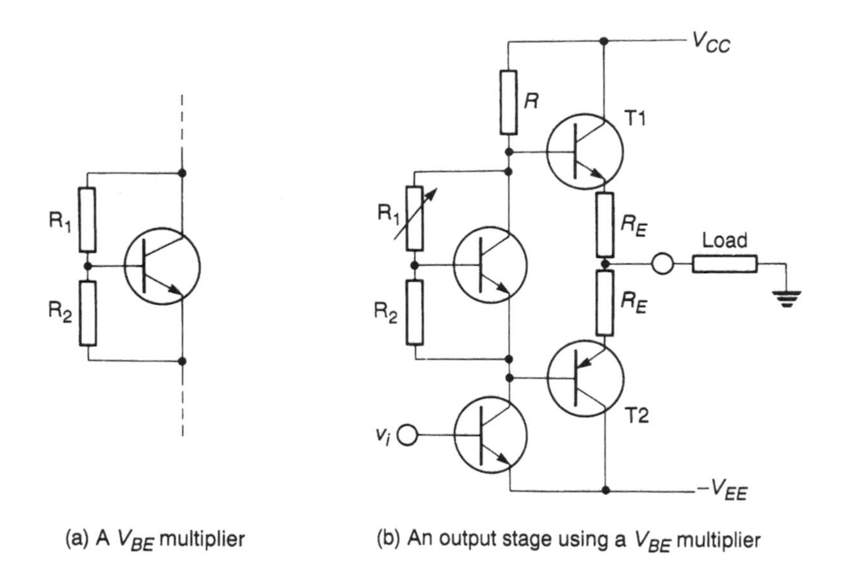

(a) A V_{BE} multiplier (b) An output stage using a V_{BE} multiplier

Figure 8.22 The V_{BE} multiplier arrangement.

The voltage across the network is thus determined by V_{BE} and the ratio of the resistors R_1 and R_2. By adjusting the relative values of the two resistors the voltage across the network can be set to equal any multiple of V_{BE}, as required. For this reason the circuit is called a V_{BE} multiplier. As with biasing diodes, it is usual to mount the transistor close to the output transistors to achieve good **temperature compensation**, as described earlier. Figure 8.22(b) shows an output stage using a V_{BE} multiplier to set the quiescent current.

8.3.3 Design for integration

When designing amplifiers for construction using discrete components (resistors, capacitors, transistors, etc.) it is natural to use low-cost passive components wherever possible to minimize the total cost of the circuit. When designing circuits for implementation in **integrated form**, the rules of economics are different. Transistors, diodes and other active components require less chip area than passive components and so are 'cheaper' to implement. This leads us to rethink the way in which we design circuits in an attempt to minimize the use of passive components and to replace them with active parts wherever possible.

Active loads

One common use of passive components within conventional circuits is the use of **load resistors**, as shown in Figures 8.23(a) and (b).

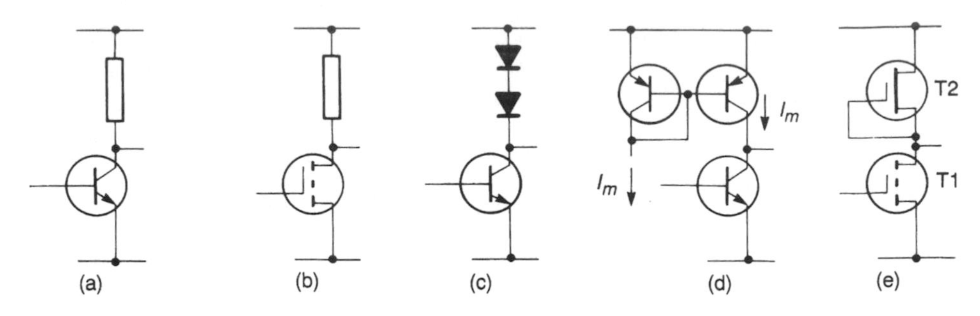

Figure 8.23 A comparison of passive and active load arrangements.

In circuits based on bipolar transistors it is possible to replace the load resistor by an active load consisting of a series of diodes, as shown in Figure 8.23(c). This produces an effective load equal to the sum of the slope resistances of the diodes. Unfortunately the resistance contributed by each diode is small and the number of diodes that can be used is limited by the fact that each contributes a voltage drop of about 0.7 V. An alternative arrangement for bipolar transistor circuits is to use a **current mirror**, as shown in Figure 8.23(d). Since a *constant* current I_m is supplied by the current mirror, any small signal variations in the collector current appear directly at the output. This arrangement is economical since a number of load transistors can be driven from one current mirror, as shown in Figure 7.32(b). A FET can also be used as an active load, as shown in Figure 8.23(e). We have already come across this circuit in Section 6.6.4 when we considered the use of a FET as a logical inverter. The arrangement of Figure 8.23(e) can be modified by replacing the DE MOSFET used for the load transistor T2 by an enhancement

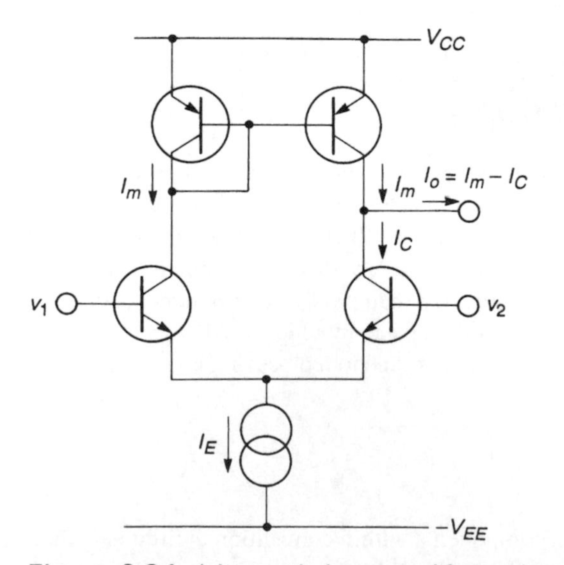

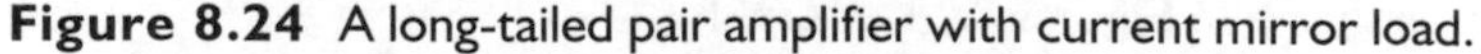
Figure 8.24 A long-tailed pair amplifier with current mirror load.

device with its gate connected to its drain rather than its source. This gives a more linear transfer characteristic which is more suitable for analogue applications. However, when used in digital circuits, such as the logical inverter of Figure 6.39(b), a DE MOSFET is preferred.

In **long-tailed pair amplifiers**, two load resistors are required. In bipolar transistor circuits this can conveniently be achieved using both arms of a **current mirror**, as shown in Figure 8.24.

This arrangement is not only very economical of space, using no resistors, but also gives a very high voltage gain of the order of several thousand.

8.4 Noise

In Section 2.2.2 we discussed thermal noise and its implications for system performance. In Section 3.8.1 we returned to the subject and looked at the representation of noise within amplifier equivalent circuits. In this section we will look at thermal noise in more detail, and at a number of other noise sources. We will then look at the measurement of noise within electronic systems and at the design of circuits to minimize its effects.

8.4.1 Noise sources

Thermal noise

We have already noted that **thermal noise** (also called **Johnson noise**) is produced by all components that exhibit resistance. Such noise is Gaussian and **white**, that is, it has components at all frequencies with equal noise power. Although the noise has an infinite bandwidth, only noise that is within the bandwidth of the system will have any effect. This leads to the concept of the **noise bandwidth** of a system. We have seen that the r.m.s. value of the thermal noise V_n generated by a resistance of value R is related to its absolute temperature T and the bandwidth B of the measuring system by the expression

$$V_n(r.m.s.) = (4kTBR)^{1/2} \tag{8.2}$$

where k is Boltzmann's constant.

It is interesting to look at the nature of this relationship. In most systems the operating temperature is close to ambient. Since this is expressed as an absolute temperature, taking an approximate value of 20 °C (68 °F) will normally suffice. We may then expand the expression as

$$V_n(r.m.s.) = (4kT)^{1/2}\,(BR)^{1/2} = 1.27 \times 10^{-10} \times (BR)^{1/2}$$

If we consider, for example, a system with a bandwidth of 20 kHz (typical of a high quality audio amplifier), we find that a resistance of 1 kΩ has an open-circuit noise voltage of about 500 nV and that a resistance of 1 MΩ has an open-circuit noise voltage of about 18 μV. Note that these are open-circuit noise voltages and that a 'noisy' resistance of R can be modelled by a voltage source V_n in series with an ideal noiseless resistance of R. This is illustrated in Figure 8.25.

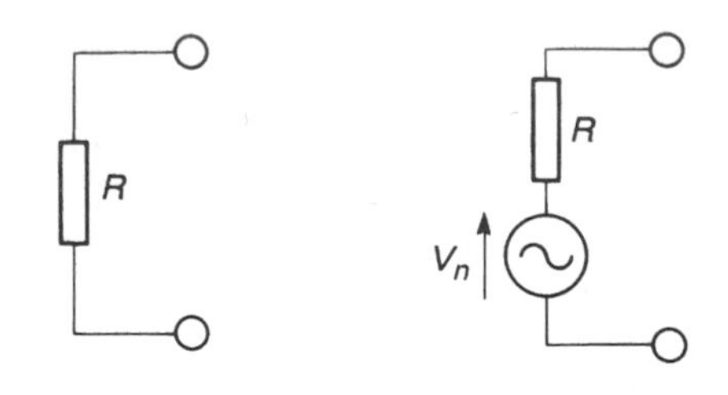

(a) A 'noisy' resistor (b) Equivalent circuit

Figure 8.25 Representation of a noisy resistor.

Since any signal source has a resistive component to its output impedance, it follows that all sources give rise to thermal noise. The noise voltage calculations above, if applied to this source resistance, thus give an indication of the ultimate limit to the **signal-to-noise ratio** of a system. In practice, thermal noise will also be produced by the various resistors within the electronic system itself and noise will be produced by other noise sources as described below. Good design techniques will reduce the noise produced by the system, but the noise produced by the source is normally beyond the control of the designer.

Shot noise

The current flowing within an electronic circuit is made up of large numbers of individual **charge carriers**. With large currents the averaging effect gives the impression of a continuous and constant stream. However, for smaller flow rates the granular nature of the current becomes more apparent. The statistical variation of the flow gives rise to a noise current, the magnitude of which is given by

$$I_n(r.m.s.) = (2qBI)^{1/2} \tag{8.3}$$

where q is the electronic charge (1.6×10^{-19} coulombs), B is the bandwidth over which the noise is measured and I is the mean value of the current. You may care to compare this expression with that of Equation 8.2. You will notice that thermal noise produces a *noise voltage*, whereas shot noise produces a *noise current*.

As with thermal noise, the magnitude of shot noise increases with the bandwidth of the measuring system. Again it is both white and Gaussian in nature. The noise current also increases with the mean value of the current (as you would expect) but since it increases with the *square root* of the mean current, the relative size of the noise decreases as the current increases. To illustrate this effect let us consider the shot noise associated with currents flowing in a circuit with a bandwidth of 20 kHz. For a current of 1 A, the r.m.s. noise current would be about 80 nA, or 0.000 008% of the mean current. With a current of 1 μA the noise current falls to 80 pA, but this now represents 0.008% of the mean current. Decreasing the current to 1 pA produces a noise current of about 8% of the mean current.

Shot noise is generated by the random flow of charge carriers across potential barriers, such as *pn* junctions. Within high gain transistors, base currents are small, making the effects of this form of noise more significant.

1/f noise

1/f noise is caused by not one but a variety of noise sources. It is so called because the power spectrum of the noise is inversely proportional to frequency. This means that the power in any octave (or decade) of frequency is the same. This represents a halving of the power for a doubling of the frequency and thus corresponds to a fall in power of 3 dB/octave. Clearly most of the power of this form of noise is concentrated at low frequencies. 1/f noise is often known as **pink noise** to distinguish it from white noise, which has a uniform spectrum.

One of the most important forms of 1/f noise is **flicker noise** which is caused by random variations in the diffusion of charge carriers within devices. Other forms of 1/f noise include the current dependent fluctuations of resistance exhibited by all real resistors. This noise is in addition to any thermal noise and is proportional to the mean current flowing through the device.

Interference

Another source of noise within an electronic system is interference from external signal sources. This can take many forms and may enter the system at any stage.

Common noise sources include: radio transmitters; AC power cables; lightning; switching transients in nearby equipment; mechanical vibration (particularly in mechanical sensors); ambient light (particularly in optical sensors); and unintentional coupling within systems (perhaps caused by stray capacitance or inductance). Interference will be discussed in more detail in Section 8.5 when we look at electromagnetic compatibility.

8.4.2 Noise in bipolar transistors

Bipolar transistors suffer from noise from a variety of sources. The semiconductor materials used to produce the devices clearly have resistance which leads to the production of **thermal noise** throughout the device. **Shot noise** is also produced by the random traversal of charge carriers across the junctions. Fluctuations in the diffusion process throughout the device lead **to flicker noise**.

At low frequencies (up to a few kilohertz) flicker noise is the dominant noise source. However, it becomes less significant at higher frequencies while the effects of thermal and shot noise become more significant. Both noise voltages and noise currents are produced. These can be represented in a **small signal equivalent circuit** as shown in Figure 8.26.

8.4.3 Noise in FETs

Noise in FETs is normally dominated by **flicker noise** (particularly at low frequencies)

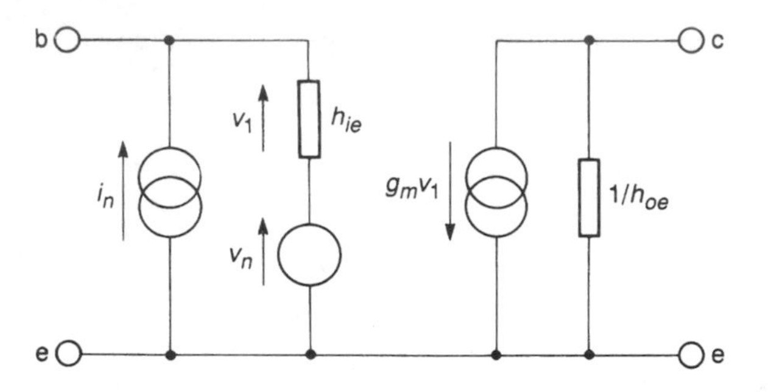

Figure 8.26 Representation of noise voltages and currents in an equivalent circuit.

and by **thermal noise** resulting from the resistance of the channel. Shot noise is generally insignificant, since in MOSFETs there is no junction and in JFETs the only currents across the gate junction are those caused by leakage.

8.4.4 Signal-to-noise ratio

From Section 3.8.2 we know that the signal-to-noise (S/N) ratio of a signal is given by the expression

$$\text{S/N ratio} = 20 \log_{10} \frac{V_s}{V_n} \text{ dB}$$

In many cases the noise voltage will be made up of a number of different components. Since noise voltages are random in nature they cannot simply be added together to obtain their combined effect. Instead we must add the squares of the r.m.s. voltages (which are related to the noise power) and then take the square root of the result to obtain the r.m.s. voltage of the combination. Thus, for two noise sources

$$V_n = \sqrt{(V_{n1}^2 + V_{n2}^2)}$$

8.4.5 Noise figure

S/N ratio calculations can be used to indicate the quality of a signal but cannot be used to describe how well an amplifier, or other circuit, performs with respect to noise. Simply measuring the S/N ratio at the output of a circuit is not an indication of its performance since this will be affected by the nature of the input signal. We have seen that any real signal source has noise associated with it, which must be taken into account when determining the noise performance of any circuit to which it is connected.

One method of describing how well an amplifier performs is to give the ratio of the noise produced at its output to that which would be present at the output of an ideal 'noiseless' amplifier of the same gain when both are connected to the same input. This comparison will give different results depending on the nature of the input signal used. A common method of comparison is to measure this ratio when the input is simply the thermal noise from a resistor of a specified value. The ratio measured under these conditions is termed the **noise figure NF** of the system, where

$$\mathrm{NF} = 10 \log_{10} \frac{\text{noise output power from amplifier}}{\text{noise output power from noiseless amplifier}}$$

$$= 20 \log_{10} \frac{\text{r.m.s. noise output voltage from amplifier}}{\text{r.m.s. noise output voltage from noiseless amplifier}}$$

In Section 3.8.1 we noted that it is often convenient to represent all the noise sources within a network by a single noise source at its input. A 'noiseless' amplifier has no noise sources and so the output noise is the same as the noise from the source. If we represent all the noise sources in our amplifier under test by a noise source of r.m.s. magnitude $V_{n(total)}$ then

$$\mathrm{NF} = 20 \log_{10} \frac{V_{n(total)}}{V_{ni}} \text{ dB}$$

where V_{ni} is the r.m.s. noise voltage of the source.

The noise figure gives a way of comparing the performance of amplifiers or other circuits in respect of noise. However, it should be remembered that the NF depends on the value of the source resistance and, since many noise sources are frequency dependent, varies with frequency. A perfect noiseless amplifier has a noise figure of 0 dB. Good low-noise amplifiers will have a noise figure of 2 to 3 dB.

8.4.6 Designing for low-noise applications

In most cases it is the noise generated by the first stage of an amplifier that determines the overall noise performance of the system. This is because noise generated near the input of the circuit is amplified, along with the signal, by all later stages whereas noise generated near the output receives relatively little amplification. Therefore, to achieve a good low-noise design particular attention must be paid to the all-important first stage.

Source resistance

From considerations of thermal noise it would seem advantageous to use a source with as low an internal resistance as possible. In fact, when the noise of the first stage is considered, this is not the best condition. It can be shown that the optimum value for the source resistance R_s is given by the expression

$$R_s = \sqrt{\frac{v_n^2}{i_n^2}}$$

where v_n and i_n are the r.m.s. noise voltage and current, respectively. These are combined in this way, rather than as a simple ratio, since they are uncorrelated quantities. The value of this optimum source resistance will vary from circuit to circuit. A typical value might be a few kilohms or a few tens of kilohms. In many cases the designer is not able to choose the resistance of the source since this is determined by another system or by a particular sensor. In such cases this must be taken into account in the design.

Bipolar transistor amplifiers

In bipolar amplifiers, both noise voltages and noise currents increase with collector current. Low-noise designs therefore use low quiescent currents of a few microamps. Suitable transistors, such as the BC212L or the BC109, combine low flicker noise and high current gain at low collector currents. Bipolar transistor amplifiers can produce good low-noise circuits with a wide range of source resistances from a few hundred ohms to a few hundred kilohms.

FET amplifiers

The dominant noise voltage source in JFETs is thermal noise caused by the resistance of the channel, which decreases with increasing drain current. Low-noise designs therefore use a fairly high drain current. MOSFETs generally have a poorer noise performance than JFETs at frequencies up to several hundred kilohertz. FETs provide a good noise performance for source resistances from a few tens of kilohms to several hundred megohms.

A comparison of bipolar and FET amplifiers

Bipolar transistors and FETs can both be used to produce excellent low-noise amplifiers with noise figures of 1 dB or better (given suitable care in design). Generally, bipolar transistors are preferable for low source resistances and will produce good results down to a few hundred ohms. FETs are superior with high source resistances and can be used with sources of 100 MΩ or more.

Interference in low-noise applications

In many applications interference plays a dominant role in determining the overall noise performance. The susceptibility of an electronic system to interference is affected not only by the circuit used, but also by its construction, location and use. Such considerations come within the very important topic of electromagnetic compatibility which is discussed in the next section.

8.5 Electromagnetic compatibility

Electromagnetic compatibility, or **EMC**, is concerned with the ability of a system to operate in the presence of interference from other electrical equipment, and not to interfere with the operation of other equipment or other parts of itself. Examples of

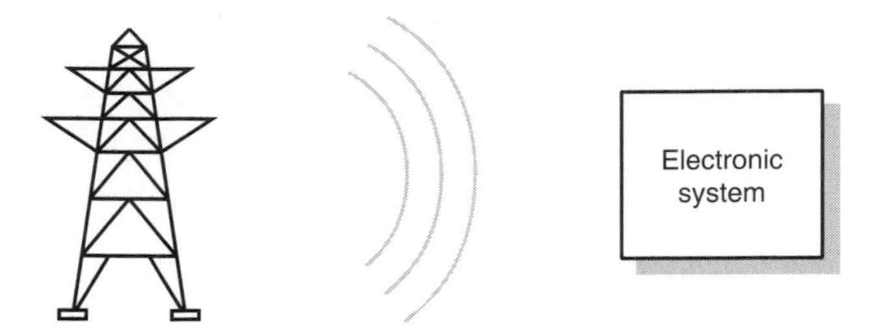

(a) An external noise source causes problems

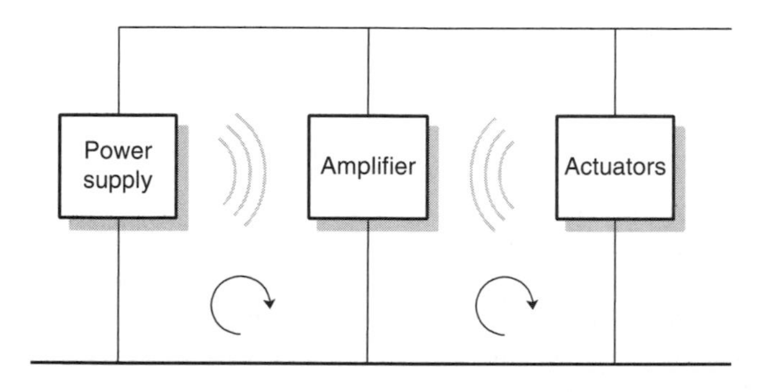

(b) One part of a system interferes with another

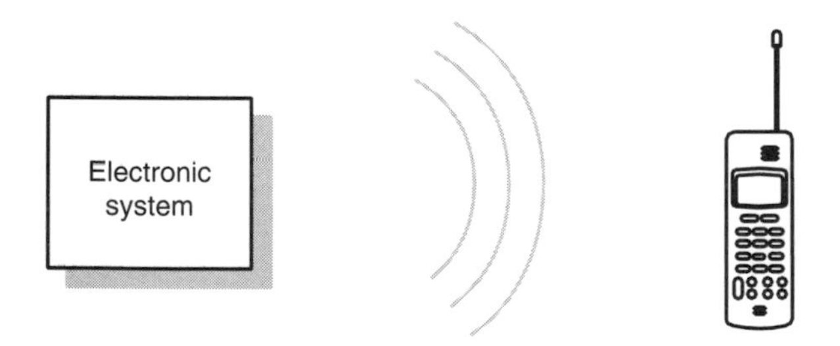

(c) A system interferes with the operation of other equipment

Figure 8.27 Examples of EMC problems.

problems associated with EMC are illustrated in Figure 8.27. Figure 8.27(a) shows the situation where an external electromagnetic noise source interferes with the operation of a system. Figure 8.27(b) shows an arrangement where the operation of one part of a system adversely affects the functioning of another part. Figure 8.27(c) illustrates the situation where a system produces radiation that interferes with other equipment.

8.5.1 Sources of electromagnetic interference

All **electromagnetic (E-M) waves** consist of an electric field and a magnetic field that are at right angles to each other. The properties of such waves vary with their frequency but all travel at the speed of light. Examples include visible light, X-rays and radio waves.

E-M waves are used as the basis of many forms of communication and information processing, and many systems rely on their production or detection. However, problems may arise when unwanted electromagnetic signals interfere with the operation of a system. Sources of electromagnetic interference may be natural or man-made.

Natural sources of interference

There are several natural phenomena that generate electromagnetic disturbances. The most significant of these are lightning, solar emissions and cosmic radiation.

The frequency of **lightning** varies dramatically between different parts of the world, being commonplace in certain areas at some times of the year. A direct lightning strike on a piece of equipment can produce voltages in the order of hundreds of thousands of volts and can deliver currents of hundreds of thousands of amps. It is therefore extremely difficult to design systems that can withstand such an incident. In most places, direct strikes are extremely rare except in the case of large exposed conductors such as overhead power cables. Unfortunately, the adverse effects of lightning are not restricted to those of a direct hit. Thunderstorms result in very intense electric fields of the order of several kilovolts per metre at ground level. At the instant of a lightning strike the intensity of the field falls rapidly as a result of the electrical discharge. This can result in large transients being induced in any nearby conductor. Lightning also produces high frequency emissions up to about 100 MHz. These can propagate over a considerable area and are a major source of atmospheric noise.

Variations in radiation from the sun cause fluctuations in the ionosphere which in turn affect the way in which it reflects or transmits radio waves. For this reason variations in **solar emissions** have a great effect on radio communications, particularly in the 2–30 MHz range, and on satellite communication at higher frequencies. **Cosmic radiation** produces significant background noise in the 100–1000 MHz range.

Man-made sources of interference

The most common sources of interference are various electrical and electronic systems which radiate energy as a direct result of their operation. Interference can also be caused by electrostatic discharges or by an electromagnetic pulse.

Electrical and electronic systems may generate interference through a number of mechanisms. For example noise may be caused by the radiation of high frequency signals used within the circuit (such as the oscillations produced within a superheterodyne radio receiver), the pulsed currents within digital circuits (such as computers) or switching transients (such as those of an electric light switch). Examples of systems that are common sources of interference include:

automotive ignition systems	switching power supplies
electric motors	power distribution systems
industrial plant	circuit breakers/contactors
mobile telephones	computers

Automotive systems in general are common sources of interference with noise being generated by the ignition system, the alternator, electrical switches, and electrostatic discharges caused by friction in the brakes (as described below).

On a smaller scale, all conductors within electronic circuits are potential sources of electromagnetic interference. The various conductors within a circuit, be they component leads, wires or printed circuit tracks, all represent small **antennas**. The mechanisms by which these antennas radiate and receive energy can be understood by considering two simplified models, namely the *short wire* and the *small loop* models. The **short wire** model is used to describe the radiation from individual wires within a circuit. When a current flows through a wire it forms a **monopole antenna** which radiates energy in all directions. This form of antenna has a very high impedance and this results in the production of an E-M field in which the electric component is much greater than the magnetic component. For this reason, such an antenna is often referred to as an **electric dipole**. Combinations of conductors and components also form loops within circuits. When a fluctuating current flows in such a loop it produces a **small loop** antenna that will again result in the generation of an E-M wave. This form of antenna has a relatively low impedance and this results in an E-M field in which the magnetic component is larger than the electric component. Such an antenna is called a **magnetic dipole**.

From the above it is clear that all useful electronic circuits have the potential to produce electromagnetic interference as a consequence of the currents within them. The extent to which a given circuit radiates energy is determined by a large number of factors including the magnitude and frequency of the signals involved and construction of the circuit. In general this form of radiation tends to be more of a problem at high frequencies (perhaps above 30 MHz) or in circuits that have transients with components at such frequencies. Unfortunately, just as a radio antenna can receive as well as transmit signals, so the electric and magnetic dipoles within circuits provide a mechanism whereby circuits are affected by E-M waves as well as producing them.

Electromagnetic interference may also be produced by equipment that would seem to have little to do with electricity. Most of us have experienced a mild electric shock as a result of touching an earthed conductor (perhaps a metal hand rail) after walking on a non-conducting carpet. This is an example of an **electrostatic discharge (ESD)**. Such discharges can also be experienced when taking off clothing made of a poorly conducting synthetic material. A build-up of static charge can be produced through friction between solids or fluids and can produce voltages of several thousand volts. In addition to giving an uncomfortable shock, the rapid discharging of the stored energy can damage electronic components and produce radiated transient signals.

Another man-made source of electromagnetic interference, though thankfully a fairly rare one, is the **electromagnetic pulse (EMP)** produced as a result of a nuclear explosion. Though of primary interest to those designing military systems that must continue to operate in the area of a nuclear explosion, it should be noted that the effects of a nuclear explosion in the upper atmosphere can affect electronic systems over an extremely large area – far greater than that directly affected by the blast.

Many electromagnetic noise sources have an extremely wide bandwidth and may produce perturbations of very large magnitudes. For example, a conventional 220 V AC domestic supply will normally have wide band noise with a bandwidth of perhaps hundreds of megahertz. It is also likely to have occasional transients of over 1000 volts. In contrast, some noise sources have a very narrow and well-defined bandwidth. An example of such a narrow-band noise source is a mobile telephone that may produce high levels of interference as a result of its transmitted signal.

8.5.2 Electromagnetic susceptibility

All electronic circuits are to some extent affected by electromagnetic interference. The **electromagnetic susceptibility** of a circuit is related to the extent to which a circuit is sensitive to such disturbances.

Interference enters a system either by **conduction** or by **radiation**. The importance of these two mechanisms varies with frequency: generally at frequencies up to 30 MHz conduction is the dominant mechanism, while at frequencies above 30 MHz radiation tends to be the more significant.

Conducted interference often enters a system through input or output cables, or through power supply leads. Radiated interference can enter a system through its casing and act directly on the internal circuitry. We shall see in Section 8.5.5 that it is common to use an earthed metal case to screen sensitive electronic equipment and so reduce the effects of radiated interference. However, radiated energy may induce noise in external cables and so enter the unit by conduction.

The electromagnetic susceptibility of a system is determined by the ease with which noise is able to enter a system, and the amount of noise that the system can tolerate before its operation is seriously affected. These factors may also be expressed in terms of the **electromagnetic immunity** of the system. This is the ability of a system to function correctly in the presence of electromagnetic interference.

8.5.3 Electromagnetic emission

EMC is concerned not only with the ability of a system to operate correctly in the presence of electromagnetic interference, but also with its ability to operate without itself generating noise that might interfere with other equipment.

As with energy entering a system it is clear that electromagnetic energy may also *leave* a system by either **conduction** or **radiation**. Again, the importance of these two mechanisms varies with frequency, and as one might expect, the 'routes' by which interference leaves a system are similar to those by which it can enter. Conducted energy tends to exit through input and output leads and through power supply lines, while radiated energy may radiate directly through the case. The use of an earthed metal case can reduce the amount of radiated energy leaving a system, although energy may radiate from noise that is conducted out of the unit through cables.

Because of the similarity between the mechanisms by which interference enters and leaves a system, it follows that many of the methods used to tackle these problems are also similar. We will look at some of these techniques in Section 8.5.5.

8.5.4 Electromagnetic coupling between stages

Another important aspect of EMC relates to the ways in which one section of a system may interfere with the operation of another. Because of the close proximity of the various parts of the system, high frequency energy can easily **radiate** from one section to another. Also, because of the interconnections between these sections, energy can be **conducted** between them through signal, power supply or ground leads. Examples of internal EMC coupling are shown in Figure 8.28.

In Section 8.5.1 we noted that all electronic circuits are potential sources of electromagnetic interference and that all circuits can be affected by such radiation. These effects produce unintentional coupling between the various sections of a system that can greatly affect its operation. High frequency circuits are particularly sensitive to this form of coupling and consequently great care is required in the design of such circuits.

Common sources of internal noise include the power supply unit. Noise on the incoming AC supply is often passed by the regulating circuitry and appears on the DC supply lines. High voltage transients are a particular problem as these will often produce spikes on the DC supply or cause the PSU to radiate noise to other parts of the system. In addition to noise that enters the system from the AC supply, the power supply unit may also produce its own interference. With simple linear supplies the fluctuating field from the transformer can induce noise currents at the supply frequency, while with **switch mode power supplies** the high frequency switching currents cause transients that can propagate throughout the system.

In addition to transmitting noise from the PSU, the power supply lines also propagate noise from one stage to another because of their impedance. When the current taken by a component or module changes, the voltage on the power line fluctuates and this

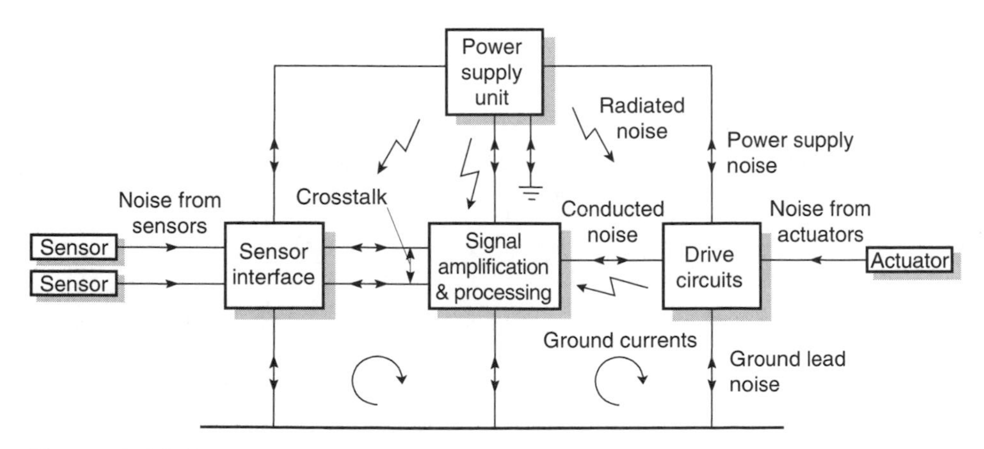

Figure 8.28 Examples of internal EMC coupling.

variation in supply voltage is coupled to other parts of the system. Similar inter-stage coupling can occur because of the impedance of the system's earth connection. Variations in the current taken by a module may cause fluctuations of the module's earth potential. If another module shares this earth connection then its earth voltage will also be affected. In this way the operation of one module can directly affect that of another.

Any lines that are routed near to each other may be susceptible to **crosstalk**, where signals on one line affect those on another. This phenomenon derives its name from the early days of the telephone industry where users would tend to overhear conversations on lines that were routed close to their own. Because of this problem the industry developed several very effective methods of reducing coupling between adjacent lines, many of which are widely used today. An example of such a technique is the use of **twisted pairs** of cables, as described in the next section.

Digital systems have particular problems with unintended coupling between stages. This topic will be discussed in more detail in Section 11.6 when we consider noise in digital systems.

8.5.5 Designing for EMC

The EMC performance of a system is affected by almost all aspects of its design. Key factors include the frequency range of the signals involved (high frequencies cause more problems than low frequencies) and the magnitudes of the voltages and currents used. Unfortunately, these factors are often dictated by the functional requirements of the system and it may not be possible for the designer to select these parameters freely. The circuits used within a system, and the components used within these circuits, also play a large part in determining EMC performance. In general the designer has much more control over the selection of these aspects of the design, and it is important that EMC issues are considered along with functional considerations within the circuit design process. Other issues of very great importance to EMC are the physical layout and the construction of the system. We have seen that conductors within a circuit act as small antennas that both radiate and receive electromagnetic interference. Keeping such conductors short and avoiding the formation of large loops can greatly improve EMC performance. Careful design of the layout of a system, and the use of appropriate grounding and shielding techniques, can reduce the emission and susceptibility of a system by several orders of magnitude.

A detailed treatment of EMC requires a study of many disciplines and is not within the scope of this book. However, it is perhaps useful to consider some aspects of design that are of particular importance.

Analogue vs digital systems

Because of the wide variety of both analogue and digital systems it is extremely difficult to make definitive statements about the characteristics of these two forms of circuitry. Analogue systems are often characterized by restricted bandwidths and small signal amplitudes. The former is generally an advantage in EMC terms since this reduces the range of frequencies that are likely to interfere with the system. However, the latter characteristic is a disadvantage, since this limits the signal-to-noise ratio that can be achieved. In analogue systems it is normal to attempt to maximize signal magnitudes in an attempt to reduce the effects of noise. We shall see in later chapters that digital circuits are affected by noise in a different way from analogue circuits. We will also see that they are generally associated with high frequency signals that require a very wide bandwidth. The use of a wide bandwidth tends to increase the problems associated with EMC since this implies sensitivity to noise over a greater range of frequencies. Fortunately, digital circuits tend to be less affected by noise than their analogue counterparts, and in general, digital systems are more immune to outside disturbances than analogue systems. Conversely, digital systems tend to produce more electrical interference than analogue systems.

Within the remainder of this section we will look at issues that are of relevance to all forms of electronic circuitry and we will leave consideration of problems that are unique to digital systems until we have looked at digital circuits in the following chapters. We will return to look at design considerations for EMC in Section 11.6 when we look in more detail at noise in digital systems.

Circuit design

We have already seen that conductors within circuits act as small unintentional antennas. In most circuits the majority of currents flow within loops formed by an outward and a return path. Such loops can be formed by any conductive route and may include wires, printed circuit board tracks and circuit components. Figure 8.29 shows examples of current loops. Each current loop forms a magnetic dipole that radiates energy with a magnitude that is proportional to the current within the loop, the loop area and the square of the frequency. Emissions can be decreased by reducing any of these factors. Emissions due to current loops are oriented such that the electric field is a maximum in the plane of the loop, and a minimum along its axis.

While most currents flow within a loop, within a given region currents may take on a unidirectional nature, to form a *short wire* antenna. This may occur, for example, in an isolated cable or a ground lead. In such cases the current flow generates a monopole or electric dipole. At a given distance from such a source the magnitude of the radiation is

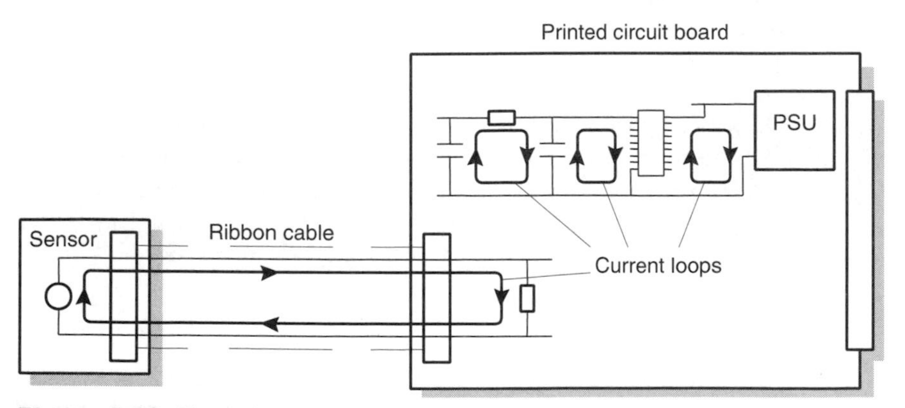

Figure 8.29 Examples of current loops within circuits.

proportional to the current, the length of the conductor and the frequency. The radiation is not oriented about the source.

It can be seen that the radiated emission of a circuit is affected by the frequency of the signals used and the magnitudes of the currents. Wherever possible these factors should be reduced in order to improve EMC performance. Current loop antennas are particularly sensitive to frequency and anything that can be done to reduce the high frequency content of a signal will be very beneficial.

Circuit layout

The radiation from both current loop and short wire antennas is greatly affected by the layout of the circuit. When laying out **printed circuit boards** (**PCB**s) every effort should be made to minimize track lengths and to minimize the areas of any current loops. These precautions are of particular importance when routing high frequency signals such as digital clock lines.

Particular attention needs to be given to the layout of the power supply lines of a PCB. Figure 8.30(a) shows a common method of laying out the power supply lines on double-sided printed circuit boards that have many digital logic devices. Here the positive supply voltage (often +5 V) is fed along one edge of one side of the board, while the 0 V return line is fed down the opposite edge of the other side of the board. Tracks are then fed from these two 'rails' to each of the logic devices. The resultant 'comb-like' arrangement has the advantage of being very easy to design, but is very poor from an EMC viewpoint. The power lines form a huge current loop that has an area almost equal to that of the board, and in which large currents flow. A better arrangement is shown in Figure 8.30(b). Here the supply and return lines are again fed to opposite sides of the board, but in this case they are placed on top of each other to minimize the area of any loops formed and to produce capacitance between the two lines to reduce the effects of transients. In this arrangement several lines are used in a grid to reduce power line and ground impedances.

Multilayer PCBs

The grid arrangement described above can be extended by the use of multilayer PCBs where separate planes are used for the power supply and the ground. This produces a very low impedance for both the supply and the ground. Tracks on alternate layers are normally arranged to be perpendicular to each other, to reduce mutual coupling (and to facilitate routing).

The parallel-plate arrangement of multilayer PCBs produces a low impedance **transmission line** effect that reduces coupling between circuits. However, when using such boards, discontinuities, such as those at right-angled bends, should be avoided since they create high fields at the corner. It is better to use curves rather than sharp corners, although in practice 90° corners are normally broken into pairs of 45° bends.

Device packaging

From an EMC standpoint surface mounted devices have several advantages over conventionally packaged parts. Many of these advantages come directly from the reduced

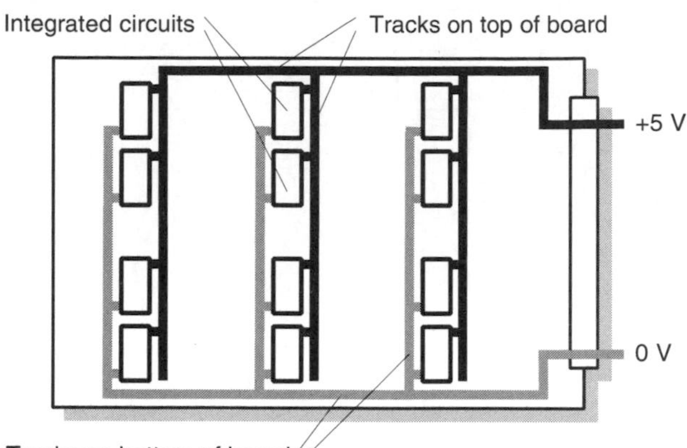

(a) A poor 'comb' layout

Tracks on top of board
+5 V
0 V
Tracks on bottom of board

(b) A better 'grid' arrangement

Figure 8.30 Examples of power supply routing methods.

size of the parts which leads to a greater board density. This in turn results in shorter lead runs and a reduction in parasitic capacitance. Most dual-in-line packages have the power connections on opposing corner pins thereby maximizing the distance between them. This not only increases the parasitic inductance of the device but tends to produce long tracks and large current loops.

Circuit partitioning and grounding

Most electrical circuits operate with one or more power supply 'rails' and a single 'ground' connection. Typically the ground is connected to a metal case or enclosure, or to the earth return of the AC power lines. Where a system has a number of components or modules, each section is normally connected to this common ground to facilitate the passing of

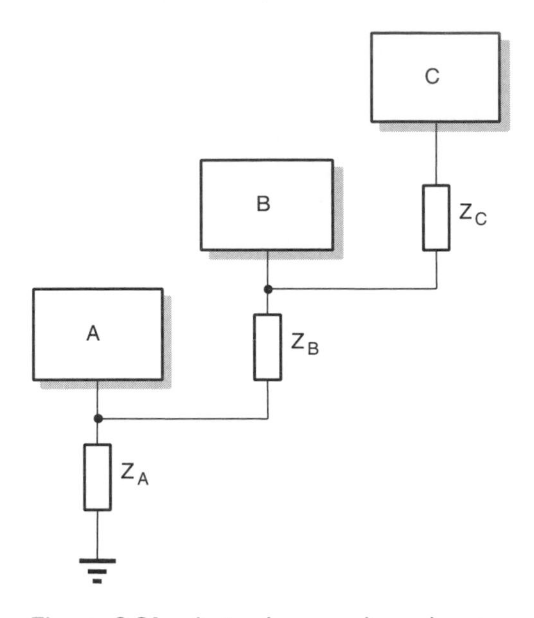

Figure 8.31 A simple grounding scheme.

signals. The way in which the various parts of a system are connected to the ground can have a large effect on its EMC performance.

The importance of grounding methods stems from the fact that any connection or conduction path has a certain impedance. The significance of this impedance can be seen by looking at the simple grounding scheme shown in Figure 8.31. This figure shows three modules A, B and C using a **series grounding arrangement**. Module A is connected to the system ground by a lead that has an impedance of Z_A. Module B has its earth connection joined not to the system ground, but to the ground of module A. The impedance of this connection is Z_B. Similarly, module C is joined to the ground connection of module B by a lead with an impedance of Z_C. This grounding scheme is often used because it simplifies wiring and is therefore cheaper. However, its disadvantage is that the common impedances produce coupling between the various modules. Variations in the currents taken by one of the modules will cause the ground potentials of the other modules to fluctuate.

An alternative grounding scheme is shown in Figure 8.32. Here each module is connected directly to the system ground. This **single-point grounding** arrangement is often referred to as a **star connection** scheme. The advantage of this method is that ground connection impedances are not shared and fluctuations in the ground potential of one module (as a result of variations in its ground current) will not be coupled to other sections of the system. This arrangement is the preferred method at relatively low frequencies and it is common for separate grounds to be used for sections where mutual interference is likely. An example of such an arrangement is shown in Figure 8.33. Here a system is partitioned into three distinct regions to minimize interaction. Analogue and digital circuitry are separated, and a third section is used to isolate particularly 'noisy' parts of the system. This last section might include drive electronics for high power actuators or electro-

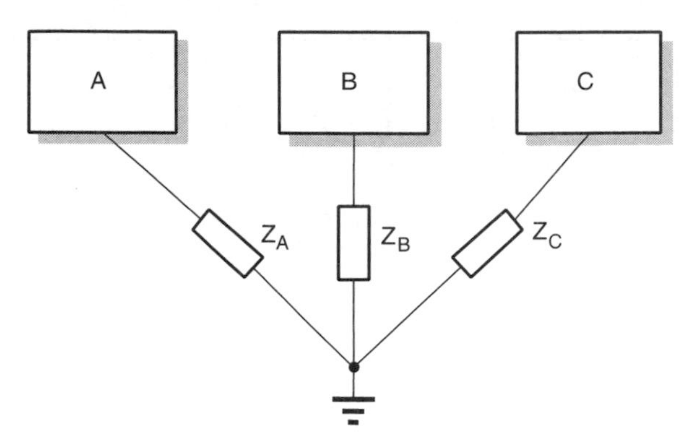

Figure 8.32 A single-point grounding scheme.

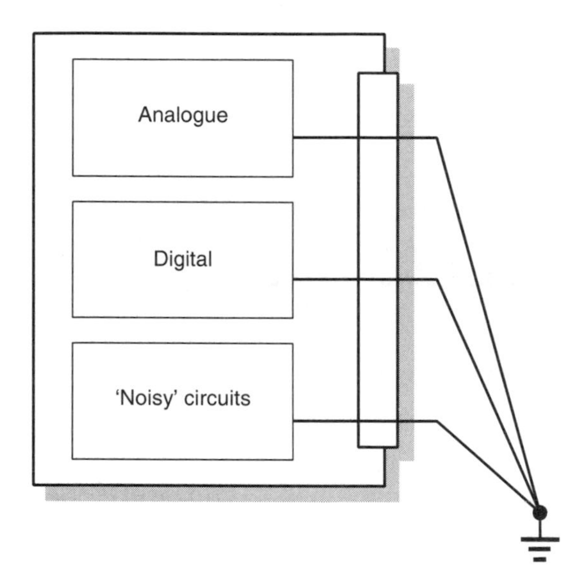

Figure 8.33 System partitioning to reduce EMC problems.

magnetic components such as relays. The ground connections of these three sections are joined off the board at the common earth point of the system.

When arranging the grounding of the various parts of a system it is important to ensure that sections do not have multiple ground paths. The presence of two or more ground connections to a given module will result in the formation of one or more **ground loops** (or **earth loops**) as shown in Figure 8.34(a). Such loops act like other forms of current loop and couple electromagnetic fields into the ground currents. Unfortunately, ground loops may also be formed by stray capacitance at high frequencies as shown in Figure 8.34(b). The single-point grounding arrangement tends to result in long ground leads for those parts of the circuit that are furthest from the common earthing point. At high frequencies stray capacitance may provide earth routes that are comparable in impedance to the

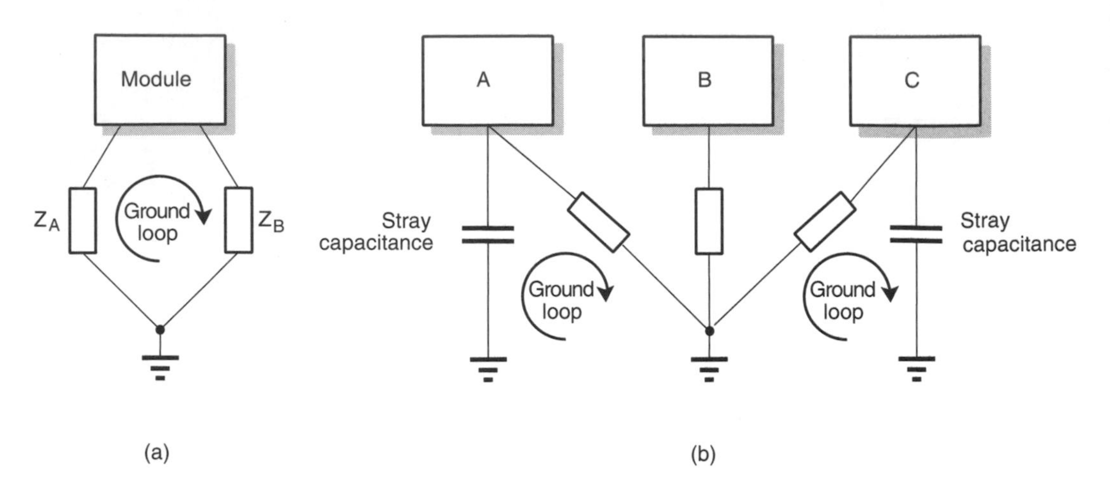

Figure 8.34 Ground loops.

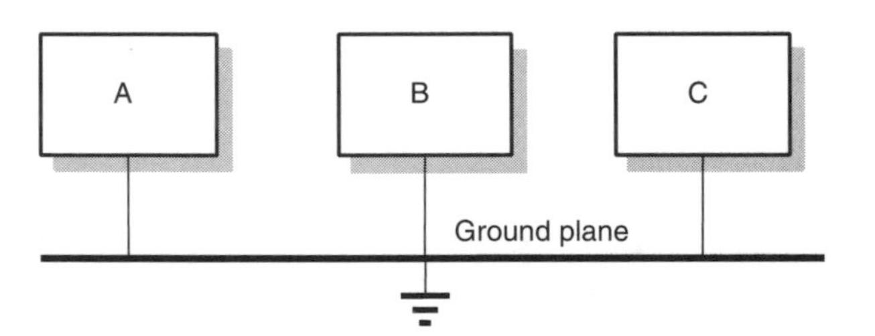

Figure 8.35 A multipoint grounding scheme.

intended grounding path forming a ground loop. For this reason it is more common to use a low impedance **ground plane** at high frequencies, as shown in Figure 8.35. In this **multipoint grounding** arrangement individual components and modules are each connected directly to the ground plane using leads as short as possible. The technique relies on having a very low impedance within the ground plane, to prevent impedance coupling problems. This is often achieved by using a plane within a multilayer PCB.

Typically single-point grounding techniques are used for systems with operating frequencies of up to 1 MHz, while systems operating at above 10 MHz will use a multipoint method. Systems operating between these values often adopt a hybrid approach.

Enclosures and cable shielding

One of the major weapons used to combat radiative coupling is the use of shielding. This will often include the use of a grounded conductive case around the system and the use of shielded connecting cables. Shielding may also be used within the system to reduce coupling between its various sections.

A shielded case acts as a form of Faraday cage and can greatly reduce the field produced inside the enclosure as a result of an external electromagnetic field. It may also dramatically reduce radiative emissions. The effectiveness of the shielding provided by a case depends on a number of factors, including the material used, its thickness, its construction, and the frequency and nature of the interference concerned. Enclosures are usually made of metal, although metal-plated plastic cases are also used. The enclosure is normally connected directly to the primary grounding point of the system in order to achieve a low ground impedance. Earthed metal cases not only provide EMC screening but also perform a safety function.

Shielding efficiency is greatly affected by any apertures in the conductive surface, such as those caused by displays, cooling fans or cable ports. Cables entering an enclosure provide a potential route for the entry of noise either by conduction through the cable itself, or by radiation through the aperture.

External signal cables have the potential to pick up radiative noise and to conduct it within the protective enclosure. Once inside it can then be distributed within the system either by conduction or by radiation. For this reason, external cables are often screened to reduce their susceptibility to noise. There are many forms of screened, cable and numerous ways of interconnecting the signal and screen conductors. A few examples are given in Figure 8.36, although these should *not* be taken as recommended configurations.

Figure 8.36(a) shows a signal being passed from one location to another using a single, unscreened conductor (wire). The return path for this signal is through the ground. Figure 8.36(b) shows an improved arrangement where a screened coaxial cable is used, the screen being joined to ground at either end. This cable has a central conductor that is surrounded by a layer of insulating dielectric and then by a ring of copper braid that forms a conducting screen along the length of the cable. The screen is then itself surrounded by an insulating layer. The effectiveness of this arrangement will depend on many factors but one might expect an improvement of several orders of magnitude in the amount of noise detected at the destination.

A further improvement might be achieved through the use of a differential arrangement as shown in Figure 8.36(c). In Section 3.10 we noted that noise pick-up could be reduced by connecting a signal source to a pair of wires and using a differential amplifier to detect the differential signal while rejecting the common-mode noise (see Figure 3.24). To make this technique as effective as possible we require the two wires to pick up exactly the same amount of noise so that the differential noise is at a minimum. It has been found that this can best be achieved by twisting the two wires around each other to form a **twisted pair cable**. One might expect that the noise picked up by this arrangement would be several orders of magnitude less than that of the coaxial cable discussed earlier. A further improvement could be achieved by using a **screened twisted pair cable** as shown in Figure 8.36(d). It is likely that the noise delivered to the destination of this final arrangement would be more than a million times less than that expected when using a single unscreened cable.

When using any cabling arrangement great attention must be paid to the connections between the cable and the rest of the system. A short length of unscreened cable where a conductor is joined to a connector can dramatically compromise screening efficiency, as can a **pigtail** (a short length of single wire used to connect onto the outer shield of a cable). For maximum effectiveness screened cables should be terminated within screened connectors that maintain the shielding integrity.

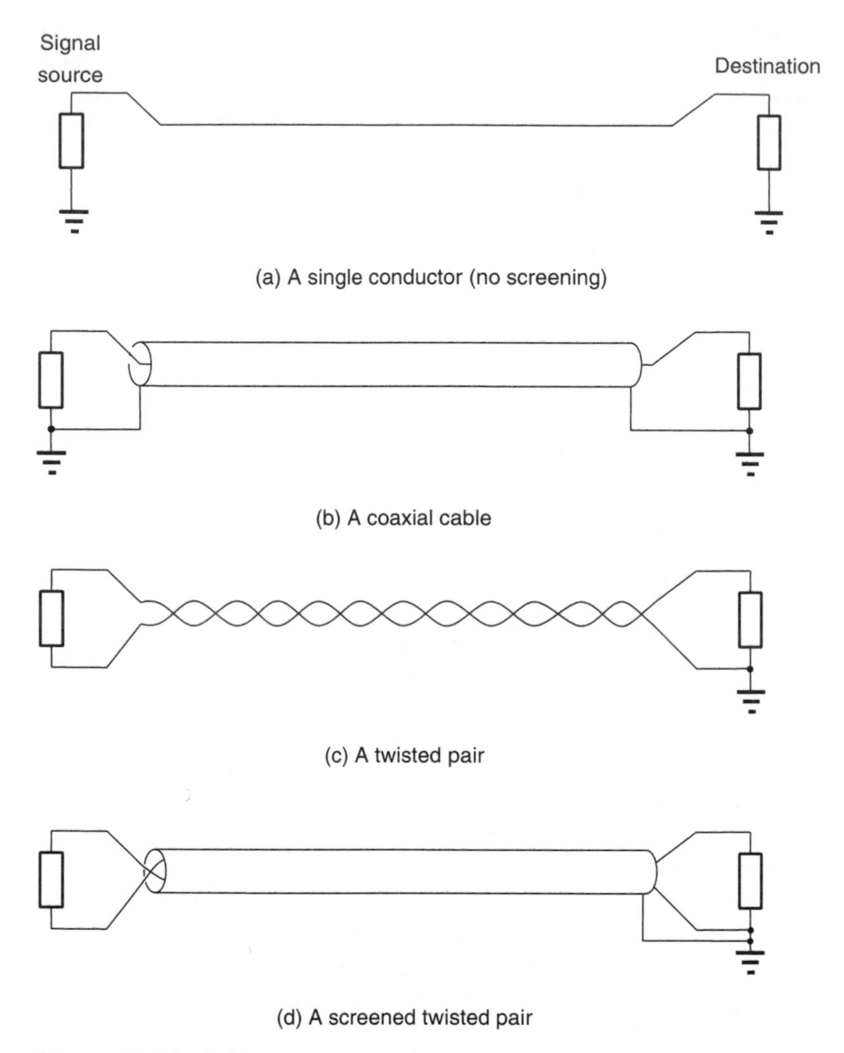

Figure 8.36 Cable screening techniques.

Supply line filtering and decoupling

We noted earlier that the domestic AC supply is a major source of electrical noise. For this reason it is normal to use a **mains filter** to remove high frequency noise from the supply in noise sensitive applications. The filter must be fitted as close as possible to the point where the supply leads enter the system's enclosure to prevent the unfiltered leads from radiating noise within the case. To reduce this problem some filters are incorporated into shielded mains sockets that can be fitted directly into an aperture in an enclosure. In this way the supply is filtered as it enters the enclosure.

Mains filters tend to reduce the amount of high frequency noise entering the system but are less effective at dealing with high voltage transients. Special **transient suppressors** can be used to counter these potentially dangerous events. Suppressors may be of several types including *pn* junction devices that resemble Zener diodes. Typically such a device would be installed across the output of a DC supply and would be chosen to have a breakdown voltage somewhat greater than the nominal supply voltage. If a transient caused the voltage to exceed the breakdown voltage of the device, the suppressor would conduct, preventing the supply voltage from rising. A typical device might respond in the order of a nanosecond and could pass tens or perhaps hundreds of amps.

Power supply line transients and noise can also be produced by the action of circuit components. This is a particular problem in digital circuits where the switching of logic gates can cause large current surges. To minimize the effects of power supply noise it is normal to employ a combination of techniques. These include the provision of a well regulated power supply with low inductance reservoir capacitors; the use of ground and power planes within a multilayer PCB to minimize the supply line impedance; and the fitting of power supply **decoupling capacitors**. Within digital circuits it is normal to fit a decoupling capacitor adjacent to every integrated circuit to supply the surge of current it requires during switching. This capacitor reduces the voltage transient produced on the supply line and minimizes the area in which the transient current flows. A typical circuit would use ceramic, or other low inductance capacitors, of between 10 and 100 nF, fitted as close as possible to each integrated circuit. Each board would then be fitted with a larger bulk decoupling capacitor of perhaps 100 μF fitted at the point where the power line enters the board. This would typically be an electrolytic or tantalum capacitor.

Isolation

In Section 7.8.1 we saw how opto-isolators could be used to safeguard electronic circuits in the presence of high voltages. These devices can also be used to pass information between circuits while preventing the passage of electrical noise. We will return to look at this use of opto-isolation in Chapter 11 when we look at the effects of noise in digital systems.

8.5.6 Achieving good EMC performance

Good EMC performance cannot be achieved simply through good design. In fact, EMC considerations affect all aspects of the design, development, construction, testing, installation, use and maintenance of a system. While a good design is essential to achieving good EMC characteristics, other issues related to **quality** are also very important. One could produce a design that was highly optimized to minimize both the sensitivity and the emission of a system, only to have one's efforts nullified by the inadvertent fitting of plastic rather than metal washers when fitting a screen, or by failing to refit an earth strap during maintenance. Good EMC performance requires commitment throughout all phases of a system's development and use.

8.5.7 EMC and the law

While there are clear commercial advantages to producing systems that have good EMC characteristics there are also legal requirements to be considered. In recent years many countries have produced legislation that places restrictions on the amount of interference that electrical equipment may produce. In Europe, legislation which came into force on 1 January 1996 imposes strict rules on the performance of electrical products and equipment in respect of EMC. The directive covers the whole range of electrical and electronic equipment that is capable of producing interference or being affected by it. The 'essential requirements' of the directive are that:

> 'The apparatus shall be so constructed that:
>
> (a) The electromagnetic disturbance it generates does not exceed a level allowing radio and telecommunications equipment and other apparatus to operate as intended.
> (b) The apparatus has an adequate level of intrinsic immunity to electromagnetic disturbance enabling it to operate as intended.'

Until recently many manufacturing companies paid very little attention to EMC issues. However, with the advent of legislation in Europe and throughout the world, all engineers need to be well acquainted with this very important area.

Key points

- Filters can be divided into passive and active types.
- Simple resistor-capacitor (*RC*) networks produce a first-order or single pole filter which achieves a maximum roll-off of 6 dB/octave.
- Multi-stage *RC* filters can produce an arbitrarily high roll-off rate, but the transition from the pass-band to the stop-band is not sufficiently sharp for many applications.
- Active filters can be used to produce a range of filter characteristics. A number of configurations are possible, of which the most significant are
  - Butterworth, which gives a maximally flat characteristic within the pass-band
  - Chebyshev, which provides the sharpest transition from the pass-band to the stop-band
  - Bessel, which has a linear phase response.
- Amplifiers can be divided into a number of classes:
  - Class A, where the output device conducts throughout the input cycle
  - Class B, where the output device conducts for only 50% of the input cycle
  - Class AB, in which the output device conducts for more than 50% but less than 100% of the input cycle
  - Class C, where the output device conducts for considerably less than 50% of the input cycle
  - Class D, which describes a switching amplifier where the output device alternates between being turned ON and being turned OFF.
- The differences between the circuitry required to produce the linear classes (that is, all except class D) are primarily concerned with the biasing arrangements.
- Power output stages requiring low distortion usually adopt class AB.

- Noise in electronic systems may be categorized into a number of classes:
  - Thermal (Johnson) noise produced by any component that possesses resistance
  - Shot noise produced by the random nature with which charge carriers cross junctions.
  - 1/f noise which is produced by a number of sources including flicker noise as a result of fluctuations in the diffusion process
  - Interference from external signals or events.
- Bipolar transistors and FETs both suffer from all these types of noise to a lesser or greater extent. High performance, low-noise amplifiers can be constructed using both forms of transistor.
- Electromagnetic compatibility (EMC) is concerned with the ability of a system to operate in the presence of interference from other electrical equipment, and not to interfere with the operation of other equipment or other parts of itself.
- There are many natural and man-made sources of electromagnetic interference.
- Interference may enter or leave a system by either conduction or radiation.
- Electromagnetic coupling between the various stages of a system can also be through conduction or radiation. The power supply is a particular problem area.
- Circuit layout plays a large part in determining EMC performance. All wires should be kept as short as possible and loop sizes should be minimized.
- Grounding is of great importance with different techniques being appropriate at high and low frequencies.
- The shielding of cables is vital and can reduce the amount of interference entering the system by many orders of magnitude.
- Good EMC performance cannot be achieved simply through good design alone. It requires attention to detail throughout all phases of the development, use and maintenance of a system.
- New legislation in this area is making EMC of increasing importance to all engineers.

Design study

At the end of Chapter 7 we considered the design of an operational amplifier and proposed a circuit based on the techniques covered at that point in the text. In this chapter we have developed our understanding of amplifier design and are in a position to improve this basic design.

Approach

Improvements to the design outlined in the design study at the end of Chapter 7 are possible in three areas: the use of emitter resistors to improve the stability of the quiescent conditions of the output transistors; an improvement in the biasing arrangements of the circuit to allow accurate setting of the quiescent current in the output devices; and the use of active, rather than passive, loads to make the circuit more suitable for integration. An improved circuit is shown below.

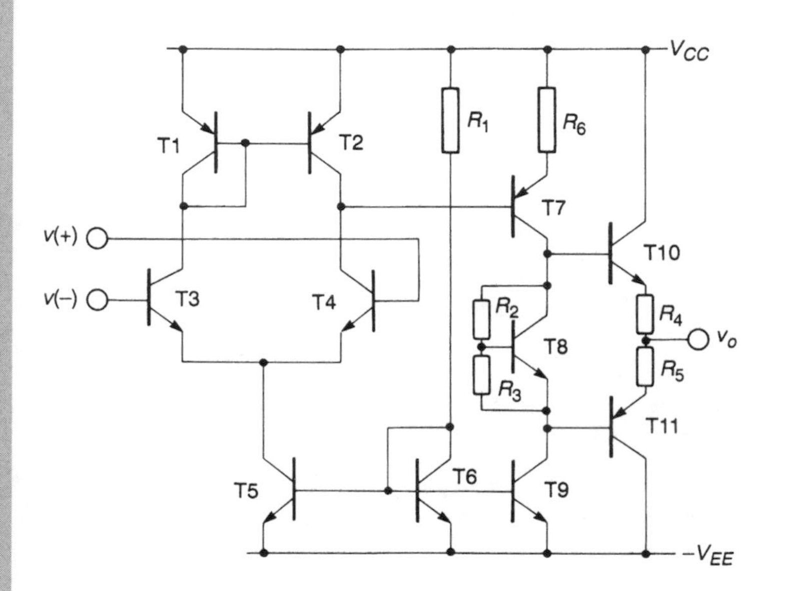

The emitter resistors R_4 and R_5 improve the stability of the quiescent conditions of the output stage by applying series negative feedback. For such a low power application the resistors would be about 25 to 50Ω.

T8 forms a V_{BE} multiplier arrangement with R_2 and R_3 to provide a stable bias voltage between the bases of T10 and T11. The values of the resistors would be of the order of a few kilohms, the ratio being chosen to select the desired quiescent current through the output transistors.

Transistors T1 and T2 form a current mirror which is used as an active load for the long-tailed pair amplifier formed by T3 and T4. The output of this load is fed to T7 which also has an active load in the form of T9. This is again part of a current mirror, which this time shares a transistor (T6) with the current mirror used to provide the constant current into the emitters of the input differential amplifier.

Real operational amplifiers are often similar in overall design to the circuit described above, but usually include certain refinements which are outside the scope of this text. These include such features as short-circuit protection, to safeguard the device against excessive output current, and offset null adjustment, to remove the effects of the input offset voltage. Below is the circuit diagram of a typical, general purpose operational amplifier, the 741.

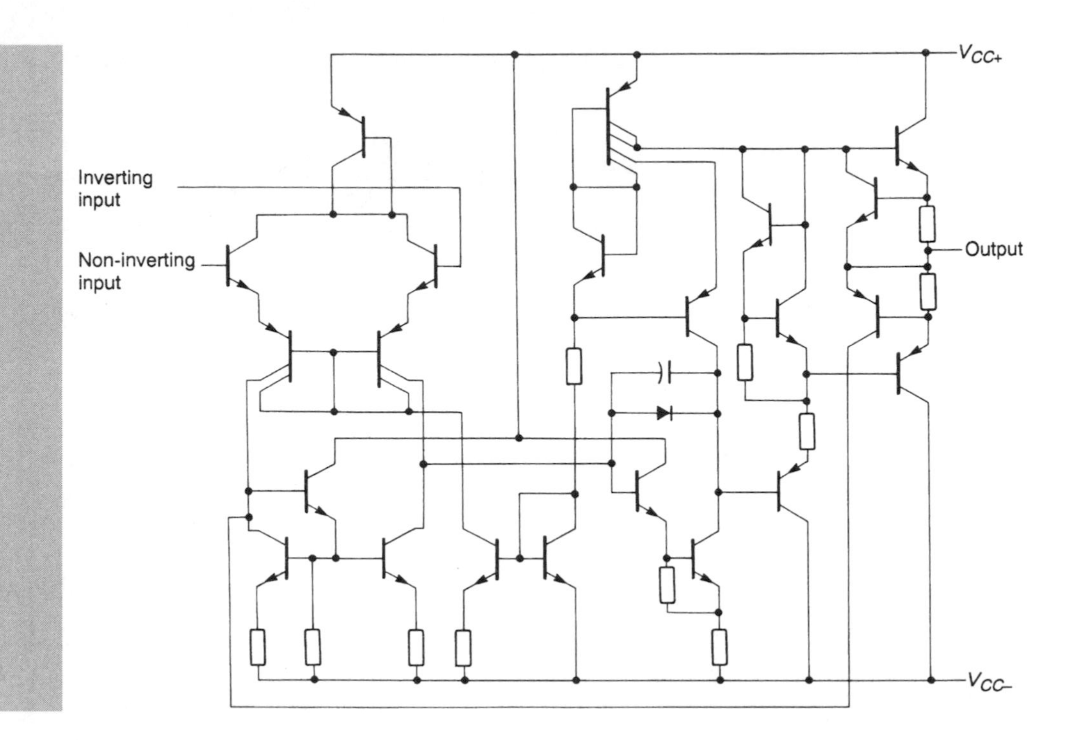

Further reading

Chatterton P. A. and Houlden M. A. (1991) *EMC: Electromagnetic Theory to Practical Design*. Chichester: John Wiley

Scot J. (1997) *Introduction to EMC*. Oxford: Newnes

Sedra A. S. and Smith K. C. (1995) *Microelectronic Circuits*, 3rd edn. New York: Oxford University Press

Exercises

8.1 Sketch the amplitude and phase responses of the following circuits, indicating the cut-off frequency in each case.

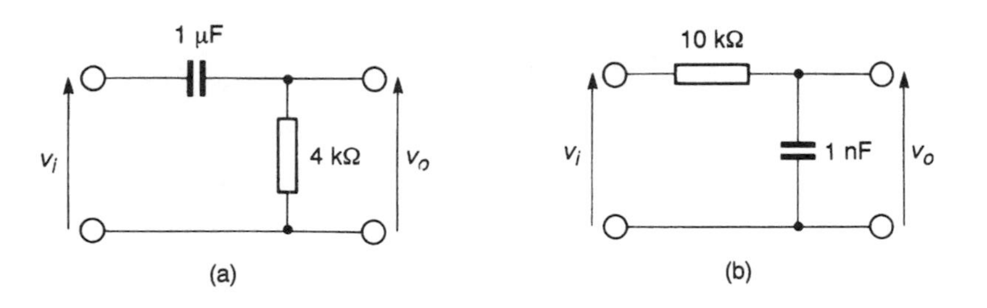

8.2 Calculate the bandwidth of the following circuit, where the amplifiers are ideal unity gain buffers.

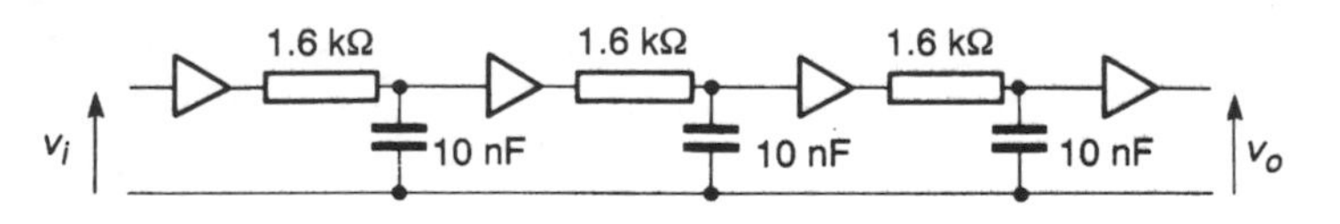

8.3 Simulate the arrangement of the previous exercise to check your calculated bandwidth. You might care to use the circuit of Computer simulation exercise 8.1 as your starting point.

8.4 Sketch the frequency response of the following circuit and estimate its resonant frequency. What is meant by the Q of the circuit?

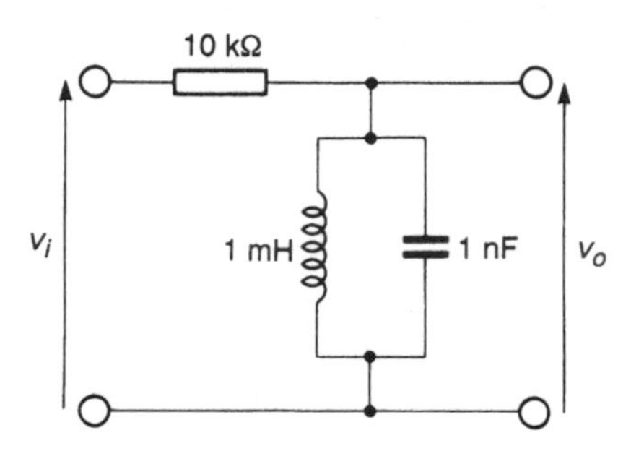

8.5 Explain why filters with different characteristics are required for different applications.

8.6 From its simulation frequency response, estimate the Q of the band-pass filter of Computer simulation exercise 8.2.

8.7 What class of amplifier (that is, class A, class B, etc.) does the following circuit represent?

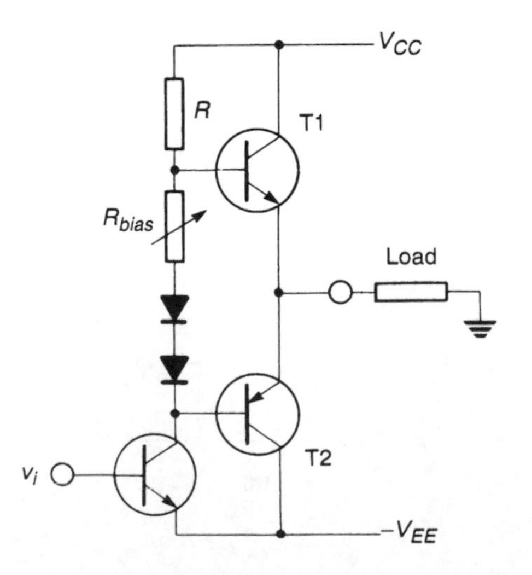

8.8 Why is class AB generally used in preference to class B in audio power amplifiers? What are the disadvantages of class AB in comparison with class B?

8.9 Explain why inductive and capacitive sensors are not free from thermal noise.

8.10 Calculate the thermal noise voltage produced by a resistor of 10 kΩ when connected to a system with a bandwidth of 5 MHz at normal ambient temperature.

8.11 Calculate the percentage fluctuation caused by shot noise current, of a 1 nA signal, when measured by a system with a bandwidth of 5 MHz at normal ambient temperature.

8.12 Explain what is meant by the terms 'white noise' and 'pink noise'.

8.13 List three natural and three man-made sources of electromagnetic interference.

8.14 Give an example of a noise source which is narrow band and one that is wide band.

8.15 How may an electronic system be protected from radiated interference?

8.16 How may an electronic system be protected from conductive interference?

8.17 What are the primary mechanisms of electromagnetic coupling between the various parts of an electronic system?

8.18 Describe the importance of the layout of a circuit in relation to its EMC performance.

8.19 Why is the choice of a system's grounding method affected by its operating frequency?

8.20 Discuss the importance of quality in relation to EMC.

Digital Systems

Objectives

When you have studied the material in this chapter you should:

- understand terms such as binary variable, logic state and logic gate, and logical operators such as AND, OR, NOT and Exclusive-OR;
- be familiar with the use of Boolean algebra in the representation and simplification of logic functions;
- be able to use Karnaugh maps to describe and minimize Boolean expressions to aid implementation;
- have an understanding of a range of number representation methods;
- be aware of coding techniques for the representation of both numeric and alphabetic data;
- be able to design combination logic circuits to fulfil functional requirements.

Contents

9.1 Introduction

Having looked in some detail at analogue systems we now turn our attention to the digital world.

We noted in Chapter 1 that some information sources produce discrete, or in other words digital, signals. That is, they produce an output signal that represents one of a fixed number of states, as opposed to an analogue signal which can take an infinite number of values. In addition to these digital information sources, there are many instances when we choose to represent an analogue quantity in a digital manner. A simple example is the use of the output of a thermostat, such as that used in a domestic heating system, to represent the temperature of a room. If the temperature is below a certain value the thermostat is in one state, whereas if it is greater than this value the thermostat switches to another state. We are effectively approximating the continuous analogue temperature of the room by a discrete digital representation, which indicates that the temperature is above or below a certain value. The thermostat produces a *binary* output, but non-binary digital examples are also readily available. A car odometer, for example, indicates the distance travelled by the car. This distance is clearly an analogue quantity, yet we choose to represent it using a digital display.

Representing an analogue quantity, which can take an infinite number of values, by a digital number, which can take only a finite number of values, must always represent an approximation. However, the process of measurement *always* represents an approximation, since all measurements are limited in accuracy as a result of errors. If a digital representation is made to sufficient resolution there is no reason why the use of digital techniques should reduce the accuracy of a measurement.

Man's use of machines of all sorts has led to an increasing dependence on digital rather than analogue techniques. One reason for this is that many binary sensors and actuators are much simpler than corresponding analogue types. It is, for example, much easier to turn a lamp ON or OFF using an electrical switch than to vary its power in a continuous manner. Another reason for the extensive use of digital techniques is that digital quantities can be more easily transmitted, processed and stored than their analogue counterparts, and are less affected by the presence of noise. Digital, and in particular binary, control is commonplace in all fields of life, from domestic heating systems, which turn heaters ON or OFF depending on whether the temperature is above or below a certain value, to sophisticated flight control computers, which take data from a range of sensors and produce output signals to determine the operation of an array of actuators.

The advent of the digital computer has had a dramatic effect on all forms of society by allowing vast amounts of data to be stored and retrieved quickly, and by allowing decisions to be made based on this information. Since the 1970s this processing capability has been available in the form of *microprocessors* which provide the calculating functions of a computer within a single integrated circuit. The low cost and great flexibility of these components has led to their use within a myriad of applications from calculators to cars and from video recorders to coffee-makers.

Microprocessors are relatively sophisticated electronic circuits containing typically 10^5 to 10^6 transistors. However, the circuits are based on combinations of much smaller elements called gates, which are themselves simple circuits using a handful of transistors and passive components. In this chapter we will start by looking at the nature of digital quantities and their representation within electronic systems. We will then investigate the various forms of logic gate and a switching algebra that is useful in describing and manipulating binary quantities. Binary arithmetic will then be discussed together with a range of applications for logic gates.

In later chapters we will consider electronic circuits to implement a number of logic elements and look at the various components of a microcomputer system.

9.2 Logic states

In this chapter we are primarily concerned with quantities that take, or may be considered to take, only two states. Examples include: a switch which can be only ON or OFF; a hydraulic valve which can be only OPEN or CLOSED; and an electric heater which is considered to be only ON or OFF. Such quantities are called **binary quantities** for obvious reasons. It is common to represent such quantities by **binary variables**, which are simply symbolic names for the quantities. Figure 9.1 illustrates a binary arrangement.

If we represent the state of the switch by the binary variable S and the state of the lamp by the binary variable L, we can represent the relationship between the two variables symbolically as follows

S	L
OPEN	OFF
CLOSED	ON

We can also use a symbolic name for the state of each variable, so that rather than using terms such as OPEN and CLOSED, or ON and OFF, we may use symbols for the states such as '0' and '1'. If we use the symbol '0' to represent the switch being OPEN and the lamp being OFF, our table becomes

S	L
0	0
1	1

The mapping between ON and OFF and '0' and '1' is arbitrary, but the user must know what the relationship is. It is common to use '1' to represent the ON state, a switch being CLOSED or a statement being TRUE. It is common to use '0' for the OFF state, a switch

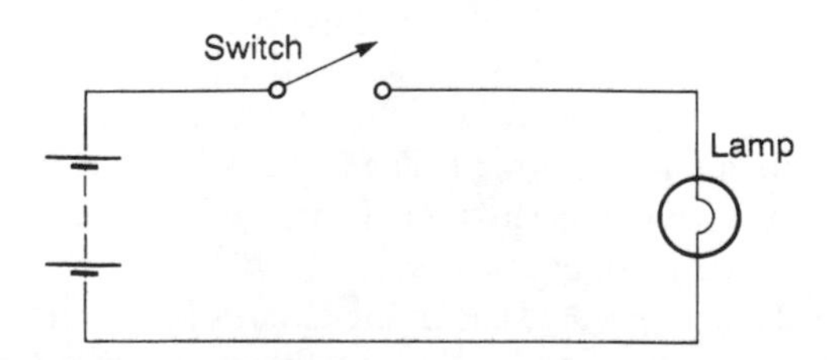

Figure 9.1 A simple binary arrangement.

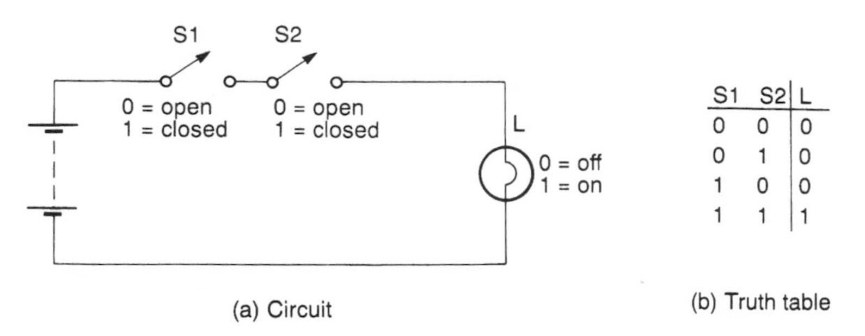

S1	S2	L
0	0	0
0	1	0
1	0	0
1	1	1

Figure 9.2 Two switches in series.

being OPEN or a statement being FALSE. The table lists on the left all the possible states of the switch and indicates, on the right, the corresponding states of the lamp. Such a table is called a **truth table** and it defines the relationship between the two variables. The order in which the possible states are listed is normally *ascending binary order*. If you are not aware of the meaning of this phrase it will become clear when we look at binary arithmetic later in this chapter.

Figure 9.2(a) shows an arrangement incorporating two switches in series. Here it is necessary for both switches to be closed in order for the lamp to light. The relationship between the positions of the switches and the state of the lamp is given in the truth table of Figure 9.2(b). Notice that the table now has four lines to represent all the possible combinations of the two switches. Alternatively, we could express this relationship in words as 'the lamp will be illuminated if, and only if, switch S1 is closed AND switch S2 is closed'. We may abbreviate this statement as

L = S1 AND S2

This **AND** relationship is very common within electronics systems and is found in a variety of everyday applications. For example, the brake lamps of an automobile are often only illuminated if the foot brake is depressed, closing a switch, AND the ignition switch is ON.

Figure 9.3(a) shows an arrangement which has two switches in parallel. In this con-

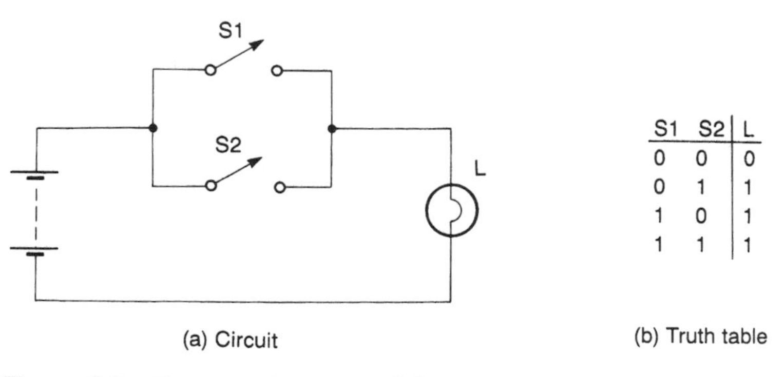

S1	S2	L
0	0	0
0	1	1
1	0	1
1	1	1

Figure 9.3 Two switches in parallel.

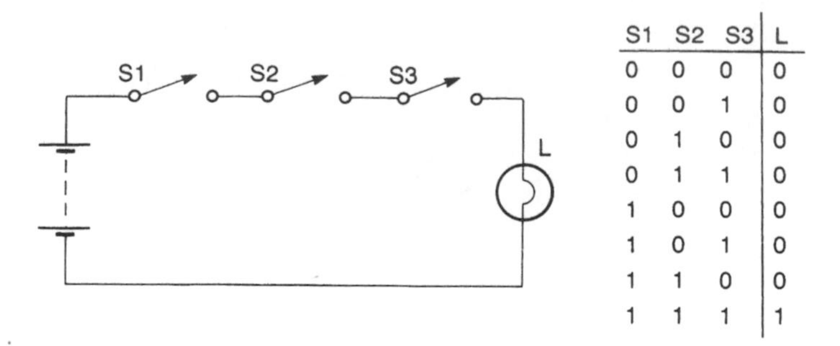

S1	S2	S3	L
0	0	0	0
0	0	1	0
0	1	0	0
0	1	1	0
1	0	0	0
1	0	1	0
1	1	0	0
1	1	1	1

Figure 9.4 Three switches in series.

figuration the lamp will light if either of the switches is closed. This function is described in the truth table of Figure 9.3(b) where the meanings of '0' and '1' are as for the previous example. As before we may express this relationship in words as 'the lamp will be illuminated if, and only if, switch S1 is closed OR switch S2 is closed (or if both are closed)' or, in the abbreviated form

L = S1 OR S2

An example of the **OR** function, again an automotive application, might be the courtesy light, which is illuminated if the driver's door is open (closing a switch) OR if the passenger's door is open. This function is sometimes called the **Inclusive-OR** function since it includes the case where both inputs are true (that is, in this case where both switches are closed).

Our examples of the AND and OR functions can be extended to the use of three or more switches, as illustrated in Figures 9.4 and 9.5 where any number of switches may be connected in series or in parallel.

Consider now the circuit of Figure 9.6. Here two switches are connected in parallel, and this combination is in series with a third switch. This produces an arrangement that can be described by the truth table as shown, or by the statement 'the lamp will be illuminated if, and only if, S1 is closed AND either S2 OR S3 is closed'. This can again be given in an abbreviated form as

L = S1 AND (S2 OR S3)

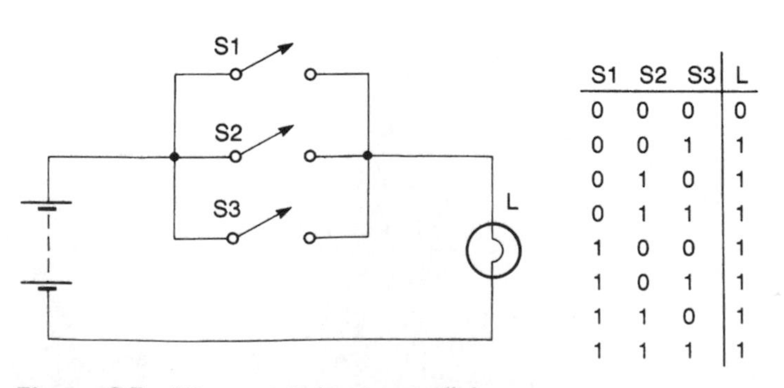

S1	S2	S3	L
0	0	0	0
0	0	1	1
0	1	0	1
0	1	1	1
1	0	0	1
1	0	1	1
1	1	0	1
1	1	1	1

Figure 9.5 Three switches in parallel.

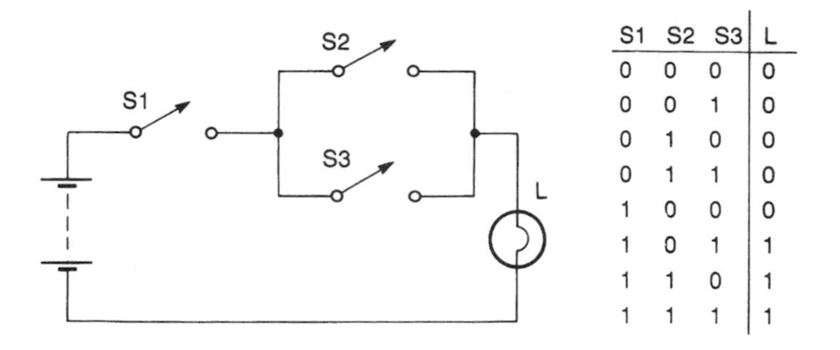

S1	S2	S3	L
0	0	0	0
0	0	1	0
0	1	0	0
0	1	1	0
1	0	0	0
1	0	1	1
1	1	0	1
1	1	1	1

Figure 9.6 A series/parallel configuration.

Notice the use of brackets to make the meaning of the expression clear and to avoid ambiguity.

In the examples so far considered we have started with a combination of switches and represented them by a truth table and a verbal description. In practice we will generally need to perform the process in reverse, being given a function and being required to devise an arrangement to produce this effect. In such cases we might be given either a truth table or a verbal description of the required system.

Consider the following truth table

S1	S2	S3	L
0	0	0	0
0	0	1	0
0	1	0	0
0	1	1	1
1	0	0	0
1	0	1	1
1	1	0	1
1	1	1	0

This represents an arrangement with three switches S1, S2 and S3, and a lamp L. We would consider the switches to be the three *inputs* to the network and the lamp to be the *output*. It is not immediately obvious what arrangement of switches would correspond to this truth table, and perhaps it is not clear whether *any* combination of the three switches can produce the desired results. In these cases it is often useful to consider the desired arrangement as a 'black box' with the various switches as inputs and the lamp as an output. Such an arrangement is shown in Figure 9.7.

The diagram of Figure 9.7 makes no assumptions concerning the method of interconnection of the switches and the lamp. It may be that in order to produce the desired function we will need some form of electronic circuitry within our 'black box'. Since the three switches and the lamp represent simple binary devices, we could produce a more general arrangement by showing these as simple binary variables, without defining their type. You will remember from Chapter 3 that when representing 'black box'

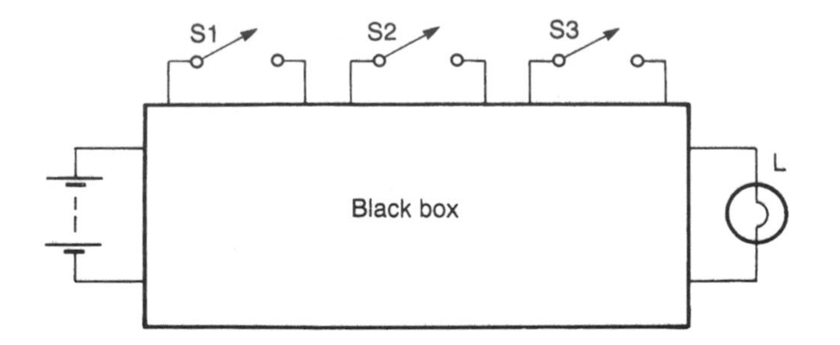

Figure 9.7 Representation of an unknown network.

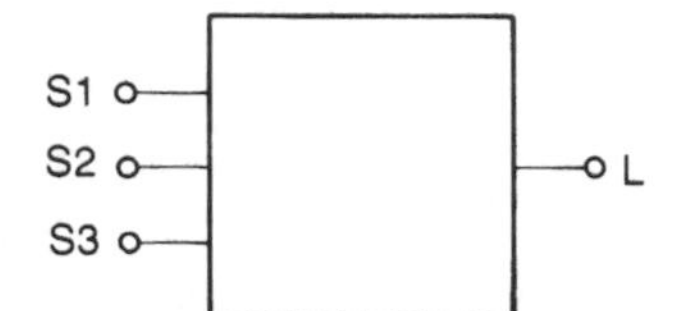

Figure 9.8 Symbolic representation of an unknown network.

amplifiers, such as operational amplifiers, we often omit the connections to the power supply. If we adopt a similar scheme here, we arrive at a diagram of the form shown in Figure 9.8.

We now have a symbolic representation of a network with three inputs and one output which makes no assumptions as to the form of the inputs or the outputs. The inputs could represent switches, as in the earlier examples, but they could equally well be signals from binary sensors such as thermostats, level sensors or proximity switches. Similarly the output device could be a lamp, but could equally well be a heater or a solenoid. We can now design the function of the 'black box' without considering its implementation. We will return to the task of actually constructing such a box in Chapter 11.

The normal building block of electronic logic circuits is the **logic gate** and in the next section we will consider a number of standard gate types.

9.3 Logic gates

A logic gate is an element that takes one or more binary input signals and produces an appropriate binary output, depending on the state(s) of the input(s). There are three elementary gate types, two of which, the AND and OR functions, we have already met. These elementary gates can be combined to form more complicated gates, which in turn may be connected to produce any required function. Each gate is given a logic symbol to allow complex functions to be represented by a **logic diagram** indicating the combination of simple gates required to implement it.

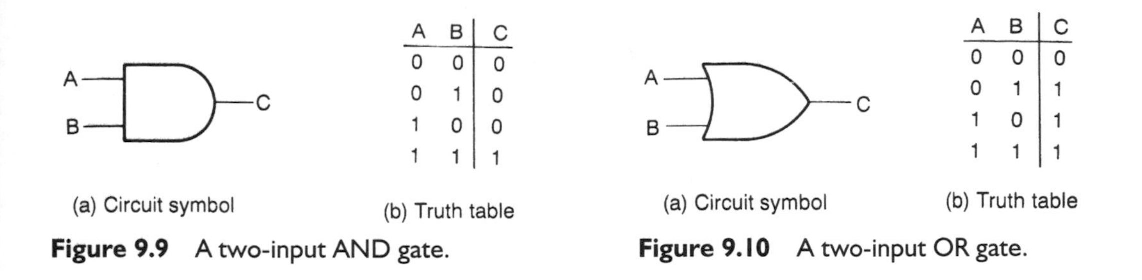

A	B	C
0	0	0
0	1	0
1	0	0
1	1	1

Figure 9.9 A two-input AND gate.

A	B	C
0	0	0
0	1	1
1	0	1
1	1	1

Figure 9.10 A two-input OR gate.

9.3.1 Elementary logic gates

The AND gate

The output of an AND gate is true (1) if, and only if, all of the inputs are true. The gate may have any number of inputs. The logic symbol and truth table for a two-input AND gate are given in Figure 9.9. The labelling of the inputs and outputs is arbitrary.

The OR gate

The output of an OR gate is true (1) if, and only if, at least one of its inputs is true. It is also called the **Inclusive-OR gate** for the reasons discussed earlier. The gate may have any number of inputs. The logic symbol and truth table for a two-input OR gate are given in Figure 9.10.

The NOT gate

The output of a NOT gate is true (1) if, and only if, its single input is false. This gate has the function of a **logical INVERTER** since the output is the **complement** of the input. The gate is sometimes referred to as an **INVERT gate** or simply as an **INVERTER**. The circuit symbol and truth table for a NOT gate are shown in Figure 9.11.

The circle in the symbol for an INVERTER represents the process of inversion. The triangular symbol without the circle would represent a function in which the output state was identical to the input. This function is called a **buffer**. The presence of a buffer does not affect the state of a logic signal. However, when we come to consider the implementation of gates using electronic circuits, we shall see that a buffer can be used to change

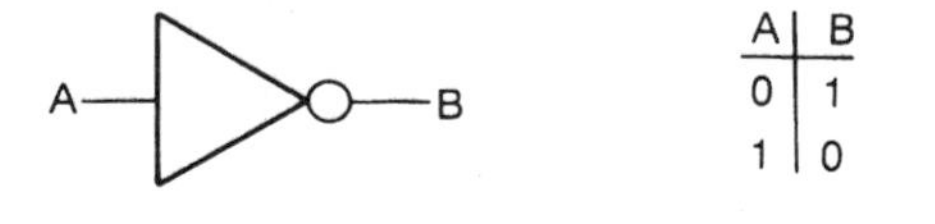

A	B
0	1
1	0

(a) Circuit symbol (b) Truth table

Figure 9.11 A NOT gate (INVERTER).

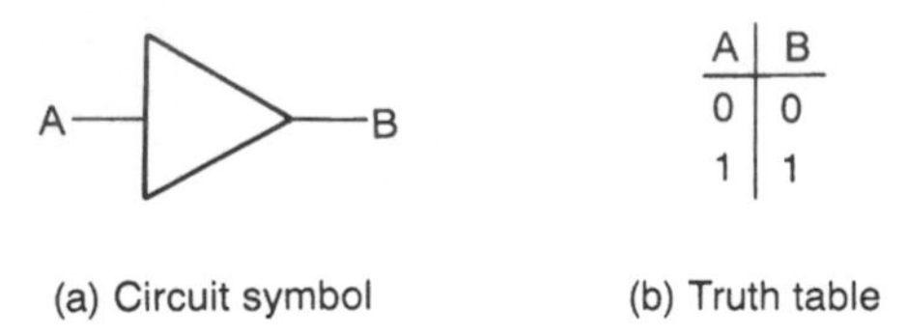

Figure 9.12 A logical buffer.

the electrical properties of a logic signal. It is interesting to note that the symbol for a buffer is similar to that used for a single-input analogue amplifier. Since the buffer does not produce any logical function it is usually not considered as an elementary gate. However, for completeness its logic symbol and truth table are given in Figure 9.12.

9.3.2 Compound gates

The elementary gates described above can be combined to form any desired logic function. However, it is often more convenient to work with slightly larger building blocks. Several compound gates are used which are simple arrangements of these elementary gates. We shall see in Chapter 11 that it is often easier to implement some of these compound gates than the simple ones described above. Consequently, the following gates include some of the most widely used types.

The NAND gate

The NAND gate is functionally equivalent to an AND gate followed by an INVERTER, the name being an abbreviation of Not-AND. Following the example set with the symbol for an INVERTER, the logic symbol for a NAND gate is simply that for an AND gate with a circle at the output. The truth table for the NAND gate is similar to that for an AND gate with the output state inverted. A NAND gate can have any number of inputs. The equivalent circuit, logic symbol and truth table for a two-input NAND gate are shown in Figure 9.13.

The NOR gate

The NOR gate is functionally equivalent to an OR gate followed by an INVERTER, the name being an abbreviation of Not-OR. Again the logic symbol is that of an OR gate with

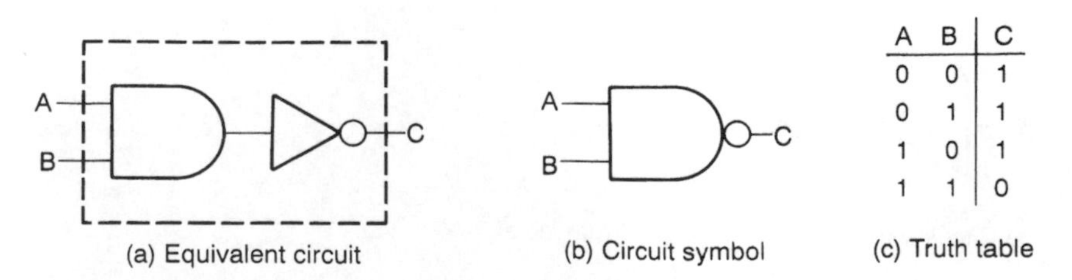

Figure 9.13 A two-input NAND gate.

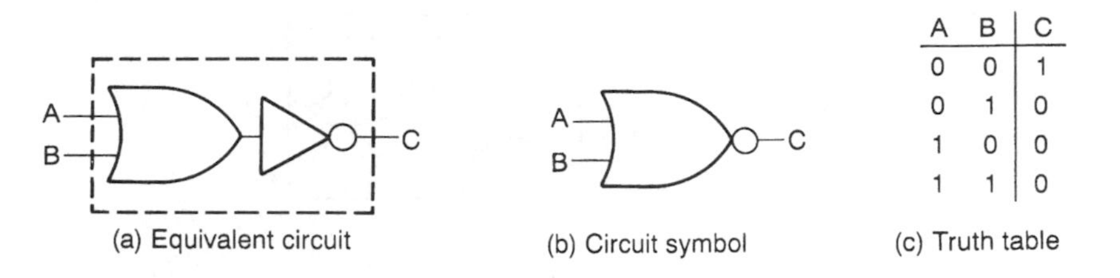

A	B	C
0	0	1
0	1	0
1	0	0
1	1	0

Figure 9.14 A two-input NOR gate.

a circle at the output to indicate an inversion. A NOR gate can have any number of inputs. Figure 9.14 shows the equivalent circuit, logic symbol and truth table of a two-input NOR gate.

The Exclusive-OR gate

The output of an Exclusive-OR gate is true (1) if, and only if, one or other of its two inputs is true, but not if both are true. The gate gets its name from the fact that it resembles the Inclusive-OR gate, except that it *excludes* the case where both inputs are true. An Exclusive-OR gate has only two inputs. An equivalent circuit, logic symbol and truth table for an Exclusive-OR gate are given in Figure 9.15.

The form of the equivalent circuit can be understood by considering the function as the combination of two conditions indicated by *X* and *Y* on the equivalent circuit. *X* represents the condition that either A OR B are true, and *Y* represents the condition that A AND B are NOT both true (that is, A NAND B). The output of the gate is true if both *X* AND *Y* are true. This combination of gates is not the only way of implementing this function and we will return to the Exclusive-OR gate later.

The Exclusive-NOR gate

The logical last member of our group of compound gates is the Exclusive-NOR gate, which, as its name suggests, is the inverse of the Exclusive-OR gate. This may be considered to be an Exclusive-OR gate followed by an INVERTER, or as the equivalent circuit of Figure 9.15 with the AND gate replaced with a NAND gate. This gate gives a true output when both the inputs are 0 or when both are 1. It therefore gives a true output when the inputs are equal. For this reason this gate is also known as an **equivalence** or

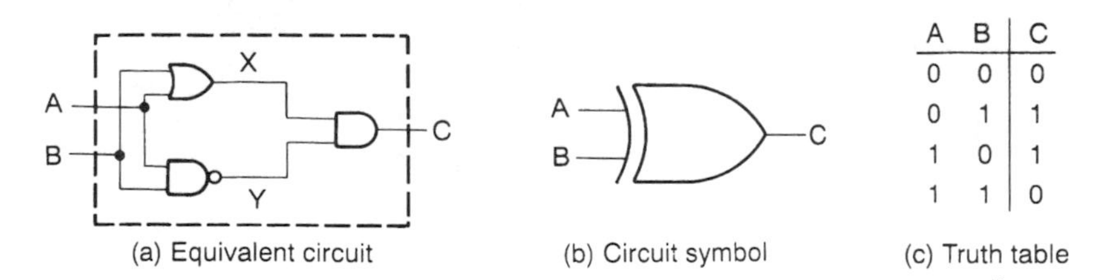

A	B	C
0	0	0
0	1	1
1	0	1
1	1	0

Figure 9.15 An Exclusive-OR gate.

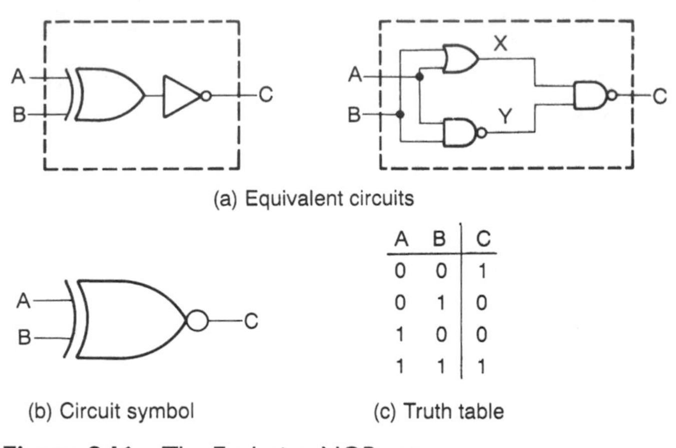

A	B	C
0	0	1
0	1	0
1	0	0
1	1	1

Figure 9.16 The Exclusive-NOR gate.

an **equality gate**. Equivalent circuits, a logic symbol and a truth table for the Exclusive-NOR gate are shown in Figure 9.16.

Example 9.1 Implementing a specified network using logic gates

Having defined our standard building blocks we may now return to the problem described at the end of the previous section, that of implementing a network from a given specification. The required network is described by the following block diagram and truth table.

The required function can be described in words by saying that the output should be '1' if exactly two of the three input signals are '1', and should be '0' otherwise. This condition is met by only three combinations of the input signals, as shown in the truth table. The function can be realized by constructing three gate networks to detect each of these combinations of the inputs and then ORing the outputs from these three networks together. The signal at point X in the circuit is '1' if S2 and S3 are true and S1 is not. Similarly Y and Z correspond to the other input combinations required to generate a true output.

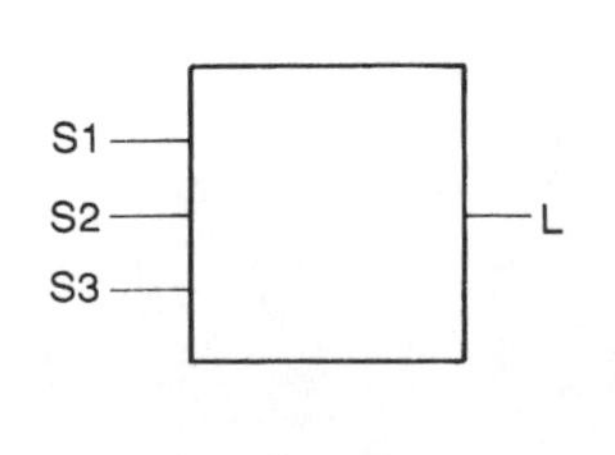

(a) Block diagram

S1	S2	S3	L
0	0	0	0
0	0	1	0
0	1	0	0
0	1	1	1
1	0	0	0
1	0	1	1
1	1	0	1
1	1	1	0

(b) Truth table

The output of the network is then made true if any of X, Y or Z is true, by ORing these signals together. You may like to confirm that this arrangement has the correct truth table by considering the input and output of each gate for each permutation of the input variables.

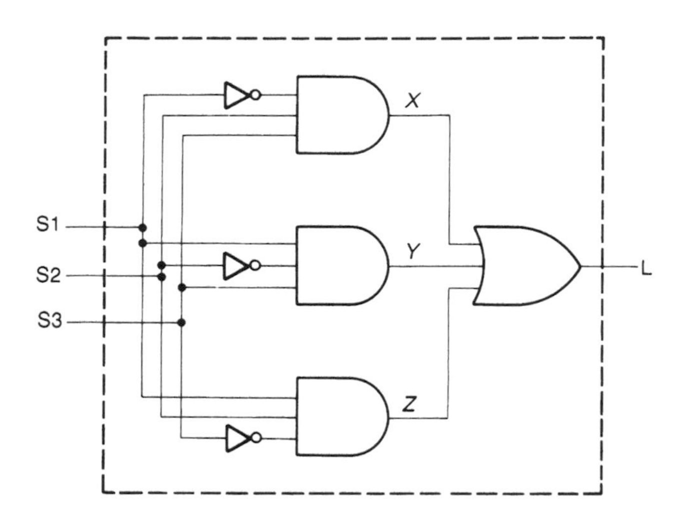

FILE 9A

Computer simulation exercise 9.1

Simulate the logic circuit of Example 9.1 and confirm that it produces the required output for all possible combinations of the inputs.

9.4 Boolean algebra

Although the technique adopted in Example 9.1 provides a solution to the problem, it is not an elegant approach. It is also not apparent whether the solution is the simplest method of achieving the desired characteristics. A more systematic and versatile method of tackling the task is to use **Boolean algebra**. This provides a notation to describe logic functions and defines a number of operations that can be performed in an attempt to simplify their form.

Boolean algebra is named after its inventor, the nineteenth century mathematician George Boole (1854), who formulated a basic set of rules concerning binary, that is true–false, logic. In fact it was not until many years after Boole's death that the importance of his work became apparent through the work of an American, Claude Shannon (1938). In 1938 Shannon, who was a postgraduate student at the Massachusetts Institute of Technology, applied *Boolean algebra* to the design of telephone switching networks. From this work it became clear that it formed a basis for the analysis of all forms of binary system.

Boolean algebra defines *constants*, *variables* and *functions* to describe binary systems. It then describes a number of *theorems* that can be used to manipulate logic expressions.

Boolean constants

Boolean constants consist of '0' and '1'. The former represents the false state and the latter the true state.

Boolean variables

Boolean variables are quantities that can take different values at different times. They may represent input, output or intermediate signals and are given names usually consisting of alphabetic characters, such as 'A', 'B', 'X' or 'Y'. Variables may only take the values '0' or '1'.

Boolean functions

Each of the elementary logic functions is represented within Boolean algebra by a unique symbol, as shown in the following table.

Function	Symbol	Example
AND	dot	$C = A \cdot B = AB$
OR	plus	$C = A + B$
NOT	bar	$C = \overline{A}$

You will notice that although the symbol for the AND operation is a '·', it is often omitted. This is similar to the omission of a dot meaning multiplication in conventional algebra, as in the expression $b^2 - 4ac$.

It may seem that the choice of symbols for AND and OR is inappropriate and that it would have been more sensible to use '+' for the AND operation. The symbols are chosen in this way because the expressions generated then follow rules which closely correspond to those of conventional algebra. For the present, simply accept that there is a good reason for this choice.

Compound gates can be simply represented by combinations of the elementary functions, as illustrated in Figure 9.17.

The Exclusive-OR function may be represented by a combination of the elementary functions, as shown in Figure 9.17(c). This is simply an algebraic statement of the circuit of Figure 9.15(a). However, the function is also given its own symbol in the form of the symbol for an Inclusive-OR surrounded by a circle (⊕).

Table 9.1 summarizes the various elementary and compound gates giving their logic symbol, Boolean algebraic representation and truth table.

Using Boolean expressions

Boolean algebra allows us to represent any logical network in a mathematical form. To make use of this technique we need to be able to generate the Boolean representation of a system from its truth table or from a circuit diagram, and to be able to take such an expression and represent it by an equivalent circuit. Fortunately both these tasks are straightforward, as is illustrated in the following examples.

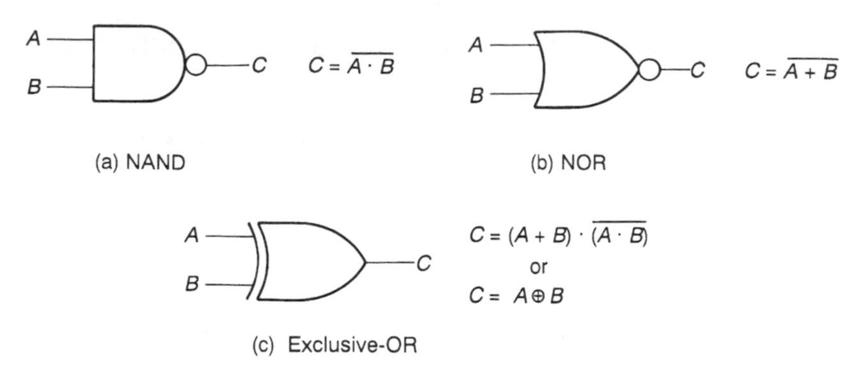

Figure 9.17 Boolean representation of compound functions.

Table 9.1 Logic gates.

Function	*Symbol*	*Boolean representation*	*Truth table*
AND		$C = A \cdot B$	A B | C 0 0 | 0 0 1 | 0 1 0 | 0 1 1 | 1
OR		$C = A + B$	A B | C 0 0 | 0 0 1 | 1 1 0 | 1 1 1 | 1
NOT		$B = \overline{A}$	A | B 0 | 1 1 | 0
NAND		$C = \overline{A \cdot B}$	A B | C 0 0 | 1 0 1 | 1 1 0 | 1 1 1 | 0
NOR		$C = \overline{A + B}$	A B | C 0 0 | 1 0 1 | 0 1 0 | 0 1 1 | 0
Exclusive-OR		$C = A \oplus B$	A B | C 0 0 | 0 0 1 | 1 1 0 | 1 1 1 | 0
Exclusive-NOR		$C = \overline{A \oplus B}$	A B | C 0 0 | 1 0 1 | 0 1 0 | 0 1 1 | 1

Example 9.2 Extracting the Boolean expression for a system from its truth table

Consider the following truth table, which you may recognize as that of the **Exclusive-OR** function.

A	*B*	*C*
0	0	0
0	1	1
1	0	1
1	1	0

We could express the information in the table in words by saying '*C* is true if *B* is true AND *A* is NOT true, OR if *A* is true AND *B* is NOT true'. This statement is extracted from the table simply by describing the states of *A* and *B* for each line on which *C* is '1' and ORing them together.

The statement can be converted into Boolean form simply by expressing the various functions using symbols, thus

$$C = (\overline{A} \cdot B) + (A \cdot \overline{B}) = \overline{A}B + A\overline{B}$$

Notice that this expression is not the same as that given in Figure 9.17 for the Exclusive-OR function. Boolean expressions are *not* unique.

A more formal description of the process might be:

(1) A **minterm** is generated for each column in which a '1' appears in the truth table.

(2) The minterm contains each input variable in turn: the input being non-inverted if it is a '1' in the truth table, and inverted if it is a '0'.

(3) The overall expression for the logic function is then the sum of the minterms.

This process can be applied to a truth table of any size, for example the table

A	*B*	*C*	*D*	*minterms*
0	0	0	0	
0	0	1	1	$\overline{A}\overline{B}C$
0	1	0	0	
0	1	1	0	
1	0	0	1	$A\overline{B}\overline{C}$
1	0	1	0	
1	1	0	0	
1	1	1	1	ABC

corresponds to the expression

$$D = \overline{A}\overline{B}C + A\overline{B}\overline{C} + ABC$$

Example 9.3 Extracting the Boolean expression for a system from its logic diagram

Consider the following logic circuit

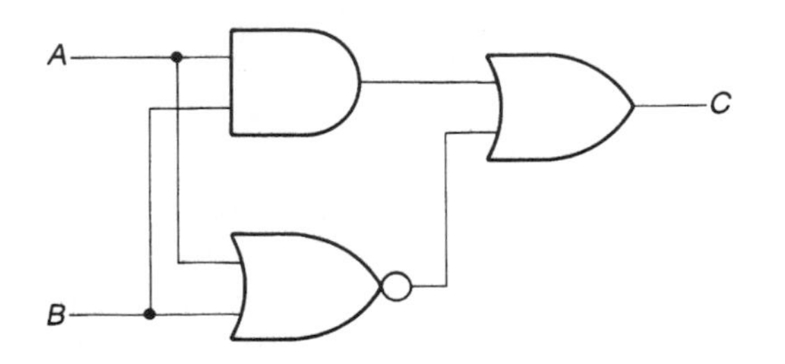

The simplest method of extracting the Boolean expression for this arrangement is simply to write on to the diagram the output of each gate in terms of its inputs, starting with gates near the inputs and working across to the output.

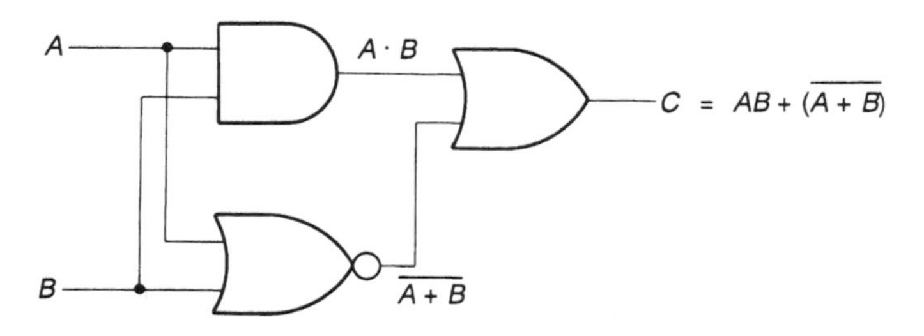

Example 9.4 Generating a logic diagram for a system from its Boolean expression

Consider the following Boolean expression

$$C = AB + \overline{\bar{A}\bar{B}} + (A+B)$$

Development of the logic diagram is the reverse of the process illustrated in Example 9.3. Here we start at the output and work back towards the inputs.

The right-hand side of the expression has three components which are ORed together. Therefore the output of our circuit will come from a three-input OR gate. The inputs to this gate will be the three components of the expression. The first, AB, comes from a two-input AND gate with inputs A and B, the second from a two-input NAND gate with inputs $\bar{A}$ and $\bar{B}$, and the third from a two-input OR gate again with inputs A and B. The complete logic diagram is shown below.

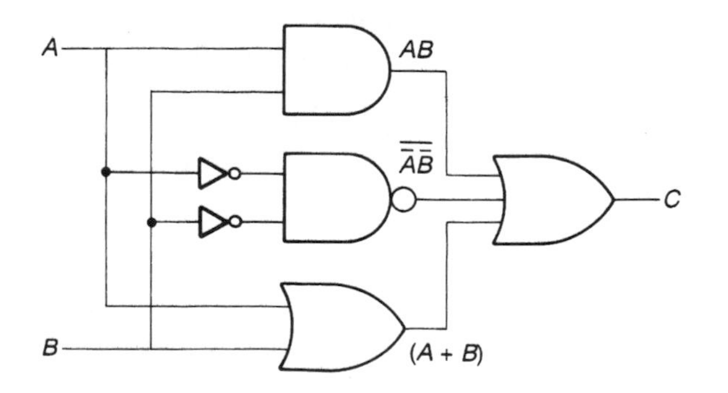

9.4.1 Boolean theorems

The techniques illustrated in the above examples enable us to generate Boolean expressions to describe a function which is specified by a truth table or a logic diagram. However, they give us no indication as to whether the expressions produced are the simplest form of the function. We noted in Example 9.2 that Boolean expressions are not unique. We therefore require some method of manipulating expressions into their simplest forms.

To enable us to perform this manipulation, Boolean algebra defines a number of **theorems** that can be used to change the form of an expression. The rules of Boolean algebra consists of a set of **identities** and a set of **laws**. Table 9.2 sets out these rules.

It is not the purpose of this text to prove the identities and laws set out in Table 9.2. It is, however, useful to have an understanding of *why* these relationships are true and this is fairly easily achieved.

When considered in isolation the Boolean identities may seem rather abstract and thus difficult to remember. It is easier to see the sense of the rules if they are converted into concrete examples when their meaning becomes clear.

Take, for example, the first identity

$$0 \cdot 0 = 0$$

The logical constant '0' represents a state that never exists and is thus an impossible condition. One could verbalize this by considering any impossible event. A common saying is 'if pigs could fly ...'. If we ignore for the moment any biological possibilities and simply accept this as an impossible event, we could illustrate the identity by the expression:

The condition will be met if, pigs could fly, AND, pigs could fly.

The effect of the two conditions is clearly the same as the single condition 'if pigs could fly' and thus $0 \cdot 0$ does indeed equal 0.

The second identity,

$$0 \cdot 1 = 0$$

requires the use of a condition that is self-evidently always true. The saying '... as sure as night follows day ...' might provide such a phrase. Thus we could verbalize identity two as

The condition will be met if, pigs could fly, AND, night follows day.

Here the second condition is true but the first is not. Since both must be true to produce a positive output from the AND function this also generates a '0'. In fact any condition ANDed with '0' will give '0', as illustrated in identity 5.

Identities 5 to 10 require the provision of a binary variable. Let us consider two binary events A and C, such that A represents the fact that it is warm outside and C that I will go out for a walk. I could thus represent the statement 'I will go out if it is warm' by the Boolean expression

$$C = A$$

We may now use these variables to investigate several of the other identities.

The statement 'I will go out if, it is warm, AND, it is warm' is clearly equivalent to saying 'I will go out if it is warm'. Thus $A \cdot A$ is equivalent to A (identity 9).

Table 9.2 Summary of Boolean algebra identities and laws.

AND function

(1) $0 \cdot 0 = 0$
(2) $0 \cdot 1 = 0$
(3) $1 \cdot 0 = 0$
(4) $1 \cdot 1 = 1$
(5) $A \cdot 0 = 0$
(6) $0 \cdot A = 0$
(7) $A \cdot 1 = A$
(8) $1 \cdot A = A$
(9) $A \cdot A = A$
(10) $A \cdot \overline{A} = 0$

OR functions

(11) $0 + 0 = 0$
(12) $0 + 1 = 1$
(13) $1 + 0 = 1$
(14) $1 + 1 = 1$
(15) $A + 0 = A$
(16) $0 + A = A$
(17) $A + 1 = 1$
(18) $1 + A = 1$
(19) $A + A = A$
(20) $A + \overline{A} = 1$

NOT function

(21) $\overline{0} = 1$
(22) $\overline{1} = 0$
(23) $\overline{\overline{A}} = A$

Commutative law

(24) $AB = BA$
(25) $A + B = B + A$

Distributive law

(26) $A(B + C) = AB + AC$
(27) $A + BC = (A + B)(A + C)$

Associative law

(28) $A(BC) = (AB)C$
(29) $A + (B + C) = (A + B) + C$

Absorption law

(30) $A + AB = A$
(31) $A(A + B) = A$

DeMorgan's law

(32) $\overline{A + B} = \overline{A} \cdot \overline{B}$
(33) $\overline{A \cdot B} = \overline{A} + \overline{B}$

Also note

(34) $A + \overline{A}B = A + B$
(35) $A(\overline{A} + B) = AB$

The statement 'I will go out if, it is warm, AND, night follows day' is equivalent to saying 'I will go out if it is warm'. Thus $A \cdot 1$ is equivalent to A (identity 7).

The same method can be applied to expressions using OR functions. For example, the statement 'I will go out if, it is warm, OR, night follows day' is equivalent to saying 'I will always go out'. Therefore $A + 1$ is equivalent to 1 (identity 17).

Identity 10 is equivalent to the statement 'I will go out if, it is warm, AND, it is cold' which is never true, and identity 20 to the statement 'I will go out if, it is warm, OR, it is cold' which is always true.

The various laws can also be understood more easily by linking them to concrete examples.

Commutative law

$AB = BA$
$A + B = B + A$

The commutative law says that the order in which terms are ANDed or ORed together is unimportant. If we return to our recreational theme and define a new binary condition B representing the fact that it is dry, then the two statements 'I will go out if, it is warm, AND, it is dry' and 'I will go out if, it is dry, AND, it is warm' are clearly equivalent.

Distributive law

$$A(B+C)=AB+AC$$
$$A+BC=(A+B)(A+C)$$

This principle can be illustrated by the statement 'I will go out if, it is dry, AND, it is either warm OR bright'. This is equivalent to saying 'I will go out if it is dry AND warm, OR, it is dry AND bright'. Similar statements can be used to illustrate the alternative form of this law.

Associative law

$$A(BC)=(AB)C$$
$$A+(B+C)=(A+B)+C$$

The law of associativity seems self-evident. It says that when many conditions are to be ANDed or ORed together, the order in which conditions are combined is unimportant. A symbolic representation of this is shown in Figure 9.18.

It should be remembered, however, that in general

$$(A \cdot B)+C \neq A \cdot (B+C)$$

as shown by the distributive law.

Absorption law

$$A+AB=A$$
$$A(A+B)=A$$

This law is exemplified by the statement 'I will go out if, it is warm, OR, if it is warm AND dry'. This statement is clearly equivalent to 'I will go out if it is warm'. A similar argument can be used to illustrate the other form of the law.

DeMorgan's law

$$\overline{A+B}=\overline{A}\cdot\overline{B}$$
$$\overline{A \cdot B}=\overline{A}+\overline{B}$$

DeMorgan's law is probably the most difficult Boolean theorem to conceptualize, since it contains both a transformation from ANDs to ORs and inversion of the state of the variables. It can be achieved, however, by noting the equivalence of the statements 'I will not go out if it is both cold AND wet' and 'I will go out if it is either NOT cold, OR, NOT wet'.

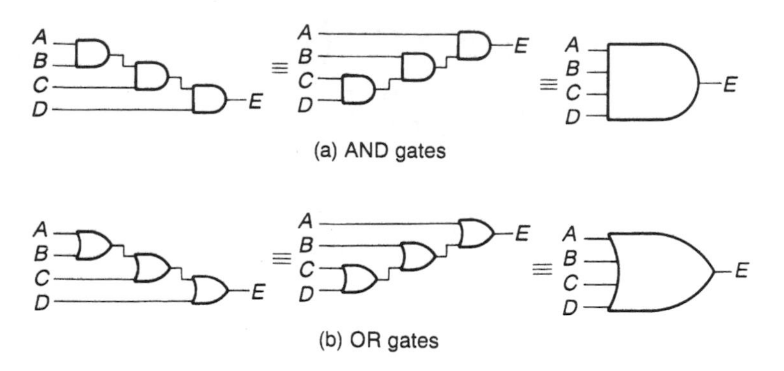

Figure 9.18 Physical significance of the associative law.

Algebraic manipulation

The various algebraic functions and laws allow us to manipulate Boolean expressions to simplify implementation. This is illustrated by the following examples.

Example 9.5 Generating a logic diagram for a system from its Boolean expression using only NAND gates.

We shall see in Chapter 11 that when using a particular device technology to implement a circuit, it is often easier, and therefore less expensive, to produce one form of gate rather than another. It is therefore important to be able to manipulate the form of a logic expression to implement it in a suitable form. In several logic families, NAND gates are the simplest form of gate and it is thus useful to be able to construct circuits using only these gates.

From Example 9.2 we know that any logic function can be expressed in a sum-of-products (sum-of-minterms) form. Consider, for example, the function

$$D = ABC + A\overline{B}\overline{C} + \overline{A}B\overline{C}$$

This can be implemented directly as

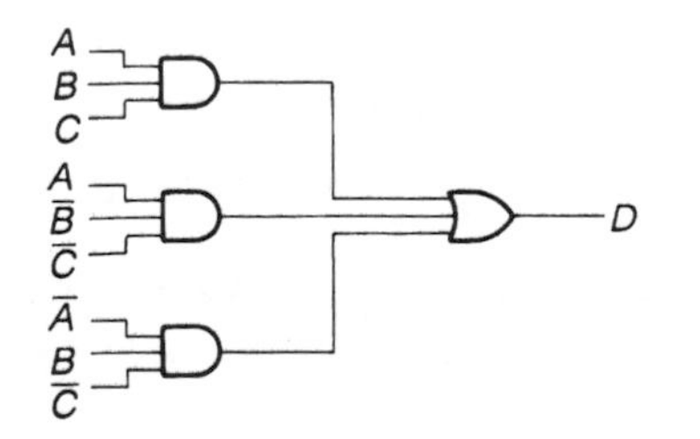

However, from DeMorgan's theorem we know that

$$A + B + C = \overline{\overline{A}\,\overline{B}\,\overline{C}}$$

and therefore

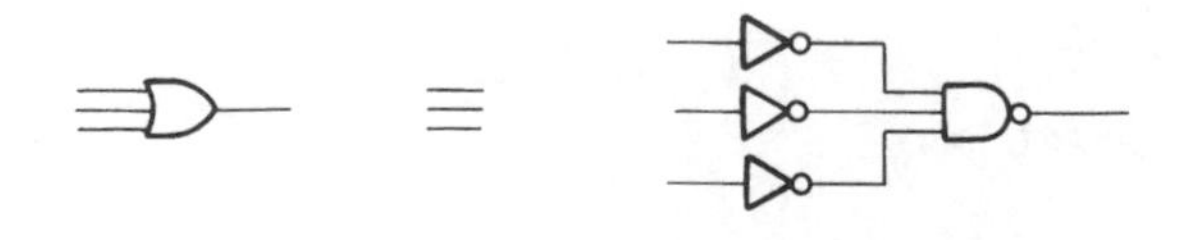

It follows that functions implemented using AND and OR gates may be modified to use only NAND gates, since

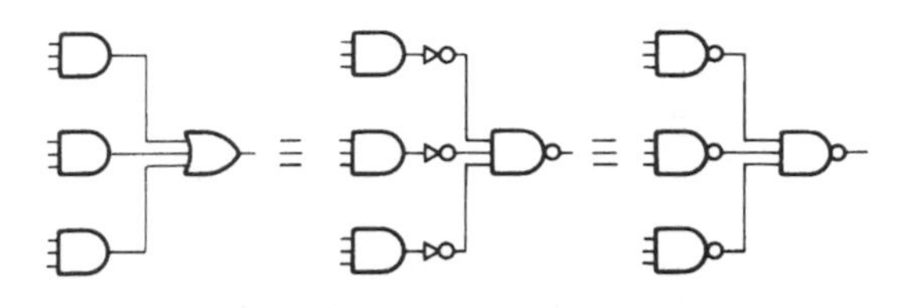

This simplification assumes that the inverses of the inputs are available. If this is not the case they can be obtained by using NAND gates as inverters. In this way any logic function can be implemented using only NAND gates, and our example is implemented as

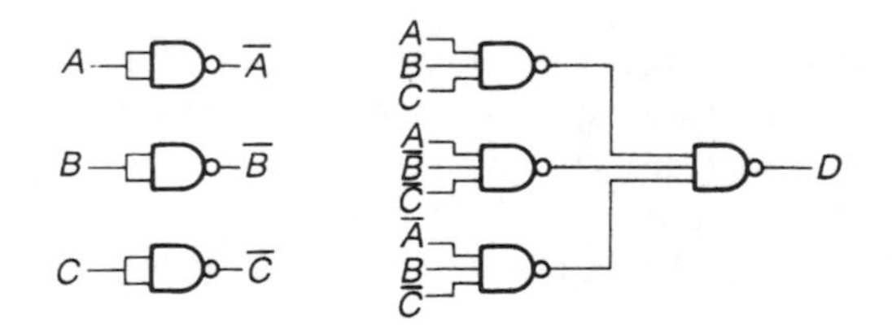

This manipulation can also be achieved using Boolean algebraic manipulation by noting that

$$D = ABC + A\overline{B}\overline{C} + \overline{A}B\overline{C} = \overline{\overline{ABC} \cdot \overline{A\overline{B}\overline{C}} \cdot \overline{\overline{A}B\overline{C}}}$$

This expression is in a form suitable for direct implementation using NAND gates.

Example 9.6 Generating a logic diagram for a system from its Boolean expression using only NOR gates

When using some logic families, NOR gates are the simplest gates. A similar manipulation to that given in the previous example can be used to implement functions using only these gates.

Again using DeMorgan's theorem we know that

$$A \cdot B \cdot C = \overline{\overline{A} + \overline{B} + \overline{C}}$$

and therefore

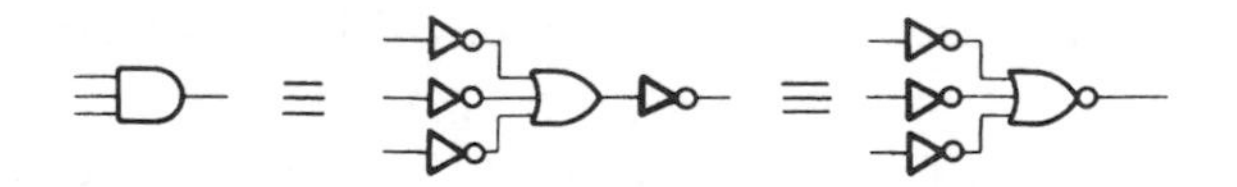

and

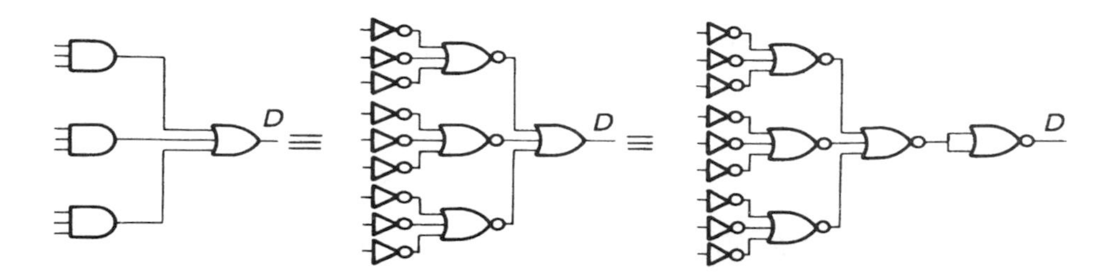

The inversions at the front end of this implementation are achieved simply by using the inverted, or non-inverted, input signal as required. Therefore, using our earlier example

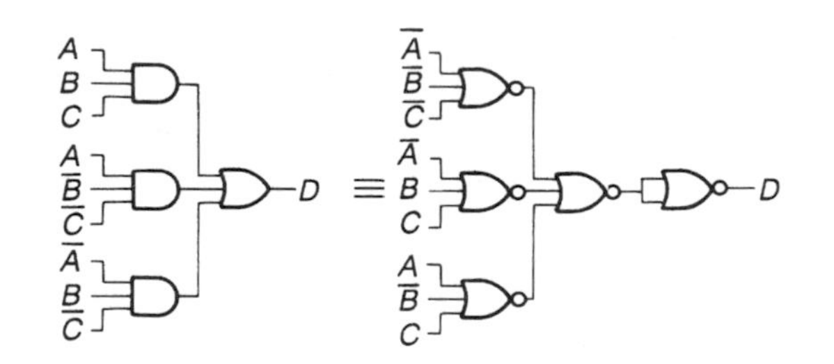

9.4.2 Logic minimization

One of the main aims of a logic designer is to produce a simple system. Logic minimization aims to take an algebraic expression and reduce it to a form that is easier to implement. In practice, circuits that contain the smallest number of gates are not necessarily the cheapest circuits to construct since other factors, such as the type of gates used, may affect the overall cost. However, reducing the complexity of a circuit is a good first step in reducing cost.

The most common form of Boolean expression is the **sum-of-products** form. This is the format obtained in Example 9.2 consisting of a number of ANDed terms (**minterms** or **products**) which are ORed together. Each product consists of a combination of some or all of the input variables, each in a complemented or uncomplemented form. Examples of sum-of-products expressions are

$$A\overline{B} + \overline{A}B$$

$$XYZ + \overline{X}Y\overline{Z} + X\overline{Y}Z$$

$$AB\overline{C}D + A\overline{B}C + \overline{A}BCD + ABC\overline{D}$$

The sum-of-products form must not include inversions of a series of terms, as in

$$A\overline{BC}D + \overline{AB}CD$$

Various techniques are available to simplify Boolean algebraic expressions to reduce the complexity of the circuitry required to implement them. Here we will look at several techniques for achieving this goal, including both manual and computer-based methods.

Algebraic simplification

Algebraic simplification is performed by taking the sum-of-products expression for a function and using the identities and rules of Boolean algebra to combine terms, progressively reducing its complexity. This is best illustrated by the use of examples.

Example 9.7 Algebraic simplification

Consider the expression

$$D = \overline{A}\overline{B}C + B\overline{C} + \overline{A}BC + ABC$$

This may be implemented directly as

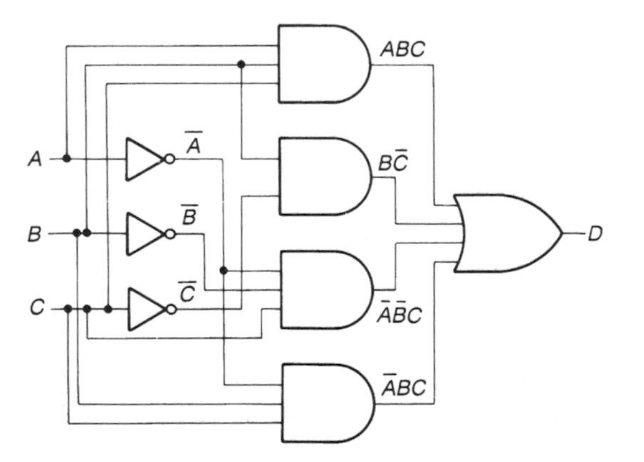

Alternatively, it can be reduced using Boolean algebra, as follows (the number following each line of the simplification represents the number of the identity or law used, as shown in Table 9.2).

$$
\begin{aligned}
D &= \overline{A}\overline{B}C + B\overline{C} + \overline{A}BC + ABC \\
&= \overline{A}\overline{B}C + B\overline{C} + BC(\overline{A} + A) && (26) \\
&= \overline{A}\overline{B}C + B\overline{C} + BC && (20) \\
&= \overline{A}\overline{B}C + B(\overline{C} + C) && (26) \\
&= \overline{A}\overline{B}C + B && (20) \\
&= \overline{A}C + B && (34)
\end{aligned}
$$

This can be implemented in only three gates

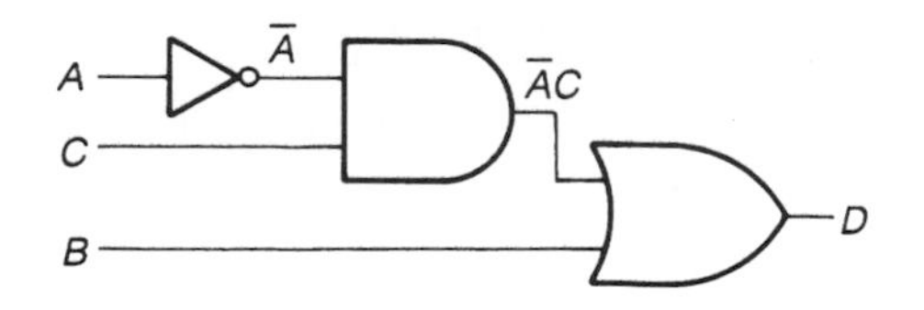

FILE 9B

Computer simulation exercise 9.2

Simulate the logic circuits of Example 9.2 and confirm that the two implementations produce the same output for all possible combinations of the inputs. Hint: construct both circuits and feed the same inputs into each. Observe both outputs and confirm that they are always the same.

Example 9.8 Algebraic simplification

Consider the expression

$$E = B\overline{C}\overline{D} + \overline{A}BD + ABD + BC\overline{D} + \overline{B}CD + \overline{A}\overline{B}\overline{C}D + A\overline{B}\overline{C}D$$

This can be implemented directly, as in the last example, but would use a large number of gates. Alternatively it may be simplified as follows.

Combining terms 2 and 3, 1 and 4, and 6 and 7

$$= BD(A + \overline{A}) + B\overline{D}(C + \overline{C}) + \overline{B}CD + \overline{B}\overline{C}D(A + \overline{A}) \quad (20)$$

$$= BD + B\overline{D} + \overline{B}CD + \overline{B}\overline{C}D \quad (26)$$

and combining the new terms 1 and 2, and 3 and 4

$$= B(D + \overline{D}) + \overline{B}D(C + \overline{C}) \quad (20)$$

$$= B + \overline{B}D \quad (26)$$

which further combines to give

$$E = B + D \quad (34)$$

This expression can be implemented in a single gate.

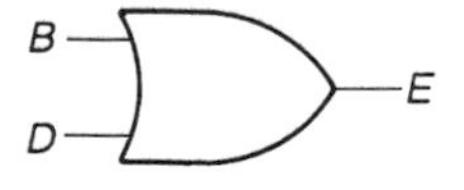

Example 9.9 Algebraic simplification

Consider the expression

$$E = A\overline{B}\,\overline{C} + \overline{A}\,\overline{C}D + ABD + BCD$$

It is not immediately obvious how to combine these terms to reduce its complexity. In fact, it is necessary to expand the terms first, as follows

$$\begin{aligned}
E &= A\overline{B}\,\overline{C} + \overline{A}\,\overline{C}D + ABD + BCD \\
&= A\overline{B}\,\overline{C}(D + \overline{D}) + \overline{A}\,\overline{C}D(B + \overline{B}) + ABD(C + \overline{C}) + BCD(A + \overline{A}) && (20) \\
&= A\overline{B}\,\overline{C}D + A\overline{B}\,\overline{C}\,\overline{D} + \overline{A}B\overline{C}D + \overline{A}\,\overline{B}\,\overline{C}D + ABCD + AB\overline{C}D \\
&\quad + ABCD + \overline{A}BCD && (26)
\end{aligned}$$

Combining the 5th and 7th terms produces

$$= A\overline{B}\,\overline{C}D + A\overline{B}\,\overline{C}\,\overline{D} + \overline{A}B\overline{C}D + \overline{A}\,\overline{B}\,\overline{C}D + ABCD + AB\overline{C}D + \overline{A}BCD \qquad (19)$$

and duplicating the 1st, 3rd and 6th terms gives

$$\begin{aligned}
&= A\overline{B}\,\overline{C}D + A\overline{B}\,\overline{C}\,\overline{D} + A\overline{B}\,\overline{C}\,\overline{D} + \overline{A}B\overline{C}D + \overline{A}B\overline{C}D \\
&\quad + \overline{A}\,\overline{B}\,\overline{C}D + ABCD + AB\overline{C}D + AB\overline{C}D + \overline{A}BCD && (19)
\end{aligned}$$

Combining terms 1 and 3, 5 and 6, 4 and 8, 2 and 9, and 7 and 10 we get

$$\begin{aligned}
&= A\overline{B}\,\overline{C}(D + \overline{D}) + \overline{A}\,\overline{C}D(B + \overline{B}) + B\overline{C}D(A + \overline{A}) \\
&\quad + A\overline{C}D(B + \overline{B}) + BCD(A + \overline{A}) && (25, 26) \\
&= A\overline{B}\,\overline{C} + \overline{A}\,\overline{C}D + B\overline{C}D + A\overline{C}D + BCD && (20)
\end{aligned}$$

and combining the 2nd and 4th, and the 3rd and 5th terms gives

$$\begin{aligned}
&= A\overline{B}\,\overline{C} + \overline{C}D(A + \overline{A}) + BD(C + \overline{C}) && (25, 26) \\
&= A\overline{B}\,\overline{C} + \overline{C}D + BD && (20)
\end{aligned}$$

This represents a considerable simplification compared with the original expression.

It is clear from Examples 9.7 to 9.9 that this form of algebraic simplification can greatly decrease the complexity of Boolean expressions, greatly reducing the cost of implementation. However, it is clear from the third of these examples that the process of simplification is not always straightforward. Manipulation of this kind is often inspired guess-work and suffers from the problem that one is never sure whether an optimum solution has been obtained. For example, can the expressions of Examples 9.7 to 9.9 be further simplified?

It is important to understand the process of algebraic simplification since it is a useful technique for simple functions. However, for more complex expressions we normally resort to more powerful methods.

Karnaugh maps

The **Karnaugh map** is a graphical method of representing the information within a truth table (Karnaugh, 1953).

In a truth table each possible combination of the inputs is represented by a unique line in the table, and the value of the output corresponding to that pattern of inputs is shown in the output column. In a Karnaugh map each possible combination of the inputs is represented by a box within a grid and the corresponding value of the output is written within that box. An example of this arrangement is shown in Figure 9.19 for a function with two inputs *A* and *B*.

The usefulness of the Karnaugh map stems from the arrangement of the boxes within the grid. For a system with two inputs the grid has four boxes, corresponding to the four combinations of the inputs. The boxes are positioned so that in going from one box to another, either horizontally or vertically, only one of the variables associated with that box changes. For example, in going from the top left-hand box to the top right-hand box, *A* changes from '0' to '1' but *B* remains unchanged. In moving from the top left-hand box to the bottom left-hand box, *B* changes from '0' to '1' but *A* remains unchanged. This principle is maintained as more boxes are added to represent systems with a greater number of inputs. Figure 9.20 shows Karnaugh maps for systems with three and four inputs. This technique can be extended to systems with a greater number of inputs, if required.

For a system with three input variables a four-by-two array is used, as in Figure 9.20(a). Two of the variables are associated with the four columns and the remaining variable with the two rows. The distribution of the variables is shown above and to the side of the map. The values of the appropriate variables are shown adjacent to each row and column. Thus box *V* corresponds to *AB* being '01', and to *C* being '0'. Box *V* is therefore $\overline{A}B\overline{C}$. Similarly, box *W* is $AB\overline{C}$. Systems with four inputs require a four-by-four array as shown in Figure 9.20(b). Here box *X* corresponds to $\overline{A}BCD$, box *Y* to *ABCD* and box *Z* to $\overline{A}BC\overline{D}$. You will notice that the grid is not labelled in simple binary order. The sequence in fact corresponds to the **Gray code** which will be discussed in the next section. For the moment it is sufficient to notice that the sequence is such that adjacent squares, both horizontally and vertically, differ in the state of only one variable. For example, *X* and *Y* differ only in the state of *A*, and *X* and *Z* differ only in the state of *D*. The importance of this

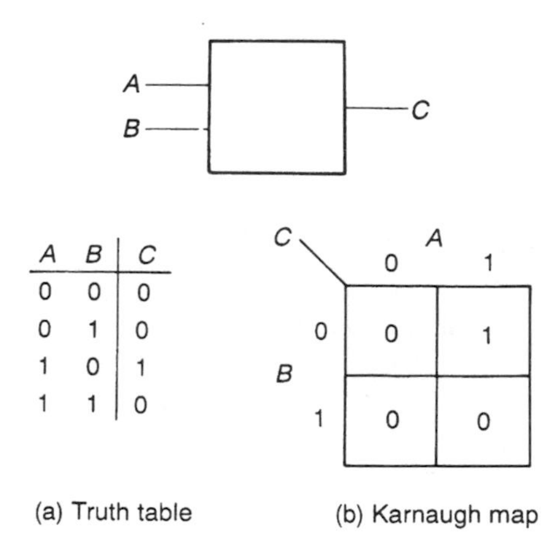

(a) Truth table (b) Karnaugh map

Figure 9.19 The truth table and Karnaugh map for a system with two inputs.

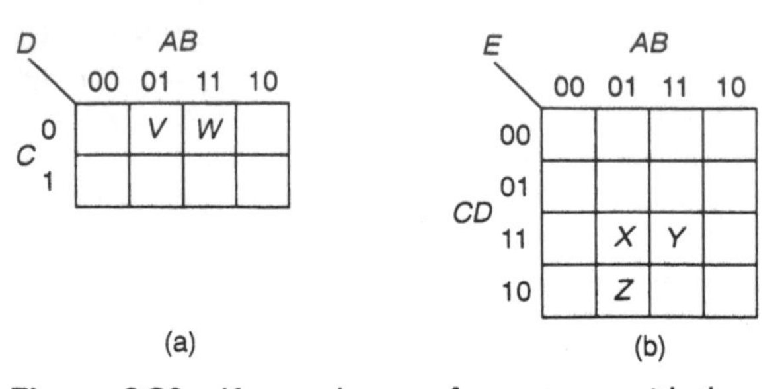

Figure 9.20 Karnaugh maps for systems with three and four inputs.

property will become apparent shortly. Note that this relationship does not apply to boxes that are linked diagonally, as with Y and Z, which differ in the states of both A and D.

To examine the use of Karnaugh maps consider the maps shown in Figure 9.21. The figure shows two Karnaugh maps for functions E and F. We can extract an algebraic expression for the functions directly from the maps in a manner similar to that used in Example 9.2 to extract this information from a truth table. Clearly E is given by the expression

$$E = \overline{A}B\overline{C}D + ABCD$$

and F is given by

$$F = \overline{A}B\overline{C}D + AB\overline{C}D$$

It is interesting at this point to attempt to simplify these expressions algebraically using the techniques described earlier. If this is done it will be found that E cannot be simplified but that F can be reduced by noting that

$$\begin{aligned} F &= \overline{A}B\overline{C}D + AB\overline{C}D \\ &= B\overline{C}D(A + \overline{A}) \\ &= B\overline{C}D \end{aligned}$$

The fact that F can be simplified while E cannot is evident if we look again at the maps of Figure 9.21. In the map for F the two '1's are adjacent. Thus, from the characteristics of the Karnaugh map, we know that these correspond to boxes representing

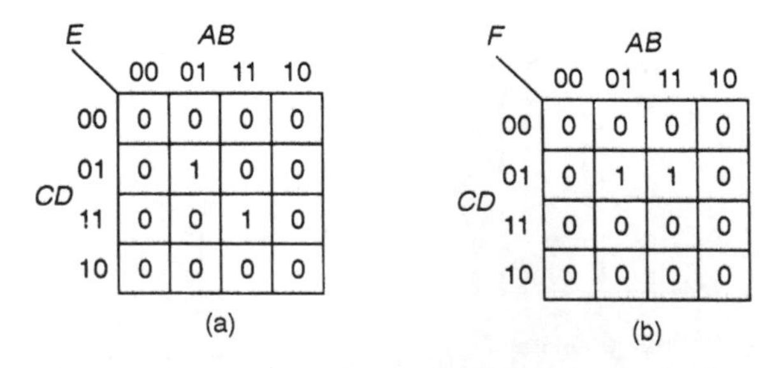

Figure 9.21 Simple Karnaugh maps.

combinations of inputs that only vary in the state of one input. Therefore *any* two adjacent elements containing '1's can be represented by an expression of the form

$$WXYZ + WXY\overline{Z}$$

where W, X, Y and Z each represent one of the input variables or its inverse. This will always simplify giving

$$WXYZ + WXY\overline{Z} = WXY(Z + \overline{Z}) = WXY$$

removing the effect of one of the variables. The variable removed from the expression is the term that is present in both its normal and its inverted form. This can be understood by noting that this means that the output will be true whether this variable is '1' or '0'. It is thus independent of this variable and it no longer appears within the expression.

Given a Karnaugh map containing two adjacent '1's, it is normal to draw a loop around the pair to indicate their union. The combination can then be represented by a term which indicates the input states that are constant for that group. In the example of Figure 9.21(b), B is '1', C is '0' and D is '1' for both elements which are '1'. However, A takes a value of '0' for one element and '1' for the other and therefore does not appear in the expression for the combination which is simply $B\overline{C}D$. Some examples of combinations of two elements are shown in Figure 9.22. Note that for the purposes of combining '1's, the top and bottom rows are considered to be adjacent, as are the right and left columns. It is as if the map were cylindrical in each plane such that the top and bottom and the two sides are touching.

Consider the map of Figure 9.23. Here four '1's are arranged in a square. The map may be described by the expression

$$E = \overline{A}B\overline{C}D + AB\overline{C}D + \overline{A}BCD + ABCD$$

This can be simplified by combining the 1st and 2nd terms, and the 3rd and 4th terms

$$\begin{aligned} E &= B\overline{C}D(A + \overline{A}) + BCD(A + \overline{A}) \\ &= B\overline{C}D + BCD \end{aligned}$$

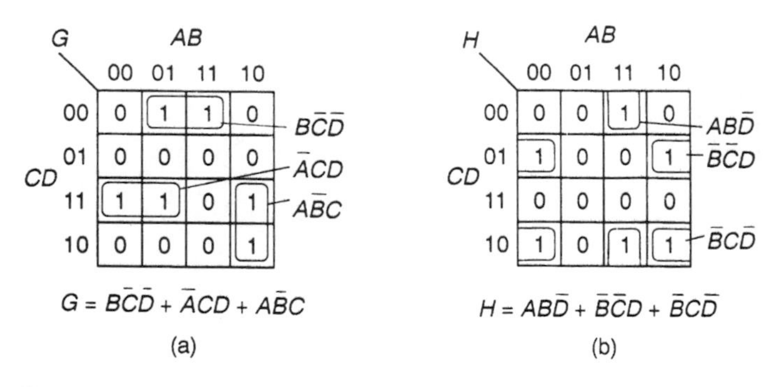

Figure 9.22 Combination of adjacent pairs of '1's in a Karnaugh map.

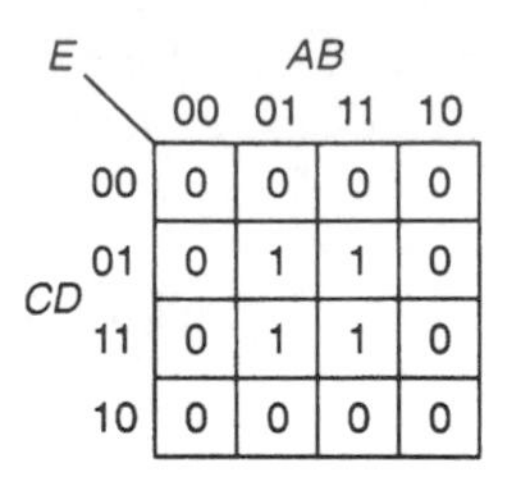

Figure 9.23 A Karnaugh map with an array of four '1's.

This simplification is evident from the map in that we could draw loops round these two pairs of elements. However, the two resulting terms may themselves be combined

$$= BD(C + \overline{C})$$
$$= BD$$

The fact that the four terms may be combined into one can be understood by noting that

$$E = \overline{A}B\overline{C}D + AB\overline{C}D + \overline{A}BCD + ABCD$$
$$= BD(\overline{A}\overline{C} + A\overline{C} + \overline{A}C + AC)$$

The term within the brackets represents all the possible combinations of the variables A and C. Thus, E is true if BD is true, independent of the values of A and C.

Other groups of four elements of a Karnaugh map can be combined provided that they represent all the possible combinations of any two variables. Some examples of allowed groupings are shown in Figure 9.24 and some illegal groupings in Figure 9.25.

Having observed that we may form groups of two or four '1's, it is perhaps not surprising to note that we may also form groups of eight elements, provided that we follow certain rules. In fact we can form groups of 2^n elements, provided that they represent all the possible states of n variables. This will be true if the groups are rectangles with each side having 2^m elements. Therefore, allowable groups will have sides of $1, 2, 4, 8, \ldots$

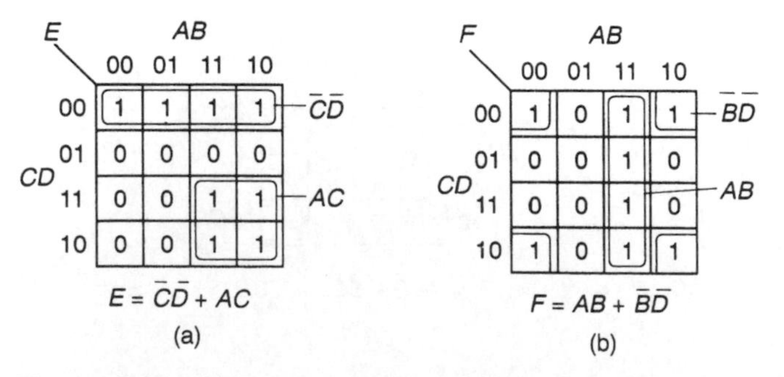

Figure 9.24 Examples of allowable groupings of four '1's.

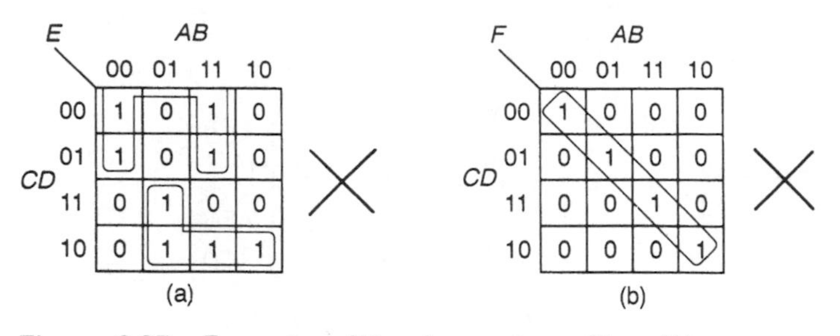

Figure 9.25 Examples of illegal groupings of four '1's.

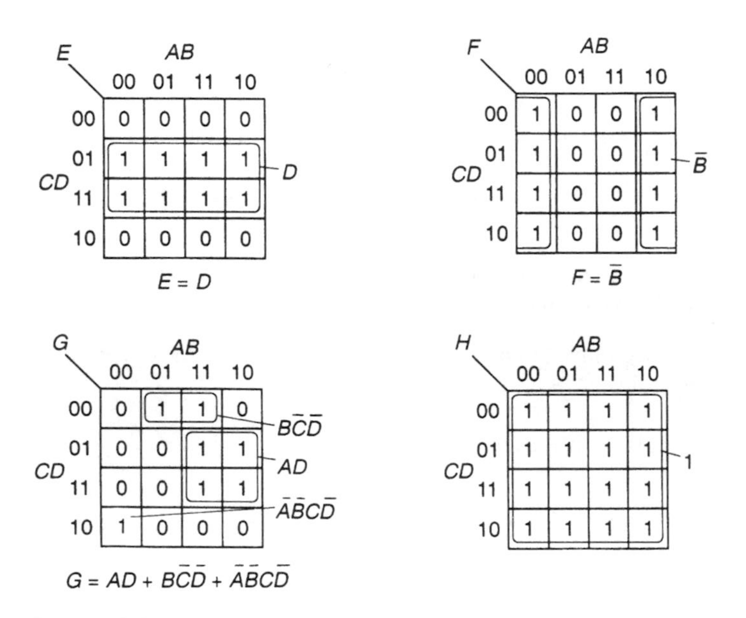

Figure 9.26 Examples of permitted groupings of 2^n elements

elements, so permitted groupings are 1×1 (a single element), $1 \times 2, 2 \times 1, 2 \times 2, 1 \times 4, 2 \times 4, 4 \times 4$, etc. The size of any rectangle will clearly be limited by the size of the map. Figure 9.26 shows some examples of such groupings. In each case the algebraic expression representing the group is obtained by noting which variable states are constant for all elements in the group.

The form of G in Figure 9.26 illustrates the fact that not all the '1's within a map may combine with other terms. It should also be noted that it is possible to combine some '1's into more than one grouping. This is shown in Figure 9.27 which shows two maps for the same function E, where the element $\overline{A}\overline{B}\overline{C}D$ is used within two groups. Also within this figure we note that the element $ABCD$ can be combined in two ways. This produces two different algebraic expressions for the function E. These two expressions are equally valid representations of E and are equally simple. This illustrates that *the simplified expressions produced by this technique are not unique.*

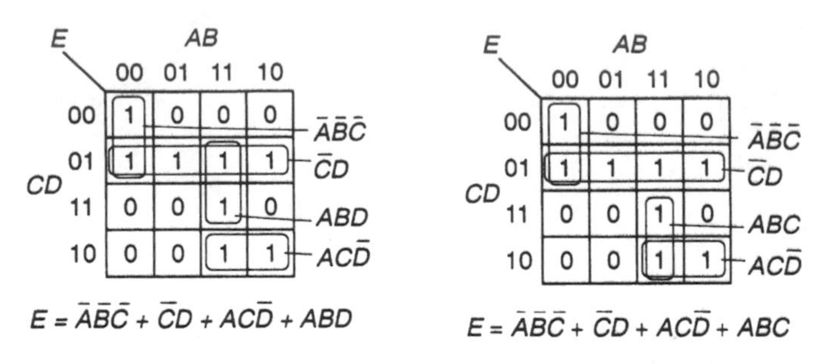

Figure 9.27 Alternative groupings of map elements.

Since it seems that there may be a number of ways of grouping the '1's of a Karnaugh map, we require some rules to ensure that we perform the groupings in such a way that the expression produced is as simple as possible. These rules may be expressed as follows.

(1) The largest possible groups of cells should be constructed first, each group containing 2^n elements.

(2) Progressively smaller groups should be added, until every cell containing a '1' has been included at least once.

(3) Any redundant groups should then be removed (even if these are large groups) to avoid duplication.

Use of these rules is illustrated in the following examples.

Example 9.10 Simplification of a logic expression using a Karnaugh map

Consider the following expression

$$D = \overline{A}B\overline{C} + AB\overline{C} + A\overline{B}\overline{C} + \overline{A}\overline{B}C + \overline{A}BC + ABC$$

This may be represented by a Karnaugh map as

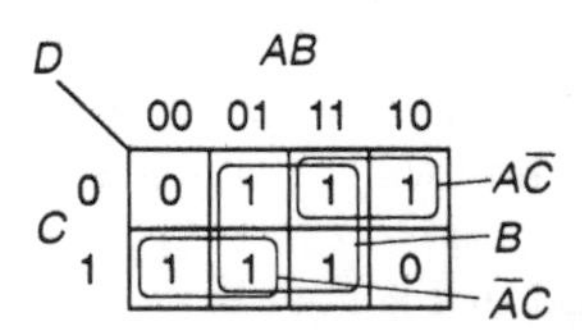

This gives an expression for D of the form

$$D = B + A\overline{C} + \overline{A}C$$

Note that the term $A\overline{B}\overline{C}$ is grouped with $AB\overline{C}$ even though the latter is contained within another group. This also applies to $\overline{A}\overline{B}C$ and $\overline{A}BC$. We always make the largest possible groups.

Example 9.11 Simplification of a logic expression using a Karnaugh map

The expression

$$E = \bar{A}B\bar{C}\bar{D} + AB\bar{C}\bar{D} + A\bar{B}\bar{C}\bar{D} + \bar{A}\bar{B}\bar{C}D + \bar{A}B\bar{C}D + AB\bar{C}D + A\bar{B}\bar{C}D + \bar{A}BCD + ABCD + \bar{A}BC\bar{D} + ABC\bar{D}$$

can be represented by a Karnaugh map as follows

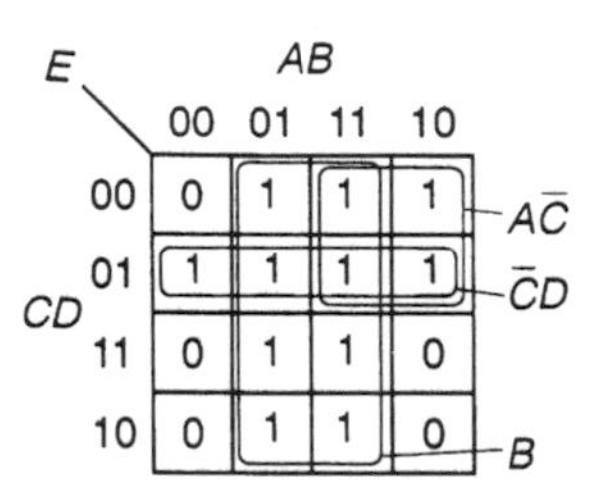

Here one element, $AB\bar{C}D$, is contained within three groups to obtain the simplest form. This is

$$E = B + A\bar{C} + \bar{C}D$$

Example 9.12 Simplification of a logic expression using a Karnaugh map

The algebraic expression

$$E = \bar{A}B\bar{C}\bar{D} + \bar{A}B\bar{C}D + AB\bar{C}D + A\bar{B}\bar{C}D + \bar{A}\bar{B}CD + \bar{A}BCD + ABCD + ABC\bar{D}$$

produces the following map

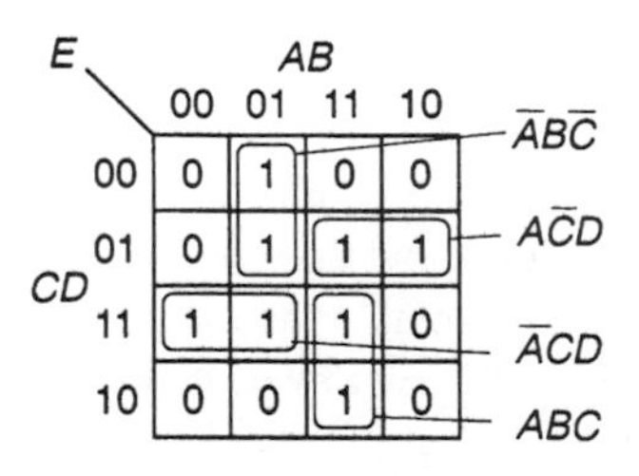

Here the largest possible grouping, formed by joining the four innermost cells, is not used. This is because the four groups shown are all essential to include the four outer '1's. This makes the inner group of four unnecessary, since all its components are already included in other groupings. This is an application of the third of our rules, which says that we must remove any redundant cells.

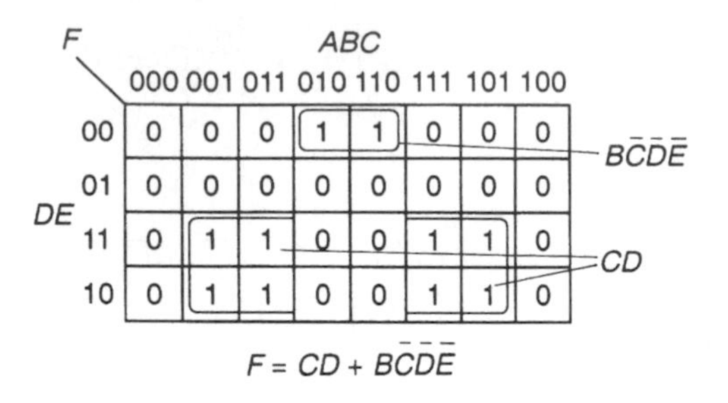

Figure 9.28 A Karnaugh map for a system with five input variables.

Karnaugh maps may also be used to simplify expressions with more than four input variables. The same basic rules apply but the topology of the map becomes more complicated. This is illustrated in Figure 9.28 which shows a five-variable map.

As the number of input variables becomes larger, this method of simplification becomes unwieldy and automated computer techniques are usually adopted. These are discussed briefly later in this section.

Don't care conditions

Don't care conditions occur when the state of an input or output variable is unimportant. In the case of an input variable this means that the output is the same whether the input is a '0' or a ' 1'. When an output variable has a *don't care* state, this means that the output state for that combination of inputs is unimportant. Often this is because that input combination will never occur. A don't care condition is represented within a truth table or a Karnaugh map by an '*X*', as illustrated in Figure 9.29.

Figures 9.29(a) and 9.29(b) show the use of don't care conditions for input variables. When used in this way they represent a *shorthand* method of representing a number of input combinations within a single line. This simplifies the truth table and aids comprehension. The truth table of Figure 9.29(a) has precisely the same meaning as that of Figure 9.29(b), but is easier to assimilate.

When used for output variables, don't care conditions are of greater significance. Here they mean that the output state is unimportant, leaving the designer with the freedom to choose which state the system should adopt. When using a truth table to represent an expression, as shown in Figure 9.29(c), it is not obvious how the don't care states should be assigned. The designer could simply choose to make them all '0' or all '1'. Alternatively, some could be set to '0' and some to '1'. However, it is not clear which choice will produce the simplest implementation.

One of the great strengths of Karnaugh maps is their ability to deal sensibly with *don't care* conditions. Figure 9.29(d) shows the Karnaugh map representation of the expression given in the truth table of Figure 9.29(c). From this it is apparent which don't care terms should be considered as '0' and which as '1', to produce the simplest expression. In this case $\overline{A}\overline{B}C$ should be chosen as a '0' and the remainder as '1's to produce the simplest grouping, as shown in Figure 9.29(e).

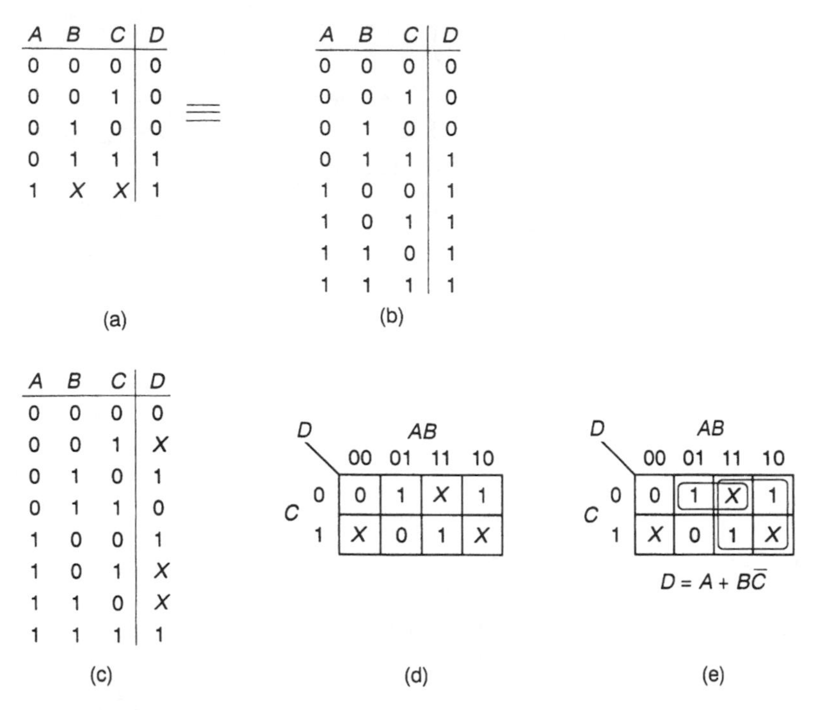

(a)

A	B	C	D
0	0	0	0
0	0	1	0
0	1	0	0
0	1	1	1
1	X	X	1

≡

(b)

A	B	C	D
0	0	0	0
0	0	1	0
0	1	0	0
0	1	1	1
1	0	0	1
1	0	1	1
1	1	0	1
1	1	1	1

(c)

A	B	C	D
0	0	0	0
0	0	1	X
0	1	0	1
0	1	1	0
1	0	0	1
1	0	1	X
1	1	0	X
1	1	1	1

Figure 9.29 The representation of 'don't care' conditions.

Automated methods of minimization

Although Karnaugh maps can be used fairly easily with up to six variables, with larger numbers it becomes impractical and other methods are required. For such problems it is normal to use a tubular method such as that developed by McCluskey (1956) from an original technique proposed by Quine (1952). This approach is universally known as **Quine–McCluskey minimization.**

The process of minimization is performed on a table of the input products (minterms) which is systematically reduced by examining each pair of terms to see if the Boolean simplification $AB + A\overline{B} = A$ can be applied. This process is applied exhaustively until a minimized expression is produced. The technique can also cater for 'don't care' conditions in a similar manner to that used with Karnaugh maps. The Quine–McCluskey method can handle systems with any number of inputs and can be performed by hand or, as is more usual nowadays, by computer. The algorithm is very simple to program and computers are now frequently used for all but the simplest minimization tasks.

Hazards

In our discussions so far we have assumed that our goal in manipulating logic functions is to implement the function with the smallest number of gates. Unfortunately this approach sometimes produces solutions that give rise to problems when they are

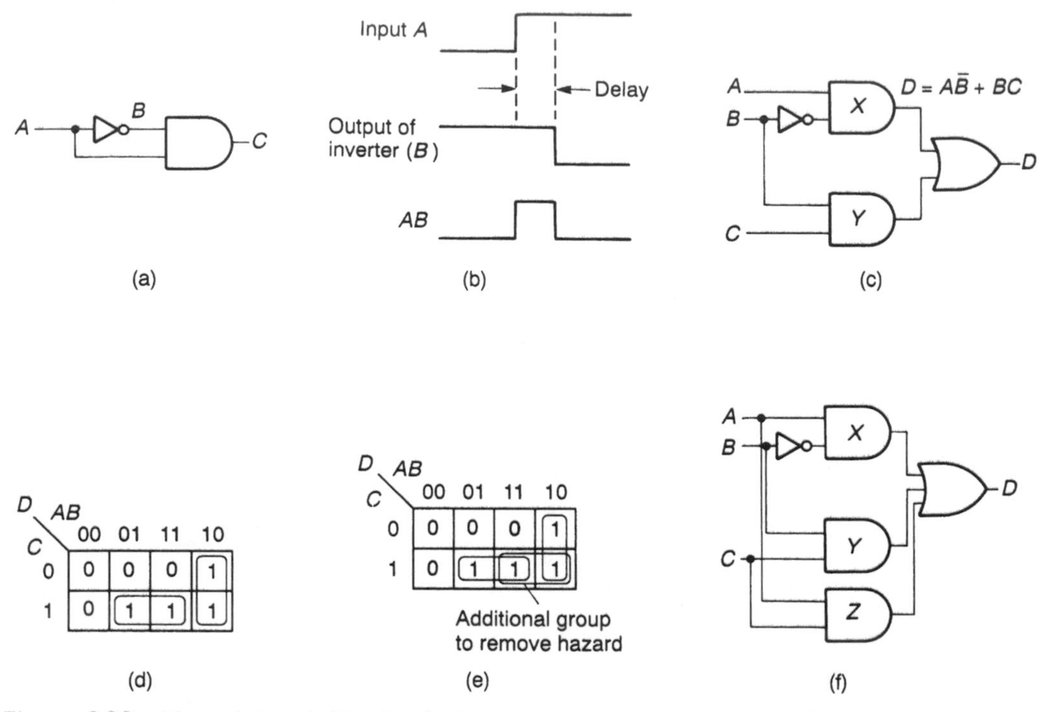

Figure 9.30 Hazards in combination logic.

constructed using physical circuits. This is because all real gates take a finite amount of time to operate and so produce a delay in the signals passing through them. The problems associated with this phenomenon are illustrated in Figure 9.30(a).

The circuit of Figure 9.30(a) represents the function

$$C = A \cdot B = A \cdot \overline{A} = 0$$

and thus the output C should always be 0. Figure 9.30(b) shows the response of the arrangement to a transition from 0 to 1 at the input. Because of the delay caused by the inverter, the AND gate sees a 1 on both inputs for a brief period and may, if it is sufficiently fast, produce an output pulse of logic 1, despite the fact that the logic function predicts that it will always be 0. This is an example of a **hazard**, a transient effect that generates unwanted transitions of the output.

The effects of hazards are further illustrated by the circuit of Figure 9.30(c). Here the output represents the function $A\overline{B} + BC$, which may be represented by the Karnaugh map of Figure 9.30(d). Let us consider the situation in which initially all three inputs are at 1. Under these circumstances both inputs to gate Y are at 1 and consequently the output D is at 1. It might be expected that taking B to 0 would have no effect on the circuit since with B at 0, both inputs to gate X are at 1, and again D is 1. However, the delay caused by the inverter results in the output of Y going to 0 before the output of X goes to 1. Consequently, for a brief period both inputs to the OR gate are 0 and the output D pulses low.

The solution to this problem can be understood by looking at the Karnaugh map of Figure 9.30(d). The conditions under which the output is high are represented by the two ringed groups. Each group is implemented by one of the AND gates in Figure 9.30(c). If either produces an output of 1, the output of the complete circuit will be high. As the pattern of inputs changes we move about the Karnaugh map producing a high or low output as appropriate. The hazard condition represents the situation in which the input combination jumps from one ring to another. While in either state the output is high, but during the brief period of transition the output is incorrect. The problem can be overcome by bridging the gap between the rings, as shown in Figure 9.30(e). The addition of an extra redundant term in no way changes the logic function, but it ensures that unwanted transitions do not occur as the inputs change between those representing the two groups. The resultant logic circuit is shown in Figure 9.30(f).

The importance of the presence of hazards within combinational logic differs between applications. In some situations they are unimportant since the system being driven is not sufficiently fast to respond to the transients produced. However, in many cases the elimination of hazards is essential, particularly when driving circuits that respond to transitions rather than levels (as in many of the sequential circuits to be discussed in the next chapter). The addition of extra gates clearly removes hazards at the expense of increased complexity. Some examples of hazard removal are given in Example 9.13.

Example 9.13 Removal of hazard conditions

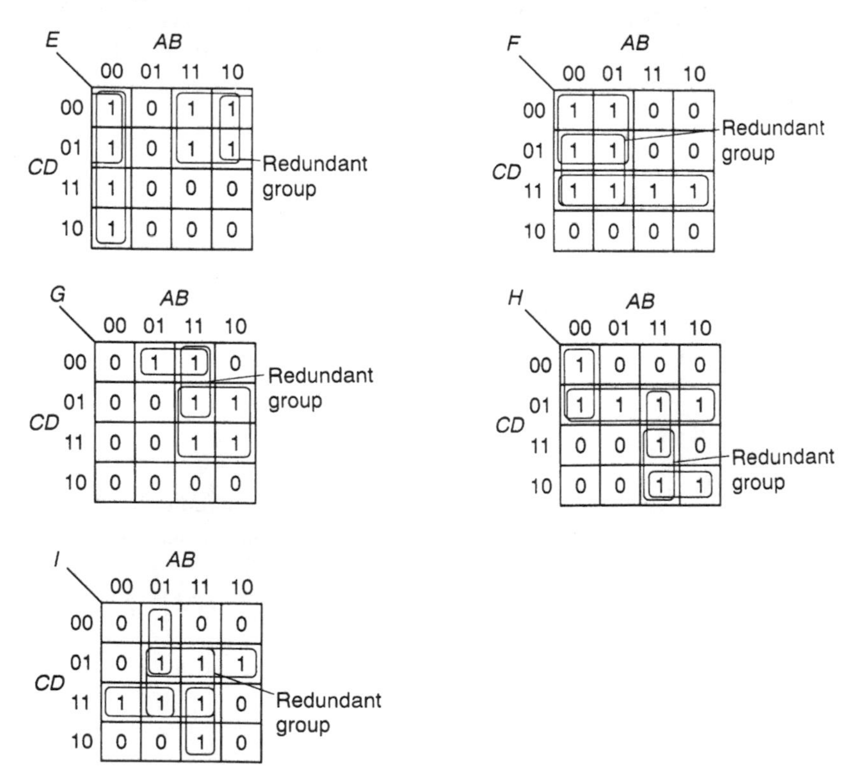

9.5 Number systems and binary arithmetic

So far we have dealt with simple binary signals, such as those produced by switches and those required to turn lamps ON or OFF. Sometimes groups of binary signals are combined to form **binary words**. These words can be used to represent various forms of information, the most common being *numeric* and *alphabetic* data. When numerical information is represented, this permits arithmetic operations to be performed on the data.

9.5.1 Number systems

The decimal number system

In everyday arithmetic, we use numbers with a base of ten, this choice being almost certainly related to the fact that we have ten fingers and thumbs. This system requires ten symbols to represent the values that each digit may take, for which we use the symbols $0, 1, 2, \ldots, 9$. Our numbering system is 'order dependent' in that the significance of a digit within a number depends on its position. For example, the number

$$1234$$

means 1 *thousand*, plus 2 *hundreds*, plus 3 *tens* plus 4 *units*. Each column of the number represents a power of ten, starting with units on the right-hand side ($10^0 = 1$), and moving to increasing powers of ten as we move to the left. Thus our number is

$$1234 = (1 \times 10^3) + (2 \times 10^2) + (3 \times 10^1) + (4 \times 10^0)$$

Digits at the left-hand side of the number are of much greater significance than those on the right-hand side. For this reason the left-hand digit is termed the **most significant digit (MSD)** while that on the right is termed the **least significant digit (LSD)**.

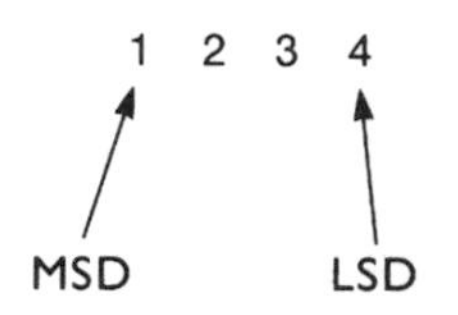

The numbering system can be extended to represent magnitudes that are not integer quantities by extending the sequence below the units column, a decimal point being placed to the right of the units column to indicate its position. Thus

$$1234.56 = (1 \times 10^3) + (2 \times 10^2) + (3 \times 10^1) + (4 \times 10^0) + (5 \times 10^{-1}) + (6 \times 10^{-2})$$

Numbers of any size can be represented by using a sufficiently large number of digits; leading zeros have no effect on the magnitude of the number, provided they are to the left of the decimal point. Similarly, trailing zeros have no effect if they are to the right of the decimal point.

The binary number system

Binary numbers have similar characteristics to decimal numbers other than that they have a base of two. Since each digit may now take only two values, only two symbols are required. These are usually 0 and 1. One advantage of this system is that digits can be represented by any binary quantity, such as a switch position or a lamp being ON or OFF.

Since binary quantities use symbols that are also used within decimal numbers, it is common to distinguish between them by adding a subscript indicating the base. Thus

$$1101_2$$

is a binary number, and

$$1101_{10}$$

is a decimal number. In many cases the base is known, or is obvious, in which case the subscript is usually omitted.

Like their decimal counterparts, the digits in a binary word are also position dependent. As before, the digits represent ascending powers of the base, such that

$$1101_2 = (1 \times 2^3) + (1 \times 2^2) + (0 \times 2^1) + (1 \times 2^0)$$

Therefore, rather than having units, tens, hundreds and thousands columns as in decimal numbers, we have 1s, 2s, 4s, 8s, 16s ... columns.

Fractional parts may also be represented, as

$$1101.01_2 = (1 \times 2^3) + (1 \times 2^2) + (0 \times 2^1) + (1 \times 2^0) + (0 \times 2^{-1}) + (1 \times 2^{-2})$$

The position of the units column is now indicated by a **binary point** (rather than a decimal point) and columns to the right of this point represent magnitudes of $1/2, 1/4, \ldots, 1/2^n$.

The term **bin**ary digi**t** is often abbreviated to **bit**. Thus a binary number consisting of eight digits would be referred to as an 8-bit number.

Other number systems

Although there are clear reasons for using both decimal and binary numbers, any integer may be used as the base of a number system. For reasons that are unimportant at this stage, common numbering systems include those using bases of 8 (octal) and 16 (hexadecimal or simply hex). Octal numbers require 8 symbols and use $0, 1, \ldots, 7$. Hexadecimal numbers require 16 symbols and use $0, 1, \ldots, 9$, A, B, C, D, E and F. From the above discussion of decimal and binary numbers it is clear that

$$123_8 = (1 \times 8^2) + (2 \times 8^1) + (3 \times 8^0)$$

and that

$$123_{16} = (1 \times 16^2) + (2 \times 16^1) + (3 \times 16^0)$$

Table 9.3 gives the numbers 0 to 20_{10} in decimal, binary, octal and hexadecimal.

Table 9.3 Number representations.

Decimal	Binary	Octal	Hexadecimal
0	0	0	0
1	1	1	1
2	10	2	2
3	11	3	3
4	100	4	4
5	101	5	5
6	110	6	6
7	111	7	7
8	1000	10	8
9	1001	11	9
10	1010	12	A
11	1011	13	B
12	1100	14	C
13	1101	15	D
14	1110	16	E
15	1111	17	F
16	10000	20	10
17	10001	21	11
18	10010	22	12
19	10011	23	13
20	10100	24	14

9.5.2 Number conversion

Conversion from binary to decimal

Converting binary numbers into decimal is straightforward. It is achieved simply by adding up the decimal values of each '1' in the number. For example

$$\begin{aligned} 11010_2 &= (1 \times 2^4) + (1 \times 2^3) + (0 \times 2^2) + (1 \times 2^1) + (0 \times 2^0) \\ &= 16 \quad + 8 \quad + 0 \quad + 2 \quad + 0 \\ &= 26_{10} \end{aligned}$$

For small numbers this conversion can be performed quite simply using mental arithmetic. Larger numbers take a little longer. Numbers with fractional parts can be converted in the same manner by adding the decimal equivalent of each term.

Conversion from decimal to binary

Conversion from decimal to binary is effectively the reverse of the above process, although the similarity is not at first apparent. It is achieved by repeatedly dividing the number by 2 and noting any remainder. This procedure is repeated until the number vanishes. This is best understood by considering an example.

Convert the number 26_{10} into binary.

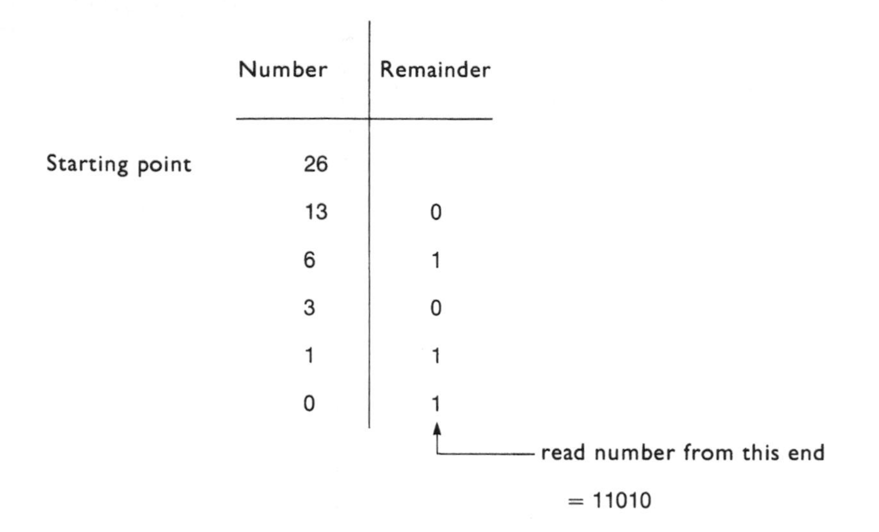

	Number	Remainder
Starting point	26	
	13	0
	6	1
	3	0
	1	1
	0	1

read number from this end

= 11010

Thus

$$26_{10} = 11010_2$$

Numbers with fractional parts are converted in parts, the integer part being converted as above and the fractional part being converted by repeated multiplication by two, noting, and then discarding, the overflow beyond the binary point after each multiplication. This is illustrated below.

Convert the number 34.6875_{10} into binary.

First 34 is converted as before

	Number	Remainder
Starting point	34	
	17	0
	8	1
	4	0
	2	0
	1	0
	0	1

number = 100010

then the fractional part 0.6875 is converted

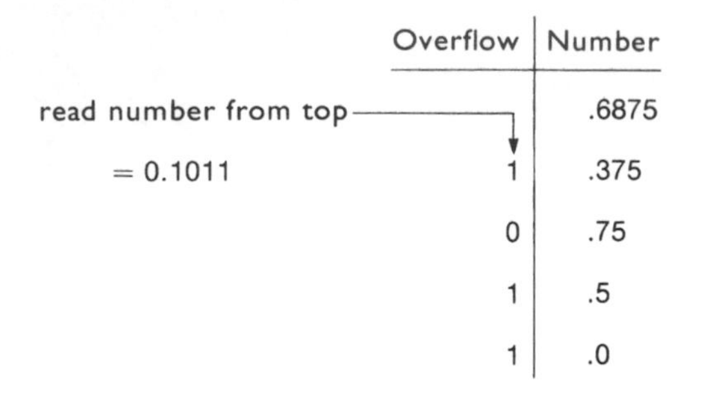

Thus

$$34.6875_{10} = 100010.1011_2$$

As with fractional parts in decimal numbers, the number of places used to the right of the binary point depends on the accuracy required.

Conversion from hexadecimal to decimal

Conversion from hexadecimal to decimal is similar to the conversion from binary to decimal, except that powers of 16 are used in place of powers of 2. For example

$$\begin{aligned} A013_{16} &= (A \times 16^3) + (0 \times 16^2) + (1 \times 16^1) + (3 \times 16^0) \\ &= (10 \times 4096) + (0 \times 256) + (1 \times 16) + (3 \times 1) \\ &= 40960 + 16 + 3 \\ &= 40979_{10} \end{aligned}$$

Conversion from decimal to hexadecimal

Conversion from decimal to hexadecimal is likewise similar to the conversion from decimal to binary, except that divisions and multiplications are by 16 rather than by 2. For example, consider the conversion of 7046_{10} into hexadecimal.

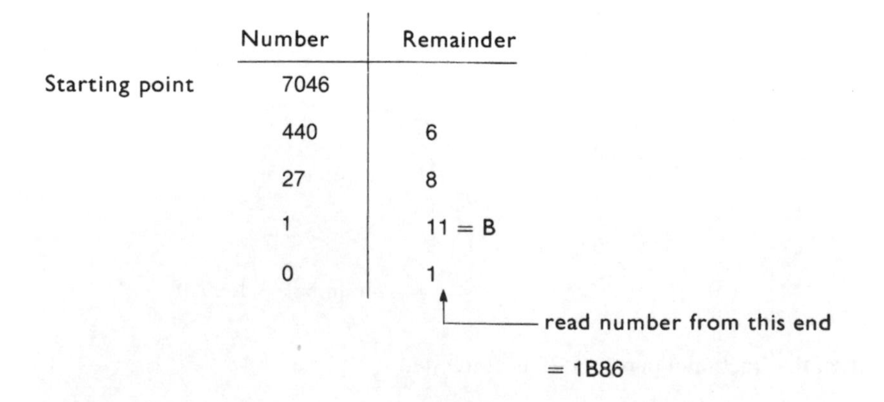

Thus

$$7046_{10} = 1B86_{16}$$

Conversions between other bases

Conversions between other number bases and decimal are similar to those described for binary and hexadecimal numbers using the appropriate power or multiplication factor. Conversion between two non-decimal number bases can be achieved using decimal as an intermediate stage. This involves converting a number in one number system into decimal form, and then converting it from decimal into the target number system.

It is also possible to convert directly between number systems, although this is sometimes tricky since most of us are strongly bound to thinking in decimal numbers. Examples of direct conversions that are easy to achieve include translations from binary to hexadecimal and vice versa. These conversions are straightforward because each hexadecimal digit corresponds to exactly four binary digits (4 bits). This allows each hexadecimal digit to be converted directly. All that is required is a knowledge of the binary equivalent of each of the 16 hexadecimal digits (as given in Table 9.3) and the translation is trivial. For example

$$\begin{aligned} F851_{16} &= (1111)(1000)(0101)(0001) \\ &= 1111100001010001_2 \end{aligned}$$

and

$$\begin{aligned} 111011011000100_2 &= (0111)(0110)(1100)(0100) \\ &= 76C4_{16} \end{aligned}$$

Note that when arranging binary numbers into groups of four for conversion into hexadecimal, the grouping begins with the right-most digit (the LSD) and extra leading zeros are added at the left-hand side as necessary.

From the above example it is clear that large binary numbers are unwieldy and difficult to remember. Because it is so easy to convert between binary and hexadecimal, it is very common to use the latter in preference to the former for large numbers. 76C4 is much easier to write and remember than 111011011000100.

9.5.3 Binary arithmetic

One of the many advantages of using binary rather than decimal representations of numbers is that arithmetic is much simpler. To see why this is true, one only has to consider that in order to perform decimal long multiplication one needs to know all the products of all possible pairs of the ten decimal digits. To perform binary long multiplication one only needs to know that $0 \times 0 = 0$, $0 \times 1 = 1 \times 0 = 0$ and that $1 \times 1 = 1$. This simplicity is a characteristic of all forms of binary arithmetic but for the moment we will consider only addition and subtraction.

Binary addition

The addition of two single-digit binary quantities is a very simple task, the rules of which may be summarized as

$$0+0=0$$
$$0+1=1$$
$$1+0=1$$
$$1+1=10$$

It can be seen that the addition of two single-digit numbers can give rise to a two-digit number. Therefore, if we construct a circuit to implement this function it must have the properties described in Figure 9.31. For reasons that will become apparent shortly, this arrangement is termed a **half adder**.

The block diagram of Figure 9.31(a) shows an arrangement with two inputs A and B and two outputs C (the **carry**) and S (the **sum**). The truth table of Figure 9.31(b) shows the states of the two outputs for all possible combinations of the inputs. This is the first example we have met of a system with more than one output. As you can see the truth table is similar to those we have used before, except that it has two output columns rather than one. The system could be described by two independent truth tables, one for each output, but it is easier to combine them.

From the truth table we can obtain Boolean expressions for the two outputs in terms of the input signals. These are

$$C=A\cdot B$$

and

$$S=\overline{A}B+A\overline{B}$$

You may recognize the expression for S as the Exclusive-OR function and therefore we could say

$$S=A\oplus B$$

We can represent the data contained in the truth table of Figure 9.31(b) by two Karnaugh maps, as shown in Figure 9.31(c), in an attempt to simplify the expressions obtained. However, as can be readily seen, the '1's cannot be combined and so no simplification is possible.

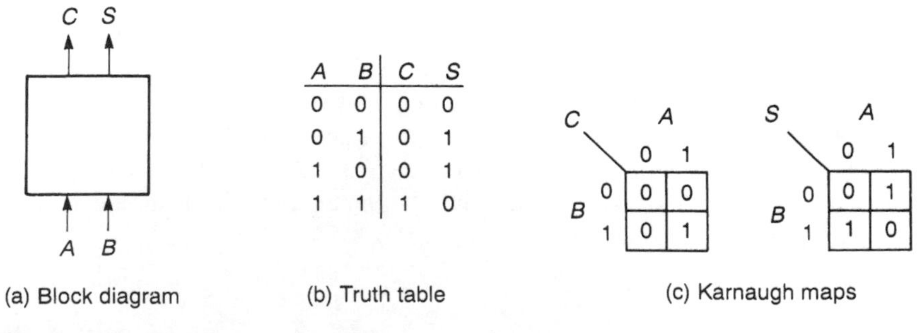

A	B	C	S
0	0	0	0
0	1	0	1
1	0	0	1
1	1	1	0

(a) Block diagram (b) Truth table (c) Karnaugh maps

Figure 9.31 A binary half adder.

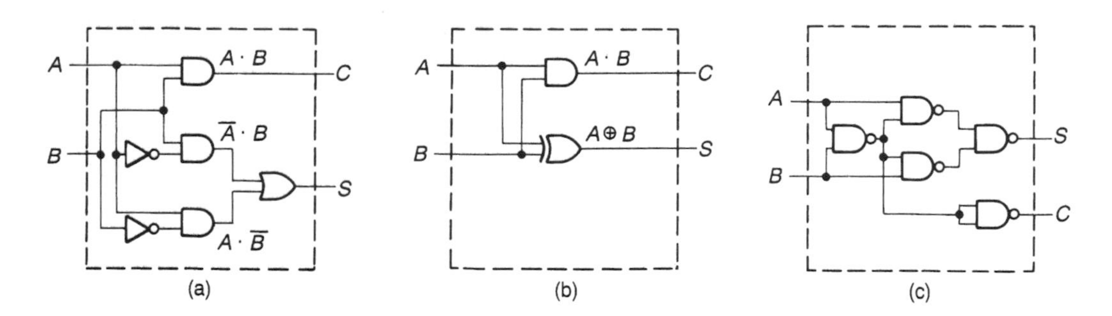

Figure 9.32 Implementation of the half adder.

The half adder can be implemented using simple gates, as shown in Figure 9.32(a), or using an Exclusive-OR gate, as shown in Figure 9.32(b).

While the circuit of Figure 9.32(a) provides the correct logic functions for a half adder, it has some problems when one considers its speed of operation. All electronic circuits take a finite time to respond to changes in their inputs. When considering logic gates, the time taken for the output to change as a result of a change in an input signal is termed the **propagation delay time** of the circuit. This delay time varies for different logic gates, but it is perhaps obvious that the time taken for a system to respond will depend, to a large extent, on the number of gates through which the signal passes on its route through the system. This leads to the concept of **logical depth**, which is the *maximum* number of simple gates through which a signal will pass between the input and the output.

If we look at the circuit of Figure 9.32(a) we see that the carry output C is separated from the inputs by only a single gate (a logical depth of 1) whereas the sum output S is controlled by signals that pass through three gates (a logical depth of 3). This difference in logical depth means that the carry output will respond before the sum output, which may cause problems in some situations. It might seem at first sight that the circuit of Figure 9.32(b) overcomes this problem since the number of gates in each path is equal. However, the S output is generated using an Exclusive-OR gate, which is not a simple gate but a collection of gates which itself has a logical depth of more than 1. The problem can be reduced by using the circuit of Figure 9.32(c) to implement the half adder. The logical depths of the C and S outputs are 2 and 3, respectively, which, although not equal, are closer than those of Figure 9.32(a). This implementation also has the advantage of using only one type of gate. You may like to convince yourself that the circuit of Figure 9.32(c) is functionally equivalent to those of Figures 9.32(a) and 9.32(b).

The design of a circuit to add together two 2-bit binary numbers could be tackled by treating it as a network with four inputs and generating an appropriate truth table and hence a logic circuit. However, this approach becomes unwieldy as we consider circuits to add together longer binary numbers. For example, an arrangement to add two 8-bit numbers would have 16 inputs, giving a truth table with over 65 000 rows! When *we* perform addition we add the digits separately and this seems a sensible approach in the design of a circuit to perform this task.

In order to add together multiple-digit numbers we need a circuit slightly more complicated than the simple half adder described above. This is because when adding all but the right-most digit, we need to cater for 'carries' from the previous digit. This can be illustrated by considering the addition of two decimal numbers

```
    1 6 5 4
+   2 6 3 7
-----------
    4 2 9 1
carry 1   carry 1
```

When adding the first (right-most) pair of digits, only these two numbers are summed. However, for all the following pairs any carry from the previous stage must also be added. In binary addition the process is similar.

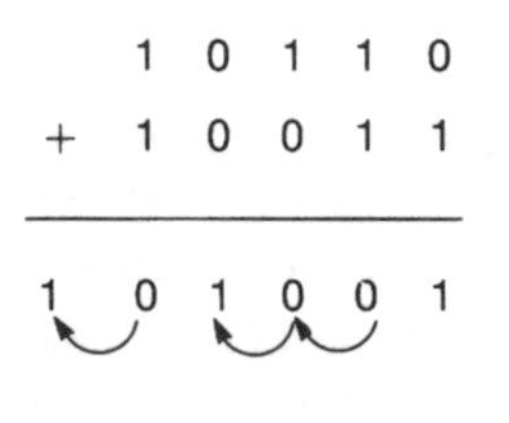

The half adder can be used to sum the right-most (least significant) digits of the numbers, but for all the following digits a circuit is required that can add together not only the two digits of the input numbers but also a carry from the previous stage. The carry from one stage to the next can be either 0 or 1. A circuit that can add together the two digits of a number and any carry input is called a **full adder** to distinguish it from the simpler half adder described earlier.

Figure 9.33 shows an arrangement for adding two 4-bit binary numbers $A_3A_2A_1A_0$ and $B_3\,B_2\,B_1\,B_0$ to give a 5-bit result $X_4X_3X_2X_1X_0$, where A_3, B_3 and X_4 represent the

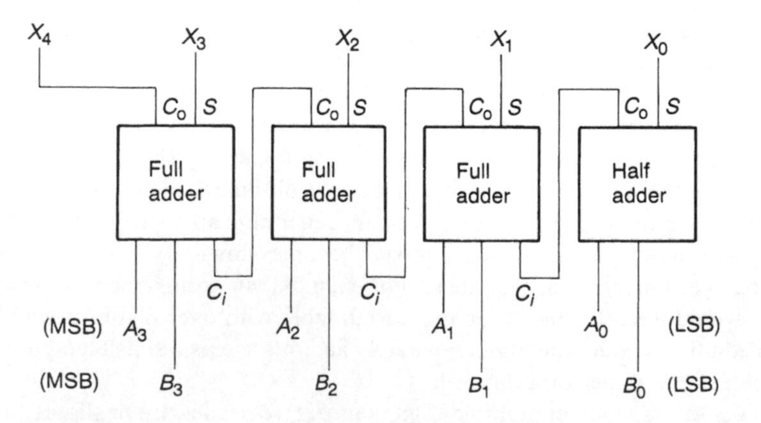

Figure 9.33 An arrangement to add two 4-bit numbers.

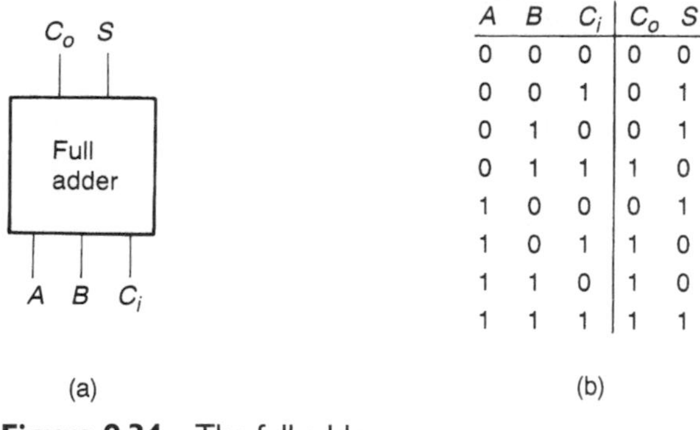

A	B	C_i	C_o	S
0	0	0	0	0
0	0	1	0	1
0	1	0	0	1
0	1	1	1	0
1	0	0	0	1
1	0	1	1	0
1	1	0	1	0
1	1	1	1	1

Figure 9.34 The full adder.

C_o: C_i \ AB	00	01	11	10
0	0	0	1	0
1	0	1	1	1

S: C_i \ AB	00	01	11	10
0	0	1	0	1
1	1	0	1	0

Figure 9.35 Representation of a full adder using Karnaugh maps.

most significant bits, and A_0, B_0 and X_0 represent the least significant bits. It is common to number the bits of binary numbers from the right and to start from 0. Thus an n-bit number has digits from 0 to $n - 1$.

It can be seen that each full adder has three inputs (A, B and the carry input C_i) and two outputs (the sum S and the carry output C_o). The function of the full adder is described by the truth table of Figure 9.34.

Boolean expressions can be obtained directly from this truth table and simplified using algebraic manipulation. Alternatively, the data can be represented using Karnaugh maps, as shown in Figure 9.35.

From either method of simplification we find that

$$C_o = AB + AC_i + BC_i$$

and

$$S = \overline{A}\overline{B}C_i + \overline{A}B\overline{C}_i + ABC_i + A\overline{B}\overline{C}_i$$

These functions can be implemented directly, as shown in Figure 9.36.

An alternative method of producing the function of a full adder is to break down its operation into two parts. Clearly adding the three components together is equivalent to adding two together and then adding the third. We can therefore produce the same effect using two half adders, as shown in Figure 9.37. The solution using two half adders requires less circuitry than that required to implement the circuit of Figure 9.36, but has a greater *logical depth* and is therefore slower to respond. The full adder has a logical depth of 3 for both outputs whereas the half adder approach has a logical depth of at least 6 for the sum output and at least 5 for the carry.

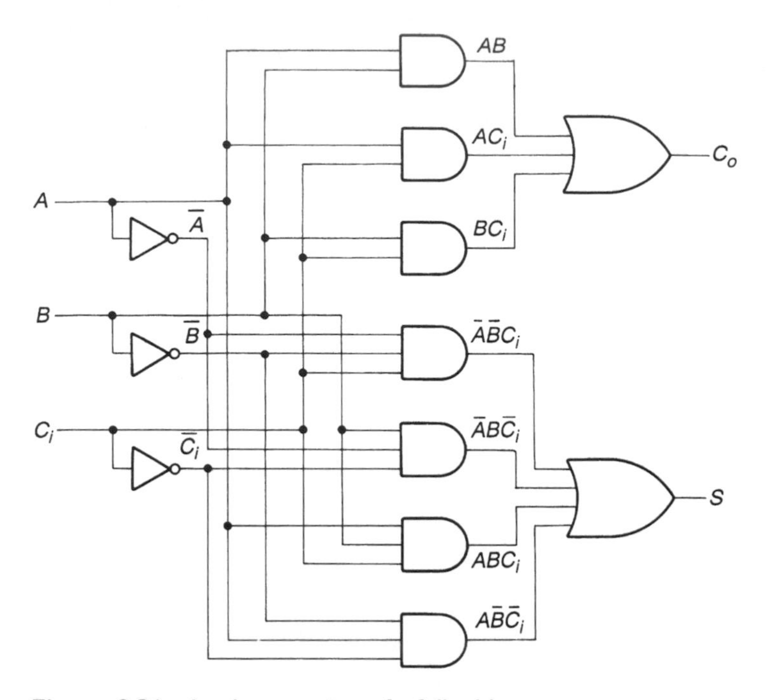

Figure 9.36 Implementation of a full adder.

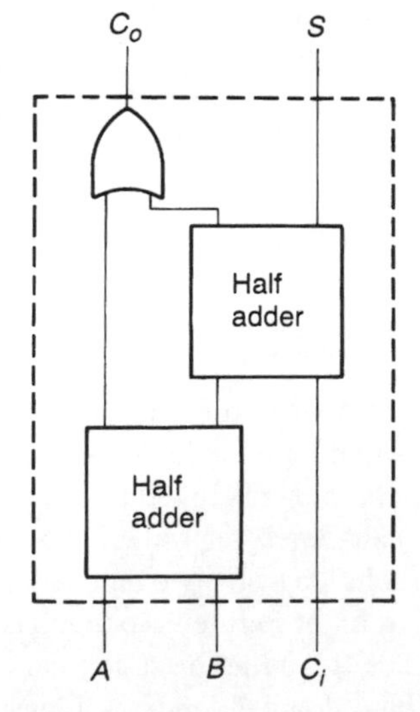

Figure 9.37 Forming a full adder from two half adders.

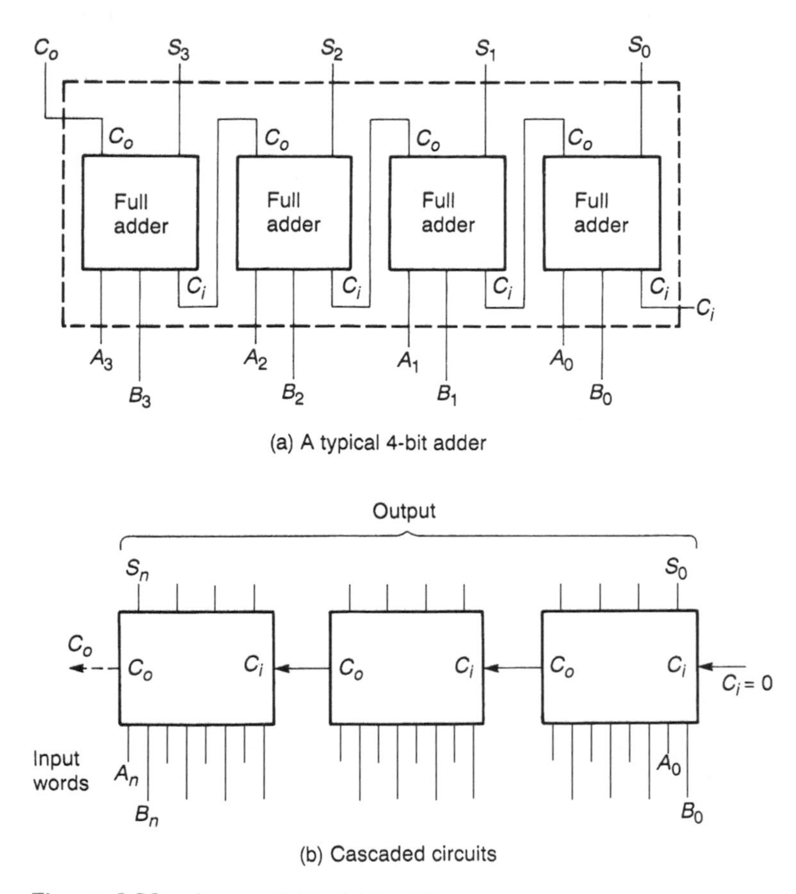

Figure 9.38 A cascadable 4-bit adder.

When adding together numbers of more than a few bits, it is unusual to construct adder networks directly from basic gates. Integrated circuits are available that provide a number of full adders within a single circuit, simplifying design and construction. A typical arrangement might incorporate circuitry to allow two 4-bit numbers to be added together. Such a circuit would be similar to that of Figure 9.33, but would normally use four full adders rather than three full adders and a half adder. This provides both a *carry in* and a *carry out* for the circuit, as shown in Figure 9.38(a). This allows a number of these circuits to be cascaded to permit binary words with any number of bits to be added. When used for the least significant (right-most) bits, the carry input is connected to '0' to indicate the absence of any carry in. This arrangement is shown in Figure 9.38(b).

Binary subtraction

The process of binary subtraction can be tackled in a similar manner to that of addition. We can construct a **half subtractor**, as shown in Figure 9.39(a), with a truth table as given in Figure 9.39(b). Since we are now concerned with subtraction rather than addition, we

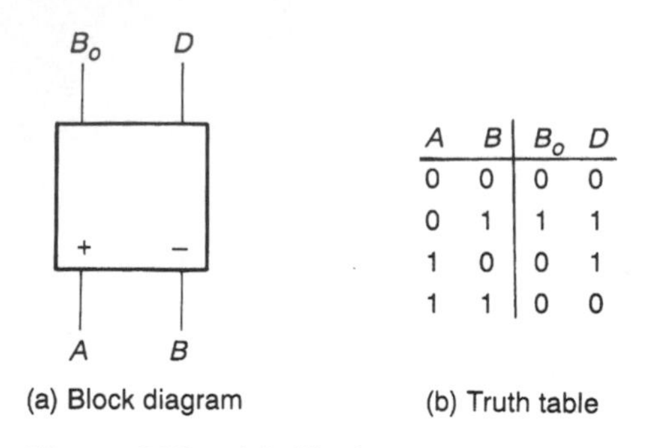

A	B	B_o	D
0	0	0	0
0	1	1	1
1	0	0	1
1	1	0	0

(a) Block diagram (b) Truth table

Figure 9.39 A half subtractor.

have *difference* (D) and *borrow* (B_o) outputs rather than *sum* and *carry*. It is also necessary to differentiate between the two inputs A and B to determine which is subtracted from which. In the example shown the output is equal to $(A-B)$.

From the truth table we can see that

$$B_o = \overline{A} \cdot B$$

and

$$D = \overline{A}B + A\overline{B} = A \oplus B$$

You will notice that D is identical to S for a half adder, but that the borrow output is not the same as the carry.

In order to perform multiple-bit subtraction we again need to consider the effect of one stage on the next. Figure 9.40 shows a 4-bit subtractor using four **full subtractors**. This circuit can be cascaded to allow larger numbers to be used.

Figure 9.41 shows the truth table for a full subtractor. The outputs can be represented by the Boolean expressions

$$B_o = \overline{A}B + \overline{A}B_i + BB_i$$

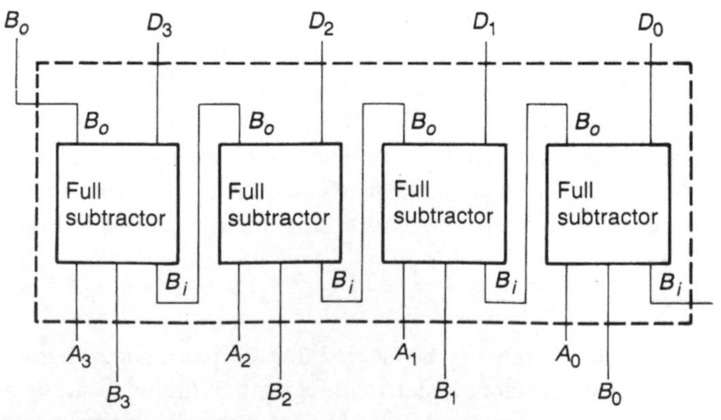

Figure 9.40 A 4-bit subtractor.

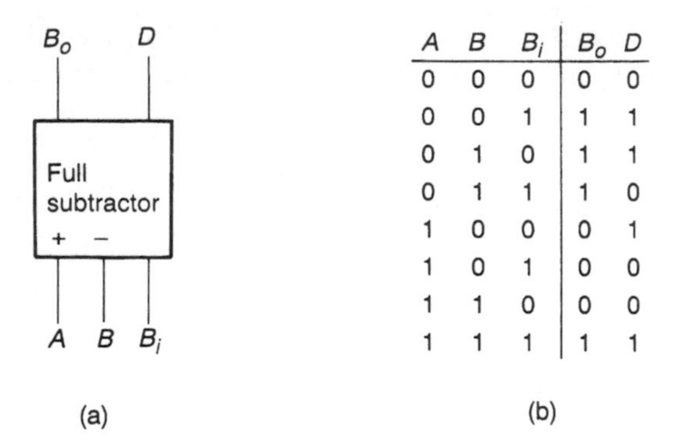

A	B	B_i	B_o	D
0	0	0	0	0
0	0	1	1	1
0	1	0	1	1
0	1	1	1	0
1	0	0	0	1
1	0	1	0	0
1	1	0	0	0
1	1	1	1	1

Figure 9.41 A full subtractor.

and

$$D = \overline{A}\overline{B}B_i + \overline{A}B\overline{B}_i + A\overline{B}\overline{B}_i + ABB_i$$

These functions can be implemented directly, as described earlier for the full adder, or constructed from two half subtractors as shown in Figure 9.42.

The subtractor circuits described above work as we would expect, provided that the result is not negative. We will leave further discussion of the representation of negative numbers until we deal with arithmetic within microprocessors in Chapter 12.

Binary multiplication and division

Although it is possible to construct circuits to perform multiplication and division using simple logic gates, it is fairly unusual as the complexity of the circuits makes them

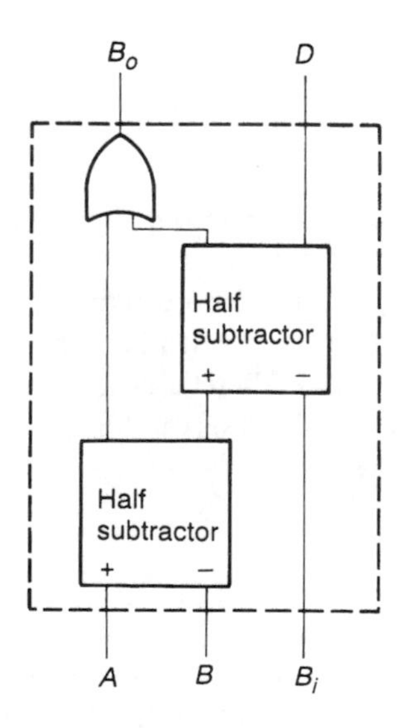

Figure 9.42 Constructing a full subtractor from two half subtractors.

impractical. It is more common to perform these functions using dedicated logic circuits containing a large number of gates, or to use a microprocessor. We will leave discussion of this topic until Chapter 12.

9.5.4 Numeric and alphabetic codes

Binary code

By far the most common method of representing numeric information within digital systems is by the use of the simple binary code described earlier. This has the advantages of simplicity of arithmetic and efficiency of storage. However, there are some applications in which other representations are used for specific purposes.

Binary coded decimal (BCD) code

Binary coded decimal (BCD) code, as its name implies, is formed by converting each digit of a decimal number individually into its binary form. For example

$$9450_{10} = (1001)(0100)(0101)(0000)_{BCD}$$

Note that this is *not* the same as the straight binary equivalent of the number, since

$$9450_{10} = 10010011101010_2$$

It is apparent from this example that the BCD form of the number requires more digits than the straight binary form and is therefore less efficient.

Conversion from decimal to BCD can be performed simply by converting each digit into decimal, as illustrated above. Conversion from BCD to decimal is just as simple and is achieved by dividing the number into groups of four, starting with the least significant digit, and then converting each digit into decimal. Additional leading zeros can be used to complete the last group, if required. For example

$$\begin{aligned} 11100001110110_{BCD} &= (0011)(1000)(0111)(0110)_{BCD} \\ &= 3876_{10} \end{aligned}$$

BCD has the advantage of very simple conversions to, and from, decimal. It is therefore widely used in situations where input and output data are in a decimal form, such as in pocket calculators. However, as we have seen, it requires more digits of storage than simple binary numbers and is therefore unattractive for applications in which large amounts of data must be stored, as within computers. Arithmetic is also more complicated in BCD than in binary. For these reasons it is usually reserved for specialist applications involving digital input and output operations.

Gray code

We have already met the Gray code when, in the last section, we discussed Karnaugh maps. There we wished to number the various elements of the map so that the codes for adjacent elements varied in only one bit. Gray code has this property, as is illustrated in Table 9.4.

Table 9.4 Gray code.

Decimal	Gray code
0	0000
1	0001
2	0011
3	0010
4	0110
5	0111
6	0101
7	0100
8	1100
9	1101
10	1111
11	1110
12	1010
13	1011
14	1001
15	1000

As with simple binary numbers, the sequence can be continued to represent arbitrarily large numbers and leading zeros have no effect. At first sight the order may seem rather strange, but there is a simple and systematic method of producing the sequence, removing the need to memorize it. This is done by first writing down the first two numbers (these are simply 0 and 1 so are not difficult to remember) and then writing these numbers down again in reverse order with a 1 in front. This gives the sequence

$$\begin{array}{r} 0 \\ 1 \\ 11 \\ 10 \end{array}$$

This sequence is then again repeated in reverse order with a 1 in front to give

$$\begin{array}{r} 0 \\ 1 \\ 11 \\ 10 \\ 110 \\ 111 \\ 101 \\ 100 \end{array}$$

The process is repeated as often as is necessary, each time repeating the previous sequence, in reverse order, with the addition of a leading 1. This method of generating the sequence gives rise to its alternative name, which is **reflected binary code**.

In addition to its use within Karnaugh maps, Gray code is found in numerous applications where changing quantities are to be read. The reason for this can be illustrated by considering an imaginary transducer which produces simple binary code as its output. Let us imagine that the output from the device at a particular time changes from 7_{10} to 8_{10}. This represents an output change from 0111_2 to 1000_2. If we now consider some external device connected to the transducer, it is interesting to note the effect of reading the output from the transducer at the exact instant that its value changes. In changing from 0111 to 1000, all four digits change. If the digits are read while they are changing, the value obtained is indeterminate. If, for example, the leading 0 happened to change to a 1 slightly faster than the '1's changed to '0's, the value could be read as 1111 (15_{10}). Alternatively, if the leading 0 turned to 1 slightly slower than the '1's changed to '0's, the number could be read as 0000 (0_{10}). This means that any combination of the four digits could be obtained, giving any number in the range 0 to 15.

If Gray code is used as the output from the transducer this problem cannot occur. In changing from 7_{10} to 8_{10} the output changes from 0100 to 1100. Since only one digit is changed, reading the output during this transition can only lead to one digit being uncertain. Thus the number read will be either 0100 or 1100. Therefore the number read will always be either the old or the new number – an ideal arrangement. Since all adjacent numbers in Gray code differ by only one digit, this property exists for all transitions.

Gray code is widely used in **counters** that must be read asynchronously (counters will be discussed in Chapter 10) as well as in absolute position encoders, as discussed in Section 2.3.4. If you look at the pattern of stripes on the encoder of Figure 2.10 you will see that this is in Gray code.

ASCII code

So far we have concentrated on codes that are used to represent numeric quantities. Often it is also necessary to store and transmit alphabetic data in digital form, for example, for storing text within a computer or word processing system. Various standard codes are used for this purpose, but by far the most widely used is the **American Standard Code for Information Interchange** which is normally abbreviated to **ASCII** (pronounced 'ass-key').

The full standard represents each character by a 7-bit code, allowing 128 possible values. Codes are defined for: both upper and lower case alphabetic characters; the digits 0 to 9; punctuation marks such as commas, full stops and question marks; and various non-printable codes which are used as control characters. These control characters include codes to produce a line feed, a carriage return and a backspace on a printer. Since codes are included for both alphabetic and numeric characters, codes of this form are often referred to as **alphanumeric** codes. It should be noted, however, that the numeric codes represent *numeric characters* – not the corresponding quantities. A partial listing of the ASCII character set is given in Table 9.5.

Error detection and correction techniques

We have seen in earlier chapters that all electronic systems suffer from noise. One possible effect of noise within digital systems is the corruption of data. This is a particular problem when data must be transmitted from one place to another.

One of the simplest methods of tackling the problem of errors is to use **parity** testing. This is done by adding a small amount of redundant information to each word of data to

Table 9.5 A partial listing of the ASCII character set.

Character	7-bit ASCII	Hex	Character	7-bit ASCII	Hex
A	100 0001	41	0	011 0000	30
B	100 0010	42	1	011 0001	31
C	100 0011	43	2	011 0010	32
D	100 0100	44	3	011 0011	33
E	100 0101	45	4	011 0100	34
F	100 0110	46	5	011 0101	35
G	100 0111	47	6	011 0110	36
H	100 1000	48	7	011 0111	37
I	100 1001	49	8	011 1000	38
J	100 1010	4A	9	011 1001	39
K	100 1011	4B	blank	010 0000	20
L	100 1100	4C	!	010 0001	21
M	100 1101	4D	"	010 0010	22
N	100 1110	4E	#	010 0011	23
O	100 1111	4F	$	010 0100	24
P	101 0000	50	%	010 0101	25
Q	101 0001	51	&	010 0110	26
R	101 0010	52	'	010 0111	27
S	101 0011	53	(	010 1000	28
T	101 0100	54	)	010 1001	29
U	101 0101	55	*	010 1010	2A
V	101 0110	56	+	010 1011	2B
W	101 0111	57	,	010 1100	2C
X	101 1000	58	-	010 1101	2D
Y	101 1001	59	.	010 1110	2E
Z	101 1010	5A	/	010 1111	2F
a	110 0001	61	:	011 1010	3A
b	110 0010	62	;	011 1011	3B
c	110 0011	63	<	011 1100	3C
d	110 0100	64	=	011 1101	3D
e	110 0101	65	>	011 1110	3E
f	110 0110	66	?	011 1111	3F
g	110 0111	67	[	101 1011	5B
h	110 1000	68	\	101 1100	5C
i	110 1001	69	]	101 1101	5D
j	110 1010	6A	^	101 1110	5E
k	110 1011	6B	_	101 1111	5F
l	110 1100	6C	{	111 1011	7B
m	110 1101	6D	|	111 1100	7C
n	110 1110	6E	}	111 1101	7D
o	110 1111	6F	~	111 1110	7E
p	111 0000	70	delete	111 1111	7F
q	111 0001	71	bell	000 0111	07
r	111 0010	72	backspace	000 1000	08
s	111 0011	73	carriage return	000 1101	0D

Continued

Table 9.5 (continued)

Character	7-bit ASCII	Hex	Character	7-bit ASCII	Hex
t	111 0100	74	escape	001 1011	1B
u	111 0101	75	form feed	000 1100	0C
v	111 0110	76	line feed	000 1010	0A
w	111 0111	77	horizontal tab	000 1001	09
x	111 1000	78	vertical tab	000 1011	0B
y	111 1001	79	start text	000 0010	02
z	111 1010	7A	end text	000 0011	03

allow it to be checked. The extra information takes the form of a parity bit which is added at the end of each data word. The polarity of the added bit is chosen so that the total number of '1's within the word (including the added parity bit) is either always even (*even parity*) or always odd (*odd parity*). For example, consider an even parity system. The ASCII character for 'S' is

1	0	1	0	0	1	1

which has an even number of '1's. Therefore a '0' parity bit is added to make an 8-bit word which still has an even number of '1's.

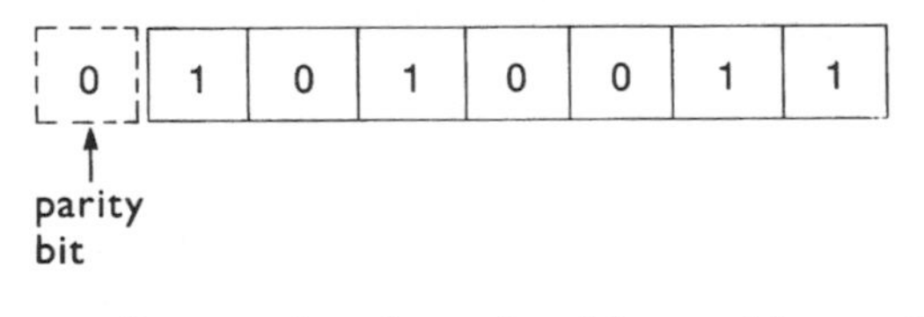

On reception, the parity of the word is tested by counting the number of '1's. If it is still even, the parity bit is removed and the original seven bits are passed to their destination. If the parity is incorrect on reception, an error has been detected and the system must take appropriate action. Although this technique indicates that an error has occurred, it cannot determine which bit or bits are incorrect. It should also be noted that if two errors are present the parity of the resultant word will again be even and the errors will not be detected. This simple error detecting technique will detect any odd number of errors but will not detect an even number of errors. Random numbers thus have a 50% chance of passing the parity test. Parity testing is often used on communications channels where it is used to give confidence that the line is working correctly. Although the reliability of testing any one word is low, when applied to a large number of words it is sure to detect errors if the line is unreliable.

An alternative method of checking the correctness of data is to use a **checksum**. This provides a test of the integrity of a block of data rather than of individual words. When a group of words is to be transmitted the words are summed at the transmitter and the sum is transmitted after the data. At the receiver the words are again summed and the result compared with the sum produced by the transmitter. If the results agree the data is probably correct. If they do not, an error has been detected. As with the parity check, the test gives no indication as to the location of the error but simply indicates that one has occurred. The action taken depends on the nature of the system. It might involve sending the data again or sounding an alarm to warn an operator.

The parity and checksum techniques both send a small amount of redundant information to allow the integrity of the data to be tested. If one is prepared to send additional redundant information, it is possible to construct codes that not only detect the presence of errors but also indicate their location within a word, allowing them to be corrected. An example of this technique is the well-known **Hamming code** (Hamming, 1950). The performance of these codes in terms of their ability to detect and correct multiple errors depends on the amount of redundant information that can be tolerated. The more redundancy that is incorporated, the greater is the rate at which data must be sent and the more complicated the system. It should also be remembered that it is not possible to construct a code that will allow an unlimited number of errors. This would imply that the system could produce the correct output with a random input – clearly an impossibility.

9.6 Combinational logic

Simple logic gates and the circuits we have so far considered can be described as **combinational logic**. By this we mean that the output is determined only by the current states of the various inputs. In other words, it is the particular combination of input signals that determines the output.

We have already looked at several design tools for use with combinational logic including Boolean algebra, truth tables and Karnaugh maps. We now look at a few examples to illustrate their use.

Example 9.14 Design a circuit to convert 3-bit binary numbers into Gray code

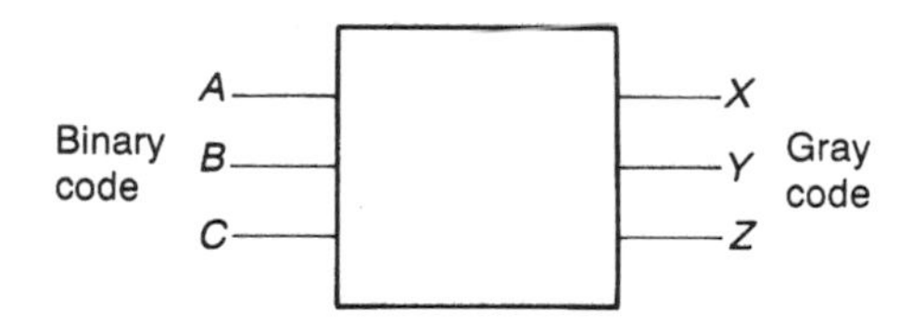

The circuit can be represented by the following truth table

A	B	C	X	Y	Z
0	0	0	0	0	0
0	0	1	0	0	1
0	1	0	0	1	1
0	1	1	0	1	0
1	0	0	1	1	0
1	0	1	1	1	1
1	1	0	1	0	1
1	1	1	1	0	0

and from this we can construct three Karnaugh maps

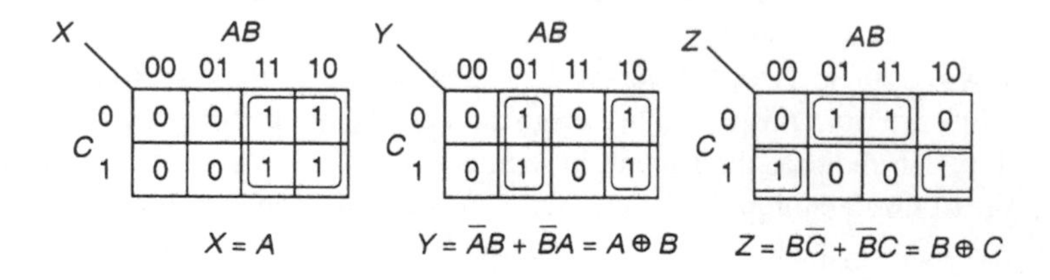

The circuit may then be implemented using standard gates

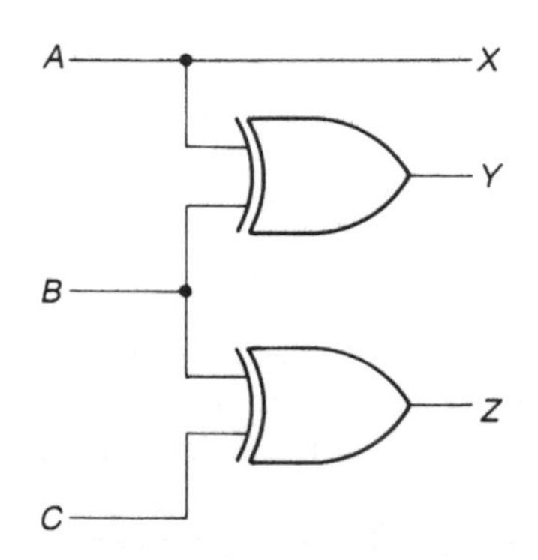

Example 9.15 Design a circuit to take a 4-bit number *ABCD* and produce a single output *Y* that is true only if the input represents a prime number

The circuit may be represented by a truth table as follows

Decimal	A	B	C	D	Y
0	0	0	0	0	1
1	0	0	0	1	1
2	0	0	1	0	1
3	0	0	1	1	1
4	0	1	0	0	0
5	0	1	0	1	1
6	0	1	1	0	0
7	0	1	1	1	1
8	1	0	0	0	0
9	1	0	0	1	0
10	1	0	1	0	0
11	1	0	1	1	1
12	1	1	0	0	0
13	1	1	0	1	1
14	1	1	1	0	0
15	1	1	1	1	0

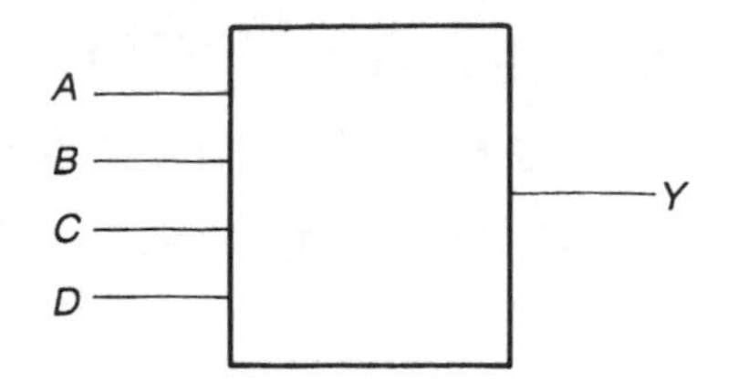

The truth table can be used to form a Karnaugh map

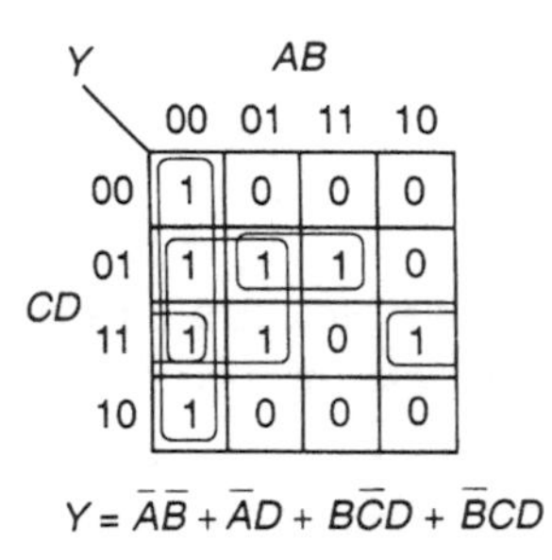

and the circuit may be implemented using standard gates.

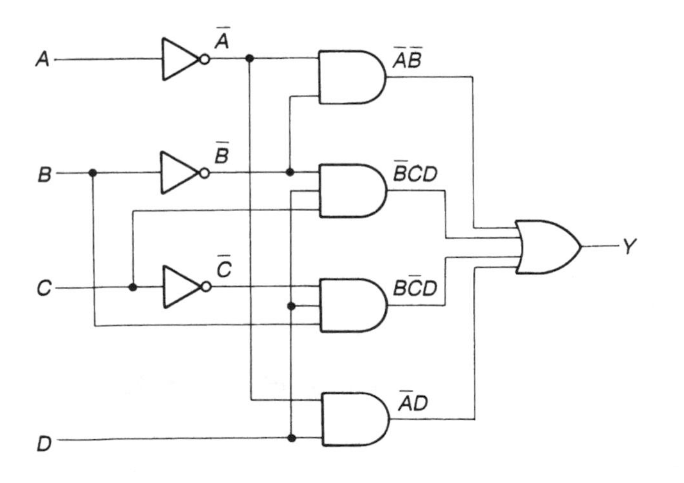

Example 9.16 Design a circuit to take a BCD number *ABCD* and produce a single output that is true only when the input corresponds to the numbers 1, 2, 5, 6 or 9

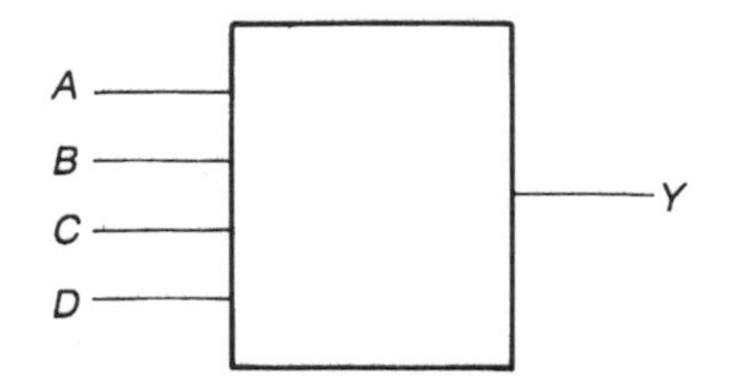

The circuit may be represented by a truth table as follows. Note that since the input is a BCD number, certain combinations of the input variables cannot occur. This means that the output state is *don't care* for these conditions.

Decimal	A	B	C	D	Y
0	0	0	0	0	0
1	0	0	0	1	1
2	0	0	1	0	1
3	0	0	1	1	0
4	0	1	0	0	0
5	0	1	0	1	1
6	0	1	1	0	1
7	0	1	1	1	0
8	1	0	0	0	0
9	1	0	0	1	1
10	1	0	1	0	X
11	1	0	1	1	X
12	1	1	0	0	X
13	1	1	0	1	X
14	1	1	1	0	X
15	1	1	1	1	X

From this a Karnaugh map can be formed and a simplified expression for *Y* obtained.

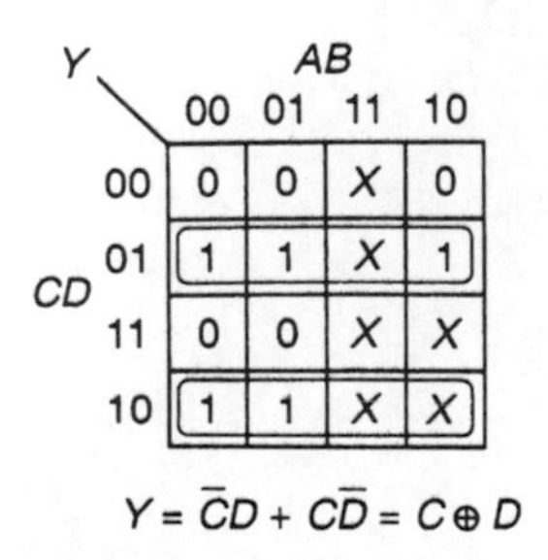

$$Y = \bar{C}D + C\bar{D} = C \oplus D$$

This can then be implemented directly.

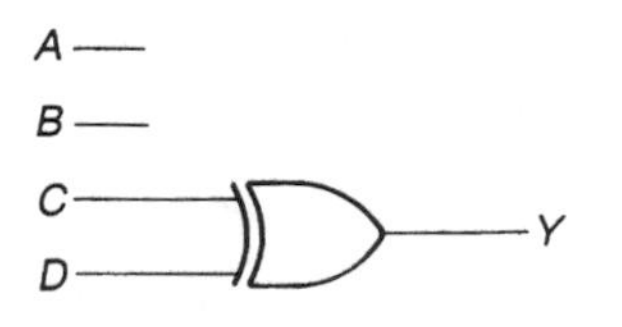

Example 9.17 Design a 4-input multiplexer

A **multiplexer** is a circuit that can perform the function of a multi-way switch by selecting one of a number of input signals and passing this to a single output line. A block diagram of a 4-input multiplexer is shown below.

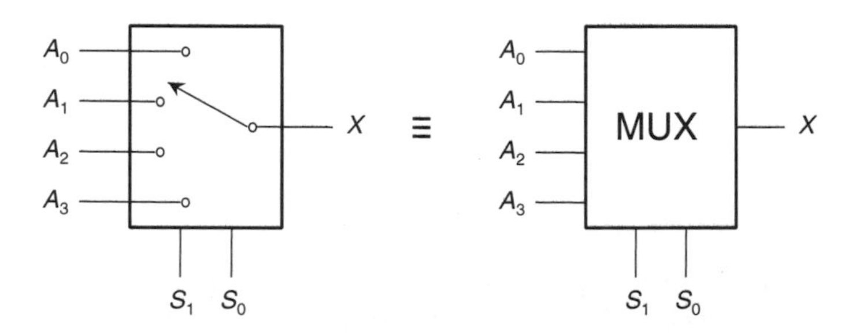

The two control inputs S_1 and S_0 are line select inputs and the signals on these inputs determine which of the four data input lines A_0 to A_3 is selected. The output X then becomes equal to the binary signal on the selected input line. The operation of the multiplexer can therefore be described by the following truth table:

S_0	S_1	X
0	0	A_0
0	1	A_1
1	0	A_2
1	1	A_3

The logic required to implement this function can be designed by treating the multiplexer as a circuit with six inputs (A_0 to A_3, plus S_0 and S_1) and drawing the necessary truth table. This can then be simplified to give the required logic. The required truth table has 64 lines and the construction of this table, and the necessary simplification, is left as an exercise for the reader.

An alternative approach to implementation is to break the design down into two components – the select logic and the gating logic.

If we look at the truth table of an AND gate we see that any signal ANDed with a '0' gives a '0', while any signal ANDed with a '1' gives the original signal. This allows us to use an AND gate as a gating network. Our multiplexer can therefore be constructed as shown below.

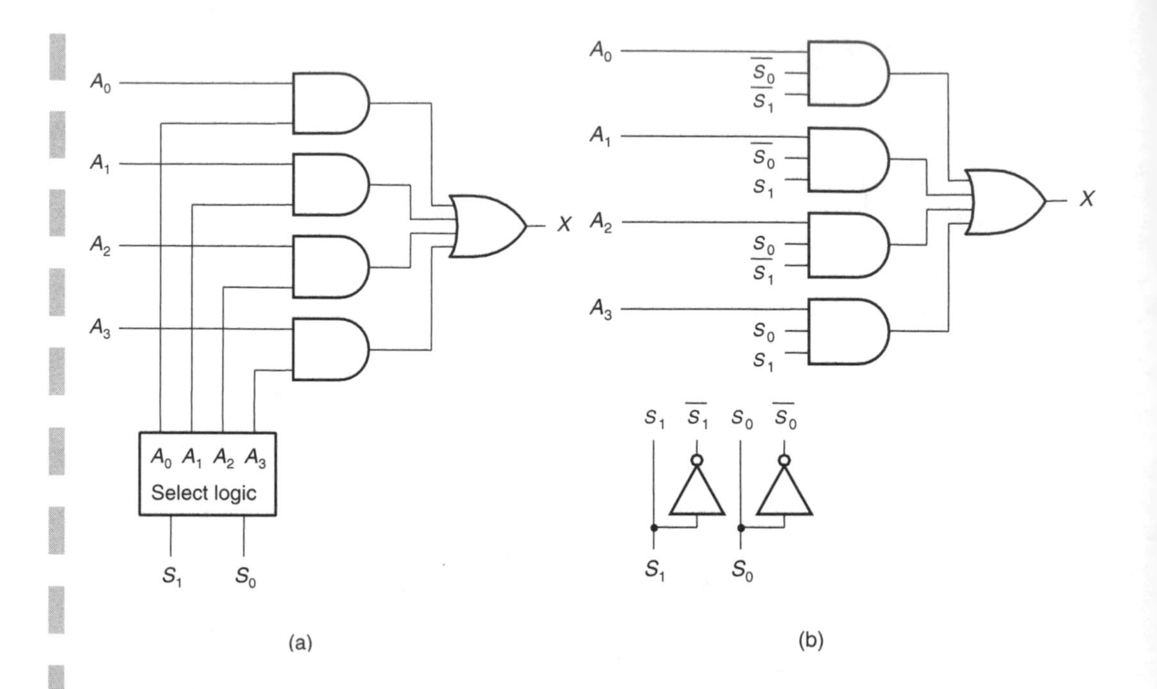

In diagram (a) the 'select logic' block produces a '1' on the appropriate select line and a '0' on the remainder. This 'enables' the AND gate connected to the appropriate input line and 'disables' the rest. The outputs from the three disabled AND gates will all be '0' and so will have no effect on the OR gate. If the output from the selected AND gate is also a '0' (because the selected input signal is '0') then the output from the OR gate will be '0'. However, if the output from the selected AND gate is a '1' (because the selected input signal is '1') then the output from the OR gate will be a '1'. Therefore the output X will always be equal to the value of the selected input signal.

If we now turn our attention to the form of the 'select logic' we see that this is very simple. Line A_0 should be selected if both S_1 and S_0 are '0' and so the appropriate select line can be formed by ANDing together the inverted form of these two signals. Similarly, the select signal for A_1 can be formed by ANDing S_1 with the inverse of S_0, and so on. Thus our four select lines may be formed by individually ANDing together the four combinations of the two select lines and their inverses. In fact, since the select lines are themselves ANDed with the data input lines, a single AND gate may be used to select each line as shown in diagram (b) above.

This implementation of the multiplexer can be obtained using a truth table and Karnaugh maps as with earlier circuits, but the example illustrates that this may not always be the simplest approach. If the problem had been to design a multiplexer with eight data lines this would have required a truth table with 11 inputs (eight data lines plus three select lines). Such a truth table has over 2000 lines. However, the approach described above can easily be extended to a circuit with eight data lines.

The circuit described in this example takes 'logical' signals on its inputs and uses them to determine the state of its output. Such a circuit is often called a **digital multiplexer**. Circuits are also available that can perform the same function with analogue as well as digital signals. Such components are described as **analogue multiplexers**.

A circuit related to the multiplexer is the **demultiplexer**. This takes a single input signal and uses it to determine the state of one of a number of output lines, under the control of an appropriate number of line select lines.

Key points

- Digital systems of one form or another are playing an increasingly important role in all aspects of everyday life.
- To simplify the description of binary variables it is common to represent their two states by the symbols '1' and '0'. These might represent ON and OFF, TRUE and FALSE or any other pair of binary conditions.
- In some simple cases it is possible to implement binary systems using switches. However, it is more generally useful to design such systems using logic gates.
- Our basic building blocks are a small number of simple gates. Three elementary forms, AND, OR and NOT, can be used to form any logic function, although it is often more useful to work with compound gates such as NAND, NOR and Exclusive-OR.
- Combinational logic circuits can be described by a truth table which lists all the possible combinations of the inputs and indicates the corresponding values of the outputs.
- It is also possible to define a logic function using Boolean algebra. This notation and set of rules and identities allows binary relationships to be described and simplified.
- Simplification may also be performed graphically using Karnaugh maps.
- In addition to binary variables, digital systems often use many-valued quantities which are represented by binary words of an appropriate length.
- In digital electronics several number systems are used, the most common being decimal, binary, octal and hexadecimal.
- Since binary numbers use only two digits, 0 and 1, arithmetic is simpler than in decimal.
- Although simple binary code is the most common way of representing numeric information it is not the only method. In some applications the use of other representations, such as Gray code, may be more appropriate.
- Codes are also used for non-numeric information, such as the ASCII code, which is used for alphanumeric data.
- Some coding techniques allow error detection and possible correction.

Design study

In Section 2.4.2 we looked at a seven-segment LED display. This device has seven inputs which can be used to illuminate the seven elements of the display. Design a BCD to seven-segment decoder which takes a 4-bit input representing a number in the range 0 to 9, and generates an appropriate 7-bit output to illuminate the appropriate elements of the display to represent this digit.

Approach

We can represent the required system by the following block diagram.

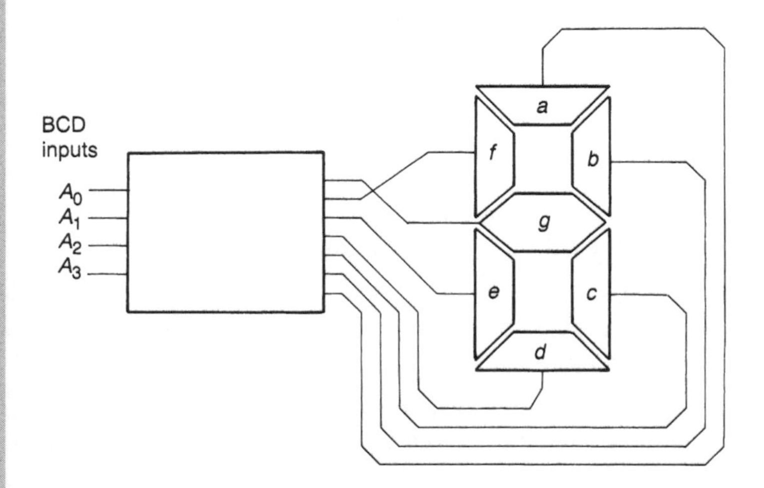

If we adopt the digit representations given in Figure 2.14(b), we can describe the required system by the following truth table.

Number	A_3	A_2	A_1	A_0	a	b	c	d	e	f	g
0	0	0	0	0	1	1	1	1	1	1	0
1	0	0	0	1	0	1	1	0	0	0	0
2	0	0	1	0	1	1	0	1	1	0	1
3	0	0	1	1	1	1	1	1	0	0	1
4	0	1	0	0	0	1	1	0	0	1	1
5	0	1	0	1	1	0	1	1	0	1	1
6	0	1	1	0	1	0	1	1	1	1	1
7	0	1	1	1	1	1	1	0	0	0	0
8	1	0	0	0	1	1	1	1	1	1	1
9	1	0	0	1	1	1	1	0	0	1	1
10	1	0	1	0	X	X	X	X	X	X	X
11	1	0	1	1	X	X	X	X	X	X	X
12	1	1	0	0	X	X	X	X	X	X	X
13	1	1	0	1	X	X	X	X	X	X	X
14	1	1	1	0	X	X	X	X	X	X	X
15	1	1	1	1	X	X	X	X	X	X	X

The outputs corresponding to the input numbers 10 to 15 are don't care conditions since these input combinations will not occur. The seven outputs can be represented by a series of Karnaugh maps, as follows.

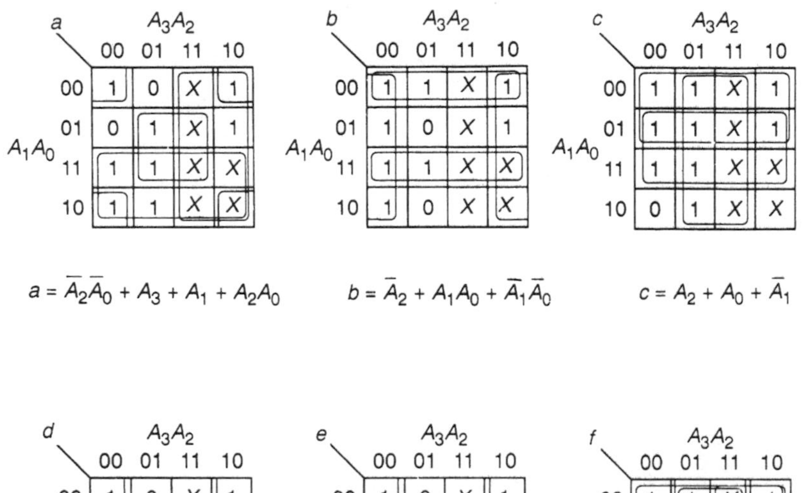

$a = \bar{A}_2\bar{A}_0 + A_3 + A_1 + A_2A_0$ $\quad b = \bar{A}_2 + A_1A_0 + \bar{A}_1\bar{A}_0$ $\quad c = A_2 + A_0 + \bar{A}_1$

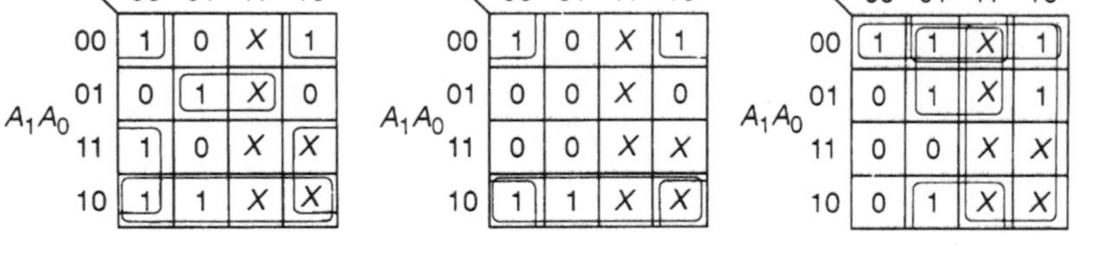

$d = \bar{A}_2\bar{A}_0 + \bar{A}_2A_1 + A_1\bar{A}_0 + A_2\bar{A}_1A_0$ $\quad e = \bar{A}_2\bar{A}_0 + A_1\bar{A}_0$ $\quad f = A_3 + \bar{A}_1\bar{A}_0 + A_2\bar{A}_1 + A_2\bar{A}_0$

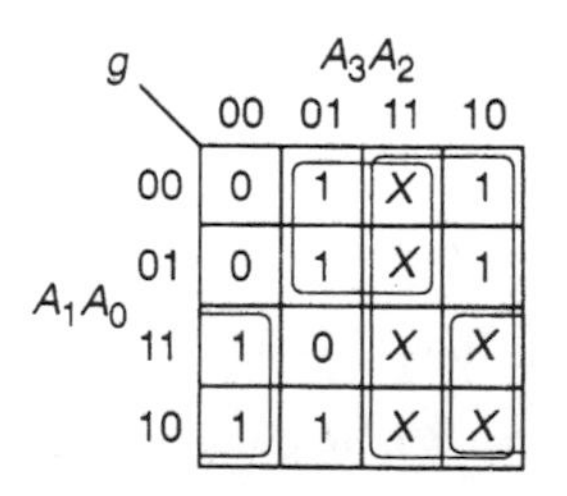

$g = A_3 + A_2\bar{A}_1 + \bar{A}_2A_1 + A_1\bar{A}_0$

The various outputs can then be generated using simple logic gates, as shown below. For clarity the interconnections between the inputs and the logic gates are not shown but are simply indicated by their functional names.

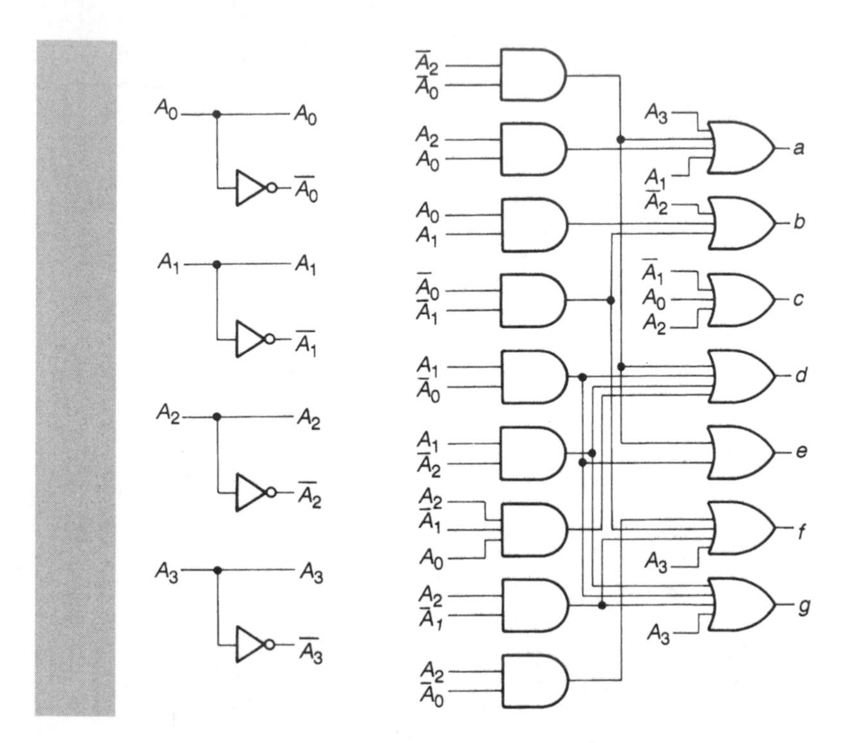

References

Boole G. (1995; originally published 1854) *An Investigation of the Laws of Thought*. New York: Dover Publishing

Hamming R. W. (1950) Error-detecting and error-correcting codes. *Bell Syst. Tech. J.* **29**, 147–60

Karnaugh M. (1953) The map method for synthesis of combination logic circuits. *Trans. Am. Inst. Elect. Engrs Comm. Electron.* **72**, 593–9

McCluskey E. (1956) Minimization of Boolean functions. *Bell Syst. Tech. J.* **35**, 1417–44

Quine W. V. (1952) The problem of simplifying truth functions. *Am. Math. Mon.* **59**, 521–31

Shannon C. E. (1938) A symbolic analysis of relay and switching circuits. *Trans. Am. Inst. Elect. Engrs.* **57**, 713–23

Further reading

Floyd T. L. (1997) *Digital Fundamentals*, 6th edn. Englewood Cliffs, NJ: Prentice-Hall

Mano M. M. (1990) *Digital Design*, 2nd edn. Englewood Cliffs, NJ: Prentice-Hall

Exercises

9.1 Show how a power source, a lamp and a number of switches can be used to represent the following logical functions

$$L = A \cdot B \cdot C$$
$$L = A + B + C$$
$$L = (A \cdot B) + (C \cdot D)$$
$$L = A \oplus B$$

9.2 Derive Boolean algebraic expressions to describe the following circuits

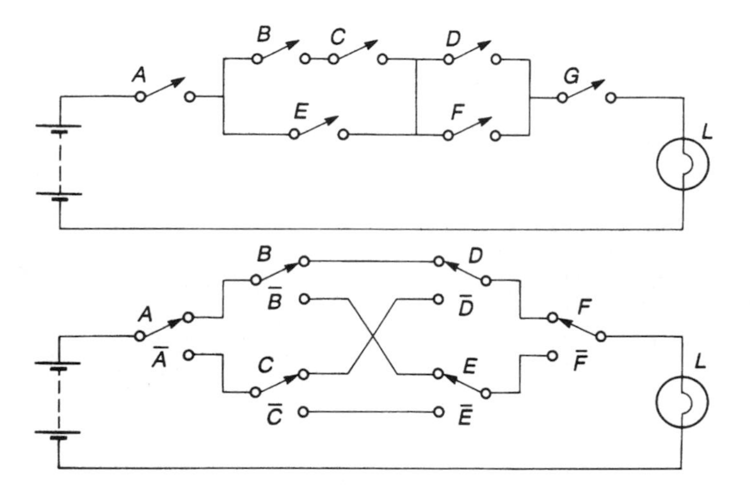

9.3 If the two circuits given in the last exercise were described by truth tables, how many lines would each table require?

9.4 Show that the two circuits (a) and (b) below are equivalent by (i) drawing truth tables for each circuit, and (ii) using Boolean algebra.

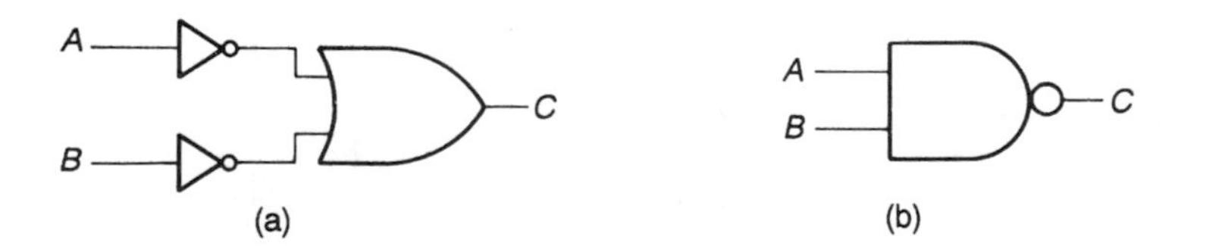

(a) (b)

9.5 Repeat the operations of Exercise 9.4 for the following circuits

(a) (b)

9.6 Derive Boolean expressions to describe the operation of the following circuit. Minimize these expressions by algebraic manipulation and hence simplify the circuit.

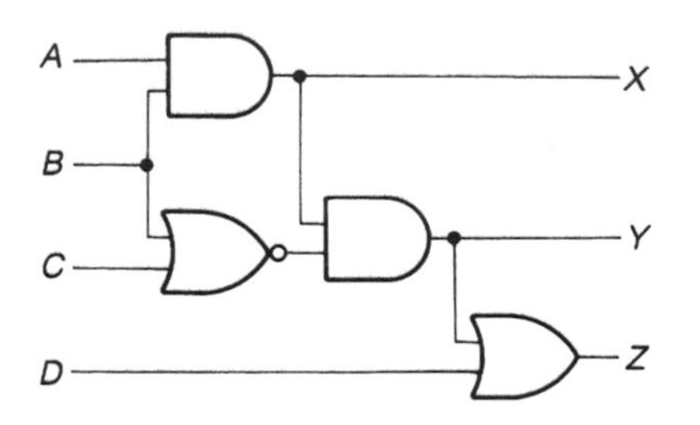

9.7 Use Karnaugh maps to obtain minimized Boolean expressions for the following functions

$$X = \overline{A}\,\overline{B} + A\overline{B}\,\overline{C} + A\overline{B}C + ABC$$

$$Y = \overline{A}\,\overline{B}\,\overline{C} + \overline{A}B\overline{C}D + A\overline{C}\,\overline{D} + A\overline{C}D + A\overline{B}C\overline{D}$$

A	B	C	D	Z
0	0	0	0	1
0	0	0	1	0
0	0	1	0	X
0	0	1	1	0
0	1	0	0	1
0	1	0	1	X
0	1	1	0	1
0	1	1	1	1
1	0	0	0	X
1	0	0	1	0
1	0	1	0	1
1	0	1	1	0
1	1	0	0	0
1	1	0	1	1
1	1	1	0	0
1	1	1	1	X

9.8 Convert the following numbers into decimal

$$110101_2,\ 754_8,\ A10E_{16}$$

9.9 Convert the following numbers into binary

$$67_{10},\ 3.625_{10},\ 635_8,\ 8FE_{16}$$

9.10 Convert the following numbers into hexadecimal

$$48602_{10},\ 307_8,\ 1100101_2$$

9.11 Perform the following binary arithmetic

$$\begin{array}{r}10111\\+1001\\\hline\end{array}\qquad\begin{array}{r}110101\\-11010\\\hline\end{array}\qquad\begin{array}{r}1011\\\times111\\\hline\end{array}\qquad\begin{array}{r}101010\\\div110\\\hline\end{array}$$

9.12 Design a circuit to convert 3-bit Gray code numbers into simple binary.

9.13 Design an 8-input digital multiplexer along the lines of the circuit described in Example 9.17. The circuit should have eight data inputs, three line select inputs and a single output.

9.14 Simulate your solution to the previous exercise to confirm that the circuit functions as expected.

9.15 Design a 4-output digital demultiplexer. The circuit should have one data input, four data outputs and two select inputs.

9.16 Simulate your solution to the previous exercise to confirm that the circuit functions as expected.

Sequential Logic

Objectives

When you have studied the material in this chapter you should:

- be familiar with a wide range of sequential logic circuits and appreciate the distinction between level sensitive, edge sensitive and pulse-triggered devices;
- be aware of the use of flip-flops in the design of a variety of registers and both asynchronous and synchronous counters;
- be familiar with the use of integrated circuit building blocks in the construction of registers and counters of various sizes;
- be able to design synchronous sequential circuits to meet specific requirements.

Contents

10.1 Introduction

We have seen that in combinational logic the outputs are determined only by the current states of the inputs. In **sequential logic**, however, the outputs are determined not only by the current inputs but also by the sequence of inputs that led to the current state. In other words, the circuit has the characteristic of **memory**.

Figure 10.1 shows a generalized sequential system which combines combinational logic with some form of memory within a feedback path. The memory elements are devices that can store binary information; the output at any time is determined by the present inputs and the data stored in the memory elements. The information stored within the memory determines the **state** of the circuit at any time, while the next state of the system is determined by the present state and the inputs. Thus the operation of the arrangement is determined by the existing inputs and the sequence of inputs that preceded them.

Sequential circuits can be divided into those which are *synchronous* and those which are *asynchronous*. In **synchronous systems** the inputs, outputs and internal states are sampled at definite instants of time, the timing of this process being controlled by a **clock** signal. In such systems it is common for many circuits to be controlled by a single clock signal such that their operations are synchronized. In **asynchronous systems** the circuitry responds to changes in the inputs at any time. The effects of input changes thus propagate throughout the system, each circuit adding its own delay. Thus in asynchronous sequential systems, internal states and output variables can be updated at any time.

Although many electronic circuits are sequential in nature, the most common sequential building blocks are the various types of **multivibrator**. This classification covers circuits of many different forms which are characterized by having two outputs which are the inverse of each other, and zero, one or more inputs. The outputs are usually given the labels Q and $\overline{Q}$. Having only these two outputs means that the circuits have only two possible output states, namely $Q=1$, $\overline{Q}=0$ and $Q=0$, $\overline{Q}=1$. Different forms of multivibrator are defined by the behaviour of the circuits in these two states. Three basic types are possible.

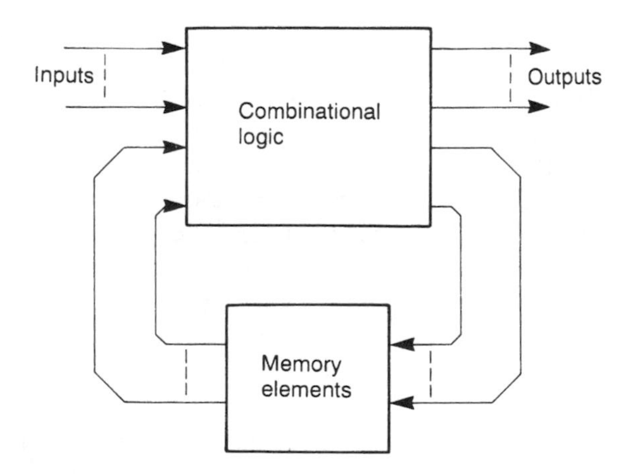

Figure 10.1 A generalized sequential system.

- **Bistable multivibrators** in which both output states are stable. When in one state the circuit will remain in that state until an input signal causes it to change state. There are several types of bistable multivibrator but unfortunately there is no general agreement on which names should be used for these classes of device. Some engineers refer to all bistable devices as *flip-flops* while others use the term *latch* for level sensitive devices and *flip-flop* for edge-triggered and pulse-triggered devices (the meanings of these terms will be explained shortly). Here we will adopt the latter terminology since it gives additional information about the form of the device.
- **Monostable multivibrators** in which one state is stable and the other is meta-stable (or quasi-stable). The circuit will remain in its stable state until acted upon by an appropriate input signal, whereupon it will change to its meta-stable state. It will remain in its meta-stable state for a fixed period of time (determined by circuit parameters) and then will automatically revert to its stable state. The circuit behaves as a single pulse generator. When *triggered* it enters its meta-stable state, causing the outputs to change for a fixed period of time. This circuit is also known as a **one-shot**.
- **Astable multivibrators** in which both states are meta-stable. The circuit stays in each state for a fixed period of time (determined by circuit parameters) before switching to its other state. This produces a circuit that continually oscillates from one state to the other – a digital oscillator.

Figure 10.2 illustrates some of the many forms of multivibrator, several of which will be discussed in this chapter. The figure shows examples of the symbols used for multivibrators and includes samples of the waveforms produced.

10.2 Level sensitive bistables or latches

Consider the circuit of Figure 10.3. The figure shows two inverters connected in a ring. If the output of the first inverter Q is equal to 1, this signal is fed to the input of the second inverter making its output P equal to 0. This in turn forms the input to the first inverter which makes its output 1. Thus the circuit is stable with $Q=1$ and $P=0$. Alternatively, if Q is equal to 0, this corresponds to a stable state with $Q=0$ and $P=1$. The circuit therefore has two stable states. It also has two outputs Q and P, where $P=\overline{Q}$. We could therefore consider the circuit to be a form of bistable multivibrator. This arrangement is an example of **regenerative switching** in which the output of one stage is amplified and fed back to reinforce that output signal, forcing the circuit into one state or the other.

Although the circuit of Figure 10.3 has the characteristics of a bistable it is of little practical use. Its state is determined when power is applied and it then remains in that state until power is removed.

10.2.1 The S–R latch

The arrangement becomes more interesting if we substitute two input NOR gates for the inverters, as shown in Figure 10.4.

We now have a circuit with two input signals R and S and two outputs which are

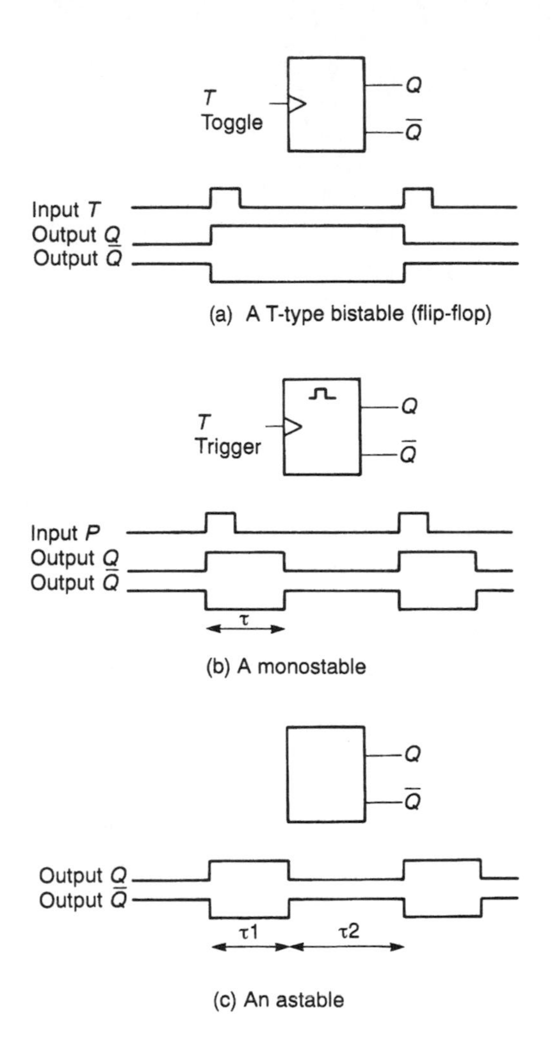

Figure 10.2 Examples of different forms of multivibrator.

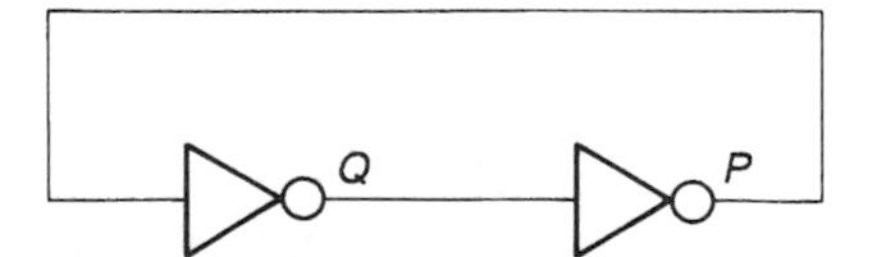

Figure 10.3 A regenerative switching circuit.

now labelled Q and $\overline{Q}$. If one input of a two-input NOR gate is held at 0, the relationship between the other input and the output is that of an inverter. Therefore, if R and S are both held at 0, the circuit behaves in the same manner as the previous circuit and will stay in whichever state it finds itself. We could call this condition the **memory**

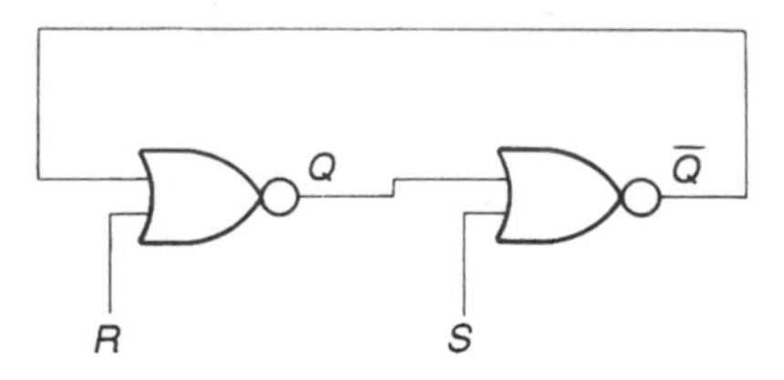

Figure 10.4 A latch formed from two NOR gates.

mode of the circuit. If now R is taken to 1 while S remains at 0, Q will be reset to 0 regardless of its previous state. In turn, this will set $\overline{Q}$ to 1. If now R is returned to 0, the circuit will re-enter its memory mode and will stay in this state. Similarly, if S is taken to 1 while R remains at 0, $\overline{Q}$ will be cleared to 0 and Q will be set to 1. Again, if S returns to 0, the circuit will re-enter its memory mode and will stay in this state. Thus the R input RESETS Q to 0 and the S input SETS Q to 1, while the other input is at 0. When both inputs are at 0 the circuit remembers the last state in which it was placed. This circuit is called a **SET–RESET latch** or simply an **S–R latch**. It should be noted that the condition $S = R = 1$ results in both outputs being 0. Under these circumstances the two outputs are no longer the inverse of each other and the circuit is not functioning as a bistable. For this reason, this combination of inputs is generally prohibited.

We can represent the action of an S–R latch using a truth table, which is now often called a **transition table** since it indicates the transitions between states.

S	R	Q_n	$\overline{Q}_n$	
0	0	Q_{n-1}	$\overline{Q_{n-1}}$	No change
0	1	0	1	RESET
1	0	1	0	SET
1	1	0	0	Ambiguous state

where Q_n represents the state of the output Q in the nth time interval, and Q_{n-1} represents the state of the output Q in the $n - 1$th time interval. The nth time interval may be taken to represent *now* and the $n - 1$th time interval then represents a time in the immediate past. What the first row of the transition table says, therefore, is that the Q output now is the same as it was a short time ago. In other words it remains unchanged. The same is true of the $\overline{Q}$ output. Therefore, when both S and R are 0 the circuit remains in its current state. The following two rows show that putting R to 1 while Q is 0 resets Q to 0, setting $\overline{Q}$ to 1, while putting S to 1 while R is at 0 sets Q to 1 and resets $\overline{Q}$ to 0. The fourth row of the transition table, corresponding to both S and R being equal to 1, gives both Q and $\overline{Q}$ equal to zero. This is the ambiguous state and is not used.

This information can also be represented by Karnaugh maps.

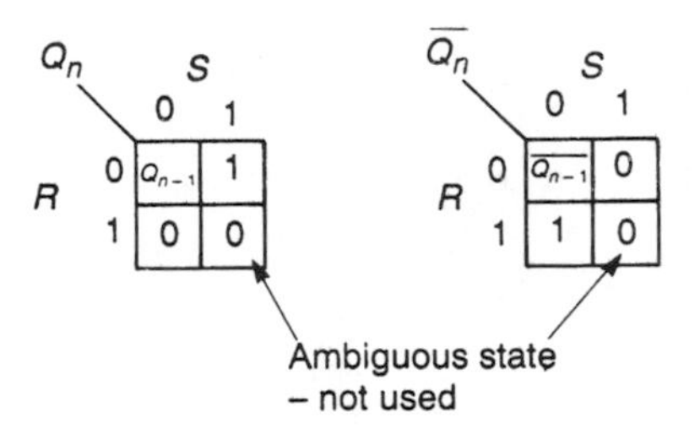

Generally the circuit diagram for the latch is redrawn as shown in Figure 10.5(a). This form emphasizes the basic symmetry of the circuit. An S–R latch can also be produced using two NAND gates, as shown in Figure 10.5(b).

Comparing the operation of a NAND gate with that of a NOR gate, we note that while the NOR gate resembles an inverter when one of its inputs is connected to 0, the NAND gate resembles an inverter when one of its inputs is connected to 1. Therefore the memory mode of the circuit of Figure 10.5(b) corresponds to both inputs being at 1. Investigation of the operation of the circuit shows that taking one input $\overline{S}$ low now SETS Q to 1, while taking the other input $\overline{R}$ low RESETs Q to 0. For this reason these inputs are called **active low inputs** and their names are given as $\overline{S}$ and $\overline{R}$ rather than S and R. In logic diagrams, any signal name that has a bar above it is active low and the function described by the name is achieved by taking the appropriate line low. Functions that are achieved by taking signals to logic 1 (such as S and R in the circuit of Figure 10.5(a)) are referred to as **active high inputs**.

The transition table of the circuit of Figure 10.5(b) differs from that of the circuit of Figure 10.5(a) because of the different polarities of the inputs. The transition table of the NAND gate circuit is as follows.

$\overline{S}$	$\overline{R}$	Q_n	$\overline{Q_n}$	
0	0	0	0	Ambiguous state
0	1	1	0	SET
1	0	0	1	RESET
1	1	Q_{n-1}	$\overline{Q_{n-1}}$	No change

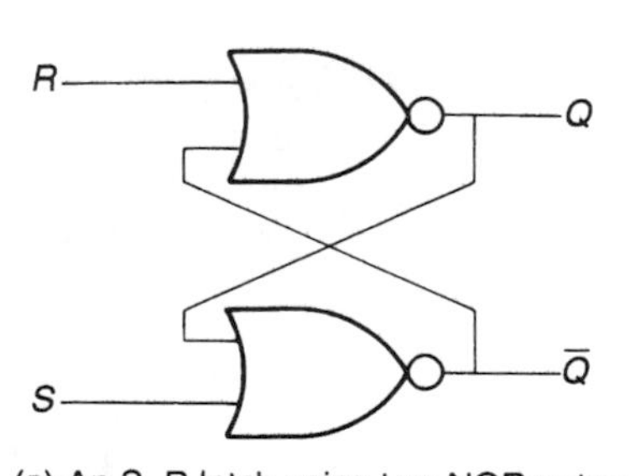

(a) An S–R latch using two NOR gates

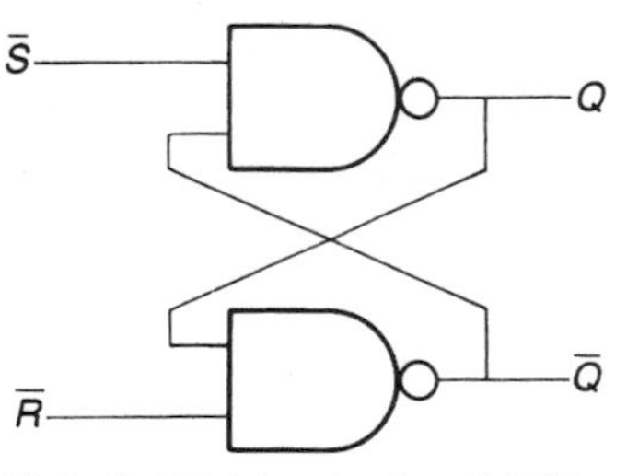

(b) An S–R latch using two NAND gates

Figure 10.5 S–R latch circuits.

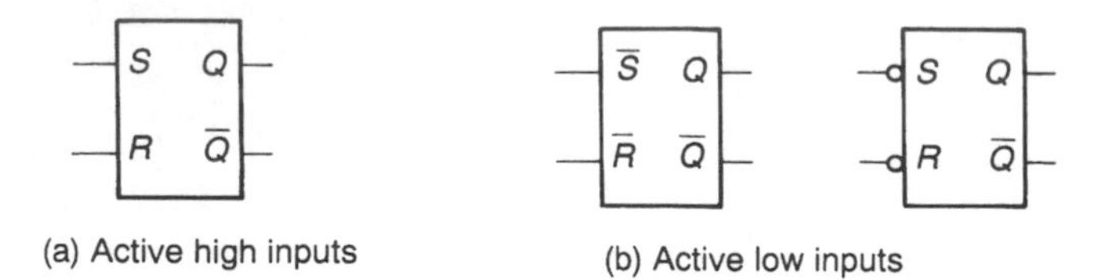

(a) Active high inputs (b) Active low inputs

Figure 10.6 S–R latch logic symbols.

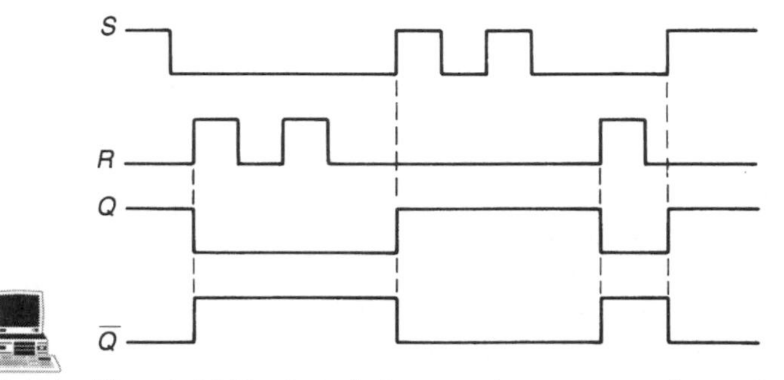

FILE 10A **Figure 10.7** Sample input and output waveforms for an S–R latch.

The symbols used for S–R latches are shown in Figure 10.6. Circuits with active low inputs can be represented in two ways. Either by labelling the inputs as $\overline{S}$ and $\overline{R}$ or by showing inverting circles at the inputs. In either case the signals applied to the input lines correspond to the active low signals $\overline{S}$ and $\overline{R}$.

Figure 10.7 illustrates the operation of an active high input S–R latch by showing its response to a series of input changes. Simulation files are available for this, and several of the other circuits described in this chapter, to allow their properties to be investigated. This diagram assumes that the latch responds immediately to changes at its input. In practice there is a slight delay which is termed the **propagation delay** time of the circuit. This topic will be discussed in the next chapter. You will notice that transitions of the S input have no effect while the Q output is already at 1. Similarly, changes in the R input are ineffective while Q is 0. The only changes that are significant are those that toggle the outputs from one state to the other.

The S–R latch may be thought of as a simple form of electronic **memory** since it remembers which of its two inputs last became active. In fact, latches of one form or another form the basis of a large proportion of the memory circuits used within all computers. However, we shall see that slight variants of this basic circuit are more convenient in many applications. Before moving on to look at these alternative forms of bistable it is perhaps worth looking at a common use of S–R latches, that of **switch debouncing**.

Use of the S–R latch for switch debouncing

We saw in Section 2.3.4 that switches play a major role in the construction of electronic sensors. We also noted that all mechanical switches suffer from **switch bounce**. The effect is illustrated in Figure 10.8.

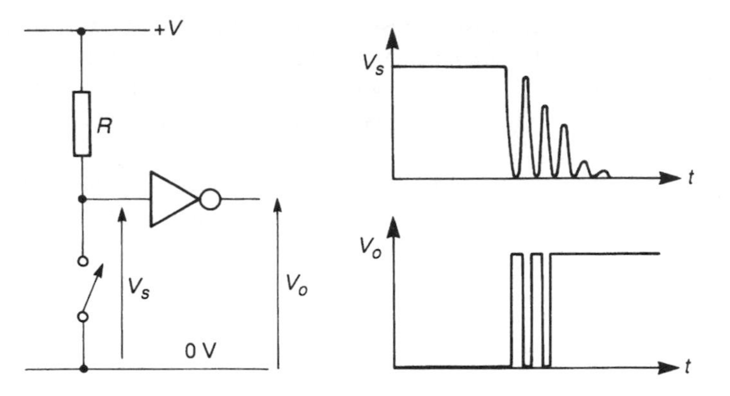

Figure 10.8 The effects of switch bounce.

Figure 10.8 shows a typical switch arrangement. When the switch is open the voltage V_s is equal to the supply voltage V, and when the switch is closed V_s is pulled to 0 V. If V corresponds to the voltage representing logical 1 and 0 V corresponds to logical 0, the voltage V_s can be used as the input to a logic gate. The figure shows that in practice there is not a clean transition from one voltage level to another. The voltage across the switch oscillates between the two voltage levels as the contacts make and break the circuit as a result of contact bounce. This oscillation generally lasts for the order of a few milliseconds for small switches and somewhat longer for larger devices. The input to the logic gate sees a signal that alternates between a voltage representing a logic 1 and that representing a logic 0, and it therefore produces an output which itself alternates between 0 and 1. Several transitions are thus produced rather than the single step expected. In the example waveform shown, the inverter produces three positive going edges rather than one. If this output were connected to circuitry designed to count the number of switch closures it would produce a count of three rather than one.

The problem can be overcome by replacing the 'make-or-break' switch with a 'changeover' switch which connects an input terminal to one of two output terminals depending on whether or not the switch is pressed. This switch is then connected to an S–R latch, as shown in Figure 10.9.

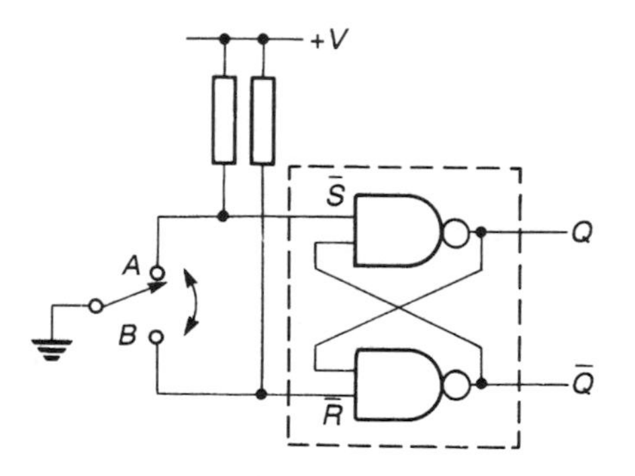

Figure 10.9 Use of an S–R latch for switch debouncing.

The switch now has three possible positions. The input terminal may be connected to terminal A, terminal B or to neither. The circuit is arranged such that when the switch is connected to neither terminal, the two lines to which they are connected are pulled high by resistors connected to V. These two signals form the $\overline{S}$ and $\overline{R}$ inputs to an active low S–R latch. When both are pulled high the circuit is in its 'memory mode' in which it simply remains in its present state. When the input terminal is connected to terminal A, $\overline{S}$ is pulled low setting Q to 1. If the contact bounces at this stage the input terminal will alternate from being connected to A to being connected to nothing. Thus the circuit alternates from the SET mode to the memory mode. Under these circumstances the circuit will simply remain with $Q = 1$ and the output will not bounce. Similarly, if the input terminal is connected to terminal B, $\overline{R}$ will be activated resetting Q to 0. Contact bounce will alternate the circuit between its RESET mode and its memory mode and again will not affect its output. Thus the circuit removes the effects of switch bounce, provided that the moving contact cannot bounce between the two output terminals. This is normally assured by its mechanical design.

Example 10.1 Design of a burglar alarm

Problem

A burglar alarm consists of a series of switches connected to doors and windows throughout a building. Opening any door or window opens the corresponding switch which should sound the alarm. It is essential that the alarm continues to sound if the door or window concerned is subsequently closed. Some method must be incorporated to silence the alarm when the building has been checked.

Solution

The following arrangement satisfies the requirements.

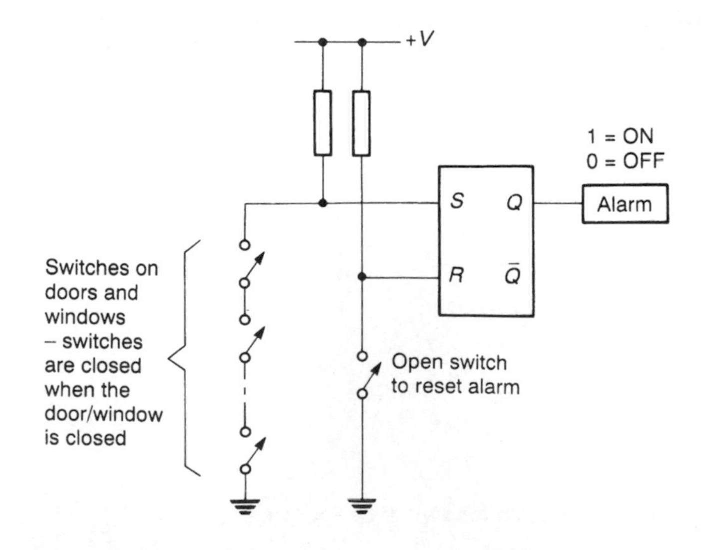

An S–R latch has two pull-up resistors connected to its S and R inputs. The various door and window switches are connected in series and wired so that when all are closed, they short the S input to ground. The R input is similarly shorted to ground by a RESET switch which is normally closed.

Initially the system is reset by momentarily opening the RESET switch with all the sensor switches closed. The latch will be reset with $Q=0$ and the alarm will be off. Once the RESET switch has been closed the system is armed. If one of the sensor switches is opened, by the opening of a door or window, the S input will go high, setting Q to 1 and sounding the alarm. If now the sensor switch is closed the system will remain in the alarm state until it is reset by opening the RESET switch.

10.2.2 The gated S–R latch

It is often useful to be able to control the operation of a latch so that the inputs can be enabled at some times and disabled at others. The circuit of Figure 10.10 provides this facility.

Two NAND gates are used to 'gate' the S and R input signals before they are applied to the latch. A third input, **latch enable** (EN), can be used to allow or inhibit the actions of the other inputs. When the enable signal is low, the signals $\overline{S'}$ and $\overline{R'}$ are both high regardless of the signals applied to the S and R inputs. This places the active-low input latch into its memory mode, preventing any change to its state. When the enable input is taken high, the S and R signals are inverted by the gating arrangement and then applied to the latch. Thus, when the enable is high the circuit acts as a conventional, active-high input S–R latch, but when the enable is low, the circuit ignores any signals applied to

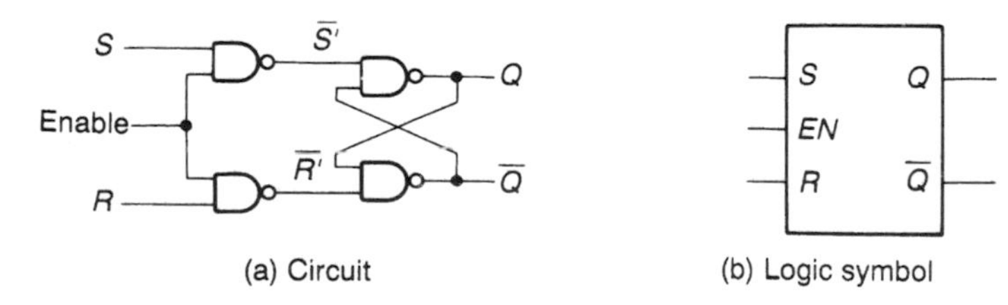

FILE 10B **Figure 10.10** A gated S–R latch.

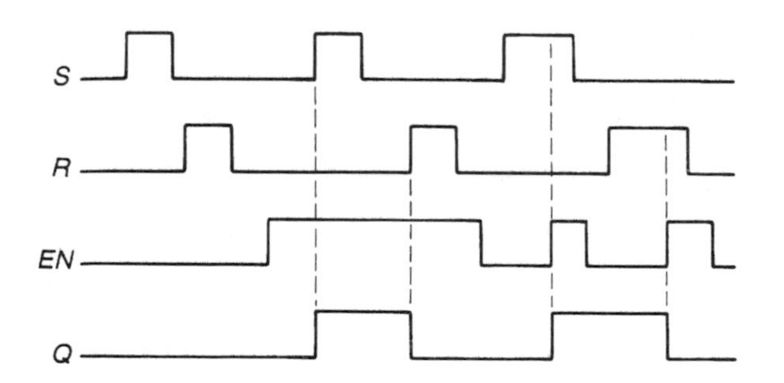

Figure 10.11 Sample input and output waveforms for a gated S–R latch.

the S and R inputs. Figure 10.11 illustrates the response of the circuit to a series of input changes. Since $\overline{Q}$ is simply the inverse of Q it is not shown in this waveform diagram.

10.2.3 The gated D latch

Another form of latch that is widely used is the **gated D latch**, which is also known as the **transparent D latch**. This circuit has two inputs D and EN, as shown in Figure 10.12.

Clearly the circuit bears a striking resemblance to that of the gated S–R latch shown in Figure 10.10, but uses a single signal D and its inverse $\overline{D}$ to act as inputs to the gating network. As before, when the enable input is low the signals fed to the latch are both high and the latch is placed in its memory mode preventing any change of state. If the enable is taken high, the D input determines the signals applied to the latch inputs $\overline{S'}$ and $\overline{R'}$. If D is high, $\overline{S'}$ will be low and $\overline{R'}$ will be high and the latch will be set with $Q=1$. If D is low, $\overline{S'}$ will be high and $\overline{R'}$ will be low which will reset the latch with $Q=0$. Thus when the enable is high the Q output takes the present value of D, and when the enable is low the Q output will remain in its present state. The D latch may therefore be thought of as a digital equivalent of the analogue sample and hold gate described in Example 6.6. When the enable is high the Q output follows the input data D and when the enable goes low the output remembers the value of D when the enable went low. This characteristic gives rise to the circuit being called a **data latch** or simply a D latch. The operation of the D latch is illustrated in Figure 10.13.

In addition to storing single bits of information, D latches are often used in groups to store words of information. It is common to combine a number of latches within a single integrated circuit to give, perhaps, four bits (a quad latch) or eight bits (an octal latch) of storage within a single device. Figure 10.14 shows an octal latch in which eight bits of data can be sampled and stored using a single enable input. When the enable input is

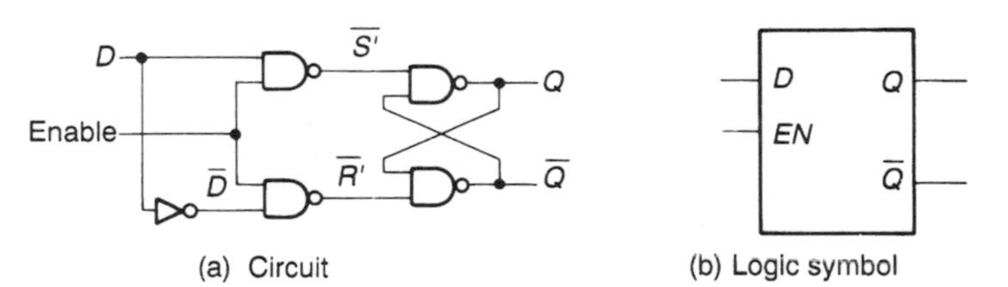

FILE 10C **Figure 10.12** A gated D latch.

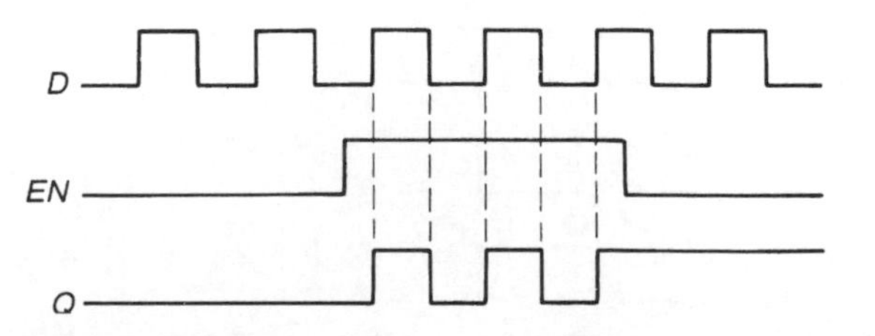

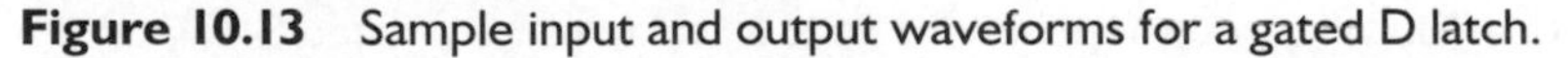

Figure 10.13 Sample input and output waveforms for a gated D latch.

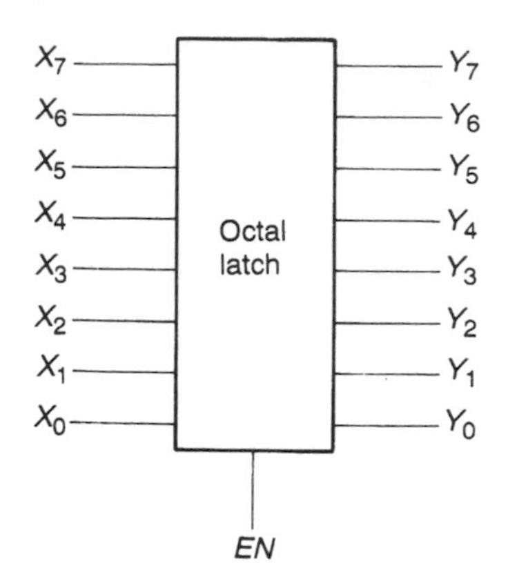

Figure 10.14 An octal data latch.

high the outputs Y_0–Y_7 are identical to the inputs X_0–X_7. When the enable input goes low the outputs are frozen at their values when the enable went low, storing this value. This technique of 'latching' data is used extensively within microcomputers and other areas of digital electronics.

10.3 Edge-triggered bistables or flip-flops

In many situations it is necessary to synchronize the operation of a number of different circuits and it is useful to be able to control precisely when a circuit will change state. Some bistable devices are constructed so that they only change state upon the application of a **trigger** signal. This trigger signal is defined as the rising or falling edge of an input signal termed the **clock**. These devices are called **edge-triggered bistables** or, more commonly, **flip-flops**. They are divided into those that are triggered by the rising edge of the clock signal (so-called positive edge-triggered devices) and those that are triggered by the falling edge of the clock (negative edge-triggered devices).

Flip-flops are available in a number of different forms, of which we will consider a selection. The logic symbols used for flip-flops resemble those for gated latches except that the enable input is replaced by a clock input. The clock line is conventionally shown by a triangle; an inverting circle is used to show a negative edge-triggered device. Examples of these symbols for S–R flip-flops are shown in Figure 10.15.

10.3.1 The edge-triggered S–R flip-flop

The operation of the S–R flip-flop resembles that of the S–R latch except that the circuit only responds to its inputs on the rising or falling edge (depending on the device) of the clock input signal. Many different forms of edge-triggered device are used. Figure 10.16(a) shows an example of a positive edge-triggered arrangement. The operation of this circuit is complex and relies in part on delays within the circuit. It is perhaps more useful to

(a) Positive edge-triggered (b) Negative edge-triggered

Figure 10.15 Logic symbols for S–R flip-flops.

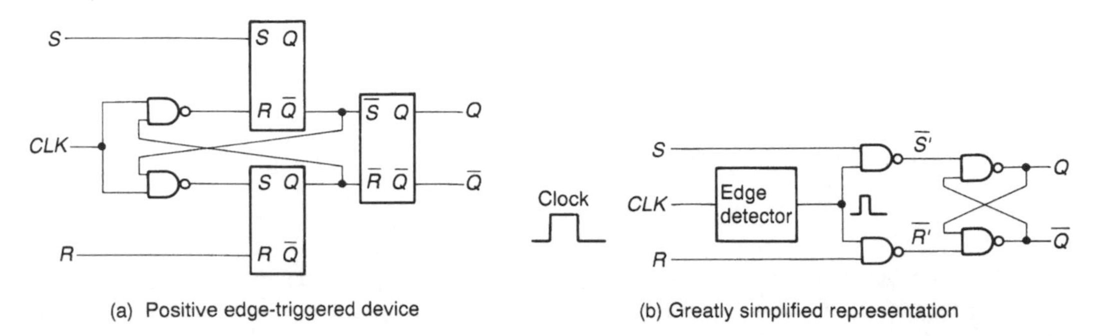

(a) Positive edge-triggered device (b) Greatly simplified representation

Figure 10.16 An edge-triggered S–R flip-flop.

consider a conceptual model of this circuit, as shown in Figure 10.16(b). This circuit is not a true representation of the operation of the earlier circuit, but gives an insight into its characteristics.

The circuit of Figure 10.16(b) is similar to that of a gated S–R latch but has additional edge-detecting circuitry. An appropriate transition of the clock input generates a short pulse which briefly enables the gating network allowing the circuit to respond to the *S* and *R* inputs. In the absence of clock transitions the circuit remains in its memory state and the outputs do not change. When an appropriate edge is detected the circuit responds to the current signals on the *S* and *R* inputs.

The operation of a positive edge-triggered S–R flip-flop can be described by a transition table.

S	R	CLK	Q_n	$\overline{Q_n}$	
0	0	↑	Q_{n-1}	$\overline{Q_{n-1}}$	No change
0	1	↑	0	1	RESET
1	0	↑	1	0	SET
1	1	↑	0	0	Ambiguous state

↑ = positive-going clock transition

Since the circuit only responds when an appropriate clock transition occurs, its operation is only shown at these times. The clock does not act like an ordinary input and thus the table has only four rows rather than eight. The transition table shows that while both

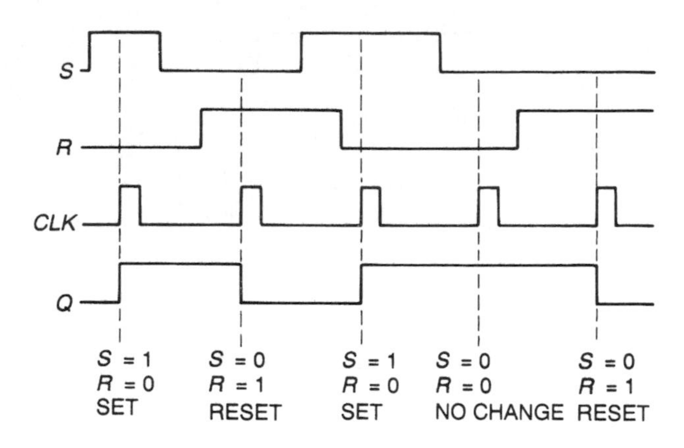

Figure 10.17 Typical waveforms for a positive edge-triggered S–R flip-flop.

S and *R* are 0, the circuit remains in its current state regardless of the signal on the clock input. If *R* is high and *S* is low when a positive-going clock transition occurs, the circuit will RESET. If *S* is high while *R* is low when a positive going clock transition occurs, the circuit will SET. As with other forms of S–R bistable, if both *S* and *R* are active the outputs are ambiguous.

The transition table for a negative edge-triggered device is identical to that given above except that it is the negative-going edge that acts as a trigger. This would be shown in the table by a downward arrow (↓).

Figure 10.17 illustrates the operation of a positive edge-triggered S–R flip-flop. You will notice that at one point during the waveforms shown in this figure, both *S* and *R* are active (high). This does not cause problems since the circuit only responds to the inputs at the rising edges of the clock and at these times only one of the inputs is active in all cases.

10.3.2 The edge-triggered D flip-flop

The D flip-flop is simply an edge-triggered version of the D latch described earlier. The logic symbol and equivalent circuit of this gate are shown in Figure 10.18.

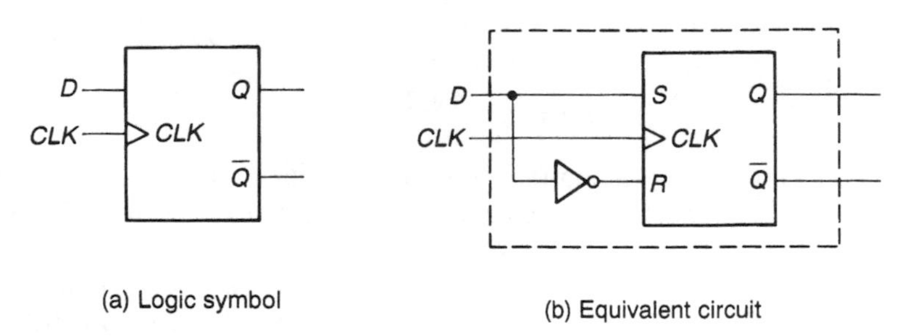

Figure 10.18 A D flip-flop.

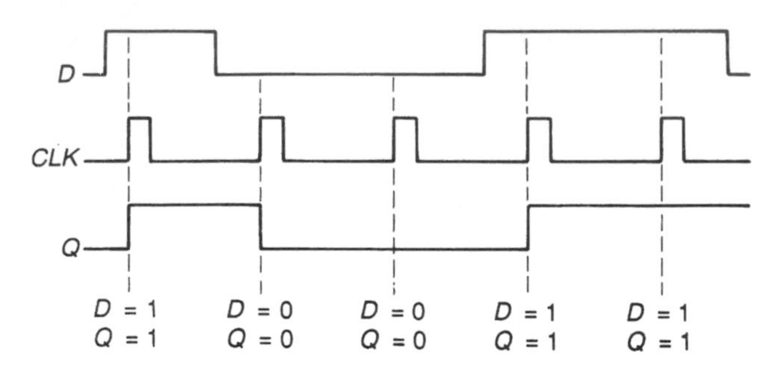

Figure 10.19 Typical waveforms for a positive edge-triggered D flip-flop.

Again it is possible to describe the operation of the D flip-flop using a transition table. Since the gate has only one input (in addition to the clock), the table has only two rows.

D	CLK	Q_n	$\overline{Q_n}$	
0	↑	0	1	RESET
1	↑	1	0	SET

Figure 10.19 illustrates the operation of the D flip-flop.

10.3.3 The edge-triggered J–K flip-flop

The S–R flip-flop has many uses but suffers from having an ambiguous input combination when both S and R are asserted simultaneously. The J–K flip-flop overcomes this deficiency by defining an operation to be performed, on active clock transitions, when both inputs are active. With the S–R device we already have input combinations corresponding to SET, RESET and No change. The obvious additional operation to give increased flexibility is TOGGLE. The transition table below shows the characteristics of a negative edge-triggered J–K flip-flop. It can be seen that the J input corresponds to the S input of an S–R flip-flop and that K corresponds to the R input.

J	K	CLK	Q_n	$\overline{Q_n}$	
0	0	↓	Q_{n-1}	$\overline{Q_{n-1}}$	No change
0	1	↓	0	1	RESET
1	0	↓	1	0	SET
1	1	↓	$\overline{Q_{n-1}}$	Q_{n-1}	TOGGLE

The function of the J–K flip-flop is achieved using circuitry similar to that of the S–R flip-flop except that the output signals Q and $\overline{Q}$ are fed back to modify the operation of the gating network. Such an arrangement is shown in Figure 10.20.

The gate is based on a negative edge-triggered S–R bistable with inputs S and R. The inputs to this arrangement come from two AND gates A1 and A2. In order for either of these AND gates to produce the logic 1 output required to activate the corresponding

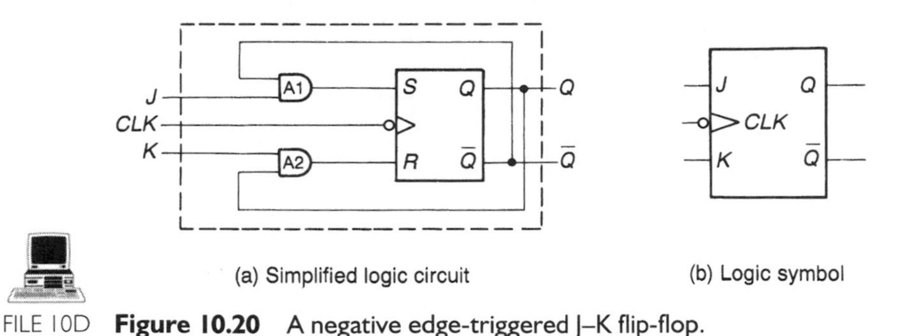

FILE 10D **Figure 10.20** A negative edge-triggered J–K flip-flop.

bistable input, both of its inputs must be at 1. From the circuit it is clear that A2 can only produce a high output when Q is high, and A1 can only produce a high output when $\overline{Q}$ is high. Let us now consider the operation of the circuit for different combinations of input conditions and states.

If both J and K are zero when a clock trigger occurs, both S and R will be low irrespective of the output state of the circuit. This is the memory mode of the bistable and so no change will occur.

If $J=1$ and $K=0$ when a clock trigger occurs, the action taken will depend on the existing output state. If $Q=0$ and $\overline{Q}=1$, both inputs to Al will be high and S will be high, so Q will be set to 1. If Q is already at 1, $\overline{Q}$ will be at 0, disabling A1 and holding S low. Thus with this combination of inputs, if Q is at 0 it is SET to 1, and if it is already at 1 it simply stays at 1.

If $J=0$ and $K=1$ when a clock trigger occurs, it is clear from the symmetry of the circuit that if Q is at 1 it will be RESET to 0 and that if it is already at 0 it will simply stay in that state.

If both J and K are at 1 when a clock trigger occurs, the action again depends on the existing output state. If Q is at 1, A2 will be enabled and A1 disabled such that R will be activated but S will not. This will result in Q being changed from 1 to 0. If, however, Q is at 0, Al will be enabled and A2 disabled so that S will be activated and R will not. This will result in Q being changed from 0 to 1. Thus in either case Q is TOGGLED between the output states.

Figure 10.21 illustrates the operation of a negative edge-triggered J–K flip-flop in response to a series of input combinations. Positive edge-triggered devices are also available, but are less common.

The great versatility of the J–K flip-flop makes it probably the most widely used type of bistable element. A number of different operating modes are possible, including using it to reproduce the functions or other types of flip-flop. This is illustrated in Figure 10.22 which shows several common configurations.

Figure 10.22(a) indicates that a J–K flip-flop can be used as a direct replacement for an S–R device since they are identical in operation for all the allowable input combinations of the latter. A J–K bistable can also be used to produce the function of a D flip-flop, as shown in Figure 10.22(b). We saw this arrangement using an S–R flip-flop in Figure 10.18. Since with this configuration S and R cannot be active simultaneously, a J–K flip-flop will produce an identical effect. Figure 10.22(c) shows a J–K flip-flop

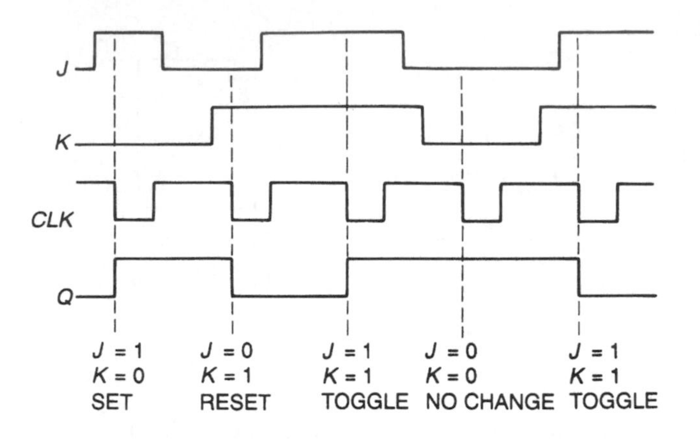

Figure 10.21 Typical waveforms for a negative edge-triggered J–K flip-flop.

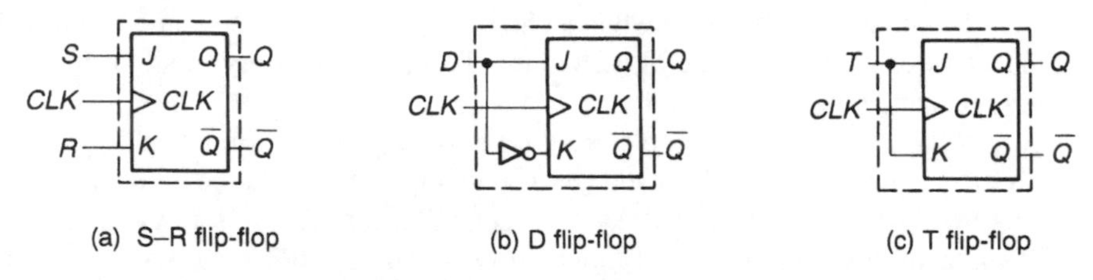

Figure 10.22 Use of a J–K flip-flop to reproduce other flip-flop functions.

configured with its *J* and *K* inputs joined to form a single input *T*. If *T* is 0 the device is in its memory mode and simply stays in its present state. If *T* is 1, both *J* and *K* are 1 and the device will TOGGLE on every clock pulse. This arrangement is called a **toggle flip-flop** or simply a **T flip-flop**. Figure 10.23 shows the operation of this form of bistable; we will look at applications for this device in the next section.

10.3.4 Asynchronous inputs

From the foregoing discussion we know that the *control* inputs of the various flip-flops (that is *S*, *R*, *J*, *K*, *D* and *T*) only have effect at the moment of an appropriate transition of the clock signal (*CLK*). We therefore refer to these control inputs as **synchronous** since their operation is synchronized by the clock input.

In many applications it is advantageous to be able to set or clear the output at other times, independent of the clock. Therefore some devices have additional inputs to perform these functions. These are termed **asynchronous inputs** since they are not bound by the state of the clock. Unfortunately, IC manufacturers are unable to agree on common names for these inputs and they may be called PRESET and CLEAR, DC SET and DC CLEAR, SET and RESET, or DIRECT SET and DIRECT CLEAR. Here we will use the names PRESET (*PRE*) and CLEAR (*CLR*). As with control inputs, these lines can be active high

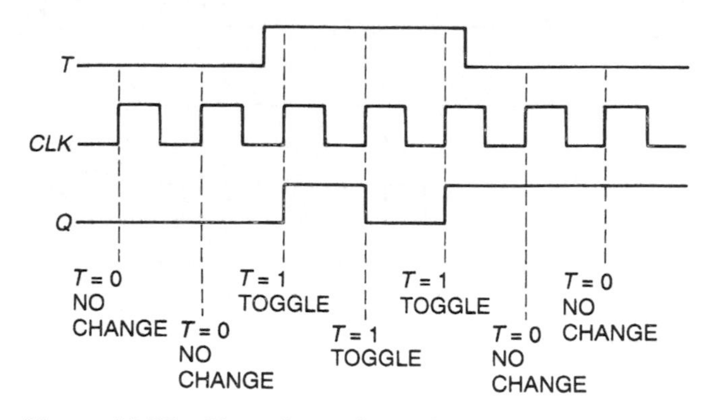

Figure 10.23 Typical waveforms for a positive edge-triggered T flip-flop.

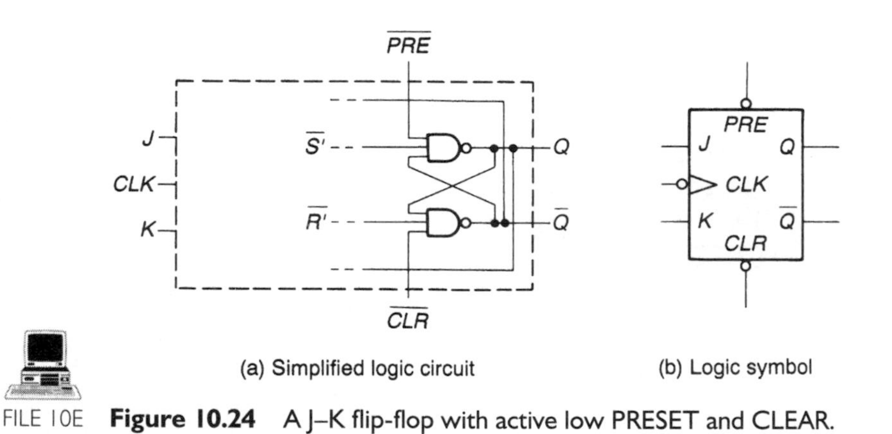

FILE 10E **Figure 10.24** A J–K flip-flop with active low PRESET and CLEAR.

or active low, although it is more common for them to be active low. Figure 10.24(a) illustrates how these inputs can be incorporated into the circuit of a J–K flip-flop while Figure 10.24(b) shows how these inputs are represented on a logic diagram.

The PRESET and CLEAR inputs override the other inputs to the circuit, acting in a similar manner to the *S* and *R* inputs in an S–R latch. Clearly, as with the S–R latch it is necessary to ensure that both asynchronous inputs are not active simultaneously, since this would make both outputs 1 and they would therefore not be complementary. Figure 10.25 illustrates the operation of these inputs on a negative edge-triggered J–K flip-flop.

Races

In many digital systems it is necessary to synchronize the operation of a large number of circuits to a single clock signal. Consider, for example, the arrangement shown in Figure 10.26.

The figure shows two edge-triggered S–R flip-flops connected such that the *Q* output from one device forms the *S* input for the next. Both bistables are driven from the same clock signal. The figure shows part of the clock waveform and indicates the relationship

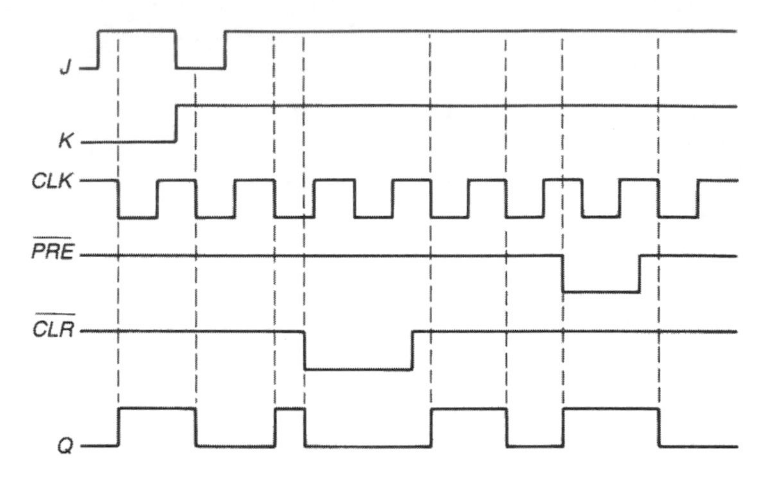

Figure 10.25 Waveforms for an edge-triggered J–K flip-flop with PRESET and CLEAR.

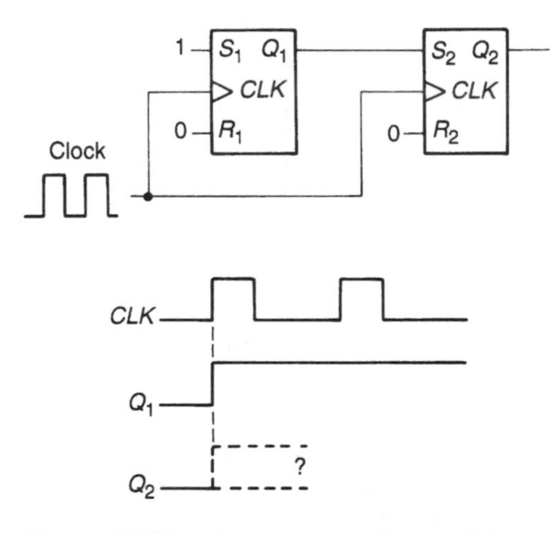

Figure 10.26 An example of a possible race condition.

between the output of the first flip-flop Q_1, and the clock signal. This waveform diagram assumes that initially both Q_1 and Q_2 are at 0. Since Q_1 forms a control input to the second device S_2, it is clear that the control input changes at the time that the clock trigger occurs. This makes the operation of the circuit uncertain. If the second flip-flop reacts quickly to the clock signal it may respond before the first device has had time to toggle. In this case it will see a 0 on the S_2 input and the output will therefore not change. If, however, the second flip-flop reacts more slowly, the first device will have had time to toggle and the second bistable will see a 1 on its S_2 input resulting in a change of its output state. This uncertainty of operation is called a **race** condition since the action of the circuit is determined by a race between two devices. Such characteristics must be avoided if circuits are to be dependable.

The problems associated with races may be tackled in a number of ways. In fact, by attention to the delays and response times of the device, the designers of edge-triggered devices ensure that a race cannot occur in the arrangement shown in Figure 10.26. For any given circuit there is a minimum time for which the inputs must be held after the clock transition to ensure correct operation. This is termed the **hold time** of the gate. In all practical edge-triggered devices the hold time is much less that the propagation delay of the gate. Therefore, in Figure 10.26 the second flip-flop will always respond to its pattern of inputs before the first flip-flop has time to update its outputs.

10.4 Pulse-triggered bistables or master/slave flip-flops

Another method of overcoming the race problems discussed in the last section is the use of **master/slave flip-flops** rather than edge-triggered devices. These bistables are also called **pulse-triggered flip-flops**.

As their name implies, master/slave flip-flops are in fact two bistables connected in series. These devices are available in S–R, D and J–K forms.

10.4.1 Master/slave S–R flip-flop

The basic S–R master/slave flip-flop is shown in Figure 10.27.

Clearly the circuit represents two S–R gated latches connected in series, with the clock input being used to form the enable for the first and its inverse forming the enable for the second.

Since the enable signals applied to the two latches are the inverse of each other, when one is enabled the other is disabled. When the clock input signal goes high the master is enabled and the slave disabled. The master reacts as an ordinary gated S–R latch and its outputs are configured accordingly. The slave, being disabled, simply stays in its previous state and the outputs remain unchanged. When the clock goes low, the master is disabled and holds its previous state. The slave is now enabled and responds to its inputs, which are the outputs of the master. Since the Q' output of the master is connected to the

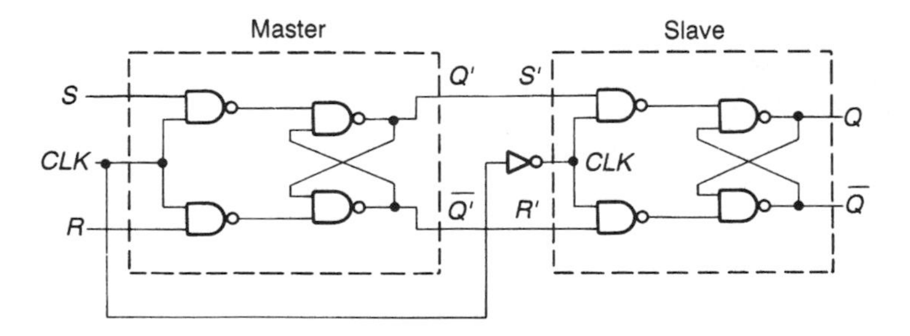

Figure 10.27 An S–R master/slave flip-flop.

S' input of the slave and the $\overline{Q'}$ output of the master is connected to the R' input of the slave, the slave will always see a 1 on one input and a 0 on the other. Thus, if the Q' output from the master is 1, the slave will be SET and its Q output will become 1. Conversely, if the $\overline{Q'}$ output from the master is 1, the slave will be RESET and its $\overline{Q}$ output will become 1. Hence, when enabled, the slave will take up the output state of the master.

The device therefore has two operating modes. When the clock is high the master responds but the slave maintains its previous outputs, and when the clock is low the master is disabled and the slave transfers the master's state to the outputs. Hence the device responds to its inputs while the clock is high, but the outputs are not updated until the clock goes low. These modes are illustrated in Figure 10.28.

It might seem that this arrangement would suffer from the 'race' problem discussed earlier, since when the clock goes high the master's outputs might change just as the slave is being disabled. In fact this is not the case. The design of the circuit ensures that the delay produced by the master latch is greater than that of the inverter used for the clock signal. This ensures that the slave is disabled shortly before the master is enabled, preventing any possible race condition.

The master/slave (M/S) device is used in a similar way to its edge-triggered cousin and has a similar transition table. The difference is that the M/S device responds at the end of an input clock pulse rather than on the rising or falling edge of the clock. This is represented in the transition table by a pulse in the clock column. The output columns show the effect on the outputs at the end of the pulse. The logic symbol used to represent an M/S flip-flop usually bears the label 'M/S' to distinguish it from other devices. The triangle used on the clock of an edge-triggered flip-flop is omitted since it is not an edge-triggered device. Figure 10.29 shows the transition table and the logic symbol for an S–R M/S flip-flop.

If we return to the arrangement of Figure 10.26, it is interesting to note the effect of

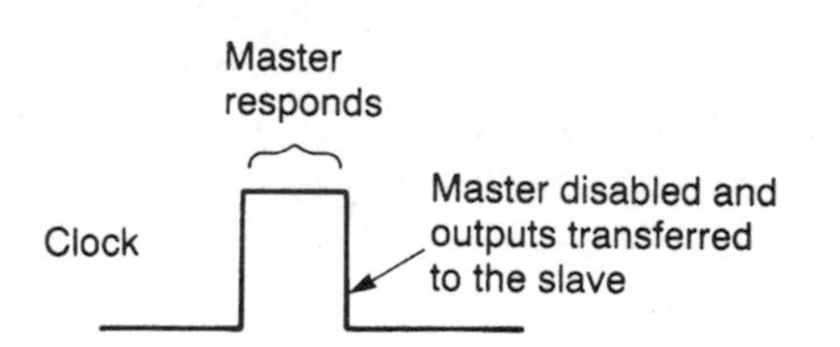

Figure 10.28 Operation of an S–R master/slave flip-flop.

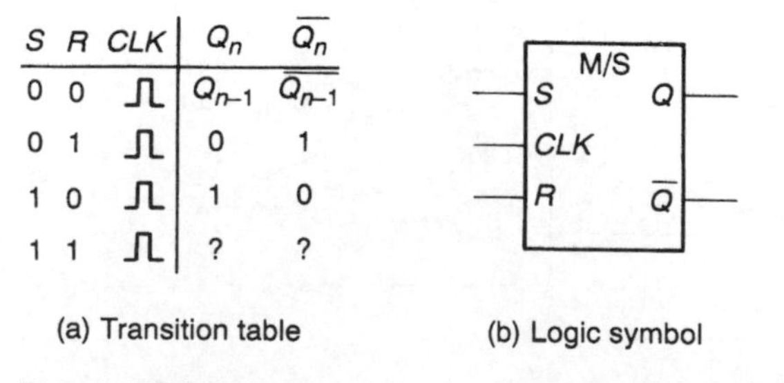

S	R	CLK	Q_n	$\overline{Q_n}$
0	0	⎍	Q_{n-1}	$\overline{Q_{n-1}}$
0	1	⎍	0	1
1	0	⎍	1	0
1	1	⎍	?	?

(a) Transition table (b) Logic symbol

Figure 10.29 Representations of an S–R master/slave flip-flop.

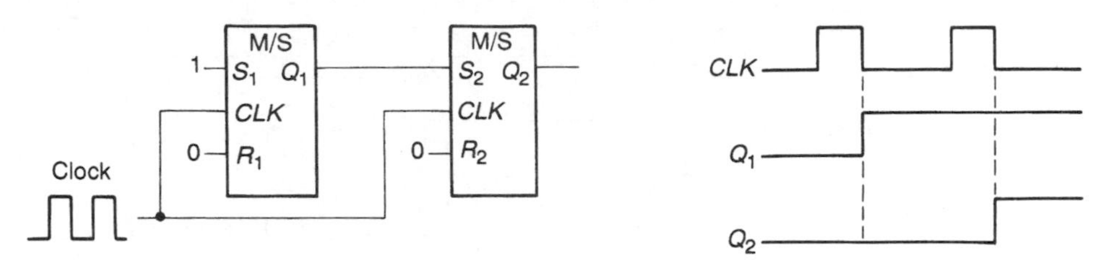

Figure 10.30 Use of M/S flip-flops in removing race conditions.

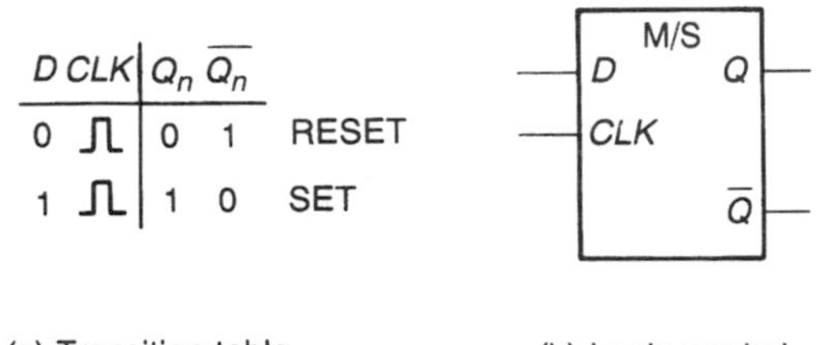

Figure 10.31 A D master/slave flip-flop.

replacing the flip-flops in that circuit by master/slave devices, as shown in Figure 10.30.

As before, the first flip-flop responds to the incoming clock pulse but this time the output changes at the falling edge of the pulse rather than the rising edge. The S_2 input to the second flip-flop sees an input of 0 while the clock is high and therefore this circuit will remain in its memory mode at the falling edge of the clock. After the first clock pulse the Q_1 output from the first flip-flop applies a 1 to the S_2 input of the second. Thus, when the second pulse arrives, the master and then the slave respond to this input. When the clock returns to 0, the Q_2 output goes to 1. Therefore, the use of master/slave devices has removed the race problem and any uncertainty as to the operation of the circuit.

10.4.2 Master/slave D flip-flop

The master/slave version of the D flip-flop is similar to the edge-triggered device except in its triggering. As with the S–R M/S flip-flop, the outputs are updated at the falling edge of the clock pulse. Thus, at the end of the clock pulse the output of the flip-flop takes up the value that was on the *D* input immediately before the rising edge of the clock. Figure 10.31 shows the transition table and logic symbol for a D master/slave flip-flop.

10.4.3 Master/slave J–K flip-flop

The J–K master/slave flip-flop is similar to the S–R M/S device with the addition of feedback to remove the ambiguous input combination (as in the J–K edge-triggered device). The arrangement is shown in Figure 10.32.

Figure 10.33 shows the transition table and logic symbol for the J–K M/S flip-flop while Figure 10.34 gives a waveform diagram which illustrates its operation.

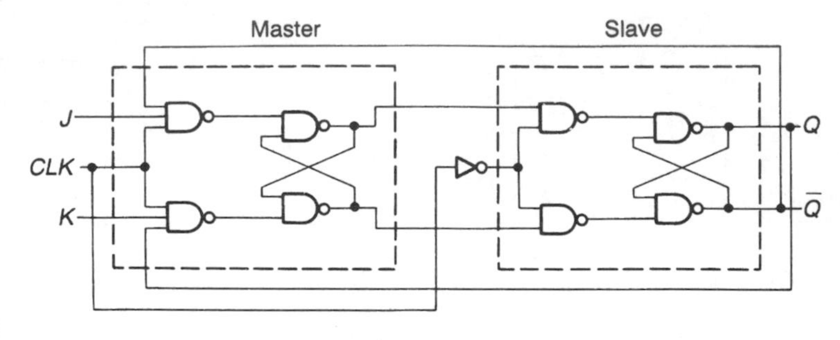

FILE 10F **Figure 10.32** A J–K master/slave flip-flop.

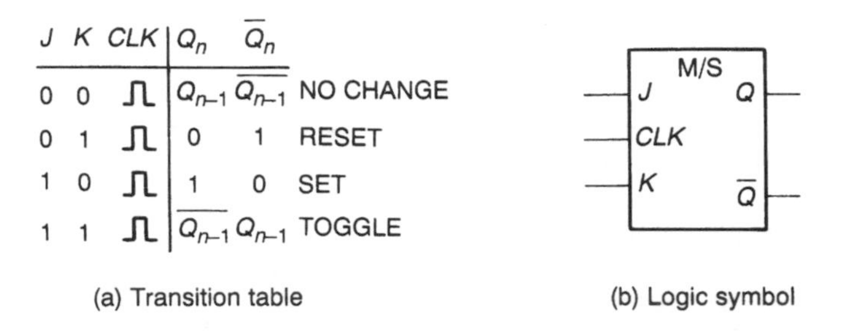

J	K	CLK	Q_n	$\overline{Q_n}$	
0	0	⎍	Q_{n-1}	$\overline{Q_{n-1}}$	NO CHANGE
0	1	⎍	0	1	RESET
1	0	⎍	1	0	SET
1	1	⎍	$\overline{Q_{n-1}}$	Q_{n-1}	TOGGLE

(a) Transition table (b) Logic symbol

Figure 10.33 A J–K master/slave flip-flop.

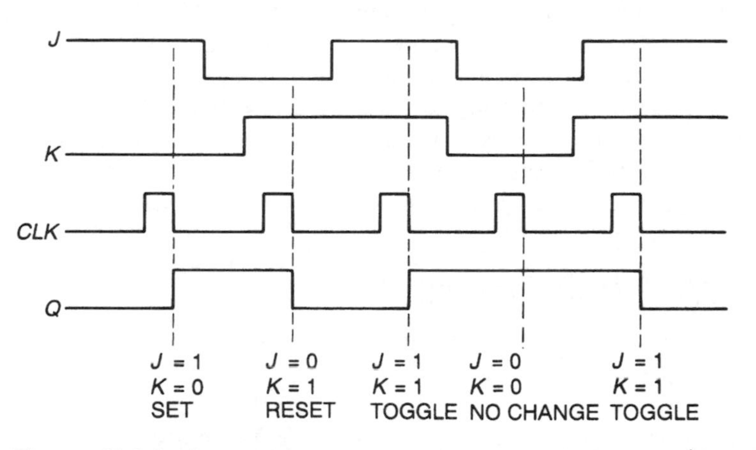

Figure 10.34 Input and output waveforms for a J–K master/slave flip-flop.

10.5 Monostables or one-shots

In Figure 10.4 we considered a bistable formed from two NOR gates connected in a ring. Now consider the circuit of Figure 10.35.

Let us assume initially that the input signal T is at 0. The resistor R, being connected to a voltage equal to logical 1, will cause the capacitor C to charge or discharge so that

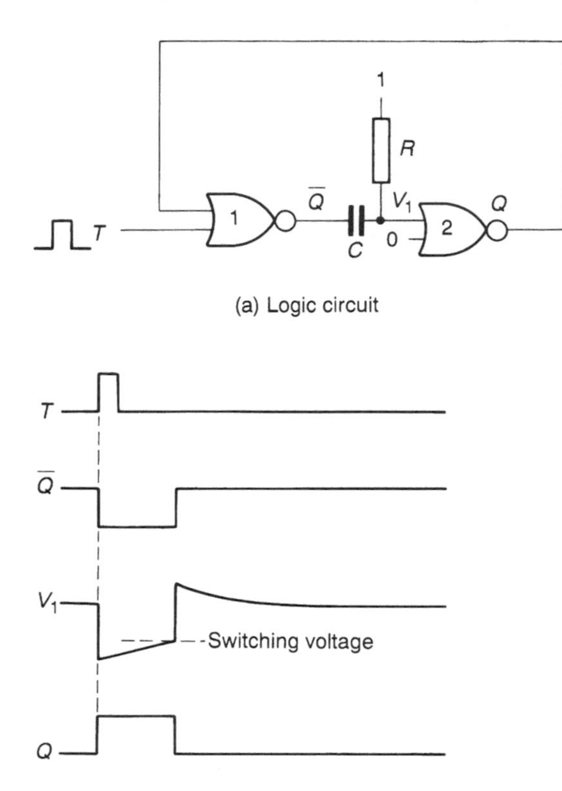

Figure 10.35 A simple monostable.

V_1 is equal to this voltage. This will cause Q to be 0, and since both inputs to gate 1 are 0, $\overline{Q}$ will be 1. Since both $\overline{Q}$ and V_1 are at a voltage corresponding to logical 1, there will be no voltage across the capacitor. This condition is **stable** and the circuit will remain in this state indefinitely unless the input changes.

Let us now consider what happens if the input T goes high for a short time. When T goes to 1, $\overline{Q}$ will go low. Since the voltage across the capacitor cannot change instantaneously, this will pull V_1 low, which in turn will take Q high. While Q is high, $\overline{Q}$ will be held low even if T reverts to 0. Thus the application of a positive pulse to T switches the circuit into a state where Q is 1, and it will remain in this state even after T returns to 0. However, this state is not stable, but only **meta-stable**. While in this state $\overline{Q}$ is at 0 and hence a voltage exists across the series combination of R and C. This produces a current through the resistor to charge the capacitor, which results in V_1 increasing exponentially towards the voltage representing logical 1. At some point it becomes sufficiently large for the input of gate 2 to interpret this voltage as a logical 1. At this point Q will return to zero, $\overline{Q}$ will go to 1 and V_1 will be pushed above logical 1 by the voltage across the capacitor. This voltage will gradually decay as the capacitor discharges under the influence of R. The circuit is now back in its stable state.

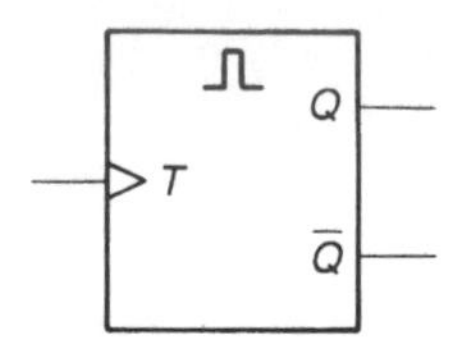

Figure 10.36 The logic symbol for a monostable.

The circuit thus has one stable state and one meta-stable state. It can be forced to enter its meta-stable state by applying an appropriate input signal. It will then stay in that state for a fixed period of time determined by the values of R, C and the switching voltages within the circuit. The circuit is therefore given the name **monostable** or **one-shot**. The label T given to the input signal stands for **trigger input**.

Although it is quite possible to construct monostables from simple logic gates, it is more common to use dedicated integrated circuits. These use more sophisticated circuit techniques than those described above but have similar characteristics. The logic symbol for a monostable is shown in Figure 10.36.

Integrated circuit monostables contain all the active components required to form the device within a single package. Generally all that is required is the addition of a single resistor and a capacitor to determine the pulse duration. Monostables can be divided into two types, namely, *nonretriggerable* and *retriggerable* monostables.

10.5.1 Nonretriggerable monostables

Nonretriggerable monostables ignore any trigger pulses that occur while the circuit is in its meta-stable state. In other words, the input is disabled during an output pulse. This is illustrated in Figure 10.37.

10.5.2 Retriggerable monostables

Retriggerable monostables are affected by the trigger input during an output pulse. These result in the output pulse being extended as shown in Figure 10.38.

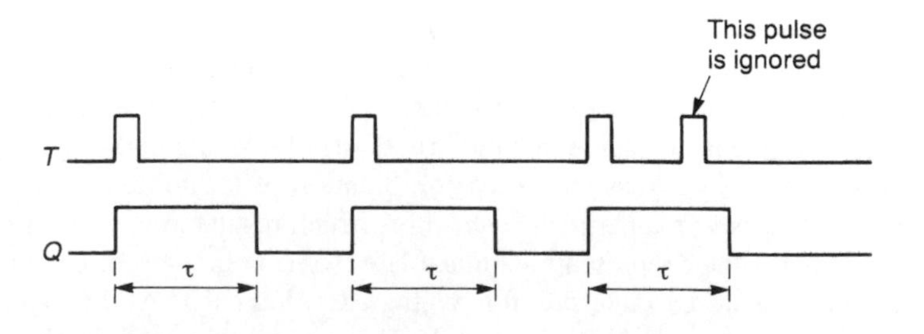

Figure 10.37 Operation of a nonretriggerable monostable.

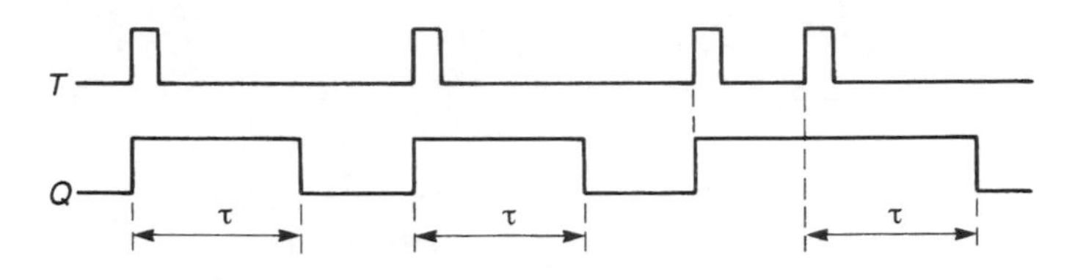

Figure 10.38 Operation of a retriggerable monostable.

10.6 Astables

By comparing the circuits of Figures 10.4 and 10.35, it is clear that the addition of a resistor-capacitor combination converts a stable state into a meta-stable state. It is then perhaps not surprising that if we add a second resistor–capacitor combination, we can generate a circuit with two meta-stable states. This circuit requires no input signals and we may therefore replace the NOR gates with inverters, as in Figure 10.3. The resultant circuit is shown in Figure 10.39.

Let us assume that initially C_1 and C_2 are discharged and that Q is 0. Since there is no voltage across C_1, V_1 will be equal to Q, that is 0. This will cause the output of the second gate to be at 1. Since C_2 is discharged, this logic level will be applied to the input of gate 1. This, in turn, will generate a logic 0 at Q which is consistent with our original assumption.

This state is meta-stable since C_1 will now charge up until V_1 is greater than the switching voltage of gate 2. At this time $\overline{Q}$ will go to 0, pushing V_2 down and making Q go to 1. C_1 will now discharge and C_2 will charge up until V_2 is greater than the switching voltage of gate 1, whereupon the circuit will again change state and the cycle will restart. The circuit will thus oscillate continuously from one state to the other with both Q and $\overline{Q}$ producing regular pulse waveforms. The length of time spent in cach state is determined by the values of the resistors and capacitors, as well as by the switching and logic voltages of the gates. If the gates are identical and the products C_1R_1 and C_2R_2 are equal, the circuit will produce a **square wave**. Figure 10.40 illustrates the waveforms produced within the circuit.

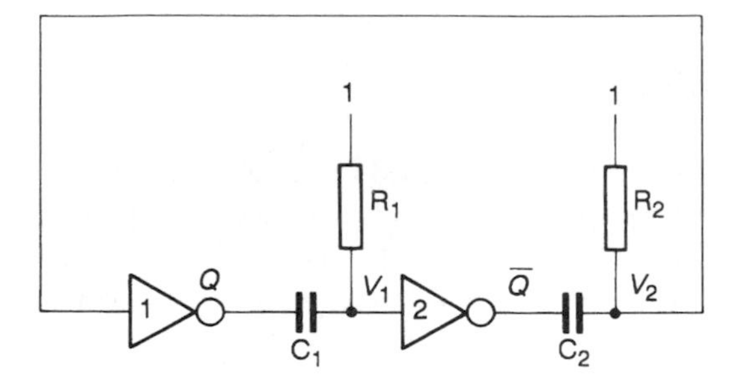

Figure 10.39 A simple astable arrangement.

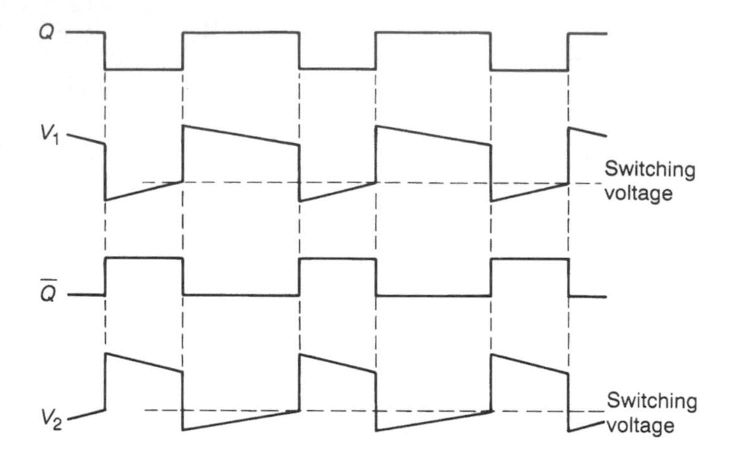

Figure 10.40 Waveforms of the simple astable circuit.

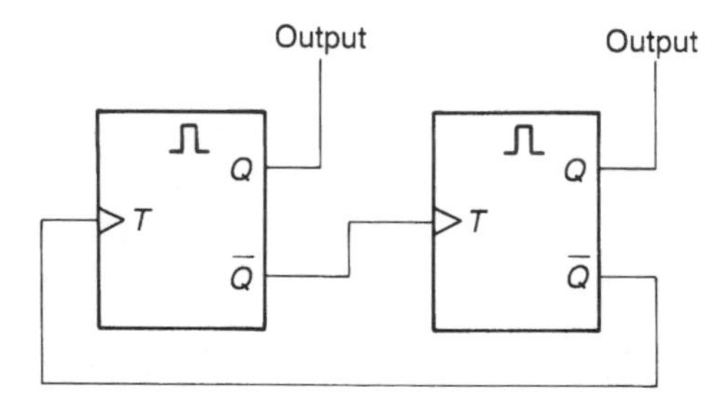

Figure 10.41 An astable formed from two monostables.

As with monostables, astables are usually used in integrated circuit form. Circuits are also available that can be configured to perform a range of functions, an example being the **555 timer**. This integrated circuit can be used as an astable or as a monostable, as well as for a number of other applications. Astables can also be constructed using two monostables connected in a ring, as shown in Figure 10.41. Here the trailing edge of the pulse generated by the first monostable triggers the second. The trailing edge of the pulse from the second monostable then triggers the first and the cycle repeats.

10.7 Memory registers

Registers of one kind or another are extensively used in almost all fields of digital electronics. One of the most widely used forms of register is that used to store words of information within computers and calculators. These registers can be used directly within calculations, as in the case of the *accumulator* within a processor or calculator, or they can be used for general memory applications where thousands, or perhaps millions, of registers are used to store programs and data.

When discussing D latches earlier in this chapter we noted that these devices could

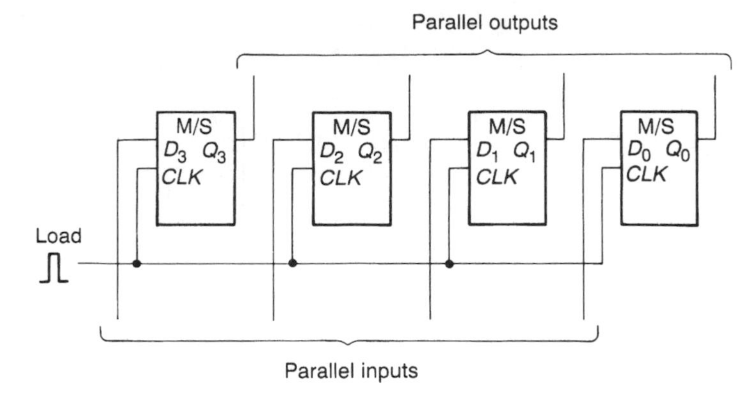

Figure 10.42 A 4-bit memory register.

be used for storing parallel words of data (see Figure 10.14). D flip-flops are also used for this purpose, as illustrated in Figure 10.42 which shows a 4-bit memory register formed using D master/slave flip-flops.

A wide range of circuit techniques are used for the storage of digital information and we will leave further discussion of this topic until later chapters.

10.8 Shift registers

Shift registers are used to convert parallel words of information into a stream of bits on a single line. This is referred to as **serial data**. A shift register can also be used to take a serial data stream and generate from it a parallel data word. This process is illustrated in Figure 10.43.

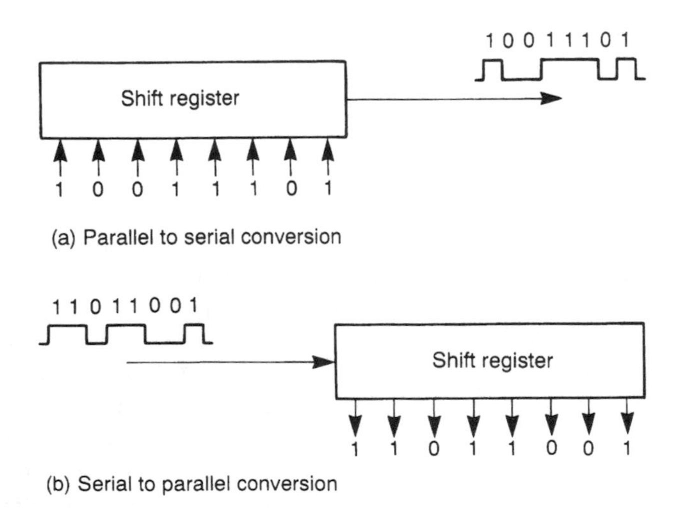

Figure 10.43 The operation of a shift register.

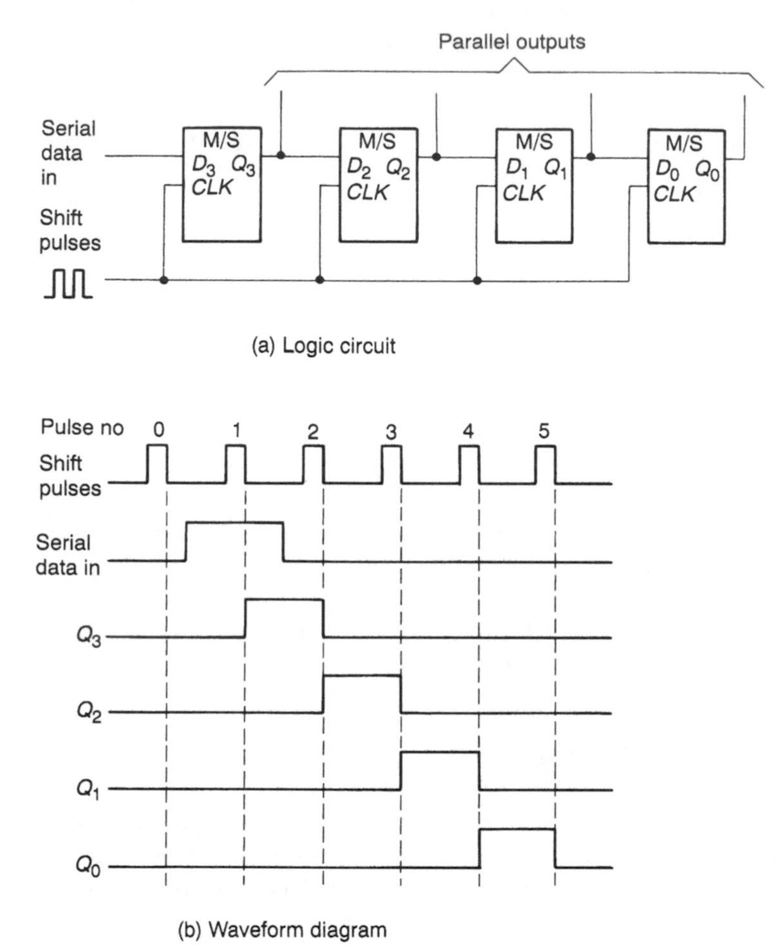

Figure 10.44 A simple shift register.

One application of this technique is in long distance communication where parallel words of information can be converted into serial form to be sent down a single wire, rather than needing a number of parallel wires.

To understand the operation of a shift register, consider the operation of the circuit of Figure 10.44(a). The circuit consists of four D master/slave flip-flops connected in series. A sequence of regular positive-going shift pulses is applied to the arrangement which forms the clock signal for each stage.

If we assume that initially the serial data input is 0 and that the output of each flip-flop is zero, then repeated shift pulses will have no effect on the circuit. If now the serial data input goes to 1 during one clock pulse and then returns to zero, the effect of this input will ripple along the register, as shown in Figure 10.44(b). During pulse number 1 the data input is 1, therefore D_3 is high immediately before the falling edge of the clock signal. Thus when the clock falls to 0, Q_3 goes to 1. None of the other flip-flops is affected by this transition, since it occurs as the clock goes low. Immediately before the end of pulse number 2, Q_3 and D_2 are high but D_3, D_1 and D_0 are low. Thus after the clock pulse Q_2 goes high, Q_3 returns to 0 while Q_1 and Q_0 are unaffected. On successive clock pulses each output goes high in turn for one clock cycle and then returns to zero. After pulse number 5 the register is back to its original state.

It should be clear from the above that any pattern of '1's and '0's within the register will move one bit to the right at the end of each shift pulse. We therefore have a **shift register**. A pattern of four bits of serial data will be shifted into the register after four shift pulses and will appear as a parallel word at the outputs $Q_3 - Q_0$. By adding more flip-flops, registers of any desired length can be formed allowing longer words to be used. The register therefore performs **serial to parallel conversion**.

In order to perform **parallel to serial conversion** it is necessary to modify the circuit of Figure 10.44(a) so that parallel data can be loaded into the shift register. We have already seen in Figure 10.42 how a series of D flip-flops can be used as a memory register. What we now require is some circuitry to allow a set of flip-flops to be switched from a memory register, which may be loaded in parallel, to a shift register which can output data in serial form. Such an arrangement is shown in Figure 10.45.

Figure 10.45(a) shows a subsystem required to implement this function. It is a **data**

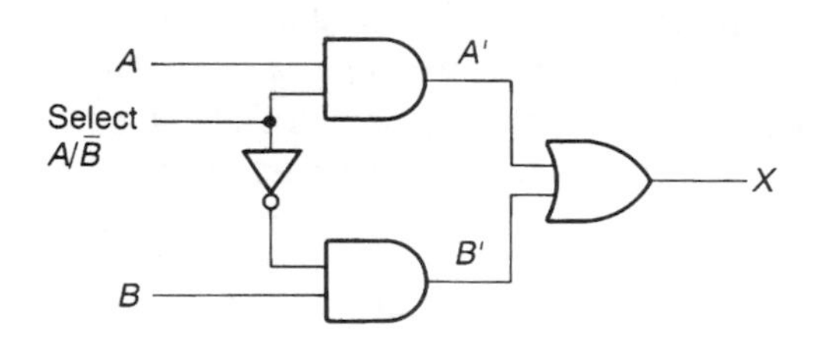

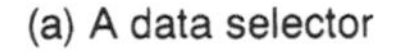
(a) A data selector

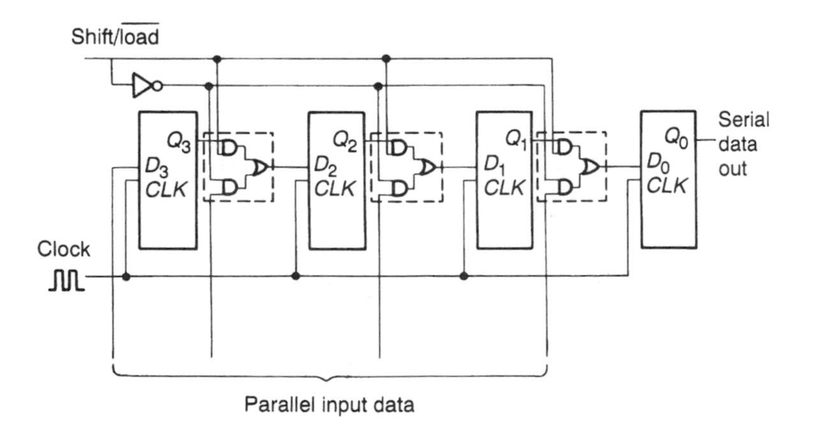

(b) Logic diagram

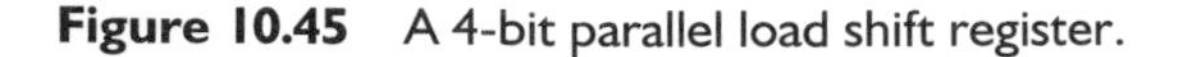
Figure 10.45 A 4-bit parallel load shift register.

selector, the function of which is to allow a control signal, SELECT, to determine which of two inputs A and B is connected to the output X. If SELECT is 1, A' will always be equal to A but B' will always be 0. Thus X will be equal to A. If SELECT is 0, A' will be equal to 0 and B' will be equal to B, making X equal to B. Thus when SELECT is high $X=A$ and when it is low $X=B$. The data selector is therefore a 2-input multiplexer. We have already met the convention of identifying active low signals by placing a 'bar' over their symbolic name. Here we describe the SELECT line by the symbol $A/\overline{B}$, which indicates that when the line is high A is selected, and when it is low B is selected.

Figure 10.45(b) shows a 4-bit parallel load shift register. The operation of the circuit is determined by a single control signal, which is inverted to drive three data selectors. When the control signal is high, the Q output from each stage is fed to the D input of the next and the arrangement is electrically equivalent to the circuit of Figure 10.44. Each time a clock pulse is applied, the contents of the register will be shifted one place to the right and will appear as serial output data at Q_0. When the control signal is low, the D input of each stage is connected to the parallel input lines. The application of a clock pulse will cause the pattern of '1's and '0's on these inputs to be loaded into the register. We therefore have a register that can be used to shift or load data under the control of a single input. This input may be labelled $shift/\overline{load}$.

The circuit of Figure 10.45 can be used for both serial to parallel, and parallel to serial conversion by applying and sensing signals appropriately. It is also possible, by slight modification of this circuit, to construct a register that will shift in either direction. This is achieved by using a data selector arrangement to determine whether the input to a particular stage comes from the output of the stage to the right or that to the left.

Although it is quite feasible to construct such circuits from standard simple gates, it is more common to use specialized integrated circuits that provide all the components of a shift register within a single package. Typical devices provide 4- or 8-bit registers with a range of features. Longer registers can be produced by connecting several devices in series. This is achieved by taking the serial output of one device to the serial input of the next.

10.8.1 Applications of shift registers

One of the most common uses of shift registers is in **serial communications** systems. This involves converting parallel data into a serial form at the *transmitter*, conveying the serial data over some distance and then converting it back into a parallel form at the *receiver*. The process is illustrated in Figure 10.46.

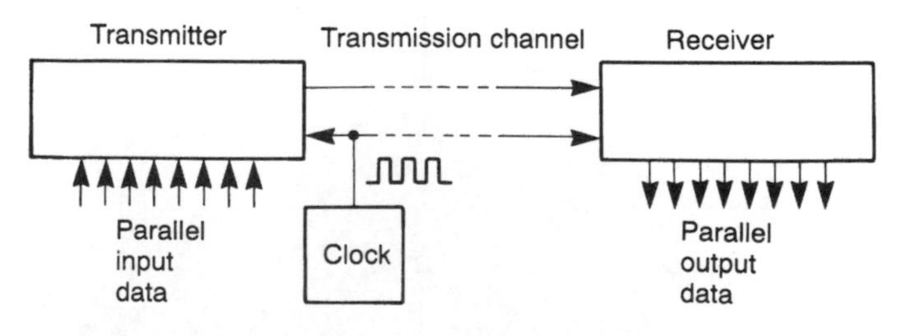

Figure 10.46 A serial communications system.

At the heart of the transmitter is a shift register which loads the input data in parallel and then outputs it in a serial form at a rate determined by a local clock signal. The serial data stream is then transmitted over some form of *transmission channel* to the receiver. This channel may take the form of a piece of wire, a radio signal, a series of laser light pulses or some other information medium. At the receiver, a second shift register loads the serial data and outputs the information in parallel form. To enable it to load the information it must receive not only the serial data but also the clock signal to allow it to *synchronize* with the transmitter.

The main advantage of this method of transmission is that it requires fewer lines for the information to be communicated, requiring only two lines (one for data and one for the clock), rather than one line for each bit of the parallel data. Serial techniques are used extensively for long distance communication. They are also used for short range applications, sometimes down to a few inches. In some systems the requirement to transmit the clock signal along with the data is removed by generating (or recovering) the clock signal at the receiver. This reduces the number of signal lines required to one. These techniques will be discussed in Section 12.5 when we look at computer input/output techniques.

10.9 Asynchronous counters

10.9.1 Ripple counters

Consider the circuit of Figure 10.47. The circuit consists of four negative-going edge-triggered J–K flip-flops configured as toggle flip-flops by connecting both J and K inputs to logical 1. The Q output of each stage forms the clock input for the next.

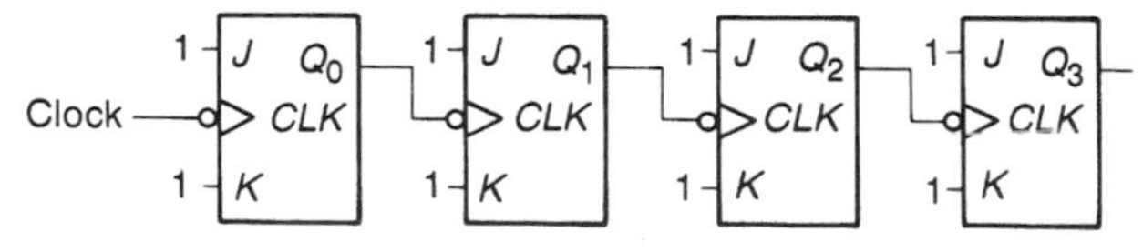

FILE 10G **Figure 10.47** A simple ripple counter.

Figure 10.48 shows the resultant waveforms within the circuit when a square-wave clock signal is applied. It can be seen that Q_0 toggles on each falling edge of the clock producing a square waveform at half the clock frequency. Q_1 toggles at each falling edge of Q_0 and therefore produces a square waveform at half the frequency of Q_0, or one-quarter the frequency of the clock. Similarly, each further stage divides the clock signal by a factor of two producing successively lower frequencies. Such a circuit may be thought of as a **frequency divider** and each stage represents a **frequency halver**.

Frequency dividers can be constructed with any number of stages to provide an appropriate division ratio. For example, digital watches often use such a divider with 15 stages to divide the oscillations of a quartz crystal at 32 768 Hz to provide a 1 Hz signal to drive a stepper motor (for watches with an analogue display using hands) or a digital display.

It is also interesting to look at the sequence of '1's and '0's which appear on the outputs

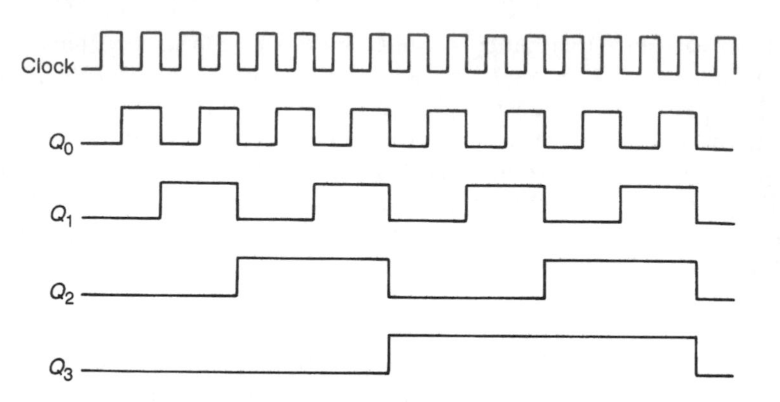

Figure 10.48 Waveform diagram for the simple ripple counter.

of the flip-flops. Figure 10.49 shows the values of the outputs after each clock pulse.

It can be seen that the pattern of the outputs represents the binary code for the number of pulses applied to the circuit. The arrangement therefore represents a *counter*. In this particular circuit the effects of the input are propagated along the series of flip-flops, the outputs changing sequentially along the line. For this reason this form of counter is called a **ripple counter**.

From Figure 10.49 it can be seen that the circuit counts up to the binary equivalent of 15 and then restarts at zero. The counter therefore takes 16 distinct values. We refer to such a counter as a **modulo-16 counter** or, sometimes, simply as a **mod-16 counter**. By using more or fewer stages we can construct a range of circuits that count modulo-2^n,

Number of clock pulses	Q_3	Q_2	Q_1	Q_0
0	0	0	0	0
1	0	0	0	1
2	0	0	1	0
3	0	0	1	1
4	0	1	0	0
5	0	1	0	1
6	0	1	1	0
7	0	1	1	1
8	1	0	0	0
9	1	0	0	1
10	1	0	1	0
11	1	0	1	1
12	1	1	0	0
13	1	1	0	1
14	1	1	1	0
15	1	1	1	1
16	0	0	0	0
17	0	0	0	1
18	0	0	1	0
19	0	0	1	1
20	0	1	0	0

Figure 10.49 The output sequence for the simple ripple counter.

where n is the number of flip-flops used. These counters will count from 0 to 2^{n-1} and then repeat.

10.9.2 Modulo-*N* counters

We have seen that by varying the number of stages we can modify our simple ripple counter to produce circuits that count up to different numbers before restarting at zero. However, varying the number of stages only allows us to choose values for the modulus of the counter which are powers of two. In many applications we wish to count up to particular numbers which may not fulfil this requirement. We therefore need a method of constructing a generalized counter that can count up to any number. Such a circuit is usually called a **modulo-*N* counter**.

In order to produce a counter with a modulus of N we simply need to ensure that on the clock pulse after the counter reaches $N-1$, it returns to zero. An example of such a circuit, for a value of $N=10$, is given in Figure 10.50. Counters that count modulo-10 are referred to as **decade counters**. Decade counters that output binary numbers in the range 0000 to 1001 are often called **binary coded decimal counters** or simply **BCD counters**.

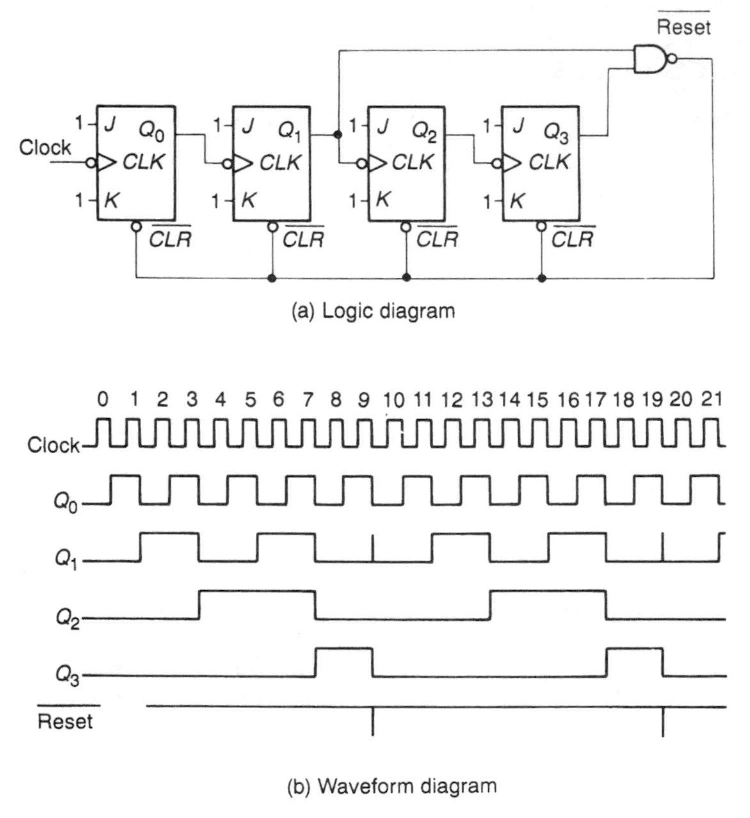

FILE 10H **Figure 10.50** A decade ripple counter.

The circuit of the decade counter is similar to that of the simple ripple counter of Figure 10.47, but has an extra reset circuit to clear all the flip-flops when they reach a count of 10. The clearing operation is achieved by using flip-flops that have **CLEAR** inputs. The signal applied to this input comes from circuitry that detects a count of 10. Since 10 is binary 1010. a simple two-input NAND gate can be used to detect the first occasion on which bits 1 and 3 are high (remember the bits start from bit 0). As soon as this is detected the $\overline{RESET}$ line goes low, clearing the counter to zero, which also sets $\overline{RESET}$ back to 1. The counter then continues from zero as before. Since the reset circuitry acts so quickly, the effects of the spike produced on Q_1 are not usually significant.

A modulo-N counter can be constructed for any value of N by forming a counter with n stages, where $2^n > N$, and then adding a reset circuit which detects a count of N.

Computer simulation exercise 10.1

FILE 10H

Design a modulo-12 asynchronous counter using negative edge-triggered J–K bistables.

Take as your starting point the circuit (and demonstration file) of Figure 10.50. Modify this circuit to produce a mod-12 counter and simulate the circuit to confirm its correct operation.

10.9.3 Down counters

In some applications it is necessary to have a counter that counts down rather than up. This can be achieved using a similar circuit to that of the earlier ripple counter by taking the clock signal for following stages from $\overline{Q}$ rather than Q. This is illustrated in Figure 10.51, in which you should note that the outputs are taken from Q_0–Q_3 and not from $\overline{Q_0}$–$\overline{Q_3}$. Table 10.1 shows the output sequence for the four-stage downcounter. This technique can be applied to counters of any modulus.

Computer simulation exercise 10.2

FILE 10G

Design a 4-bit asynchronous down counter using negative edge-triggered J–K bistables.

Take as your starting point the circuit (and demonstration file) of Figure 10.47. Modify this circuit to produce the required function and simulate the circuit to confirm its correct operation.

10.9.4 Up/down counters

Combining the circuits of Figures 10.47 and 10.51 using the **data selector** of Figure 10.45, it is possible to construct a counter that can count up or down. Such a circuit is shown in Figure 10.52.

The direction of counting is controlled by the $up/\overline{down}$ signal. When this line is high, the Q output from each stage is used to provide the clock for the next stage and the circuit counts up. When the control line is low, the $\overline{Q}$ output is fed to the next stage and the circuit counts down.

Table 10.1 The output sequence of the down counter.

Number of clock pulses	Q_3	Q_2	Q_1	Q_0	Count
0	0	0	0	0	0
1	1	1	1	1	15
2	1	1	1	0	14
3	1	1	0	1	13
4	1	1	0	0	12
5	1	0	1	1	11
6	1	0	1	0	10
7	1	0	0	1	9
8	1	0	0	0	8
9	0	1	1	1	7
10	0	1	1	0	6
11	0	1	0	1	5
12	0	1	0	0	4
13	0	0	1	1	3
14	0	0	1	0	2
15	0	0	0	1	1
16	0	0	0	0	0
17	1	1	1	1	15
18	1	1	1	0	14
19	1	1	0	1	13
20	1	1	0	0	12

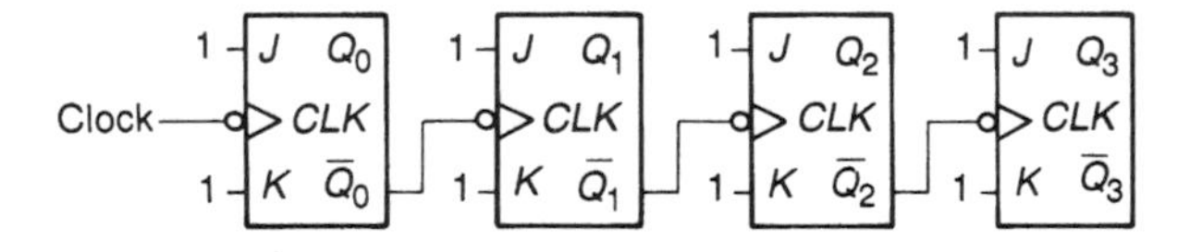

(a) Logic diagram

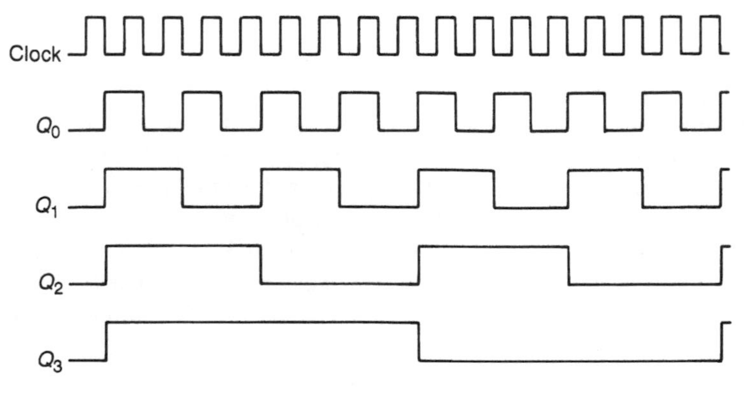

(b) Waveform diagram

Figure 10.51 A ripple down counter.

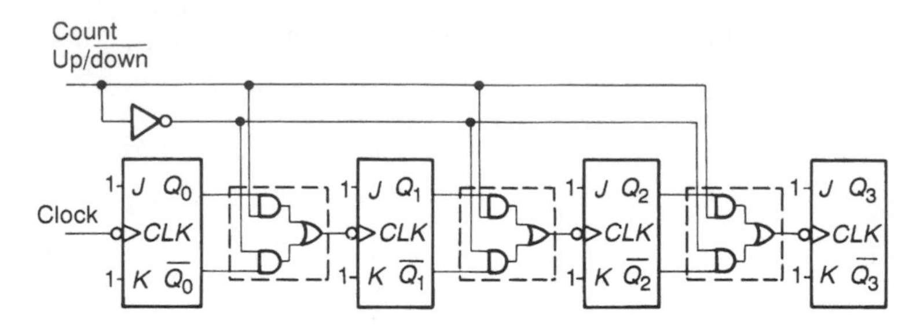

Figure 10.52 An up/down ripple counter.

Example 10.2 Converting relative motion into absolute position for an incremental position encoder or a mouse

In Section 2.3.4 we looked at the signals produced by an incremental position encoder and noted that these take the form of two square waves that are phase shifted with respect to each other. In Example 2.1 we looked at the design of a microcomputer mouse and devised an arrangement that produced a similar pair of output signals for each axis of motion. The form of the signals is shown below.

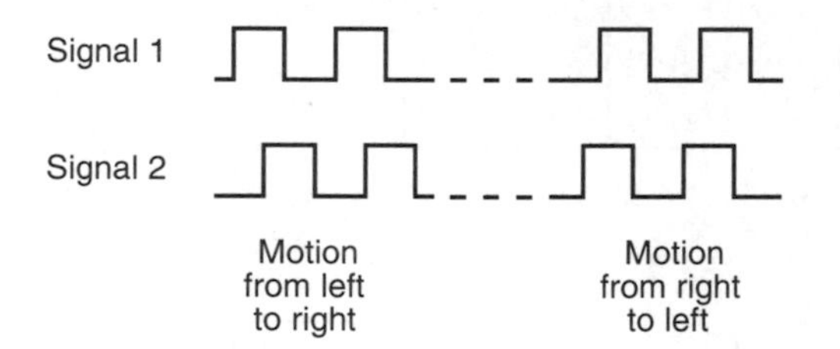

We can see that motion in one direction causes signal 2 to lead signal 1, while motion in the other direction causes signal 1 to lead signal 2. We require a mechanism to count the number of 'steps' in one direction and to subtract from it the number of steps in the other direction to give a measure of the absolute position.

To see how this objective may be met we need to look at the timing of the two signals. In particular, look at the state of signal 2 on the negative-going edge of signal 1. You will see that for motion from left to right signal 2 is high on the falling edge of signal 1. However, for motion from right to left signal 2 is low on the falling edge of signal 1. This allows us to use an up/down counter to determine the absolute position.

Signal 1 is fed to the clock of the counter and signal 2 is connected to the up/down control input. Motion from left to right will now cause the counter to count up and motion

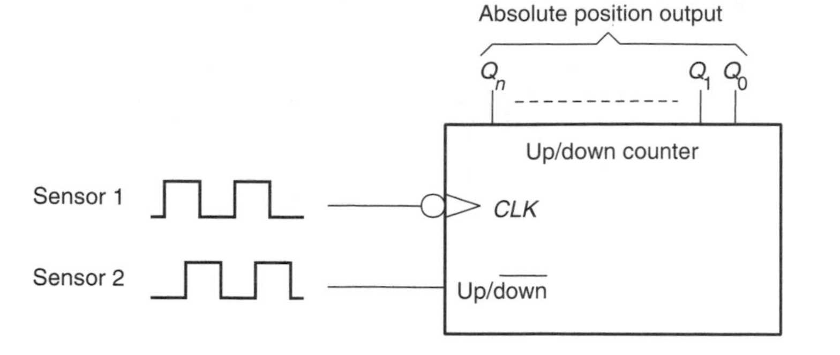

from right to left will cause it to count down. Depending on the application, some mechanism might be required to zero the counter to set its absolute value.

In practice the pulses produced by a mouse are counted within the microcomputer, rather than by an external hardware counter. However, the principles involved are the same.

10.9.5 Propagation delay in ripple counters

Though ripple counters are extremely simple to construct, they do have a major disadvantage which is particularly apparent when high-speed operation is required. Since the output of one flip-flop is triggered by the change of state of the previous stage, delays produced by each flip-flop are summed along the chain. Each flip-flop takes a finite time to respond to changes at its inputs. This is termed the **propagation delay time** or t_{pd}. In a ripple counter of n stages it will take $n \times t_{pd}$ for the counter to respond. If the counter is read during this time the value will be garbled, as some stages will have changed while others will not. This produces a fundamental limit to the maximum clock frequency that can be used with the counter. If clock pulses are received by the first stage before the last has responded, at no time will the counter read the correct value.

10.10 Synchronous counters

One solution to the problems of propagation delay within ripple counters is to use synchronous techniques. This is achieved by connecting all the flip-flops within a counter to a common clock signal so that they all change state at the same time. This ensures that a short time after the clock signal changes, the counter can be read with confidence that all stages will have responded.

Clearly, if all the stages of a counter are connected to the same clock then some method must be used to determine which stages change state and which remain the same. To investigate this, consider the required outputs of a four-stage counter, as shown in Table 10.2.

Looking at this sequence we can determine rules that define when the various outputs change. We can then produce a counter consisting of a number of edge-triggered J–K flip-flops with circuitry to enforce these rules. Since all the flip-flops are triggered simultaneously, they will respond to the values on their control inputs immediately before the

Table 10.2 The output sequence of a four-stage up counter.

Number of clock pulses	Q_3	Q_2	Q_1	Q_0
0	0	0	0	0
1	0	0	0	1
2	0	0	1	0
3	0	0	1	1
4	0	1	0	0
5	0	1	0	1
6	0	1	1	0
7	0	1	1	1
8	1	0	0	0
9	1	0	0	1
10	1	0	1	0
11	1	0	1	1
12	1	1	0	0
13	1	1	0	1
14	1	1	1	0
15	1	1	1	1
16	0	0	0	0
17	0	0	0	1
18	0	0	1	0
19	0	0	1	1
20	0	1	0	0

clock edge. The slight delay caused by the circuitry used to generate these signals will ensure that they are stable while the flip-flops respond.

Q_0 This output is the simplest to define since it changes on every clock pulse. This can be simply produced by configuring flip-flop 0 to toggle on every clock pulse by connecting both J and K to 1.

Q_1 This output changes state after each clock period where Q_0 is high. This can be achieved by connecting both J and K of flip-flop 1 to Q_0. When Q_0 is high this stage will toggle; when it is low it will remain unchanged.

Q_2 This output changes after each clock period when both Q_0 and Q_1 are high. This can be achieved by ANDing Q_0 and Q_1 together and applying the resultant signal to the J and K inputs of flip-flop 2.

Q_3 This output changes after each clock period when Q_0, Q_1 and Q_2 are all high. This can be achieved by ANDing these three outputs to produce a signal to be applied to the J and K inputs of flip-flop 3.

The resultant circuit is shown in Figure 10.53. Clearly this technique can be extended to produce synchronous counters of any length. As the number of elements in the counter

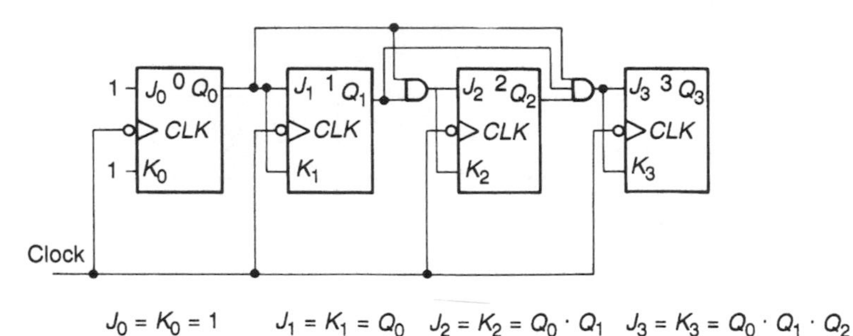

Figure 10.53 A synchronous four-stage counter.

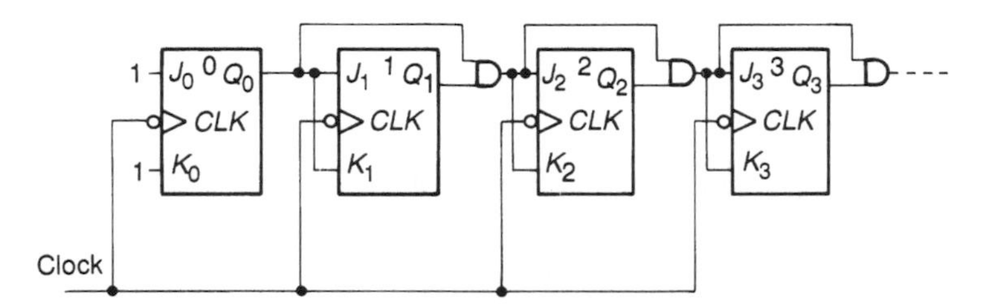

Figure 10.54 A cascadable 4-bit synchronous counter.

increases, the number of signals that must be ANDed together at each stage becomes large. This problem can be alleviated by noting that the signals required by the J and K inputs of one stage are simply those required by the previous stage ANDed with the output of the previous stage. This allows a simplification of the circuit, as shown in Figure 10.54. This circuit can be extended by adding as many identical stages as necessary.

Synchronous counters can be configured to produce up, down, up/down and modulo-N counters in a similar manner to ripple counters. They have the advantage that all outputs change at the same time, allowing the counter to be read safely a short time afterwards. They have the disadvantage of increased complexity compared with ripple counters, but can be used at higher clock frequencies.

10.11 Integrated circuit counters

Although it is feasible to construct counters using individual gates or combinations of flip-flops, it is more common to use specialized integrated circuits that contain all the functions of a counter within a single package. Both synchronous and asynchronous types are available in a number of sizes and with a range of features. Binary, decade and BCD (binary coded decimal) counters are available, as are up, down and up/down versions. Typical circuits might provide 4- to 14-bit counters, or several independent counters within a single package. Some counters allow their contents to be cleared or preset (loaded) with a particular value.

Most integrated counters are designed so that they can be **cascaded** to form counters of greater length. This is achieved with ripple counters simply by taking the most

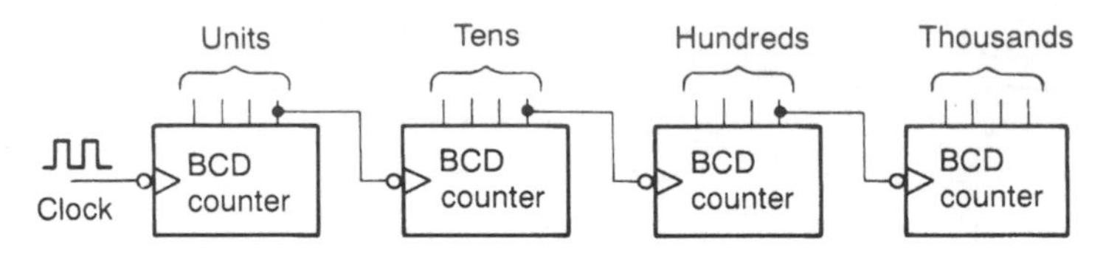

Figure 10.55 Cascading asynchronous BCD counters.

significant bit of one stage as the clock input for the next. This is illustrated in Figure 10.55, which shows a 4-decade BCD counter. The output of each digit of this counter could be used to drive a separate BCD display.

With synchronous counters, a common clock is used for all stages and appropriate signals are provided to allow a number of counters to be joined together.

10.12 Design of sequential logic circuits

At the beginning of this chapter we observed that a sequential system consists of combinational logic with some memory within a feedback path. We have looked at some examples of simple sequential elements in the form of bistables and seen how these can be used to form larger systems, such as counters. In this final section we shall look at the problem of designing sequential circuits to fulfil particular tasks.

Sequential systems may be either synchronous or asynchronous in nature. Synchronous systems are based on circuits that are controlled by a master clock; the values of the inputs and internal states are only of importance at times determined by transitions of this clock. Such arrangements often use clocked bistables as their building blocks. In asynchronous systems there is no clock. Internal states, and the outputs, may be affected at any time by changes in the inputs. Such systems are often based on unclocked bistable latches or use time delays within logic elements to represent storage within the feedback path. In general, synchronous design is more straightforward than its asynchronous equivalent, the latter being susceptible to timing problems and instability. We will therefore concentrate on the design of synchronous sequential systems.

10.12.1 Synchronous sequential systems

The approach taken to the design of a sequential system will be greatly affected by the nature of the problem and the way in which it is defined. Here we will discuss one approach to the design of synchronous systems which uses the building blocks that we met earlier in this chapter.

System states

One of the first tasks is to determine all of the discrete **system states** that exist within the system. In this context the word state refers to a combination of internal and output variables of the system.

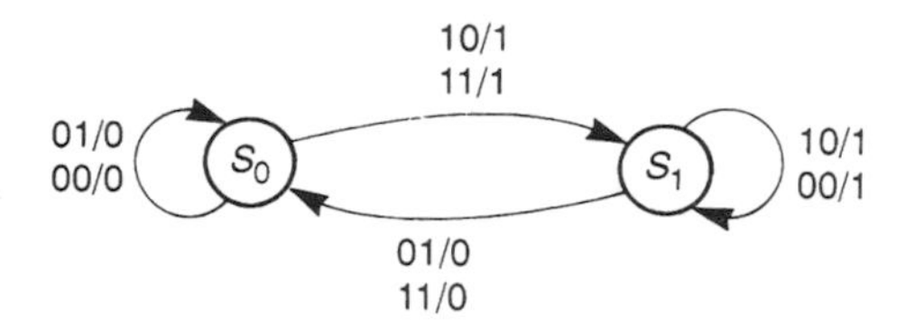

Figure 10.56 The state transition diagram of a J–K flip-flop.

State transition diagram

In simple systems the system states can be derived directly from the problem definition, but in more complex cases it may be necessary to model the system using a **state transition diagram**. Figure 10.56 shows a simple state diagram for a J–K flip-flop.

The J–K flip-flop has two inputs (J and K), one output (Q) and two states corresponding to $Q=0$ and $Q=1$. The states have been named S_0 and S_1 and are represented by the two circles in the diagram. The lines joining the two circles represent possible transitions between these states while the labels on these lines indicate the input conditions for which these transitions will occur and the resultant output. The notation used for the labels is JK/Q. The diagram also shows lines that leave each state and circle back to terminate on the same state. The labels on these lines indicate the input conditions for the system to stay in that state and the output produced while in that state.

The diagram shows that when the circuit is in S_0 the output Q is 0 and that it will remain in that state while the inputs JK are 01 or 00. A transition from S_0 to S_1 will occur for inputs 10 or 11, and if this occurs the output will go to 1. You might like to consider the other aspects of the diagram and to confirm that it corresponds to the functions of the J–K flip-flop outlined earlier in this chapter.

The names given to the two states in the above example were arbitrary and it would not affect the diagram if the labels were reversed. Most sequential systems have more than two states and a more typical diagram is shown in Figure 10.57, which shows a system with five states (S_0 to S_4). The operation of this circuit is not of paramount importance but is, in fact, that of a circuit with two inputs N (next) and R (reset) and one output Q. While the reset input is low (inactive) the circuit moves from state to state around the loop, moving on one state each time that N is high during a clock pulse. The output Q is high only in state S_0, thus the output is high for one in every five counts. Taking the R input high causes the system to jump to S_0, where the output is 1, and to remain there as long as R is held high. The arrangement is thus a form of modulo-5, resettable counter.

The state diagram indicates for each state the action (or lack of action) that will result from each of the possible combinations of the inputs. It is thus an unambiguous description of the system, unlike the verbal definition given above which is open to misinterpretation.

State transition table

From the state diagram it is possible to construct a **state transition table** (also called simply a **state table** or a **next state table**) which tabulates this information. The table lists all the possible input combinations for each system state and indicates the resultant

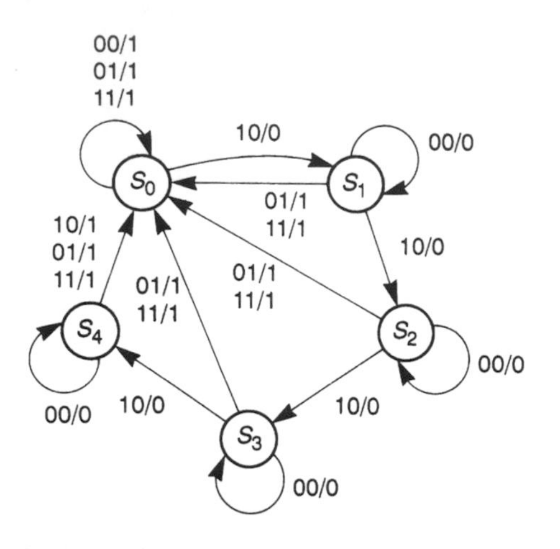

Figure 10.57 The state transition diagram of a system with five states.

state (the 'next' state). Table 10.3 shows the state transition table for the J–K flip-flop shown in Figure 10.56 and Table 10.4 shows the same table simplified by the use of 'don't care' conditions. Table 10.5 shows the more complex table for the modulo-5 counter of Figure 10.57.

Table 10.3 The state transition table for a J–K flip-flop.

Present state	Input conditions JK	Next state	Output Q
S_0	00	S_0	0
	01	S_0	0
	10	S_1	1
	11	S_1	1
S_1	00	S_1	1
	01	S_0	0
	10	S_1	1
	11	S_0	0

Table 10.4 The state transition table for a J–K flip-flop using 'don't care' conditions.

Present state	Input conditions JK	Next state	Output Q
S_0	$0X$	S_0	0
	$1X$	S_1	1
S_1	$X1$	S_0	0
	$X0$	S_1	1

State reduction

Inspection of the state table of Table 10.5 will show that all the states included in the table are unique in that they respond differently to the various input combinations. This is not always the case. Consider the state table of Table 10.6 which represents a system with one input and seven states.

Table 10.5 The state transition table far the modulo-5 counter of Figure 10.57.

Present state	Input conditions NR	Next state	Output Q
S_0	$0X$	S_0	1
	10	S_1	0
	11	S_0	1
S_1	00	S_1	0
	$X1$	S_0	1
	10	S_2	0
S_2	00	S_2	0
	$X1$	S_0	1
	10	S_3	0
S_3	00	S_3	0
	$X1$	S_0	1
	10	S_4	0
S_4	00	S_4	0
	$X1$	S_0	1
	10	S_0	1

Table 10.6 The state table for a system with redundant states.

Present state	Input conditions X	Next state	Output Q
S_0	0	S_0	0
	1	S_1	0
S_1	0	S_2	1
	1	S_3	0
S_2	0	S_0	1
	1	S_3	0
S_3	0	S_4	0
	1	S_5	1
S_4	0	S_0	0
	1	S_5	1
S_5	0	S_6	0
	1	S_5	1
S_6	0	S_0	0
	1	S_5	1

Table 10.7 The state table for a system with redundant states.

Present state	Input conditions X	Next state	Output Q
S_0	0	S_0	0
	1	S_1	0
S_1	0	S_2	1
	1	S_3	0
S_2	0	S_0	1
	1	S_3	0
S_3	0	S_4	0
	1	$S_5 \rightarrow S_3$	1
S_4	0	S_0	0
	1	$S_5 \rightarrow S_3$	1
$S_5 \rightarrow S_3$	0	$S_6 \rightarrow S_4$	0
	1	S_5	1
$S_6 \rightarrow S_4$	0	S_0	0
	1	S_5	1

Table 10.8 The state table for the system of Table 10.6 with redundant states removed.

Present state	Input conditions X	Next state	Output Q
S_0	0	S_0	0
	1	S_1	0
S_1	0	S_2	1
	1	S_3	0
S_2	0	S_0	1
	1	S_3	0
S_3	0	S_4	0
	1	S_3	1
S_4	0	S_0	0
	1	S_3	1

Close inspection of Table 10.6 will show that states S_4 and S_6 are identical, since in both cases when the input is 0 the next state is S_0 and the output is 0, and when the input is 1 the next state is S_5 and the output is 1. We can therefore remove S_6 and change all references to it to S_4. Having done this we now observe that states S_3 and S_5 are identical, and we can remove S_5 and change all references to it to S_3. This process is shown in Table 10.7.

The redundant states can now be removed leaving a system with five internal states, as shown in the state diagram of Table 10.8.

State assignment

Each of the system states must be represented in the final design by a unique combination of internal variables. Clearly the number of variables required will be determined by the number of independent states. In many cases the internal variables will be represented by the outputs of bistables and in such cases the minimum number of bistables N is related to the number of states S by the relationship

$$2^N \geqslant S$$

Thus a system with two states can be implemented using a single bistable; systems with three or four states require two bistables; systems with five to eight states require three bistables; and so on.

Our modulo-5 counter example has five states and thus needs a minimum of three bistables. Having decided on the number of bistables, it is now necessary to assign particular combinations of their outputs to each state. If we represent the Q outputs of our three bistables by A, B and C, we can describe the outputs of the bistables by the 3-digit binary number ABC. We might then choose to make the pattern 000 represent state S_0, pattern 001 represent state S_1, and so on. In synchronous systems any choice of assignments will result in a workable system, although some assignments will result in simpler circuitry than others. We will see later that in asynchronous systems this is not the case, and state assignment must take into account the stability of the resulting arrangement. Table 10.9 shows a possible state assignment for our modulo-5 counter while Table 10.10 shows a state transition table which represents the states by their pattern of internal variables.

Excitation table

The transition table indicates the action to be taken by each bistable for each combination of the input variables in each state. To design a system from such a table we need to decide on the nature of the bistables to be used, and hence to determine what the inputs to each bistable must be in order to produce the required actions.

Let us consider the modulo-5 counter and assume that we are to use J–K flip-flops to implement the system. The action of a J–K flip-flop can be deduced from the informa-

Table 10.9 The state assignment table for the modulo-5 counter of Figure 10.57.

State	Internal variables ABC
S_0	000
S_1	001
S_2	010
S_3	011
S_4	100

Table 10.10 The state transition table for the modulo-5 counter of Figure 10.57.

Present state ABC	Input conditions NR	Next state ABC	Output Q
000	0X	000	1
	10	001	0
	11	000	1
001	00	001	0
	X1	000	1
	10	010	0
010	00	010	0
	X1	000	1
	10	011	0
011	00	011	0
	X1	000	1
	10	100	0
100	00	100	0
	X1	000	1
	10	000	1

Table 10.11 The control inputs for a J–K flip-flop.

Present state Q_n	Next state Q_{n-1}	Control inputs required J	K
0	0	0	X
	1	1	X
1	0	X	1
	1	X	0

tion given in Table 10.4. This can be reorganized to give the relationship between the desired output transitions and the necessary combinations of the control inputs, as shown in Table 10.11.

Combining the data on the J–K flip-flop given in Table 10.11 with that given in the transition diagram of Table 10.10 allows us to deduce the control signals required for the three flip-flops. If we look at the first row of Table 10.10 we see that when the present state is 000 (ABC) and the inputs are 0X (NR), the next state is also 000. In other words, with this combination of internal variables and inputs, all three bistables go from 0 to 0 (in other words they remain at 0). This 'transition' is represented by the first row of Table 10.11, indicating that the J input of all three bistables must be at 0, whereas the status of the K input is unimportant (don't care). If we now turn to the second row of Table 10.10 we see that when the present state is 000 and the inputs are 10, the next state is 001. Thus bistables A and B go from 0 to 0 as before (and require $J=0$, $K=X$) but now bistable C goes from 0 to 1 and, from Table 10.11, requires $J=1$ and $K=X$. This process can be

Table 10.12 The excitation table for the modulo-5 counter of Figure 10.57.

Present state ABC	Input conditions NR	Next state ABC	Flip-flop inputs J_A	K_A	J_B	K_B	J_C	K_C	Output Q
000	0X	000	0	X	0	X	0	X	1
	10	001	0	X	0	X	1	X	0
	11	000	0	X	0	X	0	X	1
001	00	001	0	X	0	X	X	0	0
	X1	000	0	X	0	X	X	1	1
	10	010	0	X	1	X	X	1	0
010	00	010	0	X	X	0	0	X	0
	X1	000	0	X	X	1	0	X	1
	10	011	0	X	X	0	1	X	0
011	00	011	0	X	X	0	X	0	0
	X1	000	0	X	X	1	X	1	1
	10	100	1	X	X	1	X	1	0
100	00	100	X	0	0	X	0	X	0
	X1	000	X	1	0	X	0	X	1
	10	000	X	1	0	X	0	X	1
101	XX	XXX	X	X	X	X	X	X	X
110	XX	XXX	X	X	X	X	X	X	X
111	XX	XXX	X	X	X	X	X	X	X

repeated for each row of the state transition table to identify the inputs required on the J and K inputs of each flip-flop for each state and for each combination of the inputs. This information is used to create an **excitation table**, as shown in Table 10.12, where J_A is the signal on the J input of flip-flop A, K_B is the signal on the K input of flip-flop B etc. This table also indicates the output of the system for each combination.

You will notice that Table 10.12 has three more rows than Table 10.10. This is because three bistables in fact produce eight possible states and we must define what the system will do in these **unused states**. Here we have simply assumed that since these states are not used the values of the flip-flop inputs corresponding to these states are unimportant. We will return to this topic a little later.

Circuit design

The excitation table shows which signals must be applied to the inputs of each of the bistables in order to achieve the appropriate transitions between states. The final stage of the design is to define circuitry to produce these signals, and the required outputs, from the inputs and the internal variables. This process is shown in Figure 10.58.

The appropriate columns of the excitation table define the signals required for each flip-flop input in terms of the system inputs and the internal variables (the flip-flop outputs). In simple systems, Karnaugh maps can be used to produce expressions for these functions. Automated methods would normally be used in more complex arrangements, as described in Section 9.4.2.

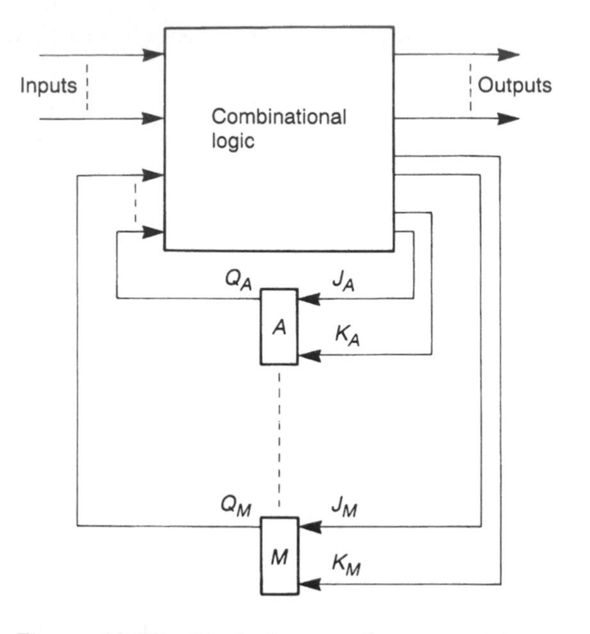

Figure 10.58 Block diagram of a sequential system.

Figure 10.59 shows Karnaugh maps for the six flip-flop input signals of our modulo-5 counter, and the resultant Boolean expressions.

The logic required to generate the output signals of a system may be defined by an **output logic table** which lists each combination of the internal states and the inputs and indicates the appropriate output. In some cases it is simpler to combine this information with that in the state transition table or excitation table, as in the earlier examples.

The excitation table of Table 10.12 indicates the required output of the modulo-5 counter for each combination of the internal states and the inputs. From this information the output logic can be designed as for the control logic for the bistables. Figure 10.60 shows a Karnaugh map that represents this information and indicates a simplified Boolean function for the required logic.

Figure 10.61 shows the complete modulo-5 counter implemented using a mixture of logic gates. Clearly this design could be adapted to use only NAND gates, or only NOR gates if required, as described in Examples 9.5 and 9.6.

The bistables used in this example are synchronous J–K devices. They may be either master/slave or edge-triggered. When using master/slave flip-flops the inputs are sampled before the outputs are allowed to change, thereby preventing any possible hazard problems. If edge-triggered devices are used, the delays within the logic gates will prevent the input signals from changing until after each flip-flop has responded to the clock transition, again preventing any hazard condition. Other forms of bistable, for example R–S flip-flops, could be used in place of the J–K devices by using appropriate control input data in place of that given in Table 10.11.

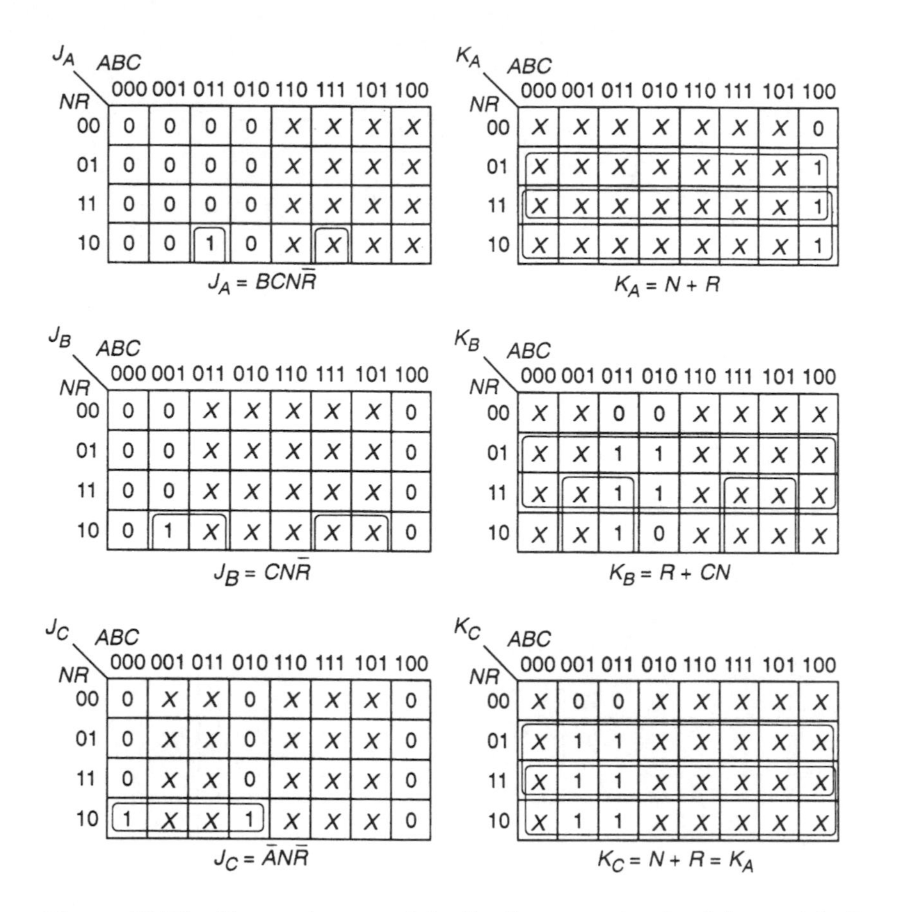

Figure 10.59 Karnaugh maps of the flip-flop input signals of the modulo-5 counter.

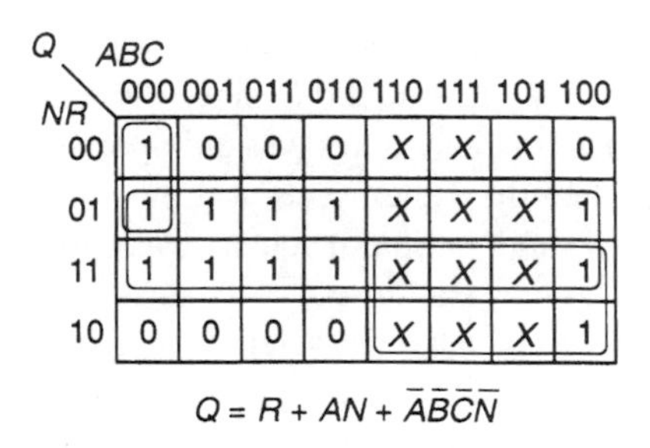

Figure 10.60 A Karnaugh map for the output logic of the modulo-5 counter.

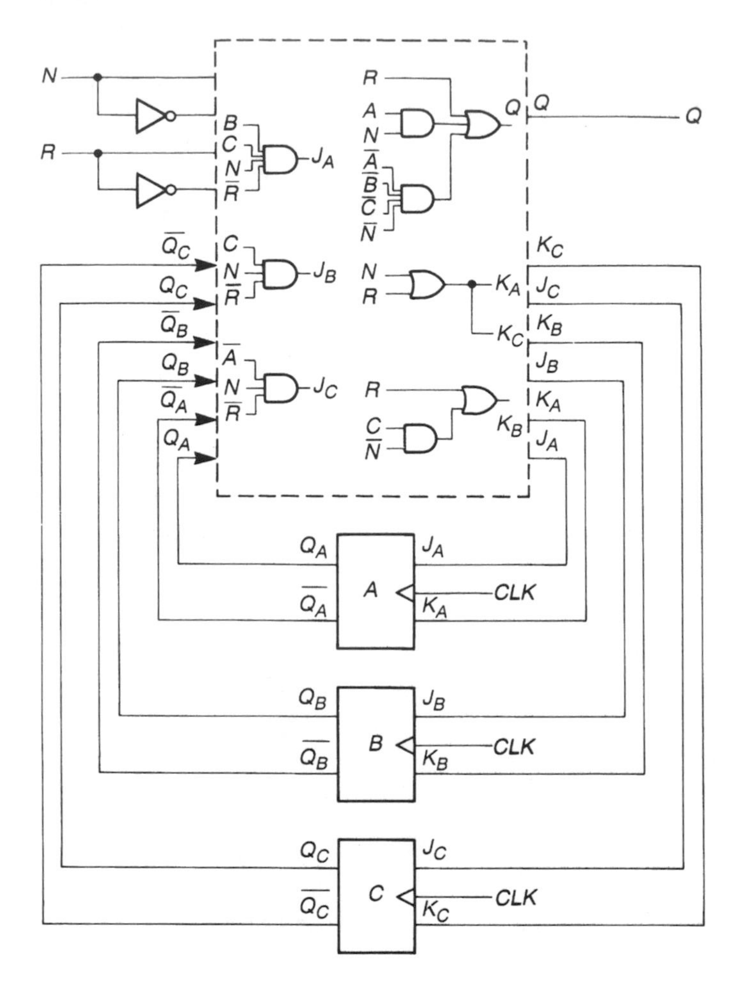

Figure 10.61 Circuit diagram of the modulo-5 counter.

Unused states

Earlier in the design of the modulo-5 counter we observed that only some of the possible internal states are used. At that time we simply assumed that since the remaining states were never used, the behaviour of the system in those states was unimportant. This leaves us with a potential problem, however, when the system is first turned on. At that time the various bistables will adopt effectively random output states, resulting in the system state being undefined. In this way the counter may start in any of the possible system states, including those which are unused. The action taken by the system in these normally unimportant states must therefore be considered.

Figure 10.62 illustrates several possible state diagrams for a simple system with five states. In (a) and (b) the unused states (S_5 to S_7) are linked to the used states, and a system

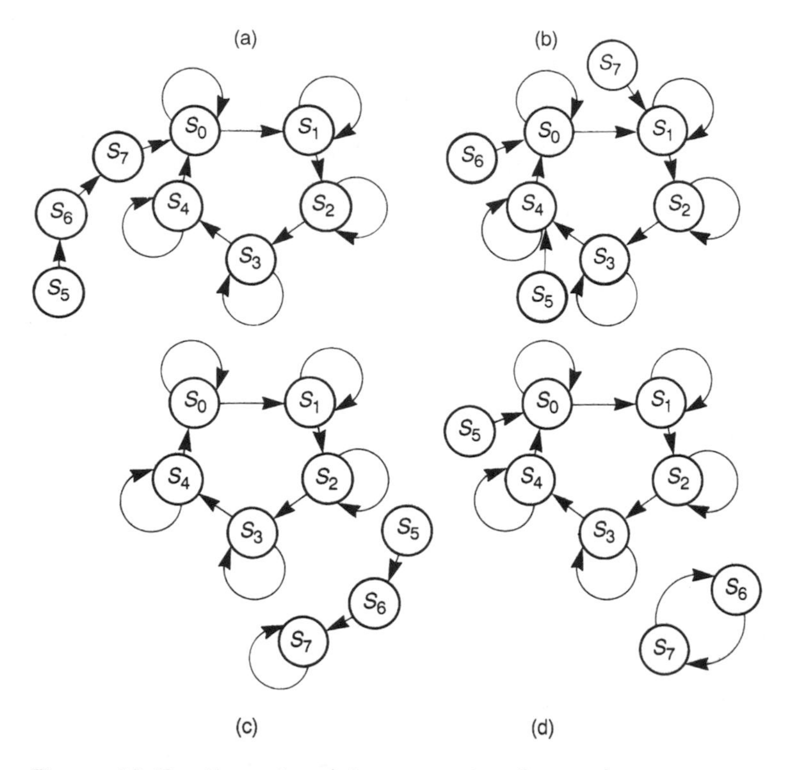

Figure 10.62 Examples of the properties of unused states.

starting in any of the unused states would soon move into the area of normal operation. In structures (c) and (d) the unused states do not all lead into the used states and in these cases it is possible for the system to become 'locked' in an undesirable mode in which it is not operating as intended. In the circuit of Figure 10.62(c), for example, the system could start in states S_5, S_6 or S_7 and would then become locked in state S_7 indefinitely.

There are several possible approaches to the problems of unused states. One is to incorporate additional circuitry to force the circuit to start in a particular state at power up. This circuitry can often be very simple, for example, a resistor and capacitor arrangement to hold the $\overline{PRESET}$ or $\overline{CLEAR}$ inputs of the bistables low for a short time after the power is applied. An alternative approach is to specify the action to be taken within the unused states, making them effectively used states. The designer could, for example, specify that the action in the unused states would be as in Figure 10.62(a), ensuring that the circuit would take up correct operation within a few clock cycles. This approach has the disadvantage that it often leads to a more complex solution than would otherwise be required since it is not usually apparent which transition routes produce the simplest system. A third approach is to adopt the technique used in the design of the modulo-5 counter earlier in this section. Here we assumed that the action of the system was unimportant within the unused states. This technique will produce simple hardware, but we must then investigate the final design to see if the behaviour in the unused states is acceptable. This is done by

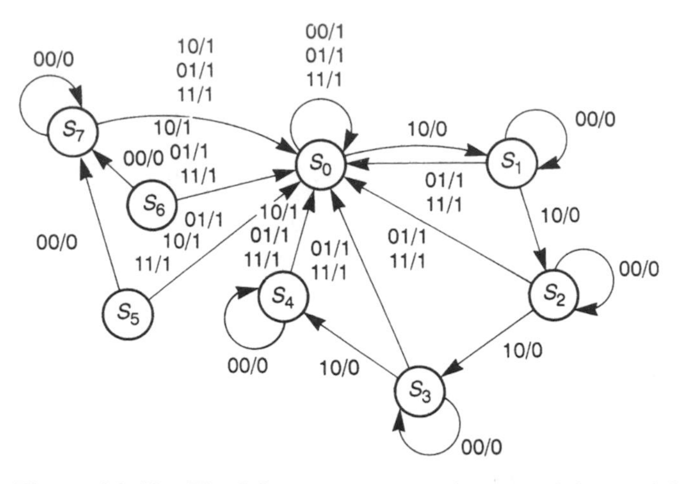

Figure 10.63 The full state transition diagram of the modulo-5 counter.

working backwards through the design replacing the 'don't care' conditions in the excitation table with the actual values used. These may be deduced from the logic functions produced or from the Karnaugh maps. From the full excitation table it is then possible to produce a complete transition diagram including the unused states.

Figure 10.63 shows the complete state transition diagram of the modulo-5 counter designed earlier. It can be seen that if the system starts in states S_5, S_6 or S_7, it will move into state S_0 as soon as either of the inputs goes to 1. If the system starts in one of the unused states, it will therefore fail to increment on the first clock pulse for which N is 1, but will then work correctly. If the system starts in any of the remaining states except S_0 it will also require a few clock pulses to get established in its correct sequence. If this characteristic is unacceptable it will be necessary to incorporate circuitry to preset the system to S_0 at power up, as described earlier.

10.12.2 Asynchronous sequential systems

In the synchronous sequential systems described in the last section, the internal states and outputs are only updated in response to transitions of a clock. In such systems a single change of state can occur each time the clock pulses; the values of the inputs and the existing internal states determine the transitions that occur at that time. Asynchronous systems have no clock and such systems will respond at any time to changes in their inputs. These changes may involve a sequence of transitions between a number of states as the effects ripple through the system. The memory elements in asynchronous systems are usually either unclocked bistables or logic gates providing a time delay. The latter act as memory devices since the logic signals are effectively stored as they propagate through the circuit.

The design of asynchronous sequential systems is more complicated than that of their synchronous counterparts since the timing of the signals plays an important role. In synchronous systems changes between states can only occur at active transitions of the clock. However, in asynchronous systems they may occur at any time and the states must

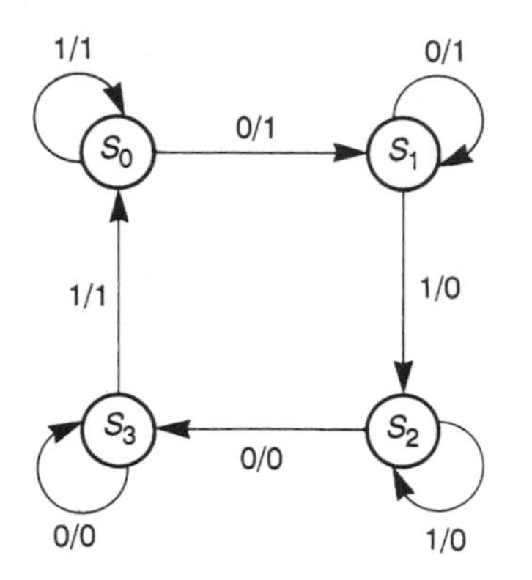

Figure 10.64 A state transition diagram for a system with stable states.

therefore be designed so that they are stable. This can be illustrated by looking back at the synchronous system of Figure 10.57. If we assume that the system is initially in state S_0 and that the input pattern is $NR = 10$, it can be seen that successive clock pulses will cause the circuit to circulate through the states, one state per clock pulse. If we now assume that this diagram represents an asynchronous system the result is quite different. As soon as the input pattern goes to $NR = 10$ the circuit will leave state S_0 and enter state S_1. However, when in state S_1 this input pattern will immediately cause it to switch to state S_2. This process will continue with the circuit circulating through the states at a speed determined by the delays within the system. A solution to this problem is to ensure that the pattern of inputs which causes the system to enter a particular state will cause it to remain in that state. A change in the inputs will then be required to move to another state. Such an arrangement is shown in Figure 10.64.

The design of asynchronous sequential systems involves many of the same components as synchronous design, although their execution is often more complicated as a result of timing and stability considerations. A detailed description of the design of asynchronous systems is not within the scope of this text and the reader is referred to one of the many books that cover this topic.

Key points

- Sequential logic circuits have the characteristic of *memory* in that their outputs are affected by the sequence of events that led up to the current set of inputs.
- Such systems may be modelled by an arrangement of combinational logic with a feedback path including some form of memory.
- Among the most important groups of sequential logic elements are the various forms of multivibrator. These circuits can be divided into three types:
  - Bistables, which have two stable outputs
  - Monostables, which have one stable and one meta-stable (quasi-stable) output
  - Astables, which have no stable states but two meta-stable states.

- The most widely used class of multivibrator is the bistable. These may be divided into
  - Latches
  - Edge-triggered flip-flops
  - Pulse-triggered (master/slave) flip-flops
- Each class of device may then be divided into a range of devices with different operating characteristics. These are often described by symbolic names, such as R–S, J–K, D or T type devices.
- Bistables are frequently used in groups to form registers or counters.
- Registers form the basis of computer memories and are also used for serial to parallel conversion.
- Counters, in their various forms, are used extensively for timing and sequencing functions. They are produced using two basic circuit techniques.
  - Asynchronous or ripple counters, in which the clock for one stage is generated from the output of the previous stage. The result is a ripple effect as each stage changes in sequence.
  - Synchronous counters, in which all stages are clocked simultaneously so that all the outputs change at the same time.
- Both techniques can be used to produce counters that can count up or down.
- Modulo-*N* counters can also be produced.
- Standard integrated circuit building blocks are available to simplify the construction of counters and registers. These can normally be cascaded to form units of any desired length.
- Sequential logic circuits can be designed using either synchronous or asynchronous techniques. The former use clock signals to control the operation of the system; the latter have no clock and will respond at any time to changes in their inputs.
- The design of synchronous systems involves a number of stages which will vary depending on the nature of the problem and the circuit techniques used. A fairly straightforward method involves the use of clocked flip-flops and appropriate combination logic. A possible design process might involve:
  - identification of the system states
  - a state transition diagram
  - a state transition table
  - state reduction
  - state assignment
  - generation of an excitation table
  - circuit design
  - investigation of unused states.
- The design of asynchronous systems is a more complicated task because of timing considerations and problems of stability.

Design study

Design a digital stop-watch that displays the time to a resolution of 1 second on four seven-segment displays showing seconds and minutes. The unit should be controlled by three push-buttons: one to start, one to stop and one to zero the timer.

Approach

Most modern digital clocks derive their time keeping from a crystal oscillator, as discussed in Section 4.7.1. This produces a timing waveform of very high stability with little temperature drift.

The resonant frequency of a crystal is related to its size, with low-frequency devices usually being larger than high-frequency types. For this, and other reasons, it would be impractical to produce crystals with resonant frequencies of the order of a few hertz. For applications that require low-frequency signals it is normal to use a higher frequency device and to divide the frequency of the output to achieve the required value. Digital watches and clocks often use a crystal with a resonant frequency of 32 768 Hz. The reason for this particular value is that it is an exact power of two ($32\,768 = 2^{15}$) which simplifies the process of frequency division. A 15-stage binary divider will generate from such a signal a 1 Hz waveform which is a suitable starting point for the clock circuit. A block diagram of a suitable arrangement is shown below.

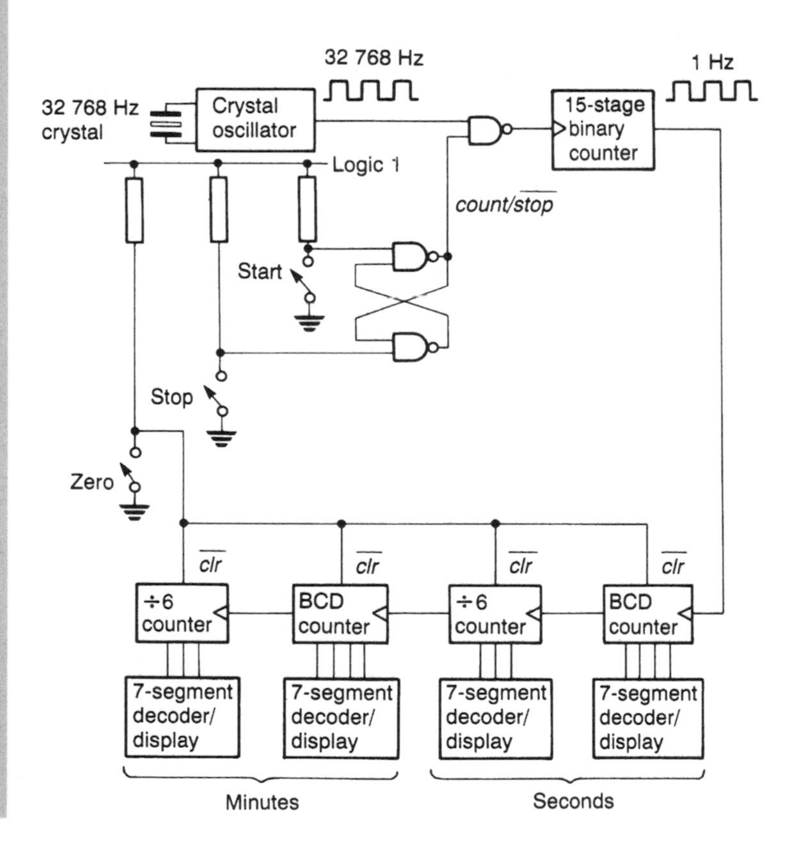

An S–R bistable is used to gate the output from a crystal oscillator running at 32 768 Hz. When the *count*/$\overline{stop}$ output from the bistable is high, the oscillator signal is inverted by the NAND gate and passed to the frequency divider, resulting in a 1 Hz signal being fed to the seconds/minutes counter chain. When the *count*/$\overline{stop}$ output is low, the output from the NAND gate is always high, preventing the frequency divider from seeing any changing input. The output from the bistable is determined by two push-buttons, *start* and *stop*.

The 1 Hz pulse train is counted by a chain of four counters. The first counter is a decade counter, the output of which is fed to a BCD to seven-segment decoder of the type described in the design study at the end of Chapter 9. This counter will count from 0 to 9 and then return to 0. When it resets to 0 it will increment the next counter which also drives a BCD to seven-segment decoder. This counter is a modulo-6 counter and thus will count from 0 to 5 before resetting. The two digits will therefore count from 0 to 59 and then return to 0. These two digits are the seconds count.

When, at the end of every minute, the second counter resets to zero, it will increment the third counter. This and the fourth counter count minutes in the same manner, displaying them on the remaining two display digits. After one hour the counter will reset to zero and continue.

At any time the third push-button *zero* may be pressed. This is connected to the direct reset input of each counter and will cause them to reset the counters. This can be done while the counter is stopped or while it is counting.

In more demanding applications the simple method adopted for gating the oscillator signal might be unsuitable since the operation of enabling or disabling the clock might itself introduce an extra half clock pulse. In such cases a slightly more sophisticated arrangement could be used using a bistable to gate the waveform. However, in this case, since the oscillator waveform is divided by 32 768 before being counted, the uncertainty introduced represents an error in timing of only about 30 μs. Since the display is being shown to a resolution of only 1 second this inaccuracy can certainly be ignored.

Further reading

Beards P. H. (1996) *Analogue and Digital Electronics*, 2nd edn. Englewood Cliffs, NJ: Prentice-Hall

Mano M. M. (1990) *Digital Design*, 2nd edn. Englewood Cliffs, NJ: Prentice-Hall

Stonham T. J. (1996) *Digital Logic Techniques*. London: Chapman and Hall

Tocci R. L. and Widmer N. S. (1994) *Digital Systems: Principles and Applications*. Englewood Cliffs, NJ: Prentice-Hall

Exercises

10.1 Define the terms bistable, monostable and astable.

10.2 Explain the origins of the labels *S* and *R* given to the inputs of an S–R bistable.

10.3 In an S–R bistable formed using two NOR gates, are the inputs active high or active low? Draw the transition table and a Karnaugh map of such a circuit.

10.4 Under what circumstances does switch bounce cause problems in digital systems? Design a circuit using two-input NOR gates to remove the effects of switch bounce from a changeover switch.

10.5 Deduce the waveform at the Q output of the following circuits

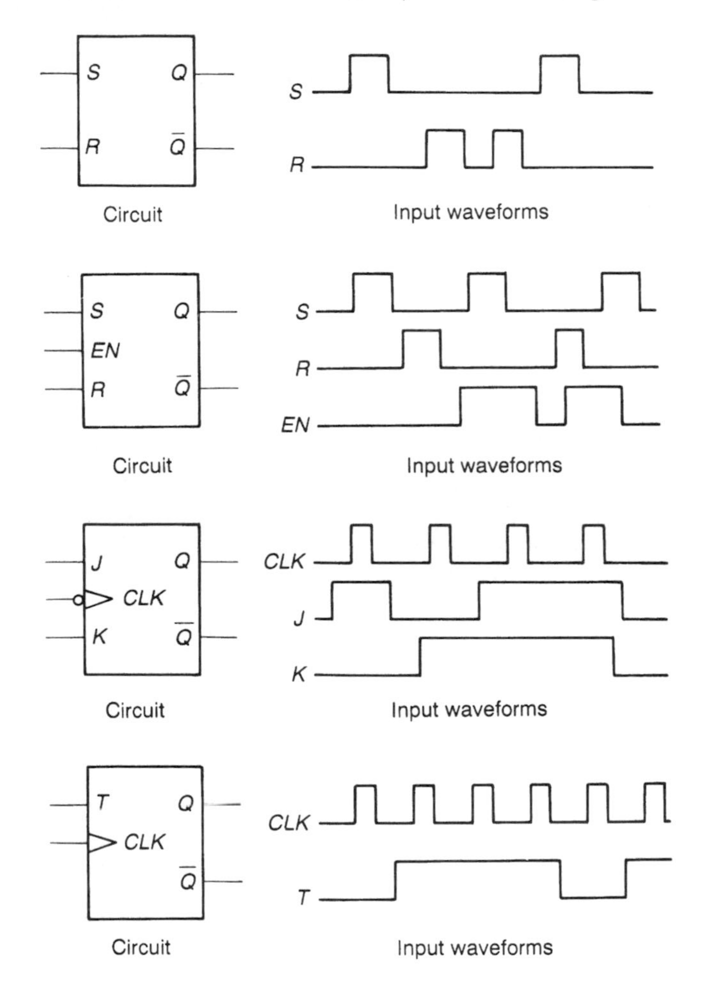

10.6 Explain the meanings of the terms 'synchronous' and 'asynchronous' when applied to counters.

10.7 Design a modulo-5 asynchronous counter using J–K negative-going edge-triggered flip-flops.

10.8 The decade counter of Figure 10.50 can be used to reduce the frequency of a clock waveform by a factor of ten by taking the output from Q_3. For some applications this has the disadvantage that the waveform produced is not a square wave. Design a 'divide by ten' counter that does produce a square wave output. Hint: the circuit from the last exercise might be useful.

10.9 Simulate your solution to the previous exercise to confirm that the circuit functions as expected.

10.10 Design a modulo-10 ripple down counter using negative edge-triggered J–K bistables. The circuit should count down to zero and then reset to 9.

10.11 Simulate your solution to the previous exercise to confirm that the circuit functions as expected.

10.12 Design a 4-bit up/down counter that does not overflow or underflow. That is, counting up is disabled when it reaches its maximum value and counting down is disabled when it reaches its minimum value. Can you think of any application for such a counter?

10.13 In Example 10.1 we looked at the design of a simple burglar alarm. Design a more sophisticated alarm that allows the house owner to leave the house after turning it ON and to re-enter the house and turn it OFF without the alarm sounding. The unit should arm itself 30 seconds after a switch is closed and allow the user 30 seconds to turn it off before sounding the alarm.

10.14 Design a digital clock that displays the time in seconds, minutes and hours on six seven-segment displays. Your circuit should take into account the fact that such clocks display hours in the range 1 to 12, not 0 to 11.

10.15 Modify your design for Exercise 10.14 to allow the time to be set by depressing buttons to increment the seconds, minutes and hours settings.

10.16 Modify your design for Exercise 10.15 to allow the circuit to display time in either a 12 hour or 24 hour format. The display mode should be controlled by the setting of a switch.

10.17 Design a synchronous modulo-5 counter with a single input D (direction) and a single output Q. The counter should count up when the D input is 1 and down when D is 0. The Q output should be 1 in one of the counter's five states and 0 in the remainder.

10.18 Explain the function of the arrangement described by the following state transition diagram

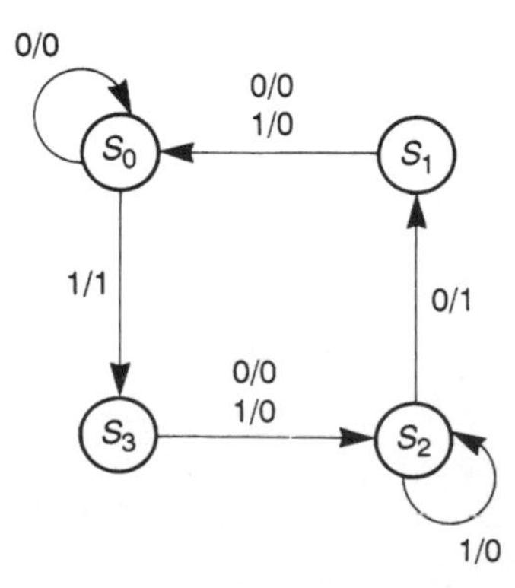

10.19 Design a synchronous circuit to implement the arrangement described by the transition diagram of Exercise 10.18.

10.20 Design a synchronous circuit to implement the following state transition diagram

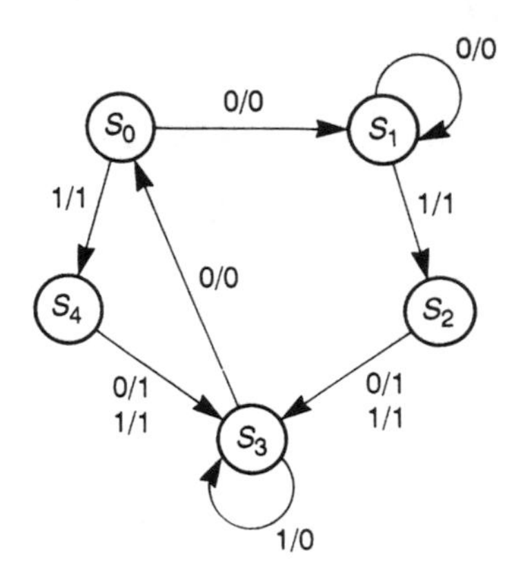

Digital Devices

Objectives

When you have studied the material in this chapter you should:

- have an understanding of the use of both bipolar and field effect transistors (FETs) as logical switches;
- be aware of a range of digital electronic 'families' and understand, in broad terms, their methods of operation and characteristics;
- be familiar with the forms and characteristics of the various members of the TTL and CMOS families of logic circuits;
- understand the basic structure and uses of programmable logic devices;
- be aware of the primary causes and effects of noise in digital systems, and be familiar with techniques that can be used to reduce it.

Contents

11.1 Introduction

In the last two chapters we looked at a number of digital applications based on the use of standard logic gates. In this chapter we will consider a range of electronic circuits that can be used to implement these gates.

In the early days of semiconductor manufacture, components were limited to single transistors. As device fabrication techniques improved it became possible to incorporate a number of active components within a single package and over the years the densities of such **integrated circuits**, or **IC**s, have steadily increased. Today it is possible to place millions of both active and passive components within a single 'chip', enabling complete computers to be constructed on a piece of silicon only a few millimetres square. Although it is possible to integrate many devices within a single package there are still applications for which single transistors are required and others that require only a handful of components. Electronic devices that contain more than one active component may be classified by their **integration level**. In the case of digital devices these are normally categorized by the number of standard gates (or circuitry of equivalent complexity) that they contain. Table 11.1 shows a common way of defining the various levels of integration.

Although modern digital electronic components are the result of many years of development and evolution, there is no single, ideal set of logic circuits that fulfils all requirements. Over the years a number of **logic families** have evolved, each offering particular advantages. Some, for example, work at very high speeds, others consume small amounts of power, while others are very tolerant of electrical noise. Part of the designer's function is to select an appropriate logic family for a given application. The information given in this chapter should help in making this choice.

Integrated circuit logic families may be divided into two main groups: those based on bipolar transistors and those that use metal oxide semiconductor (MOS) transistors. In this chapter we will start by looking at a variety of logic families and will identify common characteristics and terminologies. We will then look at the operation of both bipolar and MOS transistors, when used as logical switches, before considering in more detail two of the most important families, transistor–transistor logic (TTL) and complementary metal oxide semiconductor (CMOS) logic. Array logic is then discussed before concluding the chapter with a discussion of the effects of noise on digital systems.

Table 11.1 Integration levels for digital devices.

Integration level	Number of gates	Applications
Small scale integration (SSI)	1–11	Basic gates and flip-flops
Medium scale integration (MSI)	12–100	Counters, registers, small memories
Large scale integration (LSI)	101–1000	Memories and simple microprocessors
Very large scale integration (VLSI)	1001–100 000	Large memories, microprocessors

11.2 Logic families

Over the years a large number of logic families have evolved, the most successful being

- resistor–transistor logic (RTL)
- diode logic
- diode–transistor logic (DTL)
- transistor–transistor logic (TTL)
- emitter-coupled logic (ECL)
- metal oxide semiconductor (MOS)
- complementary metal oxide semiconductor (CMOS).

All but the second of these have been used for integrated circuit devices, although RTL and DTL are now of only historical interest. Diode logic is used mainly in discrete (that is, not integrated) circuits. Before looking at the form of these logic families, it is useful to look at the overall characteristics of logic devices and to establish a terminology to describe their characteristics.

11.2.1 Logic inverters

In Chapter 3 we looked at the characteristics of linear amplifiers and noted that all real amplifiers have an output swing that is limited by the supply voltages used. A typical inverting linear amplifier might have a characteristic as shown in Figure 11.1.

If we wish to use such a device as a linear amplifier, we must ensure that the input is restricted so that operation is maintained within the linear range of the device.

It is also possible to use the amplifier of Figure 11.1 as a logical device. If we restrict the input signal so that it is always *outside* of the linear region of the amplifier, we are left with two allowable ranges for the input voltage, as shown in Figure 11.2. We may consider these ranges as representing two possible input states, '0' and '1'.

Clearly, when the input voltage corresponds to state '0' the output voltage is at its

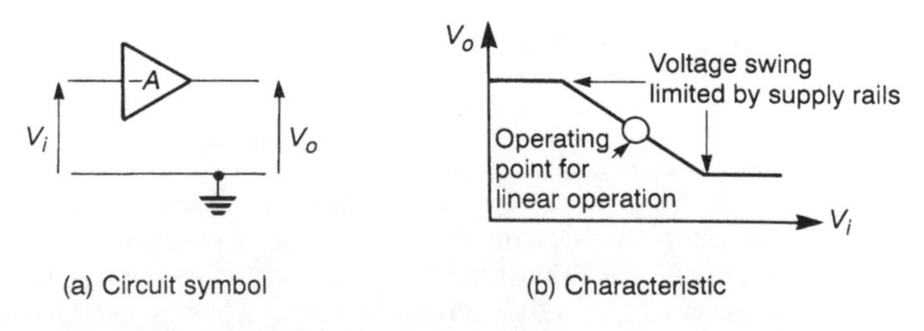

Figure 11.1 An inverting linear amplifier.

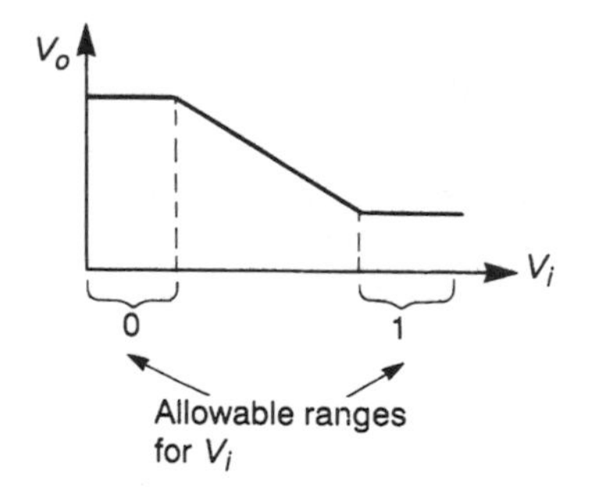

Figure 11.2 Use of an inverting amplifier as a logical device.

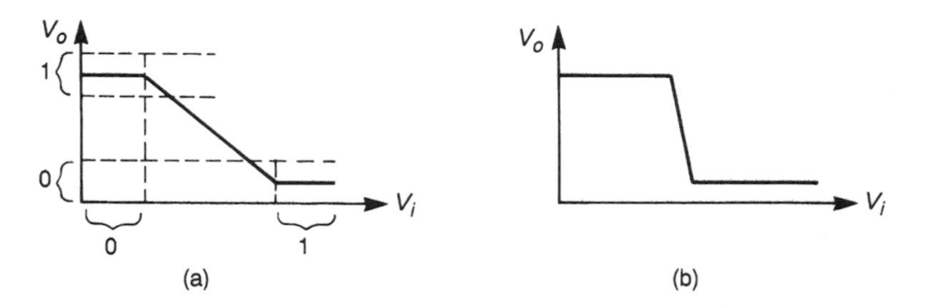

Figure 11.3 Transfer characteristics for logical inverters.

maximum value, and when the input state is '1' the output is at its minimum value. If we choose component values appropriately, we can arrange that the maximum and minimum output voltages lie within the voltage ranges defined at the input to represent '0' and '1', as shown in Figure 11.3(a).

From Figure 11.3(a) it is clear that when the input is '0' the output is '1', and vice versa. The circuit therefore has the characteristics of a **logical inverter**. Since the input and output voltages are compatible, the output of this arrangement could be fed to the input of a similar gate.

When designing linear amplifiers, we aim to produce an extended linear region to permit a large output swing. When producing a logical inverter we wish the linear portion of the characteristic to be as small as possible to reduce the region of uncertainty. Such circuits therefore have a very high gain and a rapid transition from one state to the other, as shown in Figure 11.3(b).

11.2.2 Noise immunity

Noise will be present in any real system. This will have the effect of adding random fluctuations to the voltages representing the logic levels. To enable the system to tolerate a certain amount of noise, the voltage ranges defining the '0' and '1' states at the output of a gate are more tightly constrained than those at the input. This ensures that small perturbations of an output signal caused by noise will not take the signal outside of the defined ranges of the input of another gate. Thus the circuit is effectively immune to small amounts of noise, but may be affected if the magnitude of the noise is large enough to take the logic signal, within either logic state, outside the allowable logic bands. The maximum noise voltage that can be tolerated by a circuit is termed, logically enough, the **noise immunity** V_{NI} of the circuit. It is also known as the **noise margin**.

The voltage ranges representing the two logic states at the input and output of the gate are defined by four parameters.

V_{IH} The minimum voltage required at an input for an input voltage to be interpreted as a '1' (HIGH).

V_{IL} The maximum voltage allowed at an input for an input voltage to be interpreted as a '0' (LOW).

V_{OH} The minimum voltage produced at the output of a gate to represent a '1' (HIGH).

V_{OL} The maximum voltage produced at the output of a gate to represent a '0' (LOW).

Clearly, for any real gate there will be some variation in these values and it is normal to specify maximum and minimum values (as appropriate) for each quantity. The noise immunity of a logic circuit is determined by the difference between the voltages representing the logic levels at the output of one gate, compared with those specifying the levels at the input of the next. Therefore

$$\text{Noise immunity in logic 1 (HIGH) state } V_{NIH} = V_{OH}(min) - V_{IH}(min) \qquad (11.1)$$

and

$$\text{Noise immunity in logic 0 (LOW) state } V_{NIL} = V_{IL}(max) - V_{OL}(max) \qquad (11.2)$$

This relationship is illustrated in Figure 11.4.

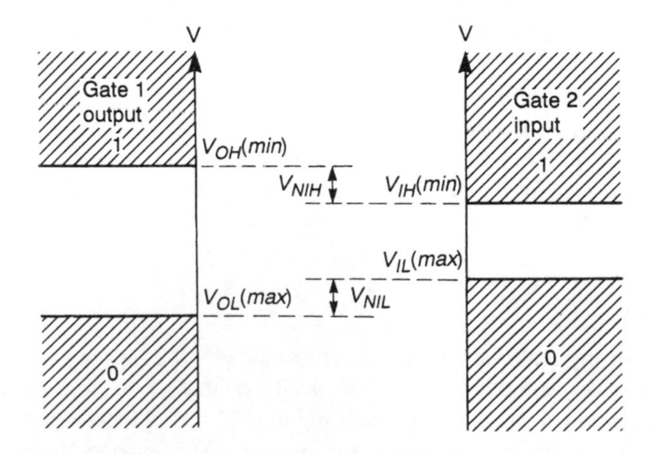

Figure 11.4 Noise immunity.

11.2.3 The bipolar transistor as a logical switch

We have seen that a logical inverter can be formed from an inverting amplifier, and in Chapter 7 we considered the use of a bipolar transistor as an inverting amplifier. When we are driving a device from one state in which it is effectively 'turned OFF' to another in which it is 'turned ON', we often liken the device to a switch that is open or closed. In Section 7.7.6 we discussed the use of a bipolar transistor as a **logical switch** and noted that it performs very well in this role.

Figure 11.5 shows a logical inverter based on a bipolar transistor. When the input voltage is close to zero the transistor is turned OFF and negligible collector current flows. The output voltage is therefore close to the supply voltage V_{CC}. When the input voltage is high the transistor is turned ON and the output voltage is equal to the **saturation voltage** of the device, which is generally about 0.1 V. Therefore the circuit acts as a logic inverter with voltages close to V_{CC} representing logical 1, and voltages close to 0 V representing logical 0.

Figure 11.5(b) shows the relationship between a pulse applied to the input of the inverter and the corresponding collector current and output voltage. The figure shows that there is a delay between a change in the input voltage and the response of the transistor. The shape of the waveforms makes it very difficult to define the point at which the device can be considered to have just turned completely ON or just turned completely OFF. For this reason it is common to measure the time taken for the device to respond to a point at which it has achieved 90% of its change in value. Thus, when determining the time taken for the transistor to turn ON, we measure the time from when the input changes to when the output changes by 90% of its full range. Since the output voltage falls when the transistor turns ON, this corresponds to the output voltage dropping to 10% of its maximum value. This is termed the **turn-on time** t_{ON}. Similarly, the time taken for the transistor to turn OFF is measured from the time that the input changes until the time when the output has risen to 90% of its maximum value. This is termed the **turn-off time** t_{OFF}.

It can be seen that the time taken for the device to turn OFF is considerably greater than that needed for it to turn ON. The primary reason for this increase in switching time

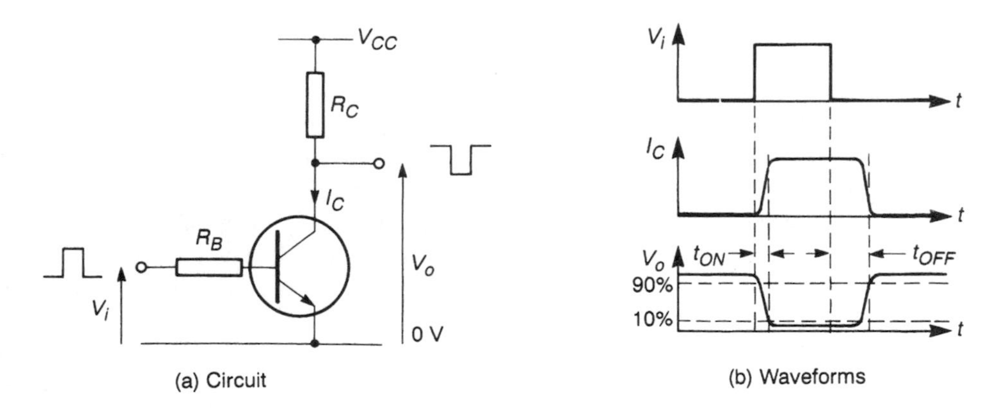

Figure 11.5 A logical inverter using a bipolar transistor.

is that when the device is turned ON it is driven into its saturation region. This results in a large quantity of **minority charge carriers** being injected into the base of the device, which in turn causes a build-up of charge within the base region. This charge is called the **saturation storage charge**. When the current into the base of the transistor is removed there is a delay while this stored charge is removed, before the collector current falls and the device turns OFF. This delay is termed the **storage time** of the device; it increases with the magnitude of the base current pumped into the base when it is turned ON. A typical value for the storage time of a general purpose transistor might be about 200 ns, several times greater than the time required for the device to turn OFF after the stored charge has been removed.

The storage time of a device can be reduced by adding impurities into the base region of the transistor to reduce the **minority carrier lifetime**. A common method of achieving this is the use of **gold doping**, which reduces the storage time but, unfortunately, decreases h_{FE} and increases leakage currents.

Some logic circuits overcome the problem of storage time by preventing the transistors from entering their saturation regions. This results in turn-off times that are comparable with the turn-on times of the transistors. We will look at examples of this technique when we consider emitter-coupled logic later in this section and when we discuss the use of Schottky diodes in the next section.

FILE 11A

Computer simulation exercise 11.1

Use simulation to investigate the characteristics of the circuit of Figure 11.5.

A suitable arrangement would use a 2N2222 transistor with $R_B = R_C = 1\ \text{k}\Omega$ and $V_{CC} = 5$ V. Apply a suitable input waveform and plot V_{in}, I_C and V_o against time. Hence measure the turn-on, turn-off and storage times for this arrangement.

Repeat this procedure using a base resistor R_B of 10 kΩ and compare your results.

11.2.4 The FET as a logical switch

In Sections 6.6.4 and 6.6.5 we looked at the use of FETs as logical switches and in Figures 6.39 and 6.40 we considered circuit examples based on these devices. MOSFETs are the dominant form of FET for digital applications. While in analogue applications it is common to describe such devices as FETs, in digital systems it is more common to talk of **MOS devices**, describing the method of construction rather than the principle of operation.

The major advantages of MOS technology over circuits based on bipolar transistors are that MOS devices are simpler and less expensive to fabricate. Each MOS gate requires a much smaller area of silicon, allowing a greater number of devices to be produced on a given chip. Power dissipation is also much less, reducing problems of heat dissipation.

The principal disadvantage of MOS devices is that they tend to be slower in operation than bipolar parts. However, other factors, such as circuit complexity, also affect operating speeds and modern advanced MOS logic families are of comparable speed to equivalent bipolar families. In many applications speed is not the dominant consideration; factors such as power consumption and circuit densities may be of more importance.

Circuitry based on *N*-channel (NMOS), *P*-channel (PMOS) and a combination of the two (CMOS) are all used, although PMOS is now rarely used in new designs. NMOS

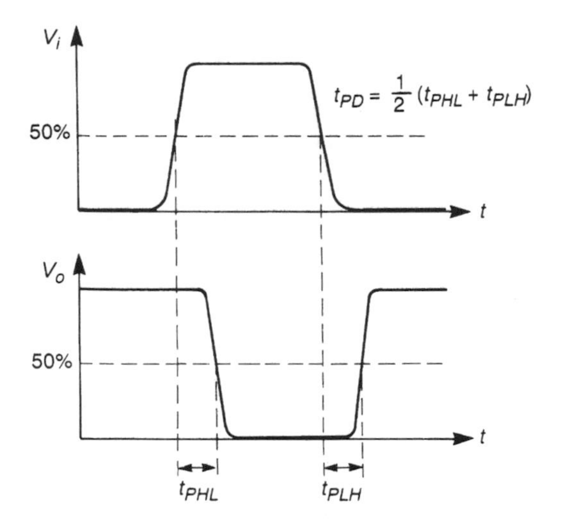

Figure 11.6 Propagation delay times.

circuits are easier, and therefore less expensive, to fabricate than CMOS devices but consume more power. CMOS is now the dominant MOS technology.

11.2.5 Timing considerations

Propagation delay time

Logic gates invariably consist of a number of transistors, each producing a slight delay as signals pass through them. Inevitably the resultant delay is different for changes in each direction, and two **propagation delay times** are used to describe the speed of response of the circuit. These are t_{PHL}, the time taken for the output to change from high to low, and t_{PLH}, the time taken for the output to change from low to high. In some cases a single value is used, corresponding to the average time for the two transitions. This average propagation delay time t_{PD} is given by

$$t_{PD} = \tfrac{1}{2}(t_{PHL} + t_{PLH})$$

Since in general the input waveform will not be a perfect square pulse, t_{PHL} and t_{PLH} are measured between the points at which the input and output signals cross a reference voltage corresponding to 50% of the voltage difference between the logic levels. This is illustrated in Figure 11.6.

Set-up time

In logic circuits that have clock input signals it is often necessary for control inputs to be applied a short while before an active transition of the clock to ensure correct operation. The time for which the control input is stable before the clock trigger occurs is termed the **set-up time** t_S. Device manufacturers normally specify a minimum value for this

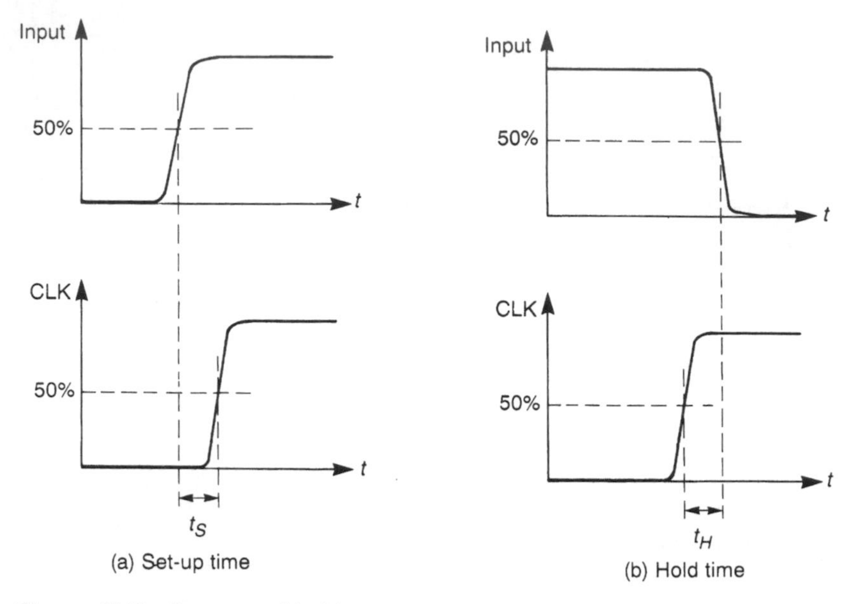

Figure 11.7 Set-up and hold times.

quantity. For typical logic circuits this minimum value might be in the range 10 to 50 ns. Figure 11.7(a) illustrates the definition of set-up time.

Hold time

It is also often required that a control signal should not change for a short interval after the active transition of a clock input. The time for which a control input is stable after a clock trigger occurs is termed the **hold time** t_H; typical minimum values range from 0 to 10 ns. This is illustrated in Figure 11.7(b).

11.2.6 Fan-out

In many cases it is necessary to connect the output of one logic gate to the input of a number of gates. Since each input draws current from this output there will be a limit to the number of gates that a single output can supply. This is termed the **fan-out** of the circuit. The fan-out is clearly determined by the output resistance of the gate and by the input resistance of the gates it is driving. Because of the very high input resistance of MOS devices, it is the input capacitance that is of importance. Circuits based on this technology generally have a greater fan-out than those based on bipolar transistors.

11.2.7 Resistor–transistor logic (RTL)

The simple inverter circuit of Figure 11.5 can be used as the basis of a family of logic devices. The addition of a second transistor forms a two-input NOR gate; extra devices

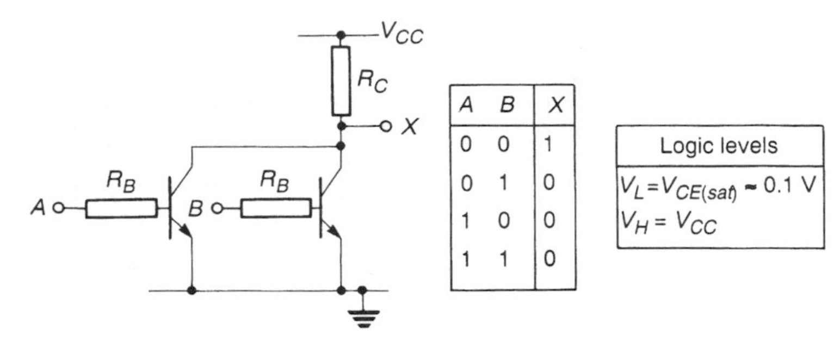

FILE 11B **Figure 11.8** An RTL NOR gate.

may be added to form gates with any number of inputs. This is illustrated in Figure 11.8. Suitable combinations of NOR gates and inverters can be used to form any required logic function.

The voltages representing logical 0 and 1 in RTL are defined by the output of the gate, since these voltages then form the input to the next gate. Logical 0 is given by the saturation voltage of the transistor, which is typically 0.1 to 0.2 V. The logic 1 voltage is simply V_{CC}. The voltages representing logical 1 and 0 are usually referred to as V_H and V_L, respectively (standing for the high and low logic states). This is preferable to the use of V_1, and V_0, which might be mistaken for V_i and V_o, the input and output voltages.

The advantages of RTL are its simplicity and compactness. Its disadvantages are that it has a relatively low **noise immunity** (even small noise spikes added to a logical 0 input voltage will start to turn on the transistor) and slow speed of operation (caused by the presence of the base resistors, which limit the base drive currents).

11.2.8 Diode logic

One of the simplest forms of logic circuit is **diode logic**, examples of which are shown in Figure 11.9.

Both AND and OR gates can be constructed using only diodes and resistors. Gates with any number of inputs can be constructed by adding additional diodes. Looking first at the AND gate of Figure 11 .9(a) it is clear that if input logic voltages of 0 V (logic 0) and V_{CC} (logic 1) are used, when either A or B is low, the output X is pulled low. Only if both A and B are high is the resistor R able to pull the output high. It should be noted that the output voltage corresponding to logic 1 is V_{CC} as at the input. However, the voltage corresponding to logic 0 at the output is not equal to that at the input since it is increased by the voltage across the diode, normally about 0.7 V.

In the OR gate circuit shown in Figure 11.9(b), X will be high if either A or B is high. In this case the output voltage corresponding to logic 0 is equal to that at the input, while that for logic 1 is reduced by about 0.7 V.

Diode logic has the advantage of simplicity but suffers from being totally passive. As signals pass through a series of gates the logic levels are gradually eroded with the voltage representing logic 0 increasing and that representing logic 1 decreasing until the distinction between them disappears. For this reason diode logic is normally used only

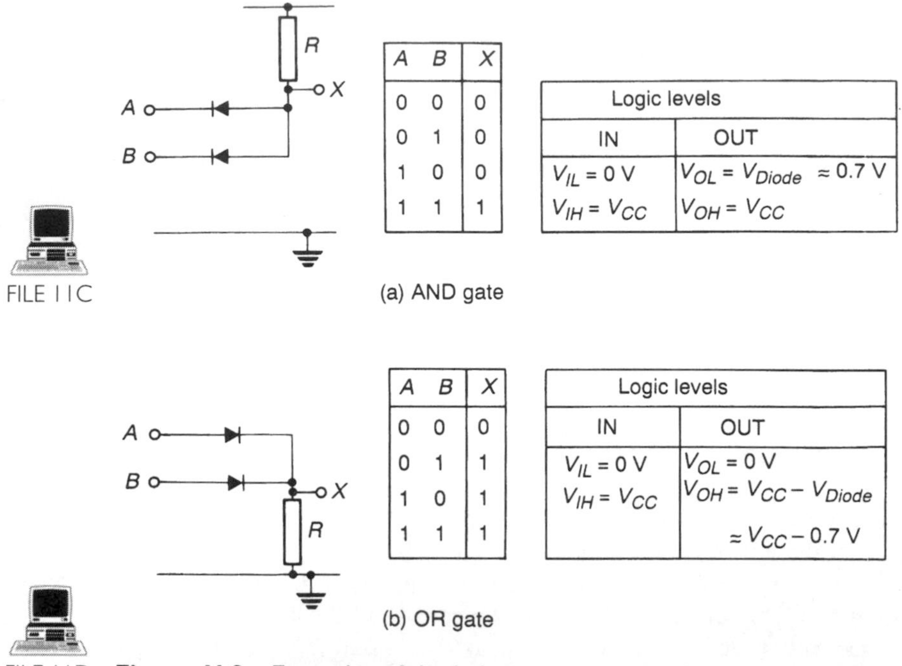

(a) AND gate

A	B	X
0	0	0
0	1	0
1	0	0
1	1	1

Logic levels	
IN	OUT
$V_{IL} = 0$ V	$V_{OL} = V_{Diode} \approx 0.7$ V
$V_{IH} = V_{CC}$	$V_{OH} = V_{CC}$

(b) OR gate

A	B	X
0	0	0
0	1	1
1	0	1
1	1	1

Logic levels	
IN	OUT
$V_{IL} = 0$ V	$V_{OL} = 0$ V
$V_{IH} = V_{CC}$	$V_{OH} = V_{CC} - V_{Diode} \approx V_{CC} - 0.7$ V

Figure 11.9 Examples of diode logic.

for simple logic operations that do not require the complexity of more sophisticated gate designs.

11.2.9 Diode–transistor logic (DTL)

Many of the problems of diode logic can be overcome by adding an amplifier at the output of the gate to re-establish the logic levels. A simple one-transistor amplifier will suffice, which provides power amplification and reduces the output resistance, increasing the fan-out of the circuit. The circuit produced contains both diodes and transistors and is thus called **diode–transistor logic** or **DTL**. Figure 11.10 illustrates such a circuit.

The circuit represents the simple AND gate of Figure 11.9(a) with a simple inverting amplifier added to its output. This produces the function of a NAND gate. In fact the circuit of Figure 11.10 is unsatisfactory since taking either A or B low takes the base of the transistor down to about 0.7 V which will not, in general, turn it off. This problem can be overcome by placing a diode in series with the base of the transistor, as shown in Figure 11.11.

In order for the transistor in this circuit to be turned ON, the voltage at point Y would need to be greater than the sum of the turn-on voltages of the transistor and the diode in series with its base. since each has a turn-on voltage of about 0.5 V, no significant current will flow in R_C unless the voltage at Y exceeds about 1 V. When a 0 logic level is applied to either input, the voltage at point Y is pulled down to about 0.7 V ensuring that the transistor is turned OFF and a logic 1 is produced at the output.

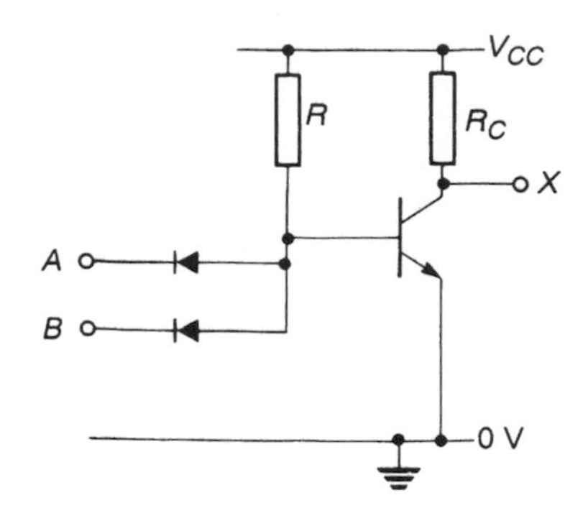

Figure 11.10 A simple DTL NAND gate.

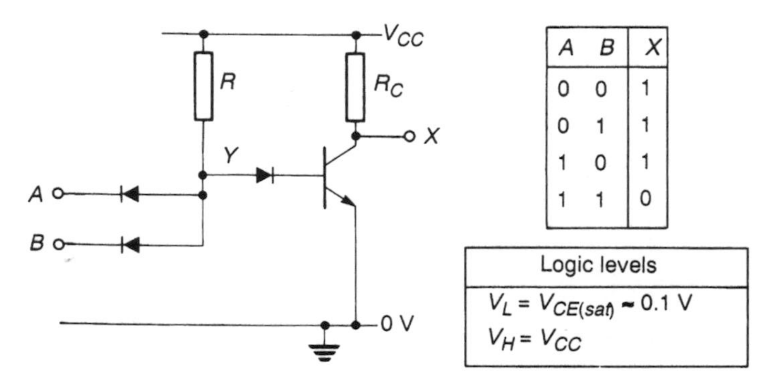

A	B	X
0	0	1
0	1	1
1	0	1
1	1	0

FILE 11E **Figure 11.11** A DTL NAND gate.

11.2.10 Transistor–transistor logic (TTL)

Looking at the circuit of the DTL gate shown in. Figure 11.11 it is clear that each input sees a series combination of two back-to-back diodes. This represents an *np–pn* structure which could be replaced by the *npn* structure of a transistor, as shown in Figure 11.12. Instead of a 'diode–transistor logic' circuit, we now have a 'transistor–transistor logic' circuit – TTL.

It can be seen that the input transistors of the circuit of Figure 11.12(b) have common connections to their bases and their collectors. When using integrated circuit techniques the circuit can be improved by combining the functions of the input transistors into a single device. This is shown in Figure 11.13(a). The multi-emitter transistor is produced by forming a number of emitter regions within a single base region, as illustrated in Figure 11.13(b).

Integrated circuit TTL gates improve this basic circuit by using additional components to increase speed and drive capabilities. Such circuits will be discussed in more detail in Section 11.3.

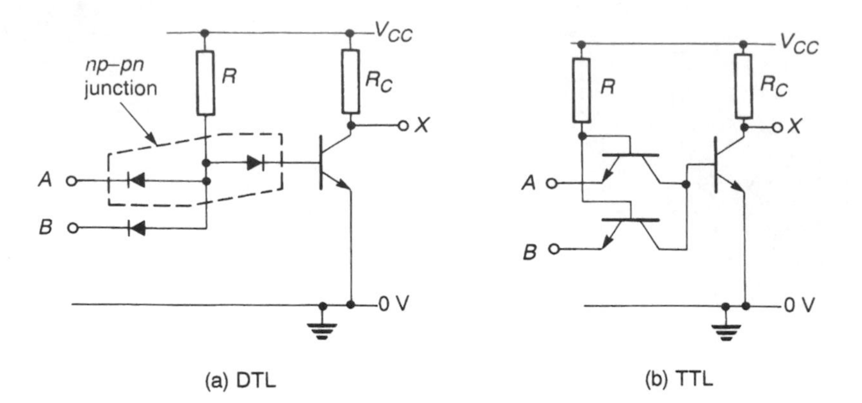

Figure 11.12 Replacing the diodes of a DTL gate with transistors.

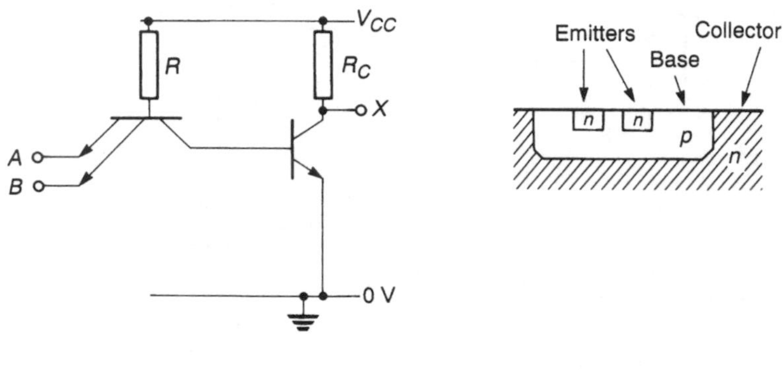

Figure 11.13 A TTL NAND gate.

11.2.11 Emitter-coupled logic (ECL)

In Section 11.2.3 we discussed the problems of **storage time**, which greatly increases the switching times of bipolar transistors that are driven into saturation. We noted that some circuits overcome this problem by keeping the transistors within their active region, preventing them from becoming saturated. Such a circuit is shown in Figure 11.14.

The circuit is similar to the long-tailed pair amplifier of Figure 7.33 with the addition of an extra transistor T1. For obvious reasons, circuits of this type are described as **emitter-coupled logic** or simply **ECL**. Further transistors can be added in parallel with T1 and T2 to form a gate with more inputs. Logic inputs are applied to the bases of transistors T1 and T2 and a constant reference voltage V_{BB} is applied to the base of T3. The logic voltages are chosen such that V_L is a little less than V_{BB} and V_H is a little greater than V_{BB}.

If A and B are both at logical 0, the voltages on these inputs will both be less than V_{BB}. T3 will be turned ON, and the voltage on the emitters of the transistors V_E will be

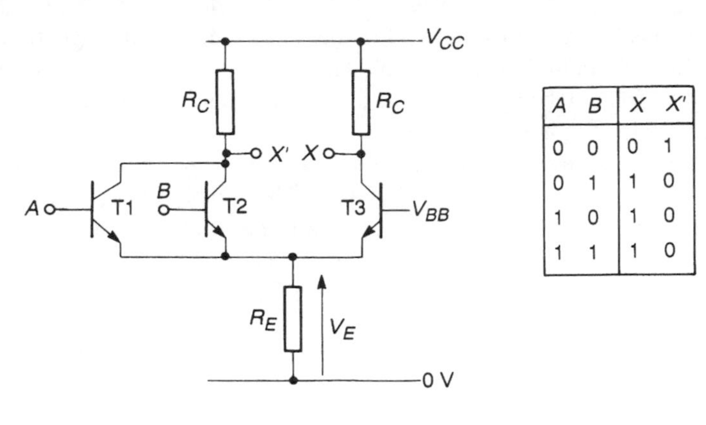

(a) Circuit

A	B	X	X'
0	0	0	1
0	1	1	0
1	0	1	0
1	1	1	0

(b) Truth table

Figure 11.14 A non-saturating logic gate.

given by

$$V_E = V_{BB} - V_{BE} \approx V_{BB} - 0.7 \text{ V}$$

Since the voltages on A and B are less than V_{BB} it follows that the base-to-emitter voltages of T1 and T2 will be less than 0.7 V and they will therefore be turned OFF. The current I_E flowing through the emitter resistor is given by the expression

$$I_E = \frac{V_E}{R_E} \approx \frac{V_{BB} - 0.7}{R_E}$$

and this flows exclusively through T3 since T1 and T2 are turned OFF. The voltage at point X is therefore

$$V_X = V_{CC} - R_C I_E = V_{CC} - R_C \frac{(V_{BB} - 0.7)}{R_E} \qquad (11.3)$$

and the voltage at X' is approximately V_{CC}. If appropriate values are chosen for V_{BB}, R_C and R_E, it can be arranged that T3 remains within its active region.

If A or B is at logic 1, V_E will be pulled up to $V_H - 0.7$ V reducing the base-to-emitter voltage of T3 to below 0.7 V. This will turn ON the corresponding input transistor and turn OFF T3. The emitter current flowing through the input transistor will now be

$$I_E = \frac{V_E}{R_E} = \frac{V_H - 0.7}{R_E}$$

This current will flow exclusively through the appropriate input transistor and so

$$V_{X'} = V_{CC} - R_C I_E = V_{CC} - R_C \frac{(V_H - 0.7)}{R_E} \qquad (11.4)$$

and V_X is approximately V_{CC}. Again, an appropriate choice of component values ensures that the input transistors are not saturated.

If we define the output logic levels such that V_{CC} represents logic 1 and the voltages given by Equations 11.3 and 11.4 fall within the range representing logic 0, the circuit has the truth table given in Figure 11.14(b). The X output therefore represents the OR function and the X' output, being the inverse of X, represents the NOR function. The gate is therefore an OR/NOR gate.

The circuit has the advantage that the transistors are never saturated but are simply switched between cut-off and some fixed current. This enables the circuit to operate at very high speeds because of the absence of the **storage time** associated with saturated transistors. The disadvantage of this circuit in its present form is that the input and output logic levels are not the same. This makes it impossible to connect the output of one gate into the input of the next. This problem is overcome by adding a **level shifting transistor amplifier** to each output. These are simple emitter followers which shift the output voltage by an amount equal to their base-to-emitter voltage (about 0.7 V). A suitable choice of component values enables this to produce equal input and output logic voltages. A typical three-input ECL OR/NOR gate is shown in Figure 11.15.

Compared with the other logic families we have considered, ECL has a relatively small output swing between its two logic levels. One effect of this is that the **noise immunity** of the device is poor at only about 0.2 to 0.25 V. Also, since the transistors are always in their active region, power dissipation is high (typically about 60 mW per gate) resulting in a great deal of waste heat which must be removed to prevent the circuits from overheating. This in turn tends to limit the amount of circuitry that can be integrated into a single chip, increasing system size and cost. However, by keeping the transistors in their active region the switching speed is greatly increased, producing propagation delays of the order of 1 ns. This is considerably faster than saturating logic such as standard TTL, allowing clock frequencies of up to 500 MHz or more.

11.2.12 Metal oxide semiconductor (MOS) logic

In Figure 6.39(a) we looked at the use of a MOSFET and a resistor in a simple logical inverter. We noted that the use of resistive loads was uneconomical of space in integrated circuits and in Figure 6.39(b) we used a second MOSFET as an **active load**. This circuit is reproduced as Figure 11.16(a), which also shows an equivalent circuit of the gate using a switch and a load resistor.

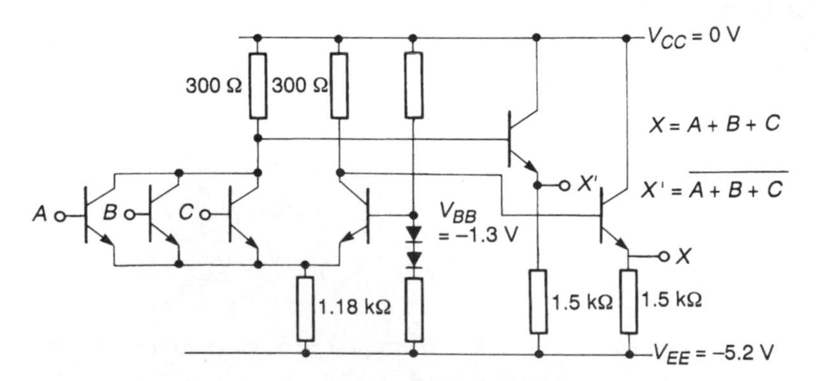

Figure 11.15 A three-input ECL OR/NOR gate.

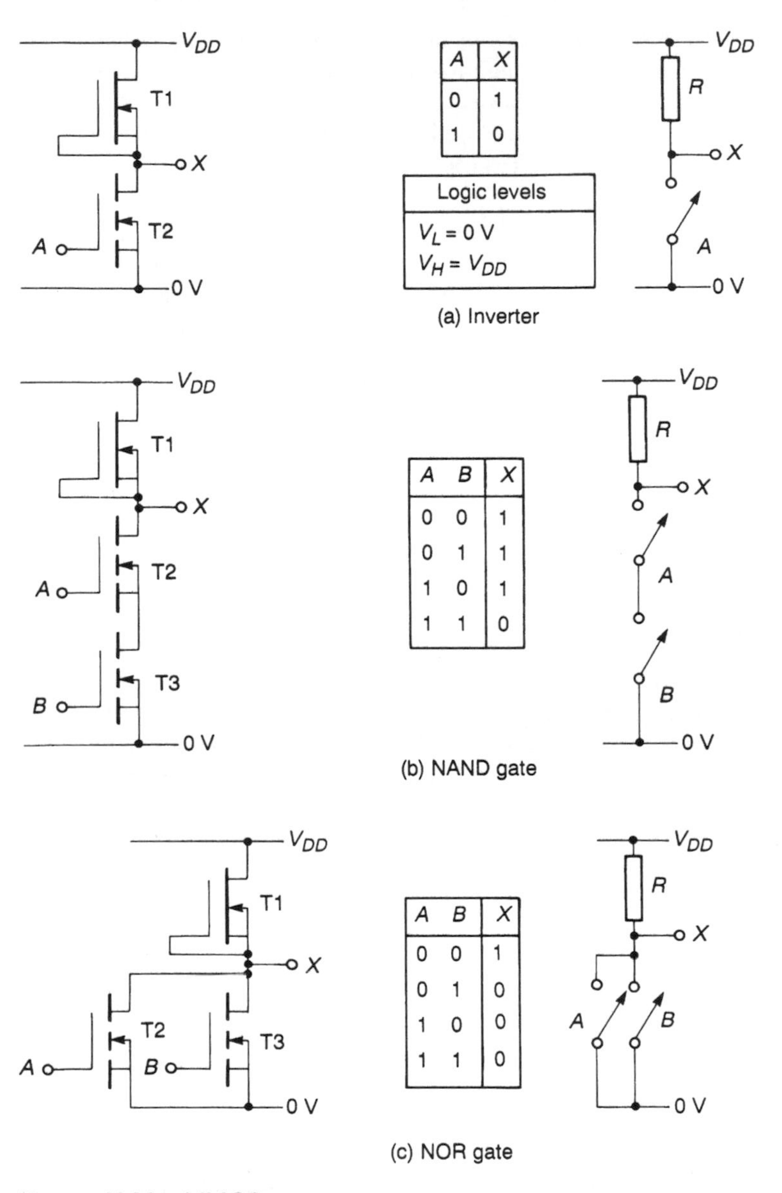

Figure 11.16 NMOS gates.

The circuit of Figure 11.16(a) uses *n*-channel MOSFETs, and is the basic form of an **NMOS inverter**. Similar circuits can be constructed using *p*-channel transistors to form PMOS devices.

One of the great attractions of MOS technology is its simplicity. The switching transistors within the circuit (as distinct from those devices which are used as active loads) act as near-ideal switches. Logic levels are equal to the supply rail voltages, giving a large output voltage swing and good **noise immunity**. Operation of the circuit of Figure 11.16(a) is very straightforward. When the input voltage is equal to the positive supply rail (logic 1), transistor T2 is turned ON and resembles a closed switch. This pulls the output down to close to 0 V. When the input is taken to 0 V, transistor T2 is turned OFF and the load transistor T1 pulls the output high to close to the positive supply V_{DD}.

This circuit technique can be expanded to give other forms of gate. Both **NAND** and **NOR** gates can be formed easily. Other functions can be produced by combining these with the basic inverter circuit. Figures 11.16(b) and 11.16(c) show circuits for two-input NAND and NOR gates. In the NAND gate it is clear that both T2 and T3 must be turned ON (*A* AND *B* high) in order for the output to be pulled low. In the NOR gate the output will be low if either T2 or T3 is turned ON (*A* OR *B* high). These circuits can be expanded to produce gates with additional inputs.

When the output of an NMOS gate is at logical 0 the output resistance is very low since the output is shorted to ground by one or more transistors which are turned ON. This enables the circuit to *sink* current efficiently. However, when the output is high the output resistance is determined by the resistance of the load MOSFET. To enable the circuit to *source* current efficiently in this state the resistance of this transistor must be low. The power dissipated by the gate is also controlled by the value of the resistance of the load MOSFET. When the output is high, current flows through the load device to the output, and when the output is low the load transistor is effectively connected directly across the supply rails. In order to minimize power consumption, the load MOSFET must have as high a resistance as possible. We therefore have conflicting requirements for the load device. To achieve a low output resistance and hence a good **fan-out**, the load MOSFET must have a low resistance, while to minimize power consumption this resistance should be as high as possible.

Fortunately, since the input resistance of NMOS gates is so high (generally greater than 10^{12} Ω), it is possible to have a good fan-out (perhaps 50) even with a relatively high output resistance (typically about 100 kΩ), enabling power consumption to be kept to low levels. The power dissipated by the gates is much greater when the output is low than when it is high, with an average value of about 0.1 mW for simple gates. However, the high output resistance combined with a relatively high input capacitance does make these devices relatively slow with a typical **propagation delay time** of about 50 ns.

11.2.13 Complementary metal oxide semiconductor (CMOS) logic

The NMOS logic gates described above suffer from a high output resistance in one of their output states which limits their speed of operation. This problem is common to all amplifiers that use a single output transistor (remember that the load MOSFET of an NMOS gate is acting as a resistor). In Section 7.7.4 we discussed a method of eliminating this problem by using a **push-pull** arrangement. This technique can also be applied to the design of logic gates and in Figure 6.40 we looked at such a circuit. For convenience this circuit is reproduced as Figure 11.17.

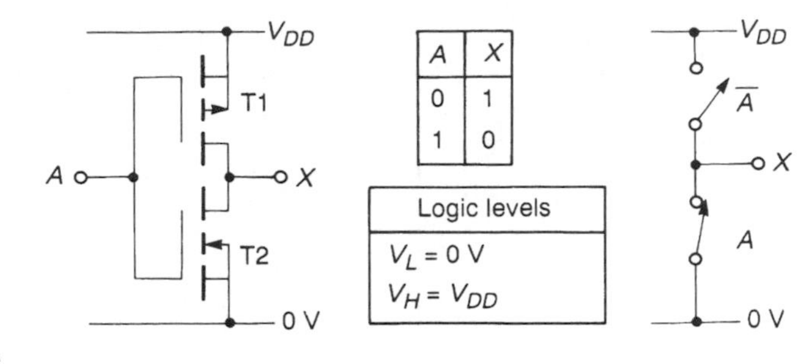

FILE 11F **Figure 11.17** A CMOS inverter.

The circuit uses both an n-channel and a p-channel device and is therefore described as **complementary MOS** logic, or simply **CMOS**. As with NMOS circuitry, V_{DD} represents logic 1 and 0 V represents logic 0. Being of different polarities, the two transistors respond in the opposite sense to voltages applied to their gates. While a gate voltage of V_{DD} will turn ON the n-channel transistor, it will turn OFF the p-channel device. Similarly a voltage of 0 V will turn OFF the n-channel transistor and turn ON the p-channel device. Since the gates of the two MOSFETs are joined, input voltages of either logic level will turn one device ON and the other OFF. This arrangement produces a low output resistance which can charge load capacitances more quickly producing a faster switching time. The low output resistance also gives a high **fan-out** of up to about 50 gates. Since one of the two transistors is always turned OFF there is no DC path between the supply rails, and the only current drawn from the supply is that which is fed to the output. The high input resistance of the gates makes this output current negligibly small except when the input capacitance of a gate is being charged or discharged after an output has changed. Power is also consumed when the circuit switches from one state to another, as, for a short period, both transistors are conducting at the same time. The resultant **power consumption** is therefore generally negligible when the circuit is static, but increases with the switching rate. Typical values for the power consumption might be about 10 nW per gate when static and about 1 mW when clocked at 1 MHz. It is clear that even when operating at high speeds the gates consume very little power. This makes them ideal for applications in which power consumption is critical, for example, where battery operation is required. Low power dissipation also reduces the amount of waste heat that must be removed, allowing more circuitry to be integrated into a single circuit.

The simple inverter of Figure 11.17 can be modified to provide additional logic functions. Examples of two-input NAND and NOR gates are shown in Figure 11.18. Like the inverter, these circuits both provide an active pull-up and active pull-down of the output, giving a low output resistance, and provide no DC path between the supply rails when in either output state.

CMOS circuitry is more difficult to fabricate than NMOS or PMOS since it requires devices of both polarities. However, its increased speed, lower power dissipation and excellent noise immunity make it the most widely used technology for new, highly integrated circuits. We will return to look in more detail at some aspects of the use of CMOS circuits in Section 11.4.

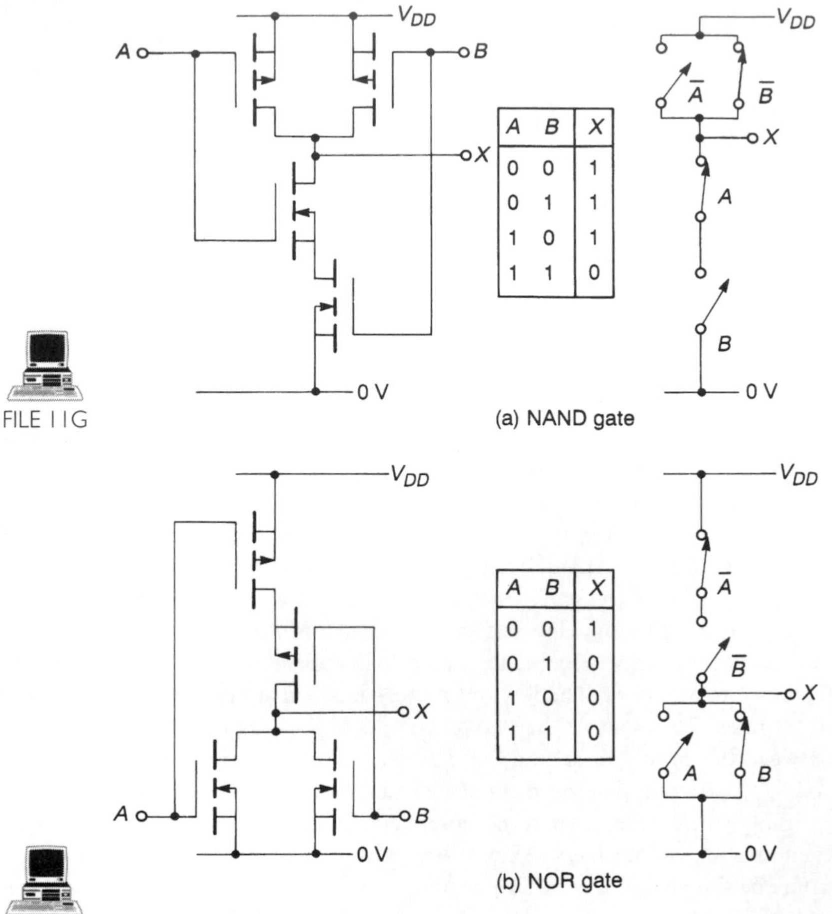

A	B	X
0	0	1
0	1	1
1	0	1
1	1	0

A	B	X
0	0	1
0	1	0
1	0	0
1	1	0

Figure 11.18 Two-input CMOS gates.

11.2.14 Integrated circuit logic gates

Most logic gates are in the form of integrated circuits which contain the functions of a number of gates within a single package. Typical ICs have 14, 16 or 20 pins, although more complicated devices may have 40 or more pins. The number of gates within a single package is generally determined by the number of pins required for each gate. Common devices include:

- octal or hex inverters or buffers
- quad 2-input gates
- triple 3-input gates
- dual 4-input gates, flip-flops, etc.
- single 8-input gates, counters, registers, etc.

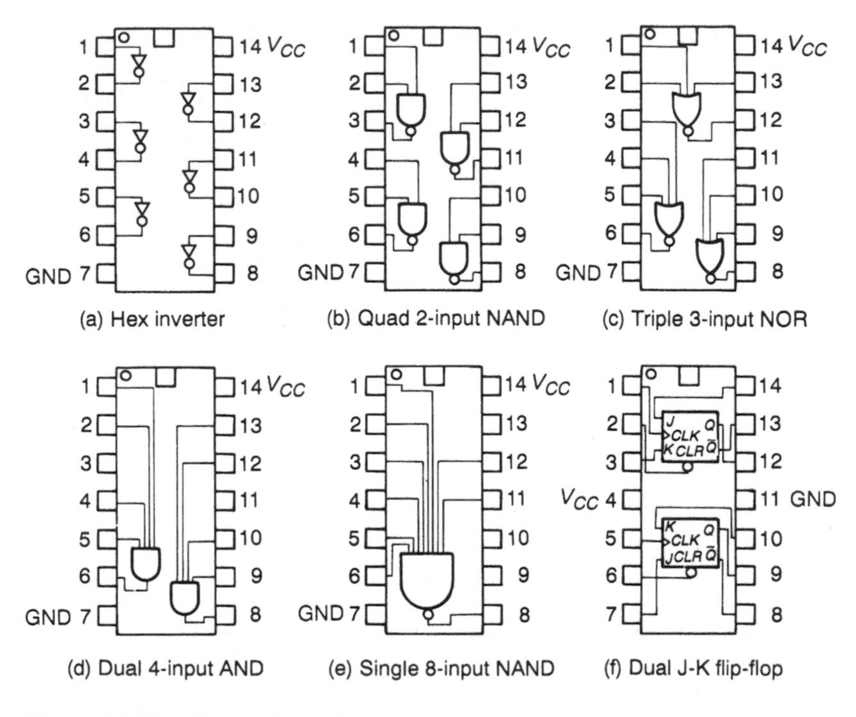

Figure 11.19 Typical logic device pin-outs.

Figure 11.19 shows typical device pin-outs for a number of gates. As with analogue ICs, pins are numbered anticlockwise when viewed from above (the side away from the pins), and the orientation of the device is indicated by a notch, or a dot, next to pin 1, or both.

11.2.15 Logic families – a summary

This section has attempted to outline various characteristics of logic gates and to introduce some of the most widely used terminology. Several semiconductor logic families have been described and broad comparisons given. It is not within the scope of this text to give detailed descriptions of the operation and characteristics of these technologies, but it is perhaps useful to summarize some of the results in tabular form. Table 11.2 gives

Table 11.2 A comparison of logic families.

Parameter	TTL	ECL	NMOS	CMOS
Basic gate	NAND	OR/NOR	NAND-NOR	NAND-NOR
Fan-out	10	25	20	>50
Power per gate (mW)	1–22	4–55	0.2–10	1@1 MHz
Noise immunity	Very good	Good	Good	Excellent
T_{PD} (ns)	1.5–33	1–4	30–300	1.5–200

a comparison of four of the most important logic families in respect of five parameters. Where a number of device families are available, the figures attempt to represent the range of the parameters across these series. Consequently, the data should be used with care. Often the values obtained will depend on other factors, and the numbers given should be taken simply as a guide for comparison and not as detailed data on a particular device family.

11.3 TTL

In Section 11.2.10 we looked at the form and operation of a simple TTL gate and traced its development from simpler logic types. TTL is one of the most widely used logic families, particularly for applications requiring small to medium scale integration (SSI and MSI).

A wide range of manufacturers produce circuits of this form and standardization has been very successful in providing a common specification for such devices. The standard TTL family of components contains a broad spectrum of circuits, each of which is specified by a generic serial number starting with the digits 54 or 74. Devices that begin with 54 are specified for operation over a temperature range from −55 to 125 °C, while those starting with 74 are restricted to a range of 0 to 70 °C. The two-digit suffix is followed by a two- or three-digit code which represents the function of the device. For example, a 7400 contains four 2-input NAND gates while a 7493 is a 4-bit binary counter. The two families are often called the 54XX and 74XX families, where the 'XX' implies some combination of two or three digits, or simply the **54/74 families**.

In addition to 'standard' 54XX and 74XX devices, there are related families with modified characteristics. These are defined by adding alphabetic characters after the 54 or 74 suffix to specify the family. For example, a 74L00 is a low-power version of the 7400, and the 74H00 is a high-speed version. In fact the 'standard' 74XX parts are used much less frequently than members of other families, such as the 74LSXX series. We will look at these variants later in this section.

Figure 11.20 shows part of a typical TTL data sheet with annotations to indicate important features.

Standard TTL

Integrated circuit TTL gates use slightly more complicated circuitry than that given in Figure 11.13. The circuit of a two-input NAND gate (one of the four gates in a 7400) is shown in Figure 11.21.

The basic operation of this gate is similar to that described in Section 11.2.10, except that a form of **push-pull output** stage has been added to reduce the output resistance, increase the current drive capability and enable the circuit to source as well as sink current. This form of output stage is known as a **totem-pole** output. We will see later that some TTL devices use alternative output circuitry. It must be said that this circuit is primarily designed to *sink* current by providing a path for current to flow into the output terminal to ground when a logic 0 is output. The circuit can also *source* current when a high logic level is output, but this current is very small. This characteristic is acceptable since the

Specification for normal commercial parts

recommended operating conditions

Input voltage limits (V_{IH}, V_{IL}); Output current limits (I_{OH}, I_{OL})

		SN5400			SN7400			UNIT
		MIN	NOM	MAX	MIN	NOM	MAX	
V_{CC}	Supply voltage	4.5	5	5.5	4.75	5	5.25	V
V_{IH}	High-level input voltage	2			2			V
V_{IL}	Low-level input voltage			0.8			0.8	V
I_{OH}	High-level output current			-0.4			0.4	mA
I_{OL}	Low-level output current			16			16	mA
T_A	Operating free-air temperature	-55		125	0		70	°C

electrical characteristics over recommended operating free-air temperature range (unless otherwise noted)

Output voltage limits (V_{OH}, V_{OL}); Input current limits (I_I, I_{IH}, I_{IL})

PARAMETER	TEST CONDITIONS †			SN5400			SN7400			UNIT
				MIN	TYP‡	MAX	MIN	TYP‡	MAX	
V_{IK}	V_{CC} = MIN,	I_I = -12 mA				-1.5			1.5	V
V_{OH}	V_{CC} = MIN,	V_{IL} = 0.8 V,	I_{OH} = -0.4 mA	2.4	3.4		2.4	3.4		V
V_{OL}	V_{CC} = MIN,	V_{IH} = 2 V,	I_{OL} = 16 mA		0.2	0.4		0.2	0.4	V
I_I	V_{CC} = MAX,	V_I = 5.5 V				1			1	mA
I_{IH}	V_{CC} = MAX,	V_I = 2.4 V				40			40	μA
I_{IL}	V_{CC} = MAX,	V_I = 0.4 V				-1.6			-1.6	mA
I_{OS}§	V_{CC} = MAX			-20		-55	-18		55	mA
I_{CCH}	V_{CC} = MAX,	V_I = 0 V			4	8		4	8	mA
I_{CCL}	V_{CC} = MAX,	V_I = 4.5 V			12	22		12	22	mA

† For conditions shown as MIN or MAX, use the appropriate value specified under recommended operating conditions.
‡ All typical values are at V_{CC} = 5 V, T_A = 25°C.
§ Not more than one output should be shorted at a time.

Conditions under which values are measured

switching characteristics, V_{CC} = 5 V, T_A = 25°C (see note 2)

Propagation delay times for different transitions

PARAMETER	FROM (INPUT)	TO (OUTPUT)	TEST CONDITIONS		MIN	TYP	MAX	UNIT
t_{PLH}	A or B	Y	R_L = 400 Ω,	C_L = 15 pF		11	22	ns
t_{PHL}						7	15	ns

NOTE 2: See General Information Section for load circuits and voltage waveforms.

Figure 11.20 Part of a typical TTL data sheet.

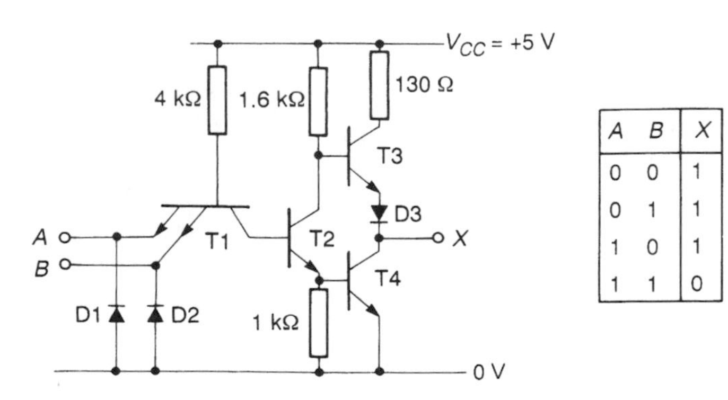

A	B	X
0	0	1
0	1	1
1	0	1
1	1	0

Figure 11.21 A TTL two-input NAND gate.

gates themselves take very little input current when a high logic signal is applied. Like all TTL devices, the circuit operates from a supply voltage of 5.0 V which must be accurate to within ±0.25 V (±0.5 V for the 54XX family). Typical power dissipation is 10 mW per gate.

If either input A or input B is low, T1 pulls the base of T2 low, turning it OFF. This causes no current to flow in the 1 kΩ emitter resistor, turning T4 OFF and allowing the base of T3 to rise, turning it ON. The output is thus pulled high, taking a value of V_{CC} less the sum of the voltage drops across the diode D3 and the emitter-to-base junction of T3. Thus the logic 1 output voltage is given by

$$\begin{aligned} V_H &= V_{CC} - V_{diode} - V_{BE} \\ &= 5.0 - 0.7 - 0.7 \\ &= 3.6\ \text{V} \end{aligned}$$

If both input A and input B are high, T1 pulls the base of T2 high turning on T2 and T4. This drives the output voltage down to the saturation voltage of T4, thus

$$V_L = V_{CE(sat)} \approx 0.2\ \text{V}$$

The voltage at the collector of T2 falls as a result of the current through its collector resistor, dropping the base of T3 to about 0.9 V (the base-to-emitter voltage of T4 plus the saturation voltage of T2). Diode D3 is present to ensure that T3 is turned OFF under these conditions. If it were absent, the voltage across the base emitter junction of T3 would be equal to the voltage on the base (0.9 V) minus the output voltage (0.2 V) which would be sufficient to turn the device ON. The presence of D3 ensures that T3 is held OFF while the output is low.

Diodes D1 and D2 are **input clamp diodes** which prevent negative-going noise spikes at the input from damaging T1. Negative-going transitions will forward bias the diodes which therefore prevent the input from going negative by more than the forward voltage of the diode (about 0.7 V).

Transfer characteristic

Figure 11.22 shows the transfer characteristic of the gate, that is, the relationship between the input voltage and the output voltage. The characteristic shows the effect of changing

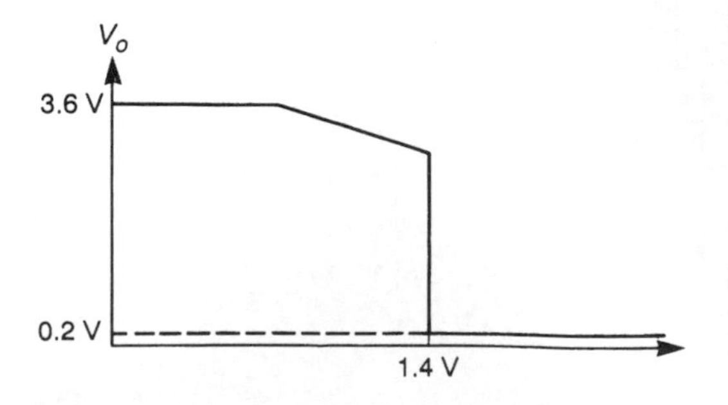

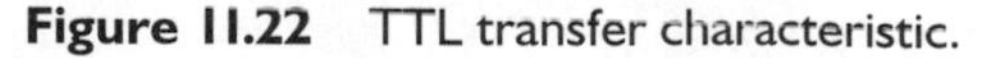
Figure 11.22 TTL transfer characteristic.

Table 11.3 Input and output voltage levels for 54/74 family devices.

	Minimum	Typical	Maximum
V_{IL}	—	—	0.8
V_{IH}	2.0	—	—
V_{OL}	—	0.2	0.4
V_{OH}	2.4	3.6	—

the voltage on one input while the other is held high (if the other input were held low, the output would not change).

The transfer characteristic shows the output logic levels of 3.6 V and 0.2 V derived above, and also indicates that the input threshold voltage is 1.4 V. That is, input voltages above 1.4 V will be interpreted as a logic 1, while voltages below 1.4 V will be taken as a logic 0.

Logic levels and noise immunity

Although the transfer characteristic indicates specific values for the input and output voltages, real devices are subject to variability and the logic levels are specified by bands of voltages. These are shown in Table 11.3.

From Figure 11.4 and Equations 11.1 and 11.2, it is clear that the noise immunities in each of the two logic states are given by

$$\begin{aligned}\text{Noise immunity in logic 1 (HIGH) state } V_{NIH} &= V_{OH}(min) - V_{IH}(min)\\ &= 2.4 - 2.0\\ &= 0.4 \text{ V}\end{aligned}$$

and

$$\begin{aligned}\text{Noise immunity in logic 0 (LOW) state } V_{NIL} &= V_{IL}(max) - V_{OL}(max)\\ &= 0.8 - 0.4\\ &= 0.4 \text{ V}\end{aligned}$$

Thus the minimum noise immunity of each logic state is 0.4 V. However, taking typical values for the logic voltages gives a noise immunity of about 1.0 V for logic 0 and 1.6 V for logic 1.

Input and output currents and fan-out

The input current taken by the gate will clearly be different depending on whether the input is high or low, and also varies between the various forms of gate. For the 7400 two-input NAND gate shown in Figure 11.21, the maximum input current for a logic 1 input signal is 40 μA, and for a logic 0 input is −1.6 mA. Since currents are conventionally measured *into* a device, the minus sign indicates that this is current flowing *out* of the device.

The specification of the 7400 states that when a logic 1 is output the circuit will source

Table 11.4 Propagation delay times for typical 54/74 family gates.

	Minimum	Typical	Maximum
t_{PHL} (ns)	—	7	15
t_{PLH} (ns)	—	11	22

an output current of at least −400 μA, and when a 0 is output it will sink at least 16 mA. Again the negative sign indicates that this current is flowing out of the device. This gives a fan-out of 10.

Switching characteristics

The switching characteristics can be described by the **propagation delay times** for transitions from high to low (t_{PHL}) and from low to high (t_{PLH}). These are shown in Table 11.4 for a typical single gate.

11.3.2 Open collector devices

Some 54/74 family devices use a different output configuration referred to as an **open collector** output stage. The 7401 contains four two-input NAND gates with open collector outputs. Figure 11.23 shows the circuit of one of these gates.

It is clear that the gate is similar to that of Figure 11.21 with the exception that the output stage has been simplified. The output is taken from the collector of an output transistor which is otherwise unconnected or *open*. In order for the circuit to function, an external **pull-up resistor** must be connected between the output and the positive supply, as shown in Figure 11.24.

The choice of a value for the pull-up resistor is a compromise between power dissipation and speed of operation. High-value resistors reduce the collector current and hence the power, but also limit the rate at which load capacitances can be charged. Even with relatively low values of resistance, the open collector circuit is not as fast as the totem-pole

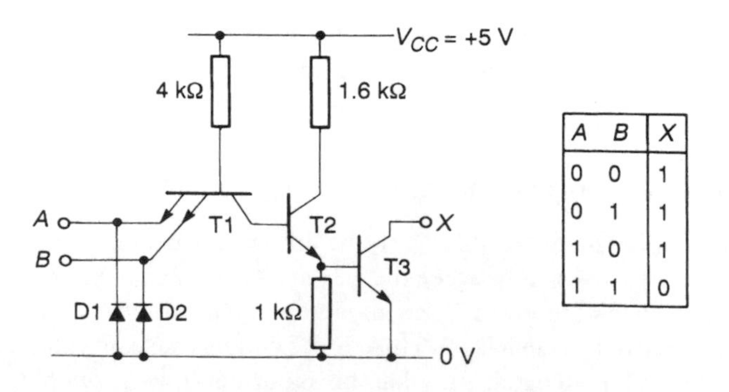

A	B	X
0	0	1
0	1	1
1	0	1
1	1	0

Figure 11.23 The 7401 two-input NAND gate with open collector output.

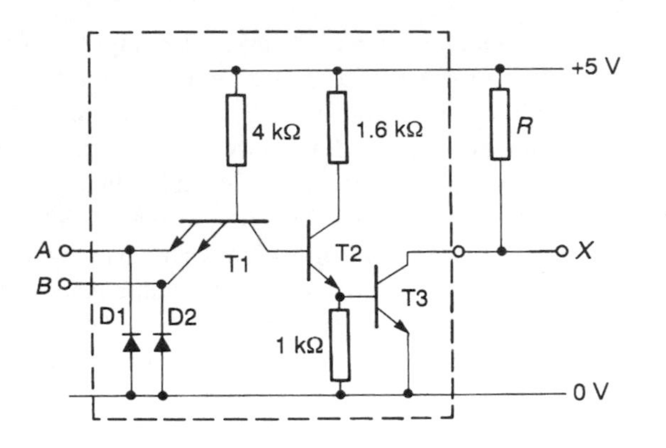

Figure 11.24 Use of an open collector gate with an external load.

arrangement. This is because the latter has an additional output transistor acting as a low-impedance emitter follower, which is able to charge load capacitances quickly.

The output logic levels of an open collector gate are not the same as those for a totem-pole arrangement. When the output transistor is turned ON the output voltage will be pulled down to the saturation voltage of the device (about 0.1–0.2 V), but when the transistor is turned OFF the output voltage will rise to the value of the supply attached to the pull-up resistor – normally V_{CC}. Thus the logic levels are close to the supply rail voltages. These voltages are completely compatible with the inputs of TTL gates and so this difference in logic levels is not usually significant.

Wired-AND operation

One of the advantages of open collector gates is that their outputs can be connected in parallel to form a **wired-AND** configuration. This is illustrated in Figure 11.25.

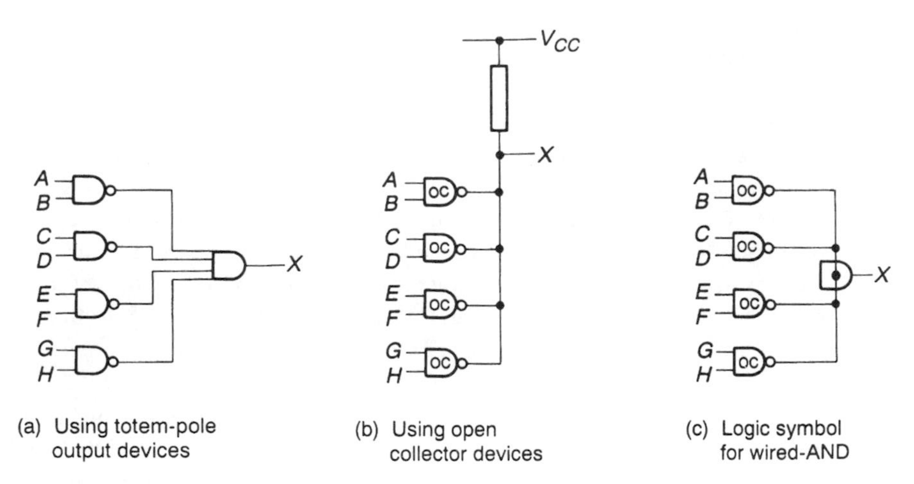

Figure 11.25 The wired-AND configuration.

Figure 11.25 shows a circuit containing four NAND gates, the outputs of which must be ANDed together to produce the function X. Using conventional, totem-pole output devices, the outputs must be combined using a four-input AND gate, as shown in Figure 11.25(a). However, if open collector gates are used the outputs can simply be connected in parallel to a single resistor to achieve the same result, as shown in Figure 11.25(b). The AND function is obtained since each gate can pull the output low, but all must be high for X to be high. The circuit symbol for the wired-AND operation is shown in Figure 11.25(c).

The wired-AND function is of particular interest when large numbers of signals must be combined, since this removes the need for gates with a large number of inputs. It is also of great importance in the production of a **bus** in which a number of devices are connected to a single line. We will consider bus systems in more detail in the next chapter when we consider microprocessor systems.

Since AND and OR functions may be interchanged (with appropriate signal inversions) by the use of **DeMorgan's theorem**, it is also possible to use the wired-AND function to produce an OR operation. For this reason the technique is sometimes referred to as a **wired-OR** configuration.

High-voltage outputs

Some open collector devices can be used with high output voltages. The device is operated from a standard 5.0 V supply, but the output is connected through a pull-up resistor to a high-voltage supply. The 7406, for example, contains six inverters with high-voltage, open collector outputs. These can switch up to 30 V at currents of up to 40 mA. The output logic levels are equal to the saturation voltage of the output transistor (about 0.1 to 0.2 V) and the high-level supply voltage.

11.3.3 Three-state devices

Conventional logic gates have two possible output states, namely 0 and 1. Under some circumstances it is convenient to have a third state corresponding to a high impedance condition, when the output is allowed to float. Under these circumstances the voltage at the output will be determined by whatever external circuitry is connected to it. Circuits with this property are called **three-state logic gates**. The output of the gate is 'enabled' or 'disabled' by a control input, which is usually given the symbol C on simple gates. In more complicated circuits this control signal is often referred to as the **output enable** line.

Figure 11.26 shows how the three-state function is represented in a circuit symbol. Figure 11.26(a) shows a non-inverting buffer with an active-high control input (that is, the output is enabled if $C = 1$); Figure 11.26(b) shows the symbol for a similar gate with an active-low control input (the output is enabled if $C = 0$). The first of these could represent one of the gates in a 74126, and the second a gate in a 74125. Both devices contain six such gates.

The output circuit of a three-state gate resembles that of a totem-pole device with the addition of extra components to turn both output transistors OFF to disable the output. This allows the output to float, independent of the other gate inputs. Since, when enabled, the output resembles the conventional totem-pole arrangement, the use of three-state techniques does not incur the speed penalty associated with open collector circuits.

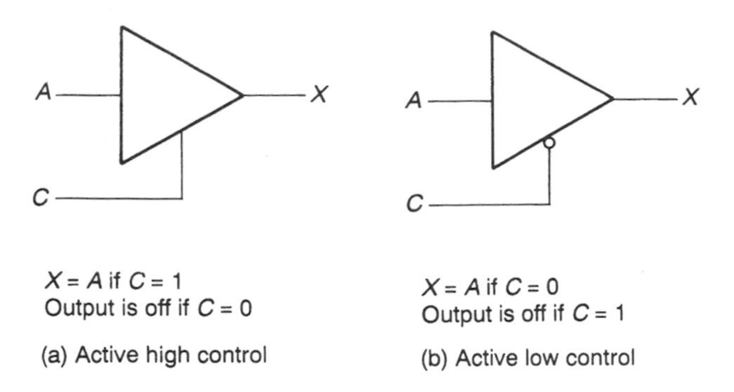

Figure 11.26 Symbolic representation of three-state logic gates.

Three-state devices can be used in the creation of bus systems in which the outputs of several devices are connected together on to a common line. Each device can then place data on the line provided that the output of only one device is enabled at any time. This arrangement differs from the use of open collector gates in the wired-AND configuration described earlier. When three-state devices are used, only one gate drives the line at any time and the outputs of disabled gates do not affect the signal on the bus.

Three-state devices are sometimes described as being **tri-state**. This term is a trade mark of the National Semiconductor Corporation.

11.3.4 TTL inputs

In certain instances a circuit may not need to use all the inputs of a particular gate. It may be convenient, for example, to OR together three signals using a four-input OR gate because such a device is already available within the circuit. If **unused inputs** are left disconnected they will act as if they were connected to a logic 1. Such inputs are said to be **floating**.

Although unused inputs will 'float' to a logic 1, it is inadvisable to leave such inputs disconnected, even if the application requires the input to be at logic 1. Unconnected inputs represent a high impedance to ground making them very sensitive to electrical noise, which could cause them to switch between states. It is much wiser to 'tie' such inputs high or low, as required. Inputs that are required to be at logic 1 should not be tied directly to the positive supply rail, but rather through a resistor. A typical value for such a resistor might be 1 kΩ. If appropriate, several inputs can be connected together to the same resistor. Inputs that are required to be low may be tied directly to ground (0 V). In some circumstances it may be appropriate to connect an input to ground through a resistance (perhaps to allow it to be pulled high through a switch to logic 1). In this case the resistor value must be sufficiently small to allow the input current to flow through the resistor without taking the input voltage above the maximum level for a logical 0 input ($V_{IL}(max)$). For standard TTL, a typical maximum value for such a resistor might be 470 Ω.

It is worth noting that when using AND or NAND gates any unused inputs should be tied high to prevent them from affecting the output state. When using OR or NOR gates any unused inputs should be tied low.

11.3.5 Other TTL families

So far we have concentrated on the 'standard' 54/74 TTL family. There are a number of related families which are optimized for particular operating characteristics.

Low-power TTL (54L/74L)

The 54L/74L families of devices are optimized for low power consumption. Figure 11.27 shows the circuit diagram of a typical gate, the 74L00; a comparison with Figure 11.21 will show that this power reduction is achieved primarily by a change of resistor values.

The average power dissipation of the low-power gate is 1 mW compared with 10 mW for a standard device, but is achieved at the expense of a reduction in speed. The average propagation delay is increased from 9 ns, for a standard gate, to 33 ns for the low-power version.

High-speed TTL (54H/74H)

The 54H/74H families are optimized for speed. Figure 11.28 shows the circuit of the 74H00 which has an average propagation delay time of only 6 ns but an average power dissipation of 22 mW.

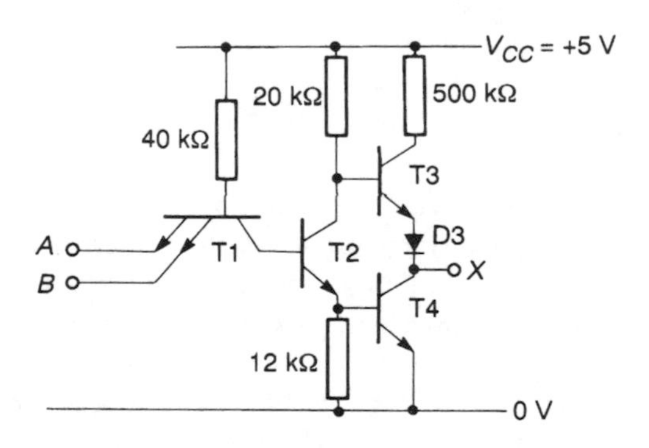

Figure 11.27 A 74L00 two-input NAND gate.

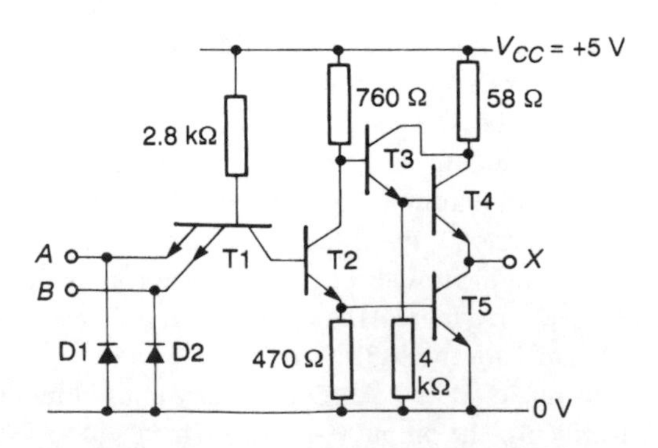

Figure 11.28 A 74H00 two-input NAND gate.

Schottky TTL (54S/74S)

The 54S/74S families use circuits similar to those of the 54H/74H devices, except that they use **Schottky** transistors and diodes in place of conventional components.

Schottky diodes are formed by the junction of a metal and a semiconductor, unlike more traditional diodes which consist of a junction between two regions of doped semiconductor (see Sections 5.5 and 5.6). The circuit symbol of a Schottky diode is shown in Figure 11.29(a).

Schottky diodes are not only very fast in operation but also have a forward voltage drop of only about 0.25 V. This allows them to be used to prevent the saturation of a transistor, as shown in Figure 11.29(b). When the transistor is operating well within its active region the collector is positive with respect to the base and the diode is reverse biased. Under these conditions the diode has no effect on the operation of the transistor. However, as the device nears its saturation region the voltage on the collector drops below that of the base and, in the absence of the diode, would ultimately fall to its saturation value of about 0.1–0.2 V above the emitter voltage. The presence of the Schottky diode across the collector–base junction prevents the transistor from entering saturation since it becomes forward biased and therefore begins to conduct before the device saturates. Once conducting, the diode robs the transistor of current, inhibiting any further drop in the collector voltage and thus preventing the device from entering its saturation region. The combination of transistor and diode is referred to as a **Schottky transistor** which has its own circuit symbol, as shown in Figure 11.29(c).

We saw in Section 11.2.3 that saturation of transistors causes a considerable increase in propagation delay because of the presence of **storage time**. The use of Schottky transistors prevents saturation and greatly increases the speed of operation while incurring only a modest increase in power consumption. Consequently, 54S/74S family gates have a propagation delay of about half that of 54H/74H devices with approximately the same power consumption. For this reason the Schottky devices have largely replaced the older high-speed family.

Figure 11.30 shows an example of a typical device, a 74S00 two-input NAND gate. This has a typical propagation delay of 3 ns and an average power dissipation per gate of about 19 mW. Note that one of the transistors T4 is not a Schottky type because it does not saturate during normal operation of the circuit.

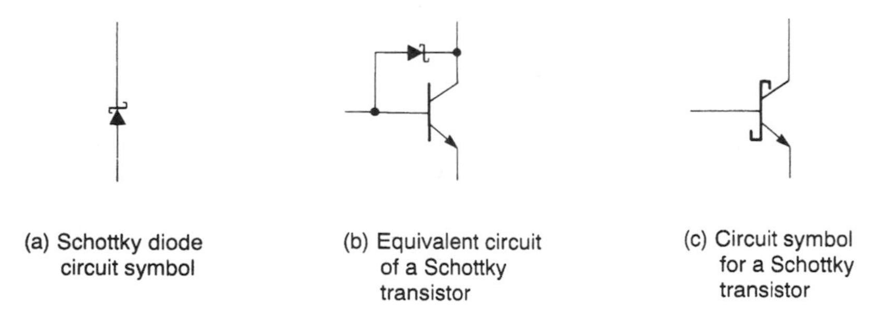

Figure 11.29 Schottky diodes and transistors.

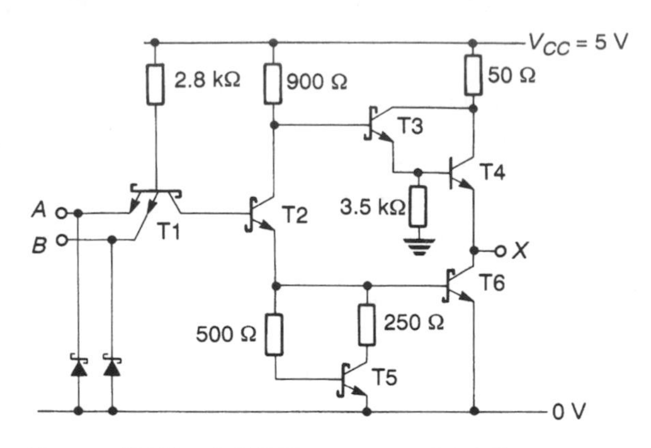

Figure 11.30 A 74S00 two-input NAND gate.

Advanced Schottky TTL (54AS/74AS)

Recent developments of TTL families have led to the production of an advanced Schottky family of devices which have a typical propagation delay of about 1.5 ns and a typical power dissipation of 8.5 mW.

Low-power Schottky TTL (54LS/74LS)

The 54LS/74LS low-power Schottky family combines considerations of speed and power consumption. Its typical propagation delay of 9.5 ns is approximately equal to that of standard TTL, but at 2 mW average dissipation per gate it consumes only one-fifth of the power. Figure 11.31 shows a 74LS00 two-input NAND gate.

Advanced low-power Schottky TTL (54ALS/74ALS)

A range of high performance low-power Schottky devices is now available as the 54ALS/74ALS family. These have a typical propagation delay of 4 ns and a power consumption of 1 mW per gate.

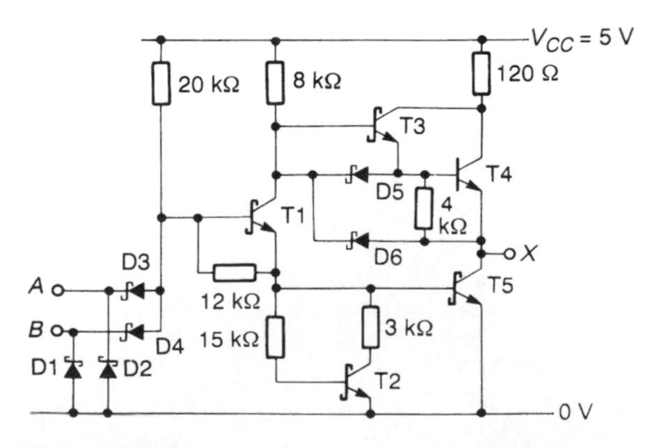

Figure 11.31 A 74LS00 two-input NAND gate.

FAST TTL (54F/74F)

The 'FAST' range of TTL logic provides high speed with low power consumption. Typical propagation delay is 2.7 ns and typical power consumption about 4 mW per gate.

74 series CMOS families (74C/74HC/74HCT/74AC/74ACT)

The 74C, 74HC, 74HCT, 74AC and 74ACT families are *not* TTL logic families, the 'C' in each name standing for CMOS. The great success of the 54/74 families, in all their forms, has led to a move to standardize the serial numbers given to the various logic devices and this has led to families of CMOS devices with the same logical functions and similar serial numbers. The 74C, 74HC and 74AC series of components are conventional CMOS devices and therefore use logic levels that are not directly compatible with TTL devices. The 74HCT and 74ACT components are also CMOS gates, but have been designed to use TTL logic levels allowing them to be interchanged with TTL devices. These families are becoming widely used in applications for which power consumption is of importance. We will leave further discussion of these CMOS gates until the next section.

11.3.6 TTL families – a summary

There are a myriad of TTL families, all with different characteristics, advantages and disadvantages. Gradually a few families are becoming dominant. The 'LS' family is used for most general applications, with standard family devices being used where extra output power is required and 'S', 'AS' or 'ALS' devices where speed is critical. Increasingly the CMOS 'HCT' and 'ACT' ranges are replacing TTL components where power consumption is of importance.

Table 11.5 gives a brief comparison between some of the TTL families in terms of propagation delay and power consumption. Figures are typical values for standard gates (usually two-input NAND gates) and will vary for other circuits.

Table 11.5 A comparison of TTL logic families.

Family	Descriptor	T_{PD} (ns)	Power per gate (mW)
Standard	74XX	9	10
Low-power	74LXX	33	1
High-speed	74HXX	6	22
Schottky	74SXX	3	19
Advanced Schottky	74ASXX	1.5	8.5
Low-power Schottky	74LSXX	9.5	2
Advanced low-power Schottky	74ALSXX	4	1
FAST	74FXX	2.7	4

11.4 CMOS

In Section 11.2.13 we looked at the basic form of CMOS gates and at the circuits of inverter, AND and OR gates. In this section we will look in a little more detail at the characteristics and use of this type of logic.

11.4.1 CMOS logic families

The first manufacturer to produce CMOS logic was RCA who described them as the **4000 series**, having numbers 4000, 4001, etc. Some other manufacturers have adopted the same numbering system, while others have devised their own related numbering schemes. Motorola, for example, produces components in the MC14000 and MC14500 series. Some years ago the original 4000 series devices were replaced by an improved 4000B series.

Some manufacturers have moved away from the original 4000 series parts and have produced a range of circuits that follow the circuit functions and pin assignments of the 74XX TTL family of devices (see Section 11.3). These are given part numbers such as 74CXX, 74HCXX, 74HCTXX, 74ACXX or 74ACTXX where in each case the 'C' stands for CMOS. The 'A' in the names of the 'AC' and 'ACT' series indicates that these are 'advanced' devices with operating speeds comparable to the fastest TTL families. The 'T' in the names of the 'HCT' and 'ACT' series indicates that these devices are unlike those of the other CMOS families in that they are designed to operate with the supply voltages and logic levels of TTL gates. This enables them to be used easily with TTL components, allowing them to act as direct, low-power replacements for the corresponding TTL parts. Conventional CMOS circuits of the 4000 series or the 74CXX types cannot normally be used directly with TTL parts since their logic levels are different. This topic will be discussed later in Section 11.4.3.

11.4.2 CMOS characteristics

CMOS gates differ in many respects from the TTL gates described in the last section. In this section we will look at the general characteristics of CMOS logic. Figure 11.32 shows part of a typical CMOS data sheet which is annotated to indicate items of importance.

Power-supply voltages

Most CMOS circuits operate using a single supply voltage of from 5 to 15 V although most gates are usable over a range of from 3 to 18 V. Common supply voltages are 5, 10 and 15 V. The speed of operation increases with the supply voltage (this will be discussed in the later section on propagation delay), as does the power dissipation.

Logic levels and noise immunity

The output logic levels of CMOS gates are very close to the supply rails and can normally be assumed to be equal to 0 (V_{SS}) and the positive supply voltage (V_{DD}). Therefore, for most purposes it is reasonable to assume that

$$V_{OL}(max) = 0$$

Maximum ratings before the chip is damaged

absolute maximum ratings over operating free-air temperature range†

Supply voltage, V_{CC} .... −0.5 V to 7 V
Input clamp current, I_{IK} ($V_I < 0$ or $V_I > V_{CC}$) .... ±20 mA
Output clamp current, I_{OK} ($V_O < 0$ or $V_O > V_{CC}$ .... ±20 mA
Continuous output current, I_O ($V_O = 0$ to V_{CC}) .... ±25 mA
Continuous current through V_{CC} or GND pins .... ±50 mA
Lead temperature 1,6 mm (1/16 in) from case for 60 s: FK or J package .... 300 °C
Lead temperature 1,6 mm (1/16 in) from case for 10 s: D or N package .... 260 °C
Storage temperature range .... −65 °C to 150 °C

† Stresses beyond those listed under "absolute maximum ratings" may cause permanent damage to the device. These are stress ratings only, and functional operation of the device at these or any other conditions beyond those indicated under "recommended operating conditions" is not implied. Exposure to absolute-maximum-rated conditions for extended periods may affect device reliability.

Specification for normal commercial parts

recommended operating conditions

Limits of input logic levels

		SN54HC00			SN74HC00			UNIT
		MIN	NOM	MAX	MIN	NOM	MAX	
V_{CC} Supply voltage		2	5	6	2	5	6	V
V_{IH} High-level input voltage	V_{CC} = 2 V	1.5			1.5			V
	V_{CC} = 4.5 V	3.15			3.15			
	V_{CC} = 6 V	4.2			4.2			
V_{IL} Low-level input voltage	V_{CC} = 2 V	0		0.3	0		0.3	V
	V_{CC} = 4.5 V	0		0.9	0		0.9	
	V_{CC} = 6 V	0		1.2	0		1.2	
V_I Input voltage		0		V_{CC}	0		V_{CC}	V
V_O Output voltage		0		V_{CC}	0		V_{CC}	V
t_t Input transition (rise and fall) times	V_{CC} = 2 V	0		1000	0		1000	ns
	V_{CC} = 4.5 V	0		500	0		500	
	V_{CC} = 6 V	0		400	0		400	
T_A Operating free-air temperature		−55		125	−40		85	°C

Conditions under which values are measured

electrical characteristics over recommended operating free-air temperature range (unless otherwise noted)

Limits of output logic levels

Maximum quiescent supply current

PARAMETER	TEST CONDITIONS	V_{CC}	T_A = 25 °C			SN54HC00		SN74HC00		UNIT
			MIN	TYP	MAX	MIN	MAX	MIN	MAX	
V_{OH}	$V_I = V_{IH}$ or V_{IL}, $I_{OH} = -20\ \mu A$	2 V	1.9	1.998		1.9		1.9		V
		4.5 V	4.4	4.499		4.4		4.4		
		6 V	5.9	5.999		5.9		5.9		
	$V_I = V_{IH}$ or V_{IL}, $I_{OH} = -4$ mA	4.5 V	3.98	4.30		3.7		3.84		
	$V_I = V_{IH}$ or V_{IL}, $I_{OH} = -5.2$ mA	6 V	5.48	5.80		5.2		5.34		
V_{OL}	$V_I = V_{IH}$ or V_{IL}, $I_{OL} = 20\ \mu A$	2 V		0.002	0.1		0.1		0.1	V
		4.5 V		0.001	0.1		0.1		0.1	
		6 V		0.001	0.1		0.1		0.1	
	$V_I = V_{IH}$ or V_{IL}, $I_{OL} = 4$ mA	4.5 V		0.17	0.26		0.4		0.33	
	$V_I = V_{IH}$ or V_{IL}, $I_{OL} = 5.2$ mA	6 V		0.15	0.26		0.4		0.33	
I_I	$V_I = V_{CC}$ or 0	6 V		±0.1	±100		±1000		±1000	nA
I_{CC}	$V_I = V_{CC}$ or 0, $I_O = 0$	6 V			2		40		20	μA
C_i		2 to 6 V		3	10		10		10	pF

Figure 11.32 Part of a typical CMOS data sheet.

and

$$V_{OH}(min) = V_{DD}$$

The input logic levels also change with the supply voltage and are defined as

$$V_{IL}(max) = 0.3 \times V_{DD}$$

and

$$V_{IH}(min) = 0.7 \times V_{DD}$$

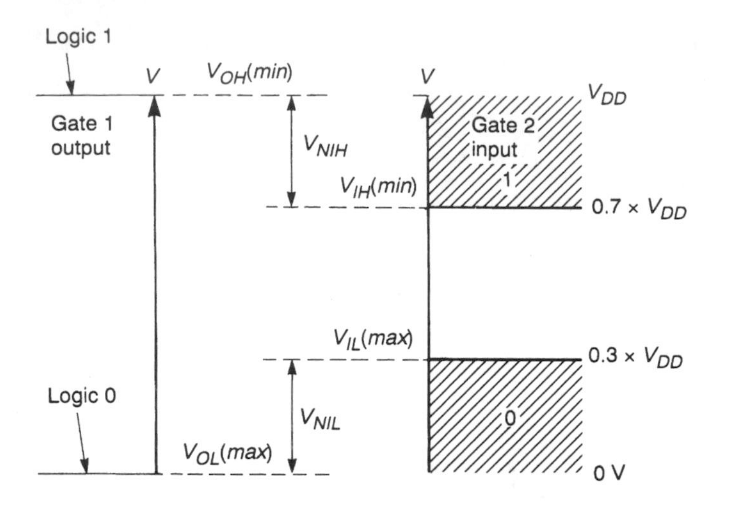

Figure 11.33 CMOS input and output logic levels.

These definitions are illustrated in Figure 11.33.

From Figure 11.4 and Equations 11.1 and 11.2, it is clear that the noise immunities in each of the two logic states are given by

$$\text{Noise immunity in logic 1 (HIGH) state } V_{NIH} = V_{OH}(min) - V_{IH}(min)$$
$$= V_{DD} - 0.7 \times V_{DD}$$
$$= 0.3 \times V_{DD}$$

and

$$\text{Noise immunity in logic 0 (LOW) state } V_{NIL} = V_{IL}(max) - V_{OL}(max)$$
$$= 0.3 \times V_{DD} - 0$$
$$= 0.3 \times V_{DD}$$

Hence the minimum noise immunity in each state is equal to 30% of the supply voltage.

When used with a supply voltage of 5 V (as for TTL), this gives a noise immunity of 1.5 V, which compares very favourably with the figure of 0.4 V for TTL devices. When used with a supply voltage of 15 V, the noise immunity rises to 4.5 V making it extremely attractive for applications in high-noise environments. Unfortunately, high noise immunity is not the only criterion for determining the susceptibility of a system to noise. CMOS gates have an output impedance between three and ten times greater than TTL, which increases their sensitivity to capacitively coupled noise. However, CMOS is generally accepted to be one of the most noise tolerant technologies, provided that appropriate design and layout rules are followed.

It might at first sight seem strange that the input logic levels are not defined such that any voltage greater than 50% of V_{DD} is interpreted as a 1, and any voltage less than this value is interpreted as a 0. This would produce a system with a noise immunity of

50% of the supply voltage. In practice this is not possible because of variations in the threshold voltages within the device. The values specified allow for this variability and assure correct operation. In operation the noise immunity will tend to be greater than the minimum value calculated above, a typical value being 45% of V_{DD}.

Where noise tolerance is a critical consideration CMOS is an obvious choice since it offers an extremely high noise immunity. It is common for such applications to use a 15 V supply voltage to improve further the noise performance, although this adversely affects the power consumption.

Power dissipation

Often one of the main reasons for adopting CMOS logic is its very low power consumption. As was observed in Section 11.2.13, the quiescent power consumption (that is, the consumption when the circuit is static) is extremely low, typically a few nanowatts for any supply voltage. However, each time the device changes state a small amount of power is used to charge capacitances within the circuit and in the load. Some power is also consumed as, for a brief period of time, both halves of the complementary pairs of transistors are ON at the same time.

Since a small amount of power is dissipated each time the gate changes state, the power consumption of the device increases steadily with the clock rate. To a good approximation it may be considered that the power consumed by the device increases linearly with frequency. Power dissipation also increases with the supply voltage.

For a supply voltage of 5 V, a typical CMOS gate consumes less than 1 μW per gate at 1 kHz, but this increases to nearly 1 mW at 1 MHz. At frequencies above 10 MHz the power consumption is greater than that of 74LSXX TTL gates. With a supply voltage of 15 V the power consumption is about 10 mW per gate at 1 MHz.

Propagation delay

The early 4000 series CMOS logic gates are generally slower in operation than all forms of TTL gate. Because of their relatively high output impedance, the propagation delay time of CMOS devices is greatly affected by the number of gates connected to their output. Their speed is also related to the supply voltage, higher voltages giving a faster response. A typical 4000 series gate operating with a 5 V supply might have a propagation delay of between 50 and 200 ns, depending on the number of gates connected to its output. A similar gate operating from a 15 V supply might have a delay of from 20 to 60 ns.

In recent years the speed of operation of CMOS logic has increased considerably. The 74AC (Advanced CMOS) and 74ACT (Advanced CMOS with TTL pin-outs) families have propagation delay times of only a few nanoseconds, making them comparable with the FAST, AS and S TTL families discussed in the last section. This increase in speed has been achieved without sacrificing the very low power consumption of CMOS.

CMOS inputs

A CMOS input looks, to the outside world, like a small capacitor of the order of 1 pF. Because of their very high input impedance such inputs are very sensitive to **static**

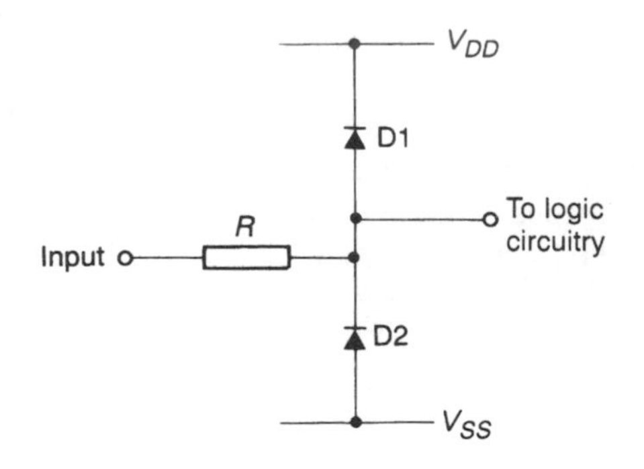

Figure 11.34 CMOS gate protection circuitry.

electricity which could easily destroy the device. To reduce these problems, most devices contain gate protection diodes, as shown in Figure 11.34.

The diodes act as **clamp diodes** preventing the voltage applied to the logic circuitry from going above, or below, safe levels. If the input goes positive with respect to V_{DD} this will tend to forward bias D1 which will clamp the voltage applied to the logic circuit to $V_{DD} + V_{diode}$. The resistor R simply limits the current passed through the diode. Similarly, if the input goes below V_{SS} it will be clamped by diode D2 to $V_{SS} - V_{diode}$ protecting the logic circuitry from damage.

Even with protection circuitry, CMOS (and other MOS circuits) are very susceptible to static charges, particularly before they are assembled into circuits. Normal precautions include storing these devices in conductive enclosures (for example, electrically conductive plastic tubes) and minimizing handling.

Unused inputs

Unused CMOS gates must not be left unconnected. Inputs that are not pulled up or down may cause problems for a number of reasons. Firstly, such inputs are prone to damage caused by static electricity, as described earlier, which could destroy the device. Secondly, unconnected inputs tend to float midway between the supply rails which means that they are liable to move above and below the threshold voltage giving rise to unpredictable behaviour. Thirdly, if inputs are not tied to either logical 1 or logical 0, the corresponding MOSFETs are not switched hard ON or OFF and the current dissipation of the device increases considerably.

All unused inputs must be tied high or low, or joined to other inputs. The choice of whether to tie an unused input high or low depends on the function of the gate. As with TTL, unused AND and NAND inputs should be tied high whereas unused OR and NOR inputs should be tied low.

CMOS outputs

CMOS gates have a typical output resistance of about 250 Ω for a V_{DD} of 5 V. Since the input resistance of the gates is so high, a large number of devices can be driven from one output, the main restriction to the **fan-out** being that the propagation delay increases with the number of gates driven (as described earlier). If high-speed operation is not required, at least 50 gates can be driven from a single output.

There is no CMOS equivalent of the TTL open collector output. However, some CMOS gates have a three-state facility which operates in exactly the same way as in TTL circuits (see Section 11.3.3).

11.4.3 Interfacing TTL and CMOS

Interfacing is the term used to describe the connecting together of two circuits or systems. Since the logic levels and characteristics of TTL and CMOS are different, an output of one cannot normally be connected directly to an input of the other.

Driving CMOS from TTL

Typical output logic levels for a TTL gate with a totem-pole output are about 3.6 V for logic 1, and about 0.2 V for logic 0. When driving other TTL gates, the high logic level will fall as current is taken from the device, and $V_{OH(min)}$ may be as low as 2.4 V. The input of a CMOS gate interprets any voltage of less than $0.3 \times V_{DD}$ as a logic 0 and any voltage greater than $0.7 \times V_{DD}$ as a 1. For a supply voltage of 5 V this gives $V_{IL(max)} = 1.5$ V, and $V_{IH(min)} = 3.5$ V. The logic 0 voltage levels clearly cause no problems, but the TTL logic 1 output is not sufficiently high to guarantee that it will be interpreted as a 1 by the input of a CMOS gate.

Fortunately the solution to this problem is very simple. We have already seen in Section 11.3.2 that the logic levels of open collector gates are approximately equal to the supply rail voltages. Therefore, an open collector gate with a pull-up resistor connected to 5 V will directly drive CMOS gates operating on a 5 V supply voltage. Moreover, since CMOS gates have a very high input resistance, they do not load TTL outputs; the addition of a pull-up resistor to a totem-pole output will cause its output voltage to rise to approximately 5 V when in the high state. This allows both open collector and conventional totem-pole devices to be interfaced to 5 V CMOS logic, simply by the addition of a pull-up resistor to 5 V. This is illustrated in Figure 11.35(a).

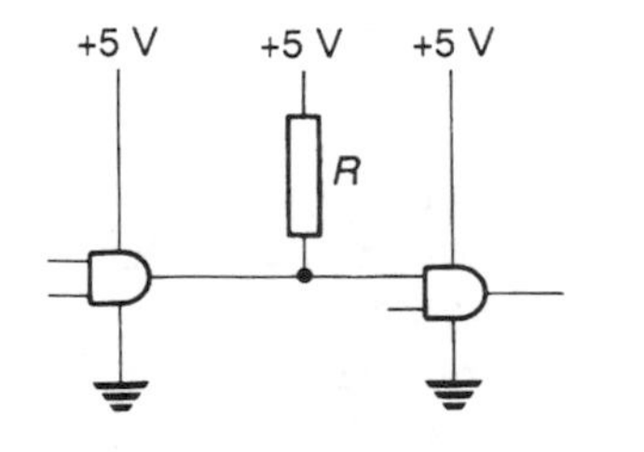

TTL gate
– open collector
or totem-pole

CMOS gate

(a) Driving 5 V CMOS logic

+5 V +15 V +15 V

R

TTL gate
– high-voltage
open collector

CMOS gate

(b) Driving 15 V CMOS logic

Figure 11.35 Driving CMOS gates from TTL.

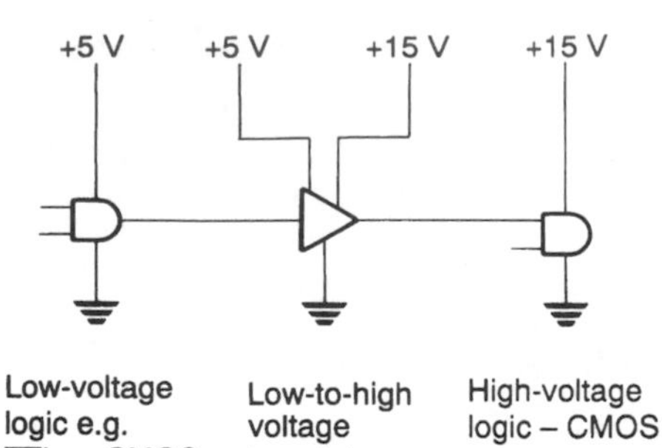

Figure 11.36 Use of a low-to-high voltage translator.

When driving CMOS gates that are operating with supply rails higher than 5 V, a high-voltage open collector TTL gate can be used. The pull-up resistor is taken to the supply rail of the CMOS logic to produce appropriate logic levels. This is shown in Figure 11.35(b) for a system using 15 V CMOS logic.

An alternative way of driving high voltage CMOS is to use a special purpose **voltage translator** such as the 4104B. This CMOS circuit is specifically designed to allow logic operating at a low supply voltage (such as TTL or low-voltage CMOS) to drive high-voltage CMOS logic. The 4104 contains eight translators in a single package. An arrangement using such a translator is shown in Figure 11.36.

Driving TTL from CMOS

The logic output levels of CMOS gates operating from a 5 V supply are approximately 0 V and 5 V and are thus compatible with the input logic levels of all forms of TTL. However, the output impedance of CMOS is too high to enable it to provide the input current required to drive standard TTL gates. Fortunately, the popular low-power Schottky family of devices (54LS/74LS) requires a much lower input current than standard TTL devices. Most CMOS gates will provide sufficient output current to drive correctly a single 'LS' gate, which can then be used to drive other 'LS' gates. Alternatively, the 'LS' gate can then be used to drive standard TTL circuits, since 'LS' gates provide sufficient output current to drive at least one standard TTL load.

High-voltage CMOS logic can be interfaced to TTL using **voltage translators**, such as the 4049B or 4050B CMOS devices. These are buffers (the 4049B is inverting and the 4050B is non-inverting) which take high-voltage input signals and generate low-voltage output signals to drive TTL or 5 V CMOS gates.

Figure 11.37 illustrates various methods of driving TTL from CMOS logic.

11.5 Array logic

In this chapter we have seen how a number of simple gates may be implemented within a single integrated circuit. Advances in technology have made it possible to produce highly

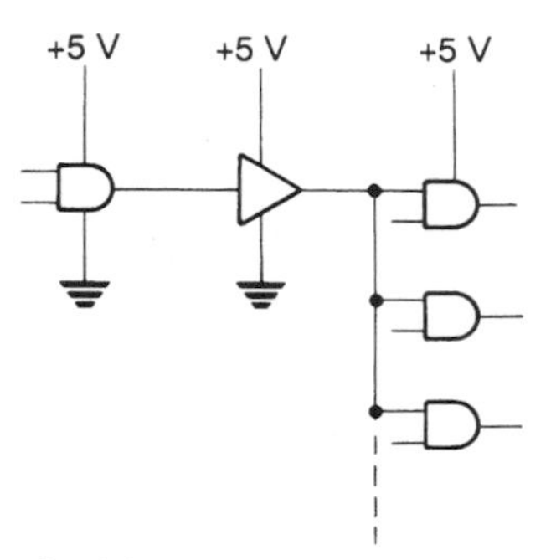

(a) Driving LS-TTL from CMOS

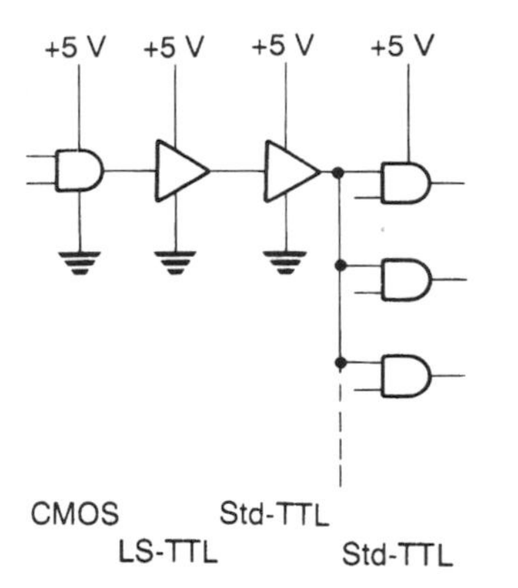

(b) Driving standard TTL from CMOS

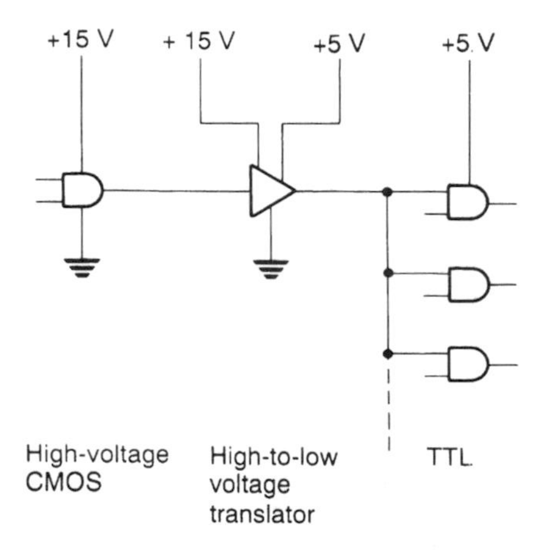

(c) Driving TTL from high-voltage CMOS

Figure 11.37 Driving TTL from CMOS.

complex circuits, but there are practical limits to the number of separate gates that may usefully be put in a single package. One of the major constraints is simply the number of pins that are required to connect to the inputs and outputs of the gates. A circuit with 1000 separate gates would require several thousand pins, and would inevitably occupy a large amount of space on a circuit board. The external interconnections between these pins would also require a large amount of board area.

To take full advantage of large scale integration it is necessary to implement not only the gates required by a circuit, but also their interconnections. If this is done then only the circuit's inputs and outputs need to be brought to the outside world, rather than connections to each node of the circuit. Internally connecting the gates within a package permits complex circuits to be implemented within a single device, but results in a device that is dedicated to a particular function.

Modern microprocessors, memories and interface circuits often contain thousands or tens of thousands of gates, within a single 'chip'. The development costs associated with such devices are very high and can only be justified for components that are used in very large numbers. Microprocessors are general purpose devices that can be used in a multitude of applications since their operation is determined by software. Similarly memory devices and several other complex components can be used in a wide range of applications, allowing them to be produced in large quantities. Unfortunately, not all electronic circuitry is manufactured in great numbers, and in many cases the circuits used are unique to a particular application. Even systems based on the use of standardized components such as microprocessors normally require a certain amount of specialized logic to 'bolt' the major components together. This circuitry is often referred to as **glue logic** for obvious reasons. While the same microprocessor may be used in thousands of designs, the glue logic varies between applications to give the system its own unique hardware characteristics. Since this circuitry tends to be specific to individual designs it is often called **random logic**, this term referring to the selection of functions rather than indicating any non-causal form of operation! This random nature makes it impossible for a manufacturer to produce a single *conventional* integrated circuit combining all the functions within a single chip.

In very small systems it may be possible to implement the required random logic using a handful of 7400 series TTL or 4000 series CMOS logic devices. However, as the complexity of the system increases the number of components required becomes prohibitive. A typical desktop computer, for example, might require only a handful of VLSI chips for the functions of the processor, memory and input/output sections, but would need more than 100 additional chips if the glue logic were implemented using simple logic circuits. What is required is a method of providing large numbers of gates within a single, mass-produced device, while allowing them to be interconnected in a manner to suit a particular application. Devices of this type come under the general heading of **programmable logic devices** or **PLD**s.

A PLD contains a large number of logic gates within a single package, but allows a user to determine how they are interconnected. This technology is also known as **uncommitted logic** since the gates are not committed to any specific function at the time of manufacture. The various gates within a device, and their interconnections, are arranged within one or more 'arrays'. For this reason this form of logic is also known as **array logic**. There are many forms of array logic and here we will look at just a few of the more important examples. Unfortunately, a study of this area is complicated by the plethora of names given to different types of programmable logic device. Here we will restrict ourselves to some of the more widely accepted terms, which include:

- PLA–programmable logic array
- PAL – programmable array logic
- GAL – generic array logic
- EPLD – erasable programmable logic device
- PROM – programmable read only memory
- CPLD – complex programmable logic device
- FPGA – field programmable gate array

11.5.1 Programmable logic array (PLA)

In Section 9.4 we saw that a combinational expression can always be represented by a series of **minterms** that may be derived directly from a truth table. For example, a system with four inputs *A*, *B*, *C* and *D* might have outputs *X*, *Y* and *Z* where

$$X = \overline{A}\overline{B}\overline{C}D + \overline{A}\overline{B}CD$$
$$Y = \overline{A}\overline{B}CD + ABC\overline{D}$$
$$Z = \overline{A}\overline{B}\overline{C}D + \overline{A}\overline{B}CD + ABC\overline{D}$$

One way of implementing such a system is to use a number of inverters to produce the inverted input signals ($\overline{A}, \overline{B}, \overline{C}$ and $\overline{D}$) and then to use a series of AND gates and OR gates to generate and combine the various minterms. A PLA has a structure that allows such functions to be produced easily.

The structure of a simple PLA is shown in Figure 11.38. This shows an arrangement with four inputs (*A*, *B*, *C* and *D*) which are inverted to produce four pairs of complementary inputs. These eight signals are then each connected to the inputs of a number of AND gates through an array of fusible links. These **fuses** are initially all intact, but they may be blown selectively to determine the pattern of connections between the input signals and the AND gates. In this way each AND gate is used to detect the input pattern corresponding to an individual minterm. A second array of fuses is used to connect the outputs

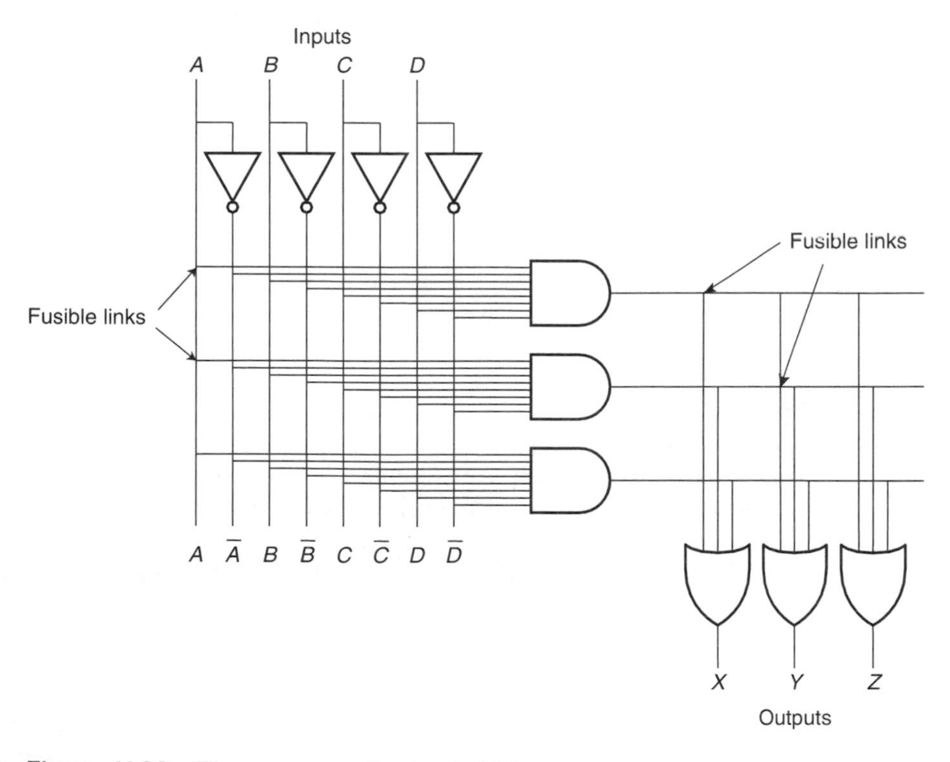

Figure 11.38 The structure of a simple PLA.

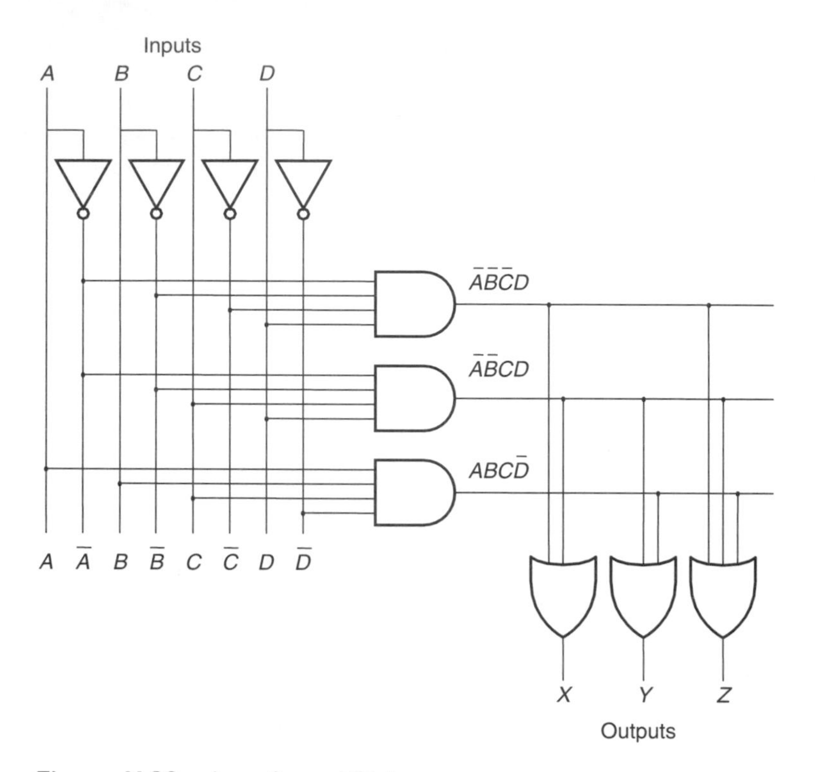

Figure 11.39 A configured PLA.

of the AND gates to a collection of OR gates. These OR gates combine the relevant minterms to produce the various outputs. An illustration of this process is shown in Figure 11.39 which shows the earlier simplified PLA configured to implement the system given in the above example. Here most of the fuses linking the inputs to the AND gates have been blown, leaving only those connecting the required signals to each gate. Similarly the fuses connected to the inputs to the OR gates have been selectively blown to produce the required three output signals.

A PLA would normally have more inputs and outputs than the simplified example shown above, and would also have a greater number of AND gates, allowing more complex functions to be implemented. In order to represent symbolically the large numbers of gates and interconnections within a typical device, it is convenient to adopt a more compact notation that reduces the large numbers of interconnecting wires within the various arrays. The symbols used when drawing logic arrays are shown in Figure 11.40. Here a single line is drawn to represent all the inputs to a gate and a cross is used to indicate those input lines that are connected to that gate. Figure 11.40(a) shows this approach applied to an array of AND gates and Figure 11.40(b) shows how it may be used with OR gates.

In the PLA shown in Figure 11.39 the input signals are inverted to produce complementary input signals using a conventional inverter as shown in Figure 11.41(a). A disadvantage of this arrangement is that the propagation delay of the gate will cause the

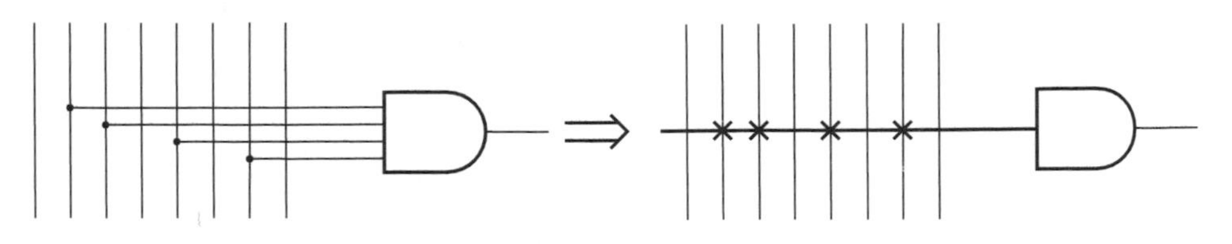

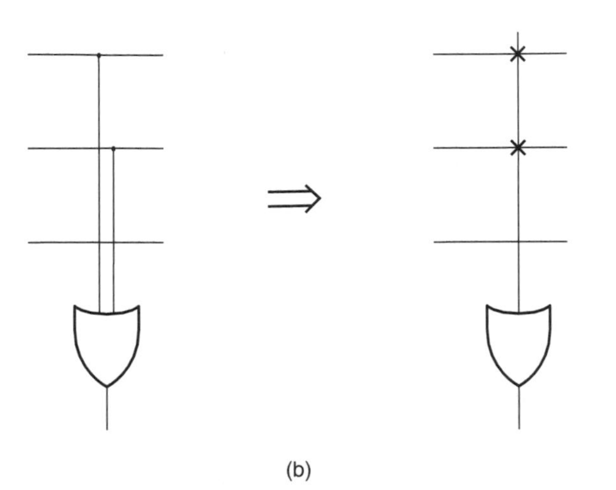

Figure 11.40 Logic array symbolic notation.

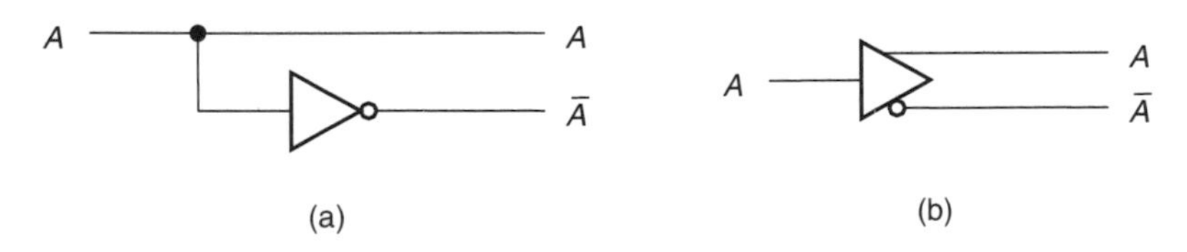

Figure 11.41 Representation of inverters within logic arrays.

inverted input to change a short time after the non-inverted signal. This is overcome by using a circuit that produces both an inverted and a non-inverted output, with equal propagation times. Such a circuit is given the symbol shown in Figure 11.41(b).

Figure 11.42 shows a PLA with six inputs, four outputs and 16 **product terms**. Devices are manufactured with all their fuses intact and this is indicated by the cross at each location in the two arrays of interconnections. In order to use the device it must be **programmed** by selectively blowing unwanted fuses. This task is performed by a **PLD programmer** which reads and interprets a **fuse map** supplied by the user. The fuse map may be produced manually, but is more often produced by a dedicated software package that deduces the required fuse pattern from a description of the desired functionality. If all the fuses connected to a particular AND gate are left intact then its output will remain

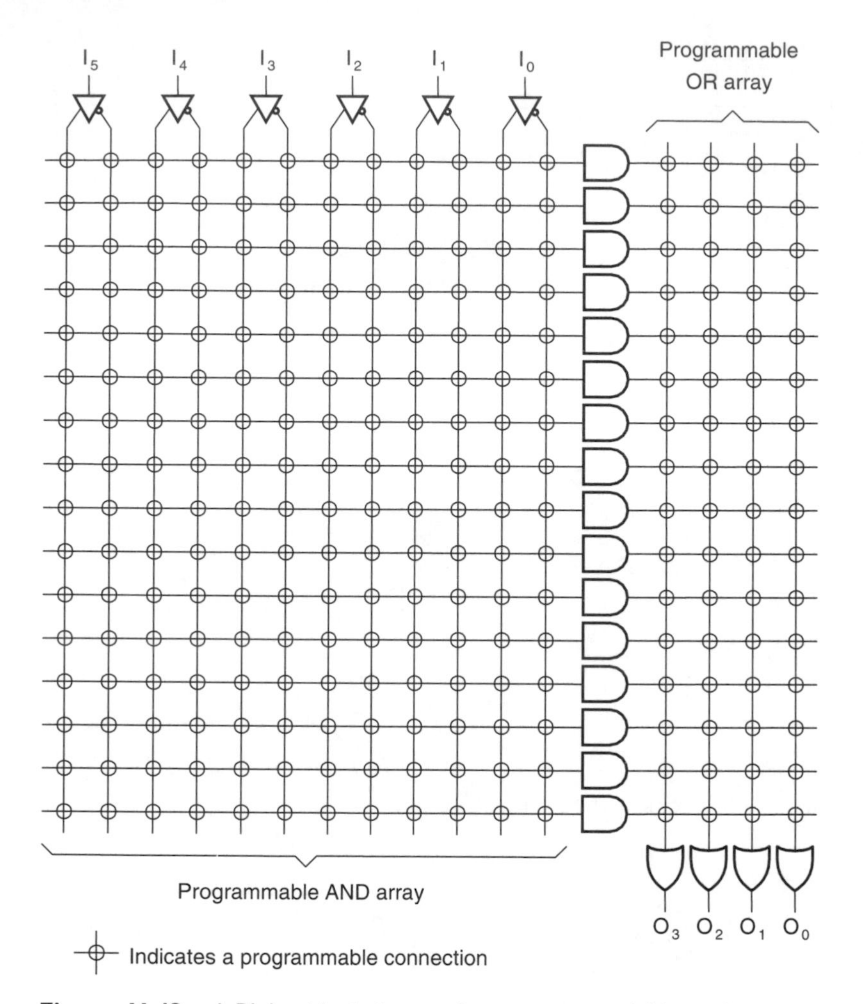

Figure 11.42 A PLA with six inputs, four outputs and 16 product terms.

low since a high output would require both the inverted and non-inverted inputs to be high simultaneously. In this way, any unnecessary AND gates may simply be left unprogrammed and ignored. If all the fuses connected to an AND gate are blown the output will be permanently high.

It can be seen that the structure of a PLA includes two arrays of interconnections. The **AND array** is used to select the components of the various minterms to be implemented and the **OR array** combines the various minterms to produce the desired output functions. Within a PLA both these arrays are programmable, giving great flexibility and the ability to make maximum use of the available product terms. However, the use of two programmable arrays makes these devices complex and relatively slow. To overcome these problems, other forms of array logic have evolved in which only one of these two arrays is programmable.

11.5.2 Programmable array logic (PAL)

Despite the flexibility of the PLA structure it became evident that there was some advantage in a less complex arrangement. One such arrangement was first developed by Monolithic Memories Inc. (MMI) in the form of **programmable array logic** or **PAL**. Figure 11.43 shows a PAL structure with six inputs, four outputs and 16 product terms. At first sight this might seem identical to the PLA shown in Figure 11.42, but it differs from the earlier circuit in that only one of its two arrays is programmable. The AND array is equivalent to that of the PLA and can be programmed to select the minterms required for a given function. However, the programmable OR array has been replaced by a fixed pattern of connections to a set of OR gates. The user now constructs the required functions

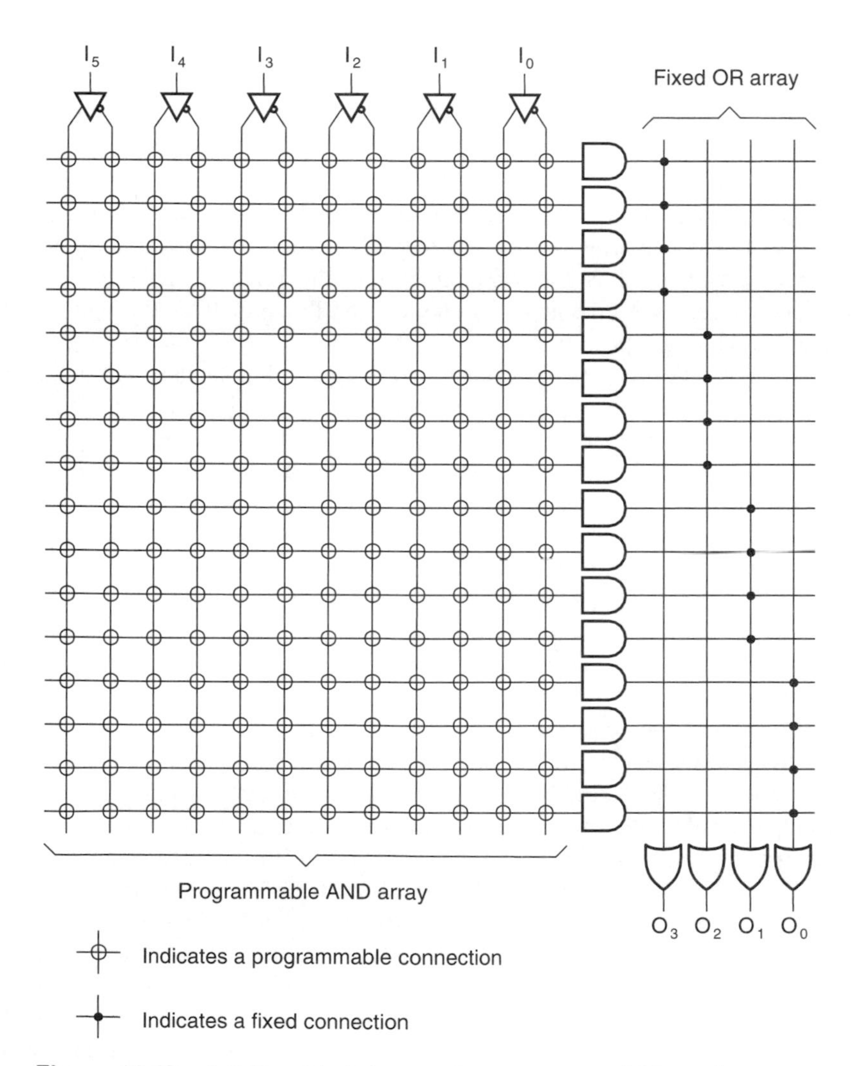

Figure 11.43 A PAL with six inputs, four outputs and 16 product terms.

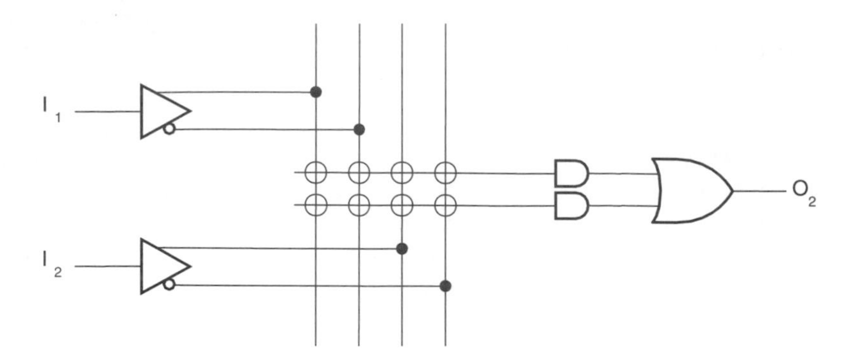

Figure 11.44 A fragment of a PAL.

by using the AND array to select the combinations of minterms that are fed into each OR gate. PAL manufacturers compensate for the absence of a programmable OR array by providing a range of devices with different numbers of OR gates and with different numbers of inputs on each OR gate. The term PAL was originally a trademark of MMI but became the property of Advanced Micro Devices (AMD) when it purchased MMI.

Because the OR array is fixed in a PAL it is common to omit this array from its symbolic representation. It is also common to move the inputs to the left of the diagram to produce a circuit with the inputs on the left and the outputs on the right, as shown in Figure 11.44.

To give increased flexibility many PALs use a technique that permits some of their pins to be used either as inputs or as outputs. Figure 11.45 shows such an arrangement. Here the output from one of the OR gates of the device is passed through a three-state inverter before being fed to the output pin. The operation of the inverter is controlled by an **output enable** signal that is derived from the AND array in a manner similar to any other minterm. If all the fuses connected to the inputs to this AND gate are blown its output will remain high, enabling the output of the inverter. This will configure this line as an output. If all the fuses connected to the inputs of this AND gate are left intact its output will remain low and the output of the inverter will be disabled, converting the pin into an input. The signal on the pin is used to generate complementary signals that are fed to lines of the input array. Depending on the state of the output enable line these complementary signals may represent either an input signal or the current state of the output. In the latter case these lines may be used to allow the output of one OR gate to be fed back to the inputs of other gates. Rather than being set continuously high or continuously low, the output enable signal can be configured to respond to the state of lines in the input array. This can be used, for example, to allow input signals to enable or disable an output.

The circuit example shown in Figure 11.45 shows the output being controlled by a three-state inverter. It would be equally possible to use a three-state non-inverting buffer in this arrangement, which would produce a functionality equivalent to that of earlier circuits. The use of an inverting or non-inverting buffer affects the fuse map that must be used; some functions are easier to implement when an inverter is used, while others are easier with a non-inverting buffer. An inverting buffer is shown in the figure since this is the more common configuration.

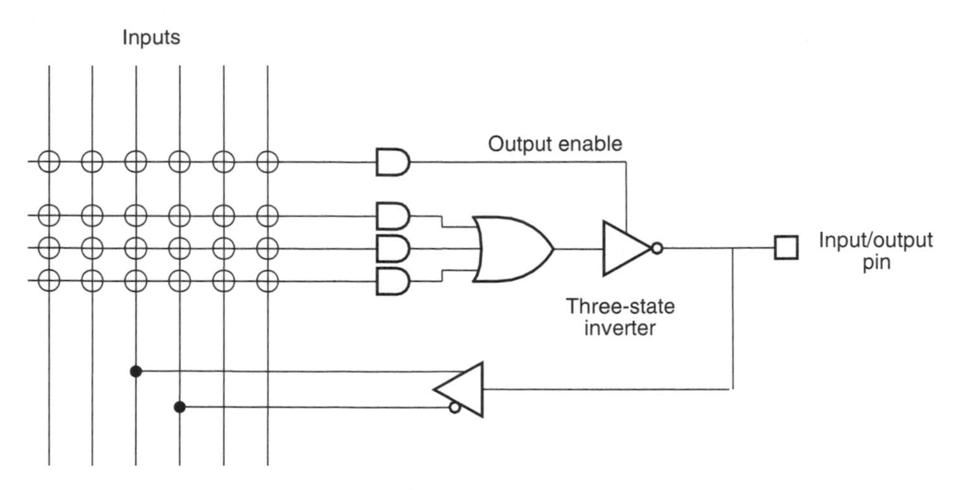

Figure 11.45 A typical PAL input/output circuit.

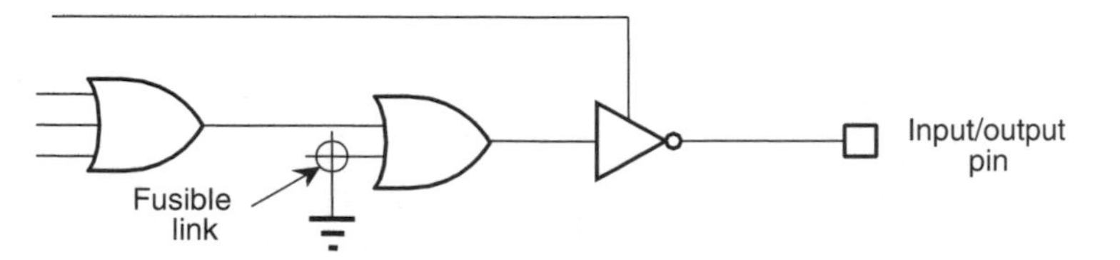

Figure 11.46 Use of an exclusive-OR gate as a programmable inverter.

Some PALs include an exclusive-OR gate in their output circuit as shown in Figure 11.46. One input to this gate is connected by a fusible link to ground. If the fuse is intact this input is pulled low and the relationship between the other input and the output is that of a non-inverting buffer. However, if the fuse is blown this input will take on a high logic level and the device will now act as an inverter. This allows each output to be individually configured to be inverting or non-inverting.

Figure 11.47 shows the functionality of a 16L8 PAL, a 20-pin device which has ten dedicated inputs, two dedicated outputs and six lines that can be used as either inputs or outputs. The device can provide up to 16 inputs and up to eight outputs (though not at the same time). Each output comes from an OR gate with seven input lines and thus the device has seven product terms for each output.

More sophisticated PALs replace the 'combinatorial' outputs used in the 16L8 with registered outputs with feedback. An example of such a device is the 16R8 PAL shown in Figure 11.48. Here each product term is stored into a D-type flip-flop on the rising edge of a clock signal. The output from this flip-flop is used to generate an output signal but is also fed back to the input array to allow this signal to be used by other parts of the PAL. The ability of the flip-flops to remember the previous state of the device permits the implementation of a range of sequential circuits such as counters, shift registers and state machines.

More advanced components remove the need to choose between devices with combinatorial and registered outputs by providing a *variable* output structure that can be made

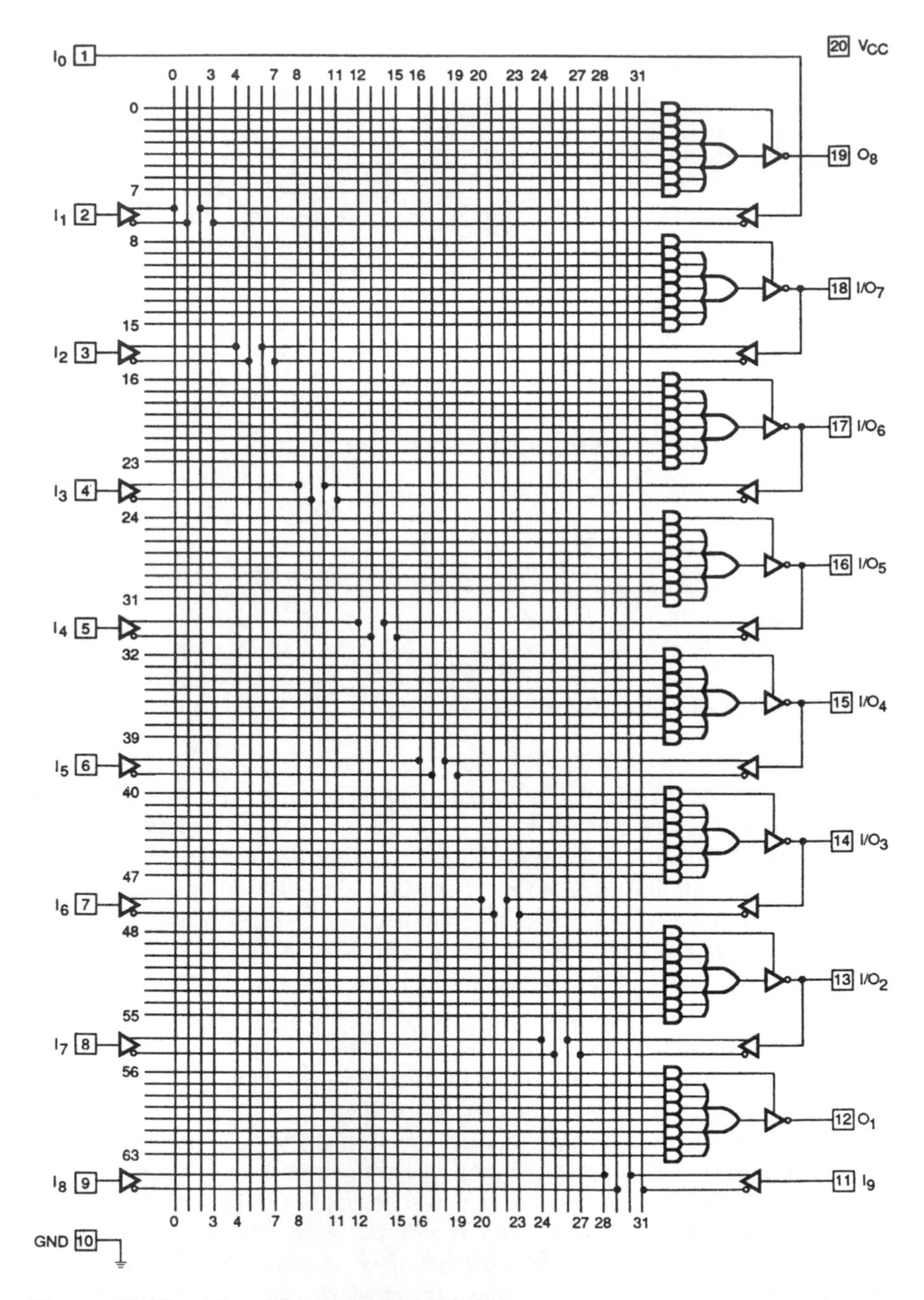

Figure 11.47 A logic diagram of a 16L8 PAL.

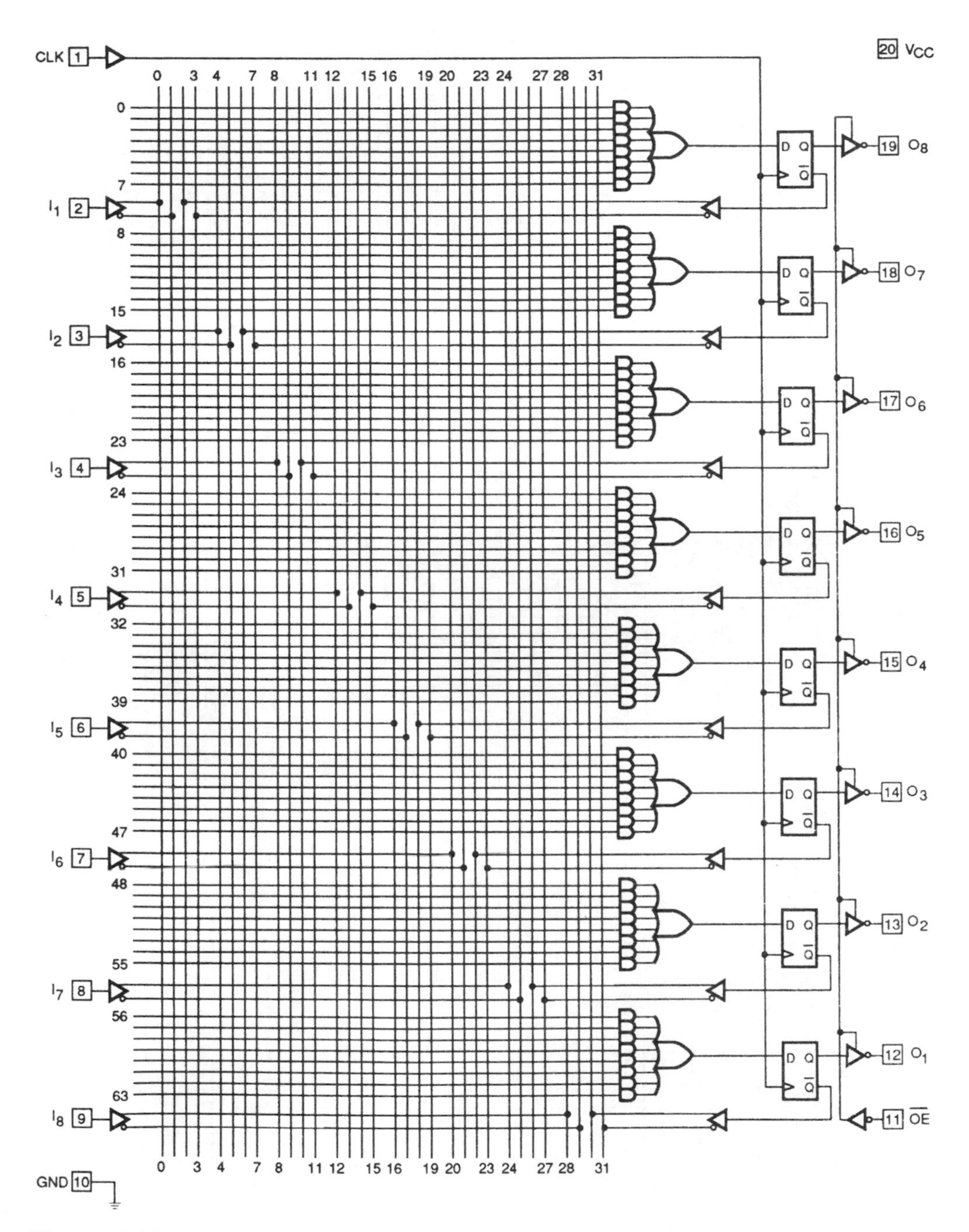

Figure 11.48 A logic diagram of the 16R8 PAL with registered outputs.

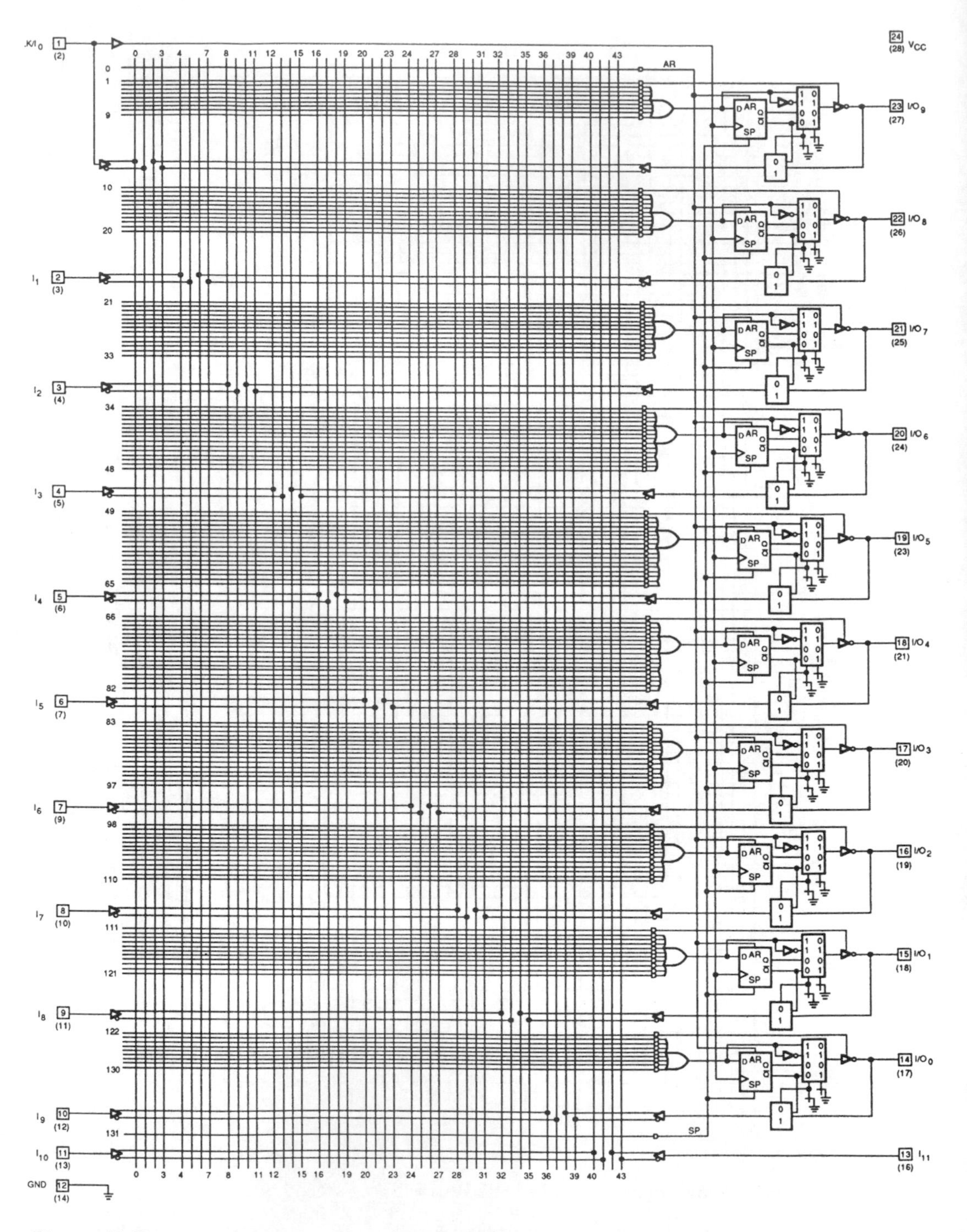

Figure 11.49 A logic diagram of the 22V10 PAL with macrocell outputs.

to emulate either form. A widely used example of this type of device is the 22V10 PAL which is shown in Figure 11.49. Here the output circuit takes the form of a **macrocell** that can be individually configured for each output. In addition to providing a combinatorial or a registered output, the macrocell allows outputs to be selectively inverted and provides an output enable function. This device has ten OR gates which have numbers of inputs ranging from eight to 16.

PALs derive their generic part name (for example 16L8 or 22V10) from their input/output characteristics:

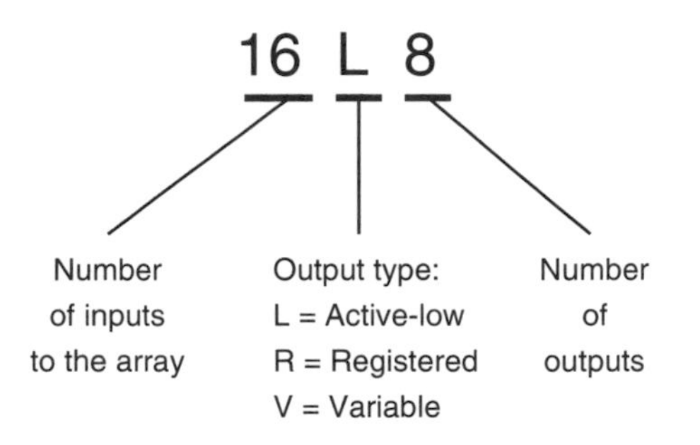

Depending on the device, the number of array inputs varies from 16 to more than 40, while the number of outputs is normally in the range four to 12. In addition to the three main output types listed above there are several variants which are given different designations. Generally the differences relate to variations in the form of the macrocell used at each output.

PALs have evolved over the years since their conception in the early 1980s, and now provide high functionality at low cost. One of the attractive characteristics of PALs is that they have a predictable propagation delay time, which in modern high-speed devices is of the order of a few nanoseconds. This allows them to be used at clock speeds of more than 100 MHz.

11.5.3 GALs and EPLDs

Because of their fuse-based construction, PALs can only be programmed once. They are therefore described as **one time programmable** (**OTP**) parts. Soon after the development of the PAL, Lattice Semiconductor produced an equivalent device that could be programmed repeatedly. These **generic array logic** (**GAL**) devices are pin compatible with conventional PALs but use **electrically erasable and programmable read only memory** (**EEPROM**) technology in place of fuses, to achieve reprogrammability (we will look at EEPROM techniques in Chapter 12 when we look at memory devices). Early GALs were much slower then PALs but more recent devices have speeds comparable with fuse-based devices.

Other reprogrammable PAL-like devices include **erasable PLD**s (**EPLD**s). This term is normally applied to parts of a form originally developed by Altera. These are similar to PALs, but use **erasable and programmable read only memory** (**EPROM**) techniques in place of fuses (we will look at EPROM in Chapter 12). This allows the devices to be erased by exposure to ultra-violet light. Once erased they can then be reprogrammed.

EPLDs generally offer more facilities than PALs and are more flexible. However, they are somewhat slower than PALs with typical delay times of between 10 and 20 ns.

11.5.4 Programmable read only memory (PROM)

At the end of Section 11.5.1 we noted that despite the flexibility of the PLA provided by its two programmable arrays, it is often more efficient to use a less complex configuration with a single programmable array. In Figure 11.42 we looked at a conventional PLA structure and in Figure 11.43 we saw how a PAL arrangement could be formed by replacing the programmable OR array with a fixed series of interconnections. An alternative method of simplifying the PLA structure is to remove the programmable AND array to form a structure as shown in Figure 11.50. This forms a **programmable read only memory** or **PROM**.

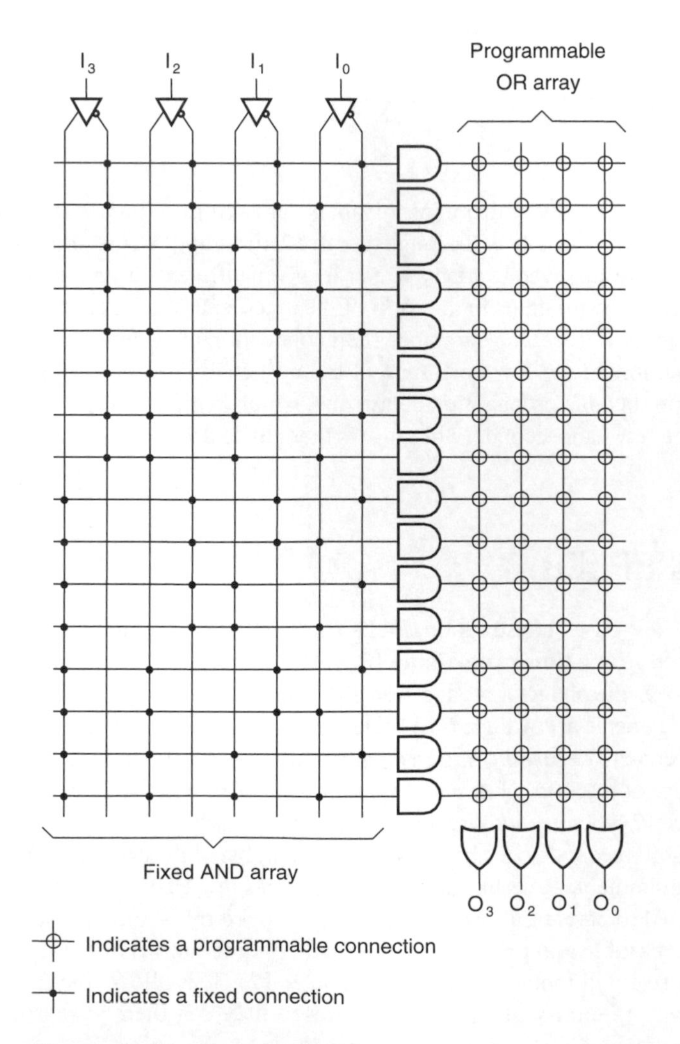

Figure 11.50 A programmable read only memory (PROM).

A PROM may be visualized as a PLA that has one fixed product term (AND gate) for every possible combination of the input variables. Thus a device with eight inputs would have 2^8 or 256 product terms. Since all possible combinations of the inputs are represented by *one* of the product terms, there is no longer any need to program the AND array and this becomes a fixed **decoder**. Each input combination selects a single AND gate and the OR array is used to determine which of the various outputs is activated (taken high) for that input combination. The pattern written into the OR array therefore determines the output pattern that will be produced for each possible set of inputs.

PROMs were one of the earliest forms of array logic and predate both PLAs and PALs. However, the use of a full decoder is inefficient for most logic applications and they are more commonly used for the storage of programs or data, rather than to implement logic functions. When used in this way the input pattern represents the address and the corresponding pattern in the OR array represents the stored data. When devices of this type are designed for program or data storage they are more often referred to as **ROM**s and we will be looking at the characteristics of these devices in the next chapter when we look at computer memory.

11.5.5 Complex programmable logic device (CPLD)

PLAs and PALs, and equivalent reprogrammable devices such as GALs and EPLDs, are often collectively referred to as **simple programmable logic devices** or **SPLD**s.

Complex PLDs can be thought of as an arrangement of several SPLDs within a single chip. In addition to providing a large number of array elements they also provide a powerful method of interconnecting inputs and outputs to allow fairly complex circuits to be implemented within a single package. A block diagram of a typical CPLD configuration is shown in Figure 11.51.

CPLDs are normally implemented using EPROM or EEPROM techniques rather than fuses, and are thus reprogrammable. EEPROM parts have the advantage of in-circuit reprogrammability, allowing their functionality to be changed without the devices being removed from the board. This is particularly useful when performing system upgrades.

CPLDs are currently the subject of a great deal of development work and the capabilities of these devices are increasing rapidly. At present, devices with several thousands of gates are available, with delay times of only a few nanoseconds. As with SPLDs (but unlike FPGAs) the propagation delay time can be predicted when performing the design.

A single CPLD might typically be used to implement a mixture of registers, decoders, multiplexers and counters, where a design using PALs or other SPLDs would require several ICs. For example, a complete 32-bit counter can be produced using a single device.

11.5.6 Field programmable gate array (FPGA)

Field programmable gate arrays take the form of a two-dimensional array of logic cells. These may be arranged in rows, or, more commonly, in a rectangular grid as shown in Figure11.52. The size of the array varies considerably, with small devices having perhaps 64 cells, while larger parts may have more than 1000. Between the cells of the array run groups of vertical and horizontal channels that can be used to route signals through the

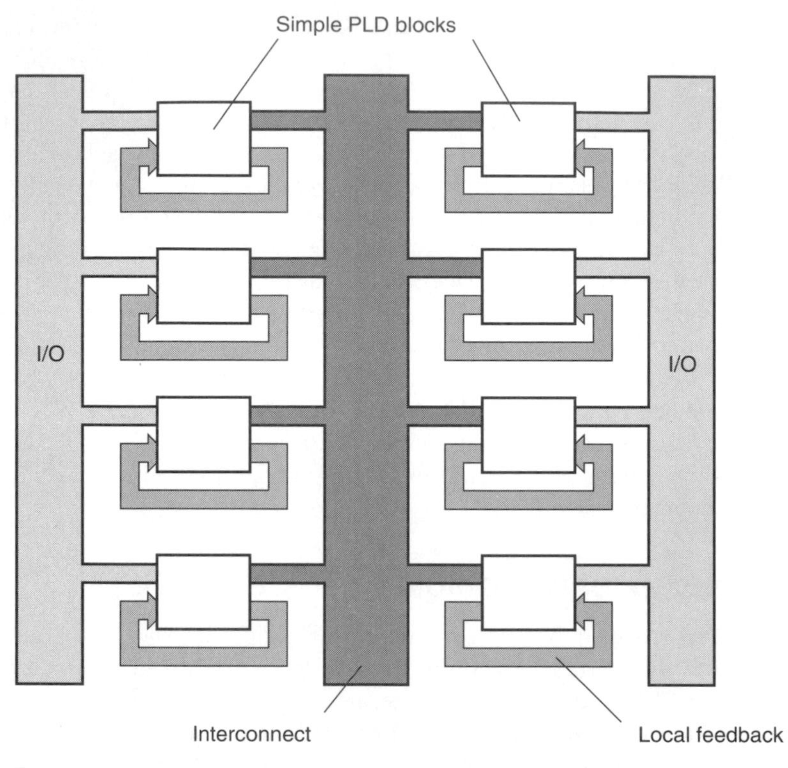

Figure 11.51 Block diagram of a typical CPLD.

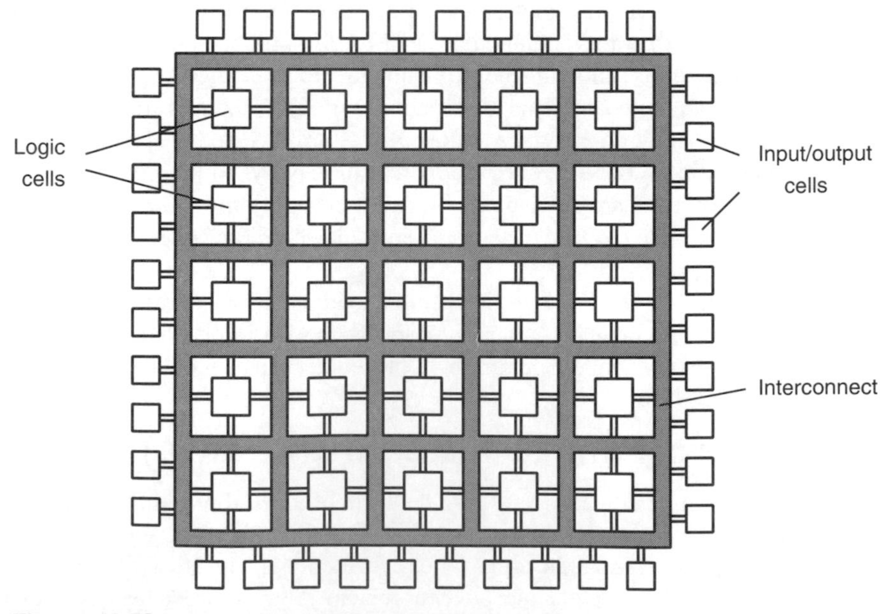

Figure 11.52 A simplified FPGA arrangement.

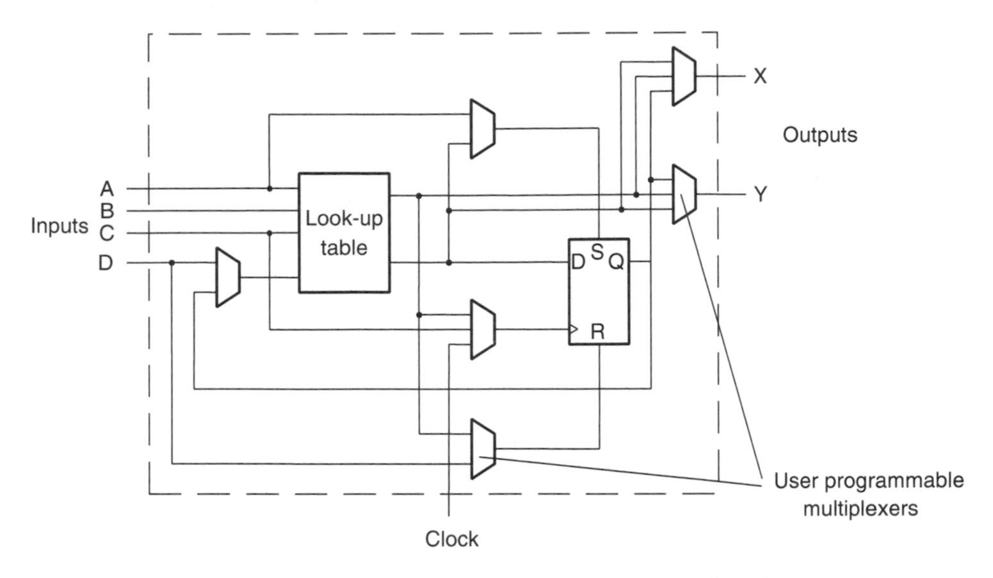

Figure 11.53 The configurable logic block of a Xilinx XC2000 FPGA.

circuit. Programmable switches are used to interconnect these conductors and so provide point-to-point connections.

The functionality of each logic cell varies considerably between manufacturers but a typical cell might contain a flip-flop, a few multiplexers and perhaps a small look-up table. Figure 11.53 shows the cell structure of the Xilinx XC2000, one of the simplest FPGAs produced by that manufacturer. The look-up table is a small user programmable memory that takes as its address the inputs to the cell. The table can be programmed to implement any logic function in a manner similar to a PROM. More sophisticated devices have much more complex logic cells that might contain several flip-flops, several look-up tables and a selection of other logic gates. The versatility of this arrangement, and the very flexible interconnection mechanism, allows complex circuits to be implemented within a single chip.

The programmable switches within FPGAs may be either one time programmable (OTP) or reprogrammable. OTP parts are based on the use of **antifuses** rather than the conventional fuses used in PALs. When manufactured, antifuses are *open-circuit* rather than closed-circuit as in an ordinary fuse. The unprogrammed antifuse resembles two back-to-back diodes which will not conduct in either direction. Programming involves forcing a large current through the diodes by breaking down the appropriate junction. Sufficient current is passed to short-out this junction permanently, removing its effect. The result is a single diode that acts as a closed switch in this circuit configuration. Since the breakdown is permanent the programming cannot be reversed and the device is **one time programmable** as in a PAL.

Reprogrammability is achieved in FPGAs by replacing each antifuse with a transistor switch (a MOSFET). The state of the switch is then determined by a memory element that can be set either to open or to close the switch. You can visualize this memory element as a bistable element. This technique is referred to as a **static random access memory**

(**SRAM**) approach and has the advantage that the contents of the memory can be changed as often as desired. The device is therefore completely reprogrammable. One disadvantage of this approach is that the content of the memory element is volatile, and is lost when power is removed. To overcome this problem the states of the various interconnections must be loaded from some non-volatile memory (typically a **read only memory** (ROM) or a computer disk) when power is first applied. In practice this procedure is not difficult to perform and the advantages of reprogrammability often outweigh this minor drawback.

FPGAs currently represent the most complex forms of programmable array logic. Small devices with perhaps 64 cells represent a functionality of about 2000 to 3000 gates while the largest parts are equivalent to tens or perhaps hundreds of thousands of gates. This represents a complexity about ten times greater than the largest CPLDs. They are therefore capable of implementing systems that are beyond the capabilities of other forms of array. However, like all forms of array logic they have their own characteristics that make them more suited to some applications than others. FPGAs are particularly useful when implementing systems that require on-chip memory, or that benefit from their distributed architecture. Modern FPGAs operate at very high speeds, but are generally not as fast as PALs or CPLDs. Also, their propagation delay times are greatly affected by the route taken by signals within the chip. This makes it very difficult to predict a circuit's performance before it is completed. This is in marked contrast to PALs and CPLDs where delay times are totally predictable.

For very high volume applications it becomes practical to consider using a **mask programmed gate array** (**MPGA**). Such devices have an architecture similar to FPGAs but are programmed during the manufacturing process rather than by the user. The configuration of the device is determined by a photolithographic mask which is used to produce direct connections between the appropriate nodes in the circuit. This removes the need for antifuses or transistor switches, and results in a device of much higher component density and greater speed. It can also produce a part with a lower unit cost than an FPGA of similar complexity. However, the disadvantage of this approach is that the mask is extremely expensive to produce. For this reason, use of MPGAs is normally only feasible in situations where tens or hundreds of thousands of similar devices are required.

11.5.7 Programming tools for array logic

The task of configuring a user programmable logic device for a particular application involves determining the appropriate pattern to be used for the various fuses, antifuses or switches, and then programming this pattern into a target component. For very simple PLDs it would be possible to derive the necessary **fuse map** manually by studying the functions required. However, this is a complicated and error-prone task, and in practice, automated tools are used for even the simplest parts. More complex components, such as CPLDs and FPGAs, are far too complex to be configured manually.

The programming of a device normally makes use of a range of computer-aided design (CAD) tools, although these may be combined within a single package. Figure 11.54 shows the main stages in the process and indicates common methods of design entry. One of the most widely used techniques for specifying the functionality

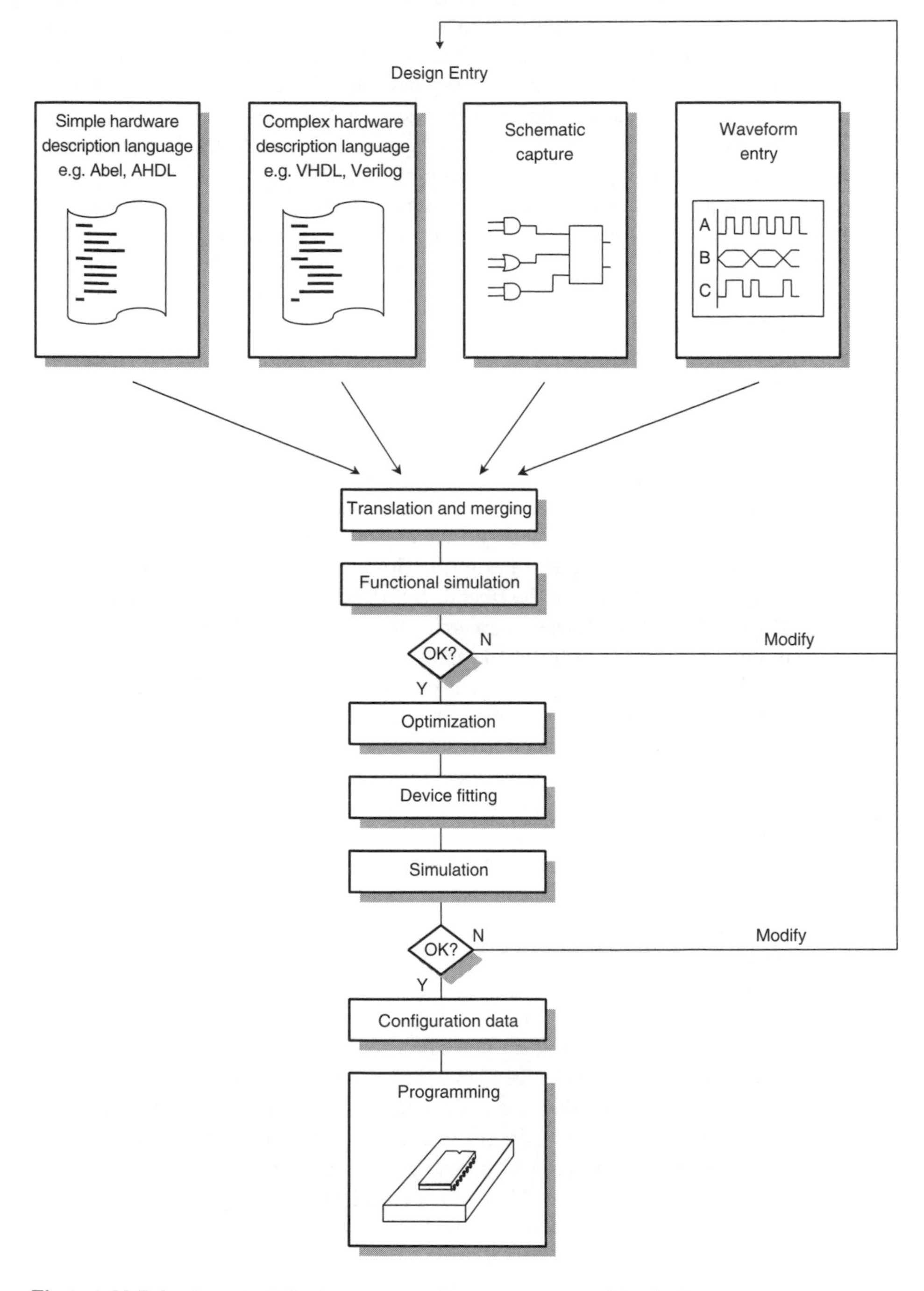

Figure 11.54 A typical design process for a programmable device.

required is the use of a simple **hardware description language** (**HDL**) such as **ABEL** or the Altera hardware description language (**AHDL**). These resemble programming languages but are used to describe the functionality of hardware rather than to define a list of instructions for a computer. They describe the required characteristics of the device using elements such as Boolean equations, truth tables and IF-THEN-ELSE constructs. The use of such an HDL is probably the most common method of design entry for simple PLDs and is also widely used for more complex devices.

More sophisticated hardware description languages such as **VHDL** and **Verilog** can also be used to specify the functionality of programmable devices. Such languages would normally be used for complex parts such as CPLDs and FPGAs, rather than for SPLDs. We shall discuss the use of hardware description languages in Chapter 14 when we look at CAD tools in more detail.

Design entry can also be achieved through the use of graphical tools. One approach is to use a schematics capture package to enter a circuit that represents the functionality of the required device. Alternatively, circuit elements can be synthesized from timing waveforms.

Many CAD packages allow the use of a range of design entry methods and permit elements defined in different ways to be merged to form a complete design. Once the design has been entered and its various components merged, the characteristics of the design can be investigated using **functional simulation**. This process does not investigate the exact timing of the final device, but is used to confirm that the circuit is logically correct. This allows design errors to be located quickly, so that time is not wasted in implementing an incorrect design.

When the design has been shown to be functionally correct, automated tools then perform optimization to simplify its implementation. There then follows a process of **fitting** the optimized design to the selected device. The difficulty of this task depends on the nature of the device concerned. For SPLDs it is a relatively straightforward process of allocating product terms to those available within the chosen part. When using CPLDs the task is somewhat more difficult because of the additional inter-block routing. The process is most demanding in the case of FPGAs because of their cellular architecture, and the vast number of possible ways of routing the interconnections. As a result the software tools required for FPGA design are perhaps 100 times more complicated than those needed for SPLDs. This difference is reflected in their speed of operation and their cost.

When the design has been fitted to its target device the resultant configuration is subjected to a detailed **timing simulation** to confirm that it will function correctly when programmed into a real device. This simulation takes into account both the functional characteristics of the design and the temporal properties of the target part. If the results of this simulation are satisfactory, a configuration file is produced which can be passed to a **PLD programmer**, which can then be used to configure any number of identical target devices. Alternatively, the file could be used to configure or reconfigure a reprogrammable device, within its circuit.

Following the programming process it is normal to read out the configuration pattern placed into a device to verify that this is correct. Once this has been done it is normal to program a **security bit**, which then prevents the contents of the component from being read. This makes it impossible for anyone to produce a copy of the part. The contents of a programmable device often represent a great deal of development effort and the ability to protect this investment against 'piracy' is a great advantage of this form of technology.

11.5.8 Custom and semi-custom ICs

When producing systems in very large quantities it may be practical to design a circuit 'from scratch' specifically for a given application. Such a **custom design** can produce a very efficient implementation by choosing an architecture to match exactly the functional requirements. Unfortunately, the cost of developing such a specialized device is very high and generally this approach is only attractive when producing components in quantities of hundreds of thousands, or millions, of components. However, for such high volume applications this may be an attractive option since custom design allows a chip to be optimized for its given function.

For more modest projects an alternative is to use a **semi-custom IC**. Such devices are also known as **application-specific integrated circuits** or **ASICs**. This term is used to refer to a range of devices, and in the past it was common for FPGAs and MPGAs to be described in this way. More recently, the term ASIC tends to be used for devices that are manufactured from a range of **standard cells**. Such components are produced by combining standard modules to produce the layout of a dedicated chip. These modules might include registers, counters, input/output circuitry and blocks of memory. This semi-custom approach is much less costly than performing a complete design from scratch, and so is practical for components that are produced in lower quantities. However, the design is also less efficient than a full custom design, resulting in a larger chip that is consequently more expensive to produce.

11.5.9 Choosing between the various forms of implementation

An important factor in determining the most suitable form of implementation for a given application is the complexity of the required functions. For systems that require only a few gates it is likely that basic TTL or CMOS logic gates will be most appropriate. However, if more than a handful of gates are needed, it will probably be more attractive to use an SPLD of some form. The choice here is likely to be between PALs, GALs or EPLDs depending on the importance of reprogrammability.

SPLDs are the preferred option for applications requiring up to 100 or so gates, even when devices are required in large quantities. However, for more complex systems, simple PLDs are insufficient and a designer will normally turn to either a CPLD or an FPGA. The choice between these two options depends on the functionality required. Some applications will fit in well with the linear structure of a CPLD, while others will be more suited to the cellular form of an FPGA. Very complex applications may be beyond the scope of CPLDs. The logic capacity of the most sophisticated FPGAs is about ten times that of the largest CPLDs.

When designing systems that are to be produced in very large numbers it may be appropriate to consider the use of devices that are programmed by the manufacturer rather than the user. MPGAs, ASICs and full custom chips come within this category. Such parts

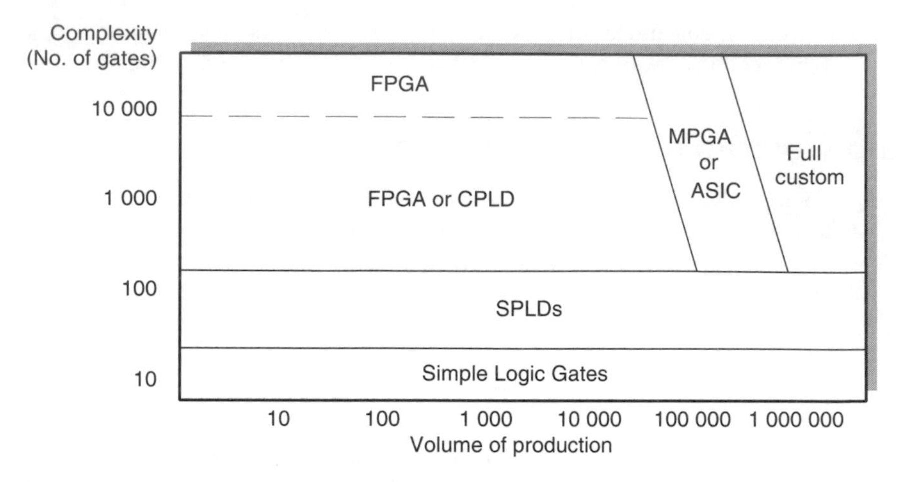

Figure 11.55 A comparison of various implementation methods.

usually have a lower unit cost than user programmable components, but have the disadvantage of high development and tooling costs. These costs can only be justified if very large numbers of devices are required. The development cost of full custom parts is considerably greater than those of MPGAs or ASICs, restricting their use to very high volume applications.

Figure 11.55 attempts to summarize the effects of system complexity and volume of production on the device selection process. This figure gives only an overview of the various technologies and the numbers given must not be taken as definitive. The capabilities of the various device families are changing constantly and the relative merits of the different techniques are subject to change.

The majority of PLDs, in common with most digital electronic components, operate with a single 5 volts supply. However, there is now a move towards parts that use a supply of 3 volts. Such components are often slightly slower than 5 volt parts, but have a much lower power consumption. They are therefore of great benefit when constructing battery powered equipment. Many PLDs are available in both 3 and 5 volt versions, and many of the associated interface components are also available in these two forms. This move towards lower operating voltages is being seen throughout the electronics industry and components such as microprocessors and memory devices are often available in low voltage versions.

11.6 Noise and EMC in digital systems

In Chapter 8 we looked at noise and EMC as they apply to analogue circuits and to electronic systems in general. In this section we will look at the implications of noise and EMC for digital circuitry.

One of the great advantages of digital systems compared to analogue equipment is their greater tolerance of noise. We have already looked at the **noise immunity** of various forms of logic gate, and noted that this is an important factor in determining the ability of a system to work correctly in the presence of noise. Unfortunately, while digital systems tend to have a high tolerance of noise, they also tend to produce more noise than analogue equipment. Therefore, from an EMC standpoint, digital systems have certain advantages and certain disadvantages when compared to analogue systems.

In the following sections we will look at sources of noise in digital systems and the ways in which noise affects their operation. We will then discuss various design techniques that can be used to improve the EMC performance of such systems.

11.6.1 Digital noise sources

Electronic noise

Electronic logic gates are constructed from electronic components. They will therefore have the same noise sources as analogue circuits, as discussed in Section 8.4. In general the designer of the gates will have taken these noise sources into account and will have chosen threshold voltages and logic levels so that the optimum noise performance is achieved.

Interference

Of less predictable nature is noise generated from interference. Nearby electrical and electronic equipment can produce large amounts of **electromagnetic radiation** which will induce voltages in any conductor. Long wires act like aerials, picking up the interference and producing noise voltages within the system. Interference can enter a system by radiation pick-up, through the power supply or through external lines to sensors and actuators.

Internal noise

Often the most important source of noise within an electronic system is the system itself! Signals at one part of a circuit can propagate throughout the system producing noise elsewhere. The **power supply** is a common source of noise. With simple linear supplies, the fluctuating field from the transformer can induce noise currents at the supply voltage frequency. With **switch mode power supplies** the high-frequency switching currents often propagate throughout the system, acting as a powerful noise source.

Power supply noise

Noise carried along the power supply lines is one of the most common forms of noise in digital systems and is one of the hardest to remove. When digital devices change state there is often a step change in the current being taken from the supply. Usually the operation of many devices within a circuit is synchronized to a master clock or oscillator, and thus large numbers of devices often change state simultaneously. When this happens extra current is taken from the supply to charge, or discharge, capacitances within the circuit. The result is that the supply rails in digital systems are usually perpetually ringing with noise spikes which are fed to all parts of the system.

Noise on the power supply rails can also originate from outside of the system, having entered the unit through the power supply. Noise from motors or other high-powered actuators propagates along the AC supply lines and is not always removed by the smoothing inside the power supply. This is normally tackled by fitting **mains filters** at the point where these lines enter the unit.

CMOS switching transients

One of the prime offenders in generating noise spikes is CMOS logic. As discussed in Section 11.4.2, CMOS takes almost no current when static but passes a surge of current when switching from one state to another. The net result is that in systems with many CMOS gates, all clocked simultaneously, this surge is many times greater than the average current. Any resistance, or inductance, within the supply lines converts this current surge into a voltage spike.

11.6.2 The effects of noise in digital systems

Small amounts of noise (less than the noise immunity of the system) usually have no effect on the operation of the system. This is in strict contrast to analogue techniques in which noise cannot normally be removed from a signal once it has been added. However, large amounts of noise can cause problems for digital systems in two ways:

- excessive noise can cause the operation of a system to be incorrect;
- excessive noise can cause permanent damage to the system.

Noise induced errors

Noise signals in excess of the noise immunity of a circuit can cause problems when added to steady logic voltages, as illustrated in Figure 11.56.

When added to slowly varying logic signals, amounts of noise considerably less than the noise immunity of the system can produce problems, as shown in Figure 11.57. Small fluctuations of the input signal cause it to cross the threshold voltage several times. This generates several transitions of the output rather than the single transition expected. If this signal were to be used as the input of a counter, an incorrect number of events would be detected.

Maximum ratings

We have seen that TTL circuitry normally operates from a supply voltage of 5 V. The manufacturers of such gates also specify absolute **maximum ratings** which define the limits of safe operation of the circuit. For TTL, the maximum supply voltage is usually 7 V and the maximum allowable voltage on any input line 5.5 V. Maximum currents are also defined. Exceeding the maximum values can destroy or permanently damage the device, sometimes in a fraction of a microsecond!

CMOS gates normally have an allowable supply voltage range of –0.5 to +18 V. The

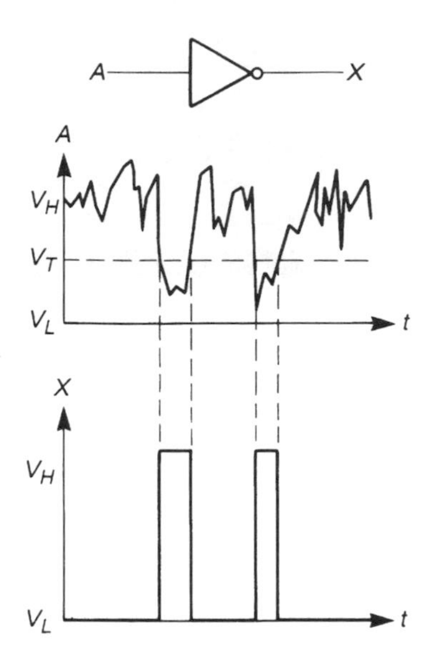

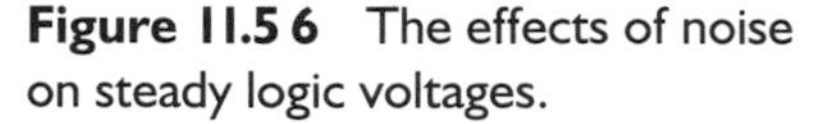
Figure 11.56 The effects of noise on steady logic voltages.

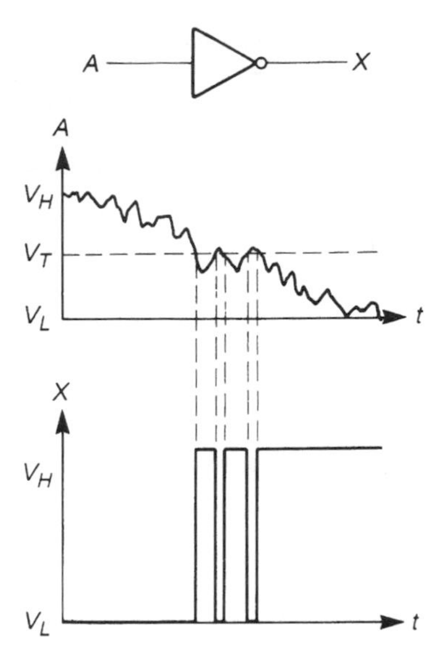

Figure 11.57 The effects of noise on slowly varying logic voltages.

allowable voltage on any input varies with the supply voltage in use and is generally −0.5 to $V_{DD}+0.5$ V. The maximum direct current into any input or output is 10 mA.

Because of the very fast response of electronic circuitry, noise spikes of only a few nanoseconds can cause serious damage. Because of the very high input resistances of many logic circuits, **static electricity** is a real problem.

11.6.3 Designing digital systems for EMC

In Section 8.5.5 we looked at various methods of designing systems in order to improve their EMC performance. In this section we will look at particular issues of relevance to the design of digital equipment.

Enclosures and cable shielding

In Chapter 8 we looked at the importance of **screening** in reducing both the susceptibility of a circuit to interference, and the amount of noise that it radiates. The use of enclosures and shielded cables is equally important, whether you are designing analogue or digital equipment.

Cables should be kept as short as possible and any long cables should be shielded. The maximum recommended distance between standard logic chips is only a few centimetres. If longer distances are required then special purpose line driver/receiver chips should be used. These are designed to cope with the increased line capacitance and to reject noise.

Clock lines cause particular problems in digital circuits since these carry fast

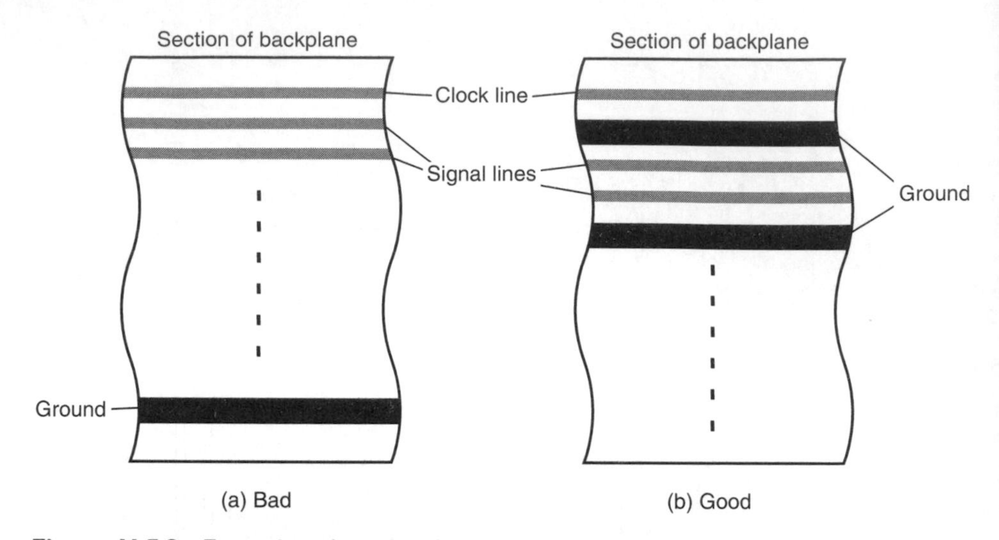

Figure 11.58 Examples of good and bad backplane arrangements.

switching signals that have very high frequency components. A Fourier analysis of a square waveform shows it to contain frequency components at odd harmonics of its fundamental frequency. Therefore even a relatively low frequency clock waveform will often contain components at very high frequencies.

Interference between conductors within a system can be a particular problem in computer **backplanes** where a large number of conductors run parallel to each other. **Crosstalk** can be reduced by placing ground conductors between clock and signal lines to provide screening. This is illustrated in Figure 11.58.

Opto-isolation

Noise from input and output lines can be reduced by **opto-isolation**, as described in Section 2.4.2. This is illustrated in Figure 11.59.

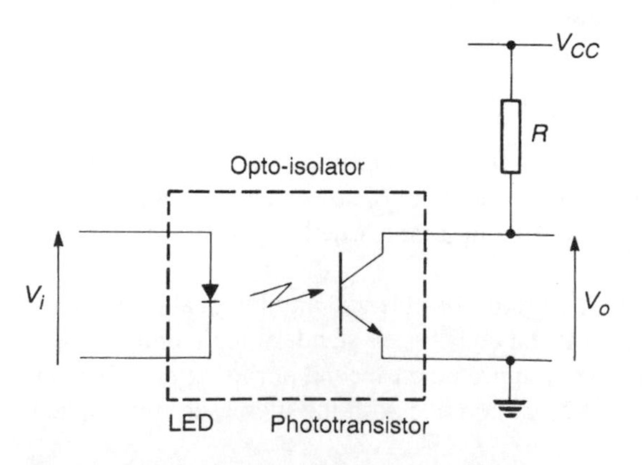

Figure 11.59 The use of opto-isolation.

The **opto-isolator** consists of an **LED** and a **phototransistor** within a single package. When zero volts (logic 0) is applied to the input of the opto-isolator the LED is turned OFF. The phototransistor receives no light and is also turned OFF. The pull-up resistor *R* therefore produces an output voltage of approximately V_{CC}. If a positive voltage (logic 1) is applied to the input of the opto-isolator, the LED will be illuminated, causing the phototransistor to turn ON. This will pull down the output voltage to close to zero volts. Thus the opto-isolator acts as a logical inverter. The important aspect of the operation of this arrangement is that there is no electrical connection between the input and the output, they are linked only by light. Typical devices produce several kilovolts of electrical isolation. Both the input and the output lines of a system can be protected in this way.

Diode clamps

In extremely noisy environments it may be prudent to provide extra **gate protection** circuitry of the form shown in Figure 11.34. Even if **clamp diodes** are fitted within the logic circuit they are of limited power handling ability and very large spikes will simply vaporize them an instant before the rest of the device is destroyed. External diodes with a fast response and capable of taking large current surges will provide improved protection for the circuit at very low cost.

Decoupling capacitors and earthing

Most problems associated with internal noise sources are tackled by the intelligent use of capacitors and by careful layout. It is normal to fit **decoupling capacitors** adjacent to *every* digital IC and to fit special low inductance capacitors across the supply lines. Earthing is of particular importance and should be as direct, and of as low resistance, as possible.

Power supply isolation

In circuits combining both analogue and digital circuits it is often necessary to use separate power supplies to prevent noise from the digital sections from interfering with low-noise analogue stages. In systems designed for use in high-noise environments, it is often advantageous to separate the power supply used for the input/output stages from that used for the remainder of the system. This, combined with opto-isolation of the control signals, prevents noise entering the system through the input/output lines.

Schmitt trigger inputs

Problems associated with slowly varying inputs can be alleviated by using logic gates with **hysteresis**. One of the most common arrangements with this characteristic is the **Schmitt trigger** circuit, which will produce a single change in logic state even for a noisy, slowly changing input. The transfer characteristic of such a circuit is shown in Figure 11.60.

From the transfer characteristic it can be seen that the output will only change from low to high if the input goes above the higher threshold voltage V_{TH}, and will only change

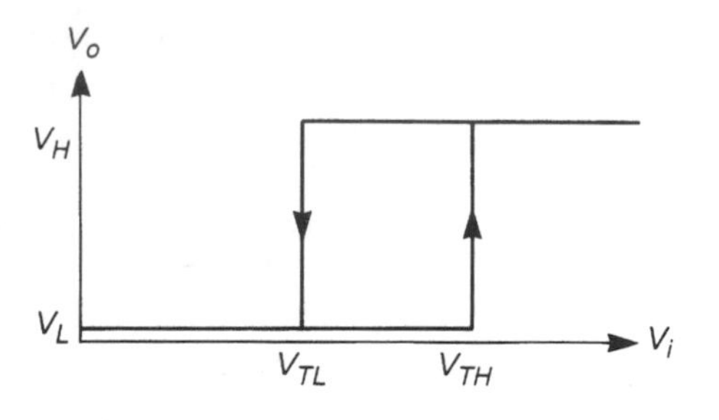

Figure 11.60 The transfer characteristic of a Schmitt trigger.

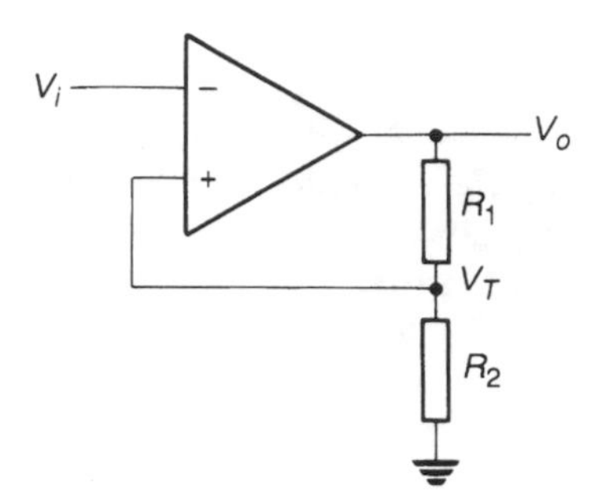

FILE 11J

Figure 11.61 A simple Schmitt trigger.

from high to low if the input goes below the lower threshold voltage V_{TL}. This prevents small amounts of noise from switching the output repeatedly between the two output states.

There are many forms of Schmitt trigger circuit. Figure 11.61 shows a simple inverting arrangement based on an operational amplifier. The combination of resistors provides **positive feedback** which tends to hold the circuit in its existing state. The output, in the absence of any negative feedback, tends to switch between voltages close to the two supply rails (+V and −V). The operational amplifier will change state when the input voltage becomes greater than, or less than, the voltage on its non-inverting input. When the output is positive, the voltage on the non-inverting input of the amplifier is given by

$$V_T = +V \times \frac{R_2}{R_1 + R_2}$$

and when the output is negative, the voltage on the non-inverting input is

$$V_T = -V \times \frac{R_2}{R_1 + R_2}$$

The switching voltage therefore changes with the output state to produce the required hysteresis. Similar circuit techniques, with appropriate resistor values and supply voltages, can be used to generate threshold voltages and output voltages suitable for particular logic devices. The various logic families include a range of gates incorporating Schmitt triggers on their inputs.

Removal of noise by filtering

We have seen that much of the noise associated with digital systems is in the form of spikes or high frequency fluctuations. If a particular logic signal is known to be only changing slowly, much of this noise can be removed using a low-pass filter, as discussed in Section 8.2. In removing high frequency components from the signal, the filter will also slow down transitions between the two logic states. It will therefore usually be necessary to follow such filtering by a **Schmitt trigger** to restore the sharp edges of the waveform.

Key points

- Integrated circuits combine the functions of many gates within a single device. Various levels of integration are available from 'small scale integration' (SSI) which combines just a handful of gates, to 'very large scale integration' (VLSI) which may provide up to 100 000 gates within a single package.
- Logic gates can be produced using either bipolar transistors or MOS devices (FETs).
- Bipolar circuits are generally faster than MOS circuits, but occupy more space and consume more power.
- Although bipolar transistors switch quickly, their speed is reduced if they are allowed to saturate, since this produces a delay, known as storage time.
- This problem can be tackled in a number of ways: gold doping can be used to reduce the lifetime of charge carriers; Schottky transistors can be used to prevent saturation; or the transistors can be kept within their active region.
- Logic circuits based on MOS transistors have the advantages of low power dissipation and very high circuit density. The input resistance is high, producing a high fan-out, but the output resistance is also high, limiting the rate at which load capacitance can be charged.
- The most popular logic families based on bipolar transistors are the various forms of transistor – transistor logic (TTL). These range from standard TTL circuits of the 54/74 families, through a spectrum of devices optimized for speed, power consumption or other characteristics.
- Conventional TTL devices have a totem-pole output producing logic levels of 3.6 V and 0.4 V. Other devices have different output configurations such as open-collector or three-state outputs.
- In recent years CMOS has become the dominant MOS logic family. Although more complicated to fabricate than other MOS technologies, CMOS offers very low power consumption coupled with a high operating speed and excellent noise tolerance.
- In many applications it is necessary to use a combination of TTL and CMOS. Since the inputs and outputs of these two families are not directly compatible, some form of interfacing is required. In fact, the circuitry needed is very simple.
- In situations where a large number of gates are required, use of standard logic gates is

impractical. In such cases it is common to use some form of programmable logic device. PLDs come in many shapes and forms, from relatively simple PALs and GALs to more complex CPLDs and FPGAs.

- All electronic circuits are susceptible to noise of one form or another.
- Within digital systems noise can cause problems in a number of ways. Firstly, noise can disrupt the correct operation of a system by causing errors in its operation. Secondly, in severe cases, it can cause physical damage to devices within the circuit.
- Various design techniques are available to reduce the effects of noise on the operation of a circuit and to provide a high level of protection. Unfortunately, complete immunity from noise is not possible.

Design study

An industrial control system is connected to two remote mechanical switches, and to an AC heater and an AC motor. The controller is required to take signals from the switches and to turn the heater and motor ON and OFF according to some control algorithm. The system is to be used in an environment with a high level of electrical noise.

Without considering the actual function of the control system, suggest how the unit could be designed to minimize the effects of noise on its operation.

Approach

The diagram shows a possible approach to this problem.

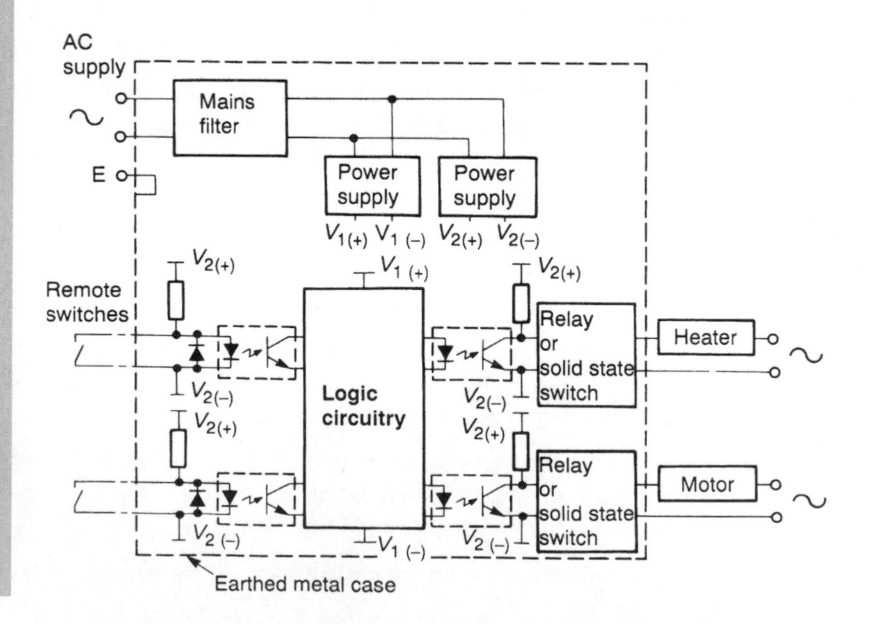

The AC supply is first filtered using a **mains filter** to remove high frequency noise, and is then fed to two isolated **power supply units** (PSUs). In practice, for reasons of economy, these two supplies may use separate windings of the same transformer, but in all other respects they are independent. The output of one of these PSUs is used to power the logic circuitry of the controller, which could use CMOS technology for improved **noise immunity**. For optimum noise rejection it may be appropriate to operate the CMOS logic from a 15 V supply.

Input signals from the remote switches will be susceptible to noise pick-up which will increase with the length of the external cables. Simple mechanical switches will produce **switch bounce** which may need to be removed by appropriate circuitry.

Since the logic circuitry will not directly drive high power AC loads, some form of **relay** or **solid state switch** is required. Mechanical relays are less expensive than solid state devices and are adequate for many applications. Their expected life is typically 10^5 operations, which is sufficient in most situations where continuous switching is not employed. Mechanical relays, being effectively electrically operated switches, produce switch bounce. If high currents are used, this can cause large amounts of noise. Solid state switches (or solid state relays, as they are sometimes called) have a much greater expected life and are preferable in applications requiring frequent changes of state. They do not suffer from switch bounce, though the transients generated when they turn ON or OFF can be considerable. Some devices reduce this problem by only changing state when the AC supply is at zero volts. This technique is called **zero voltage switching**. With any form of relay or solid state switch there is always the tendency for noise to propagate from the AC supply (which is generally very noisy) back to the control input. It is therefore necessary to provide noise protection at the outputs of the system as well as at the inputs.

All inputs and outputs to the logic circuitry are protected using **opto-isolators**. At the input, additional noise protection is provided by placing **catch diodes** in reverse parallel across the LEDs to prevent damage by negative noise spikes, which might otherwise break down the devices. LEDs have a relatively low reverse breakdown voltage of about 3 V. These diodes are not required at the output since here the LED is being fed from the comparatively low-noise logic circuitry rather than from external lines as at the input. Pull-up resistors are required to provide power to the LEDs of the opto-isolators at the input and to activate the relays or solid state switches at the output. In both cases these pull-up resistors are fed from the second PSU, so that any noise generated at the input or the output is not coupled to the logic circuitry through the power supply.

Having reduced the noise entering the system through the input and output lines, the final step is to reduce the interference entering the system due to **electro-magnetic radiation** (EMR). This is achieved by surrounding the entire circuit by an earthed metal case, which screens the system from outside influences. In extreme situations, internal screening can be used to prevent noise radiation from the power supply or the power switching devices from interfering with the logic circuitry.

Further reading

Chatterton P. A. and Houlden M. A. (1991) *EMC: Electromagnetic Theory to Practical Design*. Chichester: John Wiley

Horowitz P. and Hill W. (1989) *The Art of Electronics*, 2nd edn. Cambridge: Cambridge University Press

Millman J. and Grabel A. (1989) *Microelectronics*, 2nd edn. New York: McGraw-Hill

National Semiconductor (1996) *CMOS Logic Data Book*

Schilling D. L. and Belove C. (1989) *Electronic Circuits: Discrete and Integrated*, 3rd edn. New York: McGraw-Hill

Scot J. (1997) *Introduction to EMC*. Oxford: Newnes

Texas Instruments (1996) *TTL Data Book*

Exercises

11.1 What are the normal logic levels of TTL and CMOS gates?

11.2 Explain the meanings of the terms: noise immunity, fan-out, VLSI and programmable logic array.

11.3 Which logic families are most suitable for applications that require: (a) high speed, (b) low power consumption and (c) high noise immunity?

11.4 Define the following terms
(a) propagation delay time
(b) set-up time
(c) hold time.

11.5 What accounts for the increase in speed of Schottky TTL compared with standard devices?

11.6 Why is ECL faster in operation than TTL?

11.7 Sketch the transfer characteristic of a typical TTL gate.

11.8 Simulate a TTL inverter using the circuit of Figure 11.21 but assuming only one input. Connect a sweepable DC voltage source to the input of the circuit and a 1 kΩ resistor as a load from the output to ground. Sweep the input voltage from 0 to 5 V and observe the output voltage. Hence plot the transfer function of the gate and compare this with that given in Figure 11.22. What is the effect of removing the load resistor?

11.9 Explain the difference between a 7400 and a 5400 device.

11.10 What are the advantages and disadvantages of CMOS logic compared to NMOS?

11.11 What are the minimum noise immunities of TTL and CMOS?

11.12 What should be done with unused inputs of the following gates?
(a) TTL NOR gates
(b) TTL AND gates
(c) CMOS OR gates
(d) CMOS NAND gates

11.13 Which circuit technologies are most widely used for VLSI applications? Why?

11.14 Under what circumstances may the outputs of the following gates be connected together?
(a) totem-pole TTL gates
(b) open collector TTL gates
(c) three-state gates

11.15 How can the power consumption of CMOS gates be reduced?

11.16 Why is special consideration needed when connecting together TTL and CMOS logic gates?

11.17 Describe the operation and function of a Schmitt trigger.

11.18 What are the primary uses of open collector and three-state output gates?

11.19 Explain the function of opto-isolators in reducing noise pick-up in digital systems.

11.20 Describe the techniques used within CMOS gates to protect their inputs from static electricity. How can the device be further protected?

11.21 What distinguishes the 74HCT family of devices from the 74HC family?

11.22 How can the outputs of many logic gates be ANDed together without the expense of a multi-input logic gate?

11.23 Suggest a method of controlling a relay requiring a 24 V drive signal from TTL logic. What special precautions should be taken because of the inductive nature of the load?

11.24 What is meant by the terms 'glue logic' and 'random logic'?

11.25 Explain the differences between PLAs, PALs and PROMs.

11.26 Describe the characteristics of CPLDs and FPGAs.

11.27 Explain why CMOS circuits are a common cause of noise spikes.

Microcomputers

Objectives

When you have studied the material in this chapter you should:

- be able to describe the major components of a microcomputer system;
- have an understanding of the bus systems used to communicate between these components;
- be aware of the basic architecture of a typical microprocessor and be able to describe the functions of its main components;
- be familiar with the characteristics of the various forms of memory device employed within computer systems and the methods used to store data and programs;
- have an appreciation of the characteristics of various computer input/output methods.

Contents

12.1 Introduction

As the techniques of integrated circuit manufacture became established in the 1960s, the building of more and more sophisticated circuits became possible. However, with an increase in complexity came an increase in specialization which limited the usefulness of these techniques. Although very large numbers of devices could be combined within a single chip, it became difficult to identify circuits that could be used in sufficiently large numbers to justify the considerable development costs involved.

A solution to the problem was found in 1971 when the Intel semiconductor company developed the world's first **microprocessor**. This integrated circuit contained all the components of a computer **central processing unit** (**CPU**) within a single device. The microprocessor was of such great importance because it was **programmable**. That is, it could be made to perform a range of tasks by giving it a sequence of instructions in the form of a **computer program**. This allowed a single design of integrated circuit to be manufactured in high volume and to be programmed to perform a vast range of functions. By manufacturing the devices in large quantities, it was possible to distribute the very high development cost over many units, allowing a low component price.

At the end of the 1960s, computers were so large and so expensive that they inevitably sat in their own, air conditioned rooms with many users sharing their resources. The computer was a *tool* which was used for problem solving, for performing repetitive calculations and for storing large amounts of data. By the end of the 1970s, large mainframe computers were orders of magnitude more powerful than they had been at the start of the decade, but small **microcomputers** were also available in the form of single integrated circuits. These components were small, consumed very little power and cost less than a good meal. Some were used to build **personal computers** which performed much the same functions as mainframe machines but were cheaper to produce, allowing every user to have his or her own machine. However, the most significant development in this period was the use of microprocessors as engineering *components* within systems which performed other tasks.

Microcomputers, under the control of a program, can sense input signals, use this information within calculations and then produce relevant output signals. Therefore, given an appropriate program, they can be made to perform the functions of any combinational or sequential logic circuit. This potentially allows large amounts of logic circuitry to be replaced by a single microcomputer chip with a considerable cost saving. Moreover, since the operation of the microcomputer is controlled by a program, its operation can be modified without changing the physical structure of the system. This flexibility is invaluable in allowing products to be updated easily and cost effectively. These advantages have led to the widespread use of microcomputers in a range of applications from washing machines to aircraft autopilots, and from coffee-makers to the controllers of atomic power stations.

The physical components that make up a computer system are referred to as the **hardware** of the system, while the program which controls its operation is called the **software**. When producing a new system the designer often has the choice of performing a particular function using special purpose hardware or by employing appropriate software. In this way one can trade off the cost of the hardware against the complexity of the software. It is perhaps self-evident that when a number of identical systems are to be constructed, the hardware must be constructed for each unit to be made, but the software, being only a list of instructions, has to be written only once. This would seem to suggest that, wherever possible, one should use software to reduce the complexity of the hardware required, thus saving on construction costs. For high-volume products this is certainly true, but when only a small number of systems are required the cost of extra hardware is often outweighed by the savings in software costs. Software is time consuming, and therefore expensive, to write, and in low-volume applications it is often more sensible to perform tasks in hardware. Therefore, when designing a system an engineer must perform what is called a **hardware/software trade-off** to decide the appropriate implementation. To do this effectively requires a good understanding of electronics, programming and the application.

Microcomputers offer a host of features and possibilities which are invaluable in almost all fields of engineering, and which offer the only practical solution to many problems. However, their use also provides many challenges and a good understanding of the operation and capabilities of these components is vital to all engineers.

12.2 Microcomputer systems

All computers, from micros to mainframes, consist of three primary sections. These are shown in Figure 12.1.

The CPU is the heart of the computer. It is responsible for executing the various instructions within a program and for performing the logical and arithmetic operations that this involves. The CPU is often referred to as the **processor**.

Memory consists of a large number of registers which are used to store both **programs** (a sequence of instructions) and **data** (the information used or produced by a program). Each memory register is given its own unique identifying number or **address**. Typical small computers will have several thousands of such **memory locations**. Large computers have millions, or sometimes thousands of millions, of memory registers.

The most powerful computer would be of little practical use without some method of communicating with it. This is achieved by the input/output section of the computer, the form of which varies considerably depending on the nature of the information concerned. In a mainframe computer the input/output section may consist largely of circuitry designed to communicate with terminals (sometimes called video display units

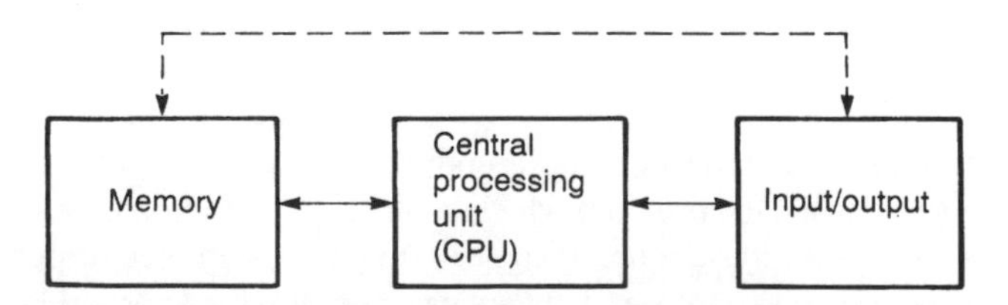

Figure 12.1 The main sections of a computer.

or VDUs), printers and links to other machines. On a small microcomputer within a washing machine, the input and output information would relate to such items as the water temperature and the motor speed, and the input/output section would be quite different.

In large computers the circuitry for each of these sections can take up several racks of equipment, while in small systems they may all be combined within a single integrated circuit. Here we will concern ourselves only with small computers in which the CPU is in the form of a single IC, a **microprocessor**. A microprocessor is *not* a complete computer but only the processor section. Additional components are required to implement the memory and input/output functions. Any computer that is constructed using a microprocessor is called a **microcomputer**. They may range from small circuits with only a handful of components to powerful systems consisting of many circuit boards.

In some cases all the three sections of the computer (processor, memory and input/output) are integrated into a single IC and this device is then called a **single-chip microcomputer**. Such devices are of obvious benefit in small, high-volume applications in which a single component can provide all the required functions. However, there is a limit to the memory and input/output facilities that can be provided within a single device, so for more demanding applications it is usually necessary to add external components.

Many microcomputers are used in situations in which their presence is far from obvious. These include such application areas as cars, domestic machines and consumer electronics. These arrangements are generally referred to as **embedded systems** since the computer is embedded out of sight, within the equipment. Microcomputers designed specifically for such applications are often called **microcontrollers** to distinguish them from desktop microcomputers.

12.2.1 Wordlength

Within a microcomputer, information is manipulated and stored in groups of bits. The size of the group which is used within a given machine is called its **wordlength**. The wordlength of a computer is one of the factors that determines its processing 'power', since it controls the amount of data that the machine manipulates at one time. The earliest microprocessors had a wordlength of 4 bits but such small machines are rarely used nowadays and then only for simple applications. Most small microcomputers have a wordlength of 8 bits, with more powerful machines using 16 or sometimes 32 bits. Machines with wordlengths of 64 and even 128 bits are available, but are normally reserved for specialized applications.

It is common to refer to a group of 8 bits as a **byte** and to a group of 4 bits as a **nibble**. The term *wordlength* has a different meaning on different machines. On an 8-bit machine (a machine with a wordlength of 8) the wordlength is equal to a byte. This is coincidental and should not be allowed to confuse the distinction between the two words.

The wordlength of a machine determines the maximum number of bits of data that can be transferred on the data bus *at one time*. It does not indicate any fundamental limit to the accuracy of any computation performed by the machine. Equivalent calculations will simply require more operations on a machine with a shorter wordlength.

12.2.2 Communication within the microcomputer

Figure 12.1 shows that information flows between the processor (CPU) and memory, and between the processor and the input/output section. In some circumstances information may pass directly between the memory and the input/output section (this is indicated by the dotted line in the figure), as will be discussed in Section 12.5.6.

Communication between the various sections of the computer takes place over a number of **buses**. These are parallel data highways which permit information to flow in one, or both, direction(s). Figure 12.2 shows the bus structure of a typical microcomputer.

The buses may be considered to be a collection of parallel conductors (wires). Three buses are used to carry data, address information and control signals. For example, if the processor wished to store a data word into a particular memory location it would place the data on the **data bus**, the address where the information was to be stored on the **address bus**, and various control signals to synchronize the storage operation on the **control bus**.

The number of lines in the data bus is equal to the wordlength of the device and thus determines the number of bits of data that can be moved about the machine at any one time. Thus in an 8-bit microprocessor the data bus would be 8 bits wide whereas in a 16-bit computer it would be 16 bits wide.

The number of lines in the address bus determines the number of memory locations that can be specified by the processor. This is called the **addressing range** of the device. An 8-bit address bus would be able to specify only 2^8 (256) addresses, which is rather limiting for the majority of applications. Most 8-bit computers use a 16-bit address bus giving an addressing range of 2^{16} or 65 536. Thus in 8-bit micros, addresses are usually represented by 2 bytes of information. Most 16-bit machines are designed for more demanding applications which often require a greater addressing range than is possible with a 16-bit address bus. Many machines use a 20-bit bus giving an addressing range of over 1 million, while others use a 24- or even a 32-bit address bus. A 32-bit bus gives

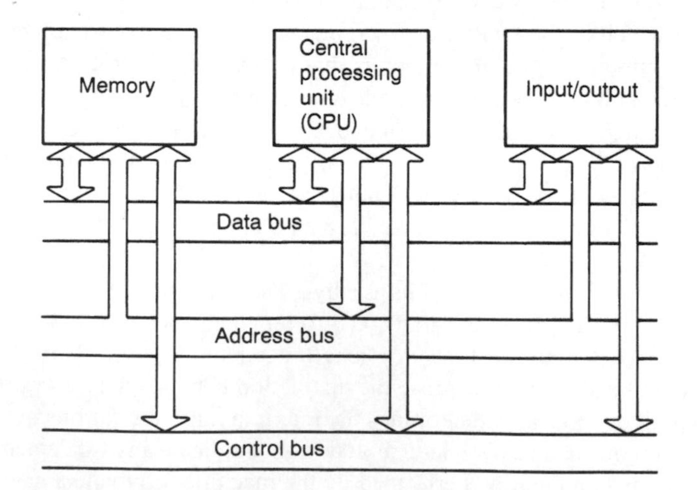

Figure 12.2 A typical microcomputer bus system.

an addressing range of over 4 000 000 000 locations which is sufficient for the vast majority of applications.

The lines of the control bus are used by the processor to produce actions in external components and to synchronize these operations. The exact nature of these lines varies between machines and we will discuss them in more detail in later sections.

12.2.3 Registers

The memory section of the computer consists of a large number of memory registers which can be used to store both data and programs. The processor and the input/output sections also contain registers for a range of purposes (these will be discussed in the following sections). The register is thus a fundamental building block within a computer system so an understanding of its operation is essential.

In Section 10.7 we looked at the use of D flip-flops in the construction of memory registers. Within a microprocessor, communication between registers takes place over a bus system, which imposes some restrictions on their design. In Section 11.3.3 we saw how gates with **three-state outputs** could be connected together, provided that the output of only one gate was enabled at any time. By using D flip-flops with three-state outputs, it is possible to produce memory registers that can be connected directly to a bus system. Figure 12.3 shows such an arrangement. Here an 8-bit register is connected to a data bus. The register has two control inputs: one to write data from the bus into the register (this corresponds to the clock input to the flip-flops) and the other to enable the outputs of the register to drive the bus (this is connected to the three-state control of the gates). The inputs and outputs of each flip-flop are connected together to the corresponding bits of the data bus.

Communication between a number of registers is achieved simply by enabling both the output of one register and the input of another, as illustrated in Figure 12.4. Since all the registers within a system are connected by the same data bus, only one piece of information can be communicated at any one time. If many pieces of data are to be transferred, this will require several operations. This arrangement is thus a **sequential** communication system.

The process of taking information from a register and placing it on to the bus is referred to as **reading**, while the process of storing information into a register is called **writing**. It is important to note that enabling the output of a register does not change its contents. Therefore the read operation is *non-destructive* and information can be read from a register many times without changing its contents. The only way in which a data word

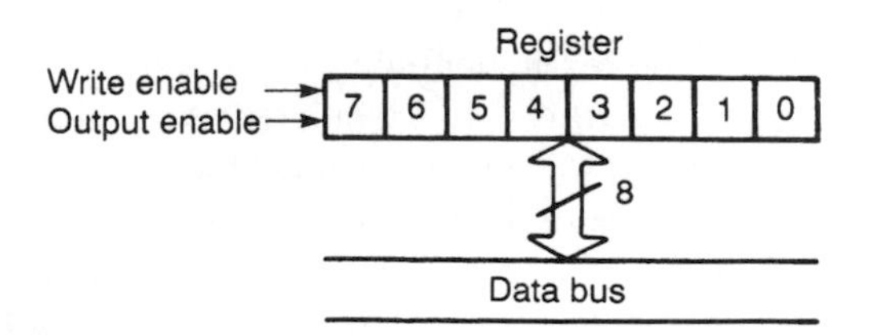

Figure 12.3 A simple 8-bit register.

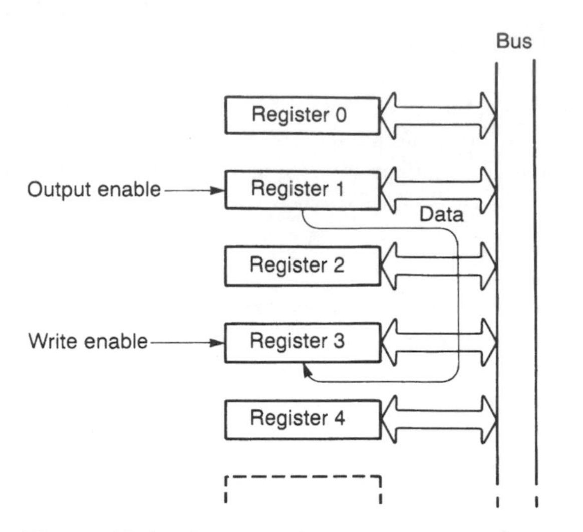

Figure 12.4 Communication between registers.

within a register is changed is if a new data word is written into it (unless, of course, the machine is turned off).

12.2.4 Data and program storage

Numeric data

The storage of numeric data within a microcomputer is straightforward. Pieces of information that can be represented by a single byte can simply be stored within a single 8-bit memory location at an address determined by the programmer. A single byte of data can store numbers in the range 0 to 255. If a piece of information requires more than a single byte, a series of memory locations can be used together. For example, two adjacent memory locations can be used to store a 16-bit number giving a range of 0 to 65 535, or four locations could store a 32-bit number with a range of 0 to 4 294 967 295. Clearly, numbers of arbitrary size can be stored by using an appropriate number of memory locations.

Negative number representation

In some applications it is important to represent not only the magnitude of a quantity but also its polarity. One method of doing this is to adopt a technique similar to that used in ordinary decimal arithmetic and to use one of the bits of an n-bit number as a sign bit. This is usually the most significant digit (MSD), which is zero if the number is positive and 1 if it is negative. Thus the MSD represents the sign of the quantity and the remainder shows its magnitude. This technique is referred to as the **sign-magnitude number representation**. The method, while simple in concept, does give rise to some problems. For example, it produces two ways of representing zero, corresponding to +0 and −0.

Most microcomputer systems use an alternative representation for positive and negative numbers which is called the **two's complement number representation**. This is most simply understood by recalling the characteristics of an up/down counter as described in Section 10.9.4. Let us imagine that we have an 8-bit counter and that we start initially from some positive value. As we count down through zero we will get the output sequence shown below.

0000011	3
0000010	2
0000001	1
0000000	0
1111111	−1
1111110	−2
1111101	−3

Clearly the pattern of all '1's represents −1, and counting down from this number gives us −2, −3, etc. This is the two's complement form.

As with the sign-magnitude notation the MSD again indicates the polarity of the number, 0 for positive and 1 for negative. The largest positive quantity that can be represented by an 8-bit number is clearly 01111111 (127) and the largest negative number is 10000000 (−128). The two's complement representation of positive numbers in this range is the same as its conventional binary form. The two's complement form of a negative quantity can be found by subtracting the binary equivalent of the quantity from 2^n, where n is the number of bits in the two's complement word. For example, the 8-bit two's complement form of −3 may be found as follows:

$$3 = 00000011$$
$$2^n = 256 = 100000000$$

Therefore

$$\begin{aligned} -3 &= 100000000 \\ &\quad -00000011 \\ \hline &= 11111101 \end{aligned}$$

Numbers that represent only magnitude are referred to as **unsigned numbers** while those that represent polarity and magnitude are called **signed numbers**. An 8-bit unsigned quantity can represent numbers in the range 0 to 255, whereas a similar signed quantity has the range −128 to +127.

Floating point numbers

In addition to the integer number representations outlined above, it is common to use a floating point format to allow both very large and very small quantities to be used. This is done by grouping a number of bytes of information together, and then using some of the resultant bits to represent a signed mantissa and the remainder a signed exponent. A typical format uses 4 bytes (32 bits) to represent each number, producing an accuracy of about seven or eight decimal places with a range of 10^{-38} to 10^{+38}.

Text

Text information can be stored using a sequence of memory locations, each character being represented by an appropriate code (for example, the ASCII code described in Section 9.5.4). This usually results in one memory location being used for each character.

Program storage

A computer program is a list of instructions to the processor. All microprocessors have a set of instructions which they can execute and these make up what is called the **instruction set** of the machine. Each type of processor has its own instruction set, and generally programs written for one machine will not operate on another.

A typical microprocessor will have within its repertoire instructions for: transferring data between registers; transferring data between registers and memory; performing various arithmetic and logical operations; performing comparisons and tests on register contents; and controlling the sequence of program execution.

Within an 8-bit microprocessor the operation to be performed by a particular instruction is usually defined by a single-byte **operation code** or **opcode**. The use of a single byte to define the operation limits the number of possible members of the instruction set to 256, though most devices use somewhat less than this. In some cases the opcode is all that is required to specify an instruction. For example, an instruction to increment (increase by one) the contents of a specified accumulator requires no further data. However, with some instructions extra information is required. For example, an instruction to store the contents of an accumulator in memory needs to include the address where the data word is to be stored. This leads to a situation in which some instructions are one byte in length (a one-byte opcode), some are two bytes (an opcode plus a single byte of data) and some are three bytes long (an opcode plus two bytes of data). The information that accompanies the opcode is called the **operand**; it might represent data used by the program or an address. A section of program stored in memory might be of the form shown in Figure 12.5.

It is worth noting that microcomputers differ in the way in which they store the bytes of a 16-bit address within memory. Some machines store the MSB (most significant byte) at the lower address of a pair of adjacent locations, while others store the MSB at the higher address. This variation between machines is called **byte sexing**, and there are arguments in favour of both methods. When considering program storage in detail, it is vital to know the orientation used by the processor in question.

Some advanced designs of microprocessor increase the number of available opcodes, and therefore the size of their instruction set, by defining one opcode which indicates that a second opcode byte should be fetched to select one of a secondary set of instructions. This allows a much larger range of instruction types and improves program efficiency.

16- and 32-bit microcomputers have correspondingly larger opcodes allowing a much more extensive range of instructions. Such computers often have very sophisticated and complex instruction sets.

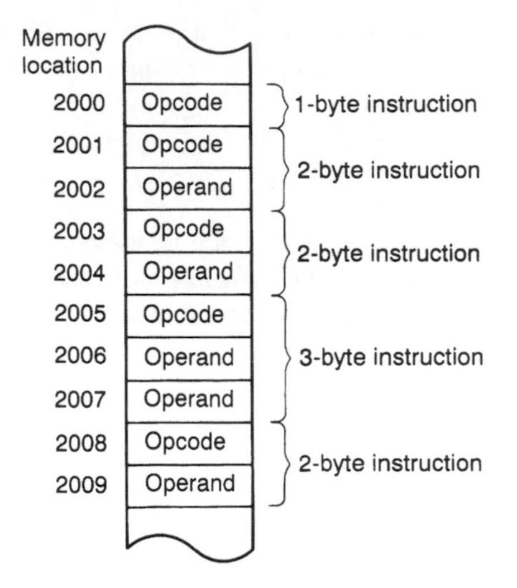

Figure 12.5 Storage of a program in an 8-bit microcomputer.

12.3 The microprocessor

The processor of a microcomputer consists of: a series of **registers**; some electronic circuitry to perform arithmetic and logical operations on the contents of these registers; some circuitry to decode and execute a sequence of instructions; buffers to interface the signals within the processor to the real world; and a series of buses join the various components together.

The structure of a microprocessor is referred to as its **architecture**. As one would expect, 8-bit devices are generally much simpler in form than 16- or 32-bit devices, although their major elements are often similar. More powerful processors tend to include features such as 'memory caches' and 'memory management units' that are not normally found in 8-bit devices.

A detailed discussion of microprocessor architectures is not within the scope of this text, but it is useful to look at the major components within a typical device. In the sections that follow we will look primarily at 8-bit devices since they are considerably simpler in construction. Much of the discussion is equally relevant to more powerful processors, although factors such as register and memory sizes will clearly be different.

12.3.1 The major components of an 8-bit microprocessor

Accumulators

An accumulator is a general purpose working register. Most operations within the processor are performed on, or are affected by, the contents of such registers. Machines vary considerably in the number of accumulators that they have. Some simple devices provide only a single 8-bit accumulator, while others provide two that can be used individually, or together to form a 16-bit register. Other processors have a bank of general purpose registers that can be used individually or in pairs.

Index registers

Most microprocessors have one or more **index registers**, which are normally used to store address information. These registers can usually be automatically incremented (increased by 1) or decremented (decreased by 1) to allow a sequence of addresses to be used in order. Since most 8-bit microprocessors use 16-bit addresses, these registers are normally 16 bits long. However, this is not always the case.

Program counter

A computer program consists of a list of instructions to the processor. These instructions are normally stored in order within a section of the memory and are fetched, one at a time, to be executed. Clearly the device must keep track of the location of the next byte of the program, and this is done using the **program counter**. This register normally contains the address of the next byte of the program. The value in the program counter is incremented each time a byte is fetched, so that it automatically points to the next. Since addresses are normally 16 bits in length, the program counter is usually a 16-bit register.

Instruction register

As a program runs, the opcode for each instruction is brought, in turn, to the **instruction register** for execution. This register is connected to the instruction decoding and control unit (see later) which produces an appropriate sequence of control signals to implement this instruction. In 8-bit microprocessors, the instruction register is 8 bits long to accommodate an 8-bit opcode.

Arithmetic and logic unit (ALU)

The ALU is responsible for performing all the arithmetic and logical operations required to implement the various opcodes. In general the ALU will be able to add, subtract, increment and decrement 8-bit quantities, and will be able to perform a range of logical operations on a pair of bytes. Arithmetic operations can be carried out on either **signed** or **unsigned** data, and some machines can also work with **binary coded decimal (BCD)** numbers. Logical functions normally include AND, OR and Exclusive-OR, and are performed bit by bit on the corresponding bits of the two data words, as illustrated in Figure 12.6.

In addition to arithmetic and logical operations, the ALU can manipulate data words by shifting them to the right or left (as in the **shift register** of Section 10.8) or by rotating them in either direction. Rotation is similar to shifting except that any data bits that fall off one end are reinserted at the other.

In simple 8-bit processors the ALU cannot directly perform **multiplication** or **division**. These functions must be carried out by a sequence of add/subtract and shift

A 1 0 1 1 0 1 1 0	A 1 0 1 1 0 1 1 0	A 1 0 1 1 0 1 1 0
B 0 1 1 1 0 0 0 1	B 0 1 1 1 0 0 0 1	B 0 1 1 1 0 0 0 1
A·B 0 0 1 1 0 0 0 0	A+B 1 1 1 1 0 1 1 1	A⊕B 1 1 0 0 0 1 1 1
(a) AND	(b) OR	(c) Exclusive-OR

Figure 12.6 Logical operations of the ALU.

instructions. More sophisticated devices do provide an 8-bit multiply operation, producing an increased speed of execution for this function.

Some 8-bit processors also provide a range of 16-bit operations, which are performed using 16-bit registers or pairs of 8-bit registers. These include 2-byte read and write operations and a range of data manipulation functions.

Processor status register (flags register)

Associated with the ALU is a **processor status register**. The individual bits of this register are set and cleared to indicate certain aspects of the machine's status and the results of ALU operations. The bits are often referred to as **flags** since they signal a particular event or condition. For this reason the register is sometimes called the **flags register**. The format of this register differs between machines, but common 'flags' are as shown in Table 12.1.

Although the flags reflect the results produced by the ALU, they are also affected by other operations such as writing data into an accumulator. Each flag has a set of rules determining when it will change and when it will not.

To make use of the information within the processor status register the processor has, within its instruction set, a number of **conditional instructions**, the operation of which depends on the status of the flags. The form of these instructions varies considerably between machines, but usually they are instructions that control the sequence of program execution. This allows the programmer to specify different sections of program which will be executed depending on the result of some calculation or manipulation.

Instruction decoding and control unit

The instruction decoding and control unit is the heart of the processor. It is responsible for sequentially *fetching* instructions from memory and then *executing* them. Attached to

Table 12.1 Common flags.

Symbol	Flag	Meaning
Z	Zero	Set if an ALU operation generates a zero result
N	Negative	Set if the result of an ALU operation is negative
C	Carry	Set if an arithmetic operation generates a carry
V	oVerflow	Set if an arithmetic operation generates an overflow into the sign bit

the control unit is a **clock generator**, which normally uses a **crystal oscillator** to produce a very accurate clock signal. This is counted to divide time into a number of **clock cycles**. The operation of the unit can be divided into two parts: the **instruction fetch cycle** and the **instruction execute cycle**. Both of these operations may take a number of clock cycles.

The control unit performs an instruction fetch cycle by generating a fixed sequence of control signals over a number of clock periods. These control signals correspond to the write enable and output enable signals associated with the various registers, as shown in Figure 12.3. The sequence produced transfers the contents of the program counter to the address buffer registers, sends a *read* command to external memory devices via the control bus, reads in data from the data bus and, if this is the first byte of an instruction, transfers this byte to the **instruction register**.

The instruction is then executed, again by the generation of a sequence of control signals over a number of clock periods. However, in this case the sequence depends on the nature of the instruction. The instruction register is connected to a decoding network, which selects the appropriate sequence of control signals for each known opcode. The selected sequence determines how many further bytes of data are required. These are then loaded by repeated fetch cycles, as above. The number of cycles required to execute the instruction differs between opcodes. When execution is complete the machine automatically commences another instruction fetch cycle to obtain the next instruction in the program.

Program execution is therefore a continuous sequence of fetch and execute cycles. Each fetch cycle is the same, but the nature of the execute cycles is determined by the opcode.

Address and data bus buffers

The address and data bus buffers are interfaces between the microprocessor and the outside world. Inside the chip, resistances are high and capacitances low, enabling very low power signals to be used. When signals are required to drive external loads, with their associated capacitances, more power is required. The buffers provide this extra power and improve the noise performance of the circuit.

Stack pointer

The stack pointer is a special purpose register found in many microprocessors. The register stores an address that *points* to a location in memory (locations that store addresses are often referred to as **pointers**). Within the instruction set of the machine are opcodes which cause information to be stored at the location defined by the stack pointer. When this is done the value in the stack pointer is automatically incremented so that it points to the next location. Instructions of this type are often given the name **push**. If several pieces of data are 'pushed' on to the stack they will be stored at sequential locations in memory forming a *stack* of data. Also provided are instructions for removing data from the stack. These are sometimes given the name **pull** and sometimes **pop**. These first decrement the address in the stack pointer and then fetch the contents of the address to which it points. The combined effect of these two instructions is that if a number of pieces of information are pushed on to the stack, they will emerge again, in reverse order, as a result of repeated pull operations. The use of these complementary instructions is best understood by an example, as shown in Figure 12.7.

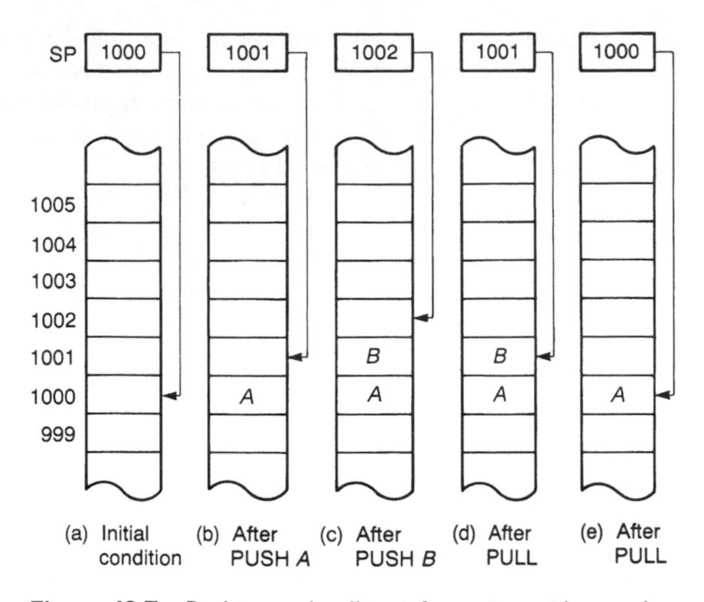

Figure 12.7 Pushing and pulling information with a stack.

The initial value in the stack pointer (SP) can be determined using a *load stack pointer* command. Let us assume that this has been set initially to 1000. This condition is illustrated in Figure 12.7(a). If now a number *A* is pushed on to the stack it will be stored at the location pointed to by the stack pointer (1000) and the stack pointer will be incremented (to 1001). This is shown in Figure 12.7(b). Pushing a second number *B* on to the stack will repeat the procedure, as shown in figure 12.7(c). If now a data word is pulled from the stack using a pull command, this will cause the stack pointer to be decremented (to 1001) and data to be read from this location. This will give the number *B*, as shown in Figure 12.7(d). If this is repeated, the next number to be obtained will be *A*, as shown in Figure 12.7(e). Thus data words are pulled from the stack in reverse order.

Although it may not seem particularly apparent at this time, this **first-in–last-out (FILO)** mechanism is of great significance. We will see why it is so useful when we come to consider the implementation of *interrupts* later in this chapter.

The number of items that can be stored and retrieved in this way is limited only by the amount of memory available and the size of the stack pointer register. Most 8-bit processors have a 16-bit stack pointer allowing the stack to occupy any portion of the address space. However, some devices use shorter registers which limit stack size. This is a considerable weakness in design and makes such processors very inefficient in the execution of many **high level languages**.

12.3.2 Communication with external components

The microprocessor communicates with external components using the bus system already described. External devices, whether used for memory or input/output, appear to the processor as registers with specific addresses. To send data to one of these components, the processor places the address of the chosen register on the address bus, the data to be transferred on the data bus, and appropriate synchronization signals on lines of the control bus. To receive information from a component, the processor places its address on the address bus, sends appropriate command signals on the control bus, and the device places the required data on the data bus which the processor reads.

Bus multiplexing

One of the constraints on the features that can be provided within a microprocessor is the number of pins that are available on the device. To reduce the number of pins required, some microprocessors use several pins to perform two functions, and many devices use eight lines to represent both the data bus and part of the address bus. This process, which is referred to as **bus multiplexing**, is illustrated in Figure 12.8.

Figure 12.8 is an example of a **timing diagram**, which shows the relationships between various signals within the system. Each signal, or group of signals, is plotted against time and its state is shown using a symbolic representation, as outlined in Figure 12.9.

Returning to Figure 12.8 we see that the high-order byte of the address bus (A_8 to A_{15}) is present throughout the period shown. However, the low-order byte of the address bus (A_0 to A_7) and the data bus are each represented for only a fraction of the period by a set of lines which is now called the **multiplexed address/data bus**. The lines of this bus are given the symbolic names AD_0 to AD_7. In order for external devices to be able to decipher this information, two control signals (or clock signals) are provided which allow external components to extract the information they require. The forms of these control signals differ between machines. In the example shown, they are called the **address strobe** and the **data strobe**. The falling edges of these signals indicate that the appropriate signals are stable and may be read by external devices. The full 16-bit address is obtained by reading the low-order byte of the address using an 8-bit latch (which is usually called an **address latch** when used in this way). The enable signal for the latch comes

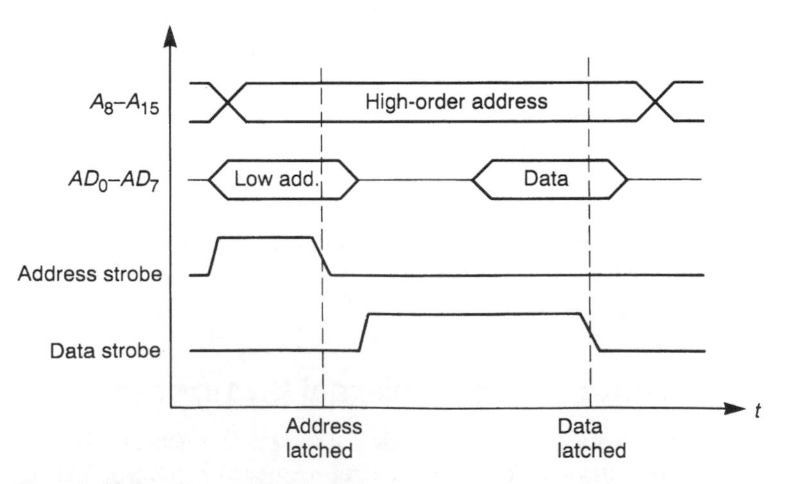

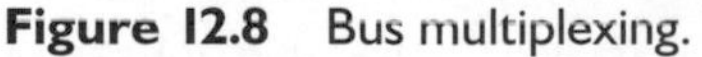
Figure 12.8 Bus multiplexing.

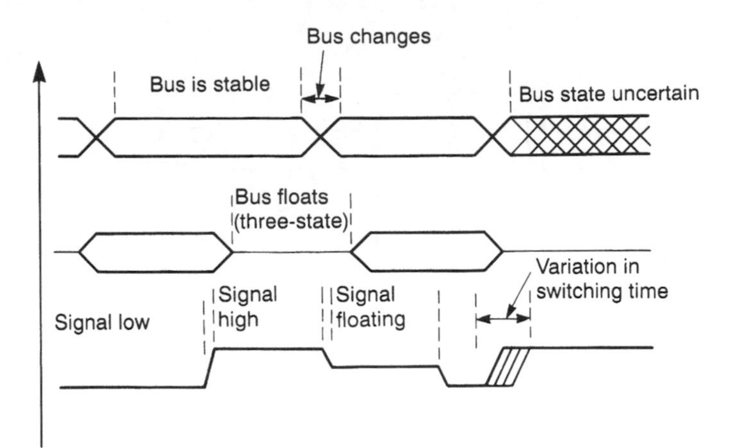

Figure 12.9 Symbolic representations used in timing diagrams.

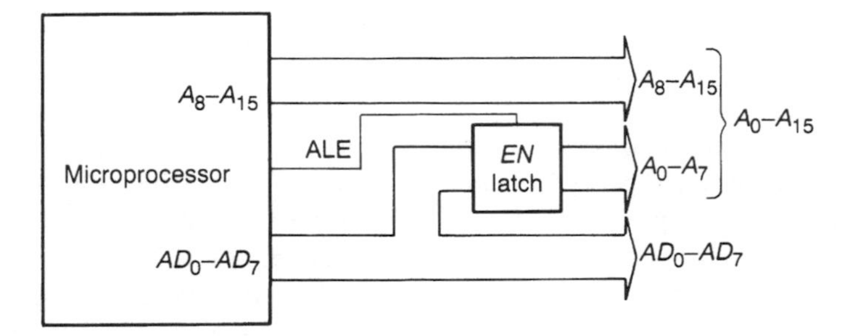

Figure 12.10 The use of an address latch.

directly from the address strobe of the microprocessor. Consequently, this signal is also called the **address latch enable** signal or **ALE**. This arrangement is shown in Figure 12.10.

Following the falling edge of the address strobe the full 16-bit address is available to external components. On the falling edge of the data strobe such components have both address and data information available.

On devices that do not use bus multiplexing the address and data buses are separate and an address latch is not required. A single data strobe can now be used to synchronize transfers between the processor and external devices.

12.4 Memory

We have seen that the memory section of a microcomputer consists of a large number of registers. Modern IC memory devices may contain thousands or even millions of memory registers. Clearly, in such circumstances it is not possible to use individual write

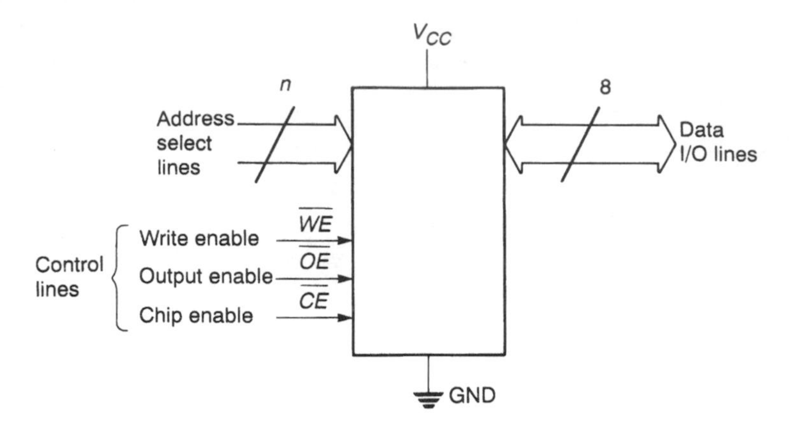

Figure 12.11 A typical memory device.

enable and output enable control lines for each register, as shown in Figure 12.3. Instead, all the registers within a device are controlled by a small number of lines and internal **address decoding** logic is used to select the appropriate location. A typical arrangement is shown in Figure 12.11.

The number of memory locations within a memory device is invariably a power of 2, since n address lines can specify 2^n memory addresses. Because the numbers concerned are large it is normal to express memory sizes in **kilobytes** (kbytes), where 1 kbyte is equal to 1024 bytes. This rather strange notation comes about because 1024 is 2^{10} which is conveniently close to 1000, which is normally represented by 'kilo' (as in kilogram and kilometre). Thus a device with 1024 8-bit memory registers would be called a 1 kbyte memory and would have 10 address lines, while a device with 4096 memory locations would be called a 4 kbyte device and would have 12 address lines. The 16 address lines of a typical 8-bit microprocessor give an addressing range of 64 kbytes. A similar notation is used for very large amounts of memory, where a block of 2^{20} (1 048 576) bytes of memory is referred to as a **megabyte** (Mbyte).

Unfortunately, the indication of memory size is made slightly more complicated by the fact that although the memory of a system is normally expressed in kilobytes, the capacity of an individual device is often given in kilobits (kbits). This is because many devices are not organized as a group of 8-bit registers. This alternative notation should not cause problems, but can occasionally lead to misunderstandings, particularly when people abbreviate the unit of memory size to 'k'. If an arrangement has 64 k of memory it makes a considerable difference whether this is 64 kbits or 64 kbytes!

In the early days of microcomputing, memory devices had typical capacities of 1 or 2 kbytes. This meant that most systems required many chips to provide their memory requirements. Now memory devices of 128 kbytes or more are commonplace, and larger devices become available year by year. The full 64 kbyte addressing range of an 8-bit microprocessor can be provided by a single chip, if required, although it is more common to use more than one chip to combine memory devices with different characteristics.

Returning to Figure 12.11 we see that the device has n address lines (where n is determined by the device capacity), eight bidirectional input/output lines and three control lines.

Two of the control lines are the **output enable** and **write enable** lines of Figure 12.3. These lines are invariably **active low inputs** and are therefore given the symbolic names $\overline{OE}$ and $\overline{WE}$. The third control input is the **chip select** line $\overline{CS}$. When this line is active (low) the device is 'selected' and will respond to the two enable signals. When the chip select line is not active (that is, it is high) the device will not respond to the other control lines thus preventing it from being accessed. The chip select line is used to determine which of a number of devices is used at any one time. The address lines of all the devices will be connected to the low-order bits of the address bus. Signals fed to the chip select line of each device will determine which is read from or written into. This arrangement is illustrated in Figure 12.12.

The figure shows an arrangement using four 16 kbyte memory devices. Each has 14 address lines ($2^{14} = 16\,384 = 16$ kbytes) which are connected to bits 0 to 13 of the address bus (A_0 to A_{13}). The most significant address lines (A_{14} and A_{15}) are connected to a '2-line to 4-line decoder'. This selects one of four output lines, depending on the combination of signals applied to its two inputs. If A_{14} and A_{15} are both 0, the first of its outputs $\overline{CS0}$ will be low and the others will be high. This will select the first of the memory devices and deselect the other three. Similarly, if A_{14} and A_{15} are both 1, memory device 3 will be selected. For each combination of the two address lines only one of the memory devices will be selected. Since A_{14} and A_{15} are the most significant lines of the address bus, addresses placed on the bus will automatically select one of the memory devices depending

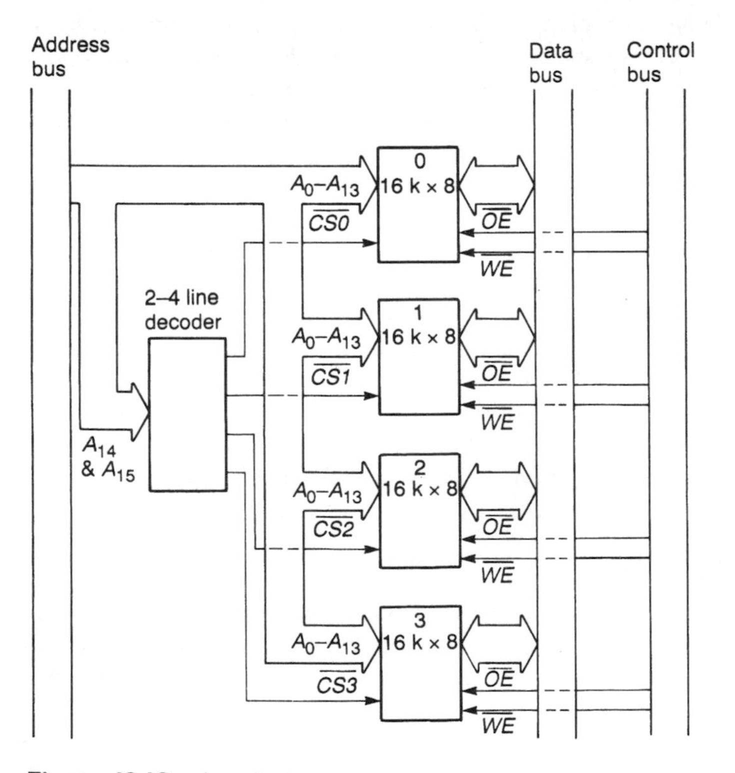

Figure 12.12 A typical memory arrangement.

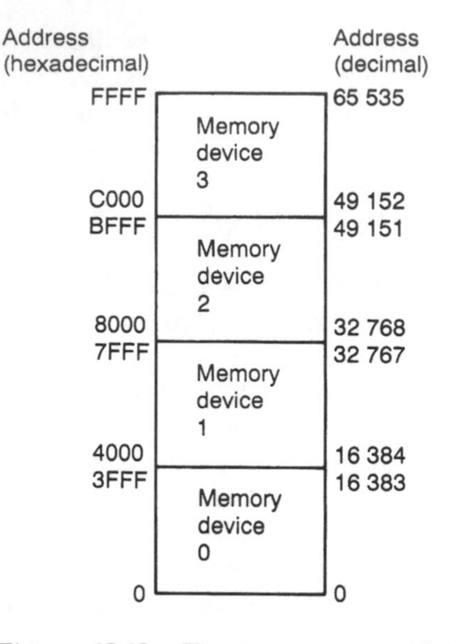

Figure 12.13 The memory map of the system of Figure 12.12.

on the two most significant bits. The location used within that device will be determined by the remaining 14 bits of the address. The address space is therefore partitioned as shown in Figure 12.13. This graphic representation of the allocation of memory within a system is called a **memory map** or sometimes an **address map**.

You will notice that the range of addresses corresponding to each memory device is shown in both decimal and **hexadecimal**. In fact the use of hexadecimal is much more common in this application, since it is more readily converted to binary to give the pattern of '1's and '0's on the bus.

The arrangement of Figure 12.12 shows a system using a non-multiplexed address and data bus. In a **multiplexed** system the circuitry is similar, except that an address latch is added, as shown in Figure 12.10. When the processor performs a *write* operation, the write enable signal $\overline{WE}$ is synchronized with the falling edge of the data strobe signal of Figure 12.8 to ensure that the data word is stable on the data/address bus when the write is performed. When the processor performs a *read* operation the output enable line $\overline{OE}$ is activated causing the selected memory device to place its information on the bus. The processor then waits for the data to become stable before reading the data from the bus.

12.4.1 Memory types

So far we have considered memory as a simple collection of registers which can be written and read at will. In fact there are several different forms of memory which have very different characteristics. Memory is used for a variety of purposes within a computer system. In some cases it stores programs that will never be changed, while in other cases it is used for data that is constantly being modified by the processor. Some systems also have **secondary memory** which is used for bulk storage of programs and data. Secondary memory is often some form of **disk drive** and is relatively slow to access. Here we will concentrate on the primary semiconductor memory of the computer which may be categorized as either RAM or ROM.

RAM

Data that must be changed during the operation of a program is normally stored in **RAM** or **random access memory**, this being the name given to memory which can be both written and read quickly. The name stems from the fact that in these devices any byte of data can be accessed in an equal time. It is tempting to suppose that this would be true of all memories. However, it is clearly not true of such storage devices as magnetic tapes since programs at the end of the tape will take much longer to access than those at the beginning. In fact, the use of the term *random access* is largely historical and is unfortunate since most forms of memory used within microcomputers (other than those used for bulk storage) are random access, even those that are *not* RAMs. A more appropriate name would be read/write memory, but the word RAM is used universally and is therefore established.

RAM in modern microcomputers is implemented using one of two circuit techniques. **Static RAM** uses a bistable arrangement similar to that described in Section 10.2.3. Information written into such a device is retained indefinitely provided power is maintained. **Dynamic RAM** stores information by charging or discharging an array of capacitors. Dynamic RAM requires far fewer components for each bit of information stored, permitting more storage elements to be integrated within a single chip. However, it suffers from the disadvantage that the charges on the capacitors tend to decay over time making it necessary to **refresh** the devices periodically by applying an appropriate sequence of control signals.

One characteristic of RAM is that it is **volatile**. That is, it loses its contents when power is removed. In many applications this is unimportant, since when power is lost the computer itself stops functioning. However, in some applications the contents of memory must not be lost when power is removed. For example the storage of the control program for an embedded system must be non-volatile since it must be present within the system when it is first turned on. For this reason programs are normally stored in ROM, as described in the next section, rather than RAM.

There are many situations where it is necessary to be able to read and write the contents of memory which must also be non-volatile. In such cases it is normal to use some form of **battery backup** to protect the contents of the RAM from power failure. Fortunately, when CMOS memory is not being clocked its power consumption is extremely low (as discussed in Section 11.4.2) allowing even a small battery to maintain its contents for extended periods. The backup battery can be mounted on the printed circuit board, or alternatively, memory devices are available that incorporate a small battery inside a CMOS integrated circuit. These devices are referred to as non-volatile RAMs and have a life of about 10 years.

ROM

ROM is **read only memory**, that is, it can be read by the processor but the processor cannot write into it. Such devices are non-volatile and are thus suitable for storing

Table 12.2 Acronyms for various types of ROM.

ROM	Read only memory
PROM	Programmable read only memory
EPROM	Erasable and programmable read only memory
EEPROM	Electrically erasable and programmable read only memory

programs or any non-changing data. There are many forms of ROM. Some must be programmed by the device manufacturer while others can be programmed by the user, often using special equipment. Table 12.2 lists the acronyms of several forms of ROM and gives their meanings.

The general term ROM is applied to all forms of read only device. Some of these are **mask programmed**, which means that the device is programmed photolithographically by the chip manufacturer as the last stage of production. A designer using this approach must supply the manufacturer with his program and pay a large fee for the production of the mask. However, when the mask has been made the unit cost of the device is low. This is the most attractive option for high-volume production but is unsuitable for development and for low-volume applications.

An alternative for small scale projects is to use one of the range of **programmable read only memories** or **PROM**s. These are available in a number of variants, but all allow the user to program the device himself, saving the high cost of mask production. In fact the term PROM is normally reserved for small, fusible link devices of the type described in Section 11.5. These were one of the earliest forms of PLD but have now been overtaken by more modern devices. A characteristic of these devices is that once programmed they cannot be modified.

For flexible system development it is advantageous to have a memory device that can be programmed, and then reprogrammed if necessary. These features are provided by **erasable and programmable read only memories** or **EPROM**s. Although the term can be applied to a number of components, usually this description is applied to memory that is erased by exposure to ultra-violet (UV) light. The chips are fitted with a transparent window which allows UV light to reach the silicon surface. Programming is normally performed using an **EPROM programmer** which provides the appropriate voltages and control signals. After use EPROMs can be erased using a UV light source in 20 to 30 minutes.

EPROMs have the disadvantage that they must normally be removed from their circuit and placed in a special eraser and a programmer to allow them to be modified. Another form of PROM, the EEPROM, can be modified electrically without the need for a UV light source. This allows a program to be modified or changed while the chip is in place.

It might seem that the EEPROM should be classified as a RAM since it can be written (programmed) as well as read. However, it should be noted that a RAM can normally be written and read in a fraction of a microsecond. An EEPROM can be read at this speed, but may require several milliseconds to write a single byte. The EEPROM is thus a read quickly but write slowly device, and is better described as a ROM than a RAM. The relatively slow programming speed of EEPROM can be rather inconvenient for memories of tens or hundreds of kilobytes. These problems are overcome by FLASH memory which provides the ability to electrically program and reprogram a device at very high speeds. This permits even the largest devices to be programmed in a few seconds.

Memory device standards

Over the years a number of standards have emerged for various forms of memory device, making it possible to purchase these components from a host of different manufacturers with confidence that there will be no compatibility problems. One of the most widely used standards is the **JEDEC** standard for byte-wide memory devices. This defines a standard pin-out for both RAMs and ROMs of a range of sizes. One advantage of this arrangement is that it allows a computer manufacturer to design a board with sockets that will accept RAM, ROM, EPROM, EEPROM or FLASH memory. These devices can be of any size up to the present maximum. This allows designers to build expandability into their products by allowing systems to be compatible with components that are not yet available.

12.5 Input/output

The input/output section of the computer is responsible for communicating with the real world. When a microcomputer is used as the heart of a desktop personal computer, its inputs and outputs will generally come from keyboards, displays, printers and communications equipment. When it is used to form an **embedded system**, its input and output devices will be far more varied and may include any of the devices discussed in Chapter 2.

In many cases the original input signal will not be directly compatible with the computer. It may, for example, be analogue in nature, or it may be too large or too small in magnitude. It is therefore usually necessary to provide some form of **interface** to make the signals produced by the sensor compatible with those required by the computer system. Similarly, the signals produced at the output of the computer may not be suitable to drive the necessary actuators directly. Again an interface may be required to overcome this incompatibility.

The operations performed by the interface circuitry will depend on the nature of the sensors and actuators within the system. This process is often referred to as **signal conditioning**. Typical functions will include: amplification; filtering (to remove noise); isolation; and analogue to digital and digital to analogue conversion. Several of these topics have already been discussed in earlier chapters: amplification in Chapter 3; filtering in Chapter 8 and isolation in Chapter 11. The conversion of signals from analogue to digital and from digital to analogue is discussed in Chapter 13.

The end result of the process of signal conditioning at the input is a clean digital signal that is compatible with the input stage of the computer. Similarly, at the output the interface is required to take signals from the output stage of the computer and to generate from them suitable voltage or current waveforms to drive the output devices. For the moment we will leave any further discussion of the interface and concentrate on the nature of the input and output stages of the computer.

12.5.1 Input/output organization

Memory mapped input/output

The simplest and most common method of computer input/output is to make external devices appear as memory registers to the processor. In this way part of the **memory map** of the system is given over to input/output devices. This technique is called, for obvious reasons, **memory mapped input/output**. As well as its basic simplicity, this approach also has the advantage that all of the instructions within the instruction set of the machine can be used to act on input/output registers in the same way that they can be used on memory locations. Disadvantages of this approach are that the full 16-bit address bus must be decoded to obtain the chip select signals for the devices, and that 16-bit addresses must be used to reference these registers, making most I/O instructions 3 bytes long. The importance of the length of the instructions is that computers spend most of their time fetching, rather than executing, instructions, so a shorter program runs more quickly.

Input/output using a separate address space

Some microprocessors have a separate input/output address space which is accessed by a range of special input/output instructions. Usually 256 input and 256 output addresses are provided, allowing the device address to be expressed in a single byte. External devices are addressed using the low-order byte of the address bus with a special control line being used to distinguish I/O from memory operations. This approach has the advantages that the special I/O instructions are only 2 bytes long (an opcode plus a single-byte device address) and that only the low-order byte of the address bus has to be decoded.

Its disadvantages are that it is limited to only 256 input and output addresses, and that only a small number of dedicated instructions can be used for I/O operations. Most microprocessors with a dedicated I/O space also support memory mapped input/output.

12.5.2 Input/output registers

One of the most common methods of input/output is the use of input and output registers. These appear to the processor as ordinary registers, each having its own unique address, but unlike conventional memory registers they have connections to the outside world. When the processor writes into an **output register** the pattern of '1's and '0's placed in the register is reproduced on a series of output lines which can be used to drive external circuitry. Similarly, when the processor reads from an **input register**, the value returned reflects the pattern of '1's and '0's on a set of input lines. Although separate input and output registers are not unheard of, it is much more common to have registers that can perform both tasks. These are thus **bidirectional input/output registers**. These devices are often referred to as **input/output ports**, the term reflecting the use of the word *port* to mean a *gateway*. They are also called **parallel I/O ports**.

12.5.3 Serial input/output

When data must be transmitted over a distance it is often advantageous to use serial rather than parallel techniques to reduce the number of wires or channels required. In Section 10.8 we discussed the use of shift registers in performing **serial to parallel** and **parallel to serial** conversions, and noted their use in serial communications. Figure 10.46 illustrates this technique and shows the need for a separate connection to carry the clock signal as well as that required for the data. The provision of a second clock channel is inconvenient and most serial communications systems use alternative methods of synchronizing the transmitter and receiver. Two basic techniques are used to tackle this problem, these being the asynchronous and synchronous approaches.

Asynchronous serial communications

The asynchronous technique provides synchronization by adding additional redundant information to each byte of data. It is used in situations in which the flow of data can be erratic, and can handle either single bytes or a stream of data. Each byte is augmented by the addition of a **start bit**, which is always 0, and by one or more **stop bits**, which are always 1. A **parity bit** may also be included. The arrangement is shown in Figure 12.14.

The data word is generated using an accurate clock signal at the transmitter (usually produced from a crystal oscillator), and is applied to the communications channel. The receiver also has an oscillator which produces its own local clock signal. Because of the high accuracy and stability of crystal oscillators, the receiver's clock will be very close in frequency to that used at the transmitter.

When the transmission line is not in use it will idle at logic 1. Thus the reception of a start bit will always represent a transition from 1 to 0. When the receiver detects this transition it will synchronize its own internal clock to that of the incoming signal. It will then sample the incoming waveform at appropriate intervals to detect each of the data bits, the parity bit (if present) and the one or more stop bits. Because of the high stability of the two oscillators used at the transmitter and receiver, the variation in phase during a single word will generally be negligible, ensuring correct reception.

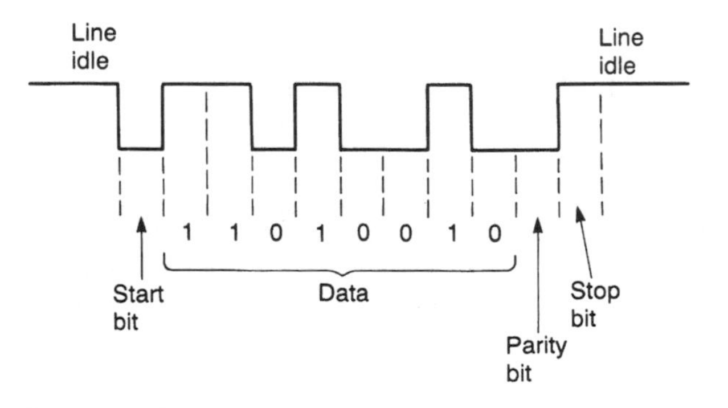

Figure 12.14 The structure of an asynchronous word.

However, the presence of the stop bit or bits, which are known to be at logical 1, allows the receiver to check that everything is correct. If the receiver's clock were running too quickly or too slowly, it would result in the parity bit, or a following start bit, being sampled in place of a stop bit. The receiver continuously monitors the polarity of the detected stop bits and, if an error is detected, flags a **framing error** to the processor. If parity is in use this will also be calculated at the receiver and any errors reported as **parity errors**.

The asynchronous system can be used with a range of data wordlengths and at a number of signalling speeds. It is, however, essential that both the transmitter and the receiver use the same arrangement. Asynchronous communications are widely used within microcomputer systems, particularly for applications producing fairly irregular data, such as **keyboards** and **visual display units** (**VDU**s).

Synchronous serial communications

Synchronous communications systems are used when a continuous stream of data is available. This makes it possible for the receiver to use **clock recovery** circuitry to reconstruct the clock signal being used at the transmitter by looking at the transitions of the incoming signal. This allows the incoming data to be directly clocked into a shift register for conversion to a parallel form. One problem remaining is the detection of the start and end of each word of data. This is achieved by periodically sending a unique **synchronizing pattern** to indicate the beginning of a block of data. Once this has been detected at the receiver, incoming bits are counted to keep track of the beginning and end of each word.

Unlike asynchronous data transfers for which error checking is generally performed using parity testing of each word, in a synchronous system it is more common to perform checking over a complete block of data. This can be achieved using a **checksum**, as described in Section 9.5.4, or by using more sophisticated techniques, such as the inclusion of **cyclic redundancy codes**.

Signalling rate

In serial communications systems it is important that both the transmitter and receiver operate at the same speed. The rate at which bits are sent down the line is described by its **baud rate** (named after Baudot) which is the number of transitions per second on the line. Common baud rates are multiples of 75 bits per second, these being: 75, 150, 300, 600, 1200, 2400, 4800, 9600, 19 200 and 38 400 bits per second.

Simplex and duplex communications

Communication channels can operate in either one or both directions. The simplest form of link can convey information in only one direction. This is termed a **simplex** channel. Systems that can communicate in both directions are termed **duplex**; they may be subdivided into **half-duplex** systems, which can communicate in only one direction at a time, and **full-duplex** systems, which can communicate in both directions simultaneously.

Serial I/O devices

Serial input/output is normally performed by one of a range of special purpose integrated circuits. As with parallel I/O devices, these are known by a variety of names. Common devices include the UART (universal asynchronous receiver/transmitter), the ACIA (asynchronous communications interface adapter), the SIO (serial I/O controller – synchronous) and the USART (universal synchronous and asynchronous receiver/transmitter). They all provide a complete full-duplex communication channel, usually with the minimum of external components.

Serial communications standards

Serial I/O devices produce and accept standard logic signals which are not themselves suitable for long distance transmission. These logic levels are normally buffered using a **line driver** or a **line receiver** to provide an increased range and improved **noise rejection**. There are several standards for serial communication signals; some of the most popular are described below.

RS-232-C

This is one of the most widely used standards and is almost always provided on computer terminals and serially driven printers. Logic 1 is represented by a voltage in the range −5 to −15 V, and logic 0 by a voltage in the range +5 to +15 V. These voltages are inconvenient in modern systems which invariably operate from a single +5V supply. The standard specifies 25 lines for a full implementation, including data lines in each direction, a set of handshaking lines, various control lines and a number of test connections. Most systems use only a handful of lines, and many use only three: transmit, receive and ground. RS-232-C was intended for short range communications and the original standard specified a maximum distance of 100 feet. The standard has now been changed to specify a maximum allowable line capacitance, allowing longer distances to be achieved.

RS-423-C

This system uses only two wires and has logic voltages of 0 and 5 V allowing operation from a single 5 V supply.

RS-422-C

Like RS-423-C this arrangement can be used with a single 5 V supply but in this case the two lines are balanced, with one or other line being taken to 5 V to select the logic level. Data rates of up to 10 Mbits per second (10 000 000 bits per second) are possible over distances of up to 4000 feet.

RS-485-C

This is a **multi-drop** version of RS-422-C. This means that several devices can be connected to a common pair of cables which acts as a **communications bus**. Usually messages sent along the bus are prefixed by a **device address** so that only the appropriate unit will respond.

20 mA current loop

Originally used in mechanical **teleprinters**, this standard is still used for a number of applications. Two wires are used for communication in each direction forming a loop, as shown in Figure 12.15(a). When the switch is closed current flows round the loop producing a voltage drop across the resistor R_2 at the receiver. No current flows when the switch is open. Normally logic 0 is represented by no current and logic 1 by a current of 20 mA. Figure 12.15(b) shows an improved arrangement using an **opto-isolator** to reduce noise problems.

12.5.4 Program controlled input/output

Input/output operations can be initiated in a number of ways. Often the programmer decides that at a particular point within a program, some data should be read from an input device or written to an output device. Logically enough, this is called **program controlled input/output**. However, in some circumstances it is not possible to determine, at the time that the program is written, when data will become available. An example of such a situation is the reading of data from a keyboard. If we assume that the device has a data reg-

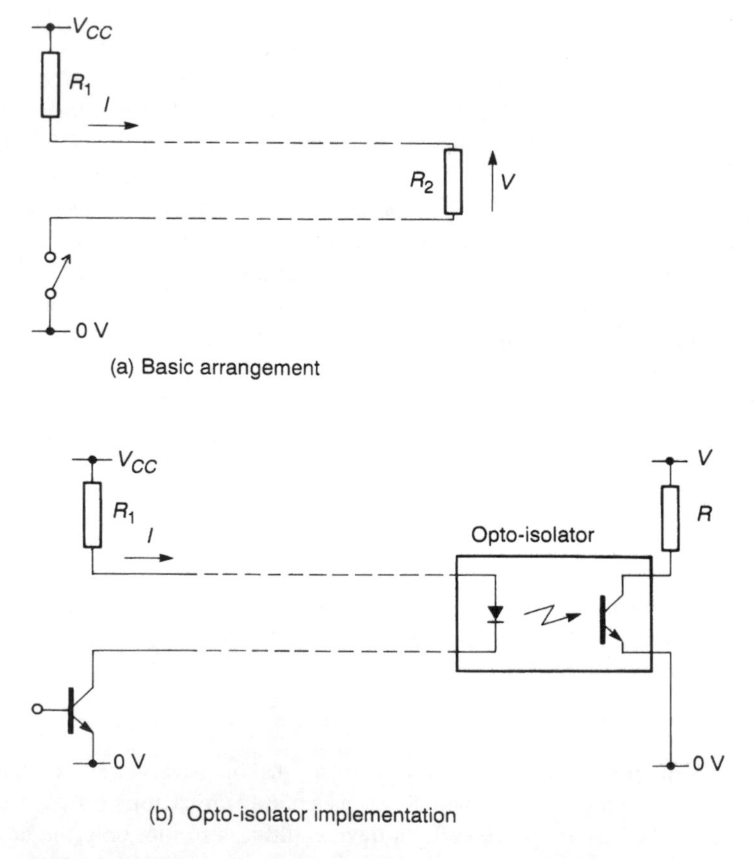

Figure 12.15 A current loop communications system.

ister associated with it, there is clearly no problem with reading the port at any time, but data will only become available when a key has been depressed. A solution to this problem, still using program controlled input/output, is to use *polling*.

Polling

Polling involves the repeated reading of an external device to determine its status. Let us initially consider the above example, involving the reading of data from a keyboard. Simply looking at the contents of the device's data register will not show whether any new data has arrived, since this new data might be identical to the previous contents of the register. We therefore require an additional register, called a **status register**, which contains a bit indicating the presence of data. This bit might be called the **data ready flag**. This flag is set when data is placed in the register and is cleared by the processor when the data has been taken. The status register can be polled in a number of ways, depending on the application. Figure 12.16 illustrates this process.

Figure 12.16(a) shows a 'straight through' program which begins, runs through a sequence of instructions and then stops. While such programs are sometimes found on personal and other general purpose computers, they are rarely found in **embedded systems** which are normally associated with control and instrumentation. Such systems usually have a program structure of the type shown in Figure 12.16(b). This has an initialization section and then a continuous loop. Once started, the program runs for ever, or until the system is turned off. This is a usual requirement of a control system.

Let us now assume that data must be read from an external device at some point during the execution of the main loop of the program. If program execution cannot continue until information is available, the system must sit and wait for it to arrive. This is achieved as shown in Figure 12.16(c). The program reads the status register of the device and looks to see if the data ready bit is set. If it is, it reads the data register, clears the data ready flag and continues with the program. If it is not, it goes back and reads the status register again, repeating this operation until it shows that data has arrived. This technique will result in the program waiting indefinitely for data to arrive, during which time it will not be performing any other task. Such an arrangement might be used when awaiting a start command from a keypad if nothing else can be done until this has been received.

An alternative arrangement is shown in Figure 12.16(d). Here the status register is tested at an appropriate point in the loop. If the data ready bit is set, the processor reads the data register and continues. If it is not, the processor simply continues and repeats the test the next time it reaches this point in the loop. This technique has the advantage that the processor can continue to do useful things until data becomes available. However, it is only practical if the processor can perform useful tasks before the data arrives.

In some applications there may be not one but a number of external devices, all of which may require action from the processor. If the processor must test them all to see if they have data available, this can waste a great deal of processor time.

12.5.5 Interrupts

The time wasted in polling large numbers of external devices can be reduced by getting these devices to inform the processor when they want attention or 'servicing'. In other

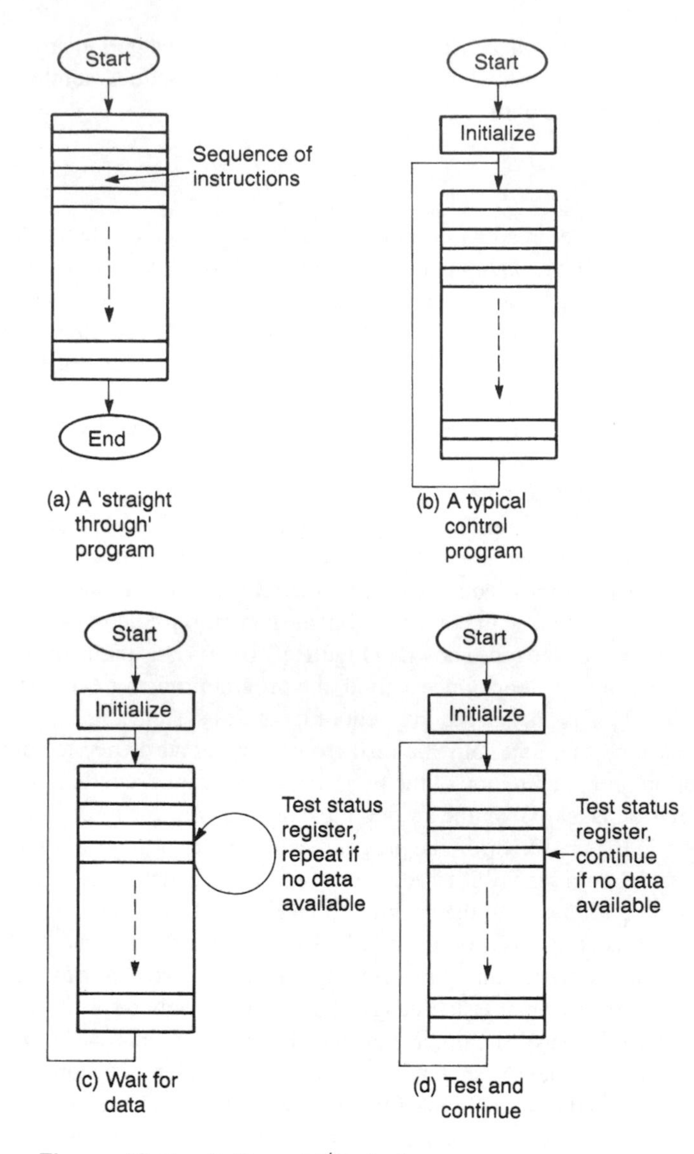

Figure 12.16 Polling of I/O devices.

words, we get external components to **interrupt** the operation of the processor. This is achieved by building into the processor hardware an input line called an **Interrupt ReQuest line (IRQ)**. When an external device activates this line the processor will stop what it is doing and service the device. The interrupt mechanism is illustrated in Figure 12.17.

Suppose that a device generates an interrupt at the time indicated in the figure. This is likely to occur during the execution of an instruction and initially no action is taken. The processor will always complete its current instruction before responding to an

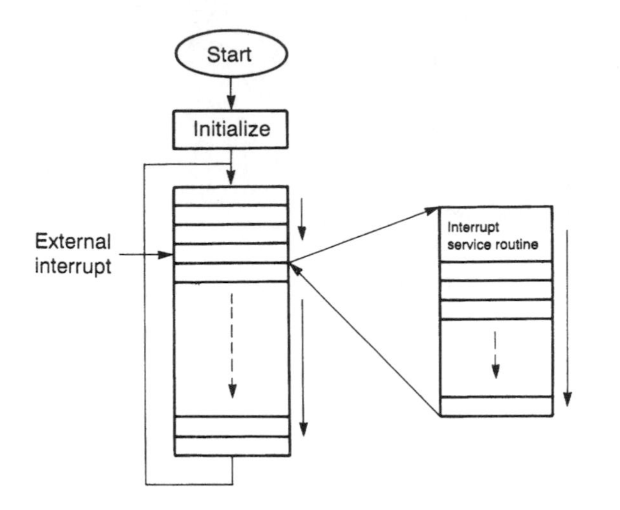

Figure 12.17 The interrupt mechanism.

interrupt, since failure to do so would leave the state of the machine uncertain. When this instruction is complete the processor leaves the main program and jumps to an **interrupt service routine**, which is simply a sequence of instructions produced by the programmer for execution when an interrupt occurs. Upon completion of the service routine the processor jumps back to the main program and continues where it left off.

Although the above description is correct, it leaves many questions unanswered. It would seem, for example, that the operation of the main program would be affected by the execution of the interrupt service routine, since this is almost certain to change the contents of the various registers within the processor. It is also not clear how the processor knows where to find the service routine, or how it remembers where to re-enter the main program. To answer these questions we need to look in a little more detail at the interrupt mechanism.

In order that the operation of the main program is not affected by the interrupt service routine, the contents of the various registers of the processor are stored before the interrupt service routine is entered. Amongst the registers saved is the **program counter** which, you will remember, contains the address of the next instruction in the main program. At the end of the interrupt service routine the original contents of these registers are restored before the main program is restarted. Since the contents of the program counter have been restored, the sequence of instructions in the main program is not disrupted. The processor detects the end of the interrupt service routine when it meets a **return from interrupt instruction** (**RTI**). This is a member of the instruction set of the machine specifically designed for this purpose. The interrupt mechanism is shown in detail in Figure 12.18.

This leaves one unexplained point. How does the processor know where to find the interrupt service routine? In fact this begs the question, how does the processor know where to find the main program? To answer these questions we need to look at the use of *vectors*.

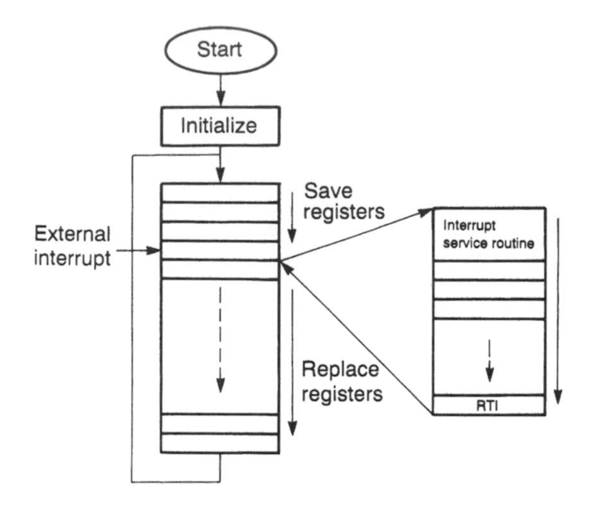

Figure 12.18 A more detailed view of the interrupt mechanism.

Vectors

A vector, as its name implies, *points* to somewhere in memory. In other words, it is an address. Within 8-bit machines, since addresses are 16 bits long, a vector is a 2-byte quantity requiring two memory locations. Microcomputers use a number of vectors to define the starting addresses of sections of program which should be executed in particular circumstances. For example, the **interrupt vector** defines the start address of the interrupt service routine and the **reset vector** (also called the **restart vector**) defines where the processor will start executing when it is first turned on, or after a particular input line (called the **reset** line) has been activated. Some processors store these vectors in fixed locations and Figure 12.19 shows a portion of the memory map of a typical processor, indicating the location of the various vectors.

The processor shown has four vectors, one corresponding to reset and the other three to different forms of interrupt. One of the interrupts is **maskable**, which means that it

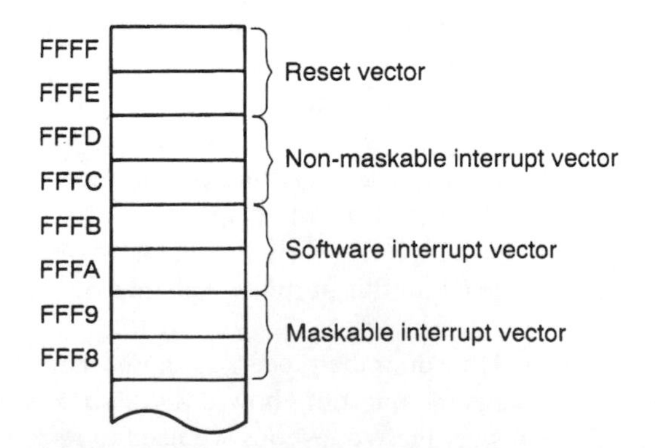

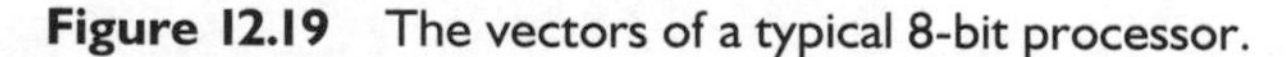
Figure 12.19 The vectors of a typical 8-bit processor.

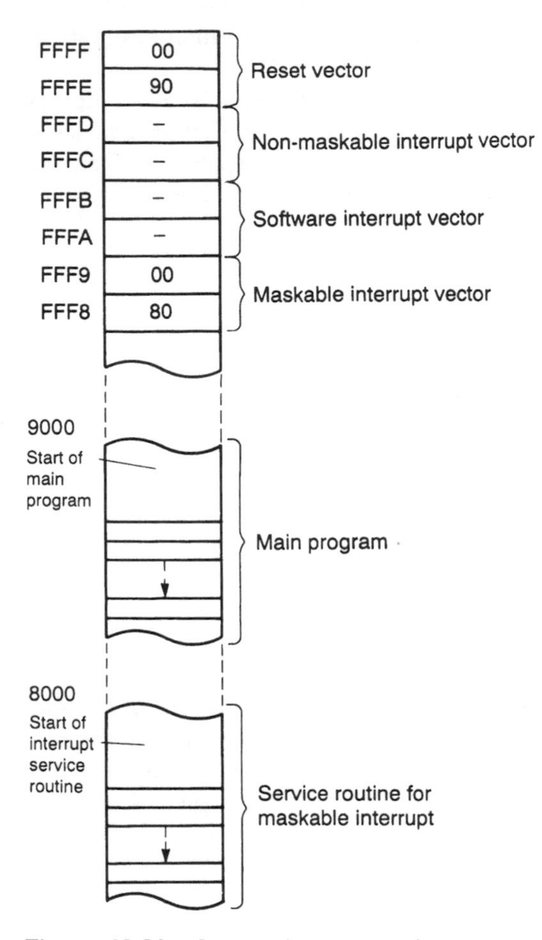

Figure 12.20 Setting the reset and interrupt vectors.

can be turned on and off under program control using dedicated instructions. There is also a **non-maskable** interrupt, which is always active. A third form of interrupt, termed a **software interrupt**, is unlike the others in that it is not activated by external hardware but as a result of the execution of a dedicated instruction. However, it derives its name from the fact that the action taken by the machine is similar to that taken in the event of a conventional hardware interrupt.

The programmer, after writing the software necessary to implement the required system functions, places the start address of his main program in the bytes corresponding to the reset vector. When the system is turned on it will fetch this address and begin executing the program from this location. If the system makes use of interrupt driven I/O, the programmer will also place the start address of his interrupt service routine(s) in the locations corresponding to the appropriate interrupt vector(s). This is illustrated in Figure 12.20.

In a general purpose computer the reset vector points to a program that requests commands from the user, and which may load further programs. In an **embedded system** the reset vector points to the start of an application program which will therefore start automatically. This self-starting arrangement is termed a **turn-key system** since, conceptually, it starts 'at the turn of a key'.

Some microprocessors use different methods to determine the starting address of the interrupt service routine. They may, for example, get the external device to supply an opcode to the processor which is then used to jump to an appropriate routine. Other processors use a variety of methods for acquiring the interrupt vector.

Interrupts and the stack

We have seen that before an interrupt service routine is executed it is necessary to store away the contents of the various registers so that the main program can be restarted successfully. In some machines this process is performed automatically while in others it is left to the programmer to save those registers that are being used. In either case it is interesting to consider where this data is stored. Clearly it would be possible to define certain fixed locations in which the contents of each register would be stored, but this has serious drawbacks. So far we have considered only the situation in which a single interrupt source is used. In many cases there may be several devices capable of generating the same, or a variety of, interrupts. In this situation it is quite likely that a second interrupt will occur while the processor is busy executing an interrupt service routine. If this happens, when the register information for the second interrupt is saved, it will overwrite the original data and the processor will be unable to return to the main program.

The solution to this problem is to save the register contents using the **stack** described earlier. This is a **first-in–last-out (FILO)** store, which means that information is retrieved in the reverse order to that in which it is saved. Using this arrangement, multiple interrupts will result in the data being stacked in order. At the end of each interrupt service routine the relevant data will be recovered from the stack, allowing the processor to return correctly to the main program. This is illustrated in Figure 12.21.

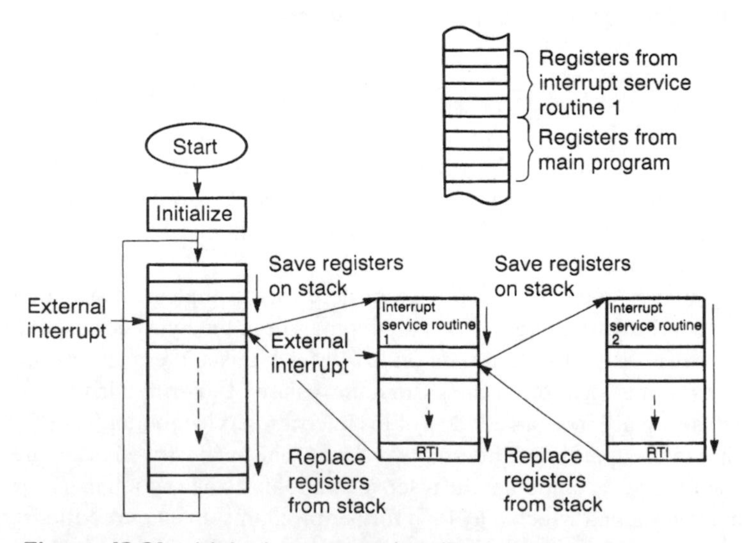

Figure 12.21 Multiple interrupt handling with a stack.

12.5.6 DMA

Although the use of interrupt driven I/O allows a rapid response to external events, for applications in which large amounts of data need to be transferred it can consume a considerable amount of processor time. Every time an external device requires servicing, the contents of the processor's registers must be stored away and then replaced. A faster alternative is to allow external devices to transfer data directly into memory without affecting the processor's registers, as shown by the dotted line in Figure 12.1. The processor is made to wait for the short time that a dedicated controller needs to write or read a byte of memory and is then allowed to continue. This is **direct memory access** or **DMA** and it is by far the fastest method of computer input/output. Unfortunately, it is also the most expensive since a considerable amount of additional hardware is required to oversee the data transfers. For this reason DMA is usually used only for applications that can justify its high cost, for example, for disk drives. This method of input/output is sometimes referred to as **cycle stealing** since the DMA unit robs the processor of memory access cycles.

12.5.7 Computer input/output – a summary

We have looked at three methods of computer input/output: program controlled, interrupt driven and direct memory access. The last of these is very fast but is also expensive. It is used primarily for transferring large quantities of data and has the advantage that it consumes very little processor time. Interrupt driven I/O can provide a fast response without wasting a great deal of processor time in polling. However, interrupt driven systems require more complicated hardware than program controlled systems, and are harder to test since their operation is essentially asynchronous. Program controlled I/O is the simplest of the three techniques and, as such, should normally be the preferred option. Interrupts should only be used if the application requires them and if the processor time saved can be used meaningfully.

12.6 Single-chip microcomputers

We have considered the three main components of a microcomputer, namely the processor (CPU), the memory and the input/output sections. For small applications it is possible to obtain all of these functions within a single integrated circuit. Such a device is termed a **single-chip microcomputer**. The combining of these components not only saves on physical size, but also reduces the number of external connections required, improving reliability and reducing cost. The single-chip microcomputer turns the computer from a 'system' into a component.

Early single-chip microcomputers provided relatively small amounts of ROM and RAM together with modest I/O facilities. A typical device might have provided a fairly conventional 8-bit CPU with 2 kbytes of mask programmable ROM, 64 kbytes of RAM, 32 bits of parallel input/output, an 8-bit counter and a single interrupt. Even with these limited capabilities such devices were sufficiently powerful to be used in a large number of applications. Unfortunately, the restricted memory meant that they generally could not be programmed using high level languages, forcing the programmer to work in assembly code. This was very time consuming and led to high development costs.

Another disadvantage of these early parts was their use of masked programmed ROM which meant that they had to be programmed by the manufacturer. This was an expensive process and as a result, use of these components was largely restricted to very high volume applications where the considerable development costs could be justified.

In recent years the capabilities of single-chip microcomputers have increased enormously. Modern devices normally have a range of memory options with large amounts of ROM, RAM and EEPROM being available if required. Both user programmed and mask programmed versions are available, simplifying development, while allowing high volume applications to take advantage of mask programmed devices if appropriate. The input/output capabilities of these devices are also extensive with typical parts providing a large number of input/output lines in addition to multiple counters, interrupts and serial channels.

Single chip 16- and 32-bit microcomputers are also available for applications that require more power than can be obtained using 8-bit components. These offer the facilities normally associated with more powerful computers such as on-board memory management and DMA.

Despite the vast capabilities of the most powerful single-chip microcomputers it is the simplest 8-bit parts that dominate the market for these components. The really high volume applications of computers are associated with small embedded systems such as those within cars and domestic appliances. In these systems relatively modest amounts of computing power are required and small 8-bit components are more than adequate.

Key points

- The development of the microprocessor allowed the great potential of very large scale integration (VLSI) to be realized.
- It did this by permitting high-volume production of components that could be tailored for a particular application by the addition of appropriate software.
- A microprocessor is a single-chip implementation of one of the main components of a computer – the CPU.
- To form a complete computer, memory and some form of input/output must be added.
- A chip that contains the processor, memory and input/output is termed a single-chip microcomputer.
- One of the most important subsections of a microprocessor is the register.
- Communications between the various registers take place over a series of buses which act as parallel data highways. There are three main buses, namely the data bus, the address bus and the control bus.
- Microcomputers obey programs by repeatedly fetching, and then executing, instructions from memory.

- All computers need some form of memory and systems will normally include both read/write memory (RAM) and read only memory (ROM).
- Typical 8-bit microcomputers have an address space of 64 kbytes. 16- and 32-bit microcomputers have much larger addressing ranges.
- The input/output section is the most application-specific part of the computer.
- In many cases the processor performs input and output operations directly under the control of the applications program. This is termed 'program controlled input/output'.
- In other instances it is an external device that initiates such actions through the use of interrupts.
- In very high-speed applications the processor may give up control of the buses temporarily to allow external devices to perform 'direct memory access' (DMA).

Design study

In the design study at the end of Chapter 1 we identified the inputs and outputs of an electronic controller for a domestic automatic washing machine. Many modern washing machines now use microcomputers to control their various functions, replacing the electromechanical controllers used in earlier models. Clearly it is not practical to consider all aspects of such a system, but it is instructive to look at some elements of the design.

At various stages of the washing operation the drum is required to rotate at different speeds. These include: a low speed of about 30 revolutions per minute (rpm) while clothes are washed; an intermediate speed of about 90 rpm while the water is pumped out; and a high speed of either 500 or 1000 rpm to spin dry the clothes. Consider how the microcomputer should control the speed of the motor.

Approach

Since a domestic washing machine is a very high-volume product our design should attempt to minimize the amount of hardware required. This necessitates a close look at the choice of actuators and sensors to select low cost items. Our first decision must be whether our system will be **open loop** or **closed loop**, as discussed in Section 4.2. In fact this decision is quite clear, since although an open-loop system is theoretically possible using a synchronous motor, the cost of such a system for high-power, variable-speed applications is prohibitive. Our system will therefore be closed loop using a motor to drive the drum and some form of sensor to measure its speed.

One of the simplest methods of speed measurement is to use a **counting technique**, as described in Section 2.3.5. A common arrangement uses a fixed inductive sensor to produce a pulse each time it is passed by a magnet which rotates with the drum. This produces one pulse per revolution of the drum, which can be used to determine its speed.

The speed of the motor will be controlled by the power dissipated in it. The simplest way of controlling this is to use a **triac**, as described in Section 7.8.2. Clearly the power could be controlled by some form of electronic circuitry, but the hardware requirement can be reduced if the microcomputer controls the power directly by firing the triac at an appropriate time during its cycle. To do this the controller must detect the zero crossing of the AC supply. This will require circuitry to detect the crossing point while protecting the processor from high voltages.

A block diagram of the resultant system is shown below.

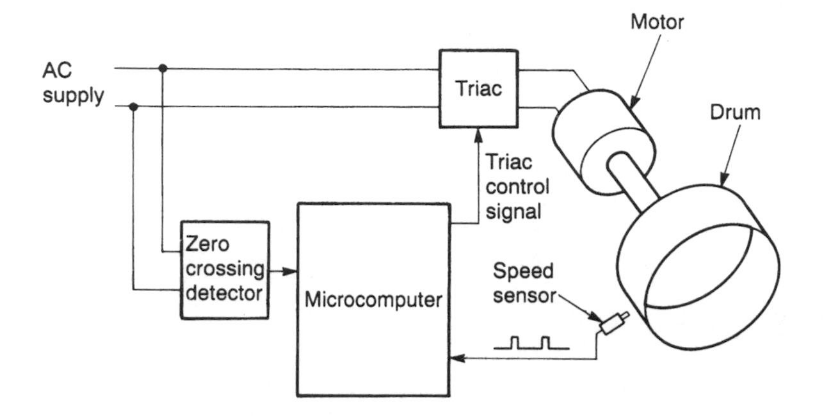

Let us initially consider the problem of determining the speed of rotation of the drum. our input is in the form of a train of pulses, the period of which is determined by the drum speed. We could simply connect this input to one line of a parallel input port and **poll** the signal to detect its state. However, this would be very time consuming. An alternative solution would be to connect this signal to an **interrupt request line** and to produce an **interrupt service routine** to increment the number in a register. The processor could then look at this register at regular intervals to see how many pulses had occurred in a fixed period of time, and hence deduce the speed. The timing could be determined by a timer which would be set to produce interrupts at regular intervals. A third alternative would be to use a hardware counter, if one were available, to count the number of pulses. Again a timer would be used to produce interrupts at regular intervals when the counter would be read and then zeroed. Using one of these methods the speed of the drum would be measured continually and the value made available to the main control program.

At any point in the washing cycle the program determines at what speed the drum should rotate. From a knowledge of the required speed and the actual speed as obtained above, the controller can determine whether to increase or decrease the power dissipated in the motor.

The motor power is determined by the timing of the triac firing pulse. If the triac is fired at the beginning of each half mains cycle it will remain on for the remainder of the half-cycle and the motor will operate at full power. The longer the processor waits before firing the triac, the less will be the motor power. The processor thus varies the delay time with respect to the zero crossing point of the mains by an appropriate amount to increase or decrease the power in the motor, as determined by the difference between the actual and required speeds.

It is clear that this method of controlling the motor speed is very 'processor intensive'. It consumes a large amount of processor time and will require a considerable amount of effort in writing and developing the software. However, this approach uses very little hardware and is thus very attractive for such a high-volume application.

Further reading

Stone H. S. (1992) *Microcomputer Interfacing*, 2nd edn. Reading, MA: Addison-Wesley

Tocci R. J., Ambrosio F. J. and Laskowski L. P. (1997) *Microprocessors and Microcomputers: Hardware and Software*, 4th edn. Englewood Cliffs, NJ: Prentice-Hall

Exercises

12.1 Draw a block diagram showing the main sections of a microcomputer and illustrating the information flows between them.

12.2 List the main components of the CPU of a typical 8-bit microprocessor and describe the function of each.

12.3 Explain the requirement for a three-state output characteristic for registers that are to be connected to a bus system.

12.4 During the operation of a program it is often useful to jump to a **subroutine** which contains a program section which is used several times. At the end of the subroutine the processor must return to the instruction following the jump to subroutine instruction. Explain how this may be achieved. Your solution should take into account the fact that subroutines can be 'nested', in that one can call another.

12.5 Explain the differences between multiplexed and non-multiplexed bus systems and describe how the full address bus is derived in the former.

12.6 Calculate the number of address lines required for the following memory devices:
(a) a 64 kbyte ROM
(b) a 2 kbyte RAM
(c) a 64 kbit (8 k × 8 bits) ROM.

12.7 Explain the meanings of the terms: memory-mapped I/O; polling; program controlled I/O; interrupt driven I/O and DMA.

12.8 Compare and contrast the use of memory-mapped I/O with the provision of a dedicated I/O space.

12.9 Explain the use of the stack in interrupt handling.

12.10 In what form of memory (ROM or RAM) would it be normal to store the system vectors? Why?

12.11 What is meant by the term 'byte sexing'?

12.12 Why is an EEPROM not considered to be a non-volatile RAM?

12.13 How does the processor distinguish between programs and data within the memory of the machine?

12.14 What methods of computer input/output would seem most appropriate for the following applications?
(a) reading push-buttons in a computer controlled washing machine
(b) interfacing a keyboard to a personal computer
(c) connecting a high-speed disk drive to a computer

12.15 Explain the meaning of the term FILO and give an example of the use of such a structure. Within computers it is often useful to produce a first-in–first-out arrangement (a FIFO). Suggest a possible use of such a structure.

12.16 What are the properties of a 'turn-key' system? How are these properties achieved?

12.17 In the system described in the design study at the end of this chapter we considered the control of the speed of a motor within a washing machine. Consider the strategy to be adopted to control the water temperature within the machine. How should the user push-buttons be sensed?

Data Acquisition and Conversion

Objectives

When you have studied the material in this chapter you should:

- recognize the need for techniques to convert analogue signals into a digital form, and vice versa;
- be aware of the major components of a typical data acquisition system;
- understand the characteristics and limitations of sampling as a method of obtaining a picture of a time varying signal;
- be familiar with common methods of digital to analogue and analogue to digital conversion;
- have an understanding of terms such as resolution, accuracy, settling time and sampling rate when applied to data conversion;
- be aware of techniques that allow a number of analogue inputs or outputs to be used with a single data converter.

Contents

13.1 Introduction

We noted in earlier chapters that the effects of noise are often less of a problem in digital systems than in those using analogue techniques. We have also seen that digital data can be easily processed, transmitted and stored. In many instances, therefore, we choose to represent analogue quantities in a digital form, which raises the question of how we translate from one form to the other.

The process of taking analogue information, often from a number of sources, and converting it into a digital form is often termed **data acquisition**. It consists of several stages. This chapter begins by looking at the process of sampling a changing analogue quantity to determine its time varying nature. It then discusses the hardware required to convert these samples into a digital form, and the reconstruction of this digital information into an analogue signal. Finally it considers the process of combining information from a number of sources into a single system input, and the converse problem of generating a number of analogue output signals from a single information source.

13.2 Sampling

In order to obtain a picture of the changes in a varying quantity it is necessary to take regular measurements. This process is referred to as **sampling**. Clearly, if a quantity is changing rapidly we will need to take samples more frequently than if it changes slowly, but how can we determine the sampling rate required to give a 'good' representation of a signal? It would seem obvious that the required sampling rate would be determined by the most rapidly changing or, in other words, the highest frequency, components within a signal, but how do we decide how fast we need to sample to get a 'good picture'?

Fortunately an answer to this question is available in the form of **Nyquist's sampling theorem**. This states that the sampling rate must be greater than twice the highest frequency present in the signal being sampled. It also states that under these circumstances none of the information within the signal is lost by sampling. In other words, it is possible to reconstruct completely the original signal from the samples.

In general the waveform to be represented will contain components of many frequencies. In order to sample it reliably we need to know the highest frequency present. Let us assume that we know that a certain signal contains no components above a frequency of F Hz. According to Nyquist's theorem, provided we sample this waveform at a rate greater than $2F$, we will get sufficient information to reconstruct completely the original signal. This minimum sampling rate is often called the **Nyquist rate**. This process is illustrated in Figure 13.1.

Figure 13.1(a) shows a sine wave of frequency F. Figure 13.1(b) shows the results of sampling this signal at a rate greater than the Nyquist rate. Given these samples it is possible to reconstruct the original waveform, since any other line drawn through the sample points would have frequency components above F. Since we know that in this case the signal has no components above this frequency, the original waveform is the only possible reconstruction. Since this sampling rate allows the reconstruction of a signal of

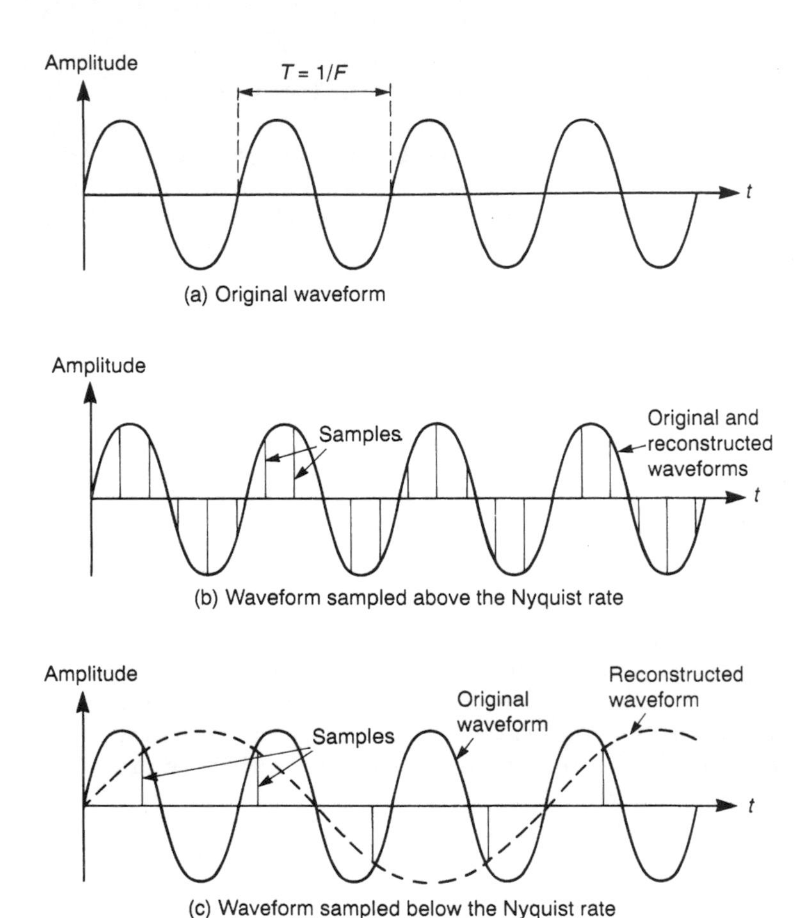

Figure 13.1 The effects of sampling at different rates.

frequency F, it will also allow reconstruction of any signal that contains no components above this frequency.

Figure 13.1(c) shows the results of sampling the waveform at a frequency below the Nyquist rate. Here the samples can be reconstructed in a number of ways, including that shown in the figure. This is clearly not the original waveform. Thus, if a signal is sampled below the Nyquist rate it will not, in general, be possible to reconstruct the original signal. The waveform generated appears to have been produced by a signal of a lower frequency than the original. This effect is known as **aliasing** and resembles a *beating* between the signal and the sampling waveform.

It should be pointed out that the Nyquist rate is determined by the highest frequencies present within a signal, *not* by the highest frequencies of interest. If a signal contains unwanted high frequency components they must be removed before sampling or they will result in spurious signals within the frequency band of interest. It is normal to use filters to remove signals that are above the range of interest to prevent this effect. These are referred to as **anti-aliasing filters**. For example, although human speech contains frequencies up to above 10 kHz, it has been found that good intelligibility may be obtained using only those components up to about 3.4 kHz. Therefore, to sample such a signal for transmission over a channel of limited bandwidth, it would be normal to filter the speech signal to remove frequencies above 3.4 kHz, and then to sample the waveform at about 8 kHz. This is somewhat above the Nyquist rate (which would be 6.8 kHz) to allow for the fact that the filters are not perfect and some frequency components will be present a little above 3.4 kHz. It is common to sample at about 20% above the Nyquist rate.

13.3 Data converters

Having decided upon the rate at which samples are to be taken, it is necessary to consider the process of generating a digital equivalent of the instantaneous value of an analogue quantity, and the converse operation of converting a digital signal back into an analogue form. These operations are performed by data converters, which can be divided into **digital to analogue converters** (**DAC**s) and **analogue to digital converters** (**ADC**s).

A range of converters is available, each providing conversion to a particular **resolution**. This determines the number of steps or **quantization levels** that are used. An n-bit converter produces or accepts an n-bit parallel word and uses 2^n discrete steps. Thus an 8-bit converter uses 256 levels and a 10-bit converter uses 1024 levels. It should be noted that the resolution of a converter may be considerably greater than its **accuracy**. The latter is a measure of the error associated with a particular level rather than simply the number of levels used.

Conversions of either form take a finite time, which is referred to as the **settling time** of the converter. We shall see that the times taken for conversion differ greatly between techniques. In general, digital to analogue conversion is faster than the inverse operation for reasons that will become apparent from the following discussion. We will start by looking at DACs since their operation is somewhat simpler, and because they are often used within ADCs.

13.3.1 Digital to analogue converters

There are two common forms of digital to analogue converter.

Binary-weighted resistor method

This form of DAC is a development of the **current to voltage converter** described in Example 4.9. A simple form of the converter is shown in Figure 13.2.

Each input controls a switch which connects a resistor to a constant reference voltage $-V_{ref}$. These switches are closed when the corresponding bit is set to 1. If the switch connected to the most significant bit (MSB) of the input digital word is closed while all the others are open, the reference voltage is connected to one side of the resistor R. The other end of this resistor is connected to the inverting input of the operational amplifier which is a **virtual earth** point and is therefore at 0 V. The voltage across the resistor is thus equal to the reference voltage and the current flowing into the virtual

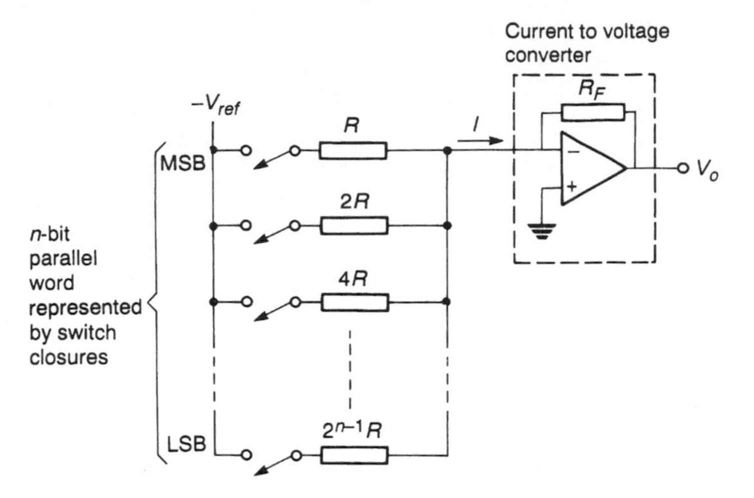

Figure 13.2 A binary-weighted resistor DAC.

earth point is given by

$$I = -\frac{V_{ref}}{R}$$

If the next most significant bit is set to 1 while all the others are at 0, the reference voltage is applied across the resistor $2R$. This will produce a current into the amplifier of

$$I = -\frac{V_{ref}}{2R}$$

this being half that produced by the MSB. The next switch, if closed while all the others are open, will produce a current of one-quarter of that produced by the MSB. The progression continues, each input in turn producing a current of one-half that of its predecessor. The inputs are thus *binary weighted.*

Since the input to the operational amplifier is a virtual earth, its voltage does not change with the current flowing into it. Thus the fact that one switch is closed will not affect the current injected by another switch. The currents therefore sum to give a value representing the combination of switches closed. The current to voltage converter then converts this input current I into an output voltage (as described in Example 4.9), obeying the expression

$$V_o = -IR_F$$

where R_F is the feedback resistor.

When the LSB alone is set to 1 the current I will be given by

$$I = -\frac{V_{ref}}{2^{n-1}R}$$

and the output voltage will therefore be

$$V_o = -IR_F = \frac{V_{ref}R_F}{2^{n-1}R}$$

This represents the output voltage for an input number of 1. For an input number of m the output will therefore be

$$V_o = m \times \frac{V_{ref}R_F}{2^{n-1}R}$$

In practice this form of DAC is implemented using electronic switches (transistors) in place of those shown in Figure 13.2. However, the principles of operation are identical to those just described.

The binary-weighted resistor method of conversion uses a small number of resistors, but requires them to have a broad spread of values (a range of R to $2^{n-1}R$). For a 10-bit converter, for example, this range will have a ratio of over 500 to 1. Unfortunately, resistors of markedly different values tend to have unequal temperature coefficients of resistance, which means that the ratios between them will change with temperature. This limits the temperature stability of this technique.

Computer simulation exercise 13.1

FILE 13A

Simulate the circuit of Figure 13.2 replacing the switches and voltage reference by logic inputs.

Use a 741 op-amp with $R_F = 1$ kΩ. Use four input resistors of values 1 kΩ, 2 kΩ, 4 kΩ and 8 kΩ. Connect the 1 kΩ resistor to a 1 kHz square wave input (use a DigStim input in PSpice). Similarly connect the 2 kΩ resistor to a 2 kHz input, the 4 kΩ resistor to a 4 kHz input and the 8 kΩ resistor to an 8 kHz input.

This arrangement models the situation where the outputs from a 4-bit binary counter are connected to the DAC. The output should therefore be a staircase waveform with 16 steps. Observe the form of the output and confirm that its magnitude and frequency are what you would expect.

Modify your circuit to model the output of an 8-bit binary counter and again observe the form of the output.

R–2*R* resistor chain method

The R–$2R$ method also makes use of the **current to voltage converter** arrangement of Example 4.9, but does not require a broad spread of resistor values. The arrangement is illustrated in Figure 13.3.

In many respects this circuit resembles that of the binary-weighted resistor arrangement. Again the binary word controls a series of switches, which generate currents in a series of resistors. The difference in this case is that all the resistors connected to the switches have the same value. The other end of the resistor in each case is joined to a chain of resistors, which goes from the inverting input of the operational amplifier to earth. The circuit is arranged such that currents flowing through each of the resistors connected

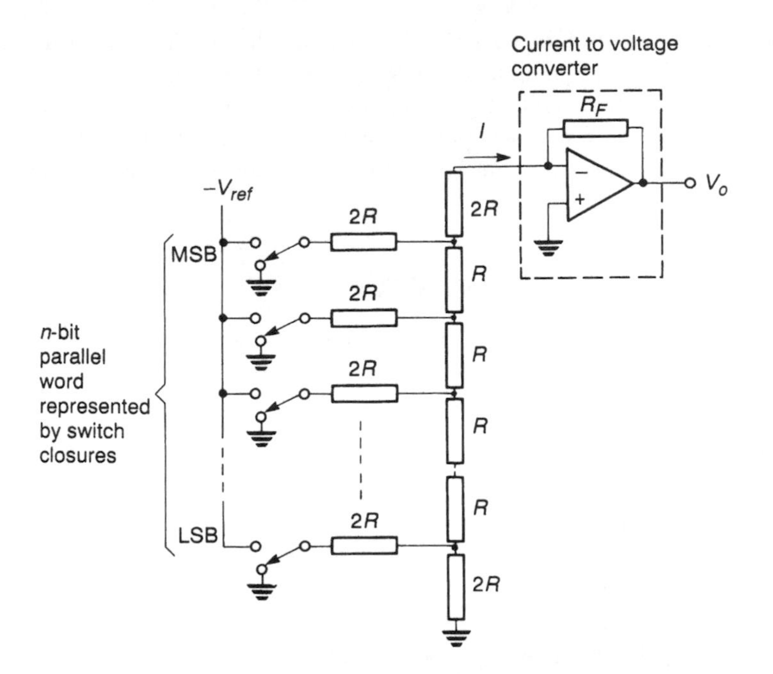

Figure 13.3 An R–$2R$ resistor chain DAC.

to the switches see a resistance of $2R$ looking in either direction along the resistor chain. Therefore, half the current will go in each direction. Similarly, currents flowing up the chain see equal resistances in either direction at each node and will again be split. Therefore, each switch contributes half as much current as the switch above, as its current is repeatedly halved at each node on its journey to the op-amp.

Therefore, the currents generated by the switches are binary weighted, as in the previous method, but without the use of a wide range of resistor values. Here only resistors of R and $2R$ are required; if appropriate these can be formed using only resistors of one value (R) by connecting two in series to form the other ($2R$). This allows temperature matched resistors to be used to provide greatly improved temperature stability.

DAC settling times

The settling times of these two methods of digital to analogue conversion are similar and are determined by the time taken for the electronic switches to operate and for the amplifier to respond. Converters are available with a range of resolutions, and in general conversion time increases with resolution. A typical, general purpose 8-bit DAC would have a settling time of between 100 ns and 1 μs, while a 16-bit device might have a settling time of a few microseconds. However, for specialist applications high-speed converters might have settling times down to a few nanoseconds. It is sometimes more convenient to specify the number of samples that can be converted in a second rather than the settling time. Converters used for generating the video signals used in graphics display systems might have a resolution of only 4 bits, but may have a maximum **sampling rate** of above 100 MHz, corresponding to a settling time of less than 10 ns.

It is normal to low-pass filter the output of a DAC to smooth the resulting waveform and thus remove the effects of sampling. The cut-off frequency of such a **reconstruction filter** would be chosen to remove components at the sampling frequency without attenuating the required signal.

13.3.2 Analogue to digital converters

There are a number of techniques available for analogue to digital conversion. Of these, four types are the most widely used.

Counter or servo

The counter method of conversion gives one of the simplest forms of ADC. The principle is illustrated in Figure 13.4.

At the heart of the converter is a DAC connected to the parallel outputs of an **up counter**. The output of the DAC is compared with the analogue input signal using a **comparator** (a comparator is a device which produces an output of 0 or 1 depending on which of its two inputs is most positive). The output of the comparator is used to generate a 'stop' control for the counter. Initially the counter is zeroed and the counter starts to count up. As it does so, the output from the DAC increases. When the DAC voltage becomes equal to the analogue input signal, the output from the comparator will change state, stopping the counter. This signal is also used to generate a 'conversion complete' control signal. At this stage the digital equivalent of the analogue input signal can be found by reading the parallel output from the counter. When external equipment has read this value, the counter is set to zero and the process begins again.

The counter ADC is one of the simplest forms of converter, but is relatively slow in operation. For each conversion the counter must increment from zero, allowing sufficient

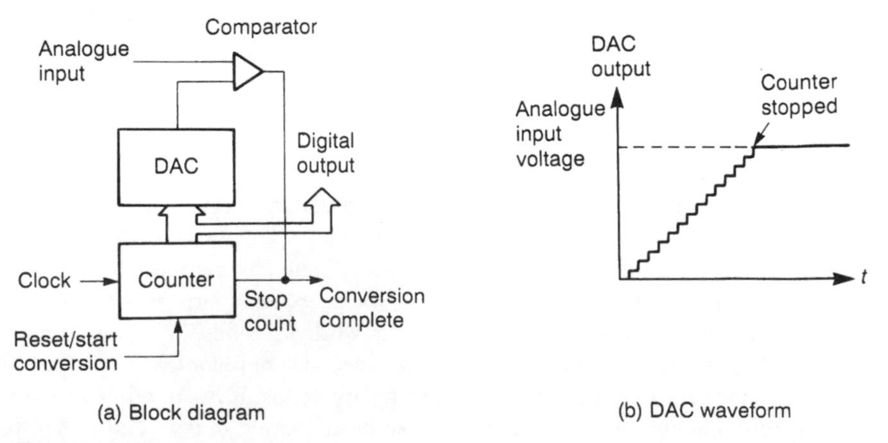

Figure 13.4 A counter ADC.

time after each count for the DAC and the comparator to settle. The conversion time will therefore be at least m times the settling time of the DAC and the comparator, where m is the final digital output value of the converter. For an n-bit conversion this could take as long as 2^n times this settling time. Settling times of the order of a few milliseconds are typical.

A modification of the counter ADC is formed by replacing the up counter with an **up/down counter**. The output from the comparator is now used as an up/down control signal, forcing the counter to track the analogue input signal. This arrangement is called a **servo ADC**.

Successive approximation

The counter ADC is slow in operation since it uses a very inefficient method of searching for the correct value. This is perhaps best illustrated by taking an analogy. Let us suppose that we wish to determine which of the 1000 pages of a dictionary contains a particular word. We could perform this task by looking at the first page, seeing if the word was on this page, and if not moving on to the next. This process would involve us searching progressively through the book until we found the correct page; this technique is similar to that adopted by the counter ADC. A more efficient technique would be to open the book half-way through (at page 500) and looking to see whether the appropriate page was before or after this point. This will locate the page within either the first or the second half of the book and eliminate 500 pages from our search. Let us suppose that we discover that the page we require is before page 500. We would then open the book at page 250 (half-way through the first half of the book) and again see if the required page was before or after this point. In this way we would 'home in' on the required page, reducing the region of uncertainty by 50% each time we open the book. Since 2^{10} is 1024, it will take at most ten attempts to locate the correct page, which is considerably faster than looking at each page.

The successive approximation ADC is similar in many respects to the counter ADC except that the simple counter is replaced by logic circuitry which operates in a manner similar to that described in our dictionary searching analogy. The arrangement is shown in Figure 13.5.

The DAC is driven from a digital word produced by the successive approximation logic. Initially all the bits of this word are set to 0 and then the most significant bit (MSB) is set to 1. This input word is converted by the DAC into an analogue signal corresponding to half of the full range of the DAC. This value is compared with the analogue input signal using a comparator and the result is fed back to the control logic. If the comparison shows that the DAC output is less than the analogue input, the MSB will be left at 1; if not it will be reset to 0. In any event the logic then sets the next most significant bit and again compares the output of the DAC with the input signal. In this way each bit of the input to the DAC is set in turn and its correct state determined. The conversion is completed when all the bits of the DAC input have been set correctly. Therefore, for an n-bit conversion this will take approximately n times the settling time of the DAC and the comparator. This compares favourably with the counter type, which requires up to 2^n times the settling time of the DAC and comparator.

Typical successive approximation converters might have settling times of from 1 to 10 μs for an 8-bit conversion increasing to perhaps 10 to 100 μs for a 12-bit device.

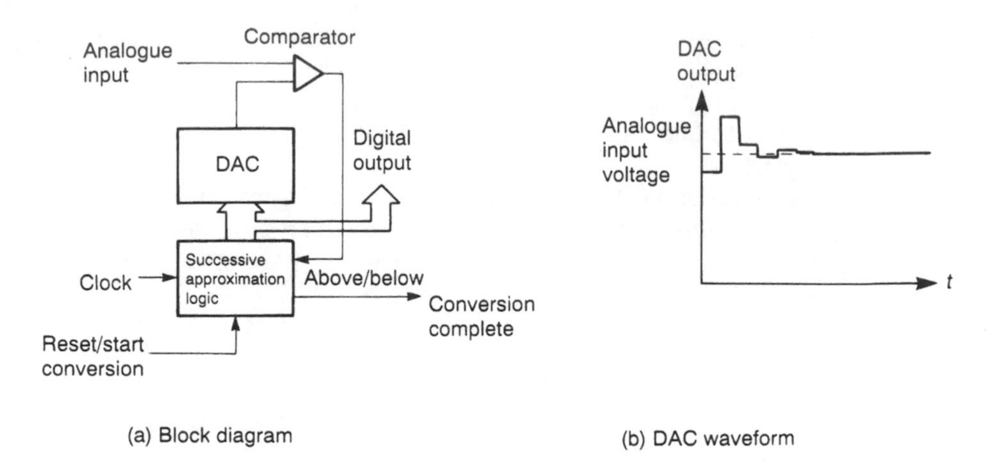

Figure 13.5 A successive approximation ADC.

High-speed variants are available with considerably improved conversion times. The complexity of this form of converter is somewhat greater than that of the counter type. However, its superior speed of operation make it one of the most common arrangements for integrated circuit converters.

Dual-slope

The basic form of this ADC is shown in Figure 13.6.

An operational amplifier is used to integrate the input signal for a fixed period of time, producing a charge on the integrator's capacitor which is proportional to the input

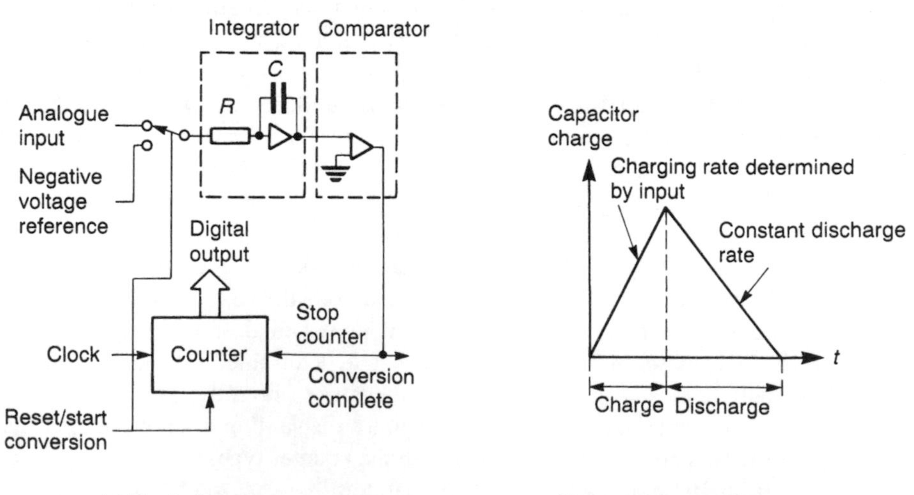

(b) Capacitor waveform

Figure 13.6 A dual-slope integrator ADC.

voltage. The integrator is then connected to a constant current source which discharges the capacitor at a constant rate. The time taken to reduce the charge to zero is measured by counting the cycles of a stable clock oscillator, this time being proportional to the charge on the capacitor and hence the input voltage.

The dual-slope technique has the advantages of high accuracy and low cost, and is often used in such applications as digital panel meters. It is also used where a very high resolution is required and can give a resolution of better than 20 bits, if necessary (a 20-bit conversion represents a resolution of better than one part in one million). The speed of conversion is relatively slow, with high resolution devices producing only perhaps 10 to 100 conversions per second.

Parallel or flash

The parallel, or flash, converter is the fastest of the various forms of ADC. It operates by having a separate comparator to compare the input voltage with every discernible voltage step within the converter's range. The arrangement is illustrated in Figure 13.7.

The various voltage steps are produced using a precision resistor chain from a reference voltage source. Each voltage increment is connected to a separate comparator which compares it with the input voltage. The result is that all of the comparators

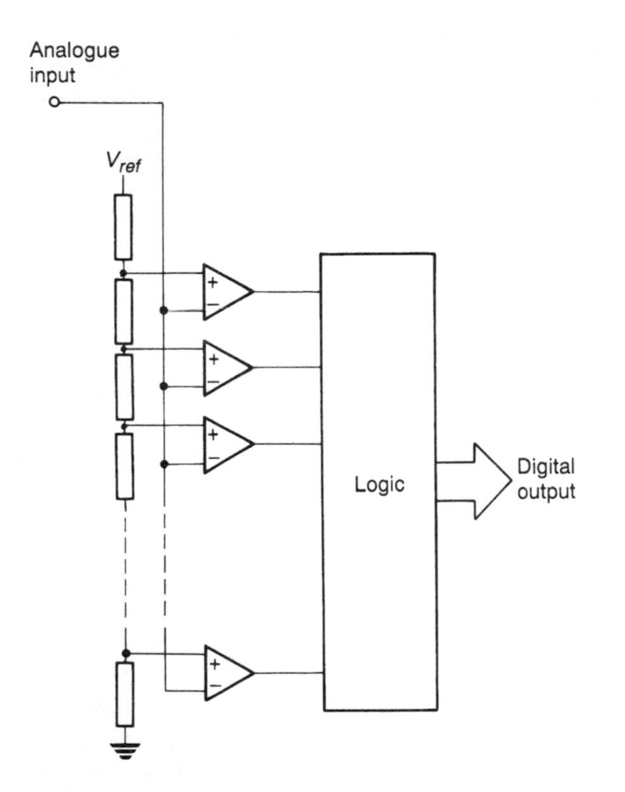

Figure 13.7 A parallel, or flash, ADC.

connected to points along the resistor chain which have voltages greater than the input voltage will produce an output of one polarity, whereas those connected to voltages below the input voltage will produce voltages in the opposite sense. Combinational logic is then used to determine the value of the input voltage from this pattern.

The great advantage of this method is its high speed of conversion, all the comparisons being performed simultaneously. This allows sample rates in excess of 150 million conversions per second with conversion times of only a few nanoseconds. However, since an n-bit converter requires 2^n comparators, the hardware is significantly more complex, and hence expensive, than other techniques.

13.3.3 Sample and hold gates

With rapidly changing quantities it is often useful to be able to *sample* a signal and then *hold* its value constant. This may be required when performing analogue to digital conversion, so that the input signal does not change during the conversion process, upsetting the operation of the converter. It may also be necessary when performing digital to analogue conversion, to maintain a constant output voltage during the conversion period of the DAC.

We have already come across a circuit to perform this function in the form of the **sample and hold gate** described in Example 6.6. These gates can be constructed using discrete components or, more commonly, in integrated circuit form. Typical integrated components require a few microseconds to sample the incoming waveform, which then decays (or **droops**) at a rate of a few millivolts per millisecond. Higher speed devices, such as those used for video applications, can sample an input signal in a few nanoseconds, but are designed to hold the signal for a shorter time. Such high-speed devices may experience a droop rate of a few millivolts per microsecond.

13.4 Multiplexing

Although it is quite possible to have a system with a single analogue input or a single analogue output, it is more common to have multiple inputs and outputs. Clearly, one solution to this problem is to use a separate converter for each input and output signal, but often a more economical solution is to use some form of **multiplexing**. The principle of signal multiplexing is illustrated in Figure 13.8.

A number of analogue input signals can be connected to a single ADC using an **analogue multiplexer**. This is a form of electrically controlled switch based on the use of **analogue switches**, as discussed in Section 6.6.3. Each analogue signal is connected in turn to the ADC for conversion, the sequence and timing being determined by control signals from the system. This is illustrated in Figure 13.8(a).

For certain applications the arrangement of Figure 13.8(a) is unsuitable, since each analogue input signal is sampled at a different time. This may make it impossible to obtain detailed information as to the relationship between the signals, such as their phase difference. The problem can be overcome by sampling all the inputs simultaneously using a number of sample and hold gates, as shown in Figure 13.8(b). Once the input signals

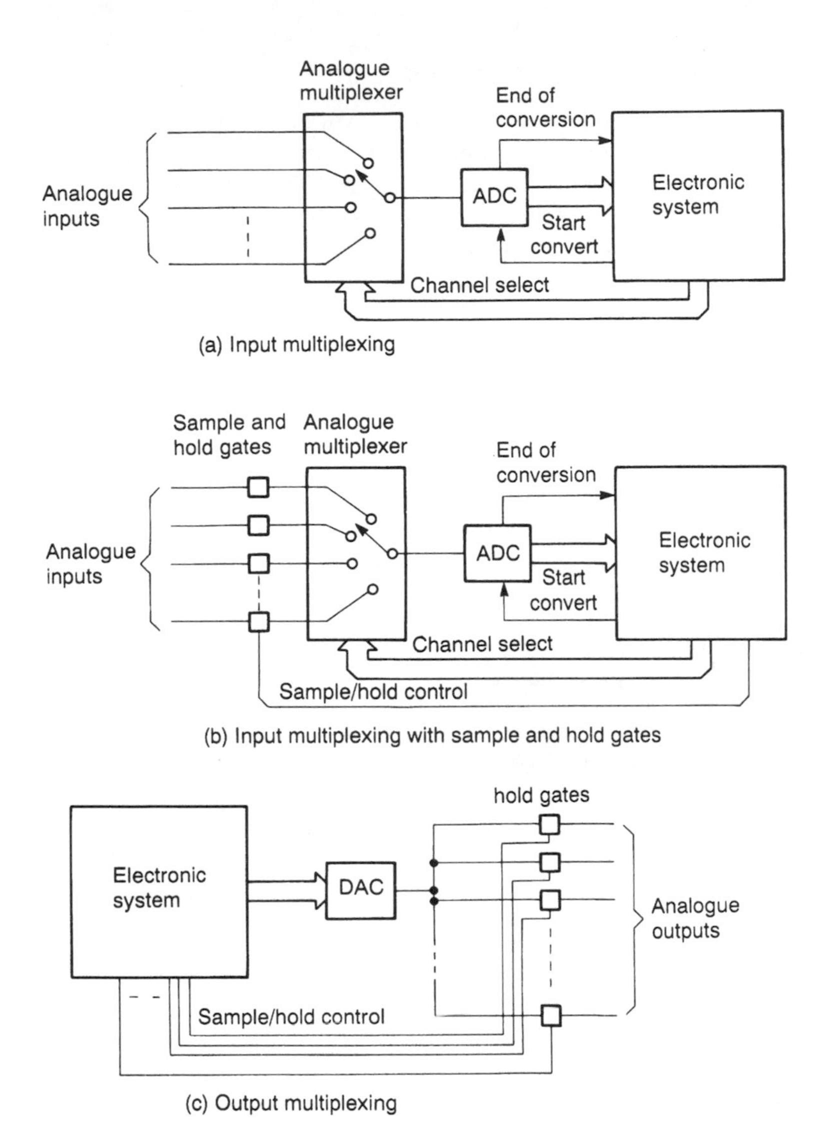

Figure 13.8 Input and output multiplexing.

have been sampled they can be read sequentially without losing the time relationship between the channels.

Figure 13.8(c) shows an arrangement whereby a number of output channels are produced from a single DAC. Here the converter is sent data relating to each channel in turn. When the conversion is complete, a control signal is used to activate the appropriate sample and hold gate. The gate samples the output from the DAC and reproduces this value at its output. The system sets the values of each output channel in turn, updating the values as necessary. Filters would normally be used to smooth the output signals to remove the effects of sampling.

Single-chip data acquisition systems

The combination of an ADC and a multiplexer within a single integrated circuit is often termed a **single-chip data acquisition system**. Although a slight exaggeration, the combination of the two functions is often convenient. These components, which are often specifically designed for use with microprocessor based systems, provide all the control lines necessary for easy interfacing.

Key points

- The extensive use of digital techniques for information processing, storage and transmission means that conversions between digital and analogue quantities are widely used.
- Converting an analogue signal into a digital form is achieved by sampling the waveform and then performing analogue to digital conversion.
- As long as the signal is sampled at a rate above the Nyquist rate, no information is lost as a result of the sampling operation.
- Digitization is always an approximation and so effectively adds noise to the signal. This noise is related to the resolution of the conversion performed.
- When sampling signals that have a broad frequency spectrum it is necessary to use anti-aliasing filters to remove components that are at frequencies above half the sampling rate.
- Several forms of DAC and ADC are available. These differ in their resolution, accuracy, speed and cost.
- Sample and hold gates may be required to hold an input signal constant while it is sampled, or to hold an output signal constant between the times at which it is updated.
- In systems with a number of analogue inputs or outputs, multiplexing can be used to reduce the number of data converters required.

Design study

A microcomputer system is required to sense eight analogue input signals and to produce eight analogue outputs. The inputs come from sensors that produce useful signals with a bandwidth of 1 kHz, but which are known to pick up higher frequency noise. These signals are to be measured to an accuracy of at least 1%. The output signals are to drive actuators with a maximum operating bandwidth of 100 Hz but which are affected by higher frequency signals. The actuators require signals to an accuracy of at least 1%.

Ignoring the need for any amplification or isolation, design an analogue input/output arrangement suitable for this system.

Approach

A block diagram of a suitable system is shown below.

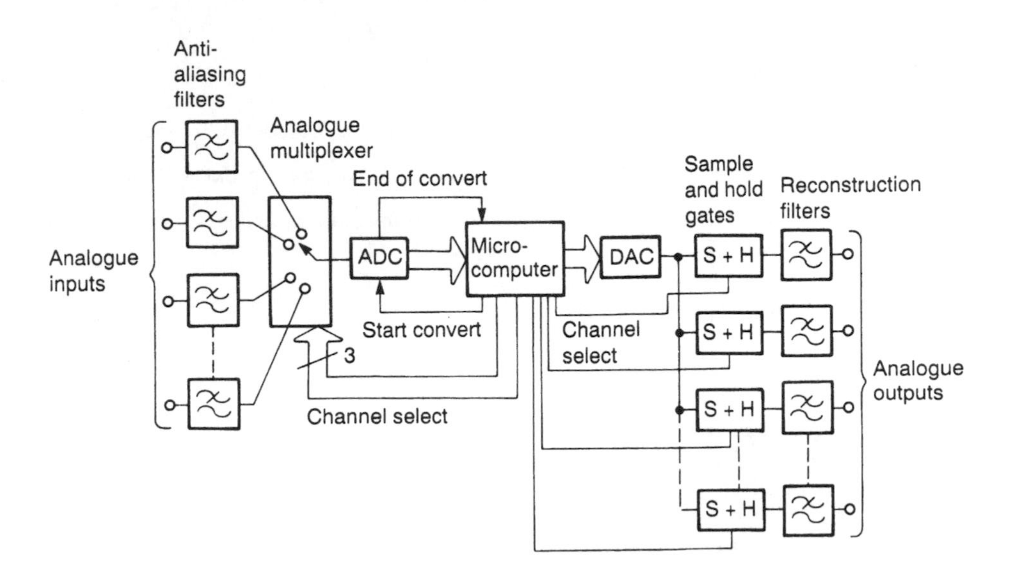

An analogue multiplexer selects one of the eight input signals and applies it to an ADC. The converter needs two control lines from the microcomputer. One of these is an output from the computer which is used to instruct the ADC to begin a conversion (the **start converting signal**) and the other is an input which allows the ADC to tell the micro when it has completed the conversion and is ready for more data (the **end of conversion signal**). The multiplexer requires three control lines from the processor to select one of the eight input channels.

Since the sensors have a useful signal bandwidth of 1 kHz the converter will need to sample each channel with a sampling rate of at least twice this frequency. However, the presence of high frequency noise necessitates the use of **anti-aliasing** filters to remove this noise. Assuming that a slight attenuation near 1 kHz is acceptable, it would be appropriate to use **low-pass filters** with a cut-off frequency of 1 kHz. Sixth order **Butterworth filters** would be a typical choice. Since these are not ideal filters it is necessary to sample at somewhat above the **Nyquist rate**. An increase of 20% gives a sampling rate of 2.4 kHz. If each channel is to be sampled at this rate, the ADC must be capable of 8×2.4 kHz = 19.2 kHz, which corresponds to a conversion time of 52 μs. This is well within the capability of a general purpose successive approximation ADC.

In order to achieve an accuracy of 1%, 7-bit resolution is required. In fact, most general purpose ADCs and DACs have a resolution of at least eight bits, so it would seem sensible to use such devices for the input and output.

The eight output channels are obtained from a single DAC using a series of **sample and hold gates**. These are individually controlled by lines from the processor, although if input/output lines were in demand, these could be produced using a '3-line to 8-line decoder'. The outputs from the sample and hold gates would have step changes of voltage as the outputs were updated. These fast transitions represent high frequency components in the output which may upset the operation of the actuator. It is therefore necessary to **low-pass filter** the output signals to remove these unwanted components. As with the input filters, these **reconstruction filters** would typically be sixth order **Butterworth filters**, in this case with a cut-off frequency of 100 Hz.

Further reading

Analog Devices (1986) *Analog Devices Conversion Handbook*, 3rd edn.

Analog Devices (1989) *Analog Devices Data Conversion Products Databook*

de Sa A. (1990) *Principles of Electronic Instrumentation*, 2nd edn. London: Edward Arnold

Exercises

13.1 Identify the main components of a typical data acquisition system.

13.2 What is meant by the term 'Nyquist rate'? A signal has a frequency spectrum which extends as high as 2 kHz. What is the minimum rate at which this signal may be sampled to obtain a good representation of its form? What would be the effect of sampling below this rate?

13.3 Explain the terms resolution and accuracy as applied to data converters.

13.4 Discuss the relative advantages and disadvantages of the two methods of digital to analogue conversion outlined in Section 13.3.1.

13.5 Modify the circuit used in Computer simulation exercise 13.1 to investigate the behaviour of the circuit of Figure 13.3. Apply similar input signals and confirm that the circuit functions as expected.

13.6 Explain why successive approximation converters have a superior speed performance to those using a simple 'counter' technique.

13.7 Describe the advantages and disadvantages of parallel analogue to digital conversion.

13.8 Explain the use of sample and hold gates in analogue input/output systems. What is meant by the term 'droop' when applied to such gates?

13.9 What is meant by the term 'single-chip data acquisition system'?

13.10 To achieve good intelligibility, speech transmissions require a bandwidth of about 3.4 kHz. If such a signal is sampled with an accuracy of eight bits, what is the minimum number of bits that must be transmitted every second to communicate a single speech channel?

13.11 Sketch a system to multiplex and demultiplex eight of the voice channels described in Exercise 13.10 on to a single channel.

System Design

Objectives

When you have studied the material in this chapter you should:

- be able to identify the major tasks associated with the design of an electronic system;
- be familiar with a range of alternative system implementation methods, including both analogue and digital techniques;
- be aware of the principal characteristics of the various device technologies and be able to identify the appropriate techniques for a variety of applications;
- recognize the relative merits of programmable and non-programmable systems and be able to suggest an appropriate strategy for a given task;
- have a knowledge of a wide range of electronic design tools for both analogue and digital systems;
- be aware of the uses of system description languages and formal methods for system specification and design.

Contents

14.1 Introduction

In Chapter 1 we looked in very simple terms at the process of system design and outlined some of its major components. In this chapter we will look in more detail at the topic and, in the light of the material covered in the intervening chapters, attempt to chart a course through this rather ill-defined area.

A good design is one that solves a particular problem in the most appropriate and efficient manner. To achieve this the designer requires not only a good understanding of the problem but also a wide ranging knowledge of available techniques and technologies. Inevitably this means that design ability increases with experience, but this should not be seen as reducing the importance of a systematic and methodical approach. Design is a creative process but must be based on sound engineering principles to achieve a result that is both cost effective and efficient.

A range of automated tools is available to aid the designer at each stage of a project. These include packages that allow circuit diagrams to be drawn easily and quickly; simulate the operation of the circuit for testing; produce layouts for either printed circuit or VLSI implementations; and verify that these layouts follow a series of design rules.

14.2 Design methodology

In Chapter 1 we identified a number of components of the process of system design. We will now return to expand on these areas and pull together strands from other chapters.

Customer requirements

The customer requirements represent the problem which the system is to solve. In some circumstances the customer and the designer may be one and the same person, but the principle remains the same. The requirements of the system are usually, and correctly, expressed in terms of the problem rather than in terms of the solution. It should be noted, however, that the customer requirements represent the system's *actual* requirements, rather than any verbal or written description of them.

Top-level specification

The top-level specification is an attempt to define a system that will satisfy the customer's requirements. The definition is usually in the form of a written description in a natural language (the term 'natural' is used here to distinguish it from a programming or other computer language), but may include appropriate mathematical equations or expressions. We shall see later, when we come to look at electronic design tools, that there are other more precise methods of defining a system. The problem with a written specification in a natural language is that it is extremely difficult to write in a way that is not open to misinterpretation.

As discussed in Chapter 1, it is important to ensure that the specification defines *what* the system is to do, not *how* it is to do it. Such topics as the appropriate device family, or the use of analogue or digital techniques, do not fall within the realms of the specification.

In large companies it is normal for the specification to be produced by a different team from the one that will ultimately perform the design. This is done to preserve the independence of these two functions. When complete the specification should be agreed by all concerned, including, if appropriate, the customer. Once the specification has been finalized, work may begin on the next stage of the project. It should be noted that occasionally it may be necessary to make modifications to the specification during the project, in the light of new information. There is no reason why the specification should not be revised, provided that it is done with the agreement of all interested parties. However, it must not be done unilaterally by the designer simply to make his job easier.

In addition to providing a specification for the system, it is common at this stage to define a series of tests which the resultant system must perform to prove its suitability for the task in hand. These tests will then form the basis of system testing and the ultimate demonstration of the system.

Top-level design

Once the specification of the system is complete, the design stage may be commenced in a **top-down** manner. For large projects the first task is usually to divide the system into a number of manageable modules. A specification is then produced for each module, enabling it to be designed and tested independently.

One of the earliest design decisions to be made concerns the choice of technology. Invariably it will be possible to implement a given function in a number of ways, using, for example, analogue, digital or software techniques. Often large systems will include elements of each of these methods. Until these decisions have been made it is usually impossible to progress to more detailed aspects of the design. This topic is covered in more detail in Section 14.3.

For systems that are to include programmable devices, such as microprocessors or microcontrollers, it is also necessary to perform a **hardware/software trade-off** to decide which functions are to be performed in hardware and which in software. It was noted in Chapter 12 that this procedure requires a knowledge of the predicted volume of production of the system.

The top-level design results in a block-diagram form of description of the system and a specification of each block of hardware and software. Based on this information, work can move to progressively greater levels of detail.

Detailed design

If the top-level design has been performed efficiently, the detailed design stage of the project should be relatively straightforward. Each hardware section will consist of a series of functions which can usually be assembled from standard circuit building blocks, as described in earlier chapters. Each software section will require the provision of a segment of computer program. Again standard functions and structures are available to simplify this task.

A range of automated tools is available to help the designer at all stages of the project. These are discussed later in this chapter.

Module testing

When the design of a system is complete, the various modules must be constructed and tested to ensure that they perform their required functions correctly.

Unlike design which is performed in a top-down direction, testing is *usually* performed using a **bottom-up** approach. This involves first verifying the operation of each individual circuit element, and then investigating the functioning of progressively larger subsystems. Testing is performed in this way since both error detection and fault location are easier when dealing with small sections of circuitry rather than a complete system. Any faults found at this stage must be rectified before continuing to system testing.

System testing

When each module has been tested and any corrective work performed, the complete system may be assembled and tested. Only at this stage is it possible to see whether the system meets its top-level specification, and to confirm that it does indeed fulfil the customer's requirements. It will normally be necessary to demonstrate the system to the customer and to perform any prescribed system proving tests.

In systems where incorrect operation could have serious safety or financial implications, the tests required for the system will often be very stringent, and it will often be necessary to demonstrate that the system is 'safe' or is of 'high integrity'.

14.3 Technological choice

One of the crucial decisions to be made in the design of any system is the choice of technology. In the broadest sense this could involve implementing a system using mechanical, hydraulic, pneumatic, electrical or electronic means, but in this text we are primarily interested in the choice between various forms of electronic circuit. The major options include the use of analogue or digital techniques, programmable or non-programmable methods, and bipolar or FET devices. We have already noted that it is normal to subdivide a large project into a number of more manageable modules. Clearly, there is no reason why each module should adopt the same approach, and indeed it is common for a combination of techniques to be used even within the same section of a system.

Figure 14.1 indicates some of the alternative approaches to the design of a particular circuit. One of the most fundamental decisions is whether the solution is to be of an analogue or a digital form. If an analogue solution is chosen it is then necessary to decide on either a discrete or an integrated approach. If a digital system seems more appropriate, then it is necessary to select either a programmable technique, as in the case of a microcomputer system, or a non-programmable implementation. In either case, several options are available to the designer.

Unfortunately it is difficult to provide hard and fast rules as to when one solution is more appropriate than another. A more useful approach is perhaps to identify some of

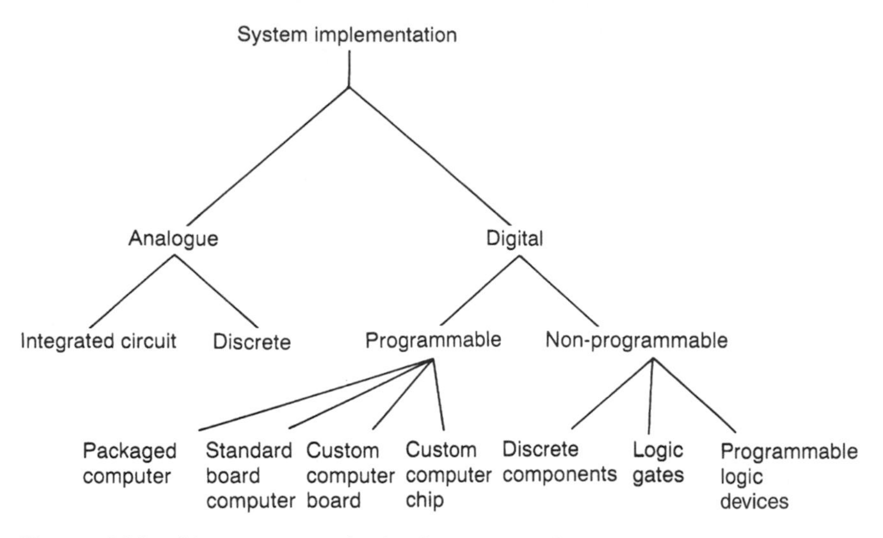

Figure 14.1 Alternative methods of system implementation.

the characteristics associated with the various options to allow the designer to assess which method would be appropriate for a given application.

Analogue vs digital

Often the choice between an analogue and a digital solution to a particular problem is indicated by the nature of the input and output signals. Clearly, if a system uses only binary sensors and actuators then a digital approach is indicated, whereas a system with analogue inputs and outputs would suggest an analogue technique. However, although it is uncommon to use an analogue method for a system with purely digital inputs and outputs, it is quite common to use digital techniques in applications that use analogue signals. A disadvantage of this latter approach is that **data converters** must be incorporated to translate signals between analogue and digital forms. However, in some cases the advantages of digital processing make this worthwhile.

In general terms one could say that the potential advantages of a digital system are that it offers improved consistency between units and a better noise performance. The storage and transmission of digital signals is easier than for analogue quantities and complex signal processing can be performed more easily. Because of the improved predictability of digital systems, design is also easier and there is a greater range of automated design tools available.

If we compare an analogue implementation of an application with a programmable system based on a microcomputer, then there are further aspects to be considered. One might identify the advantages of a computer based system as:

- improved consistency
- greater flexibility through programmability

- standard hardware
- reduced component count
- lower unit cost (sometimes)
- improved testing and the ability to include self-testing
- the opportunity to add more features
- greater reliability
- the possibility of providing automatic, or simplified, calibration.

Against these advantages we should list the potential disadvantages of the computer based approach, which are:

- higher development cost
- higher investment required in development equipment
- greater system complexity.

Integrated vs discrete

The choice between the use of integrated circuits and discrete components when producing analogue circuits must be based on a number of factors, including function, noise, power consumption, cost, size and design effort. In general using ICs is easier, since the chip designer has done much of the hard work for you. However in some very simple, or very specialized applications, discrete components may be more appropriate. Often high power circuitry must be implemented using discrete transistors.

Programmable vs non-programmable

In simple applications, non-programmable solutions are preferable since they require no software development and therefore have a lower development cost. However, as the complexity of the system increases, the potential advantages of the use of a programmable (that is, computer-based) approach become considerable. One of the greatest advantages of computer-based systems is their flexibility. This allows a single standard computer board to be used for a range of applications, reducing the range of subsystems that must be produced. This saves on both design time and inventory costs (the cost of holding stocks of components). It also allows the operation of a system to be updated simply by changing its operating program without having to redesign the hardware. Set against these advantages are the high cost of software development.

As the performance and capabilities of PLDs increase, much of the flexibility of computer-based systems is becoming available within non-programmable systems. In systems that are based on PLDs much of the design complexity is implemented within the logic device. This enables the functionality of the system to be modified simply by changing the PLD configuration. This allows such systems to be upgraded in much the same way as a computer-based system – often without modifications to the hardware. When complex PLDs are used in this way the task of configuring the device is actually very similar to that of the production of software. The process uses an array of complex development tools and can be very time consuming and expensive. One can view microprocessors and PLDs as similar devices that each implement a potentially complex set of instructions defined by the programmer. The primary difference between the two approaches is that in a computer they are executed in a serial manner, while in a PLD they are executed in parallel.

Beyond a certain level of complexity the cost of implementing systems using non-programmable logic becomes prohibitively expensive. In such cases a computer based system is the only practical solution, allowing very complex control algorithms to be constructed in software, rather than by adding more complicated hardware.

Implementing programmable systems

Computer based systems can be implemented in a number of ways, the strategy adopted depending to a great extent on the volume of production. Low-volume projects will tend to favour the use of ready made systems, reducing the need for a large amount of expensive design work. Higher volume applications, on the other hand, will tend to demand a custom approach with a considerably greater amount of design effort being used to produce specialized circuit boards. For very high-volume projects it may be appropriate to have **custom integrated circuits** produced. This approach yields a very low unit cost, but is associated with extremely high development costs.

Implementing non-programmable systems

With non-programmable digital systems the method of implementation is likely to be determined by the complexity of the functions to be produced. Where very limited logical operations are required it may be possible to produce these using simple discrete circuits based on diode logic or the use of a small number of transistors. However, for functions of all but the simplest form, the use of more conventional logic circuits is normal. Applications that require only a handful of gates will normally use standard TTL or CMOS logic devices. Unfortunately, even relatively simple logic arrangements can produce circuits requiring many devices and it soon becomes economical, in terms of both cost and space, to use some form of array logic.

14.3.1 Device technologies

Having decided on the method of implementation for a particular system, or module, it is then necessary to consider the **device technology** that will be used to produce it. Figure 14.2 outlines some of the major choices to be made.

In both analogue and digital systems one of the major decisions to be made is the choice between circuits using bipolar transistors and those based on FETs.

Bipolar vs FET in analogue systems

The characteristics of FETs and bipolar transistors (as discussed in Chapters 6 and 7) are sufficiently different for them to be used in many circuits where they are not

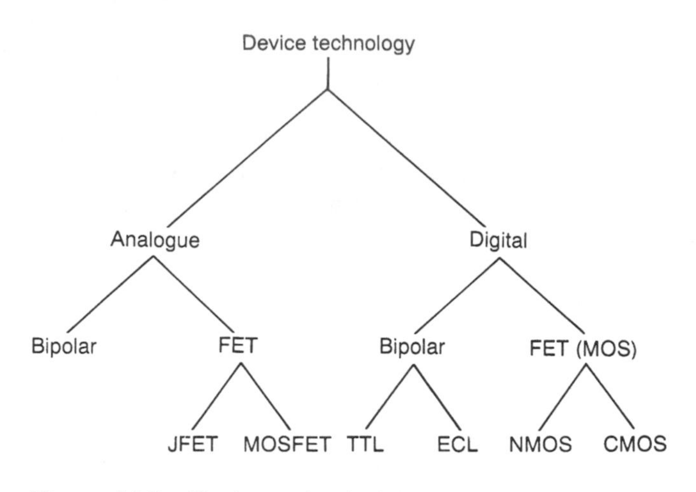

Figure 14.2 Device technologies.

interchangeable. In such cases the choice of component is clear. However, in some applications it is possible to use either device technology and so a decision must be made between them. In general FETs are used in applications where a high input impedance is required, and they can provide a good noise performance in such cases. Bipolar transistors have a lower input resistance but can often produce a much higher gain, and have superior noise performance when used with low impedance sources.

Bipolar vs FET (MOS) in digital systems

In digital systems the choice between bipolar circuitry and that based on FETs (the latter normally being described as MOS devices when considering digital components) is made largely on considerations of speed, power consumption, noise immunity and component density. These considerations have been discussed in some detail in Chapter 11.

The bipolar logic families, such as TTL and ECL, are usually faster in operation than MOS devices, with the non-saturating families, such as ECL and Schottky TTL, being faster than saturating types.

The price paid for the high speed of operation of bipolar devices is a very much greater power consumption than is associated with their MOS counterparts. Both NMOS and CMOS circuits have low power consumptions with CMOS being extremely low and being the natural choice for any battery-operated system.

MOS circuits also have a better noise immunity than bipolar devices, with CMOS gates being able to tolerate noise of at least 30% of the supply voltage. Since these circuits can also be used with a 15 V supply rail, this gives a noise immunity of about 4.5 V compared with about 0.4 V for TTL and somewhat less for ECL. It should be noted, however, that the high input impedance of MOS logic makes it less impressive in terms of its noise performance than these figures might suggest.

Another important characteristic of the MOS technologies is their very high circuit densities when implemented in integrated circuit form. This allows far more circuitry to be combined within a single device than is possible within a bipolar part. For this reason most microcomputers and their associated memory and support devices are constructed using NMOS or CMOS techniques. For high-speed applications bipolar microprocessors are available, but these are generally very expensive and are not as sophisticated as MOS types.

14.4 Electronic design tools

A vast range of computer based electronic aids are available to simplify and speed the process of design. These come under the general title of **computer aided design** (CAD) **tools**, but may also be referred to as **electronic computer aided design** (ECAD), **computer aided engineering** (CAE) or **computer aided software engineering** (CASE) utilities. It is not within the scope of this text to describe the use of these packages, or even to discuss in detail their functions. It is, however, relevant to look in general terms at the overall characteristics of the various forms of tools.

Although many of the functions provided by individual packages are distinct, it is often advantageous to use a number of tools in succession. For this reason it is normal to adopt certain standard forms for the data produced and accepted by the packages, in an attempt to make their data interchangeable.

14.4.1 Schematic capture

A schematic capture package is to the circuit designer what a word processor is to the writer. It allows circuit diagrams to be drawn quickly on to a computer screen using both keyboard commands and input from a pointing device such as a mouse or graphics tablet. Components may be added, deleted or moved, and may be joined as appropriate to produce the desired circuit. However, the system would be of only limited use if it were simply a graphics package capable of drawing objects on the screen. The power of the technique comes from the use of **component libraries** within the package. These store the technical details of each component including its **pin-out** and its circuit symbol. The designer can select components from the library as required, and position them within a circuit, without having to draw each one laboriously. Most packages come with an extensive range of standard components which will meet most needs and which are updated as new components become available. It is also normally possible to add new components by drawing the required circuit symbol and adding the appropriate information to the library.

The output from the schematics capture package may be in a number of forms. One of the most obvious is that it can produce a 'hard-copy' of the finished diagram on a printer or plotter. It can also provide a **components list** and a **net list** which defines their interconnection. This information can be used as input to some of the other packages described below.

An example of a schematic capture package is the 'Schematics' utility within the PSpice design suite. This provides a huge library of standard components, and also allows new parts to be defined. The program can produce a standard net list file that is compatible with other Spice programs, and circuit files that can be used by other programs within the PSpice design suite.

14.4.2 Circuit simulation

Having designed a circuit and produced a computer representation of it using a schematic capture package, it is then necessary to find out if it works as required. Obviously one approach to achieving this is simply to build the circuit and test it. This is, however, a time-consuming, and often inaccurate, method, and can result in a waste of time and effort. A more attractive solution is to **simulate** the circuit, so that any required modifications can be made before construction.

A wide range of circuit simulation packages is available, although probably the best known is **SPICE** (Tuinenga, 1995). This is a computer aided simulation program, the name of which stands for **S**imulation **P**rogram with **I**ntegrated **C**ircuit **E**mphasis. It was originally developed by the Electronic Research Laboratory of the University of California and became available in 1975. The original version of the program produces only numeric and textual output and it is now common to use more modern packages which operate in a similar manner but present the output in a more meaningful graphic format. A number of such tools are available and they often include the word SPICE within their name. Perhaps the best known is the PSpice package produced by MicroSim.

SPICE, and its derivatives, can be used for both analogue and digital circuits, for AC and DC analysis, and for both continuous and transient conditions. As with the schematic capture packages described above, these use a **component library** to store the characteristics of each device. As before, new components can be added by the user. The standard libraries include details of a range of active and passive components, including transistors and integrated circuits.

When the circuit has been defined the user can then specify a set of initial conditions and input signals, and the program will simulate the operation of the circuit and display the results. Simulation is an invaluable tool for the circuit designer and is often used interactively during the development process rather than simply to test a completed design.

14.4.3 PCB layout

When the design has been completed, the next stage is to build it. In the case of circuits that are to be implemented in the form of a **printed circuit board (PCB)** this requires the production of a photographic mask which is used to define the positions of the copper tracks and holes on the board. Before the advent of electronic tools, the production of these masks was performed manually by placing strips of opaque tape on to a transparent film. Today all but the simplest boards are laid out using one of a number of **PCB layout packages**.

Layout packages take as their input the **component list** and **net list** produced by a schematic capture package. The program then allows the user to define the dimensions and shape of the board and to position the components. Again a **component library** is used to store the dimensions and **pin-out** of each component, and, as before, the user can specify custom components. Some systems can perform **automatic placement** of components, taking into account their sizes and connections.

Most packages then perform **automatic routing** to join up the required pins on the various components. Until recently the automatic routing utilities on these packages were very primitive and could not be relied on to perform the complete task unaided. Using such a system it was normal to position the earth and power supply tracks manually, and then to allow the program to attempt to route the remaining tracks automatically. The user would then often need to complete the task manually to route any connections that the machine could not manage itself. Newer systems have much more powerful automatic routing algorithms which can often perform the complete layout task without human intervention. This can reduce the time taken to lay out a complicated board from perhaps 50 hours to a few minutes.

For initial checking of the design, the layout package can produce its output on a printer or plotter. For final production it is normal to use the output from the package to drive a **photoplotter** which produces its output directly on to transparent film, to form a printing mask for the board.

14.4.4 PLD design and programming packages

In Chapter 11 we looked at the use of **array logic** in the construction of digital electronic systems. Many of these devices can be programmed by the user, these components being given the general term of **programmable logic devices** or **PLD**s. Since these arrays may contain several thousands of gates, the task of selecting an appropriate interconnection pattern is not trivial. Consequently this task is normally delegated to one of a number of automated tools which simplify this task. These packages take as their input a description of the required functions of the device, written in a specification language. This description sets out which of the pins of the PLD will be inputs and which will be outputs, and defines the relationships between them. It also defines a set of **test vectors** which consist of a set of input combinations and the expected outputs. From this data the package generates a **fuse map** which can be loaded into a **PAL programmer** which writes this pattern into a target device. After programming the PLD is automatically tested using the specified test vectors to ensure that it is functioning as required. The process of programming and verifying a device generally takes a few seconds.

14.4.5 VLSI layout

If a circuit is to be implemented as a VLSI component, an alternative form of layout package is required. These have certain similarities with PCB layout programs, but work in dimensions measured in microns (1 micron = 10^{-6} metres) rather than in millimetres. At this level, individual components are assembled from regions of different forms of semiconductor and it is necessary to position these regions, and the layers of metallization which connect them together, with great accuracy. Several distinct regions will be required to form a single transistor, but once this has been designed it may be stored away as a library component to be used again. A number of transistors might then be connected to form a gate, which could also be stored for future use. In this way a **component library** is assembled. In digital designs there is often a great deal of repetition of circuit components, and the use of standard circuit 'cells' greatly simplifies design.

14.4.6 Design verification

In both PCB and VLSI layout there are a number of **design rules** which must be obeyed to produce a circuit that will work reliably. These will govern, for example, the minimum separation between conductors, the minimum thickness of tracks and the relative positions of semiconductor regions. Various packages exist for checking that designs do not breach any of these rules. Sometimes these utilities are provided within a particular layout package. In other cases they are a separate piece of software.

Another aspect of design verification relates to EMC. Conventional circuit simulation packages allow the functional characteristics of a system to be investigated but do not consider its EMC behaviour. Specialist packages allow the EMC performance of a system to be predicted before construction, to verify that this is acceptable. This can save a great deal of time and effort in fine tuning the design.

14.4.7 System specification and description

In addition to the CAD packages outlined above there are a number of computer languages for use in the specification and design of electronic systems. Examples include such languages as **VDM** (Jones, 1991), **ELLA** (Morison and Clarke, 1993), **HILO** (Blundell *et al.*, 1987), **HOL** (Gordon and Melham, 1993) and **Z** (Spivey, 1992). These languages are not used to produce software that will run on a target system to perform a specified task, but rather they are used to define the nature of the system itself. VDM, for example, is essentially a **system specification language** which can be used to describe any system in terms of its inputs, outputs and the relationship between them. This description forms a very precise specification of the system which is independent of the eventual method of implementation. It does not, for example, define whether a particular function will be achieved in hardware or in software.

Once a system has been defined in this way, tools are available to simplify the task of implementing it in hardware, software or a combination of the two. Languages are available to specify circuitry at the logic gate level, or to interface directly to VLSI design packages. Techniques also exist for generating conventional programs to implement the required software aspects of the target system. Of paramount importance in this process is the need to confirm that the circuits and software produced correspond directly to the original top-level specification. Various software tools are available to facilitate this process, although they rely heavily on the intellectual abilities of the user.

Because of their rigorous mathematical basis, these forms of specification and design are often described as **formal methods**. The use of these techniques greatly increases the reliability of the design process and gives much greater confidence in its correctness.

Perhaps the most important **hardware description language** is **VHDL** (Hunter and Johnson, 1996). The name is an acronym standing for VHSIC Hardware Description Language, where VHSIC is another acronym for Very High Speed Integrated Circuits. VHDL can be used for many purposes including the documentation, verification, synthesis, simulation and testing of circuits. It can be used to describe the structure, data flow or behaviour of a system, and can be used in a number of ways within the development process. For example, VHDL can be used to specify a system by defining the functionality of each section, the interactions between these sections and even acceptance criteria for use in testing. Following specification VHDL can be used for design capture, as an alternative to a schematic representation. The use of VHDL in this way allows the system to be simulated using some of the very powerful simulation tools that are available. An increasingly important use of VHDL is in the specification of complex PLDs, as discussed in Section 11.5.7.

Key points

- A good design produces an appropriate solution to a problem.
- The normal method of achieving such a design is the use of a top-down approach.
- This begins with the requirements of a system which are formalized to produce a specification.
- The specification forms the basis of the top-level design. This may involve subdividing the problem into a number of sections.
- When this has been done the detailed design of the hardware and software can begin.
- Top-down design is normally followed by bottom-up testing. This culminates in the testing of the complete system to ensure that it meets its original specification.
- At an early stage it is necessary to decide on the form of implementation to be used. This will involve choosing between a range of technologies.
- A wide range of automated tools are available to simplify the design of both analogue and digital systems. These include packages to perform schematic capture, circuit simulation, PCB layout, PLD design and programming, VLSI layout and design verification.
- A number of specification languages are also available. Of these, perhaps the most important is VHDL.

Design study

The production of safety critical systems is normally the province of highly skilled and experienced engineers, whose work is meticulously checked by independent workers. However, some aspects of fairly conventional designs may have safety implications which must be treated seriously to ensure that they operate correctly.

Consider how an emergency stop button should be interfaced to a microcomputer based machine control system, to ensure its correct operation.

Approach

The following diagram shows four possible approaches to the problem.

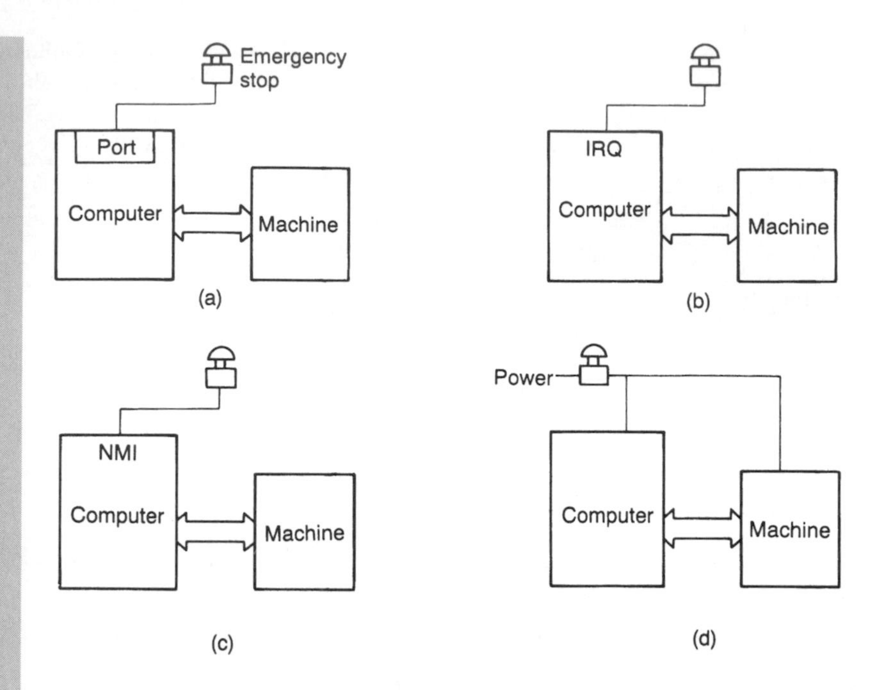

One of the simplest methods of interfacing a switch to a microcomputer is simply to connect it to one of the lines of a parallel input port and to **poll** it periodically. This arrangement is shown in diagram (a). Each time the processor reaches the appropriate point in its control loop, it senses the switch and can take action, if required. In most cases the speed of response of this arrangement will be adequate, since it will normally be short compared with the time taken by the user to detect a problem and to press the button. However, this method is *not* suitable for use with an emergency stop button because of the effects of a **system crash**.

During the operation of a microcomputer system there is always the possibility that a noise spike may cause an error within the execution of the program. This can result in an address being misread, and in some cases can result in the processor jumping to an effectively random address. Because of the large numbers of jump and branch instructions within the instruction set of the machine, it is not uncommon for this to result in the processor becoming stuck in a loop, executing random code. This results in a system crash, since the processor is no longer executing its control program. In this situation any emergency stop button which relied on polling by the processor would become inoperative, since the polling routine would not be executed.

An alternative arrangement is shown in diagram (b). Here the switch is connected to the **interrupt request line** (**IRQ**). The programmer now provides an interrupt service routine which is designed to take appropriate action in the event of the emergency stop button being pressed. The advantage of this method, compared to that given in (a), is that the interrupt mechanism should still function correctly even in the event of a system crash. However, the use of the IRQ line is *not* appropriate since this is usually a **maskable interrupt**, which may be enabled or disabled by machine instructions. Clearly, if this interrupt were being used for this purpose, the programmer would ensure that it was enabled at the beginning of the main program, but a problem again arises in the event of a system crash. Following a crash the processor may fetch bytes from anywhere in memory, and if it happens to fetch a byte corresponding to the instruction to disable interrupts, this will turn off the emergency stop function. This possibility may seem remote, but remember that there are only 256 possible values for each byte fetched.

The problems associated with the masking of interrupts may be overcome by the use of the **non-maskable interrupt** (**NMI**) as illustrated in (c). This operates in a similar manner to IRQ but cannot be turned off, thus providing a mechanism which is sure to operate, even in the event of a system crash. This arrangement provides a solution which is far more acceptable than those outlined in (a) and (b), but is still unattractive from a number of viewpoints.

Microprocessors are extremely reliable components but are still subject to random hardware failures. Unfortunately, because of the complexity of these circuits it is not possible to predict the effects of such failures. To ensure reliable operation it is often necessary to provide hardware and software redundancy, and to employ rigorous design and testing methods (Storey, 1996). In fact, in applications where failure could cause injury, it is not normally acceptable to use only a single processor in a situation where it performs a primary safety function. For these reasons it is normal to ensure that the microprocessor is not directly responsible for the safety of the system. An example of how this may be achieved is illustrated in diagram (d). Here the emergency stop button is connected into the main power supply lead to the system, such that hitting the button removes power from the machine. Under these circumstances it is the emergency stop button which is responsible for providing the safety function, *not* the processor. This removes the need for the controller to be a high integrity system. This last approach is by far the most attractive of the four options described and is the normal method adopted. However, there are some applications where it is not possible simply to remove power from the system in the event of an emergency. In such cases it may be necessary for the microprocessor to play an active role in the shutdown procedure. Under these circumstances great demands are placed on the design and construction of the control system.

References

Blundell G. *et al.* (1987) *An Introduction to Silvar Lisco and HILO Simulators*. London: Macmillan

Gordon M. and Melham T. F. (1993) *Introduction to HOL: Theorem-Proving Environment for Higher-Order Logic*. Cambridge: Cambridge University Press

Hunter R. D. M. and Johnson T. T. (1996) *Introduction to VHDL*. London: Chapman and Hall

Jones C. B. (1991) *Systematic Software Development Using VDM*. Englewood Cliffs, NJ: Prentice-Hall

Morison J. and Clarke A. (1993) *ELLA 2000: A Language for Electronic System Design*. Maidenhead: McGraw-Hill

Spivey J. M. (1992) *The Z Notation – A Reference Manual*, 2nd edn. Englewood Cliffs, NJ: Prentice-Hall

Storey N. (1996) *Safety-Critical Computer Systems*. Harlow, UK: Addison-Wesley

Tuinenga P. W. (1995) *SPICE – A Guide to Circuit Simulation Using PSPICE*. Englewood Cliffs, NJ: Prentice-Hall

Further reading

Jones P. L. and Buckley A. (1989) *Electronic Computer Aided Design.* Manchester: Manchester University Press

Exercises

14.1 List the major tasks associated with the design of an electronic system.

14.2 What is the difference between the customer's requirements and the specification of a system?

14.3 Why is it important that the specification describes *what* a system must do, rather than *how* it must do it?

14.4 What factors affect the decisions made in the hardware/software trade-off?

14.5 Define the terms 'top-down' and 'bottom-up'. Which of these methods is appropriate for design and which for testing?

14.6 Why are individual modules tested separately before a complete system is assembled?

14.7 Compare the characteristics of systems constructed using analogue techniques as opposed to those using a digital approach.

14.8 In digital systems, what factors determine whether a microprocessor should be used rather than a circuit based on non-programmable techniques?

14.9 Explain the use of a schematic capture package and describe how the output from this package may be used in association with other software tools.

14.10 Explain the function and characteristics of the following ECAD tools:
(a) circuit simulation
(b) PCB layout
(c) PLD design and programming
(d) design verification.

14.11 Describe the function of a system specification language. How does this differ from that of a computer programming language?

Appendices

Appendix A

Appendix B

Appendix C

Appendix A Symbols

Below are the principal symbols used within the text and their meanings.

Symbol	Meaning
A_i, A_p, A_v	current, power and voltage gains
β	bipolar transistor DC current gain (equivalent to h_{FE})
$\overline{CE}$	chip enable
f_{co}	cut-off frequency
f_T	transition frequency
G	overall gain
g_m	transconductance
h_{FE}	bipolar transistor DC current gain in common-emitter configuration
h_{fe}	bipolar transistor small signal current gain in common-emitter configuration
h_{ie}	bipolar transistor small signal input resistance in common-emitter configuration
h_{oe}	bipolar transistor small signal output conductance in common-emitter configuration
h_{re}	bipolar transistor small signal reverse voltage gain in common-emitter configuration
I_{CBO}	leakage current from the collector to the base with the emitter open circuit
I_{CEO}	leakage current from the collector to the emitter with the base open circuit
i_b, i_c, i_e	small signal base, collector and emitter currents
I_B, I_C, I_E	DC base, collector and emitter currents
i_d, i_g, i_s	small signal drain, gate and source currents
I_D, I_G, I_S	DC drain, gate and source currents
I_{DSS}	drain to source saturation current
I_n	noise current
I_s	reverse saturation current
$\overline{OE}$	output enable
P_n, P_o, P_s	noise, output and signal powers
r_e	emitter resistance
r_i, r_o	small signal input and output resistances
R_i, R_o	input and output resistances

R_L	load resistance
t_H	hold time
t_{OFF}, t_{ON}	turn-off and turn-on times
t_{PD}	propagation delay time
t_{PLH}	propagation delay for transitions from low to high
t_{PHL}	propagation delay for transitions from high to low
t_S	set-up time
T	time constant
V_A	Early voltage
V_{br}	breakdown voltage
v_{be}, v_{ce}	small signal base–emitter and collector–emitter voltages
V_{BE}, V_{CE}	DC base–emitter and collector–emitter voltages
V_{BB}, V_{CC}, V_{EE}	base, collector and emitter supply voltages
V_{DD}, V_{GG}, V_{SS}	drain, gate and source supply voltages
v_{ds}, v_{gs}	small signal drain–source and gate–source voltages
V_{DS}, V_{GS}	DC drain–source and gate–source voltages
V_H, V_L	voltage representing logic 1 and logic 0
v_i, v_o	small signal input and output voltages
V_i, V_o	input and output voltages
V_{IH}, V_{IL}	input voltage representing logic 1 and logic 0
V_{ios}	input offset voltage
V_n	noise voltage
V_{NI}	noise immunity
V_{OH}, V_{OL}	output voltage representing logic 1 and logic 0
V_{ref}	reference voltage
V_T	threshold voltage
V_Z	Zener breakdown voltage
V_+, V_-	voltages on the non-inverting and inverting inputs of an op-amp
ω_{co}	angular cut-off frequency
$\overline{WE}$	write enable

Appendix B IEC 617 symbols for logic elements

In recent years the **International Electrotechnical Commission** (**IEC**) has published a series of documents defining the ways in which electrical and electronic components should be represented within circuit diagrams. These documents collectively form the **IEC 617** standard, which is divided into several parts covering different circuit elements. Most of the symbols used within the 'analogue' sections of this book are of a form similar to those described within the IEC documents. However, the symbols used for the various logic elements are markedly different from those defined by the standard.

The reason for ignoring the dictates of the international standard is that the recommended symbols do not aid the understanding of simple circuits. Therefore this text uses a range of widely used symbols that are much easier to recognize and understand. Most electronics text books use these symbols.

While this book does not make use of the standardized logic symbols it is useful to be able to understand diagrams based on them. This appendix therefore gives a brief overview of the recommended notation. The data concerned comes from **IEC 617**: Graphical Symbols for Diagrams, Part 12: Binary Logic Elements. In the UK this is also published as **BS 3939** Graphical symbols for electrical power, telecommunications and electronics diagrams, Part 12: Guide for binary logic elements. The symbols used are similar to those given in the American standard **IEEE 91-1984**.

Rather than the distinctive shapes of the symbols used within this book, the standard

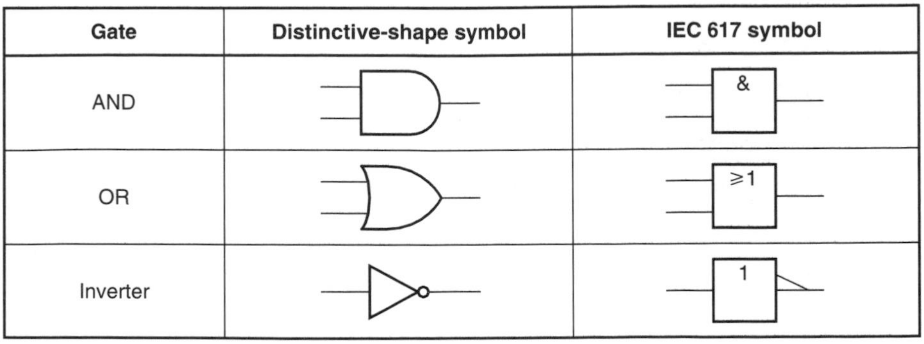

Gate	Distinctive-shape symbol	IEC 617 symbol
AND		&
OR		≥1
Inverter		1

Figure B.1 Examples of symbols for simple gates.

defines symbols that have outlines based on a rectangular grid. Figure B.1 shows a comparison of the symbols used for a few simple gates.

Simple gates normally have a square outline with an inner **qualifying symbol** to define their function. The qualifying symbol indicates the number of inputs that must be at logical 1 in order for the *output variable* to take on a value of logical 1. In an AND gate all the inputs must be at logical 1 in order for the output variable to become logical 1 and so the '&' symbol is used. In an OR gate the output variable will be at logical 1 if one or more of its inputs is at logical 1, and so the qualifying symbol is '⩾1'. Similar application of this rule defines that the qualifying symbol for a non-inverting buffer is '1' and that for an exclusive-OR is '=1'.

An inversion of the polarity of an output signal may be indicated using a 'half arrow' as shown in the symbol for an inverter in Figure B.1. This indicates that the signal that appears on the output pin is the inverse of the value of the output variable as defined by the qualifying symbol. This allows NAND, NOR, inverter and exclusive-NOR gates to be formed by adding half arrows to the symbols for AND, OR, buffer and exclusive-OR gates. Figure B.2 shows the resulting symbols for a range of simple gates.

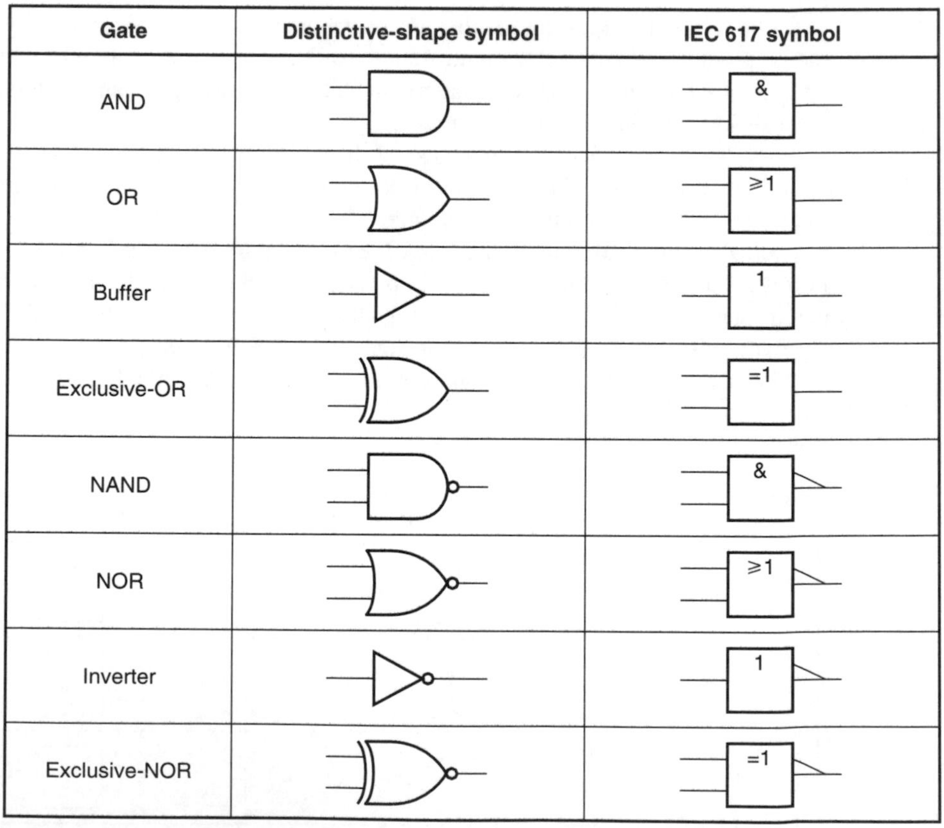

Gate	Distinctive-shape symbol	IEC 617 symbol
AND		&
OR		⩾1
Buffer		1
Exclusive-OR		=1
NAND		&
NOR		⩾1
Inverter		1
Exclusive-NOR		=1

Figure B.2 IEC 617 symbols for simple gates.

The great strength of the IEC symbols is their ability to accurately define the form of the devices concerned. For example, it is possible to indicate not only the logical

Symbol	Device characteristic	Notes
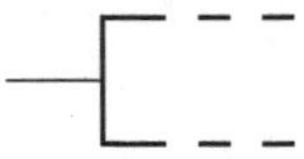	Ordinary input	The internal input variable takes a value of 1 when the input is high
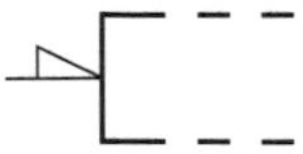	Inverted input	The internal input variable takes a value of 1 when the input is low
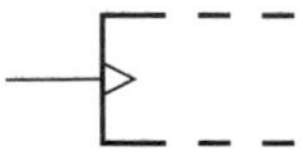	Dynamic input	A positive-going edge sensitive input
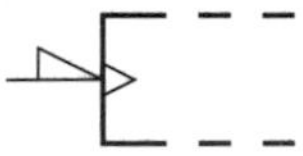	Dynamic input	A negative-going edge sensitive input
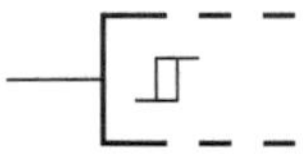	Hysteresis input	Bi-threshold input with hysteresis
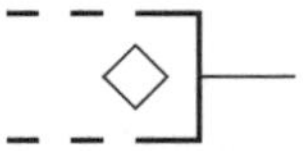	Open-circuit output	e.g. Open-collector, open-drain
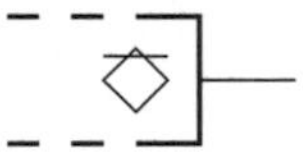	Open-circuit output	Open circuit output (H-type) e.g. open-collector PNP transistor. When not in its high impedance state, gives a low impedance H-state
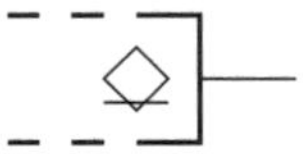	Open-circuit output	Open circuit output (L-type) e.g. open-collector NPN transistor. When not in its high impedance state, gives a low impedance L-state
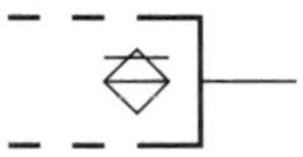	Passive pull-down	Similar to open-circuit (H-type) but does not need external components
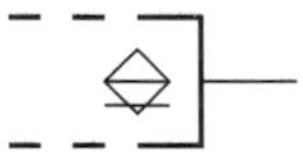	Passive pull-up	Similar to open-circuit (L-type) but does not need external components
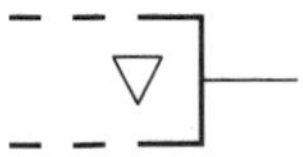	Three-state output	Output can take a high impedance state
	Postponed output	Output change is delayed until the initiating input returns to its initial state (as in pulse-triggered bistables)
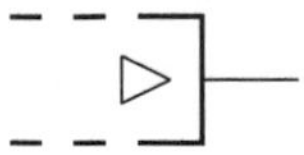	Special amplification	Output with special drive capabilities

Figure B.3 Representation of input and output types.

function of a gate but also the form of the input and output circuitry used. This makes it possible to read and understand the nature of the interactions between devices directly from a circuit diagram without having to refer to device data sheets. The standard gives the symbols for a wide variety of possible features and it would not be appropriate to list all of these within this text. However, Figure B.3 illustrates the symbols used to represent some of the more important device characteristics.

Representation of repeated elements and control blocks

Many components have several identical sections. To simplify the representation of such devices these sections are shown as a stack of elements with only the top element labelled. This is illustrated by the first symbol in Figure B.4. Here two pairs of inputs are first ANDed together, and then the resultant signals are ORed to produce the output variable. This is then inverted to produce the output signal. Such a circuit is called an AND-OR-invert gate.

In some cases, parts have a series of identical sections that are all controlled by one or more common control inputs. This is represented by the use of a common control block

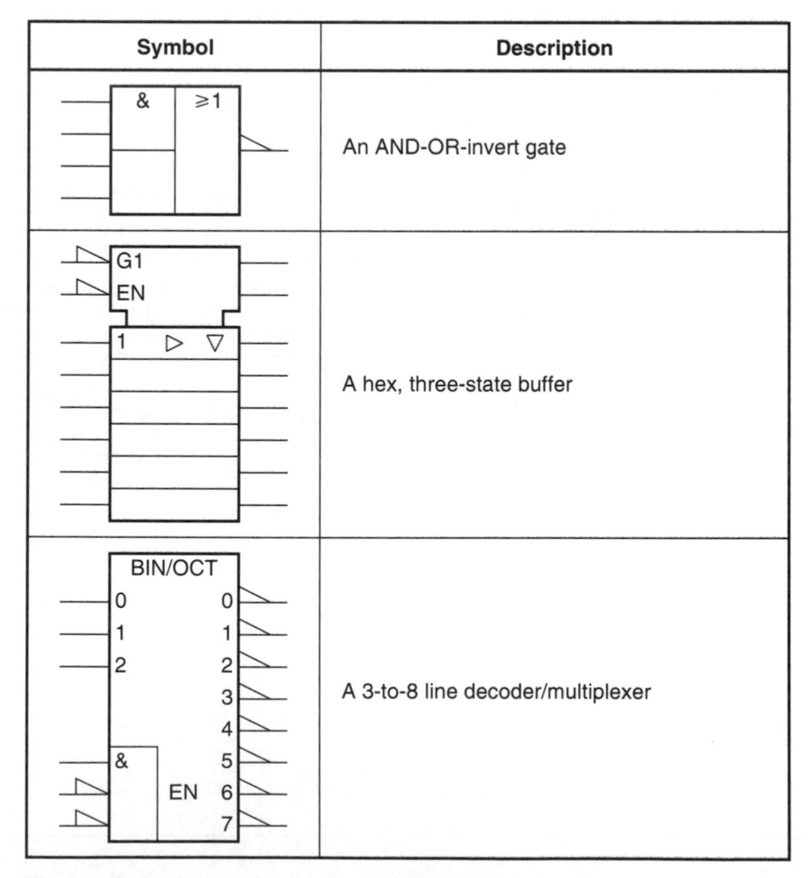

Figure B.4 Representation of repeated elements.

as shown in the second symbol of Figure B.4. This shows a hex buffer which has six identical elements each with a single input and a single output. The topmost element includes the symbol for a buffer and shows that the output is of a three-state type. Each of the six elements is controlled by two control inputs – EN and G1. These are shown connected to a common control block. The EN input controls each of the outputs and these will be enabled when the *internal variable* corresponding to EN is high. Since this signal shows a polarity inversion at its input, the outputs will be enabled when the input *signal* is low. The label of the second control input, G1, includes the number '1' which allows the diagram to indicate which of the other inputs are controlled by this signal. In this case the input to the first buffer element is labelled '1', indicating that this input is controlled by the G1 control input. Since the six buffer elements are identical, this implies that the inputs to all six channels are gated by the G1 control input. We shall see other examples of the use of numbered control inputs in later figures.

Often a number of inputs are used together to specify an address or to select one of a number of input or output lines. An example of the notation used to represent such an arrangement is shown in the third symbol of Figure B.4. Here the outputs are numbered from 0 to 7, and three inputs are labelled 0, 1 and 2. The numbers of the inputs represent powers of two and indicate their weight when selecting one of the outputs. This symbol also shows how internal variables may be defined in term of inputs or other terms. Here an enable signal EN is derived by ANDing together three input terms. The internal variable EN is then used as an enabling signal for the outputs.

Representation of bistables

The various forms of bistable input are represented by labelling the input with a qualifying letter (R, S, J, K, D or T). This, combined with the use of appropriate notations for the clock input and the outputs, allows a wide range of bistables to be defined unambiguously. Some examples of bistable devices are shown in Figure B.5.

Symbol	Description	Notes
S R	RS bistable (RS latch)	Both inputs active high
S 1D >C1 R	Edge-triggered D-type bistable	Clock input C1 is positive edge triggered and controls the D input since this is labelled '1D'. Labels on S and R are not preceded by a 1 and so are not controlled by the clock. Hence these are direct set and clear inputs.
1J >C1 1K R	Edge-triggered J-K bistable	Clock input C1 is negative edge triggered and controls both the J and K inputs since these are labelled 1J and 1K. Input R is active low and is not controlled by the clock. It is a direct reset input.

Figure B.5 Representation of bistable devices.

Symbol	Description
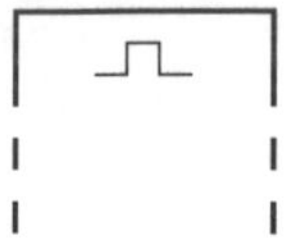	A retriggerable monostable
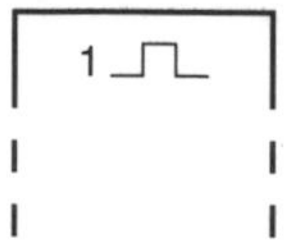 1	A non-retriggerable monostable
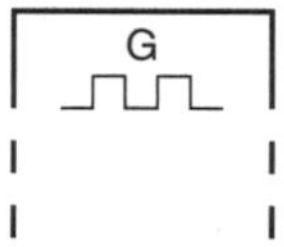 G	An astable

Figure B.6 Representation of monostables and astables.

Representation of monostables and astables

The qualifying symbols used to represent monostables and astables are shown in Figure B.6. The letter 'G' in the symbol for an astable is the qualifying symbol for a generator.

More complex devices

The notation described within the standard may be used to represent even quite complex devices such as PLDs, microprocessors and memory chips. The resultant symbols are often complex and use elements of the standard that have not been discussed within this brief description. Figure B.7 shows a couple of examples of the representation of complex parts. These symbols are included without further explanation simply to illustrate the scope of the notation.

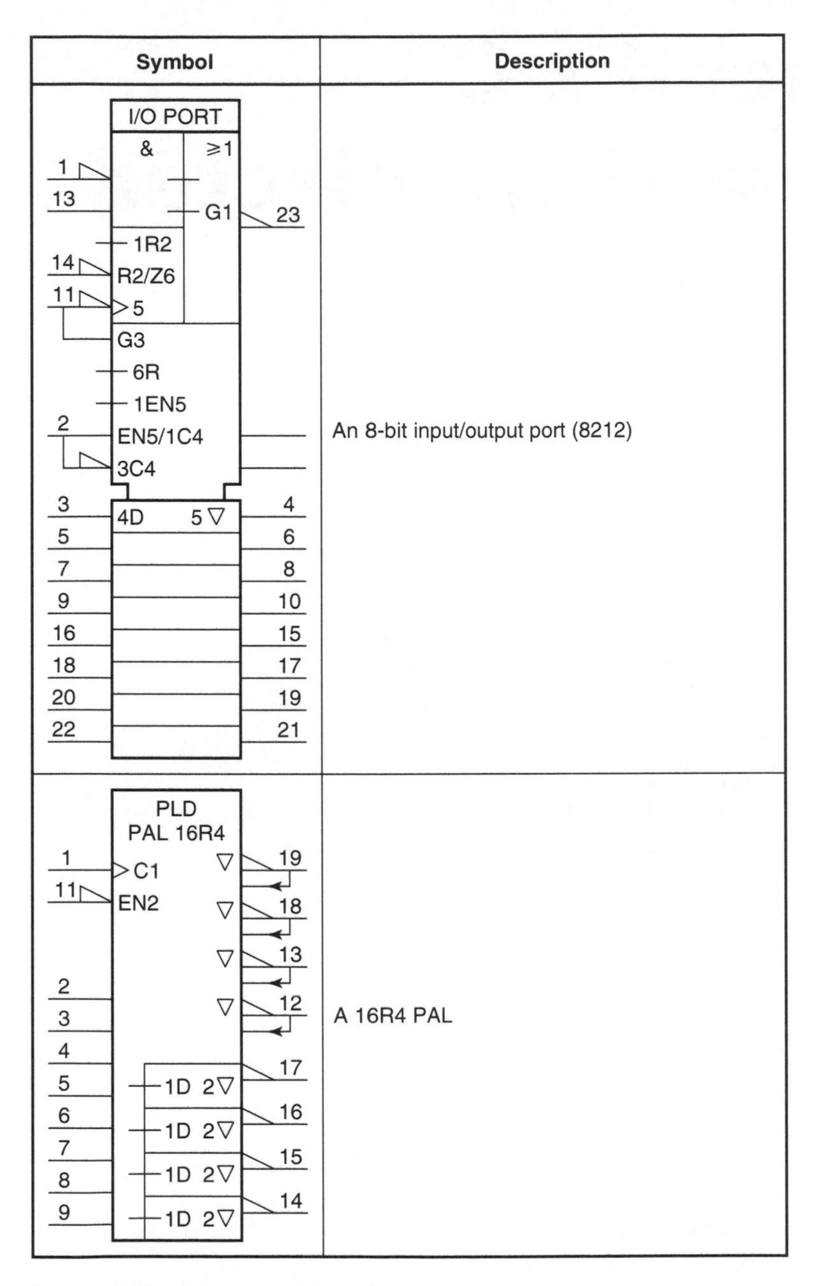

Symbol	Description
I/O PORT	An 8-bit input/output port (8212)
PLD PAL 16R4	A 16R4 PAL

Figure B.7 Representation of complex components.

Appendix C Answers to selected exercises

1.11	14 V
2.5	3.9 μV
2.9	138.5 Ω, 1.385 V, 2.77 °C
3.5	15 Ω
3.6	18.2 V r.m.s.
3.8	(a) 10 dB, (b) 0 dB, (c) −3 dB, (d) 120 dB
3.9	(a) 100, (b) 10, (c) 0.03, (d) 0.18
3.10	(a) 20 Hz, (b) 8 Hz, (c) 250 Hz, (d) 1 kHz, (e) 100 Hz
3.13	145 Hz, lower
3.15	58 dB
4.11	0.014, 15
4.15	65 Hz
4.17	(a) 215, 1×10^9 Ω, 80 mΩ
	(b) −33, 10 kΩ, 10 mΩ
	(c) 11, 2×10^{10} Ω, 4 mΩ
	(d) −67, 1.5 kΩ, 20 mΩ
4.19	160 Hz
4.21	0.000 04%
5.6	~0.4 V
6.7	1.2 mS, 1.7 mS, 2.4 mS
7.1	930 μA, 4.8 V
7.3	~ −240, 730 Ω, 985 Ω
7.4	~ −180, 730 Ω, 750 Ω
7.5	1.5 mA, 6.15 V, ~ −3.9
7.6	1.8 kΩ, 3.9 kΩ
7.7	$f_{co} = 88$ Hz
7.11	1.8 kHz, >18 μF
7.12	5.2 mA, 5.2 V, 1, 2.3 kΩ 4.8 Ω
7.13	6.8 V, 90, 11 Ω, 1 kΩ
8.10	28 μV
8.11	4%
9.3	128, 64
9.8	53_{10}, 492_{10}, $41\,230_{10}$
9.9	$1\,000\,011_2$, 11.101_2, $110\,011\,101_2$, $1000\,1111\,1110_2$
9.10	$BDDA_{16}$, $C7_{16}$, 65_{16}

9.11 $100\ 000_2$, $11\ 011_2$, $1\ 001\ 101_2$, 111_2
12.6 (a) 16, (b) 11, (c) 13
13.2 4 kHz
13.10 ~65 600 bits per second

Index